A Companion to Feminist Geography

A Companion to Feminist Geography

Edited by

Lise Nelson & Joni Seager

WILEY Blackwell

This edition first published 2005
©2005 John Wiley & Sons Ltd
except for editorial material and organization © 2005 by Lise Nelson and Joni Seager

Registered Office(s)
John Wiley & Sons, Inc., 111 River Street, Hoboken, NJ 07030, USA
John Wiley & Sons Ltd, New Era House, 8 Oldlands Way, Bognor Regis, West Sussex, PO22 9NQ, UK

Library of Congress Cataloging-in-Publication Data

A companion to feminist geography / edited by Lise Nelson & Joni Seager.
    p. cm. — (Blackwell companions to geography ; 6)
  Includes bibliographical references and index.
  ISBN 1–4051–0186–5 (hardback : alk. paper) ISBN 978–1–394–34952–4 (paperback)
    1. Feminist geography. 2. Women—Social conditions. 3. Women—Employment. 4. Women
and city planning. 5. Women and the environment. I. Nelson, Lise. II. Seager, Joni. III. Series.

HQ1233.C53 2005
305.42—dc22

                                                                              2004007288

A catalogue record for this title is available from the British Library.

The publisher's policy is to use permanent paper from mills that operate a sustainable forestry policy,
and which has been manufactured from pulp processed using acid-free and elementary chlorine-free
practices. Furthermore, the publisher ensures that the text paper and cover board used have met
acceptable environmental accreditation standards.

For further information on
Blackwell Publishing, visit our website:
www.blackwellpublishing.com

Contents

Contributors

Kawango Agot received her PhD in medical geography, and a master's in public health, from the University of Washington in 2001. Until August 2003, she was a Lecturer in Geography at Moi University in Kenya. Presently, she is the Director of Impact Research and Development Organization and the Vice-Chairperson of Nyanza Reproductive Health Society. Her current research focuses on culture, gender, and HIV in Kenya, specifically examining the association between widow inheritance and the risk for HIV acquisition. She founded a non-profit organization, Ushindi Children's Support Services, to assist children orphaned by AIDS in Kenya.

Megan Blake is a Lecturer in Human Geography at the University of Sheffield. Her research interests focus on the intersections of home, work, and community; job satisfaction; and family migration. She has published research on care responsibility and migration decision-making, gendered notions of innovation, employment change during industrial restructuring, and links between domestic lead poisoning and industrialization.

Liz Bondi is Professor of Social Geography at the University of Edinburgh. She has published extensively in feminist geography and is founding editor of *Gender, Place and Culture*. Her most recent publications include the co-authored volume *Subjectivities, Knowledges and Feminist Geographies* (2002). Her current research focuses on counseling, psychotherapy, and emotional life.

Kate Boyer received her PhD in geography from McGill University in 2001. Her work has appeared in the *Journal of Urban Geography* and *Gender, Place and Culture*. She is presently serving as a Visiting Assistant Professor of Geography in the Department of Science and Technology Studies at Rensselaer Polytechnic Institute in Troy, New York. Her current research focuses on issues relating to gender, place, and technological change.

Judith Carney is Professor of Geography at the University of California, Los Angeles. She has conducted field research in West Africa, Brazil, Mexico, and Central America. Her research interests include gender and development, political ecology, and cultural ecological issues surrounding the African Diaspora. She is the author of *Black Rice: The African Origins of Rice Cultivation in the Americas* (2001).

Altha Cravey is Associate Professor of Geography and Directory of Undergraduate Studies at the University of North Carolina at Chapel Hill and former construction electrician and member of Local 481 in Indianapolis. She is author of *Women and Work in Mexico's Maquiladoras* and articles on globalization, work, and gender. Working in the building trades led her to seek understanding of the ways in which notions of work and gender become deeply entwined with identity and embedded in particular places.

Joyce Davidson is Assistant Professor of Geography at Queen's University, Kingston, Ontario. She completed her PhD at the University of Edinburgh and took up a post-doctoral research fellowship at the Institute for Health Research, Lancaster University. Her research and teaching focus on geographies of emotion and embodiment, and her publications include the forthcoming *Phobic Geographies: The Phenomenology and Spatiality of Identity* (2003).

Giovanna Di Chiro teaches women's studies and environmental studies at Mount Holyoke College. She has published widely on the topics of community-based knowledge production and environmental justice. She is a co-editor of the forthcoming collection *Appropriating Technology: Vernacular Science and Social Power*.

Teresa Dirsuweit received her PhD in geography from the University of the Witwatersrand in Johannesburg in 2003. She currently holds a lecturing position at the University of the Witwatersrand. She is primarily concerned with theorizing space within an African context and her current research focuses on institutional and urban spaces. Recently she guest edited a special edition of the African journal *Urban Forum*, entitled "Safety, Fear of Crime and Social Exclusion." Her work has also appeared in *Geoforum* and *Development Update*.

Mona Domosh is a Professor of Geography at Dartmouth College. She is the co-editor (with Kay Anderson, Steve Pile, and Nigel Thrift) of *Handbook of Cultural Geography* (2003), co-author (with Joni Seager) of *Putting Women in Place: Feminist Geographers Make Sense of the World* (2001), and author of *Invented Cities: The Creation of Landscape in 19th-Century New York and Boston* (1996).

Glen Elder is Associate Professor of Geography at the University of Vermont. He has researched and written about geographies of race, sex, and gender, with a particular emphasis on southern Africa. He is the author of *Hostels, Sexuality, and the Apartheid Legacy: Malevolent Geographies* (2003). He teaches classes in political and feminist geography.

Marlène Elias is a graduate student in the Department of Geography, University of California, Los Angeles. A French-Canadian citizen, she has worked on agroforestry issues in Panama and Burkina Faso. Her master's thesis (2003) examines the burgeoning international demand for shea butter on female producers in Burkina Faso.

Jody (Jacque) Emel is a Professor of Geography and the Director of Women's Studies at Clark University in Worcester, Massachusetts. Her research focuses on the geopolitics of natural resources and on animal geographies. She is beginning a new project on the implications of the new "livestock revolution" (which promotes intensification of livestock production) for women and the rural poor in developing countries. Her most recent publications are on the effects of environmental NGO campaigns against the gold mining industry.

Kim England is Associate Professor of Geography and Director of the Canadian Studies Center at the University of Washington. Her research interests focus on North America and revolve around the intersection of labor markets, families, and the state; feminist and social theories; and the politics and ethics of "doing" research.

Ayda Eraydın is Professor in the Department of Urban and Regional Planning at the Middle East Technical University, Turkey. She served as visiting scholar in INTAN (Malaysia) the and Institute of Developing Economies (Japan). Her areas of interests are regional/local development, gender, and the spatial organization of production and labor processes. She has published more than ten books and fifty papers in Turkish and other languages, including "Building up competence, institutions and networks in order to catch up in the knowledge economy," in R. Hayter and R. LeHeron (eds), *Knowledge, Territory and Industrial Space* (2002).

Tovi Fenster is a Senior Lecturer at the Department of Geography and Human Environment, Tel Aviv University. She has published articles and book chapters on ethnicity, citizenship, and gender in planning and development. She is the editor of *Gender, Planning and Human Rights* (1999) and the author of *The Global City and the Holy City: Narratives on Knowledge, Planning and Diversity* (2004). She is the founder and the first chair (2000–3) of Bimkom – Planners for Planning Rights in Israel, and has been on the board of directors of the Association for Civil Rights in Israel (1994–9).

Melissa R. Gilbert is Associate Professor of Geography and Urban Studies at Temple University. Her current research interests are in the areas of theoretical and policy debates concerning: urban poverty; labor and community organizing; the social and economic causes and consequences of "the digital divide"; and university–community partnerships. She has published in the *Annals of the Association of American Geographers, Urban Geography*, and the *Professional Geographer* and chapters in various edited volumes.

Matthew Hannah is Associate Professor and Chair of the Geography Department at the University of Vermont. His book *Governmentality and the Mastery of*

Territory in Nineteenth Century America (2000) explores the racial and gender politics involved in the evolution of the US censuses of the post-Civil War decades. He is currently researching anti-census protest movements in 1980s West Germany.

Susan Hanson is Professor of Geography at Clark University. She is an urban geographer with interests in feminist geography, urban transportation, urban labor markets, and gender issues. Her current research delves into geography, gender, and entrepreneurship in two metro areas – Worcester, Massachusetts and Colorado Springs, Colorado.

Maureen Hays-Mitchell is Associate Professor in the Department of Geography at Colgate University. Her research interests include the gendered dimensions of economic restructuring in Latin America, the role of women's grassroots associations in post-conflict societies, and broader issues of human rights and development. She has conducted field research in Peru, Mexico, Chile, and Spain.

Phil Hubbard is Reader in Urban Social Geography at Loughborough University. His research focuses on the social life of cities, and his previous publications include *Sex and the City: Geographies of Prostitution in the Urban West* (1999), *People and Place* (with Lewis Holloway, 2001) and *Thinking Geographically* (with Rob Kitchin, Brendan Bartley, and Duncan Fuller, 2002).

Jennifer Hyndman is Associate Professor of Geography at Simon Fraser University, Vancouver. Her work focuses on the implications of feminist politics for analyses of migration, mobility, and security. She is the author of *Managing Displacement: The Politics of Humanitarianism* (2000) and co-editor of *Sites of Violence: Gender and Conflict Zones* with Wenona Giles (2004).

Hiroo Kamiya is a professor in the Department of Geography, Kanazawa University, Japan. His research interests lie in geographic aspects of welfare provision in east Asian cities.

Audrey Kobayashi is a Professor of Geography and Women's Studies at Queen's University, Kingston, Canada. She has published extensively in the areas of gender and racism, immigration, work, and human rights, and works as an activist in the areas of anti-racism and disability rights. Her most recent publication, co-edited with David Goldberg and Philomena Essed, is *A Companion to Gender Studies*, forthcoming. She is currently Editor of *People, Place and Region, Annals of the Association of American Geographers*.

Eleonore Kofman is Professor of Human Geography at Nottingham Trent University. She has co-edited *Globalization: Theory and Practice* (2003) and *Mapping Gender, Making Politics: Feminist Perspectives on Political Geography* (2004). She has published widely on gender, migration, and citizenship, and co-authored *Gender and International Migration in Europe* (2000).

Hille Koskela is a senior lecturer in planning geography in the Department of Geography in the University of Helsinki, where she has conducted research projects on video surveillance, street prostitution, sexual harassment, fear of violence, and multiculturalism. Her research interests cover the interrelations between security, surveillance, and the politics of looking, as well as geographies of gender. She is currently working for the Academy of Finland on a research project titled "Urban Security Policy, Control and Space." She has published several book chapters in international multidisciplinary anthologies and articles in journals such as *Gender, Place and Culture, Progress in Human Geography* and *Urban Geography.*

Victoria A. Lawson is Professor of Geography, University of Washington. Her research and teaching interests include: critical development theory in the Americas; migration and identity formation; the social and economic effects of global economic restructuring; poststructural analyses of power dynamics structuring discourses and practices of race, class, gender, sexuality, and nationality. She has published in journals such as *Annals of the Association of American Geographers, The Professional Geographer, Economic Geography, Environment and Planning A, Environment and Planning D: Society and Space, Antipode, Regional Studies, World Development, Progress in Human Geography, International Journal of Population Geography,* and *International Regional Science Review.*

Robyn Longhurst is Associate Professor and Chair of the Department of Geography at the University of Waikato in Hamilton. She is author of *Bodies: Exploring Fluid Boundaries* (2001) and a co-author of *Pleasure Zones: Bodies, Cities, Spaces* (2001). Her research coalesces around themes of bodies, spaces, and power. She is currently researching social and spatial aspects of body size.

Michele Masucci is an Associate Professor of Geography and Urban Studies and the Director of the Information Technology and Society Research Group at Temple University. Her current research focuses on the impacts of digital divide barriers on access to health and educational services among inner city populations in north Philadelphia. Her most recent publications are on the relationship between institutional partnerships and the development and use of information technology resources by environmental and economic justice organizations in Philadelphia and Brazil.

Sara McLafferty is Professor of Geography at the University of Illinois at Urbana-Champaign. Her research interests include geographies of health, urban geography, and GIS/spatial analysis. She has examined the intersections between spatial and social inequalities in the New York metropolitan region, focusing on issues of health and geographic access to healthcare services and employment opportunities.

Robina Mohammad is currently working as a postdoctoral fellow in the South Asian Studies Programme at the National University of Singapore. Her research interests include political transformations and their impact on women in different national contexts, and the marketing and consumption of the gendered, sexed body. Her current project is on the transnationalization of Bollywood, to and beyond the South Asian diaspora.

Pamela Moss is professor in the Faculty of Human and Social Development, University of Victoria. Her research coalesces around themes of power and body emerging from women's experiences of changing environments – women with chronic illness, women's collective autobiographies, and the organization of support services for women in crisis. She edited *Placing Autobiography in Geography* (2001) and *Feminist Geography in Practice* (2002). She is co-author with Isabel Dyck of *Women, Body, Illness* (2002). She is also active in local feminist community politics around issues in women's housing and invisible disabilities.

Richa Nagar is Associate Professor of Women's Studies at the University of Minnesota. Her research has focused on the politics of space, identity, and community among South Asians in Tanzania, and more recently, on the contradictions of empowerment in women's organizations in North India. Her new work seeks to reconceptualize transnational feminist theory and praxis through collaborations with academic and non-academic actors across institutional, geographic, and socioeconomic borders.

Lise Nelson is an Assistant Professor of Geography at the University of Oregon. Her research interests include gender, globalization, democracy, and critical development studies. She is currently finishing a book entitled *Defending the Plaza: Gender, Citizenship and the Politics of Place*, and is initiating research on immigrant settlement processes and struggles over race, place, and citizenship in the Pacific Northwest.

Geraldine Pratt teaches in the Geography Department of the University of British Columbia. She is author of *Working Feminism* (forthcoming), co-author of *Gender, Work and Space* (1995), and editor of *Environment and Planning D: Society and Space*. The Philippine Women Centre was launched in February 1990 and officially registered as a non-profit society in January 1991. The PWC (pwc@telus.net) aims to empower women of Filipino origin in Canada to understand the roots of their challenges as immigrants, as women of color, and as low-income earners; and to collectively assert their struggle for their rights and welfare.

Valerie Preston is Professor and Graduate Director in the Department of Geography at York University, where she teaches social geography. She has published in numerous journals including *The Canadian Geographer*, *Urban Geography*, *Gender, Place and Culture*, and *Economic Geography*. Her current research examines how transnationalism is transforming citizenship practices in various metropolitan areas, and the integration of immigrant women in urban labor markets.

Jasbir K. Puar is Assistant Professor of Women's and Gender Studies and Geography at Rutgers University. Her work on topics such as queer tourism, transnational sexualities, and South Asian cultural production appears in *SIGNS*, *GLQ*, *Antipode: A Radical Journal of Geography*, *Society and Space*, and *Social Text*.

Saraswati Raju is Professor of Social Geography at the Centre for the Study of Regional Development at Jawaharlal Nehru University, New Delhi. Her research has focused on gendered geographies, with social development as a main concern, in which she has published extensively in national and international journals of repute. Her recent publications include the co-edited *Atlas of Women and Men in India* (1999). She is consultant to many national and international agencies trying to forge partnerships between academia, policy implementers, and members of civil society on gender, empowerment, and development issues in South Asia.

Dianne Rocheleau is an Associate Professor of Geography at Clark University, where she also teaches in the Women's Studies Program, the Global Environmental Studies Program, and the International Development, Commmunity and Environment Program. She has worked with community-based organizations on land use and conservation research for the past 25 years, on questions of applied ecology, political ecology, landscape change, and environmental justice, primarily in Kenya, the Dominican Republic, and the United States. She has served on the boards of the Land Tenure Center, the Center for International Forestry Research, and the International Association for the Study of Common Property, as well as the editorial boards of *Forests, Trees and Livelihoods*, *Gender, Place and Culture*, and the *Geographical Journal*. She received a fellowship at Radcliffe Institute for Advanced Study at Harvard University in 2002–3 to work on a book in progress: *The Invisible Ecologies of Machakos: Landscapes and Life Stories, 1900–2000*.

Vidyamali Samarasinghe obtained her PhD from Cambridge University in economic geography. She is an Associate Professor in International Development in the School of International Service at American University in Washington, DC. She has also taught at the University of Sri Lanka and has been a research/teaching fellow at Swarthmore College and the University of Maryland at College Park. Her research is focused on gender in developing countries, with a particular emphasis on South Asia. Her writings cover female sex trafficking, the impact of globalization on women's work, the impact of conflict on women, women and politics, and feminist methodology. She is currently working on a book project on sex trafficking in South and South East Asia, based on fieldwork in Nepal, Cambodia, and the Philippines. She has co-authored *A Historical Dictionary of Sri Lanka* (1998), and co-edited *Women at the Center: Gender issues for the 1990s* (1994) and *Women at the Crossroads: A Sri Lankan Perspective* (1990).

Joni Seager is Dean of the Faculty of Environmental Studies at York University in Toronto, Canada. She is the author of several books, including recently *The Penguin Atlas of Women in the World* (2003), and co-author, with Mona Domosh, of *Putting Women in Place: Feminist Geographers Make Sense of the World* (2001). Her research interests include feminist environmentalism, militarism and the environment, and global political economy. She is also co-founder of the "Center for New Words" project, a Boston-based non-profit organization committed to creating space for women's voices to be heard in civil society.

Rachel Silvey is Assistant Professor of Geography and Research Associate at the Population Program, Institute of Behavioral Science, at the University of Colorado, Boulder. Her research interests include gender and feminist geography, migration studies, social activism, critical development studies, and Indonesia. Her work has appeared in a range of journals, including the *Annals of the Association of American Geographers*, *Political Geography*, *Progress in Human Geography*, *Gender, Place, and Culture*, and *World Development*.

Amanda Lock Swarr is a Mellon Postdoctoral Fellow at Barnard College of Columbia University, teaching in the department of Women's Studies. She holds a PhD in feminist studies from the University of Minnesota. Her work is forthcoming in *Signs* (with Richa Nagar) and the *Journal of Homosexuality*, and she is currently completing a book about South African transgenderisms, race, and apartheid.

Asuman Turkun-Erendil is Assistant Professor at the Department of Landscape Architecture and Urban Design at Bilkent University, Turkey. Her areas of interest are local development, industrial districts, urban regimes, and methodological issues in social sciences. With Zuhal Ulusoy she published "Re-invention of tradition as an urban image: the case of Ankara Citadel" (2002), *Environment and Planning B: Planning and Design*, 29, 655–72.

Julie Urbanik is a doctoral student in the Department of Geography at Clark University. She also holds a master's in women's studies from the University of Arizona, where her thesis focused on the integration of ecofeminist theory into women's studies curriculums. Her current research explores the material and theoretical challenges animal biotechnology poses for feminism, animal rights activism, and geographies of place.

Ebru Ustundag is a doctoral student in the Department of Geography at York University, Toronto. She also has an MSc and and a BSc from the Middle East Technical University, Ankara. Currently she is working on her dissertation, titled "Building a Nation, Contesting the City: Articulations of Turkish Republican Citizen." Her research interests include theories of the city, philosophical constructions of space, citizenship studies, and feminist geographies. Her current publications are "Theorizing Turkish urban studies with 'rights to the city'," in *Rights to the City* (2004) and "Women, Gender and Poverty: Turkey," in *Encyclopedia of Women and Islamic Cultures* (2004).

Anoja Wickramasinghe is Senior Professor and Head of Geography, Chair, Postgraduate Unit, Faculty of Arts at the University of Peradeniya, Sri Lanka and Co-ordinator of the Collaborative Regional Research Network in South Asia (CORRENSA). Her research focus has been on environment, forestry, rural development, indigenous knowledge, common property management, community development, and gender. She is the author of several books related to these areas: *Deforestation, Women and Forestry* (1994), *People and the Forest: Management of the Adam's Peak Wilderness* (1995), *Land and Forestry: Women's Local Resource-based Occupation for Sustainable Survival in South Asia* (1997), *Gender Aspects in Woodfuel Flows in Sri Lanka: A Case Study Conducted in Kandy District* (1999). Anoja is also a trainer in participatory planning, gender analysis, and environmental management.

Jennifer Wolch is Professor of Geography and Director of the Center for Sustainable Cities at the University of Southern California, where she teaches and conducts research on Los Angeles, social problems, urban sustainability, and human–animal relations. She is co-editor of *Animal Geographies: Place, Politics and Identity in the Nature/Culture Borderlands* (1998), and *Up Against the Sprawl: Public Policy and the Making of Southern California* (2004).

Brenda S. A. Yeoh is Associate Professor, Department of Geography, National University of Singapore. Her research foci include the politics of space in colonial and postcolonial cities; and gender, migration, and transnational communities. She has published a range of scholarly journal papers and several books including *Gender and Migration* (2000, with Katie Willis), and *Gender Politics in the Asia-Pacific Region* (2002, with Peggy Teo and Shirlena Huang).

Jin Zhang is a PhD candidate in psychology at the University of Southern California. Born in Xi'an, PR China, Jin Zhang earned his MA in psychology at Beijing University in 1989, where he went on to serve as a research fellow at the National Center for Education Development Research in China, and to conduct market research for the Gallup Organization from 1995 to 1998. His specialties include positive psychology, quantitative psychology, research design, survey methodology, and marketing research. His work has appeared in book chapters, journals, and reports, and his current research on subjective well-being, culture, personality, and demographics won the International Positive Psychology Fellow Award.

Acknowledgments

The editor and publisher gratefully acknowledge the permission granted to reproduce the copyright material in this book:

Esmeralda Bernal, "My Womb," from *The Americas Review* (Houston: Arte Publico –University of Houston, 1986). Reprinted with permission from the publisher of *The Americas Review* (Houston: Arte Publico Press – University of Houston, 1986).

Kishwar Naheed, extract from *Woman of Pakistan: Two Steps Forward One Step Back*, Mumtaz Khawar and Shaheed Farida, p. 77. (London: Zed Books, 1987). Reprinted with permission from the publisher.

"A Punjabi suit for evening wear," from *Women's Own* (Pakistan), vol. 13(11), February 2001. Reprinted with permission from the publisher.

"Policy Department *c.*1920" (Sun Life Archives). Reprinted with permission from Sun Life Financial.

"Montreal Head Office Building and Tulip Festival, *c.*1947" (Sun Life Archives). Reprinted with permission from Sun Life Financial.

"1945 Corporate Annual Report Cover" (Sun Life Archives). Reprinted with permission from Sun Life Financial.

"Pink Map 2000," Reprinted with permission of A & C Maps, Cape Town.

"Pubs, Clubs and Entertainment," from *Out and About*, 1998. Reprinted with permission from the publisher.

"Cartoon depicting connections between violent crime in Cape Town and tourism," *The Sowetan Newspaper*, November 30, 1999. Reprinted with permission.

Every effort has been made to trace copyright holders and to obtain their permission for the use of copyright material. The publisher apologizes for any errors or omissions in the above list and would be grateful if notified of any corrections that should be incorporated in future reprints or editions of this book.

Chapter 1

Introduction

Lise Nelson and Joni Seager

The Poetics of Bodies, Spaces, Place, and Politics

> my womb
> a public domain
> erotica a doormat
> trampled on by
> birthright
>
> my womb a
> legislated periphery
> no longer mine
> but public space
>
> my womb
> a palestinian front
> fighting for
> the right to be
> a private space

"My Womb" by Esmeralda Bernal (reprinted with permission from the publisher of *The Americas Review*, Houston: Arte Publico Press–University of Houston, 1986)

The ideas and materialities woven into Bernal's poem evoke some of the key insights and sites of feminist geography. Locating her poem within the gendered body, Bernal weaves together the politics of public and private space, the state and nationalism. While no single poem can represent the diverse issues and questions in feminist geography, the centrality of the body in her poem is significant. Only a few decades after feminists levered "woman" and "gender" into the lexicon of geographic thought, it is "the body" and the multidimensionality of embodied experience(s) that continue to anchor feminist geography at the dawn of the twenty-first century.

The body is the touchstone of feminist theory. Within contemporary feminist theory "the body" does not have a single location or scale; rather it is a concept that disrupts naturalized dichotomies and embraces a multiplicity of material and symbolic sites, ones located at the interstices of power exercised under various guises. From the pivotal second-wave feminist understanding that "the personal is political" to the postmodern decentering of a singular notion of gendered experience, feminist theory draws on understandings of embodied experience to fundamentally challenge bedrocks of Western social and political thought. Feminist geography, anchored in the body, moves across scale, linking the personal and quotidian to urban cultural landscapes, deforestation, ethno-nationalist struggles, and global political economies.

But what does it mean analytically, theoretically, and methodologically to center the body? What contours define the map of feminist geography? Where is feminism on the map of geography? What do feminist geographers *do*? This companion to feminist geography approaches these questions by assembling the work of a wide range of contemporary feminist scholars in geography, ones located in Anglo-American as well as global contexts. It examines historiographies of feminist thinking and charts emerging research trajectories that continue to transform not only feminist geography as a field, but the discipline of geography itself.

Changing Terrains of Feminist Geography

Most chroniclers mark the emergence of feminist geography in North America and the UK in the early 1970s, sparked by movements both within and outside the academy. Within geography, feminist critiques emerged as part of the ferment of "new" radical geographies – especially Marxism – that was raising challenges in the 1970s to the hegemonies of positivistic and corporatist geography (for discussion see Mackenzie, 1984). This was a productive, but also thorny, convergence. The degree of synergy between Marxist and feminist frameworks varied considerably across subfields within the discipline and from country to country; the socialist–feminist intellectual link was much stronger in the UK than in the USA, for example, and remains so today. Clearly feminist geography draws on radical intellectual traditions in the discipline; nevertheless, as feminist geography matured it *also* served as a corrective – and to some extent as a rebuke – to its radical counterparts which, at least into the 1990s, remained as stubbornly androcentric as mainstream geography (and in some instances just as openly hostile to feminist approaches). The contested relationship between Marxism and feminism continues to shape epistemology and intellectual debates within both subfields.

From its earliest inception, a defining characteristic of feminist geography was its intellectual cross-fertilization and multidisciplinarity; this remains one of its strengths today. In comparison to other cognate social science fields, geography as a discipline was slower in developing and embracing feminist scholarship; this delayed engagement meant that the critical work already under way in disciplines such as anthropology, sociology, history, political philosophy, and economics was available to the early cohort of geographers who were pioneering feminist geography. Economist Esther Boserup's *Women's Role in Economic Development* (1970) and Barbara Rogers's *The Domestication of Women* (1979), for example, were instrumental in the emergence of the "women and development" subfield in geography. Early geographical analyses of women's perceptions of and relationships to new landscapes drew heavily from then-extant feminist research in historical disciplines, including key works such as Annette Kolodny's *Lay of the Land* (1975). Similarly, feminist geographical urban and built-environment research was infused by historical, sociological, and architectural work (including, prominently, now-classics such as Jane Jacobs's (1961) *Death and Life of Great American Cities* and Dolores Hayden's (1984) *Redesigning the American Dream*). Carol Pateman's *Participation and Democratic Theory* (1970) helped inspire feminist geographers interested in gendered divisions between public and private, the state and democracy.

The exuberance and vitality of women's movements outside the academy during the 1970s also strongly influenced the emergence of feminist geography. The women's liberation movement demanded accountability, visibility, equality. Within feminist geography, this translated first into a project to "add women" to the field, both as producers of knowledge and as subjects of analysis (for discussion see Monk and Hanson, 1982; Mayer, 1989). Starting from the newly legitimated interest in the lives of "real women," the earliest feminist geographical work focused on mapping (literally and metaphorically) the spatial constraints facing women (for examples see Davies and Fowler, 1971; Hayford, 1974; Tivers, 1977; Ardener, 1981; Seager and Olson, 1986; Seager, 2003; for reviews see Bowlby et al., 1981; Zelinsky et al., 1982). The work of making – and keeping – women's lives visible is far from complete, and such projects remain at the heart of feminist geography.

The efflorescence of feminist geography in the 1980s laid the foundation for many of the subfields and interests that define the contemporary field (for excellent overviews see McDowell, 1992a, b; Domosh, 1999; Longhurst, 2002). In tandem with ground-breaking research on the material realities of women's lives, feminist geographers in the 1980s adopted and introduced theoretical constructs about the role of gender as an instrumental force and as a category of explanation in geographical processes. Extending work of the previous decade, feminist geographers sought to document and bring into geographical inquiry the analytical significance of gendered spatial divisions between public and private, particularly as they shape work (paid and unpaid), and urban processes (for examples see Christopherson, 1983; Rossini, 1983; Mackenzie, 1986; Nelson, 1986; Pratt and Hanson, 1988). Relatedly, feminist geographers turned to an examination of the spatial and gendered dimensions of industrial restructuring, and in the process challenged gendered assumptions within Marxist geography (see Massey, 1984; Murgatroyd et al., 1985). Other scholars sought to make visible women's roles as actors in built and natural landscapes (for discussion see Monk, 1984).

The reverberation of research agendas in the 1980s is still being felt today: feminist geographical work in that decade on ecology and social constructions of nature (see, for example, Fitzsimmons, 1989) are at the heart of contemporary work in feminist political ecology. An expanding literature on "women and development" and women's work in the Global South (see, for example, Momsen and Townsend 1987; Chant and Brydon 1989; Carney and Watts 1990) laid the foundation for a robust subfield, one that today infuses much of the feminist work on globalization and transnational processes. Finally, early feminist forays into political geography (such as Drake and Horton 1983; Peake 1986) led to a 1990 special issue of

Political Geography that charted emerging feminist agendas that are still under debate today (Kofman and Peake 1990; see Kofman, chapter 34 in this volume). Key theoretical insights from this period might be summarized by phrases that are now part of our ordinary geographical conventional wisdom: "space is gendered"; "place doesn't just reflect gender, it produces it"; "sexuality is constructed in place and spatially."

By the 1990s, feminist geographers were actively contributing to broader feminist debates that questioned both the unity/singularity of knowledge *and* the very subject of "woman" that once occupied the central position in feminist thought. While issues such as reproductive rights or the epidemic of violence against women might suggest a unifying gendered experience – perhaps validating a sense of a common "sisterhood" – textured (and textual) analysis of women's experiences acted out in particular lives and particular places reveals deep cleavages in the notion of what it means to be gendered as a woman. Writers such as bell hooks (1984) and Chandra Talpade Mohanty (1984) powerfully argued that feminist thinking implicitly and explicitly centered a white, First World and middle-class female subject, downplaying (or ignoring) power relations and differences between women (see also the collection edited by Moraga and Anzaldúa, 1981).

This critique within feminism by women of color and Third World women dovetailed with emerging epistemological debates to transform the nature of feminist thinking. Taken-for-granted assumptions about knowledge as singular (represented as a quest for the "grand narrative") and "science" as a neutral, disembodied endeavor gave way to feminist reconsiderations of knowledge as situated and embodied – produced by concrete subjects and shaped specific histories and geographies (Harding, 1986; Haraway, 1991; for disscussion in Geography, see Bondi 1990a, b; Rose 1993). As a result of these philosophical and political debates, "the body" and "the subject" theorized in feminist thinking became more nuanced and explicitly located within various geographic and historical contexts beyond the horizon of white, middle-class and Western spaces.

Emphasis on the politics of knowledge and the intersectionality of multiple oppressions and identities invigorated feminist geography by providing a wide array of new theoretical and methodological tools for feminist geographical work. Developments in critical race theory, pscychoanalysis, poststructuralism, postcolonialism, and queer theory led feminist geographers to develop more nuanced approaches to identity, power, and difference. These perspectives destabilized sexual and gender categories, shifted understandings of space and place, and led to new methodological approaches and understandings in the field.[1] These debates were not merely theoretical: this period witnessed a growing number of students and scholars from diverse geographic and social backgrounds contributing to, and enriching, feminist geography. Just as (usually white) women in geography demanded a seat at the academic table in the 1970s, women from a multiplicity of racialized, sexualized, classed, and transnational experiences have been exploding the canon of feminist geography. Nothwithstanding these important advances, diversifying the professional ranks of geography remains a crucial and ongoing project (see Al Hindi, 2000).

Theoretical perspectives and epistemological critiques increasingly visible by the early to mid-1990s not only strengthened feminist geography "internally," but deepened the contribution of feminist geographers to geographic thought and analysis more broadly. On one hand, during the past fifteen years a distinct set of research agendas has emerged "within" and in close relation to feminist geography. Examples include a rich literature on the spatialized performance of sexuality, gender, and race as well as spaces of embodiment (Bell et al., 1994; Binnie 1997; Nast and Pile, 1998; Skelton, 1998; Mahtani, 2002; in this volume see Dirsuweit, Longhurst, Mohammad, and Puar), on the geographies of masculinity (Morrell, 1998; Myers, 2002; Bye, 2003; for review see Longhurst, 2002; in this volume see Hannah), and on geographies of (dis)ability, illness, and health (Asthana, 1996; Moss and Dyck, 1996; Butler and Parr 1999; Dyck et al., 2001; in this volume see Agot). These contributions do not exclusively "belong" to feminist geography; instead they represent work that cross-cuts feminist geography and more recently formed geographical literatures that closely engage queer theory, critical race studies, and the social construction of (dis)ability.

On the other hand, feminist geographers have continued to bring new and important perspectives to a variety of "traditional" subfields in geography. While a comprehensive review is beyond the scope of this introduction, examples of contributions by feminist geographers to various fields during the past decade include: economic and labor geography (McDowell and Court, 1994; Gibson-Graham, 1996; Stiell and England, 1997; Freidberg, 2001; in this volume see chapters by England and Lawson, Eraydin and Erendil, Hanson and Blake, Samarasinghe); political geography (Staeheli, 1996; Secor, 2001; in this volume see Domosh, Hyndman, Elder, Hays-Mitchell); urban geography (Gilbert, 1997; Bondi, 1998; in this volume see Boyer, Fenster, Hubbard, Kamiya, Koskela, Preston and Ustundag); cultural geography (Katz and Monk, 1993; Anderson, 1996; Jacobs and Nash, 2003); critical development geography (Ulluwishewa, 1992; Momsen and Kinnaird, 1993; Radcliffe, 1996; in this volume see Agot, Cravey, Elias and Carney, Hays-Mitchell, Nagar and Swarr, Raju); environmental geography (Rocheleau et al., 1996; Okono, 1999; Gururani, 2002; in this volume see Di Chiro, Emel and Urbanik, Rocheleau, Wickramasinghe, Wolch and Zhang); geographies of migration (Tyner, 1994; Wright et al., 2002; Yeoh and Willis, 2002; in this volume see Pratt, Silvey, Yeoh); and geographic information science (see Kwan, 2002; in this volume see Gilbert and Masucci, and McLafferty).

Just as work of the 1970s arose in tandem with the women's movement, current debates and topics in feminist geography are profoundly influenced by social movements and geo-historical dynamics. In particular, feminist geography is currently being shaped by (and responding to) contemporary globalization processes and neoliberal discourses, including but not limited to distinct political, economic, and cultural connections that are transnational and translocal. These influences are relevant to a range of the work showcased in this volume because contemporary global/local connections, processes, and movements are profoundly gendered, whether in the context of labor on the global assembly line (see the chapter by Cravey), markets (see chapters by Elder, Elias and Carney, Eraydin and Erendil, Puar), global health policy and risk (see the chapter by Agot), transnational politics and social movements (see chapters by Di Chiro, Nagar and Swarr, Raju), the globalization of low-wage service work (see chapters by England and Lawson, Pratt), local/global performances of identity (see chapters by Dirsuweit, Puar), regional and transnational migration flows (see chapters by Preston and Ustundag, Mohammad, Silvey), "anti-terrorism" discourses and uneven manifestations of state violence (see chapters by Hannah, Hays-Mitchell, Hyndman), or the astonishing acceleration of the global sex trafficking (see the chapter by Samarashinghe). We mention these not to straitjacket every contribution in terms of the ubiquitous "global," but to point out a thread of analysis that connects many chapters in this volume, as well as feminist geography more broadly.

Finally, over the past ten to fifteen years close attention to the politics of knowledge production and epistemology in feminist theory has inspired careful explorations of methodology by feminist geographers (see, for example, Katz, 1992; Kobayashi, 1994; Lawson, 1995; in this volume see Moss). These debates represent the cutting edge of methodological thinking and practice in the discipline of geography: questions of positionality, power, and embodied knowledge production permeate discussions of methodology within various fields of geographic thought.

Given the diversity of questions, approaches, and methodologies in feminist geography, the contours of which we only touch upon above, it is unsurprising that the work of feminist geographers is becoming more visible than ever in feminist thinking and social theory broadly defined. Examples of this are numerous. Feminist political ecology, cited above, has reshaped thinking about women, development and environment throughout the social sciences, in policy circles, and among grassroots development organizations. Work on the spatialized performance of identity and power, also cited above, complements textual approaches to performativity in the humanities. Finally, feminist geographers' nuanced theorizations of place and scale are shaping broader feminist and social theoretical debates about identity, transnationality, and globalization (see, for example, Massey, 1993, 1999; Katz, 2001; Nagar et al., 2002; in this volume see Yeoh).

Given the complexity of the field, it is difficult to generalize about approaches and ongoing conversations within contemporary feminist geography. In our view, nevertheless, four themes of feminist geography emerge and re-emerge to distinguish it as an area of study.

First, feminist geography is closely allied with diverse political movements and commitments; this invigorates it as an arena of analysis and broadens its appeal both within and outside the academy. Feminism is defined by explicit political commitments (against oppressions, or to making visible the workings of social power), and feminist geography is unapologetically marked by this agenda. Some traditionalists suggest that this delegitimizes feminist geography as an *academic* enterprise; feminists argue that the myth of "objective neutrality" has been debunked long ago and that it is only the explicitness of its ideological commitments that distinguishes feminist from mainstream social science.

Second, it is an innately interdisciplinary subfield. This too reflects a politicized intellectual stance as much as it reflects the historical emergence of the field. As feminist geography grew from its early materially grounded and radical roots, it has engaged a feminist re-reading of key theoretical approaches throughout the discipline – in political and cultural geography, in urban and environmental research, in economic and migration literatures, and in methodological engagements. In the process, feminist geographers drew inspiration from, and contributed to, work in fields far beyond the domain of conventional social sciences – particularly engaging with poststructural, psychoanalytic, critical race, postcolonial, and queer theory.

Third, to the extent that feminism is at the forefront of theorizing the intersectionality of multiple oppressions, feminist geographers demonstrate how these oppressions are embedded in, and produced through, material and symbolic space and place. Place matters. The particularities of *where* social processes unfold, and how they unfold *in relation to* other social, political, and economic processes, shape the *way* in which they do so. In this context, feminist geographers (like other geographers) often underscore the importance of asking *where*. The focus of this question can range from so-called mundane spaces – the kitchen, urban park, or forest – to more clearly ideological spaces of territory, nation, and place. Whatever the focus, feminist geographers insist that asking *where* is not a secondary question, an afterthought, but instead represents a crucial entree into understanding the world in which we live, particularly a world marked by difference including but not limited to gender. Asking "where" forces us to map the complex relationships between bodies, identities, places, and power, and represents an arena in which feminist geographers are making their most important contributions to feminist theory.

And finally, feminist geography asserts the importance and salience of foregrounding women as a subject of study and "gendering" as a social and spatial process. This seems obvious, perhaps, but nonetheless worth repeating. Women's lives are so easily and so often trivialized and "disappeared" that a commitment to taking women seriously needs conscious and continuous reassertion. But "gender" does not only read as "women." One of the exciting theoretical turns in feminist geography is in grappling with the spatialized construction of femininity *and* masculinity – as ideology, materiality, and practice.

A Companion to Feminist Geography

Clearly the landscape of geographical inquiry and knowledge has been irrevocably challenged – and changed – by feminist geography. A substantive field in its own right, feminist geography has also reframed fundamental approaches across the discipline, reconceptualizing core subjects, concepts, epistemologies, and methodologies of geographic investigation.

Producing an anthology is, as colleagues once wrote, a "terrifying experience" (McDowell and Sharp, 1997). As editors, the problems of "selecting in" and "leaving out" produced in us constant moments of indecision and doubt. In the end, we resigned ourselves to the reality that we could never produce a definitive or even comprehensive volume – the field of feminist geography is too rich, diverse, and expansive to be contained in any single volume. Rather, we offer this *Companion* as a mosaic: in close view, each chapter can be read alone for its own distinctive contribution, but stepping back from each contribution we see an assembled portrait of a vibrant field. The primary purpose of the *Companion* is to showcase cutting-edge research by feminist geographers for both scholars and students – research that charts emerging issues in feminist geography while remaining grounded in the historiography of work in the field.

The first part of the volume, "Contexts," offers in four chapters a broad-brush assessment of contemporary feminist geography in its biggest frame – from core concepts of space and place, to intersectionalities of power and difference, to methodological engagements, and finally to the challenges of transnationality. The remaining thirty-four chapters are organized into five parts that reflect key spaces and scales central to feminist geography: work, city, body, nature, and the state/nation. Introductory chapters to these five parts frame each subset of chapters, providing an overview of the arc of intellectual developments in feminist geography as they relate to each part's particular theme. In developing these five parts we intended to create a loose-fitting organizational frame without corseting the field into a tight-fitting thematic structure. In the end we realized that even this loose thematic structure is capricious: most of the chapters in this volume could easily have been "placed" in two or more parts. As editors, we have come to appreciate the extent to which feminist geography, appropriately, resists constraint.

NOTE

1 Nevertheless, as in other fields, this engagement by feminist geographers with the "post-modern turn" of social theory was contested and continues to be subject to ongoing examination (see McDowell, 1991; Bondi and Domosh, 1992).

BIBLIOGRAPHY

Al-Hindi, K. F. (2000) Women in geography in the 21st century. Introductory remarks: structure, agency, and women geographers in academia at the end of the long twentieth century. *Professional Geographer*, 52(4), 697–702.

Anderson, K. (1996) Engendering race research: unsettling the self–other dichotomy. In N. Duncan (ed.), *Bodyspace Destabilizing Geographies of Gender and Sexuality*. London: Routledge.

Ardener, S. (1981) *Women and Space Ground Rules and Social Maps*. New York: St Martin's Press.

Asthana, S. (1996) The relevance of place in HIV transmission and prevention: geographical perspectives on the commercial sex industry in Madras. In R. Kearns and W. Gesler (eds), *Putting Health into Place: Landscape, Identity and Well-being*. Syracuse, NY: Syracuse University Press.

Bell, D., Binnie, J., Cream, J. and Valentine, G. (1994) All hyped up and no place to go. *Gender, Place and Culture*, 1(1), 31–47.

Binnie, J. (1997) Invisible Europeans: sexual citizenship in the new Europe. *Environment and Planning A*, 29(2), 237–48.

Bondi, L. (1988) Gender, class and urban space: public and private space in contemporary urban landscapes. *Urban Geography*, 19(2), 160–85.

Bondi, L. (1990a) Progress in geography and gender: feminism and difference. *Progress in Human Geography*, 14, 438–45.

Bondi, L. (1990b) Feminism, postmodernism and geography: space for women? *Antipode*, 22, 156–7.

Bondi, L. and Domosh, M. (1992) Other figures in other places: on feminism, postmodernism and geography. *Environment and Planning D: Society and Space*, 10(2), 199–213.

Boserup, E. (1970) *Woman's Role in Economic Development*. New York: St Martin's Press.

Bowlby, S., Foord, J. and Mackenzie, S. (1981) Feminism and geography. *Area*, 13(4), 711–16.

Butler, R. and Parr, H. (1999) *Mind and Body Spaces: Geographies of Illness, Impairment, and Disability*. London: Routledge.

Bye, L. (2003) Masculinity and rurality at play in stories about hunting. *Norsk Geografisk Tidsskrift*, 57(3), 145–53.

Carney, J. and Watts, M. J. (1990) Manufacturing dissent: work, gender and the politics of meaning in a peasant society. *Africa*, 60(2), 207–41.

Chant, S. and Brydon, L. (eds) (1989) *Women in the Third World: Gender, Issues in Rural and Urban Areas*. New Brunswick, NJ: Rutgers University Press.

Christopherson, S. (1983) Female labor force participation and urban structure: the case of Ciudad Juarez, Mexico. *Revista Geografica*, 97, 83–5.

Davies, C. S. and Fowler, G. L. (1971) The disadvantaged black female household head: migrants to Indianapolis. *Southeastern Geographer*, 11, 113–20.

Domosh, M. (1999) Sexing feminist geography. *Progress in Human Geography*, 23(3), 429–37.

Drake, C. and Horton, J. (1983) Comment on editorial essay: sexist bias in political geography. *Political Geography Quarterly*, 2, 329–35.

Dyck, I., Lewis, N. D. and McLafferty, S. (2001) *Geographies of Women's Health*. London: Routledge.

Fitzsimmons, M. (1989) The matter of nature. *Antipode*, 21, 106–20.

Friedberg, S. (2001) To garden, to market: gendered meanings of work on an African urban periphery. *Gender, Place and Culture*, 7(4), 341–62.

Gibson-Graham J. K. (1996) *The End of Capitalism (as We Knew It): A Feminist Critique of Political Economy*. Oxford: Blackwell.

Gilbert, M. R. (1997) Feminism and difference in urban geography. *Urban Geography*, 18(2), 166–79.

Gururani, S. (2002) Forests of pleasure and pain: gendered practices of labor and livelihood in the forests of the Kumaon Himalayas, India. *Gender, Place and Culture*, 9(3), 229–43.

Haraway, D. (1991) *Simians, Cyborgs and Women: The Reinvention of Nature*. New York: Routledge.

Harding, S. (1986) *The Science Question in Feminism*. Ithaca, NY: Cornell University Press.

Hayden, D. (1984) *Redesigning the American Dream*. New York: Norton.

Hayford, A. M. (1974) The geography of women: an historical introduction. *Antipode*, 6(2), 26–33.

hooks, bell (1984) *Feminist Theory from Margin to Center*. Boston: South End Press.

Hyndman, J. (2001). Towards a feminist geopolitics. *Canadian Geographer*, 45(2), 210–22.

Jacobs, J. (1961) *The Death and Life of Great American Cities*. New York: Vintage Books.

Jacobs, J. M. and Nash, C. (2003) Too little, too much: cultural feminist geographies. *Gender, Place and Culture*, 10(3), 265–79.

Katz, C. (1992) All the world is staged: intellectuals and the projects of ethnography. *Environment and Planning D: Society and Space*, 10(5), 495–510

Katz, C. (2001) On the grounds of globalization: a topography for feminist political engagement. *Signs*, 26(4), 1213–34.

Katz, C. and Monk, J. (1993) *Full Circles: Geographies of Women over the Life Course*. London: Routledge.

Kobayashi, A. (1994) Coloring the field: gender, "race" and the politics of fieldwork. *Professional Geographer*, 46, 73–80.

Kofman, E. and Peake, L. (1990) Into the 1990s: a gendered agenda for political. *Political Geography Quarterly*, 9(4), 313–36.

Kolodny, A. (1975) *The Lay of the Land: Metaphor as Experience and History in American Life*. Chapel Hill: University of North Carolina Press.

Kwan, M.-P. (2002) Feminist geography and GIS. *Gender, Place and Culture*, 9(3), 261–2.

Lawson, V. A. (1995) The politics of difference: examining the quantitative/qualitative dualism in post-structuralist feminist research. *Professional Geographer*, 47(4), 449–57.

Longhurst, R. (2000) Geography and gender: masculinities, male identity and men. *Progress in Human Geography*, 24(3), 439–44.

Longhurst, R. (2002) Geography and gender: a "critical" time? *Progress in Human Geography*, 26(4), 544–52.

McDowell, L. (1991) The baby and the bathwater: deconstruction, diversity and feminist theory in geography. *Geoforum*, 22, 123–34.

McDowell, L. (1992a). Space, place and gender relations: part 1. *Progress in Human Geography*, 17, 157–79.

McDowell, L. (1992b). Space, place and gender relations: part 2. *Progress in Human Geography*, 17, 305–18.

McDowell, L. and Court, G. (1994) Performing work: bodily representations in merchant banks. *Environment and Planning D: Society and Space*, 12, 727–50.

McDowell, L. and Sharp, J. (eds) (1997) *Space, Gender, Knowledge: Feminist Readings*. London: Arnold.

Mackenzie, S. (1984) Editorial introduction: women and the environment. *Antipode*, 16(3), 3–10.

Mackenzie, S. (1986) Women's responses to economic restructuring: Changing gender, changing space. In R. Hamilton and M. Barrett (eds), *The Politics of Diversity: Feminism, Marxism and Nationalism*. London: Verso.

Mahtani, M. (2002) Tricking the border guards: performing race. *Environment and Planning D: Society and Space*, 20(4), 425–40.

Massey, D. (1984) *Spatial Divisions of Labor: Social Structures and the Geography of Production*. New York: Methuen.

Massey, D. (1993) *Space, Place and Gender*. Minneapolis: University of Minnesota Press.

Massey, D. (1999) Imagining globalization: Power-geometries of time-space. In A. Brah, M. J. Hickman and M. Mac an Ghaill (eds), *Global Futures: Migration, Environment and Globalization*. New York: St Martin's Press.

Mayer, T. (1989) Consensus and invisibility: the representation of women in human geography textbooks. *Professional Geographer*, 41, 397–409.

Mohanty, C. T. (1984) Under western eyes: feminist scholarship and colonial discourses. *Boundary*, 2(12), 333–58.

Momsen, J. H. and Kinnaird, V. (1993) *Different Places, Different Voices: Gender and Development in Africa, Asia, and Latin America*. London: Routledge.

Momsen, J. H. and Townsend, J. G. (eds) (1987) *Geography of Gender in the Third World*. Albany: State University of New York Press.

Monk, J. (1984) Approaches to the study of women and landscape. *Environmental Review*, 13(1), 23–33.

Monk, J. and Hanson, S. (1982) On not excluding half of the human in human geography. *Professional Geographer*, 34, 11–23.

Moraga, C. and Anzaldúa, G. (eds) (1981) *This Bridge Called My Back: Writings by Radical Women of Color*. Watertown, MA: Persephone Press.

Morrell, R. (1998) Of boys and men: masculinity and gender in Southern African studies. *Journal of Southern African Studies*, 24, 605–30.

Moss, P. and Dyck, I. (1996) Inquiry into environment and body: women, work, and chronic illness. *Environment and Planning D: Society and Space*, 14(6), 737–53.

Murgatroyd, L. et al. (the Lancaster Regionalism Group) (1985) *Localities, Class and Gender*. London: Pion.

Myers, G. A. (2002) Colonial geography and masculinity in Eric Dutton's Kenya Mountain. *Gender, Place and Culture*, 9(1), 23–38.

Nagar, R., Lawson, V., McDowell, L. and Hanson, S. (2002) Locating globalization: feminist (re)readings of the subjects and spaces of globalization. *Economic Geography*, 78(3), 257–84.

Nast, H. and Pile, S. (1998) *Places through the Body*. London: Routledge.

Nelson, K. (1986) Labor demand, labor supply and the suburbanization of low-wage office work. In A. J. Scott and M. Storper (eds), *Work, Production, Territory*. Boston: Allen and Unwin.

Okono, E. (1999) Women and environmental change in the Niger Delta, Nigeria: Evidence from Ibeno. *Gender, Place and Culture*, 6(4), 373–8.

Pateman, C. (1970) *Participation and Democratic Theory*. Cambridge: Cambridge University Press.

Peake, L. (1986) A conceptual enquiry into urban politics and gender. In K. Hoggard and E. Kofman (eds), *Politics, Geography and Social Stratification*. Beckenham: Croom Helm.

Pratt, G. and Hanson, S. (1988) Gender, class and space. *Environment and Planning D: Society and Space*, 1, 15–35.

Radcliffe, S. (1996) Gender nations: nostalgia, development and territory in Ecuador. *Gender, Place and Culture*, 3(1), 5–22.

Rocheleau, D., Thomas-Slayter, B. and Wangari, E. (1996) *Feminist Political Ecology: Global Issues and Local Experiences*. London: Routledge.

Rogers, B. (1984) *The Domestication of Women: Discrimination in Developing Societies*. London and New York: Tavistock Publications.

Rose, G. (1993) *Feminism and Geography: The Limits of Geographical Knowledge*. Minneapolis: University of Minnesota Press.

Rossini, R. E. (1983) Women as labor force in agriculture: the case of the state of São Paulo, Brazil. *Revista Geografica*, 97, 91–5.

Seager, J. (2003) *The Penguin Atlas of Women in the World*. New York: Penguin.

Seager, J. and Olson, A. (1986) *Women in the World: An International Atlas*. New York: Simon and Schuster.

Secor, A. (2002) The veil and urban space in Istanbul: women's dress, mobility and Islamic knowledge. *Gender, Place and Culture*, 9(1), 5–22.

Skelton, T. (1998) Ghetto girls/urban music: Jamaican ragga music and female performance. In R. Ainley (ed.), *New Frontiers of Space, Bodies and Gender*. London: Routledge.

Staeheli, L. A. (1996) Publicity, privacy, and women's political action. *Environment and Planning D: Society and Space*, 14(5), 601–19.

Stiell, B. and England, K. (1997) Domestic distinctions: constructing difference among paid domestic workers in Toronto. *Gender, Place and Culture*, 4(3), 339–59.

Tivers, J. (1977) *Constraints on Spatial Activity Patterns: Women with Young Children*. Occasional Paper No. 6. London: Department of Geography, King's College.

Tyner, J. (1994) The social construction of gendered migration for the Philippines. *Asian and Pacific Migration Journal*, 3, 589–617.

Ulluwishewa, R. (1992) Development planning, environmental degradation, and women's fuelwood crisis: a case study. *Abstracts, 27th International Geographical Congress*, 629–31.

Wright, R., Bailey, A. J., Miyares, I. and Mountz, A. (2000) Legal status, gender and employment among Salvadorans in the US. *International Journal of Population Geography*, 6(4), 273–86.

Yeoh, B. and Willis, K. (2002) Gendering transnational communities: a comparison of Singaporean and British migrants in China. *Geoforum*, 33(4), 553–65.

Zelinsky, W., Monk, J. and Hanson, S. (1982) Women and geography: a review and prospectus. *Progress in Human Geography*, 6(3), 317–66.

Part I   Contexts

Chapter 2

Situating Gender

Liz Bondi and Joyce Davidson

Introduction

Over the past three decades, feminist geographers have challenged entrenched processes through which the discipline of geography has produced and reproduced inequalities between women and men. Importantly, this work has considered how geography is taught, and practices of academic labor, as well as the substance of research. Thus, for the pioneers of feminist geography, key tasks have included redressing the neglect of women, challenging misrepresentations of women, and insisting on the salience of gender as an axis of social differentiation and inequality, in teaching materials, within the academic workforce, and in geographic research (Hayford, 1974; Tivers, 1978; Monk and Hanson, 1982; McDowell, 1983, 1992; Women and Geography Study Group, 1984).[1]

One of the most important effects of feminist geography has been to unsettle taken-for-granted assumptions about women's and men's "places" in the societies, communities, organizations, and relationships within which we live and work. Thus, feminist geography has opened up questions about ways in which spaces and places – from bathrooms to call centres, from urban parks to teaching spaces – are experienced differently by different people, and come to be associated with the presence or absence of different groups of people (Nairn, 1997; Burgess, 1998; Longhurst, 2001; Belt et al., 2002). While gender is one salient dimension in these experiences and associations, so too are age, class, ethnicity, and many other factors. Consequently, bringing issues about gender into geography has entailed much more than attending to differences and inequalities between women and men. It has also prompted much reflection on what the categories "women" and "men" mean, and on the concept of gender, in the context of social identities and social relations more generally. One expression of this has been growing interest in a diversity of "masculinities" and "femininities"; that is, in different ways of being men and women (Valentine, 1996; Ainley, 1998; Laurie et al., 1999; Longhurst, 2000).

As feminist geographers have worked to enable and ensure the inclusion of women, and of gender issues, in the literal and figurative maps geographers use and produce, so deficiencies in these maps and the associated map-making techniques – tools of the trade – have become apparent. Indeed, what might appear to be the discipline's most fundamental concepts have tended to remain at least as unquestioned as gender. As Massey (1992, p. 66) has put it, conceptual discussion of "space" "never surfaces [within geography] because everyone assumes we already know what the term means." Consequently the project of feminist geography has been as much about rethinking core geographical concepts like space and place as about rethinking gender and its relationship to space and place.

It is important to emphasize that this chapter does not provide stable, singular definitions of the concepts space, place, and gender. Instead we show how feminist geography has enabled new ways of thinking about, and thinking with, interconnections between these terms. Nor do we offer an exhaustive review or history of feminist geographical usage of the concepts of space, place, and gender. Instead we outline and illustrate two approaches evident within the rich literature of feminist geography. It is our hope that readers will find that this selective way of reflecting on key concepts in feminist geography provokes fresh thinking on themes and issues within and beyond the subdiscipline.

The first approach to space, place, and gender we discuss has its origins in Doreen Massey's (1984) highly influential analysis of the dynamic interplay between social relations of class and the spatial organization of production. This approach conceptualizes space, place, and gender as interrelated, mutually constitutive processes. Neither gender identities nor places are stable, fixed, or given. However, neither are they freely chosen or easily transformed. Instead the dynamic interplay between space, place, and gender is subject to inertia and "stickiness."

The second approach to space, place, and gender we discuss shifts the focus from the persistence of dominant versions of masculinity and femininity to points at which limits may be reached and breached. It has its origins in Gillian Rose's (1993) description of paradoxical space. This approach analyses the relationship between space, place, and gender in terms of contradictions that can render everyday experiences – especially women's experiences – fraught and even tortuous, but through which radically different possibilities can also be glimpsed.

Sedimenting Gender

The term "place" is often thought of as referring to a bounded entity, containing a unique assemblage of characteristics, and within which people forge profound attachments and identities. This definition links "place" to subjective, meaningful, and emotional experiences, whether at the scale of a home, a neighborhood, or a country. It has provided a point of departure for a good deal of humanistic geography, including studies of people's "sense of place" (Relph, 1976; Eyles, 1985). Standing in implicit contrast to this, space is often thought of as abstract, objective, and defined by geometric and locational properties such as distance, latitude, and longitude. This conceptualization has been foundational for the "spatial science" and "spatial analysis" associated with quantitative perspectives in geography (Berry and Marble, 1968; Harvey, 1969).

These definitions of space and place have been subject to extensive challenge and criticism by feminist (among other) geographers. The notion of places as producing shared experiences and the notion of space as abstract geometry both conveniently ignore the myriad ways in which differences of gender, age, class, "race," and other forms of social differentiation shape people's lives. To cite one example, the places and spaces of homes and streets are often experienced very differently by gays and lesbians compared to heterosexuals (Valentine, 1993, 1996; Johnston and Valentine, 1995). These conceptualizations of space and place also fail to recognize how social relations shape geography. Interconnections between places do not necessarily lead to homogeneity, but instead produce what Doreen Massey (1994, p. 23) has described as "social relations stretched over space." Put another way, "geography matters" in the constitution of gender, class, and other social relations (Massey and Allen, 1984). According to Massey (1984) the uniqueness of a place can be understood as the expression of a particular mix of social relations which stretch far beyond that place. She argued that places could be understood in terms of "the combination of their succession of roles within a series of wider, national and international, spatial divisions of labour" (Massey, 1994, p. 14). This interpretation conceptualizes social relations, places, and their interconnections as layers deposited one on top of another, and, because of parallels with the way in which sedimentary rocks are laid down, it is often described as a "geological metaphor."

Massey's original formulation focused overwhelmingly on class relations. An early and influential extension of the approach to matters of gender was presented by Linda McDowell and Doreen Massey (1984), who showed how nineteenth-century economic change in Britain helped to produce a mosaic of regions differentiated in terms of gender as well as class, and how these patterns have influenced and been reshaped by twentieth-century economic restructuring. For example, in nineteenth-century colliery villages in northeast England women were almost completely excluded from employment in coal mining (but see John, 1980) and from a wider culture of mining work associated with labor organizations and social (working men's) clubs. Women, including wives and often daughters too, undertook very heavy burdens of unpaid domestic labor, which was essential to reproduce coal miners' labor power on a daily basis (until the middle of the twentieth century when proper shower and laundry facilities were finally provided by employers). In the context of men's and women's different relation to waged labor, nineteenth-century pit villages became intensely patriarchal, with households overtly ruled by men, and women seriously disempowered in all realms of their lives. Women and men led very different lives, generating highly contrasting versions of masculinity and femininity, in which, McDowell and Massey (1984, p. 132) argue, "male supremacy . . . became an established, and almost unchallenged, fact."

The cotton towns of northwest England present a very different story. Here women became paid workers in the new nineteenth-century textile factories. Although excluded from paid work as spinners soon after cotton production moved from domestic settings into factories, significant numbers of women secured waged work as weavers in the cotton mills (Hall, 1982). A strong tradition of women's involvement in waged labor was associated with working-class women's political organization through trades unions and suffrage campaigns (Liddington, 1979). Notwithstanding unequal pay and other forms of gender inequality, women's relatively high degree of autonomy and financial independence generated different versions of femininity and masculinity from those of the pit villages of northeast England. In a variety of ways, women challenged traditional assumptions about the gender order, prompting some commentators to fear that the "home [was being] turned upside down" (Hall, 1982, p. 17).

McDowell and Massey (1984) examined the legacies of these nineteenth-century patterns in relation to twentieth-century economic restructuring. They noted how, by the 1970s, it was in the former mining areas of northeast England that "homes [were] being 'turned upside down'" (McDowell and Massey 1984, p. 139). A steep decline in employment in coal mining prompted the introduction of government incentives for employers to relocate to the region. Although it was men who had lost their jobs, many of the incoming industries employed mostly women. Research by Jane Lewis (1984) demonstrated how women in this region had become an attractive pool of unskilled labour, often for assembly-line work, in part because of the very absence of any tradition of waged labor or union organizing. Prospective employers viewed them as cheap, flexible, and, perhaps above all, "docile" workers. Thus, gender relations in which women had been seriously disempowered relative to men had produced images of femininity that became a significant factor in the changing economic geography of the UK. A long established, dominant pattern in which men were wage-earners and women were unpaid domestic workers began to be reversed because employment opportunities for men were subject to relentless decline, while for women wage-earning opportunities increased. The impact on gender relations within and beyond households has been complex. Men's involvement in childcare and other domestic responsibilities has not increased in proportion to the loss of employment (Wheelock, 1990). Instead, as in many other areas, women combine paid and unpaid work, while the lack of job prospects for young men from educationally and materially disadvantaged backgrounds has prompted concern about a "crisis in masculinity" (Campbell, 1993; McDowell, 2001).

The decline of the cotton industry in northwest England resulted in job losses of a similar magnitude to those suffered in northeast England but did not prompt the same scale of government intervention to attract new employers into the region. One important reason for this was that many of the jobs lost were lost to married women who did not appear in the unemployment statistics because the tax and benefit system in place at the time defined such women as dependent rather than unemployed. Moreover, McDowell and Massey (1984) argued that the long tradition of women's employment, and of their involvement in political movements, made them significantly less attractive as a pool of labor for the "footloose" industries relocating during the 1960s and 1970s (also see Glucksman, 2000). In such ways, gender divisions and gender ideologies in different places contribute to the dynamic processes through which "social relations [are] stretched over space."

McDowell and Massey's (1984) analysis emphasized the dynamic effects of the development of different forms of gender relations and femininities in different places. They pointed to the influence of perceptions, assumptions, and stereotypes about women in particular places (such as their "docility"), as well as to geographically differentiated histories of gender divisions of labor. During the second half of the twentieth century, regional variations in women's rates of labor market participation in the UK converged towards levels similar to men's participation rates

(Lewis, 1984). However, as Simon Duncan and Darren Smith (2002) have shown, spatial differences in how women combine wage-earning and domestic responsibilities persist, demonstrating the influence of local gender cultures in the production of masculinities and femininities (also see McDowell, 2003).

Analyses such as these have demonstrated the uniqueness of particular places. However, this does not mean that places or place-based identities are stable, singular, or essential. On the contrary, the accounts we have discussed argue that places and place-based identities are unboundaried, open, porous, and fluid entities that are always "in process" in relation to numerous other places (Massey, 1994). Parallel arguments have been advanced in relation to gender identities; indeed, we have already drawn attention to the multiplicity and mutability of femininities and masculinities.

The notion that identities are fluid and mutable is powerfully and problematically reinforced by consumer cultures, which frequently market goods and services in ways that suggest consumption will confer desirable identity attributes on the consumer. Social theorists from a range of disciplines, including geography, have argued that consumers buy products in an attempt to "buy into" a particular lifestyle or "look," that consumers are, in effect, "identity shoppers" (Goss, 1992; Langman, 1994; Gabriel and Lang, 1995). Zygmunt Bauman (1988, p. 808), for example, has described the consumer as someone who is engaged in "self construction by a process of acquiring commodities of distinction and difference"; that is to say, buying an identity. We *interact* with consumer products, as part of what has been called a reflexive project of self, asking questions of ourselves such as "Am I like that? Could that be (part of) me?" (Falk and Campbell, 1997, p. 4). Consumer cultures have, therefore, fostered the idea that identities can be more or less freely chosen, and that people are individually responsible for making themselves the people they are. Feminists have, of course, drawn attention to the highly gendered nature of the marketing processes associated with consumerism, which often propagate deeply masculinist ideals of femininity, and we return to this theme in the next section. However, the point we wish to emphasize here is that conceptualizing identities in these terms is diametrically opposite to the view of identities as essential, natural, persistent, and immutable attributes of places, gender categories, age groups, and so on. Feminist geographers have been equally critical of both points of view, and the argument that places and gender are fluid and mutable therefore needs to be tempered by an appreciation of the "stickiness" of gendered and place-based identities.

In this context, Geraldine Pratt and Susan Hanson (1994, p. 25) have provided a useful analysis of what they call the "geography of placement," in which they show how geography constructs differences, often generating "a stickiness to identity that is grounded in the fact that many women's lives are lived locally." They examined the experiences of women living in four different neighborhoods in the metropolitan area of Worcester, Massachusetts, which vary between city and suburban locations, as well as in terms of income levels, racial composition, occupational histories, and employment opportunities. They found a high degree of localism, as well as racial and class segregation, in women's employment, linked to employers' stereotypical views of potential labor pools. As well as contributing to the maintenance of distinct ethnic communities, the recruitment strategies adopted by employers perpetuated assumptions about household gender relations; for example, through the establishment of "women's shifts," which in some neighborhoods were designed to fit between school hours, and in other neighborhoods meant night hours when children were assumed to be asleep and partners home from day shifts. Thus, in one neighborhood employers construed the "normal" household to be one in which the care of children falls entirely to women except during school hours, whereas in the other they implicitly encouraged men to take responsibility for childcare at night. Pratt and Hanson (1994, p. 19) argued that these local labor market patterns therefore helped to produce "different gender identities [that] congeal around women living in different areas." Thus, while "people may 'travel' globally (for example, when they turn on their televisions or when they go out to eat)," this apparent stretching across space needs to be held in tension with the intensely local character of other aspects of women's lives (Pratt and Hanson, 1994, p. 25). Consequently, while places and gender are mutually constitutive processes that exist in dynamic relationships across space and time, the "sedimenting" of these dynamics as geographical layers produces considerable inertia or "stickiness." Put another way, this perspective recasts attachments between people and places as both powerful and contingent, persistent and mutable.

In summary, in this section we have discussed an influential approach to the concepts of space, place, and gender in which they are theorized as interrelated processes, rather than as fixed entities. Thus, gender relations and gender identities are constructed in and through space and place, and, conversely, space and place construct gender (Bondi and Rose, 2003; also see Mackenzie, 1988). The examples we have chosen illustrate this argument in relation to regions, neighborhoods, labor markets and households. While we have drawn on studies that pay particular attention to women's employment in order to elucidate the interplay between gender, these studies also demonstrate that employment is closely bound up with the reproduction of labor power through domestic work and consumption. We have, however, restricted our attention to studies concerned with quite broad contours of gender. For example, although we have insisted on the multiplicity of femininities and masculinities, the story we have told has not problematized issues of sexuality, but has presumed a predominance of heterosexual households. While this is not inevitable within the approach to space, place, and gender discussed in this section, one of the key limitations of this perspective is its tendency to focus on dominant patterns rather than disruptive possibilities. In the next section, therefore, we turn our attention to an approach that is more concerned with imagining radically different versions.

Reimagining Gendered Space

In *Feminism and Geography*, Gillian Rose (1993) advanced a groundbreaking account of the gendered subject, conceived in terms of its positioning in what she calls "paradoxical space." This concept entails a radical rethinking of gender, space, and the relations between, in a way that helps to explain some of the complexities and contradictions involved in women's diverse experiences, and, importantly, to suggest new imaginative geographies. For Rose (1993, p. 159), paradoxical space "is a space imagined in order to articulate a troubled relation to the hegemonic discourses of masculinism." In other words, paradoxical space is a dynamic and truly different sense of space through which to unsettle and displace key assumptions underlying predominant ways of thinking about and experiencing gender. Above all, paradoxical space opens up possibilities radically different from the traditional and "transparent" spatialities associated with patriarchal accounts of gender as a stable, natural, mutually exclusive binary distinction between "Man" and "Woman." Consequently, this approach examines the ordinary, taken-for-granted spatial operation of the gender inequalities for an explicitly subversive purpose: by disclosing the gendering of dominant concepts, feminist geographers can also engage in a radical rethinking that goes beyond the limits of geographical knowledge.

Rose (1993) argued that dominant and dominating (or hegemonic) conceptualizations of space limit our ability to experience and express difference except in terms of a dualistic distinction between "Man" and "Woman." By excluding all that cannot be pressed into a gender binary, this system represents us as behaving and believing in accordance with long established patriarchal principles, and as experiencing social space exclusively in terms of a simple gender divide.[2] Potentially subversive and emancipatory differences among women (and men) are kept out of the spatial equation, and we are stuck within a restrictive and hierarchical social space. Moreover, the real differences generated by other axes of social identity, such as age, "race," sexuality, and class, are reduced to superficial variations on a theme.

Following accounts of gender developed by feminist theorists such as Marilyn Frye (1983), Teresa de Lauretis (1989), and Iris Marion Young (1990), Rose argued that being defined as a woman is likely to entail feeling confined in and constrained by space. Being a woman means living largely according to a geographical imagination that is masculinist in nature, that privileges and makes room for male subjects to express and impose themselves in and on their environs. In contrast, women "do not often gesture and stride, stretch and push to the limits of our physical capabilities" (Rose, 1993, p. 144). One explanation for this constricted sense of spatiality relates to women's heightened awareness of embodiment, associated with a sense of being the object of other people's (potentially evaluative) gaze that creates and strengthens the notion that space is not our own. Women, on this account, rarely claim or control space but instead are caught and confined by it.

One of the ways in which this theorization of the relationship between gender and space has been used to develop insightful accounts of gendered experience is in feminist analyses of agoraphobia. Agoraphobia is a disorder characterized by intense fear (panic attacks) and avoidance of social situations and spaces (Bankey, 2002; Davidson, 2003), and is suffered primarily by women. Agoraphobics' spatial experience can be understood as an exaggerated example of the restricted, excluded spatiality that Rose (1993) argued is typical of women's lives in hegemonic, masculinist space (also see da Costa Meyer, 1996). In a study conducted by one of the co-authors of this chapter, Joyce Davidson (2000, 2001a) has argued that the panic attacks described by agoraphobics can be understood as "boundary crises" in which sufferers cease to sense themselves as separately and securely bounded from what would "normally" be considered their "surroundings," and which therefore fundamentally threaten their sense of self and ontological security. Typically experienced in social space, these intensely disturbing experiences prompt sufferers to retreat to the perceived safety of their homes, whose walls serve to reinforce their own weakened boundaries and fragile sense of identity.

Agoraphobic respondents frequently referred to consumer spaces – shops, streets, supermarkets, and especially shopping malls – as sites most strongly associated with panic attacks. Reflecting on why shopping spaces present such acute difficulties, sufferers described the variety and intensity of the multisensory stimuli characteristic of these environments, which they argued are liable to confuse and displace any prior sense of calm for phobics and non-phobics alike (Davidson, 2001b). Shopping, especially in the mall, has been described as contradictory, even paradoxical: "it is an experience that yields both pleasure and anxiety, a 'delightful experience' that can quickly become a 'nightmare'" (Falk and Campbell, 1997, p. 12). Consumer spaces are designed to capture our attention and persuade us to buy. As we argued in the preceding section, this persuasion often entails the marketing of identities. Shoppers are, in effect, bombarded with images and messages about how we might "improve" or "enhance" ourselves and our lives, playing on insecurities, actively unsettling our sense of secure boundaries in order to encourage us to "fix" ourselves up with appropriate goods and services (Longhurst, 1998). Moreover, consumer spaces are also strongly gendered. As Nicky Gregson has pointed out, "it is still overwhelmingly women who shop . . . just as much as it is women who form the majority of retail sales workers" (Gregson, 1995, p. 137; also see Lowe and Crewe, 1991). Shopping spaces are thus very much part of women's spatial experience, and marketing strategies often make extensive use of ideas about gender, simultaneously addressing women as consumers and objectifying women's bodies. It is hardly surprising, therefore, that consumer spaces stir up troubling emotions and associations for many women. For women whose sense of self is fragile or impaired, they can be incomparably difficult places to be. Thinking about space as produced by, and saturated with, traditional and often contradictory conceptions of gender within which women are caught and confined sheds new light on women's everyday experiences, whether or not we suffer from agoraphobia.

Returning to Gillian Rose's (1993) account of paradoxical space, she argued that as well as recognizing the powerful effects of hegemonic spatialities, women need to find ways in which to exercise our particular positioning for personal and political advantage. In other words, women must "insist on the possibility of resistance" (Rose, 1993, p. 155). Dominant geographies need to be challenged, and, although women are often caught in oppressive spaces, Rose argued that this capture is only ever partial. Thus, while women are trapped within oppressive, hegemonic spaces, we are at the same time excluded from these spaces on the grounds of illegitimacy, of not being a fully fledged – that is, masculine, "master" – subject. Women are thus simultaneously both "prisoners and exiles": "No wonder," according to Rose (1993, p. 150), "space is so tortuous for so many women." However, this positioning – as simultaneously trapped and excluded – is paradoxical in the sense of invoking an apparently impossible combination of positions. Rose (1993) insisted that this contradictoriness is generative in the sense that some aspects of our identities always exceed the constitutive framing of hegemonic, masculinist space. It is in this excess that the radical, emancipatory potential of paradoxical space lies.

To illustrate the possibility of resisting dominant spatialities, we return to experiences of agoraphobia. As well as highlighting the oppressiveness and constricting effects of dominant understandings and enactments of space, place, and gender, accounts of agoraphobia contain suggestions of radically different, potentially liberating, spatialities. In a chapter co-authored with her sisters, feminist theorist Susan Bordo offered an insightful analysis of her own relatively brief but nevertheless debilitating experience of agoraphobia (Bordo et al., 1998). She described a panic attack in a crowded supermarket, when everything felt alien and frightening to her, "the noise, the crying children, the pushing and shoving" (Bordo et al., 1998, p. 84). As with other agoraphobics, her experience made her long for something solid to grab hold of: "I was stranded on the edge of an enormous [iceberg], rising high out of the sea, perched, precarious, desperate for walls to plant my hands against" (Bordo et al., 1998, p. 80). However, Bordo also explained how, when she was recovering, spaces that had once radically threatened her sense of herself came to enhance her feeling of embodied identity. "Being outside, which when I was agoraphobic had left me feeling substanceless, a medium through which body, breath, and world would rush, squeezing my heart and dotting my vision, now gave me definition, body, focused my gaze" (Bordo et al., 1998, p. 83).

The fragile relationship between self and space associated with panic can thus be transformed into excitement. Like the child who spins in circles, manufacturing dizziness for the delightful disorientation it creates, temporary shifts in "normal" perspective can be liberating. To open oneself up to excitement, to learn to endure and even enjoy the radical otherness of "disorderly" spatial experiences, is to glimpse the potentialities of paradoxical space. As Davidson (2002, pp. 31–2) has noted

> the person who is "ill" will most likely want nothing more than to "fit in", to conform with society's expectations of normality . . . [but] the language used by [agoraphobic] women to express their experience is at times an attempt to transcend the dualisms implied by, and inherent to, our specular symbolic order (Irigaray, 1985). Consequently, it may not always be helpful . . . to (op)press this experience back into dualistic discourse.

Accounts of agoraphobia provide one example of how the idea of paradoxical space has helped feminist geographers to make sense of women's emotional, embodied, spatial experience, and to develop new understandings of the spatial constitution of gender. But it is important to note that agoraphobics are just one of many "groups" confronted with a sense of space as restrictive, ill-fitting, and prejudicial, *and* whose experiences can be understood as calling into question hegemonic versions of spatial subjectivity. Indeed, we are all paradoxically positioned, albeit in many different ways, which are often closely bound up with major axes of differentiation such as age, class, sexuality, and "race," as well as gender, and the means by which we might seek to contest dominant spatialities therefore differ in numerous ways. To illustrate this further we draw on discussions of sexuality, which we use to show how dominant and restrictive spatialities can be challenged and even "breached" by subversive actions, but also how uncertain and contestable the effects of such actions can be.

Feminist geographers have shown that, like agoraphobics, although for different reasons and in different ways, gays and lesbians also often experience feelings of restriction in, and exclusion from, dominant, taken-for-granted, everyday spaces. Following arguments developed by feminist and "queer" theorist Judith Butler

(1990), Gill Valentine (1996, p. 147), for example, has suggested that space is produced as "naturally" heterosexual through repeated, regulated, performative acts, ranging from:

> heterosexual couples kissing and holding hands as they make their way down the street, to advertisements and window displays which present images of contented "nuclear" families; and from heterosexualized conversations that permeate queues at bus stops and banks, to the piped music articulating heterosexual desires that fill shops, bars and restaurants.

Thus, everyday spaces are powerfully infused with traditional, normative versions of sexuality, which impact upon all those who enter them, even if just fleetingly by "passing through." Such spaces, which include many workplaces and home spaces as well as supposed "public" spaces of streets, plazas, and malls, are constructed as normatively heterosexual, and therefore deny and negate the very existence of "other" sexual identities. In addition to the profound and pervasive oppression of gays, lesbians, and others who do not "fit in," and who may, literally, find that they have nowhere to go (Valentine and Skelton, 2002), the denial and negation of other sexualities leaves hegemonic conceptualizations of space undisturbed.

Of course, sexual dissidents have challenged the normative heterosexuality of space in a variety of ways. Perhaps most well known of these strategies is the carving out of alternative gay spaces in the form of gay and/or lesbian residential neighborhoods (see for example Castells, 1983; Lauria and Knopp, 1985; Adler and Brenner, 1992; Peake, 1993; Forest, 1995; Kramer, 1995; Rothenberg, 1995; Valentine, 1997). Such neighborhoods create zones in which heterosexuality is no longer normative. Spaces are produced and performed in ways that accept and celebrate alternative sexual identities, and increase the visibility of alternative sexualities. Notwithstanding the enormous importance of these effects, there is, nevertheless, a risk that these "alternative" spaces become ghettos or enclosures within which "other" sexual identities are contained, leaving the heterosexuality of all other spaces untouched. Indeed, in some instances the very existence of gay and lesbian neighborhoods may intensify the vulnerability of individuals to homophobic violence (Myslik, 1996; Namaste, 1996). Clearly, the difference between "gay and lesbian space" and "straight space" is very far from equal, exemplifying the unequal, hierarchical operation typical of many dualistic concepts, including "man/woman," "white/not white," as well as "straight/not straight."

As Valentine (1996) and others (for example, Davis, 1995; Rothenberg, 1995) have shown, gays and lesbians have struggled against and refused the restrictions of heterosexist social space in other ways that are perhaps more subtly subversive of dominant conceptions of everyday space. For example, by kissing in public, gays and lesbians actively challenge the heterosexuality of space, and begin to "rearticulate the very fabric of that social space" (Probyn, 1995, p. 81). These actions help to ensure that sexual dissidents are seen outside of gay and lesbian "enclosures," and such tactics have been taken further by activists such as the "Lesbian Avengers" (Valentine, 1996, p. 153), who set out to "bend" and to "queer" everyday social space, performing in ways that: "rupture the taken-for-granted heterosexuality of these spaces by disrupting the repetitive performances of the mall and the shopping street as heterosexual places and (re)imagining/(re)producing them as queer sites." By making other (non-hetero-) sexualities both visible and unavoidable, activists undermine the supposedly natural, "neutral," and timeless heterosexuality of social space. Conceived as paradoxical – and, crucially, open to change – gay and lesbian positioning in such space can be seen as positively provocative, simultaneously demonstrating and undoing its presumed heterosexuality.

Gay pride parades and similar "spectacles" can also disrupt the status quo in complex and multiple ways, enacting what Sally Munt (1995, p. 124) has called a "politics of dislocation." In consequence, the views of the "viewing public," together with politicians and policy-makers, are at least challenged, if not straightforwardly changed (Johnston, 2002). Moreover, by "challeng[ing] the production of everyday spaces as heterosexual" (Valentine, 1996, p. 152), pride parades empower participants to (at least partially) break out of and reach beyond the restrictive dualistic framework of hegemonic space: participation can be individually liberating for those who have labored and learned to *revel* in their paradoxical experience of space by stretching its definitions and limits. Gay and lesbian pride parades are thus, at least potentially, transgressive and transformative. "By numerically appropriating the streets (and surrounding transport system, car parks, pubs, parks, shops, McDonalds and so on) and filling them with lesbian and gay meaning for one day, marchers pierce the complacency of heterosexual space" (Valentine, 1996, p. 152).

Pride parades are short-lived, spectacular events. As with the creation of gay and lesbian neighborhoods, their subversive potential can be limited, defused and even co-opted through containment strategies. For example, when they are harnessed as tourist attractions, spectators are often securely separated from participants (for example behind barriers), implicitly ensuring that the former cannot be confused with, or "contaminated" by, the latter (Johnston, 2002). Thus, the paradoxical possibilities of public celebrations of sexual dissidence are not easily or straightforwardly harnessed to emancipatory ends (also see Bell et al., 1994). But, nevertheless, the repeated reimposition of dominant norms does not automatically recapture or undo the unpredictable effects of these temporary "breaches" in hegemonic spatialities. On the contrary, through interventions that exploit vulnerabilities in hegemonic space it is possible to show that "space teems with many other possibilities" (Valentine, 1996, p. 154). The challenge we all face entails using the tension of our contradictory positionings to critique and undermine hegemonic space, and to reveal what lies beyond it, elsewhere. By speaking out about the complex and multiple spaces in which we live, and about the ways in which we experience space differently from each other, all of us can seek to disrupt, rupture, and perhaps partially transform the masculinist and heterosexist status quo.

In concluding this section, it is important to acknowledge that paradoxical space is a very slippery concept that risks reinscribing dualistic frameworks – for example, between "here/within," "beyond/outside," and "women/men" – as it unsettles them (Desbiens, 1999). One aspect of this slipperiness arises because the concept of paradoxical space refers to, and attempts to holds together, spaces of experience and spaces of imagination. Consequently, while we can perhaps glimpse the emancipatory potential of paradoxical space – as we have illustrated in relation to agoraphobia and sexual dissidence – we also have to work, and work hard, to realize its potential.

Conclusion

In the preceding sections we have outlined and illustrated two different ways in which feminist geographers have conceptualized gender, space, and place as inter-related terms. In conclusion we briefly discuss the consequences of these two approaches for the discipline of geography and for feminist politics.

The approach to space, place, and gender described in the section on "sedimenting gender" exemplifies materialist analyses of geographies of gender, class, "race," and other forms of social differentiation. That is, it attends to the material conditions of people's lives, and elaborates how space, place, and gender are interwoven means through which inequalities and oppressions are forged and perpetuated. In so doing, this perspective presents important challenges to the discipline of geography in (at least) two ways. First, it demonstrates the vital importance of attending to gender as a key dimension through which spaces and places are produced, reproduced and transformed. Second, just as it helps to make visible the reality of women's lives within a discipline that has traditionally neglected or misrepresented "half the human in human geography" (Monk and Hanson, 1982, p. 11), so too it helps to illuminate the taken-for-grantedness of dominant conceptualizations of space. More specifically, this approach advocates and develops an understanding of the dynamic, mutual constitution of the spatial and the social, thereby challenging traditional distinctions between places as discrete assemblages of attributes, and space as abstract three-dimensional geometry.

This materialist approach to space, place, and gender offers important resources to feminist politics as well as to the discipline of geography. By attending to the geographical construction of class, gender, and other forms of social differentiation, this approach speaks to a feminist political strategy of building alliances across differences. This is of particular importance in the context of what are sometimes called "horizontal hostilities" among women – conflicts generated by affirmations of distinctive identities articulated in terms of dis-identifications, exclusions, and even, on occasion, hatreds (Fraser, 1995, 2000; Ahmed, 2002). In this context, Geraldine Pratt and Susan Hanson (1994, pp. 25–6) expressed the scope for building alliances in appropriately tentative terms:

> by writing about the ways in which differences are constructed, we hope to allow groups of women, who conceive themselves as different from one another, to gain some mutual understanding; we hope, not for cross-cultural identification, but an informed knowledge of how the conditions of others' lives are shaped by local opportunities and in relation to each other.

However modest such hopes may be, they illustrate the profound importance of feminist geographical analyses for feminist politics.

The approach to space, place, and gender described in the section on "reimagining gendered space" exemplifies the application by feminist geographers of theories of subjectivity, including especially phenomenological and psychoanalytic theories, which in different ways problematize and analyze taken-for-granted features of everyday life (for feminist renditions see, for example, Grosz, 1994; Butler, 1997; Battersby, 1998). Although phenomenology has been drawn upon by geographers over many years (for example Relph, 1976; Tuan, 1977; Pickles, 1985;

Strohmayer, 1998) these uses have largely failed to consider (gendered) embodiment as integral to human existence. In the case of psychoanalytic theories, feminist geography has been a key route through which such ideas have been brought to bear on geographic concerns. Overall, therefore, this approach to space, place, and gender has enriched the discipline of geography by opening up and extending geographical analyses of human subjectivity (Bondi et al., 2002). It has also elaborated and invited subversive ways of working within and against disciplinary conventions by simultaneously questioning the authority of geographic knowledge, and reaching beyond that authority to generate new geographies (Rose, 1996; Bondi, 1997).

The project of reimagining space beyond the limits of a gender binary has important and enabling implications for feminist politics. As we have shown, this approach helps to explain why keenly felt aspects of women's experience are often difficult to represent by bringing into view ways in which what constitutes "knowledge" is limited. By insisting on the productive possibilities of paradoxical space, this approach invites feminist interventions at numerous sites, whether construed as central or marginal. In this context it is important to stress that the two approaches we have discussed are complementary and overlapping. By challenging the constraints of binary thinking, paradoxical space is an approach that seeks to disrupt the opposition between similarity and difference that underpins "horizontal hostilities" among women. This perspective seeks to dislodge the idea that identities (or places) are ever coherent or stable enough to provide a basis for collective action, and invites instead a creative engagement with the uncertainties, fractures, and differences that are integral aspects of all of us.

In this chapter we have introduced and illustrated two of the ways in which feminist geographers have conceptualized space, place, and gender. In so doing, we have shown how feminist geography has unleashed new creative energy for thinking about the world we inhabit. The remaining chapters in this volume elaborate some of the rich possibilities that ensue.

NOTES

1 In this chapter we focus on feminist geography as it has developed in the anglophone literature, primarily within Anglo-American contexts. There are important issues to consider about the multiple genealogies of related concepts across languages and contexts, but these lie beyond the scope of this chapter.
2 Our discussion slips between the third person ("women" as "they") and the first person ("women" as "us") in a way that reflects our argument about the contradictory positions invoked by the idea of paradoxical space. While we belong to the broad category of "women" about whom we write, we also take up a position apart from that category as we question its limits and meanings.

BIBLIOGRAPHY

Adler, S. and Brenner, J. (1992) Gender and space: lesbians and gay men in the city. *International Journal of Urban and Regional Research*, 16, 24–34.
Ahmed, S. (2002) Affective economies. Paper presented to the Emotional Geographies Conference, Lancaster University, September.

Ainley, R. (ed.) (1998) *New Frontiers of Space, Bodies, Gender*. London: Routledge.

Bankey, R. (2002) Embodying agoraphobia: rethinking geographies of women's fear. In L. Bondi, H. Avis, A. Bingley, J. Davidson, R. Duffy, V. I. Einagel, A.-M. Green, L. Johnston, S. Lilley, C. Listerborn, M. Marshy, S. McEwan, N. O'Connor, G. Rose, B. Vivat and N. Wood, *Subjectivities, Knowledges, and Feminist Geographies*. Lanham, MD: Rowman and Littlefield.

Battersby, C. (1998) *The Phenomenal Woman*. Cambridge: Polity Press.

Bauman, Z. (1988) Sociology and postmodernity. *Sociological Review*, 36, 790–813.

Bell, D., Binnie, J., Cream, J. and Valentine, G. (1994) All hyped up and nowhere to go. *Gender, Place and Culture: A Journal of Feminist Geography*, 1, 31–47.

Belt, V., Richardson, R. and Webster, J. (2002) Women, skill and interactive service work in telephone call centres. *New Technology, Work and Employment*, 17, 20–34.

Berry, B. J. L. and Marble, D. F. (eds) (1968) *Spatial Analysis*. Englewood Cliffs, NJ: Prentice Hall.

Bondi, L. (1997) In whose words? On gender identities and writing practices. *Transactions of the Institute of British Geographers*, 22, 245–58.

Bondi, L. and Rose, D. (2003) Constructing gender, constructing the urban: a review of Anglo-American feminist urban geography. *Gender, Place and Culture*, 10, 241–57.

Bondi, L., Avis, H., Bingley, A., Davidson, J., Duffy, R., Einagel, V. I., Green, A.-M., Johnston, L., Lilley, S., Listerborn, C., Marshy, M., McEwan, S., O'Connor, N., Rose, G., Vivat, B. and Wood, N. (2002) *Subjectivities, Knowledges, and Feminist Geographies*, Lanham, MD: Rowman and Littlefield.

Bordo, S., Klein B. and Silverman, M. K. (1998) Missing kitchens. In H. Nast and S. Pile (eds), *Places through the Body*. London: Routledge.

Burgess, J. (1998) "But is it worth taking the risk?" How women negotiate access to urban woodland: a case study. In R. Ainley (ed.), *New Frontiers of Spaces, Bodies and Gender*. London: Routledge.

Butler, J. (1990) *Gender Trouble*. London: Routledge.

Butler, J. (1997) *The Psychic Life of Power*. Stanford, CA: Stanford University Press.

Campbell, B. (1993) *Goliath: Britain's Dangerous Places*. London: Methuen.

Castells, M. (1983) *The City and the Grassroots*. Berkeley: California University Press.

da Costa Meyer, E. (1996) La donna è mobile: Agoraphobia, woman and urban space. In D. Agrest, P. Conway and L. Weisman (eds), *The Sex of Architecture*. New York: Abrams.

Davidson, J. (2000) ". . . the world was getting smaller": women, agoraphobia and bodily boundaries. *Area*, 32, 31–40.

Davidson, J. (2001a) Pregnant pauses: agoraphobic embodiment and the limits of (im) pregnability. *Gender, Place and Culture*, 8, 283–97.

Davidson, J. (2001b) Fear and trembling in the mall: women, agoraphobia and body boundaries. In I. Dyck, N. D. Lewis and S. McLafferty (eds), *Geographies of Women's Health*. London: Routledge.

Davidson, J. (2002) All in the mind? Women, agoraphobia and the subject of self-help. In L. Bondi, H. Avis, A. Bingley, J. Davidson, R. Duffy, V. I. Einagel, A.-M. Green, L. Johnston, S. Lilley, C. Listerborn, M. Marshy, S. McEwan, N. O'Connor, G. Rose, B. Vivat and N. Wood, *Subjectivities, Knowledges, and Feminist Geographies*. Lanham, MD: Rowman and Littlefield.

Davidson, J. (2003) *Phobic Geographies: The Phenomenology and Spatiality of Identity*. Aldershot: Ashgate.

Davis, T. (1995) The diversity of queer politics and the redefinition of sexual identity and community in urban spaces. In D. Bell and G. Valentine (eds), *Mapping Desire*. London: Routledge.

de Lauretis, T. (1989) *Technologies of Gender*. Basingstoke: Macmillan.

Desbiens, C. (1999) Feminism "in" geography. Elsewhere, beyond and the politics of para-doxical space. *Gender, Place and Culture*, 6, 179–85.

Duncan, S. and Smith, D. (2002) Geographies of family formations: spatial differences and gender cultures in Britain. *Transactions of the Institute of British Geographers*, 27, 471–93.

Eyles, J. (1985) *Senses of Place*. Warrington: Silverbrook.

Falk, P. and Campbell, C. (eds) (1997) *The Shopping Experience*. London: Sage.

Forest, B. (1995) West Hollywood as symbol: the significance of place in the construction of a gay identity. *Environment and Planning D: Society and Space*, 13, 133–57.

Fraser, N. (1995) From redistribution to recognition? Dilemmas of justice in a "post-socialist" age. *New Left Review*, 212, 68–93.

Fraser, N. (2000) Rethinking recognition. *New Left Review*, 3, 107–20.

Frye, M. (1983) *The Politics of Reality*. New York: Crossing Press.

Gabriel, Y. and Lang, T. (1995) *The Unmanageable Consumer: Contemporary Consumption and Its Fragmentations*. London: Sage.

Glucksman, M. (2000) *Cottons and Casuals: The Gendered Organisation of Labour in Time and Space*. Durham: Sociology Press.

Goss, J. (1992) Modernity and post-modernity in the retail landscape. In K. Anderson and F. Gale (eds), *Inventing Places: Studies in Cultural Geography*. Harlow: Longman.

Gregson, N. (1995) And now it's all consumption? *Progress in Human Geography*, 19, 135–41.

Grosz, E. (1994) *Volatile Bodies*. Bloomington: Indiana University Press.

Hall, C. (1982) The home turned upside down? The working class family in cotton textiles 1780–1850. In E. Whitelegg, M. Arnot, E. Bartels, V. Beechey, L. Birke, S. Himmelwit, D. Leonard, S. Ruehl and M. A. Speakman (eds), *The Changing Experience of Women*. Oxford: Basil Blackwell in Association with the Open University.

Harvey, D. (1969) *Explanation in Geography*. London: Edward Arnold.

Hayford, A. (1974) The geography of women: an historical introduction. *Antipode: A Radical Journal of Geography*, 6, 1–19.

Irigarary, L. (1985) *Speculum of the Other Woman*. Ithaca, NY: Cornell University Press.

John, A. V. (1980) *By the Sweat of Their Brow: Women Workers at Victorian Coal Mines*. London: Croom Helm.

Johnston, L. (2002) Borderline bodies. In L. Bondi, H. Avis, A. Bingley, J. Davidson, R. Duffy, V. I. Einagel, A.-M. Green, L. Johnston, S. Lilley, C. Listerborn, M. Marshy, S. McEwan, N. O'Connor, G. Rose, B. Vivat and N. Wood, *Subjectivities, Knowledges, and Feminist Geographies*. Lanham, MD: Rowman and Littlefield.

Johnston, L. and Valentine, G. (1995) Wherever I lay me girlfriend, that's my home: the per-formance and surveillance of lesbian identities in domestic environments. In D. Bell and G. Valentine (eds), *Mapping Desire*. London: Routledge.

Kramer, J. L. (1995) Bachelor farmers and spinsters: gay and lesbian identities and commu-nities in rural North Dakota. In D. Bell and G. Valentine (eds), *Mapping Desire*. London: Routledge.

Langman, L. (1994) Neon cages: shopping for subjectivity. In R. Shields (ed.), *Lifestyle Shop-ping: The Subject of Consumption*. London: Routledge.

Lauria, M. and Knopp, L. (1985) Towards an analysis of the role of gay communities in the urban renaissance. *Urban Geography*, 6, 152–69.

Laurie, N., Dwyer, C., Holloway, S. and Smith, F. (1999) *New Geographies of New Femi-ninities*. Harlow: Pearson.

Lewis, J. (1984) The role of female employment in the industrial restructuring and regional development of the UK. *Antipode: A Radical Journal of Geography*, 16, 47–60.

Liddington, J. (1979) Women cotton workers and the suffrage campaign: the radical suffragists in Lancashire 1893–1914. In S. Burman (ed.), *Fit Work for Women*. London: Croom Helm.

Longhurst, R. (1998) (Re)presenting shopping centres and bodies: questions of pregnancy. In R. Ainley (ed.), *New Frontiers of Space, Bodies, Gender*. London: Routledge.

Longhurst, R. (2000) Geography: masculinities, male identity and men. *Progress in Human Geography*, 24, 439–44.

Longhurst, R. (2001) *Bodies: Exploring Fluid Boundaries*. London: Routledge.

Lowe, M. and Crewe, L. (1991) Lollipop jobs for pin money? Retail employment explored. *Area*, 23, 344–7.

McDowell, L. (1983) Towards an understanding of the gender division of urban space. *Environment and Planning D: Society and Space*, 1, 59–72.

McDowell, L. (1992) Doing gender: feminism, feminists and research methods in human geography. *Transactions, Institute of British Geographers*, 17, 399–416.

McDowell, L. (2001) The trouble with men? Young people, gender transformations and the crisis of masculinity. *International Journal of Urban and Regional Research*, 24, 201–9.

McDowell, L. (2003) The particularities of place: geographies of gendered moral responsibilities among Latvian migrant workers in 1950s Britain. *Transactions of the Institute of British Geographers*, 28, 19–34.

McDowell, L. and Massey, D. (1984) A woman's place? In D. Massey and J. Allen (eds), *Geography Matters*. Cambridge: Cambridge University Press in Association with the Open University.

Mackenzie, S. (1988) Building women, building cities: toward gender sensitive theory in the environmental disciplines. In C. Andrew and B. Moore Milroy (eds), *Life Spaces*. Vancouver: University of British Columbia Press.

Massey, D. (1984) *Spatial Divisions of Labour*. Basingstoke: Macmillan.

Massey, D. (1992) Politics and space/time. *New Left Review*, 196, 65–84.

Massey, D. (1994) *Space, Place and Gender*. Cambridge: Polity Press.

Massey, D. and Allen, J. (eds) (1984) *Geography Matters*. Cambridge: Cambridge University Press in Association with the Open University.

Monk, J. and Hanson, S. (1982) On not excluding half the human in human geography. *Professional Geographer*, 34, 11–23.

Munt, S. (1995) The lesbian flâneur. In D. Bell and G. Valentine (eds), *Mapping Desire*. London: Routledge.

Myslik, W. (1996) Renegotiating the social/sexual identities of places. Gay communities as safe havens or sites of resistance. In N. Duncan (ed.), *BodySpace: Destabilizing Geographies of Gender and Sexuality*. London: Routledge.

Nairn, K. (1997) Hearing from quiet students: the politics of silence and voice in geography classrooms. In J. P. Jones III, H. J. Nast and S. M. Roberts (eds), *Thresholds in Feminist Geography*. Lanham, MD: Rowman and Littlefield.

Namaste, K. (1996) Genderbashing: sexuality, gender and the regulation of public space. *Environment and Planning D: Society and Space*, 14, 221–40.

Peake, L. (1993) Race and sexuality: challenging the patriarchal structuring of urban social space. *Environment and Planning D: Society and Space*, 11, 415–32.

Pickles, J. (1985) *Phenomenology, Science and Geography*. Cambridge: Cambridge University Press.

Pratt, G. and Hanson, S. (1994) Geography and the construction of difference. *Gender, Place and Culture*, 1, 5–29.

Probyn, E. (1995) Lesbians in space: Gender, sex and the structure of missing. *Gender, Place and Culture*, 2, 77–84.

Relph, E. (1976) *Place and Placelessness*. London: Pion.

Rose, G. (1993) *Feminism and Geography*. Cambridge: Polity Press.

Rose, G. (1996) As if the mirrors had bled. Masculine dwelling, masculinist theory and feminist masquerade. In N. Duncan (ed.), *BodySpace: Destabilizing Geographies of Gender and Sexuality*. London: Routledge.

Rothenberg, T. (1995) "And she told two friends": lesbians creating urban social space. In D. Bell and G. Valentine (eds), *Mapping Desire*. London: Routledge.

Strohmayer, U. (1998) The event of space: geographic allusions in the phenomenological tradition. *Environment and Planning D: Society and Space*, 16, 105–22.

Tivers, J. (1978) How the other half lives: the geographical study of women. *Area*, 10, 302–6.

Tuan, Y.-F. (1977) *Space and Place*. Minneapolis, Minnesota University Press.

Valentine, G. (1993) (Hetero)sexing space: lesbian perceptions and experiences of everyday spaces. *Environment and Planning D: Society and Space*, 11, 395–413.

Valentine, G. (1996) (Re)negotiating the heterosexual street. In N. Duncan (ed.), *BodySpace: Destabilizing Geographies of Gender and Sexuality*. London: Routledge.

Valentine, G. (1997) Making space: separatism and difference. In J. P. Jones III, H. J. Nast and S. M. Roberts (eds), *Thresholds in Feminist Geography*. Lanham, MD: Rowman and Littlefield.

Valentine, G. and Skelton, T. (2002) "I just felt different though for a long time I didn't understand why": emotional geographies of coming out as lesbian or gay. Paper presented to the Emotional Geographies Conference, Lancaster University, September.

Wheelock, J. (1990) *Husbands at Home: The Domestic Economy in a Post-industrial Society*. London: Routledge.

Women and Geography Study Group (1984) *Geography and Gender*. London: Hutchinson.

Young, I. M. (1990) *Justice and the Politics of Difference*. Princeton, NJ: Princeton University Press.

Chapter 3

Anti-racist Feminism in Geography: An Agenda for Social Action

Audrey Kobayashi

We act from within the social relations and subject positions we seek to change. (Frankenberg, 1993, p. 5)

Setting the Frame: Anti-racist Feminist Geography in Historical Context

For at least the past decade, critical feminist geography has operated on the principle that gendered constructions of "race" are fundamental to social life as we know it. Indeed, theorizing about the mutually constitutive qualities of "race" and gender, in addition to a much older tradition of class analysis, has provided much of the zing to our discipline. Poststructuralist accounts of social construction have provided a conceptual paradigm that, while constantly developed and embroidered, is seldom if ever now refuted by feminist or anti-racist scholars. As Kay Anderson (2002, p. 25) suggests, the "most useful insight of the constructivist perspectives on race . . . was to position race within a *social* (as opposed to *natural*) area of contestation" (italics original). This statement applies more broadly to the discipline as a whole, where the shift from the natural to the social is all the more profound given the historical roots of our discipline sunk deeply in naturalist soil.

It was not always so. Until the early 1990s, by which time feminist scholarship was well established in geography, the connections between gender and "race" were almost entirely unexplored by human geographers. One of the earliest references is in the introduction to feminist geography compiled by the Women and Geography Study Group of the Institute of British Geographers (1984, pp. 25, 101, 145), which signals the fact that in other disciplines there was at the time emerging attention to the intersections of "race," class, and gender, most notably by such pioneer anti-racist feminists as Collette Guillaumin (1995) and Angela Davis (1981) (for a review see Collins, 1990). Over the next decade, it became common for feminist geographers to recognize the significance of "race" in understanding both the geography

of every life (Massey, 1994, p. 148) and the theoretical emergence of modern feminism – and its conflicts – as disciplined by whiteness (Rose, 1993, p. 13). Fundamental to the inscription of a mindset that links the processes of gender and
racialization was the recognition that analyses of both are predicated upon seeing
human relations as the construction of *difference* (Bondi, 1990) and resulting *differentials* of power, prestige, identity, wealth, and other markers of human life.
Bondi's (1990) review, however, while it provides a considerable list of works that
address women's differences at an international scale, identifies only two contributions by feminist geographers in which racial difference is the dominant theme
(Petras, 1989; Kobayashi, 1994b).[1]

For geographers in general a decade ago, a critical development was the extension of poststructuralist theories to recognize the crucial role of spatial mediation
in the formation of social difference, identity, and political struggles. For Keith and
Pile (1993, p. 22) the project of placing political struggles depended upon place-
specific processes of racialization, class differentiation, and multiple positionalities.
The emergent fields of diaspora studies, and the emphasis on social memory and
imagination as the basis for differentiated, and dislocating, life experiences, cried
out for a spatial analysis (Hesse, 1993), building upon a growing body of literature
in which geographers were making a significant contribution to understanding the
lived realities of racialization (Jackson, 1987; for reviews see Bonnett and Nayak,
2003; Kobayashi, 2003).

By the mid-1990s, there emerged the first set of geographical writings in which
the intersection of "race" and gender was the focus rather than a topic mentioned
in passing, as feminist geographers began to publish the results of empirical work
that took such questions seriously. The methodological basis for gendering anti-
racist geography was established at this time. Sanders (1990) called for an "integration" of gender and "race," considering the double jeopardy of racist and
gendered marginalization. Kobayashi and Peake (1994) provided a basis for
activist/advocacy-informed research, calling for a form of "unnatural discourse"
that would both take seriously the theoretical contribution of anti-essentialist, but
place-based, research and provide a means of disrupting the whiteness of both the
academy and the streets. Linda Peake (1993) provided a template for understanding the ways in which urban space is simultaneously raced and sexualized, while
Kobayashi (1994b) presented the paradox of identity politics and reflexivity as a
basis for deconstructing the essentialized category of "race."

Publications that resulted from a workshop at the University of Kentucky in
January 1995 (Jones et al., 1997) highlight several in-depth studies of the geographical impact of gendered racialization. Laura Pulido uses a case study of low
income women of color in environmental justice movements to show that the
concept of "woman" needs to be understood as racialized and simultaneously *placed*
within specific community concepts. She also challenges, however, the act of *placing*
on the part of the academic: "how do we represent those who have been historically invisible, especially in light of geography's legacy of colonization? . . . without
imposing a form of 'epistemic violence' on the subject of our research" (Pulido,
1997, pp. 12–13; quoting Jackson, 1993, p. 208). Pulido's work is important not
only because it is part of the first significant wave of empirical work by geographers
that actually takes seriously the lives of women of color (see Kodras and Jones,

1991), but because it recognizes the role of the researcher in *situating* both herself and her subjects, thus constituting a relationship through which the identity of both is transformed (see also Pulido, 1996, 2000). Lamenting the lack of attention to racialized women by feminist geographers, she asks:

> why do we persist in such presentations? Even more fundamentally, what are the implications of this practice for feminist geography as a whole? Who is our subject and what are our goals? On what bases are various subjects incorporated into the feminist research agenda? And what happens when they resist? (Pulido, 1997, p. 24)

I shall return below to some of the after effects of this reflexive turn.

In a similar vein, Melissa Gilbert links naturalized concepts of "race," gender, and space in the academic and popular imaginations in her examination of debates over American poverty. She too engages in detailed empirical research among Black women living in poverty, emphasizing the spatiality of their lives (as opposed to the essentialist concept of lives within a spatial container). Her analysis is enriched by taking to task the inadequacy of essentialist explanatory frameworks. Academics have a significant social responsibility because what is at stake is not simply their intellectual understanding of the world, but the very construction of the world in which people are required to live their lives. Essentialist works (and by this she also includes works that fail to problematize the concepts, such as "race," class, and gender, used as variables to describe human beings) at worst contribute to conditions of racialized poverty, and at best obscure the "connections that could form the basis for political action" (Gilbert, 1997, p. 33; see also Gilbert, 1993, 1998).

I wish to use this forum to celebrate these contributions from Pulido and Gilbert, for they provide a frame – in a book most appropriately entitled *Thresholds* – for an anti-racist and feminist geography. To my mind, such a geography has three essential elements. First, it is directly engaged with the world, taking seriously the lives of racialized subjects. Second, it recognizes not only the positionality, the epistemic presence, of the researcher, but also the responsibility of the researcher to make a difference. Third, it takes seriously the issue of representation as a discourse that (potentially at least) alters material reality. In the above historical narrative, the difference between the feminist geographies of the 1980s and the 1990s is that while earlier works made only passing recognition of the *theoretical* significance of recognizing the simultaneous construction of "race," class, and gender, in effect gazing out from a white academy as though through an ideological window, the work that began in the 1990s takes on the challenge of engaged and activist research. After all, what good does it do to recognize the construction of difference if you do not thereby make a difference? In the balance of this chapter, I shall expand upon these three challenges.

Anti-racist Feminist Engagement: Activism, Activism, Activism

For Ruth Gilmore, racism is a "practice of abstraction" that

> produces all kinds of fetishes: states, races, normative views of how people fit into and make places in the world. A geographical imperative lies at the heart of every struggle

for social justice; if justice is embodied, it is then therefore always spatial, which is to say part of a process of making a place. (Gilmore, 2002, p. 16)

If it seems unusual that the concept of abstraction should be introduced to describe a process whose results are anything but abstract for those who live them, Gilmore's point is that such abstractions are both powerful and fatal. So much for any lingering thoughts that social constructions lie *only* in the imagination! The role of the academic activist is therefore twofold: to take on the abstractions both on the ground and on paper, to "change aspects of both the forces and the relations of knowledge production in order to produce new and useful knowledges" (Gilmore, 2002, p. 22). For Gilmore that change involves a dizzyingly complex analysis that moves from the global to the local (see Gilmore, 1999).

A key concept here is "useful knowledges." The knowledges that constructed and fetishized racism and gender as the dominant conditions of social life have been tremendously useful to those for whom the power to invent, abstract, and impose normative conditions have remained relatively unchecked for centuries. If in fact we are, as I believe, in the midst of a major shift of that power, then of course the challenge is to identify which forms of knowledge are most useful, as inscribed through and upon human bodies, in the streets and in the pages of geographical journals. But such knowledges cannot be useful if imposed upon the world without the direct involvement of the subjects about whom we profess knowledge.

Taking seriously the lives of our subjects is no simple task, and much has been written about how to do so (e.g. Katz, 1992; Limb and Dwyer, 2001). The challenge is to go beyond the methodological strategies of giving respect and of negotiating the relative effects of ethnographic difference and positionality, to a recognition of the transformative nature of the research process. Too often, issues of reflexivity become fixed upon the experiences of the researcher and upon the process through which the researcher is herself transformed, which can have the ironic effect of creating a kind of double abstraction, multiplying the privilege of the researcher with the privilege to be self-reflective (Kobayashi, 2004).

To cite an example, I recently became involved in a project on neighborhood development in Old Havana, Cuba.[2] Old Havana is a vibrant but very poor neighborhood, predominantly inhabited by descendants of plantation slaves. There is a high degree of revolutionary spirit, in which traditional elements of western and African religions, racialized identities, and *machista* gender relations are accommodated and constantly reinscribed across a socialist political economy. As in any society, the contradictions are rife, but accommodated and compromised in everyday discourse. For two Northern researchers trained to have a sensitive eye for the ways in which difference is socially constructed, however, these gendered and racialized dimensions of life seemed at first strikingly evident. But the women with whom we are working made it very clear from the start that they were not interested in having us there simply to observe and try to "understand" them. They have nothing but disdain for a relativist position in which the researcher tries to set aside her own values and judgments. They agreed to work with us only on the condition that we join *their* project, of transforming life on their terms. We cannot be inhabitants of Old Havana (and for that reason I have always found the term "participant observation" a misnomer); but we can be part of the project.

Being "part of the project" is of course an old theme among feminist geographers. During the 1980s we engaged in a very significant debate over who was in or out of the feminist circle, depending on one's analytical position with respect to "race" and class. We have long transcended that debate, largely because social constructionist approaches have established the contingent possibility of multiple, sometimes competing, constitutive elements in modern social relations. Attention to processes of racialization was a key element in that transcendence. But for all our analytical acuity and, indeed, professed commitment to social change, feminist geography remains predominantly white (Kobayashi 2002; Pulido, 2002). This fact does not in itself mean that progress cannot be made towards anti-racist feminist geography, but it does signal a significant gap between recognizing difference and making a difference within the social world of geography, and it contributes to what Pulido (2002, p. 46) describes as the "marginalized and fragmented" approach to racism that occurs in feminist geography. We have some way to go before the lives of women of color are seriously engaged in the academic world. Echoing Pulido's and Gilmore's questions, what are the abstractions that continually reinscibe the face of whiteness upon our disciplinary gaze? As geographers, we live our own abstractions.

Positionality and the Reflexive Turn: A Double Bind

There is no doubt that recognition of the positionality of the researcher is one of the most "useful" analytical concepts of the past decade or so. This recognition echoes the work of feminist standpoint theorists, for whom a standpoint is:

> a hard-won product of consciousness-raising and social-political engagement that exposes the false presuppositions upon which patterns of domination and subordination are built and sustained. Standpoint theorists contend that the minute detailed, strategic knowledge that the oppressed have had to acquire of the workings of the social order just so as to be able to function within it can be brought to serve as a resource for undermining that very order. (Code, 1998, p. 180)

Positionality is thus more than simply the space that one occupies: it implies a recognition that such occupation is active, engaged, and contested. It makes a difference that I am a woman of color, in part because as such I am assigned a particular place in society, but more significantly because I choose that identity as a focus for anti-racist struggle, and I do so with the full recognition of how I might thus participate in transforming the social field of which I am a part. Such recognition depends strongly upon an understanding of the thoroughly geographical concept of *situation* (see Haraway, 1991). Situational thinking:

> owes a debt to Foucault's thinking about "local knowledges." It resists homogenizing under one unified model the people, artefacts, material objects, or events that become the "objects" of knowledge, to work by analogy (and disanalogy) from one knowledge-making situation to another. (Code, 1998, p. 182)

To recognize the mutually transformative nature of one's positionality in relation to a particular material situation does not in itself, however, set the terms of that situation. The current emphasis on reflexivity among feminists is a good example of this problem (Kobayashi, 2004). No matter how difficult the struggle to reach a particular position, to occupy a particular space, the power to situate always represents a form of privileging. The very act, therefore, of casting the spotlight upon the academic "I" reinforces her position and makes more difficult the transition from one knowledge-making situation to another. There is a tendency, in other words, to hold on to standpoints once they are established, even if to do so goes against the concept of transformative situation. Much as we have learned, therefore, from the concept of positionality, with all its rich appeal to geographical metaphors, it is social transformation, not reflexivity, that holds the key to overcoming racism and lessening the strength of whiteness that occupies the center ground of feminism.

Shifting the Discourse

Finally, I should like to take up the question of how we might shift the discourse of whiteness, both within the discipline and in society more generally. In a series of recent articles that link the concept of human "race" with the concept of the "beast," Kay Anderson (1998, 2000, 2002) holds that the power of the race idea owes much to the construction of the animal-as-nature as a reference point for civilization and, by extension, whiteness. She contends that exposing this connection entails not only "enlarging the epistemic field within which cultures of race are conceived to have evolved," but also "erod[ing] the conceits of a transcendent, unmarked white self" (Anderson, 2002, p. 29). In other words, we need to engage in discourse that disrupts the discursive fields in which race is continually reconstructed. I would call this process a form of "unnatural discourse" (Kobayashi and Peake, 1994).

The challenge to anti-racist feminists today is to gain more skill, both analytically and socially, at disruptive discourse. To do so, we need to ask not only the questions posed in the above sections about who we are and how we might engage in transformative social action, but how we might influence the process through which racial and gendered imaginations are stimulated. Such questions push us beyond familiar territory, to engage that which is thought but unarticulated, spoken but publicly unspeakable, to lay bare not only the public architecture of social construction, but to approach the thought processes of its architects. It means to challenge whiteness not only as a form of othering and exclusion, but also as a form of inclusion, as an emotional safety zone, a set of cultural practices that are not only normative, but through which the appeal of normativity is constituted.

Heidi Nast (2000) posits eight "mappings" through which the black – and especially the black male – body gains social embodiment. She employs an "excavation of imaginary-symbolic stories connecting the white maternal to the black son," contending that in so doing she uncovers some of the "unconscious" meanings of "race" that exist "between the lines" of landscape, thus exposing "the disavowing racist force of the oedipal drama . . . [making] each body a place of geographical scrutiny" (Nast, 2000, p. 243). Nast's approach runs the significant risk that her own imagination may prove more powerful than the repository of imaginary-symbolic images from which her maps are drawn. Any concept of the unconscious, moreover, runs the risk of being essentialized as an area of precognitive knowledge, upon whose origins we can only speculate. These caveats aside, what is most striking about Nast's work is her attempt to reach that area of human activity in which the products of human imagination are constructed through telling stories that evoke not only familiar images, but familiar and self-validating emotions and desires. To understand the construction of "race," gender, or any other process that gives significant meaning to the world, it is necessary to understand the process of meaning making itself. And if we are to influence that meaning, then we need to tell the geographic counter-stories that will do much more than explain the world, but will engage sufficient imagination to change it.

The Challenges Before Us

I have identified three ways in which anti-racist feminist geographers during the past decade have advanced an agenda for overcoming racialization both within the discipline and in the world at large. They have dedicated themselves to engaged, activist research that takes seriously the lives of the women with whom they work. They have recognized the importance, but also the pitfalls, of positionality as a starting point for social transformation. And they have engaged the shady and ineffable realm of geographical imagination to uncover the discursive field in which racism is constructed and, potentially, to shift the human imagination beyond its racialized boundaries. All three endeavors are works in progress that deserve much more discussion than I have been able to give them here.

The current agenda calls for continued activism, both to engage with the lives of people of color and to change the face of the discipline. At no time in history has this agenda been more important, for in the ideological turmoil of the post-September 11 world, the discursive field has become much more fraught, violence has once again become an easier and more acceptable solution to combat those who disrupt normative practice, and fear makes it all too easy to reimagine difference as a threat to cultural stability and safety. Anti-racist feminists hold much potential to counter the tendencies of our times.

NOTES

1 The Kobayashi article was forthcoming in 1990.
2 C. Krull and A. Kobayashi, Shared memories, common visions: neighbourhood, generation and social organization among women in Havana, Cuba. Social Sciences and Humanities Research Council of Canada Grant Number 410-2003-0742.

BIBLIOGRAPHY

Anderson, K. (1998) Science, the savage, and the Linnean Society of New South Wales. *Ecumene*, 5, 125–43.

Anderson, K. (2000) The beast within: race, humanity, and animality. *Environment and Planning D: Society and Space*, 18, 301–20.

Anderson, K. (2002) The racialization of difference: enlarging the story-field. *Professional Geographer*, 54(1), 25–30.

Andersen, M. L. and Collins, P. H. (1995) *Race, Class, and Gender*. Belmont, CA: Wadsworth.

Bondi, L. (1990) Progress in geography and gender: feminism and difference. *Progress in Human Geography*, 14, 438–45.

Bonnett, A. (2000) *White Identities: Historical and International Perspectives*. London: Routledge.

Bonnett, A. and Nayak, A. (2003) Cultural geographies of racialization – the territory of race. In K. Anderson, M. Domosh, S. Pile and N. Thrift (eds), *Handbook of Cultural Geography*. Thousand Oaks, CA: Sage.

Code, L. (1998) Epistemology. In A. M. Jaggar and I. M. Young (eds), *A Companion to Feminist Philosophy*. Oxford: Blackwell.

Collins, P. H. (1990) *Black Feminist Thought: Knowledge, Consciousness, and the Politics of Empowerment*. London: Routledge.

Davis, A. (1981) *Women, Race and Class*. New York: Random House.

Domosh, M. (1997) Geography and gender: the personal and the political. *Progress in Human Geography*, 21, 81–7.

Eisenstein, Z. R. (1994) *The Color of Gender: Reimaging Democracy*. Berkeley: University of California Press.

Frankenberg, R. (1993) *White Women, Race Matters: The Social Construction of Whiteness*. Minneapolis: University of Minnesota Press.

Frankenberg, R. (1997a) *Displacing Whiteness*. Durham, NC: Duke University Press.

Frankenberg, R. (1997b) Growing up white: feminism, racism and the social geography of childhood. In L. McDowell and J. P. Sharp (eds), *Space, Gender, Knowledge: Feminist Readings*. London: Arnold.

Gilbert, M. R. (1993) Ties to people, bonds to place: The urban geography of low-income women's survival strategies. PhD dissertation, Graduate School of Geography, Clark University, Worcester, MA.

Gilbert, M. R. (1997) Identity, space and politics: A critique of the poverty debates. In J. P. Jones III, H. J. Nast and S. M. Roberts (eds), *Thresholds in Feminist Geography*. Lanham, MD: Rowman and Littlefield.

Gilbert, M. R. (1998) "Race," space, and power: the survival strategies of working poor women. *Annals of the Association of American Geographers*, 88, 595–621.

Gilmore, R. W. (1999) "You have dislodged a boulder": mothers and prisoners in the post-Keynesian California landscape. *Transforming Anthropology*, 8(1/2), 12–38.

Gilmore, R. W. (2002) Fatal couplings of power and difference: notes on racism and geography. *Professional Geographer*, 54(1), 15–24.

Guillaumin, C. (1995) *Racism, Sexism, Power and Ideology*. London: Routledge.

Haraway, D. (1991) A manifesto for cyborgs: science, technology and socialist feminism for the 1980s. In D. Haraway (ed.), *Simians, Cybords, and Women: The Reinvention of Nature*. New York: Routledge.

Hesse, B. (1993) Black to front and black again: racialization through contested times. In M. Keith and S. Pile (eds), *Place and the Politics of Identity*. London: Routledge.

Jackson, P. (1987) *Race and Racism: Essays in Social Geography*. London: Allen and Unwin.

Jackson, P. (1991) The cultural policies of masculinity: towards a social geography. *Transactions of the Institute of British Geographers*, 16, 199–213.

Jackson, P. (1993) Changing ourselves: a geography of position. In R. J. Johnston (ed.), *The Challenge of Geography: A Changing World, A Changing Discipline*. Oxford: Blackwell.

Johnson, L. C. (1994a) Occupying the suburban frontier: accommodating difference on Melbourne's urban fringe. In A. Blunt and G. Rose (eds), *Writing Women and Space: Colonial and Postcolonial Geographies*. New York: Guilford Press.

Johnson, L. C. (1994b) Colonising the suburban frontier: place-making on Melbourne's urban fringe. In S. Watson and K. Gibson (eds), *Metropolis Now: Planning and the Urban in Contemporary Australia*. Sydney: Pluto Press.

Jones, J. P., Nast, H. J. and Roberts, S. M. (eds) (1997) *Thresholds in Feminist Geography: Difference, Methodology, Representation*. Lanham, MD: Rowman and Littlefield.

Katz, C. (1992) All the world is staged: intellectuals and the project of ethnography. *Environment and Planning D: Society and Space*, 10, 495–510.

Keith, M. and Pile, S. (eds) (1993) *Place and the Politics of Identity*. London: Routledge.

Kobayashi, A. (1994) For the sake of the children: Japanese/Canadian workers/mothers. In A. Kobayashi (ed.), *Women, Work, and Place*. Montreal and Kingston: McGill–Queen's University Press.

Kobayashi, A. (1994b) Coloring the field – gender, race, and the politics of fieldwork. *Professional Geographer*, 46(1), 73–80.

Kobayashi, A. (2002) A generation later and still two percent: changing the culture of Canadian geography. *The Canadian Geographer/le géographe canadien*, 46(3), 245–8, 262–5.

Kobayashi, A. (2003) The construction of geographical knowledge – racialization, spatialization. In K. Anderson, M. Domosh, S. Pile and N. Thrift (eds), *Handbook of Cultural Geography*. Thousand Oaks, CA: Sage.

Kobayashi, A. (2004) GPC ten years on: is self-reflexivity enough? *Gender, Place and Culture*, 10(3), 345–9.

Kobayashi, A. and Peake, L. (1994) Unnatural discourse: "race" and gender in geography. *Gender, Place and Culture*, 1, 225–44.

Kobayashi, A. and Peake, L. (2000) Racism out of place: thoughts on whiteness and an anti-racist geography in the new millennium. *Annals of the Association of American Geographers*, 90(2), 392–403.

Kodras, J. and Jones, J. P. (1991) A contextual examination of the feminization of poverty. *Geoforum*, 22(2), 159–71.

Limb, M. and Dwyer, C. (eds) (2001) *Qualitative Methods for Geographers*. Oxford: Oxford University Press.

McDowell, L. and Sharp, J. P. (eds) (1997) *Space, Gender, Knowledge: Feminist Readings*. London: Arnold.

Massey, D. (1994) *Space, Place and Gender*. Minneapolis: University of Minnesota Press.

Moss, P. (2002) *Feminist Geography in Practice: Research and Methods*. Oxford: Blackwell.

Nast, H. J. (2000) Mapping the unconscious: racism and the Oedipal family. *Annals of the Association of American Geographers*, 90, 215–55.

Peake, L. (1993) "Race" and sexuality: challenging the patriarchal structuring of urban social space. *Environment and Planning D: Society and Space*, 11, 415–32.

Peake, L. and Kobayashi, A. (2002) Policies and practices for an anti-racist geography at the millennium. *Professional Geographer*, 54(1), 50–61.

Petras, E. M. (1989) Jamaican women in the US health industry: caring, cooking and cleaning. *International Journal of Urban and Regional Research*, 13, 304–23.

Pulido, L. (1996) A critical review of the methodology of environmental racism research. *Antipode*, 28(2), 142–9.

Pulido, L. (1997) Community, place, and identity. In J. P. Jones III, H. J. Nast and S. M. Roberts (eds), *Thresholds in Feminist Geography*. Lanham, MD: Rowman and Littlefield.

Pulido, L. (2000) Rethinking environmental racism: white privilege and urban development in southern California. *Annals of the Association of American Geographers*, 90(1), 12–40.

Pulido, L. (2002) Reflections on a white discipline. *Professional Geographer*, 54(1), 42–9.

Rose, G. (1993) *Feminism and Geography: The Limits of Geographical Knowledge*. Cambridge: Polity Press.

Ruddick, S. (1996) Constructing difference in public spaces: race, class and gender as interlocking systems. *Urban Geography*, 17, 132–51.

Sanders, R. (1990) Integrating race and ethnicity into geographic gender studies. *Professional Geographer*, 24, 228–30.

Women and Geography Study Group (1984) *Geography and Gender: An Introduction to Feminist Geography*. London: Hutchinson.

Chapter 4

A Bodily Notion of Research: Power, Difference, and Specificity in Feminist Methodology

Pamela Moss

Feminist methodology in geography has had what could be called a conventional history. It is conventional in the sense that it emerged in the late 1980s through the interstices of the ways in which geography as a discipline took up dominant discourses of science. In other words, feminists in geography began by addressing the question that DeVault (1999, pp. 22–3)[1] suggests defines feminist methodology out of existence: "is there a feminist method?" (*Canadian Geographer*, 1993).[2]

This approach parallels how feminist methodology emerged in other fields of study. Feminists were engaged with challenging the prominent methodological paradigms of the day: modernist conceptions of research, including positivist notions of truth, experiment, and knowledge. As a result, much of the methodological work emerging in white, Anglo-centered academic feminism focused on rejecting "malestream" scientific modes of inquiry (Roberts, 1981; Mies, 1983; Bowles and Duelli Klein, 1984; Harding, 1987a).

Defining feminist methodology by deciding on and then agreeing to a set of parameters that locates research within a particular politics and ethics is not a task to be taken lightly. As DeVault (1999, p. 23) says, to define feminist methodology risks distorting what feminist methodologists actually do. Nevertheless, earlier pieces in feminist geography did try to set parameters around feminist methodology (e.g. McDowell, 1992a, 1993a, b; Moss, 1993). These works were useful in that they grappled extensively with issues that feminists outside geography were writing about – positionality and reflexivity, fieldwork, quantitative methods – as well as issues salient to an ever-burgeoning critical and poststructural geography – multilocality, subjectivity, colonialism, praxis (e.g. *Professional Geographer*, 1994, 1995; *Antipode*, 1995). Methodological works in the latter half of the 1990s developed these themes even further, positioning feminism as a leading influence of criticism in geography as a whole (e.g. Women and Geography Study Group, 1997; Nast, 1998; Pratt, 1998; McDowell, 1999).

Even as feminist methodology converges with (and leads) other broadly critical methodological approaches, it still makes sense to describe in some manner what it is that makes feminist methodology *feminist*. Such a discussion would of course be different from the earlier focus about whether or not a distinctive feminist methodology exists. Indeed, it would draw attention to the multiplicity of feminisms and how each contributes to methodological debates in feminism, the framing of feminist research questions, and the actual practice of feminist research. The rest of this chapter is a discussion of one way to sort through this issue of showing how a methodology is "feminist." I first offer a framework for conceptualizing prominent concepts in feminist theory – power, difference, and specificity – that are helpful in understanding feminist methodology. I then show how these concepts have been taken up methodologically by feminist geographers. I next discuss how specificity can be useful in tilling new ground in feminist methodology. Following from this discussion, I put forward a set of topics that could become part of feminist methodological debates in geography. I close by urging feminist geographers to take seriously the principles of a feminist methodology they choose and to be critical about the context within which research takes place.

Power, Difference, and Specificity

After the clash with poststructuralism, with its insistence that analytical categories be deconstructed, feminism is still struggling for a collective presence, searching for an identifying characteristic setting feminism apart from other radical social movements. This task translates methodologically into the question of what differentiates feminism from other approaches to research. Is it that the subject of the research is women? Is it that the central construct in the analytical framework is patriarchy, sex, or the intersection of class, race, and gender? Is it that the object of analysis is gender? Sex? Women? Oppression? Resistance? Is it that differences among women are brought to the fore of the research? Or is it that none of these on its own is enough to make an inquiry feminist because all of these things are important and need to be included in any critical inquiry?

I venture to claim that the distinguishing characteristic of feminism is that it deals with power in some way – whether conceived as something to be held, exerted, deployed, mobilized, sought after, or refused, or as something structural and inevitable, despotic and concentrated, or dispersed and everywhere. The interactive dynamics of various feminisms since the emergence of second wave feminism in the 1960s has shown that sensitivity to power *within* feminism and the women's movement was not initially incorporated into feminist inquiry. Throughout the 1970s and 1980s, women of color critiqued white, Anglo-centered feminism, maintaining it was racist because it universalized white women's experiences and because many white feminists dismissed women of color politics as divisive and "anti-feminist" (e.g. Beale, 1971; Mirikitani, 1973; Davis, 1981; Moraga and Anzaldkúa, 1981; Smith, 1983; Allen Gunn, 1986). Despite a women of color politics in North America gaining widespread popularity in activist circles as well as gaining an increasingly commanding academic presence, it was not until feminism's encounters with poststructuralism from white, French feminists in the 1980s that the white, Anglo-centered feminism's self-image as uniquely radical and all-inclusive fractured (Sandoval, 1991).[3]

Although poststructuralism was a welcomed influence by some feminists (e.g. Butler, 1990, 1993; Butler and Scott, 1992; Gibson-Graham, 1996), others were less enthusiastic and more strategic in taking up poststructural arguments (e.g. Fraser, 1989; Nicholson, 1990; Pratt, 1993; Ramazanoğlu, 1993). Nonetheless, the imperative to address issues involving differences among women became a central task to all feminisms. The shattering of the once monolithic category of "woman" offered space for new politics – ones sensitive to power through difference. Sandoval (2000, p. 54) provides one of the strongest cases for a new politics with what she calls an *oppositional consciousness*, a differential mode of opposition that moves between and among ideological positionings.[4] Through what she calls US Third World feminism, she brings together insights from writings by women of color and poststructural thought in a unique way. She argues that under modernist forms of capitalist production, four tactics of resistance were mobilized: equal rights, revolutionary, supremacist, and separatist. These roughly correspond to four types of feminism: liberal, Marxist, cultural, and radical, respectively. With the transformation of capital in the late twentieth century, it is only through a differential mode of opposition that one can resist such forceful neocolonizing processes of postmodernity. With an oppositional consciousness, sensitive to differences among women specifically and to the differentiation of identities more generally, resistance emerges differently, strategically, immediately, and appropriately – all at the same time. Oppositional consciousness provides a way to sort through how power when wielded can bring about differentiated strategies of resistance. This reconfiguration of how power differentiates identities highlights the significance of the role that "the local" plays in effecting social change. The local is a site of mediation of power that sets the stage for how resistance, protests, and contestations take place.

One way to gain insight into how the local mediates power is to look at feminist theorizations of the body. For some time the body has been "a key site of contention" between feminists and poststructuralists for interrogating issues of self, subjectivity, identity, and difference (Canning, 1999, p. 57). The surfacing of the body as a vital subject of analysis brought with it intense engagement with an ongoing feminist dilemma about the core source of sex and gender identities; this time it manifested as a debate between biological and social constructionist conceptualizations of the body (Shilling, 1993). Williams and Bendelow (1998) try to overcome this dichotomous approach to understanding the body by distinguishing theories of the body that address bodily order (regulation and restriction of bodies through social functions) and bodily control (influences and choices of how individuals live life and express themselves). As an alternative to both, they prefer theories that address *real* bodies – ones that sleep, eat, menstruate, experience pleasure, are in pain – that are at one time both biological and social. They argue that with an integrated analytical framework of the social and the biological, the body can be theorized *as a body*. For example, rather than leaving sleeping bodies in the hands of psychologists (for sleep studies), psychoanalysts (for dreams), and the emerging sleep industrialists (for the production of sleeping aids), sociologists could investigate sleep as a social phenomenon. Understanding that sleep for humans is located at the intersection of "physiological *need*, environmental *constraint*, and sociocultural *elaboration*" (Williams and Bendelow, 1998, p. 211, emphasis in original) can open up research around sleep as an activity that is socially and spatially regulated (e.g. in institutions, in homes, in parks, and in battlefields), discursively and materially (e.g. using sleep clinics in the medicalization of sleep, implementing policy designating that social housing must have one bedroom for each child, enacting vagrancy laws criminalizing sleeping in public space, stipulating that soldiers take uppers to stay awake for extended periods of time).

Another concept that attempts to overcome the binary of the biological and the social is embodiment. Although notions of embodiment abound in the feminist literature (e.g. Benhabib, 1992; Grosz, 1994; in geography, see essays in Nast and Pile, 1998), one work in particular stands out as a useful way to conceive the relationship between ideas and acts. To define embodiment, Bray and Colebrook (1998) draw on Deleuze's notions of positive ontology, the idea that being is based on what is, rather than on a deficiency, lack, or absence. For example, they argue that anorexia nervosa, rather than a manifestation of the negation of the body,[5] is actually a set of activities and behaviors that actively produce the body as an entity. These activities along with discourse, thought, and reason shaping these activities comprise bodily events. In their conceptualization, then, embodiment is an ethics through which "bodies become, intersect and affirm their existence" (Bray and Colebrook, 1998, p. 36).

Combining the notion of a *real* body of Williams and Bendelow and the notion of bodily events from Bray and Colebrook produces what I call *specific* bodies – actively produced concrete entities constitutive of subjectivity, emotion, pain, and body parts textured through/by/in everyday life. Spaces that bodies inhabit are material and temporal juncture points, dripping with the minutiae of the immediate environment, the mediation of power through multiscale processes, and the culmination of historical moments. So, like bodies, spaces are *specific*, at any scale – the local, the regional, the national, the global. Specificity then describes the process through which bodies come to be made *specific* through interactions with other bodies and spaces. Through constitutive interaction of bodily events, *specific* bodies come into being. A particular expression of a body changes depending on the various constitutive interactions. So, for example, the body of a woman with an unpredictable chronic illness like lupus may take on different meanings in different contexts – in the doctor's office, at the workplace, and in the home – as well as in relation to co-workers, employers, doctors, disability insurance agents, children, neighbors, and partners. Because of the constitutive relationship bodies have with spaces, and vice versa, bodies always already have with them traces of *specific* spaces, just as spaces have traces of *specific* bodies. So, for example, the same woman with lupus takes with her traces of the doctor's office visit in the form of perhaps a diagnosis, which is then taken up differently in her daily living and working spaces, where in the former she may be permitted to be openly ill and in the latter she may not. In her social relationships too there are the same traces resulting in some people being supportive and others not. From elaborating the specificity of a body, insights into various social processes can be gained – identity formation, constitution of subjectivity, and notions of self. Using these insights to sort out connections among *specific* bodies and linking them in/through/by *specific* spaces can facilitate the development of liberatory acts associated with social change.

Attending to the specificity of bodies and spaces is a possible strategy that can mobilize an oppositional consciousness. Instead of looking at the circumstances under which such a consciousness emerges (e.g. Meoño-Picado, 1997; Pulido,

1997), I focus on how this bodily notion of specificity as an oppositional strategy can be incorporated into an approach to research that facilitates the development of an oppositional consciousness in the research process.

Feminist Geography and Approaches to Feminist Methodology

Sorting through questions regarding what makes a research project feminist methodologically still makes sense even while using a bodily notion of research. Some of the parameters feminists use to show that feminism is a unique approach to research have been outlined in the numerous collections of essays and "how-to" guidebooks in feminist methodology (e.g. Kirby and McKenna, 1989; Fonow and Cook, 1991; Reinharz, 1992; Wolf, 1996). Outside geography, feminist methodology tends to be defined as the constellation of the political positionings of the researcher with some action for social change that improves women's positions in society (Gottfried, 1996, p. 13; DeVault, 1996, pp. 32–4; Hesse-Biber et al., 1999, p. 6; Ramazanoğlu and Holland, 2002, pp. 15–16). In feminist geography, ruminations on feminist methodology have taken a somewhat different path, influenced by both feminist theory and geographic knowledge. Most of this debate has taken place in discussions about epistemology, fieldwork, and choice of method.

Epistemology

Feminist geographers took the lead from feminist philosophers in asking questions that would shape their inquiries: who are knowers, what can be known, and how do we know what we know (following, for example, Harding, 1987b; Haraway, 1988; Hawkesworth, 1990). They explicitly addressed these issues through discussions about reflexivity and positionality. In earlier renditions, reflexivity was taken up as a task of the researcher in coming to terms with her own position among a web of power relations constituting the research process (McDowell, 1992b; England, 1994; Moss, 1995). The idea of being introspective and self-critical soon gave way to a different notion of what can possibly be known. Rose (1997) claimed that the reflexivity feminist geographers desired was unattainable because positionings are not transparent. For her, it is the analytical uncertainty of interactions, interpretations, and partial understandings within sets of webbed relations – both of the researcher and of the research participants – that provides room for another type of reflexivity, that which is performed as an uncertain piece of interpretive authority and that queries the role a feminist researcher has in the creation of knowledge. Although the point about transparency was absorbed into the feminist methodological literature quickly, there are still feminist geographers trying to create space for other conceptions of reflexivity, ones not rooted in performativity. For example, Falconer Al-Hindi and Kawabata (2002) suggest that feminist research in geography has the potential to become more fully reflexive. Falconer Al-Hindi[6] argues that an open reflexivity on the part of the researcher, the research participant and, in the case discussed, a researcher's research advisor produces an interactive knowledge base conducive to multiple interpretations. Moss and Dyck (2002, pp. 70, 76–8) decline to engage with debates on reflexivity and shift the discussion toward the "interpretive act" or analysis. They go on to contextualize their analysis by identifying partially their own social positionings *vis-à-vis* the research participants and issues they identify as important for the analysis. Dyck (2002, p. 244) extends this line of thinking, claiming that although the research process can never be transparent, "it is important to continue to make our best efforts to uncover the mechanisms of truth claims we produce" through the "social and political processes of academic knowledge construction."

Fieldwork

Cultural anthropology has probably been the strongest outside influence in geographic debates about the "field" as a site of research. Geertz's (1973) cross-cultural studies shaped the way in which researchers – both feminists and non-feminists – undertook ethnographic work. Following Geertz, some anthropologists expressed their unease about using ethnographic concepts like being both an insider and an outsider, "thick description," and difference, and initiated a flood of postmodern interpretations of the "field" (e.g. Clifford and Marcus, 1986; Clifford, 1997; Gupta and Ferguson, 1997; Marcus, 1998). Feminists engaged these debates and contributed extensively in the area of representation of difference in ethnographic writing (e.g. Kirby, 1989, 1993; Gordon, 1993; Visweswaran, 1994; Behar and Gordon, 1995; Behar, 1996). As much of the feminist anthropological literature incorporated space between women and between cultures (Kirby, 1991, p. 400), the masculinity of the act of "going to the field" was not brought out as a pressing issue. In geography, "going to the field" has been shown to have connections to masculinities deeply rooted in a colonialism and imperialism (G. Rose, 1993; Sparke, 1996). These masculinities play out in subtle and unsuspected ways, even in field trips and geography camps (Nairn, 1999, 2002). Parallel with these critiques of "the field," feminist geographers have created a sensitivity to spaces of interaction – in interview settings (Gilbert, 1994; Oberhauser, 1997), in ethnographic field settings (Nast, 1998; Marshall, 2002), and in constructing difference (Kobayashi, 1994; Nast, 1994; Dyck et al., 1995; Valentine, 2002). Within these writings, there is a recognition that the boundaries of what constitutes the field are in flux, for as Katz (1994, p. 72) proclaims, she is "always, everywhere, in the 'field'." Bondi (2003) offers yet another dimension of interaction to consider while undertaking research in the "field." Drawing on psychoanalysis, she specifies psychic space as the predominately unconscious communication that sorts through similarities and differences among researchers and research participants. Such conceptualizations of the "field" and the interactions that take place within it draw into question the various truth claims feminists make in the process of collecting and translating "data" or information for analyses, as well as the way interaction and relations are represented in the write-up of the research.

Choice of method

Feminist geographers have been deeply influenced by the longstanding divisiveness between quantitative and qualitative methods in the social sciences. Debates about feminist methodology, within and outside feminist geography, initially focused on rejecting positivist claims of objectivity, value-free inquiry, and rational logic, and on embracing research methods that refused to break the living connections of women's lives (after Mies, 1991, p. 67). Qualitative methods appeared to be more congruent with understanding women's lives, less intrusive, and more sympathetic (following Oakley, 1981), until another feminist pointed out that qualitative methods could be just as exploitative as quantitative ones (Stacey, 1988).[7] In the mid-1990s, Mattingly and Falconer Al-Hindi (1995) brought together feminist geographers who were interested in contemplating whether women could "count." The framing of the question as women counting was a clever way to bring together concerns about the epistemic claims being made at the time of the possibility of feminists committed to making visible women's geographies in non-positivistic ways through the use of numbers. Lawson (1995) argues that the ontological status of counting was not in itself masculine. What was problematic was the use of the counting that had come to support masculinist science. Rocheleau (1995) demonstrates Lawson's claims in showing how in her own work the strategy of mixing methods was useful in understanding the ways people come to make sense of their world while remaining liberatory.

Although feminist geographers accepted that women could indeed "count," the rationale for choice of method remained unresolved. If there was nothing intrinsic about qualitative (or quantitative) methods that made them feminist (or masculinist), what was the basis upon which a feminist would choose a method? These discussions about choice of method seemed to be moving toward advocating a practical strategy of having the research question determine the choice of method.[8] Yet, just as it was not enough to spurn numbers because they were masculinist, it was also not enough to choose a method solely based on the type of information needed to address the (feminist's) research question. This shift in conceptualizing what is at issue in feminist geographical analysis is illustrated by Shuurman and Pratt (2002, p. 291) as they embrace feminist critique as a means to examine critically "assumptions, ideas, statements and theories." By introducing the notion of the "care of the subject" as part of a feminist critique of geographic information systems (GIS), they transform the debate from one focused on choice of method for data collection to one focused on method for analysis – a critique that interrogates the processes through which certain truth claims are produced as knowledge systems.

Tilling New Ground in Feminist Methodology

This review indicates that there is strong agreement among feminist geographers (along with feminists in other disciplines) that there is something about the positioning of the researcher that matters in feminist methodology. In addition to the feminist researcher's positioning *vis-à-vis* research participating in the research projects, other things matter too, such as politicizing the research process, effecting social change through participatory and emancipatory research, being an activist scholar, allying with specific theories, and dealing with allegations of hegemony (e.g. Humphries, 1997; Thorne and Varcoe, 1998; Byrne and Letnin, 2000; Gatenby and Humphries, 2000; Arat-Koç 2002; Parker and Lynn, 2002). Feminist geographers have not always taken up these types of issues as methodological queries. For example, feminist geographers' discussions about activism are not predominantly located within feminist methodology, even though many of the feminists writing in the methodological literature make calls for social change (e.g. D. Rose, 1993; Kobayashi, 1994; Moss and Matwychuk, 2000). Likewise, the focus on women is not a particularly well developed topic methodologically for feminist geographers. For the most part the subject of women-centered research is either assumed or diffused by the focus on difference (see Pulido, 1997). It appears that the "woman question" has been supplanted by queries about how masculinist truth claims permeate epistemologies and research practices that feminists either adopt or engage in, even in the more conventional inquiry of the possibility of men being feminists (see Butz and Berg, 2002).

If we take at face value the overlap of issues raised by feminists in other disciplines with feminist geographers, then the only thing that would seem to matter in defining what is feminist about methodology is the positionality of the researcher. But this clearly is not the case if debate in feminist geography is to be taken seriously. Such debate seems to have actually transformed feminist methodology *from* a woman's experientially based approach to social change *through* research based in the academy *into* an academic inquiry of truth claims as the basis for the production of knowledge systems, and to have located feminism outside *specific* bodies. If this is so, where does this leave non-epistemological considerations in feminist research approaches?

It may leave us more convinced than ever about the importance of a specified bodily notion of research. Using (bodily) specificity in research is a way both to locate research participants in multiple sets of social relations and to contextualize this location *vis-à-vis* other participants and research settings. Conceived this way, specificity fits into Sandoval's (2000) general framework of a methodology of the oppressed, which she puts forward as a force of dissidence within a neocolonizing postmodern world. She identifies ways to decolonize spaces that can then be reconstructed to mobilize an oppositional consciousness to resist globalization, oppression, and authority and to find new visions, alliances, and possibilities for transformation. Decolonization, in a simple form, means an undermining or unsettling of the power relations constituting a particular space, access to which comes about through contradictions, paradoxes, and gaps in the way power plays out in constitutive interaction among multiple sets of social relations. Within these technologies, there are various strategies that promote decolonization and assist in the creation of an oppositional consciousness. Because Sandoval focuses primarily on theory, she does not go into detail about specific strategies for engaging in research processes.

But this does not mean that they do not exist.[9] As a strategy for decolonizing *research* spaces, specificity can be used not just to interpret the everyday lives of research participants but also to open up diverse spaces within feminist methodology. Discursive spaces such as those written up and frozen in text in the pages of academic journals can be opened up just like the ones emerging through other means of communication – aural, verbal, visual, and unconscious (e.g. Rose, 2001; Besio, 2002; Bondi, 2003). Specificity can decolonize material spaces too – classrooms, interview settings, and private flats – and transform them for radical ends (e.g. McKay, 2002; Nagar, 2002; Oberhauser, 2002; Pratt, 2002).

For example, feminist geography graduate students as newly emerging scholars are beginning to publish their longstanding thoughts about being feminist and doing feminist research. Publishing their work is a specific practice that opens up the discourse congealed in previous publications about feminist methodology and feminist research. Feminist methodology emerged in a decade when it was politically important to claim intellectually that feminists actually do research differently. Throughout the late 1980s, and into the 1990s in geography, feminists undertaking research in the academy needed to differentiate their efforts from their colleagues. In masculine-dominated disciplines like geography, other strategies may be apropos in defining what is and is not feminist research. Newly emerging scholars who consider themselves feminist may not engage in what has been conventionally defined as feminist geography research, that which has been rooted in the debates struggling for a definition of feminist methodology (see collection about female graduates students in the *Great Lakes Geographer*, 2002). Drawing on their own material and embodied experiences, Murphy and Cloutier-Fisher's (2002) discussion of how being mothers in graduate school influenced their graduate training is an example of researchers taking up their own specificity. Although neither is engaged in traditional feminist research, they draw on feminist analyses of gender to make sense of their experiences. Hall's (2002) work is a little different in that she wonders about being feminist without studying women, and whether that makes her a postfeminist somehow. What she comes to in sorting through her positioning within the discipline is that she is a feminist first, and that her study of landscapes is permeated by her being a feminist even if it doesn't show in conventional ways.[10]

What the points of discussion in *The Great Lakes Geographer* special issue insinuate is that feminist methodology as a topic can be opened up further to make room for sundry discursive understandings and material undertakings of what is feminist about feminist methodology. The endeavor of tilling new ground for methodological engagement has potential to reconfigure challenges facing feminists doing research. So rather than asking head-on what is feminist about feminist methodology, perhaps asking what issues are significant to feminist researchers taking on, thinking about, and doing feminist research is a way to unsettle the stability of feminist methodology as a field of inquiry.[11]

Specificity matters here because feminist methodology itself can be "decolonized" as a field of inquiry. Through the multiple expressions of feminist methodology, power and difference, text and action, discourse and materiality (e)merge together to create *specific* research bodies and research spaces. Without attention to specificity, access to the social practices that materially and discursively shape the research process would be limited. By limited, I do not mean that with specificity a researcher can get beyond the partiality of any one view; I mean that with specificity the partial view generated should not be taken to be the only one. It is not enough just to accept partiality as a state of being; partiality itself needs to come under scrutiny. Through specificity, kaleidoscopic views of the research process could be generated, troubling the clarity of any one (partial) view.

Occasions for Specificity

As part of elaborating the specificity of feminist methodology as a field of inquiry, there are issues that feminists in geography may want to look at in more detail, including several that have not yet received much attention in the extant literature.

I offer the following suggestions merely as departure points for discussion and not as an authoritative agenda for debate about feminist methodology in feminist geography.

Context

Geographers like to talk about "context," but what actually does context mean? Is context always already "out there," poised for interrogation? What does it mean to place something "in context"? And why would that matter? A colleague and I undertook a study of women with chronic illness and in the write up, we created a conceptual framework that seeks to contextualize women's experiences of being ill (Moss and Dyck, 2002). We attend to the specificity of bodies and how they structure and restructure their environments as congruent with, in response to, and in spite of their ill bodies. One way we deal with context involves the concept of "bodies in context," shorthand for describing socially constructed bodies and their material presences. " 'Bodies in context' focuses on the placement of *specific* bodies in relation to each other through recursive constitutive processes – socially and spatially" (Moss and Dyck, 2002, p. 56).

But bodies were not the only thing that we could have set "into context." For women in the study, using the telephone, for example, was not simply a matter of picking up a phone and calling someone on the support group's list. What was significant about using the telephone was that the phone list created helped women with chronic fatigue syndrome to cope with their fluctuating bodily sensations *and* the social and biomedical rejection of their bodies as legitimately ill. In order for the phone list to exist in the first place, some women with chronic illness had already negotiated the power within these social relations and created access to a specific space (a virtual space through telephone wires) where they could find support with one another. For this kind of contextualization, an approach like institutional ethnography (IE) might be useful in figuring out the implications of "placing in context" specific acts (see Campbell and Gregor, 2002). Given that IE is an approach to research rooted in feminist theory (Smith, 1990, 1998), and depending on the ethics and politics feeding into the interpretation, IE could be adapted to different research situations investigating places where power and difference constitute *specific* bodies – universities, hospitals, bureaucracies, homes, and workplaces.

Information as data

As researchers, we ask ourselves: what can you say with the information you collected? If partiality is an issue, then do we need different words to make the points we can make? Instead of talking in conventional terms like validity, accuracy, and triangulation, it might make sense to think about, for example, the momentary-ness of how and when information gets collected, the intention of the participants in being part of a project, or the fleeting moments of negotiating meaning in any research interaction. Dyck (2002, p. 244) makes mention of how snapshots of people's lives become "fixed" in ethnographic moments through the process through which feminist researchers construct knowledge with consequences we may not be aware of. She also claims that it is probably worthwhile to "set into context," even if only partially, these snapshots, for without doing so, they become unanchored narratives that have little use in sorting out the mechanisms of truth claims. For what if on that day the research participant were tired, hungry, or in a hurry? What if the research participant were making assumptions that you didn't know about?

Valentine (2002) takes up these issues by sorting through mutual constitutive subjectivities of herself and the research participants during the interview. There were some instances in her research about sexuality and parenting where research participants established some sense of connection with her by assuming commonality or differentiated themselves from her by assuming differences in sexuality, age, class, employment, and status as a parent. She draws the conclusion that "we can never really know what is going on in any given research encounter and therefore how the knowledge we take from it is being produced, nor how the information we use might have been different if our performances, those of our interviewees, or interactions between us, had been different" (Valentine, 2002, p. 125). In looking for such occurrences, an approach like "conversational analysis" (a specific type of discourse analysis) might be useful. Historically, feminists have engaged in discourse analysis primarily with regard to understanding power and resistance. Finding places in the data that question the stability of a particular reading of power and resistance might point toward instances of the momentary-ness of the information gathered in an effort to decolonize the snapshot of an underscrutinized space of analysis.

Praxis

If we engage in an ongoing deconstruction of each research moment, to the point where we cannot know what a research encounter is, then why do research at all? To what end does one engage in research as a feminist? How does activism fit into the research process? Is research about finding a truth or some truth-correspondence? How does praxis work methodologically? Although some of these questions seem to be basic, perhaps even pedantic in light of the history of feminist thought, they still matter. They matter not so much in the sense that they need an answer; they matter in that the answers themselves need to be decolonized. Decolonization would favor a system of answers that are flexible enough to be reincarnated in various contexts that would continually spark new ways of reading cultural artifacts, thinking new thoughts, acting in opposition, and imagining liberatory spaces. Some elements of participatory research have made their way into the "norm" of doing qualitative research among feminists. For example, there has been an ethic in feminist geography that the only choice feminists have available to them, if they want to be inclusive in their data collection and analysis, is to share the write-up of their stories with the research participants.[12]

Naples (1996), a sociologist, questions the extent to which participatory research strategies can be maintained given the complexities of politicized social positionings that are part of the research project. Feminist geographers are increasingly interrogating this practice: that is, which research participants get to participate in the data gathering and in the analysis. England (2002, p. 211), for instance, in her research with bank managers, notes that she does not use feedback from the managers about the transcription of their interview to make her text polyvocal; rather she uses it, at least in one case, to gain insight into why a manager might return a grammatically corrected version of the transcription text. McKay (2002) questions her self-silencing in the excerpts she chose to analyze – she kept out details of her life and included details from one of the women in the study whom she also saw as a friend. These examples provoke thought about what are considered to be liberatory strategies. What does participation mean? And who gets to decide? Conventions within feminism need to be rethought and reconstituted ongoingly in order to be fastidious in the commitment to decolonizing research methods.

Multiscale analyses

What about a bodily notion of research at scales other than the body? Would the same principle of specificity apply? There are works that address this issue, at least tangentially, mostly in the field of economic restructuring. Price (1999), for example, finds support for her interpretation of global socioeconomic processes by reading the body as the "very local." She argues that the women in her study embodied the very processes restructuring the region, evidenced by the way the women held themselves (comportment), the bruises on their bodies, and their poor mental health. Callard (1998) rescales body studies theoretically from the corporeal body to the abstraction of the laboring body. By focusing on the body as a laboring entity, she traces the connections between the body and that which gives rise to particular meanings of the body. And Leslie and Butz (1998) argue that restructuring occurs at multiple geographical scales at the same time. Scrutinizing injured bodies in an auto plant, they argue that the injuries take on specific meanings depending on the context within which the body appears – on the assembly line, in union meetings, or in non-working places like ski resorts. This contextual reading of injuries (smaller scale than the body) permits insight into how the contradictions of the labor process play out for workers (larger scale than the body). These three examples show how a study of the body can move through multiple scales within the same analysis. This notion of multiscale analysis opens up the possibility of undertaking nested research projects, ones that investigate the same topic, with the same information, at various scales. These examples also show that specificity is not restricted to the scale of the body; rather, specificity as a bodily notion of research can be taken up at different scales.

Parting Comments

I close with a simple message, but by simple I don't want to imply easy to abide by. Researchers using feminist methodology need to take seriously the principles of the feminist methodology they piece together from the fragments available to them and the ones they create themselves. To get to this point there needs to be a willingness to "shift" while remaining "rooted" (after Pryse, 2000), to engage with new ideas while sorting through what is still useful from previous debate, to be topical while nourishing your own interests, and to be critical of the context of the research process, the research knowledge, and the "products" produced.

Unlike the early feminists who were trying to determine a set of principles that guide the use of feminist methodology, feminist geographers today need no such thing. For now, feminist researchers in geography need to open up discussions about more and more issues – even the field of feminist methodology itself! By posing questions differently about feminist approaches to research, the various dimensions of the research process, data collection, and analytical methods, feminist geographers will be able to draw out connections, gaps, entanglements, ravelings, paradoxes, contradictions, congruences, and simultaneities among the topics that interest them, the concepts that assist in making sense of what information is available to them, and the actions that will effect social and political change. These efforts in pulling together a specific ethics and politics grounded in a commensurable set of truth claims will contribute to constructing a knowledge that, at lest for the interim, defines what is feminist about a feminist methodology.

NOTES

1   "To focus prematurely on the question of whether feminist methodology 'exists' or is 'really different,' before making a serious attempt to study the field, seems at heart an insistence on constructing questions about feminist method that define it out of existence. When I am pushed to define feminist methodology simply and completely in the terms of mainstream social science, I risk distorting what feminist methodologists do. Instead of rushing to answer, it may be more useful to notice that the question comes from a discourse that is not eager to make room for us" (DeVault 1999, p. 22–3).

2   In that collection, I offered what I saw as constituting feminism as an approach to research (Moss, 1993). Two feminists, D. Rose (1993) and Dyck (1993), wrote about issues that they were dealing with in their own work – how weaknesses in some research methods can be compensated for by other methods and how to resolve insider–outsider tensions as a woman researching women, respectively. A fourth contributor, Eyles (1993), who is not a feminist, wrote from a humanist perspective questioning the existence of a particularly feminist viewpoint in research.

3   This is a debatable point. Barrett and Phillips (1992) maintain that the biggest influence on white feminism has been the continual critique by black women of the racist and ethnocentrist assumptions in white feminism. Given the engagement of white, Anglo-centered, academic feminist literature with writings by women of color, including feminist geography, it appears that poststructural thought by white men dominates both the citations and the quotations of feminists attempting to change the face of feminism. This debate in itself demonstrates the relative ease with which academic feminists fit into their privileged positions intellectually.

4   Sandoval's (2000) work *The Methodology of the Oppressed* includes pieces published as early as 1991. In this sense "new" does not mean new in time; by "new," I mean an alternative in comparison to existing sets of political practices. Sandoval (2000, p. 68) argues that her notion of oppositional consciousness or "varying dimensions of its differential form" are akin to several other concepts in the literature: " 'hybridity,' 'nomad thought,' 'marginalization,' *'la conciencia de la mestiza,'* 'trickster consciousness,' 'masquerade,' 'eccentric subjectivity,' 'situated knowledges,' 'schizophrenia,' *'la facultad,'* 'signifin',' 'the outsider/within,' 'strategic essentialism,' *'differance,'* *'rasquache,'* 'performativity,' *'coatlicue,'* and 'the third meaning.' "

5   This is the prevalent view of anorexia nervosa (e.g. Bordo, 1993).

6   Falconer Al-Hindi and Kawabata make different arguments and are clear about who maintains which views.

7   Oakley (2000) has gone on to work toward fracturing this divide and to provide a rethinking of how experimental ways of knowing, although central to social science work, have been rejected by most feminists.

8   This debate was also taking place in population geography (see *Professional Geographer*, 1999a) and in medical and health geography at the time (*Professional Geographer*, 1999b; see especially Elliott, 1999). Dyck (1999) provides insight into both the feminist debate over choice of method and the use of qualitative methods into the relatively highly quantitative medical geography. She argues that the use of qualitative methods does not transform an inquiry; the use of theory does.

9   See especially Smith (1999).

10 The authors of the other contributions in this collection, Mahtani (2002), Frohlick (2002), and Nash (2002), locate themselves within feminism to provide ways of reading their work and experiences in the "field." Mahtani (2002) uses feminist theory to report on her study of women of color graduate students in the UK, Canada, and the USA. Frohlick (2002) argues that the presence of her spouse and two children shattered the masculine, geographic convention of a lone, detached male in the field. Her query then comes in the form of figuring out how the shattering of this convention is to be represented in subsequent write-ups. Nash (2002) draws parallels between the masculine environments of geography and law as she changed her career path.

11 The potentially decolonizing acts of taking on, thinking about, and doing feminist research is the structure I use in the textbook *Feminist Geography in Practice: Research and Methods* (Moss, 2002).

12 This is common across social research disciplines (e.g. Riessman, 1993) and has been questioned in terms of its usefulness in terms of checking truth claims (Taylor, 2001, p. 322).

BIBLIOGRAPHY

Allen Gunn, P. (1986) *The Sacred Hoop: Recovering the Feminine American Indian Traditional*. Boston: Beacon Press.

Antipode (1995) Discussion and debate: symposium on feminist participatory research. *Antipode*, 27, 71–101.

Arat-Koç, S. (2002) Imperial wars or benevolent interventions? Reflections on "global feminism" post September 11th. *Atlantis*, 26(2), 53–65.

Barrett, M. and Phillips, A. (1992) *Destabilizing Theory: Contemporary Feminist Debates*. Cambridge: Polity Press.

Beale, F. (1971) Double jeopardy: to be black and female. In R. Morgan (ed.), *Sisterhood is Powerful: An Anthology of Writing from the Women's Liberation Movement*. New York: Random House.

Behar, R. (1996) *The Vulnerable Observer: Anthropology that Breaks Your Heart*. Boston: Beacon Press.

Behar, R. and Gordon, D. (1995) *Women Writing Culture*. Berkeley: University of California.

Benhabib, S. (1992) *Situating the Self: Gender, Community and Postmodernism in Contemporary Ethics*. London: Routledge.

Besio, K. (2002) "I think we may need a divorce": spatial stories of weddings and postcolonial geographies. Paper presented at the annual meeting of the Association of American Geographers, Los Angeles, March 20–24.

Bondi, L. (2003) Empathy and identification in fieldwork. *ACME*, 2(1), 64–76.

Bordo, S. (1993) *Unbearable Weight: Feminism, Western Culture, and the Body*. Berkeley: University of California Press.

Bowles, G. and Duelli Klein, R. (eds) (1983) *Theories of Women's Studies*. Boston: Routledge & Kegan Paul.

Bray, A. and Colebrook, C. (1998) The haunted flesh: corporeal feminism and the politics of (dis)embodiment. *Signs*, 24(1), 35–67.

Butler, J. (1990) *Gender Trouble: Feminism and the Subversion of Identity*. London: Routledge.

Butler, J. (1993) *Bodies that Matter: On the Discursive Limits of "Sex"*. London: Routledge.

Butler, J. and Scott, J. W. (1992) *Feminists Theorize the Political*. London: Routledge.

Butz, D. and Berg, L. D. (2002) Paradoxical space: geography, men, and duppy feminism. In P. Moss (ed.), *Feminist Geography in Practice: Research and Methods*. Oxford: Blackwell.

Byrne, A. and Letnin, R. (eds) (2000) *(Re)Searching Women: Feminist Research Methodologies in the Social Sciences in Ireland*. Dublin: Institute of Public Administration.

Callard, F. J. (1998) The body in theory. *Environment and Planning D: Society and Space*, 16, 387–400.

Campbell, M. L. and Gregor, F. A. (2002) *Mapping Social Relations: A Primer in Doing Institutional Ethnography*. Aurora, Ont: Garamond Press.

Canadian Geographer (1993) Feminism as method. *Canadian Geographer*, 37, 48–61.

Canning, K. (1999) Feminist history after the linguistic turn: historicizing discourse and experience. In S. J. Hesse-Biber, C. K. Gilmartin and R. Lydenberg (eds), *Feminist Approaches to Theory and Methodology: An Interdisciplinary Reader*. New York: Oxford University Press.

Clifford, J. (1997) *Routes: Travel and Translation in the Late Twentieth Century*. Cambridge, MA: Harvard University Press.

Clifford, J. and Marcus, G. E. (eds) (1986) *Writing Culture: The Poetics and Politics of Ethnography*. Berkeley: University of California Press.

Davis, A. (1981) *Women, Race, and Class*. New York: Random House.

DeVault, M. (1996) Talking back to sociology: distinctive contributions of feminist methodology. *Annual Review of Sociology*, 22, 29–50.

DeVault, M. (1999) *Liberating Method: Feminism and Social Research*. Philadelphia: Temple University Press.

Dyck, I. (1993) Ethnography: a feminist method. *Canadian Geographer*, 37(1), 52–7.

Dyck, I. (1999) Using qualitative methods in medical geography: deconstructive moments in a subdiscipline? *Professional Geographer*, 51, 243–53.

Dyck, I. (2002) Further notes on feminist research: embodied knowledge in place. In P. Moss (ed.), *Feminist Geography in Practice: Research and Methods*. Oxford: Blackwell.

Dyck, I., Lynam, J. M. and Anderson, J. M. (1995) Women talking: creating knowledge through difference in cross-cultural research. *Women's Studies International Forum*, 18, 611–26.

Elliott, S. J. (1999) Introduction: and the question shall determine the method. *Professional Geographer*, 51, 240–3.

England, K. V. L. (1994) Getting personal: reflexivity, positionality and feminist research. *Professional Geographer*, 46, 80–9.

England, K. V. L. (2002) Interviewing elites: cautionary tales about researching women managers in Canada's banking industry. In P. Moss (ed.), *Feminist Geography in Practice: Research and Methods*. Oxford: Blackwell.

Eyles, J. (1993) Feminist and interpretive methods: how different? *Canadian Geographer*, 37(1), 50–1.

Falconer Al-Hindi, K. and Kawabata, H. (2002) Toward a more fully reflexive feminist geography. In P. Moss (ed.), *Feminist Geography in Practice: Research and Methods*. Oxford: Blackwell.

Fonow, M. M. and Cook, J. A. (eds) (1991) *Beyond Methodology: Feminist Scholarship as Lived Research*. Bloomington: Indiana University Press.

Fraser, N. (1989) *Unruly Practices: Power, Discourse and Gender in Contemporary Social Theory*. Minneapolis: University of Minnesota Press.

Frohlick, S. E. (2002) "You brought your baby to base camp?" Families and field sites. *Great Lakes Geographer*, 9(1), 49–58.

Gatenby, B. and Humphries, M. (2000) Feminist participatory action research: methodological and ethical issues. *Women's Studies International Forum*, 23, 89–105.

Geertz, C. (1973) *The Interpretation of Cultures*. New York: Basic Books.

Gibson-Graham, J.-K. (1996) *The End of Capitalism (As We Knew It)*. Oxford: Blackwell.

Gilbert, M. (1994) The politics of location: doing feminist research at "home". *Professional Geographer*, 46, 90–6.

Gordon, D. (1993) Worlds of consequences: feminist ethnography as social action. *Critique of Anthropology*, 13(4), 429–43.

Gottfried, H. (ed.) (1996) *Feminism and Social Change: Bridging Theory and Practice*. Urbana: University of Illinois Press.

Great Lakes Geographer (2002) Feminism and the academy: the experiences of women graduate students in geography. *Great Lakes Geographer*, 9(1), 1–58.

Grosz, E. (1994) *Volatile Bodies: Toward a Corporeal Feminism*. Bloomington: Indiana University Press.

Gupta, A. and Ferguson, J. (eds) (1997) *Anthropological Locations: Boundaries and Grounds of a Field Science*. Berkeley: University of California Press.

Hall, J. (2002) The next generation: can there be a feminist geography without gender? *Great Lakes Geographer*, 9(1), 19–27.

Haraway, D. (1988) Situated knowledges: the science question in feminism and the privilege of partial perspective. *Feminist Studies*, 14(3), 575–99.

Harding, S. (ed.) (1987a) *Feminist Methodology*. Bloomington: Indiana University Press.

Harding, S. (1987b) Introduction: is there a feminist method? In S. Harding (ed.), *Feminist Methodology*. Bloomington: Indiana University Press.

Hawkesworth, M. E. (1990) *Beyond Oppression: Feminist Theory and Political Strategy*. New York: Continuum.

Hesse-Biber, S. J., Gilmartin, C. K. and Lydenberg, R. (eds) (1999) *Feminist Approaches to Theory and Methodology: An Interdisciplinary Reader*. New York: Oxford University Press.

Humphries, B. (1997) From critical thought to emancipatory action: contradictory research goals? *Sociological Research Online*, 2(1) (http://www.socresonline.org.uk/socresonline/2/1/3.html).

Katz, C. (1994) Playing the field: questions of fieldwork in geography. *Professional Geographer*, 46, 67–72.

Kirby S. and McKenna, K. (1989) *Experience Research Social Change: Methods from the Margins*. Toronto: Garamond Press.

Kirby, V. (1989) Corporeographies. *Inscriptions*, 5, 103–19.

Kirby, V. (1991) Corporeal habits: addressing essentialism differently. *Hypatia*, 6(3), 4–24.

Kirby, V. (1993) Feminisms and postmodernisms: anthropology and the management of difference. *Anthropological Quarterly*, 66, 127–33.

Kobayashi, A. (1994) Coloring the field: gender, "race" and the politics of fieldwork. *Professional Geographer*, 46, 73–80.

Lawson, V. (1995) The politics of difference: examining the quantitative/qualitative dualism in post-structuralist feminist research. *Professional Geographer*, 47, 449–57.

Leslie, D. and Butz, D. (1998) "GM suicide": flexibility, space, and the injured body. *Economic Geography*, 74, 360–78.

McDowell, L. (1992a) Doing gender: feminism, feminists and research methods in human geography. *Transactions, Institute of British Geographers*, 17, 399–416.

McDowell, L. (1992b) Valid games? A response to Erica Schoenberger. *Professional Geographer*, 44, 212–15.

McDowell, L. (1993a) Space, place and gender relations: part 1. Feminist empiricism and the geography of social relations. *Progress in Human Geography*, 17, 157–79.

McDowell, L. (1993b) Space, place and gender relations: part 2. Identity, difference, feminist geometries and geographies. *Progress in Human Geography*, 17, 305–18.

McDowell, L. (ed.) (1999) *Gender, Identity, and Space: Understanding Feminist Geographies*. Cambridge, Polity Press.

McKay, D. (2002) Negotiating positionings: exchanging life stories in research interviews. In P. Moss (ed.), *Feminist Geography in Practice: Research and Methods*. Oxford: Blackwell.

Mahtani, M. (2002) Women graduate students of colour in geography: increased ethnic and racial diversity, or maintenance of the status quo? *Great Lakes Geographer*, 9(1), 11–18.

Marcus, G. E. (1998) *Ethnography through Thick and Thin*. Princeton, NJ: Princeton University Press.

Marshall, J. (2002) Borderlands and feminist ethnography. In P. Moss (ed.), *Feminist Geography in Practice: Research and Methods*. Oxford: Blackwell.

Mattingly, D. and Falconer Al-Hindi, K. (1995) Should women count? A context for the debate. *Professional Geographer*, 47, 427–35.

Meoño-Picado, P. (1997) Redefining the barricades: latina lesbian politics and the creation of an oppositional public sphere. In J. P. Jones, H. J. Nast and S. M. Roberts (eds), *Thresholds in Feminist Geography: Difference, Methodology, Representation*. Lanham, MD: Rowman & Littlefield.

Mies, M. (1983) Towards a methodology for feminist research. In G. Bowles and R. Duelli Klein (eds), *Theories of Women's Studies*. London, Routledge & Kegan Paul.

Mies, M. (1991) Women's research or feminist research? The debate surrounding feminist science and methodology. In M. M. Fonow and J. A. Cook (eds), *Beyond Methodology: Feminist Scholarship as Lived Research*. Bloomington: Indiana University Press.

Mirikitani, J. (ed.) (1973) *Third World Women*. San Francisco: Third World Communications.

Moraga, C. and Anzaldúa, G. (eds) (1981) *This Bridge Called My Back: Writings by Radical Women of Color*. Watertown, MA: Persephone.

Moss, P. (1993) Introductory comments. *Canadian Geographer*, 37, 48–9.

Moss, P. (1995) Embeddedness in practice, numbers in context: the politics of knowing and doing. *Professional Geographer*, 47, 442–9.

Moss, P. (ed.) (2002) *Feminist Geography in Practice: Research and Methods*. Oxford: Blackwell.

Moss, P. and Dyck, I. (2002) *Women, Body, Illness: Space and Identity in the Everyday Lives of Women with Chronic Illness*. Lanham, MD: Rowman & Littlefield.

Moss, P. and Matwychuk, M. L. (2000) Beyond speaking as an "as a" and stating the "etc." Toward a praxis of difference. *Frontiers*, 21(3), 82–104.

Murphy, B. L. and Cloutier-Fisher, D. (2002) A balancing act: motherhood and graduate school. *Great Lakes Geographer*, 9(1), 37–47.

Nagar, R. (2002) Women's theatre and the redefinitions of public, private and politics in North India. *ACME*, 1, 55–72.

Nairn, K. (1999) Embodied fieldwork. *Journal of Geography*, 98, 272–82.

Nairn, K. (2002) Doing feminist fieldwork about geography fieldwork. In P. Moss (ed.), *Feminist Geography in Practice: Research and Methods*. Oxford: Blackwell.

Naples, N. with Clark, E. (1996) Feminist participatory research and empowerment: going public as survivors of child sexual abuse. In H. Gottfried (ed.), *Feminism and Social Change: Bridging Theory and Practice*. Urbana: University of Illinois Press.

Nash, C. J. (2002) Law and the discipline of geography: a parallel universe. *Great Lakes Geographer*, 9(1), 29–36.

Nast, H. J. (1994) Opening remarks on "women in the field." *Professional Geographer*, 46, 54–66.

Nast, H. J. (1998) The body as "place": reflexivity and fieldwork in Kano, Nigeria. In H. J. Nast and S. Pile (ed.), *Places through the Body*. London: Routledge.

Nast, H. J. and Pile, S. (eds) (1998) *Places through the Body*. London: Routledge.

Nicholson, L. J. (ed.) (1990) *Feminism/Postmodernism*. New York: Routledge.

Oakley, A. (1981) Interviewing women: a contradiction in terms. In H. Roberts (ed.), *Doing Feminist Research*. New York: Routledge & Kegan Paul.

Oakley, A. (2000) *Experiments in Knowing: Gender and Method in the Social Sciences*. New York: New Press.

Oberhauser, A. (1997) The home as "field": households and homework in rural Appalachia. In J. P. Jones, H. J. Nast and S. M. Roberts (eds), *Thresholds in Feminist Geography: Difference, Methodology, Representation*. Lanham, MD: Rowman & Littlefield.

Oberhauser, A. (2002) Examining gender and community through critical pedagogy. *Journal of Geography in Higher Education*, 26(1), 19–31.

Parker, L. and Lynn, M. (2002) What's race got to do with it? Critical Race Theory's conflicts with and connections to qualitative research methodology and epistemology. *Qualitative Inquiry*, 8(1), 7–22.

Pratt, G. (1993) Reflections on poststructuralism and feminist empirics, theory and practice. *Antipode*, 25, 51–63.

Pratt, G. (2002) Studying immigrants in focus groups. In P. Moss (ed.), *Feminist Geography in Practice: Research and Methods*. Oxford: Blackwell.

Pratt, G. in collaboration with the Philippine Women Centre (1998) Inscribing domestic work on Filipina bodies. In H. J. Nast and S. Pile (eds), *Places through the Body*. London: Routledge.

Price, P. L. (1999) Bodies, faith, and inner landscapes: rethinking change from the very local. *Latin American Perspectives*, 26(3), 37–59.

Professional Geographer (1994) Women in the field: critical feminist methodologies and theoretical perspectives. *Professional Geographer*, 46, 54–102.

Professional Geographer (1995) Should women count? The role of quantitative methodology in feminist geographic research. *Professional Geographer*, 47, 426–66.

Professional Geographer (1999a) Multi-method research: an introduction to its application in population geography. *Professional Geographer*, 51, 40–89.

Professional Geographer (1999b) And the question shall determine the method. *Professional Geographer*, 51, 240–320.

Pryse, M. (2000) Trans/feminist methodology: bridges to interdisciplinary thinking. *NWSA Journal*, 12(2), 105–18.

Pulido, L. (1997) Community, place, and identity. In J. P. Jones, H. J. Nast and S. M. Roberts (eds), *Thresholds in Feminist Geography: Difference, Methodology, Representation*. Lanham, MD: Rowman & Littlefield.

Ramazanoğlu, C. (1993) *Up against Foucault: Explorations of Some Tensions between Foucault and Feminism*. London: Routledge.

Ramazanoğlu, C. and Holland, J. (2002) *Feminist Methodology: Challenges and Choices*. New York: Sage.

Reinharz, S. (1992) *Feminist Methods in Social Research*. Oxford: Oxford University Press.

Riessman, C. K. (1993) *Narrative Analysis*. Newbury Park, CA: Sage.

Roberts, H. (ed.) (1981) *Doing Feminist Research*. New York: Routledge & Kegan Paul.

Rocheleau, D. (1995) Maps, numbers, text and context: mixing methods in feminist political ecology. *Professional Geographer*, 47, 458–66.

Rose, D. (1993) On feminism, method and methods in human geography: an idiosyncratic overview. *Canadian Geographer*, 37, 57–61.

Rose, G. (1993) *Feminism and Geography: The Limits of Geographical Knowledge*. Minneapolis, University of Minnesota Press.

Rose, G. (1997) Situating knowledges: positionality, reflexivities and other tactics. *Progress in Human Geography*, 21, 305–20.

Rose, G. (2001) *Visualizing Methodologies: An Introduction to the Interpretation of Visual Materials*. London: Sage.

Sandoval, C. (1991) US Third World feminism: the theory and method of oppositional consciousness in the postmodern world. *Genders*, 10, 2–24.

Sandoval, C. (2000) *Methodology of the Oppressed*. Minneapolis: University of Minnesota Press.

Shilling, C. (1993) *The Body and Social Theory*. Thousand Oaks, CA: Sage.

Shuurman, N. and Pratt, G. (2002) Care of the subject: feminism and critiques of GIS. *Gender, Place and Culture*, 9(3), 291–9.

Smith, B. (1983) *Home Girls: A Black Feminist Anthology*. New York: Kitchen Table, Women of Color Press.

Smith, D. (1990) *Texts, Facts, and Femininity: Exploring the Relations of Ruling*. New York: Routledge.

Smith, D. (1998) *Writing the Social: Critique, Theory and Investigations*. Toronto: University of Toronto Press.

Smith, L. T. (ed.) (1999) *Decolonizing Methodologies: Research and Indigenous Peoples*. London: Zed Books.

Sparke, M. (1996) Displacing the field in fieldwork: masculinity, metaphor and space. In N. Duncan (ed.), *BodySpace: Destabilizing Geographies of Gender and Sexuality*. London: Routledge.

Stacey, J. (1988) Can there be a feminist ethnography? *Women's Studies International Forum*, 11, 21–7.

Taylor, S. (2001) Evaluating and applying discourse analytic research. In M. Wehterell, S. Taylor and S. J. Yates (eds), *Discourse as Data*. London: Sage.

Thorne, S. and Varcoe, C. (1998) The tyranny of feminist methodology in women's health research. *Health Care for Women International*, 19, 481–93.

Valentine, G. (2002) People like us: negotiating sameness and difference in the research process. In P. Moss (ed.), *Feminist Geography in Practice: Research and Methods*. Oxford: Blackwell.

Visweswaran, K. (1994) *Fictions of Feminist Ethnography*. Minneapolis: University of Minnesota Press.

Williams, S. J. and Bendelow, G. A. (1998) *The Lived Body: Sociological Themes, Embodied Issues*. London: Routledge.

Winchester, H. P. M. (1999) Interviews and questionnaires as mixed methods in population geography: the case of lone fathers in Newcastle, Australia. *Professional Geographer*, 51(1), 60–7.

Wolf, D. (ed.) (1996) *Feminist Dilemmas in Fieldwork*. Boulder, CO: Westview Press.

Women and Geography Study Group (1997) *Feminist Geographies: Explorations in Diversity and Difference*. Harlow: Longman.

Chapter 5

Transnational Mobilities and Challenges

Brenda S. A. Yeoh

Introduction

Feminists have generally approached the emergent field of transnationalism with a certain degree of ambivalence (Pratt and Yeoh, 2003). On the one hand, it is clear that new conceptual maps are needed to capture a range of migrations today, as mobilities often no longer take the form of permanent ruptures, uprooting, and settlement, but are more likely to be transient and complex, ridden with disruptions, detours, multi-destinations, and founded on interconnections and multiple chains of movement. The terrain sketched by transnationalism, with its emphasis on "to-ing and fro-ing" and multistranded linkages across space, appears promising. It has also been noted that not only have mobilities across borders intensified in complex ways, but there has been a feminization of many of these flows as a result of changing production and reproduction processes worldwide. Globalization processes, in which production activities are relocated from core economies to those of the periphery to take advantage of cheaper input costs, have drawn on the pre-existing gender relations and targeted cheap and flexible female workers – many from countries with declining employment opportunities – to enter factory employment in export processing zones and industrial parks in rapidly industrializing countries. The other numerically more important form of female labor migration is linked to reproductive activities, such as domestic service and the sex industry. The increasing demand for women's paid domestic work on an international scale is due to: (a) the decline in state support for childcare and care for the elderly and infirm; (b) an increase in women's labor force participation in industrializing countries, resulting in a crisis on the domestic front; and (c) the expansion of hospitality and sexual services as male executives and entrepreneurs become more mobile. The increasingly ramifying web of transnational flows where women are key participants in a wide range of capacities clamors for the attention of feminist scholars.

On the other hand, there are several interrelated reasons for retaining a sense of ambivalence about the value of transnational perspectives. First, against the claims that the transnational subject (along with the diasporean subject) embodies liberatory potential which could challenge "oppressive nationalism, repressive state structures, and capitalism," Ong (1999, p. 15) reminds us of the "diverse forms of interdependencies and entanglements between transnational phenomena and the nation-states" which trouble any easy assumptions equating mobility with emancipatory impulses. Indeed, the hypermobility and the easy transgression of national borders in today's globalizing world may well be liberating or emancipatory for the individuals involved, but may also reinforce existing social ideologies, including those of the nation-state. It is important in a feminist approach to transnationality[1] not to somehow attribute a sense of inevitability in thinking about transnational processes but to look them squarely in the eye and ask the all important question as to who reaps the gains and who bears the costs of rapidly developing transnational social practices. Is it the transnational subject himself or herself, in subjecting his or her body to the constant state of being "neither here nor there"? Or are transnational families at risk, and does the cost fall mainly on women who often bear the emotional costs of holding the family stretched across countries together? Or do those outside the transnational project, such as the poor and those with little access to networks, bear the brunt of growing transnationality as a privileged discourse?

There is also the suspicion that major accounts of cultural globalization to which the transnational framework is linked, such as the work of Appadurai (1990), are masculinist, for they "make no attempt to identify the processes that increasingly differentiate the power of mobile and non-mobile subjects" (Ong, 1999, p. 11). In traversing transnational space, men often feature as entrepreneurs, career-builders, adventurers, and breadwinners who navigate transnational circuits with fluidity and ease, while women are alternatively taken to be missing in action from globalized economic webs, stereotyped as exotic, subservient, or victimized, or relegated to playing supporting roles, usually in the domestic sphere (Yeoh et al., 2000). As Freeman (2001, p. 1018) observes, "travel, with its embodiments of worldliness, adventure, physical prowess, and cultural mastery, is widely constructed as a male pursuit." In short, that "women stay home and men go abroad" is a taken-for-granted expectation (Clifford, 1997, p. 6) and as such, "when diasporic experience is viewed in terms of displacement rather than placement, traveling rather than dwelling, and disarticulation rather than rearticulation, then the experiences of men will tend to predominate" (Clifford, 1997, p. 259).

Third, the "transnational optic" (Smith, 2001) emerges out of an untidy terrain of overlapping similar-but-different concepts – from "diaspora" to more "inclusive" accounts of globalization – which have already been proposed, evaluated, critiqued, "tried and tested" for their value and power in producing gendered geographies which could cut through layers of the accumulated fat of masculinist thinking to reveal the vital sinews of the discipline. For example, Anthias (1998, p. 572) has called for the need to "gender the diaspora," so as to acknowledge the different "ways in which men and women of the diaspora are inserted into the social relations of the country of settlement, within their own self-defined 'diasporic communities' and within the transnational networks of the diaspora across national borders." From a different perspective, in her attempt to rethink "the gender of globalization," Freeman (2001, pp. 1008, 1010, 1032) argues that because "not only has globalization *theory* been gendered masculine but the very processes defining globalization itself . . . are implicitly ascribed a masculine gender," the task before us is to "attune our critical gaze to the range of actors and practices on the global stage" such that "producers, consumers, and bystanders of globalization are not generic bodies or invisible practitioners of labor and desire but are situated within social and economic processes and cultural meanings that are central to globalization itself."

Amidst these ongoing endeavors, does shifting the focus of feminist critique toward transnationalism offer us a new set of lenses with a better chance of rendering the hitherto-invisible visible? Freeman (2001, p. 1017) is of the view that the more important task lies with challenging the constitution of hegemonically persistent structuralist models such as globalization theory, rather than redirecting attention towards the new vocabulary of the "transnational." Also giving primary attention to "locating globalization," Nagar et al. (2002) advocate the use of feminist understandings of global processes to challenge conceptions of economic globalization, insisting on greater attention to informal spheres of economies, cultures, and politics (and not just the formal and public spheres of globalization), emphasizing the community, household, and bodily scales (and the way they interconnect with the national and supranational) and including neglected subjects and subjectivities of marginalized groups (and therefore broadening "actors" of economic globalization beyond formal political and economic institutions). These feminist scholars do go on to note that a major key to creative and critical engagement with globalization theory is to place "questions of mobility and subjectivity at the center," as "such feminist interventions emphasize how economic dimensions of contemporary globalization processes are thoroughly entangled with the ways that these same processes reconstitute and reinvigorate preexisting social hierarchies" (Nagar et al., 2002).

The importance of grappling with the way "embeddedness" and "mobility" work with and against each other in specific contexts is also championed by Smith (2001), who finds the use of the "transnational optic" as a counter to the globalization perspective promising, as it is capable of acting like a bifocal lens which brings into view ordinary people on the move and at the same time frames them within contested historical and geographical contexts as socially and spatially situated subjects. Given its inherent accents on mobility, and at the same time interconnectedness, of people and products through space and across lines, the transnational perspective throws into sharp relief the different ways in which men and women are mapped into transnational flows, and their differential positionalities in taking advantage of, or resisting, schemes of mobility across international borders. The terrain opened up by the transnational "optic," while uneven and fragmented, offers a salient opportunity to rethink key concepts underpinning contemporary social life, from notions which serve to "ground" social life such as "family," "community," "place," "home," "nation," and "identity," to those which "transgress" or "unmoor," including "mobility," "migrancy," and "transience." Indeed, what constitute "groundings," which locate people in particular places, as opposed to "unmoorings," which destabilize these localizations, have become inextricably linked. The interplay between "groundings" and "unmoorings," and the accompanying possibilities, challenges, and constraints, are of interest to feminist geographers working in at least three interrelated spheres: the formation of transnational families and women's identity negotiations; the relationship between transnational migrant women and the nation-state; and women's agencies and activisms and the challenges of constructing transnational civil society.

Transnationalism and the Family

Scholars have shown that the relationship between "transnational migration" and "family" are mutually constitutive, arguing on the one hand that migration as a life-changing decision and process is deeply embedded, and must be understood, in the context of family norms, relations, and politics; and on the other that transmigratory moves often reconstitute the "family" in ways which are sometimes destabilizing, sometimes affirming. It is increasingly acknowledged that the transnational family – generally regarded as a family where one or more constituent core members are distributed in two or more nation-states, but which continue to share strong bonds of collective welfare and unity – is a strategic response to the changing social, economic, and political conditions of a globalizing world (Parreñas, 2001; Bryceson and Vuorela, 2002). As Chan (1997, p. 195) puts it, dispersal is "often a rational family decision to preserve the family, a resourceful and resilient way of strengthening it: families split in order to be together translocally."

Understanding transnationalism as a "family" strategy, however, does not imply an uncritical acceptance of a unitary and unproblematic notion of the "family" – especially common when considering certain strains such as the "Asian family" – as one where decision-making is hierarchical and therefore consensual, where individual desires are usurped by the "greater good" of the family, and which is based on a "nostalgic vision of femininity" (Stivens, 1998, p. 17) and accepted as inherently (read "culturally") "Asian." Instead, the increasing participation of Asian women in migration and their absences from not only the home but also the home-nation have the potential of both perpetuating and challenging the ideology of Asian familialism advocated by the state as a means of "manufactur[ing] consent and contain[ing] dissent" (Stivens, 1998, p. 17; see also Elmhirst, 2000; Huang et al., 2000). On the one hand, the principle of "for the sake of the family" or "all in the family" – a principle which mobilizes family members to work toward common interests – is clearly part of the discourse that many would-be female labor migrants sustain in seeking contract work abroad. The notion of the migrant as a "martyr mother," "dutiful daughter," or "sacrificial sister" holds considerable currency (Yeoh and Huang, 2000; Asis, 2002) in fueling the desire to help the family, whether this takes the form of bettering the family's material well-being, supporting elderly parents, or furthering their children's or siblings' education. In fact, Hugo (2002) notes that in the case of Indonesian migration to neighboring countries, the deployment of women family members to overseas labor markets "may be part of a strategy to maintain the [patriarchal] status quo" and that "family control of women migrants can still be exercised via the strong social networks which have developed between origin and destination." Similarly, Ong (1999, p. 118) describes the Chinese family regime as one where "an individual's sense of moral worth is based on endurance and diligence in income-making activities, compliance with parental wishes, and the making of sacrifices and the deferral of gratification, especially on the part of women and children." The resilience of such a regime and its underlying ideology does not necessarily dissipate in the face of migrating family members, for dispersal may ironically be seen as a resourceful way of strengthening the family, as well as (patriarchal) control over the family.

On the other hand, the impulse to migrate for particular individuals, particularly women, may be rooted in a desire to escape from the strictures of the traditional patriarchal family and construed as "flight" from home. The use of "flight" from one's natal community to "fight" constricting gender norms and identities has been noted among young single women who take the route of migration as a means of freeing themselves from parental control and making more independent choices about marriage (for example, marrying foreigners becomes a distinct possibility) (Elmhirst, 2002; le Espritu, 2002; Suzuki, 2002). In Suzuki's (2002) account, for example, Filipino women sever affective ties to the family and resort to marriage and/or labor migration to escape gendered surveillance and sexual violence. Trapped in a masculinist regime with double standards – women's sexual purity and chastity is tightly controlled while men's overstepping of sexual boundaries through extra-marital affairs and rape is often overlooked – these women decide to "gamble" with the risks involved in unmooring themselves from "home" to "navigate a potentially 'suicidal voyage' on the unfamiliar sea of migration." While marriage to Japanese men or working as "entertainers" in Japan may not always offer the emancipation that the women seek (and in fact brings with it a whole new set of gender negotiations), Suzuki (2002, p. 116) shows that they find empowerment through breaking "hegemonic constructions of the unwed daughter as virginal and dutiful" and finding "alternative moralities" to counter patriarchal control and remake their agency in striving for "new life chances in an otherwise unkind world." Such a breaking away from the family may be permanent, but is often provisional, for family ties and obligations may in time reassert themselves (Tyner, 2002).

The notion of a "transnational family" where family members are separated over long distances for prolonged stints is not new. Historically, the multilocal residence of Chinese migrant families in different parts of the world, for example, was a common phenomenon – male migrants (known as *huaqiao*) seeking work as gold or tin miners, traders, and railway, plantation, and manual laborers in Southeast Asia, North America, Australia, and New Zealand in the nineteenth and early twentieth centuries often had wives and children left behind in the ancestral homeland. In fact, some eventually set up at least two "homes," and had more than one wife, often a "first" wife in China, and a "second" (and possibly "third" and "fourth" as well) wife elsewhere. Today, what may be new, however – apart from the feminization of labor migration in Asia – is the fact that family relations stretched across distance have to be managed in the context of the nation-state framework, which is predicated on "the idea of a primary belonging to one society and a loyalty to just one nation-state" (Castles, 2000, p. 8). Individual family members who find themselves located in different nation-states wrestle with different policies governing the politics of incorporation, including basic questions such as: What constitutes a citizen? What rights and responsibilities are conferred on migrants in different categories and what regulations are they subjected to? How is ethnic or cultural difference at the boundaries of the nation-state managed? Sustaining the "family" transnationally as a social, emotional, and economic unit hence also depends on making sense of, capitalizing upon, or simply living with the different sets of conditions confronting individual members within the framework dictated by the nation-state.

In this context, the economic, social, and emotional costs of transnationalizing the family fall unevenly on different individual members of the family, as long-term and long-distance separation across national boundaries may require a hardening of the prescribed division of labor assigned to each member in order to maintain the "flexibility" of the family as a whole. This is, for example, indicated in the "astronaut family," a particular form of the transnational family which originally arose as a result of the increasing pace of skilled migration from Hong Kong in the run-up to the return of the British colony to China in 1997. Fearing the financial and social instability which might be set in train with the new Chinese government, Hong Kong businessmen and professionals have sought to relocate their families in countries such as Australia, New Zealand, Canada, and the United States, while they return to work or manage their businesses in Hong Kong. For these "reluctant exiles" (Skeldon, 1994, p. 2), the rationality behind such an arrangement is to spread economic risk and maximize social benefits for their families. Different permutations of the transnational middle-class family have emerged: Chee (2003) writes about Taiwanese women who migrate in their capacity as "mothers" accompanying their children to study in the United States, while financial support is provided by the father who remains in Taiwan as the breadwinner; Yeoh and Willis (2004) argue that transnational entrepreneurialism encouraged by the state as a means of building Singapore's "second-wing" economy in China "extends and crystallizes male roles as exemplary fighters, breadwinners, adventurers, husbands, fathers, and lovers, against a background where women somehow remain sequestered in a supportive, domestic sphere." While much of the work (Man, 1997; Ong, 1999; Chee, 2003) has indicated that middle-class women belonging to such transnational households bear the brunt of social and personal adjustments needed to make the transnational family work – primarily from having to relinquish careers and being assigned to the domestic sphere – Waters's (2002) analysis of "astronaut" households in Vancouver showed that these conclusions need to be taken as provisional. In her words, "traditional conceptions of settlement and adaptation over time retain some explanatory power when confronting the experiences of particular family members. . . . Women's sense of oppression was gradually transformed over time, . . . as they consequently gained in confidence" (Waters, 2002, p. 130). In time, the "transnational family" may assume the contours of the more established "immigrant family" seeking to adapt to a new homeland, or may indeed transmute further.[2]

Migrant women hence continue to shoulder the "pains and gains" of simultaneous embeddedness in "home" and "host" countries, and the need for constant mobility in linking the two in order to sustain the family. While there can be no easy generalizations as to whether the "gains" outweigh the "pains," it is likely that migration involving the movement of women away from the immediate control of traditional, often patriarchal, families is usually, on the whole, an empowering process for these women (Hugo, 2002). Women who are "left behind" in the source area from which menfolk migrate may also find themselves taking on a wider range of roles and responsibilities and may become more autonomous and involved in decision-making within the family and community. In contrast, le Espritu (2002) argues that while being "left behind" might give Filipino women more authority over family governance, it also increases their domestic burdens and overall work-loads. It is also unclear whether inversions of the traditional gender division of labor and male privilege resulting from transnational migration is a transient phenome-non. Le Espritu (2002) shows that male Filipino migrants who became navy stew-ards doing feminized work may bring back to the family fold a host of domestic skills; at the same time, it is also possible that the "stripping of male privilege" expe-rienced by these migrants may result in their attempting to reclaim their masculin-ity by denigrating women and children. When it is the women who migrate to seek employment opportunities beyond national borders, a reversal of gender roles is also not automatic, as traditional patterns and expectations of what men and women do tend to be "sticky." It is hence important to underscore the fact that any gains in gender equality in transnationalizing the family are uneven and hard fought for, often entailing conflict and confrontation.

Transnationalism and the Nation-state

Nonini (2002, p. 5) argues that transnational labor movements cannot be construed in purely economic terms. Instead, "when labor migrates transnationally, this is not an incidental fact but signals crossing from one territorially and culturally specific regime of power and knowledge to another. . . . In moving across national bound-aries laborers are moving in and out of range of regulation by these territorially based regimes." As transnationalism draws attention to what it negates – that is, the continued significance of the "national" in defining territorially based regimes of power and knowledge – it continues to remind us that we have far from reached a postnationalist state of affairs. In the face of the "transgressive fact" (van der Veer, 1995) of increasing transnational migration, nationalisms on the part of both sending and receiving states have in fact been reignited to strengthen or reconfigure the nation-state. Often, in coming to grips with the relationship between transna-tional communities such as ethnic "minorities" and "guest"-workers and their "host," "mainstream" societies, "the presence of the migrant 'other' is used . . . in the nationalist discourse of the established" (van der Veer, 1995, p. 7).

In exploring the relationship between transnational flows and nation-building projects, feminist scholars have raised crucial issues of nation, state, and citizenship, arguing that gendering and racializing of such migrant flows have further compli-cated definitions of citizenship and the constitution of civil society. Stasiulis and Bakan (1997, pp. 118–19) explain that today citizenship is no longer a "linear, static, thing-like status" bestowed by a single state on an individual but "a terrain of struggle" shaped by ideologies of gender, race, and class operating within and beyond national levels. They argue that citizenship should be considered a "nego-tiated relationship" between a particular nation-state and individual. Using a case study of Filipina and West Indian domestic workers in Toronto, they argue that foreign domestic workers are exploited, as they have little protection from the Canadian government, and their national governments are unable to intervene in the affairs of another nation. In addition, they highlight the differential incorporation of Filipinas and West Indian domestic servants into the employment and citizenship framework. These differentials represent variations in Canadians' gendered and racialized conceptions of these women, and also in the strategies adopted by the women themselves. Women transmigrants during their sojourn as labor migrants often find themselves accorded few citizen's and civil rights in destination countries, trapped within patriarchal notions of "women's work" and "women's place" (see also Chin, 1997; Pratt, 1999).

The possibilities of gradual incorporation into receiving nation-states and societies available to middle-class women (and the consequent diminution of transnationality as a family strategy) belonging to "astronaut" families (discussed previously) are not open to women who move transnationally as low-skilled labor migrants. In the rapidly industrializing nations of East and Southeast Asia (e.g. Hong Kong, Taiwan, Malaysia, and Singapore), for example, female transnational migration is concentrated in domestic services (also entertainment work) to fill the reproductive labor gap as local women are increasingly absorbed in paid work outside the home. Brought in to perform what is perceived to be low-skilled work, these women are admitted as transient contract workers embodying cheap labor, not as social or political subjects, and as such, are accorded few rights of participation in wider civil society. In Singapore, for example, a range of policy measures have been put in place to ensure surveillance of migrant bodies and that they gain no permanent structural foothold in the geobody of the nation; these include framing migration conditions in such a way as to tie the validity of the work permit to employment under a specific employer at a specific address, and preventing family formation and settlement by disallowing accompanying dependants, prohibiting marriage to Singapore citizens and permanent residents, and immediately repatriating the worker if she is found to be pregnant (Yeoh and Huang, 1999). In short, "by virtue of being a woman, a foreigner, a domestic, and a menial, not only is the 'maid' in Singapore significantly excluded from the material spaces in the public sphere but also her physical invisibility signals the lack of a foothold on the metaphorical spaces opened up in recent public discourse on potentially more inclusive notions of citizenship and civil society" (Yeoh and Huang, 1999, p. 1164).

It is precisely these policies ensuring the transience of female foreign bodies in the receiving nation-state that, in turn, reinforce the permanence of transnational family forms among unskilled labor migrants. With little chance of sustainable employment in her home country where she is a citizen and even less likelihood of becoming an immigrant-turned-citizen in the country where she is employed, the migrant domestic worker is locked into unending circuits of transnational care, affection, money, and material goods in order to sustain the family in its transnational form. In Asia, where the notion of turning immigrants (apart from the highly skilled) into citizens is unthinkable, marriage to a national of the host country thus presents the unskilled contract worker the only opportunity (at least available in some countries) of "achieving secure residential permits before embarking on the still bumpy road to gain full citizenship rights" (Piper and Roces, 2003, p. 16), as well as "a way of addressing the 'shame' of domestic work and affiliating herself to her host society, while distancing herself from some of the stereotypes of her ethnic group" (McKay, 2003, p. 49).

The rules of marginality and "otherness" which operate to keep transmigrant contract workers in their place are often refracted through the gendered lenses of the host nation; as a result, women transmigrants when put into a comparative frame with their male counterparts find their bodies subject to a more oppressive disciplinary framework, their skills further devalorised, and their spaces even more circumscribed (Mackie, 2002; Huang and Yeoh, 2003). This further deepens, and naturalizes, the lines of gender inequalities already etched on the body of the nation (even as the entry of transmigrants no doubt produces new inequalities), for such transnationalisms fail to transcend, or trouble, the ideological gender bases upon which nationalisms are built. In Japan, for example, Mackie and Taylor (1994) note that while male migrant workers in construction and manufacturing are often discussed in terms of labor policy, immigrant women (mainly from the Philippines) who work in the entertainment industry are more often than not discussed in terms of morality and policing. Thus, while male immigrant workers engaged in productive labor are given space in discussions about policy matters, immigrant women workers remain beyond the pale where citizenship discourses are concerned. Taking a different perspective, Aguilar (1996) revisits the centre–marginality relationship by considering how the "export" of citizens to become "exilic communities" elsewhere reshapes the national consciousness at "home" as well as among the global Filipino elite. He argues that the transnational diaspora of Filipinas, many of whom enter host economies as contract domestic workers, serves to conflate the "Filipina" with a "transnational servant class," an image which generates a politics of "transnational shame" from the point of view of the Filipino professional elites, also in the business of navigating the globalizing world.

Precisely because the e/immigrant other is a gendered subject and precisely because the state articulates nationalism by employing "genderic" modes (offering women and men specific positions as "mothers of the nation" and "defenders of the nation" and making specific appeals to women and men as gendered subjects: see Yuval-Davis and Anthias, 1989; Radcliffe, 1990; Westwood, 1995), the potentially disruptive absence/presence of female e/immigrant others needs to be carefully managed by the state through the politics of inclusion and exclusion. Deconstructing state migration and labor policies relating to the import and export of labor to reveal their ethnicized, classed, and gendered connotations continues to form an important research arena for understanding the relationship between transnationalism and the nation-state.

Transnationalism and Civil Society

Feminist scholars have been generally wary about overstating the potential of constructing transnational civil society and opening "new" interconnected political spaces of emancipatory value. It is first of all important to note that activisms – even if intended to be transnational – are often rooted in particular local contexts. As Mahler (1999, p. 710) notes, "transnational processes may produce new spaces, but this does not mean that actors within these spaces are set completely loose from their social moorings. The tether [including that of place] may be loosened, redirected, and perhaps frayed but not lost into an imaginary 'third space'." It should be remembered that while women as transmigrants may built for themselves new roles and new political spaces, it is also clear that because such women continue to "maintain connections with homelands, with kinship networks, and with religious and cultural traditions" in complex and strategic ways, they may renew patriarchal structures in their new homelands which are at times empowering, and at other times limiting (Clifford, 1997, pp. 258–9). Thus, Dwyer (2000) argues that the young girls in her study are expected to maintain "appropriate attire" (i.e. traditional South Asian dress) as a reflection that they have not succumbed to English values; on the other hand, the "rhetoric of the veil" may be drawn upon by some as a means of gaining greater freedom, because those who don Islamic dress are perceived as worthy of trust. Kibria (1990) uses the case of Vietnamese migrants in Philadelphia to show that the "patriarchal bargain" can be renegotiated on arrival in the United States, but that many elements of gender relations remain the same. Women transmigrants seem prepared to uphold the patriarchal system within the Vietnamese community, in return for economic and social protection by men in an unfamiliar environment.

The difficult challenge of constructing transnational civil society lies in conceptualizing gender politics as a set of embodied practices played out in the context of a specific nation-state, and at the same time tracing transnational connections between activisms and across different histories and geographies. Mackie (2002), for example, argues that the seeds of oppression against which women mobilize are often "created through the histories that connect people in different nations." In her analysis of the "spaces of difference" formed as a result of the insertion of immigrant others into the once-thought-to-be homogeneous fabric of Japanese society, Mackie (2002) reminds us that "the relationships between immigrants and their relatively privileged hosts in Japan have been shaped by a history of imperialism and colonialism and the features of the contemporary political economy of East Asia." The Japanese nation-state has to move from a time in the 1970s when it could assume that embodied encounters with "difference" in the form of Southeast Asian women could be safely displaced offshore (as played out in sex tourism and other sexualized practices of "gazing" on the rest of Asia), to having to confront the presence of these "others" within its own boundaries. Immigrant women who enter Japan through labor and marriage migration are often marked by sexualized images, a construction with which Japanese immigration policy is complicit, for immigrant female workers are barred from being employed as domestics and are limited to entering the country under the legal category of "entertainer," which is often a mask for the provision of sexualized activities, from singing and dancing, waitressing and hostessing, to prostitution.

In response, Japanese activist groups such as the Asian Women's Association situate their aims of activism within the broader context of Asia, seeking "narratives of liberation in the histories of Third World women" in order to imagine "a form of resistance they could not find in their own history." Mackie (2002) shows that in dealing with "spaces of differences" resulting from the placing of immigrant women within the nation, these activist groups have not only attempted to counter negative representations of immigrant workers and provided material assistance (such as the setting up of shelters for immigrant workers), but also "taken their struggles back into international circles through such forums as the United Nations Conference on Women in Beijing in 1995." This has helped to place the living conditions of immigrant labor in Japan on the international agenda, alongside the issue of military prostitution during the Second World War and the problem of militarized sexual violence in Okinawa.

Also concerned with similar questions as to whether advocacy work among women's non-government organizations (NGOs) on migration issues can remain nation-bound, but from a different perspective, Law (2002) examines what she calls "the emerging spaces of transnational cultural production in the realm of political activism." Unlike Mackie (2002), who focuses on the activist groups constituted by host society working for immigrant others, Law (2002) locates the activist discourses and networks forged among NGOs for migrant women of different nationalities present in Hong Kong to illustrate "the tensions between nationalism and transnationalism in the global cultural economy." Drawing on several campaigns spearheaded by a coalition of NGOs in Hong Kong to address the protection of migrant workers' rights, Law (2002) shows how NGOs drew connections between the specificities of labor migration issues and government policies in both the Philippines (the sending country where the majority of female domestic workers in Hong Kong originate) and Hong Kong (the receiving country), as well as global discussions about human rights at the Beijing Conference. In this sense, NGOs operate as "transforming terrains which expand the discursive field of their activities," and this in turn connects up with and potentially influences global discourses on human rights. In a globalizing world where the nation is "no longer the key arbiter of important social change" (Appadurai, quoted by Law, 2002), such a conception points to multiple sites of transnational activism within "a broader social space where new alliances between migrant, feminist and workers' organizations" may be envisaged. At the same time, it has to be remembered that while these "terrains" allow for coalition-building – for example, when domestic worker groups of five different nationalities (Filipino, Thai, Indonesian, Sri Lankan, and Nepali) came together to protest against minimum wage cuts – they also decenter other nationally bound issues (such as differential wages among domestic workers) and may themselves be fraught with inequalities. If indeed these transnational terrains and advocacy networks give shape to a "postnational," "diasporic public sphere" as emergent forms of political spaces, it is likely that they will continue to reflect the tensions between national and transnational politics.

In an intensely interconnected world, globalization processes at the supranational scale have "strengthened the claims of powerful actors (such as global corporations, global financial markets, international institutions and treaties)," often at the expense of the vulnerable, including the poor, women, and minorities who find themselves pushed into casualized and flexibilized work (Nagar et al., 2002, p. 7). At the same time, the possibilities of interconnected linkages have also given rise to transnational civil society energized by non-governmental organizations and networks of activists working to create new political spaces from which to act. Transnational activism to improve the lot of women depends for its success on the ability to negotiate and cross social and spatial boundaries, and to take full account of the multiplicity of scales at which processes operate. Feminist geographers are potentially well placed to come to grips with the intersecting politics of space and scale. Going beyond Hall's view (quoted in Nagar, 2002) that the goal of research and the purpose of theory is "to understand the situation you started out with better than before," the litmus test as to whether transnationalism constitutes a productive field for feminist engagement must lie with whether it leads geographers not only to "understand" but to frame disciplinary practices differently to construct "counter topographies" (Katz's term, quoted in Pratt and Yeoh, 2003) which hold out the possibility of remapping the globalized world to recover a more humanized face.

NOTES

1 Aihwa Ong (1999, p. 4) notes that the term "*trans* denotes both moving through space or across lines, as well as changing the nature of something," preferring to use the term "transnationality" to refer to "the condition of cultural interconnectedness and mobility across space," while reserving "transnationalism" to refer to "the cultural specificities of global processes." She usefully reminds us that "transnationality also alludes to the *trans*versal, the *trans*actional, the *trans*lational, and the *trans*gressive aspects of contemporary behavior and imagination that are incited, enabled, and regulated by the changing logics of states and capitalism."

2 Ho (2002) also argues that transnational family arrangements are fluid and in fact may be highly volatile. Using longitudinal survey data involving a resurvey of Hong Kong Chinese youths first surveyed as children accompanying parents who transplanted to New Zealand, Ho documents the strategic transnationality of such family arrangements. In the space of eight years, families which have originally utilized the astronauting strategy have reconfigured through reunion of the whole family in New Zealand, reverse relocation of the whole family back to Hong Kong, continuation of the strategy of having members spanning both countries, or an extension of their transnational networks beyond the source and destination to include other countries. The transnational family is hence a highly fluid formation which transmutes in response to each family's life-cycle changes, family members' personal aspirations, and the wider socio-economic context.

BIBLIOGRAPHY

Aguilar, F. V. Jr (1996) The dialectics of transnational shame and national identity. *Philippine Sociological Review*, 44(1–4), 101–36.

Anthias, F. (1998) Evaluating "diaspora": beyond ethnicity? *Sociology*, 32(3), 557–80.

Appadurai, A. (1990) Disjuncture and difference in the global cultural economy. *Public Culture*, 2(3), 1–24.

Asis, M. M. B. (2002) From the life stories of Filipino women: personal and family agendas in migration. *Asian and Pacific Migration Journal*, 11(1), 67–94.

Bryceson, D. and Vuorela, U. (2002) Transnational families in the twenty-first century. In D. Bryceson and U. Vuorela (eds), *The Transnational Family: New European Frontiers and Global Networks*. Oxford: Berg.

Castles, S. (2000) *Ethnicity and Globalization: From Migrant Worker to Transnational Citizen*. London: Sage.

Chan, K. B. (1997) A family affair: migration, dispersal, and the emergent identity of the Chinese cosmopolitan. *Diaspora*, 6(2), 195–213.

Chee, M. W. L. (2003) Migrating for the children: Taiwanese American women in transnational families. In N. Piper and M. Roces (eds), *Wife or Worker? Asian Women and Migration*. Lanham, MD: Rowman & Littlefield.

Chin, C. B. N. (1998) *In Service and Servitude: Foreign Domestic Workers and the Malaysian Modernity Project*. New York: Columbia University Press.

Clifford, J. (1997) *Routes: Travel and Translation in the Late Twentieth Century*. Cambridge, MA: Harvard University Press.

Dwyer, C. (2000) Negotiating diaspora identities: young British South Asian Muslim women. *Women's Studies International Forum*, 23(4), 475–86.

Elmhirst, R. (2000) A Javanese diaspora? Gender and identity politics in Indonesia's transmigration resettlement program. *Women's Studies International Forum*, 23(4), 487–500.

Freeman, C. (2001) Is local:global as feminine:masculine? Rethinking the gender of globalization. *Signs: Journal of Women in Culture and Society*, 26(4), 1007–37.

Ho, E. S. (2002) Multi-local residence, transnational networks: Chinese astronaut families in New Zealand. *Asian and Pacific Migration Journal*, 11(1), 145–64.

Huang, S., Teo, P. and Yeoh, B. S. A. (2000) Diasporic subjects and identity negotiations: women in and from Asia. *Women's Studies International Forum*, 23(4), 391–8.

Huang, S. and Yeoh, B. S. A. (2003) The difference gender makes: state policy and contract migrant workers in Singapore. *Asian and Pacific Migration Journal*, 12(1/2), 75–98.

Hugo, G. (2002) Effects of international migration on the family in Indonesia. *Asian and Pacific Migration Journal*, 11(1), 13–46.

Kibria, N. (1990) Power, patriarchy and gender conflict in the Vietnamese immigrant community. *Gender and Society*, 4(1), 9–24.

Law, L. (2002) Sites of transnational activism: Filipino NGOs in Hong Kong. In B. S. A. Yeoh, P. Teo and S. Huang (eds), *Gender Politics in the Asian-Pacific Region*. London: Routledge.

Le Espritu, Y. (2002) Filipino navy stewards and Filipino health care professionals: immigration, work and family relations. *Asian and Pacific Migration Journal*, 11(1), 47–66.

McKay, D. (2003) Filipinas in Canada – De-skilling as a push towards marriage. In N. Piper and M. Roces (eds), *Wife or Worker? Asian Women and Migration*, Lanham: Rowman & Littlefield, pp. 23–52.

Mackie, V. (2002) "Asia" in everyday life: dealing with difference in contemporary Japan. In B. S. A. Yeoh, P. Teo and S. Huang (eds), *Gender Politics in the Asian-Pacific Region*. London: Routledge.

Mackie, V. and Taylor, V. (1994) Ethnicity on trial: foreign workers in Japan. Paper presented at the Conference on Identities, Ethnicities and Nationalities, July, La Trobe University, Melbourne, July.

Mahler, S. J. (1999) Engendering transnational migration. *American Behavioral Scientist*, 42(4), 690–719.

Man, G. (1997) Women's work is never done: social organization of work and the experience of women in middle-class Hong Kong Chinese immigrant families in Canada. *Advances in Gender Research*, 2, 183–226.

Nagar, R., Lawson, V., McDowell, L. and Hanson, S. (2002) Locating globalization: feminist (re)readings of the subjects and spaces of globalization. *Economic Geography*, 78(3), 257–84.

Nonini, D. M. (2002) Transnational migrants, globalization processes, and regimes of power and knowledge. *Critical Asian Studies*, 34(1), 3–17.

Ong, A. (1999) *Flexible Citizenship: The Cultural Logics of Transnationality*. Durham, NC: Duke University Press.

Parreñas, R. S. (2001) *Servants of Globalization: Women, Migration and Domestic Work*. Stanford, CA: Stanford University Press.

Piper, N. and Roces M. (2003) Introduction: marriage and migration in an age of globalization. In N. Piper and M. Roces (eds), *Wife or Worker? Asian Women and Migration*. Lanham, MD: Rowman & Littlefield.

Pratt, G. in collaboration with the Philippine Women Centre (1999) Is this Canada? Domestic workers' experiences in Vancouver, BC. In J. Momsen (ed.), *Gender, Migration and Domestic Service*. London: Routledge.

Pratt, G. and Yeoh, B. S. A. (2003) Transnational (counter) topographies. *Gender, Place and Culture*, 10(2), 156–66.

Radcliffe, S. (1990) Ethnicity, patriarchy and incorporation into the nation: female migrants as domestic servants in Peru. *Environment and Planning D: Society and Space*, 8, 379–93.

Skeldon, R. (1994) Reluctant exiles or bold pioneers: an introduction to migration from Hong Kong. In R. Skeldon (ed.), *Reluctant Exiles? Migration from Hong Kong and the New Overseas Chinese*. Hong Kong: Hong Kong University Press.

Smith, M. P. (2001) *Transnational Urbanism: Locating Globalization*. Oxford: Blackwell.

Stasiulis, D. and Bakan, A. B. (1997) Negotiating citizenship: the case of foreign domestic workers in Canada, *Feminist Review*, 57, 112–39.

Stivens, M. (1998) Sex, gender and the making of the new Malay middle classes. In K. Sen and M. Stivens (eds), *Gender and Power in Affluent Asia*. London: Routledge.

Suzuki, N. (2002) Gendered surveillance and sexual violence in Filipina pre-migration experiences to Japan. In B. S. A. Yeoh, P. Teo and S. Huang (eds), *Gender Politics in the Asian-Pacific Region*. London: Routledge.

Tyner, J. A. (2002) The globalization of transnational labor migration and the Filipino family: a narrative. *Asian and Pacific Migration Journal*, 11(1), 95–116.

van der Veer, P. (1995) Introduction: the diasporic imagination. In P. van der Veer (ed.), *Nation and Migration: The Politics of Space in the South Asian Diaspora*. Philadelphia: University of Pennsylvania Press.

Waters, J. L. (2002) Flexible families? "Astronaut" households and the experiences of lone mothers in Vancouver, British Columbia. *Social and Cultural Geography*, 3(2), 117–34.

Westwood, S. (1995) Gendering diaspora: space, politics and South Asian masculinities in Britain. In P. van der Veer (ed.), *Nation and Migration: The Politics of Space in the South Asian Diaspora*. Philadelphia: University of Pennsylvania Press.

Yeoh, B. S. A. and Huang, S. (1999) Spaces at the margin: migrant domestic workers and the development of civil society in Singapore. *Environment and Planning A*, 31, 1149–67.

Yeoh, B. S. A. and Huang, S. (2000) "Home" and "away": foreign domestic workers and negotiations of diasporic identity in Singapore. *Women's Studies International Forum*, 23, 413–29.

Yeoh, B. S. A., Huang, S. and Willis, K. (2000) Global cities, transnational flows and gender dimensions: the view from Singapore. *Tijdschrift voor Economische en Social Geografie (Journal of Economic and Social Geography)*, 2(2), 147–58.

Yeoh, B. S. A. and Willis, K. (2004) Constructing masculinities in transnational space: Singapore men on the regional beat. In P. Jackson (ed.), *Transnational Spaces*. London: Routledge.

Yuval-Davis, N. and Anthias, F. (1989) *Woman-Nation-State*. Basingstoke: Macmillan.

Part II Work

Chapter 6

Feminist Analyses of Work: Rethinking the Boundaries, Gendering, and Spatiality of Work

Kim England and Victoria Lawson

The social organization of work is a longstanding theme for feminist social scientists. Employment and home–work linkages are enduring themes in feminist geography scholarship and were pivotal in the early development of the field. Feminist geographers, including both of us, are greatly influenced by analyses of work by feminist scholars outside geography. However, in this chapter we have chosen to emphasize geographers' research. Our discussion is not an exhaustive review of the rich and now voluminous feminist literature on work; instead we draw on examples from English language feminist geography literature to illustrate the epistemological, theoretical, and empirical scope of research. We frame feminist research broadly, reflecting an emphasis in geography on the ways in which gender systems operate in relation to the processes and politics of class, "race," ethnicity, (dis)ability, sexuality, and postcoloniality.

Our overview chapter charts the central conceptualizations, approaches to, and emerging questions in feminist geographers' analyses of work. Our discussion is organized around three themes. First, we discuss the ways feminist research explodes the concept of work. A fundamental contribution of feminist research is to trouble the boundaries of the category "work" (a concept that has been largely framed in the West). Specifically, feminists have expanded "work" beyond solely waged work to encompass unpaid reproductive work and the work of politics and community activism. Second, we discuss feminist research identifying the multitude of ways work is gendered. Although many focus exclusively on women, others have expanded the conceptualization of worker identities and subjectivities to include both women and men and how those categories are internally fractured by difference. Finally, we discuss feminist researchers' important contributions to understanding the spatiality of work in the process of economic restructuring. Recently geographers in several subfields have theorized the social production of scale and scalar politics, and we draw on that literature to describe the experience of work from the body to the global scale.

Exploding "Work"

A critical contribution of feminist theorizing is the interrogation of the central and powerful concepts in social science research for their exclusions and absences. Thus, feminist (re)conceptualizations of "work" reveal its historically and geographically constructed boundaries, along with the exclusions set up by Eurocentric notions of the public–private divide and waged work. Feminists have long argued that the boundaries constructed around "work" are part of an ongoing process with its ideological, material, and historical roots in the separation of home and work. This separation was bound up with the rise of a domestic ideology which established the home as refuge supported by the unpaid domestic work of women. Feminist scholars argue that historically the social and spatial separation of waged work from sites of reproduction (like communities and homes) occurred throughout Europe and subsequently North America alongside the rise of patriarchal capitalist systems (Hayford, 1974; Mackenzie and Rose, 1983; Mies, 1999). In the process, "work" became socially constructed as waged activity in the "public" sphere of capitalist production and "non-work" as the activities in the feminized "private" household sphere (like providing a comforting refuge for waged workers). Basically, the concepts of "work" are shaped by the gendered sites where it is performed.

The discursive and material consequences of these binaries linger and still shape the lives of women (and men and children) around the globe. Mainstream definitions of "work" continue to ignore or undervalue informal, domestic sphere, and reproductive work. For instance, most national accounting systems and official censuses ignore unpaid and reproductive labor. As Mona Domosh and Joni Seager (2001, p. 43) note, "Household work, volunteer activities, reproductive activities, subsistence activities, and bartering are entirely absent from national accounts. Of course, all of these arenas of activity are dominated by women." Thus feminist geographers continue to challenge limited conceptualizations of "work" and articulate the ways gendered spatial divisions of labor between "productive" and "reproductive" spaces produce a series of exclusions and devaluations of the many types of work that women do.

Feminist expansions of "work" opened up analysis of diverse sites of work, beyond formal workplaces, to include homework and other unregulated sites (such as sweatshops, domestic, and community work). Contemporary research demonstrates that industrial homework persisted in the "Global North" and "Global South" throughout the twentieth century and that production is shifting *back* into homes in sites as diverse as West Virginia (Oberhauser, 2002), Spain (Baylina and Garcia-Ramon, 1998), Ecuador (Lawson, 1995), and Mexico (Beneria and Roldan, 1987). In cases where dependent subcontract workers (rather than independent producers) undertake production, they are often women and children performing under precarious employment relations and piece rate systems. Researchers first made this home-based waged work visible as a way of critiquing mainstream concepts of work. Then they illustrated how these precarious working conditions result from ideological processes of feminization, positioning the home as a site of domestic "non-work" where waged work is secondary or supplemental. These gendered constructions of home as a *domestic, feminized space* mean that women's homework is devalued, unregulated, and without political representation. Such constructions are meeting with resistance, as Rowbotham (1998) demonstrates in her account of homeworkers organizing and networking internationally. Another key element of these debates examines how the separation (and blurring) of home and work in the West plays out differently for women and men across differences of class, race, nationality, and particular urban contexts (compare England, 1993; Hanson and Pratt, 1995; Massey, 1995; McLafferty and Preston, 1997; Law, 1999).

One feminist strategy for resisting hegemonic concepts of "work" is making visible the variety of work performed in homes and communities, particularly caring and reproductive labor both paid and unpaid. Empirical research in Britain (Gregson and Lowe, 1994), Canada (England and Stiell, 1997; Pratt, 1997), Peru (Radcliffe, 1990), the USA (Mattingley, 2001), and other sites around the globe (see Momsen, 1999, on Africa, Asia, Europe, and the Americas) demonstrate that paid domestic work complicates cherished assumptions in mainstream economics and liberal feminism (like troubling the production/reproduction dichotomy). Paid housecleaners and nannies disrupt the economistic assumption that domestic work is "non-work" performed for love, and (to the extent that this work is recorded) insert these forms of work into national accounting systems. Paid domestic work is often performed by lower income women for middle-class families. Kim's work with Bernadette Stiell (England and Stiell, 1997; Stiell and England, 1999) shows that in the context of women's migration from the Global South to the West, "immigrant" women of color often perform this work for white Western women. The revelations about domestic workers draw attention to class and race/ethnicity divisions *among* women, as well as women's diverse experiences of work.

Other examples relate to unpaid work within the home and community, research that challenges hegemonic concepts of "work" by making visible volunteering, community activism, and mothering. Women have long been recognized as pivotal in building communities (Moore-Milroy and Wisner, 1994; Staeheli, 2003). For instance, lesbians have worked hard to develop "safe spaces" and "out" lesbian territories by building social networks and community organizations in rural settings (Forsyth, 1997) and cities in the USA and Britain (Valentine, 1995), as well as internationally, such as within the Europe Union (Valentine, 1996; Binnie, 1997). Feminist research on community-based organizations providing social services also reveals how unpaid work subsidizes neoliberal reforms by stepping in where government services have withdrawn; for instance, in Ecuador (Moser, 1987; Lind, 2000), Australia (Fincher and Panelli, 2001), and the USA (Staeheli, 2003). Also see chapters on political and caring work in this volume.

Mothering work is a frequent topic of feminist scholarship, and the subject of bitter debate. Mothering is demanding and is steeped in normative discourses – a "labor of love," "the family," and "good mothers." This affective labor is also "care work" and is central to social reproduction. Geographers' work here mainly explores childcare arrangements in the Global North (Rose, 1993; England, 1996; Gilbert, 1998; Holloway, 1998, Preston et al., 2001; but see Yeoh et al., 2000, on Singapore). Some of this work considers class and "race"/ethnicity differences in child care arrangements (Rose, 1993; Yeoh and Huang, 1995; Gilbert, 1998), while other research focuses on how mothers use local networks to negotiate mothering strategies, and find jobs and child care (England, 1993, 1996; Holloway, 1998). Finally, some argue that concepts of "mother" and "mothering" and the home–work boundary be opened up to include men and fathers (Pulido, 1997; Smith and Winchester, 1998), as well as extending the concept of family care to eldercare (Joseph and Hallman, 1998).

Feminist international research is uncovering diverse informal activities combining caring and productive work in ways considered empowering by workers (Lawson, 1995, 1999; George, 1997; Oberhauser, 2002). For example, consider Vicky's (Lawson, 1995, 1999) research with poor women in Quito: while they were indeed a vulnerable workforce, politically constructed as cheap (by industrialists and the Ecuadorian state), their own interpretations were much richer. Several women insisted that homeworking allowed them to be independent and powerful as household heads; to avoid oppressive household situations; or to keep families together. Women entrepreneurs in Worcester, Massachusetts, often begin their own businesses to create greater time–space flexibility and a better balance between paid work and caring work (Hanson and Blake, this volume).

Feminist analyses of work also examine individuals', firms', and states' efforts to address the work–life balance. Dominant themes in European-focused analyses of work are about ways that home–work relations shape, and are shaped by, social policy. A growing transdiciplinary, comparative literature considers the social and spatial implications of welfare state regimes and gender contracts across the EU (Perrons, 1995; García-Ramon and Monk, 1996; Bühler, 1998; Duncan and Pfau-Efflinger, 2001). Indeed, gender equity in the family–labor market–state relation is pivotal to recent efforts at "gender mainstreaming" economic and social policy in the EU (Perrons, 1995; Aufhauser and Hafner, 2002) and in the transition economies of the former Soviet bloc (Smith, 1999; van Hoven, 2001; Timar, 2002).

Research on the home–work balance in the UK and North America tends to focus on the household strategies of dual earner and dual career households (Rose and Villeneuve, 1998; Preston et al., 2000; Clark and Withers, 2002), although such households have also garnered attention in Europe (see, for example, Fagnani, 1993; McDowell, 1997; Jarvis, 1999; Hardill, 2002). These bodies of research all analyze everyday practices that confound the public–private, work–home binaries, and examine the difficulties of negotiating the divide in ways that enable gender equality. At the same time there is the potent narrative of "super woman" constructed around masculinist models of career and success. Yet here too are spaces of transgression. For instance, George (1997) describes leaving a secure academic position in India to focus on being a wife and mother in the Netherlands. She continues to write and do research, simply not in a formal career-building manner. George does not consider her decision a step backwards from a feminist agenda, but a "step sideways to explore new choices and combinations in addition to familiar ones" (George, 1997, p. 67). Drawing on spatial metaphors, George challenges the discourse of individual achievement through workplace competitiveness, and captures the complex interweaving of "work," home, and everyday life.

Gendered Work

Feminist research identifies the myriad ways "work" is socially constructed as "men's work" or "women's work." This gendering of work is rooted in the discursive and material separation of home and work. The ideological construction of caring and reproductive work as "non-work" lowered the value society assigns to women's paid work, while simultaneously constructing women as suitable for only certain types of jobs. For example, in Kim's work with Bernadette Stiell about nannies in Toronto one of the placement agency owners remarked, "I have now come to realize that what the government thinks of as a nanny is just any female – you know, if you are a woman, you must be able to cook, clean and look after children" (Stiell and England, 1999, p. 55).

These ideological constructions also help to produce the gender wage gap, an issue that has received considerable attention in feminist social science research. The social construction of gender valuations in paid work is evidenced in gender wage gaps at a variety of scales. Wage differentials are measured in several ways, including comparisons of women and men with the same educational level, and for women and men working in the same economic sector. Studies look at the wage gaps across a wide variety of countries using national data (Beneria, 2001). Despite improvements in the gender wage gap between 1980 and 1997 (UNIFEM, 2000), women around the world still earn substantially less than their male counterparts. For instance, within the same economic sector, women's wages in Honduras are 78 percent of men's, in Singapore and the UK they are 74 and 72 percent respectively (Beneria, 2001). Research investigates wage differences at the regional scale (Duncan and Pfau-Efflinger, 2001), the subnational scale (McCall, 1998), and the intraurban scale (Carlson and Persky, 1999). However, it remains unclear how much of the declining wage gap results from increases in women's incomes, versus declines (or stagnation) in men's incomes. The declining gender wage gap is also offset by increasing evidence of growing wage polarization *among* women in the Global North (Perrons, 1995; Armstrong, 1997). Wage gaps between women and men, and among women, become especially acute when "race," ethnicity, class, and other social characteristics are added to the equation (McCall, 1998; Rose and Villeneuve, 1998). Clearly a complex mix of factors explains these wage patterns, and feminist geographers should continue to enrich our understanding of how space and place play a role.

The gendering of work is also evident in occupational segmentation. Explanations include sexist practices and gender stereotypes in workplaces that restrict women's access to male-dominated occupations (Reskin and Roos, 1990; Hanson and Pratt, 1995). This research examines the gendering of occupations, of workers in occupations (Pringle, 1989; McDowell, 1997; Lawson, 1999), and gendering of access – including sticky floors and glass ceilings (McDowell, 1997; England and Gad, 2004). For instance, Kim's research with Gunter Gad shows that even in a female-dominated industry like banking, very few women occupy senior management positions.

Geographers raise important questions about the spatiality of occupational segmentation. Some research points to an intraurban geography of gendered occupations in Tel Aviv (Blumen, 1998), Worcester (Hanson and Pratt, 1995), Montréal (Rose and Villeneuve, 1998), and Los Angeles (Wright and Ellis, 2000). Another dimension of the geographies of occupational segmentation concerns the gendering of occupations in particular workplaces (Hahner, 1976; Wolfe, 1993; McDowell, 1997; Wright, 1997; Lawson, 1999). Others analyze social policies put in place to address gendered occupational segmentation (Armstrong, 1997; Aufhauser and Hafner, 2002). There are also growing efforts to recover the diverse positionings of women (and men) across a range of social and spatial settings. Although much research on gendered work is focused on national or urban spaces, *fields* are still important places of work for women and men (Whatmore, 1990; Mackenzie, 1992). Feminist research on agricultural production indicates that while particular gender divisions of labor may vary across settings, studies in India (Raghuram, 1993) and West Africa (Carney, 1992; Schroeder, 1999) illustrate women's enormous contributions to agricultural and subsistence household production.

Feminist geographers have also demonstrated that women who work in feminized occupations are often spatially constrained, both in the distance traveled to work, and in access to transportation (see England, 1993; Hanson and Pratt, 1995; Blumen, 1998; Law, 1999). Recent work on commuting complexifies our understandings around the effects of "race" and ethnicity (Cooke, 1997; Johnston-Anumonwo, 1997; McLafferty and Preston, 1997), family structure, and non-paid employment responsibilities (Fagnani, 1993; Joseph and Hallman, 1998; Kwan, 1999). Complementing longstanding attention to how spaces and occupations gender workers, feminist geographers conceptualize worker identities and subjectivities through the ways in which women and men transgress gendered boundaries of work. Research has typically focused on women moving into "men's jobs" in banking management (McDowell, 1997; England and Gad, 2004), steel mills (Tonkin, 2000), construction (Laurie, 1997), dockyards (Bartham and Shobrook, 1998), and fishing (Munk-Madsen, 2000; Nadel-Klein, 2000; Hapke, 2001). Recent work also considers how masculinities are constructed at work. For instance, McDowell (1997) found that managers in merchant banking define potential for success in terms of competitiveness, toughness, and self-confidence, none of which is a gender-neutral characteristic. Similarly, Johnson (1990) found that as the Australian textile industry shifted from predominantly female to majority male workers, authority and technical skill were invoked to prioritize men for new high technology production processes.

In addition, feminist researchers are examining multifaceted processes of accommodation and resistance to capitalist production in various parts of the globe: for example, Wolf's (1992) work on the contradictory lives of Javanese "factory daughters"; Ong's (1987) study of high technology labor control systems and indigenous cultural forms in Malaysia; Mullings's (1999) research on the subversive workplace strategies of Jamaican data-entry operators; Wright's (1997) work on the ways in which one woman subverts the historical representation of the "Mexican Woman" within a US-owned maquiladora in order to gain skills and be promoted. These studies capture the diverse ways managers construct mobile or docile workers through the deployment of cultural and racial notions of "appropriate" femininity and sexuality. At the same time, this research does not simplistically represent women (or men) as passive victims of capitalist restructuring, but focuses on how workers challenge the feminization of work, reformulate their political identities, and organize through unions or social movements. For instance, Smith (1999) evaluates discourses of citizenship in women's urban-based activism in East Germany. Van Eyck (2002) in collaboration with union organizers, examines the ways in which Colombian bank workers experience and resist deskilling, subcontracting, and feminization of banking work in the context of liberalization and global restructuring of bank work. Mackenzie (1992) argues that local, provincial, and federal political successes of the "Women for the Survival of Agriculture" in Eastern

Ontario bring attention to the plight of the "family farm" and make visible farm women's labor. More generally, rural women's groups are challenging male hegemony in farm work and masculinist farm property rights in Australia (Liepens, 1998; Fincher and Panelli, 2001), Canada (Mackenzie, 1992), and the Gambia (Carney, 1992; Schroeder, 1999).

Scaling Work

Recently geographers have been deeply engaged with theorizing the social production of scale and scalar politics; we draw on those ideas to examine experiences of work from the body to the global. Feminist geographers have long argued that identities and spaces are recursively (re)created, and they have considered the myriad ways in which power and knowledge are reproduced through identity, representation, and space. Recent shifts in feminist analyses of work include a focus on the embedded nature of economic processes in cultural contexts. Linda McDowell and Doreen Massey (1984) published their pioneering work on women's historically and spatially differentiated experiences of work in diverse localities and economic sectors across England. They looked at differences in gender "contracts" through which women and men enter/exit the labor force in distinct localities such as East Anglia, Durham, and Lancashire. That same year Massey also published *Spatial Divisions of Labor*. These influential ideas about spatializing gendered divisions of labor inspired research on the relations between industrial restructuring and both spatialized and gendered divisions of labor in the "Global South" and the "Global North." In the following paragraphs, we touch on a range of questions, raised by feminist geographers, about gendered spatial divisions of labor across a range of scales and settings.

Feminist geographers are interested in the body as a surface of inscription and in how different "bodies" are produced under distinct discursive regimes. The embodied nature of work influences the gendered construction of spatial (and social) divisions of labor, and vice versa. Recent studies examine the clothing, comportment, and gendered performances of managers' bodies in Britain and New Zealand (McDowell, 1997; Longhurst, 2001); while other research examines how sexuality is deployed by both workers and managers as a strategy of control in Mexican maquiladoras (Wright, 1997; Salzinger, 2001). In the Caribbean, Western sex tourists (both men and women), subject Caribbean bodies to sexualizing fantasies and in the process these bodies become "the sites for the construction of (white) North American and Western European power, wealth and well-being" (Kempadoo, 2001, p. 58). Other research considers the "strong male bodies" associated with agricultural work in Australia (Liepens, 1998), while Leslie and Butz (1998) focus on the injured body among Canadian GM car workers. Some bodies disrupt accepted notions of "appropriate" embodied employment and are constructed as "out of place" in their workplace. Examples include women in investment banking in contemporary Britain (McDowell, 1997), Francophone women in early twentieth-century Anglophone Montréal banks (Boyer, 1998), women in Australian steel mills (Tonkin, 2000), people with disabilities in most workplaces (Dyck, 1999), and sex workers in English neighborhoods (Hubbard, 1998).

Workplaces are now under scrutiny as sites where power and knowledge are discursively (re)produced, and as sites implicated in the reconstitution of workers'

subjectivities. Dyck (1999, p. 133) argues that "(w)orkplaces are recursively implicated in the reconstitution of subjectivity as women become defined as 'disabled',"
and Valentine (1996) makes a similar point about lesbians and gay men in EU workplaces. Feminist geographers also explore the cultural dimensions of workplace
dynamics and the role of gender in the organization of firms (Hanson and Pratt
1995; Halford and Savage, 1997; McDowell, 1997). Attention is focused on the
ways in which masculinity and femininity are constructed in the workplace, and
how the embodied nature of work influences spatial divisions of labor within offices,
factories, or fields (for instance, Massey, 1995; Mez and Bühler, 1998; Mullings,
1999; Munk-Madsen, 2000; Tonkin, 2000).

The rapid restructuring of work in the Global South has prompted researchers
to examine how gendered identities and relations get contested at the scale of workplaces. For instance, Vicky's (Lawson, 1999) interviews with garment producers in
Quito revealed struggles over narratives of women as "seamstresses," not "skilled
tailors," which exclude them from tailoring guilds dominated by men. Wright
(1997, p. 278) highlighted "one woman's journey through the ideological representation of her as a 'typical Mexican woman' " in a factory along the USA–Mexico
border, as she works to disrupt this representation and to be promoted. And Laurie
(1997) contrasts the Peruvian state's representation of women as receiving welfare
(through the emergency employment program) with those of women seeing themselves as engaging in "masculine" work (i.e. construction of roads, schools, etc.)
and their concomitant renegotiations of gender identities.

At a larger spatial scale, especially in the USA and Europe, there are geographies
of gender work at the regional scale (Kodras and Padavic, 1993; Perrons, 1995;
García-Ramon and Monk, 1996; Duncan and Pfau-Efflinger, 2001) and intermet-
ropolitan scales (García-Ramon and Monk, 1996; McCall, 1998; Odland and Ellis,
1998; Duncan and Pfau-Efflinger, 2001). Complementing that literature, numerous
studies examine the spatiality of gendered work at the intrametropolitan scale,
perhaps because women have more localized job searches and shorter commuting distances than men (Hanson and Pratt, 1995; Johnston-Anumonwo, 1997;
McLafferty and Preston, 1997). Hanson and Pratt's (1995) classic study of (predominantly white) women in Worcester showed that recruitment practices of
employers and job searches of employees were embedded in localized sociospatial
relations of everyday life. Other work has emphasized highly localized female clerical labor supplies as a locational factor in the suburbanization of service sector jobs
(England, 1993; Mez and Bühler, 1998; Carlson and Persky, 1999; England and Gad,
2003). Finally, a parallel set of inquiries suggest a spatial unevenness to the occupational prestige associated with different workplace locations within cities (Hanson
and Pratt, 1995; Blumen, 1998; Mez and Bühler, 1998; England and Gad, 2003).

The spatiality of work in cities of the "Global South" is quite different from that
in the "Global North." These differences are often expressed in the greater dispersal and/or invisibility of informal work such as street trading or home-based work
(Chant, 1989; Hays-Mitchell, 1993; Aspaas, 1998), in the containment of women's
work in export processing zones (Wright, 1997; Cravey, 1998), or through the intersections between transnational export work and informal economies (Freeman,
2001; Nagar et al., 2002). Feminist geographers could usefully attend to these
spatial dynamics of invisibility, containment, and dispersal in cities of the "Global

South" to more fully understand how people live geographically bounded lives in places constructed through various complex, intersecting social relations operating within and across scales.

Feminists have examined the national scale in terms of how the state mobilizes gendered representations and political identities in their economic policy. Mullings (1999) focuses on the Jamaican government's strategy of developing an export industry based on data entry and telemarketing. Meier (1999) questions whether the Columbian government's promotion of its globalized cut-flower industry based on cheap female labor really is a development success story. Raynolds (1998) shows how the Dominican Republic government's collaboration with corporations draws women into non-traditional agriculture and export manufacturing, while simultaneously devaluing their labor to enhance profits. However, these practices have met with resistance, as demonstrated by research on feminist political organizing in the Global North and South. Peake and Trotz (1999) highlight Guyanese women's political activism to show how constructions of racialized and classed differences (such as Afro-Guyanese and Indo-Guyanese) are located in a particular nationalized politics. Mackenzie (1992) demonstrates how women farmers in Ontario developed discourses of resistance at federal as well as local and provincial scales (see Liepens, 1998, on Australia and New Zealand).

At the global scale, feminist research is contributing to critical globalization studies by starting from the positions of those feminized by globalization. Feminists examine multiple and intersecting scales through which work is redefined and transformed through globalization. Examples include the ways in which processes of economic governance, social regulation, and workforce formation are jumping from national to global scales (including the implications of these processes for the rise of new workforces around the globe); the emergence of non-standard work; the reworking of peasant and agricultural economies; the intensification of homeworking and sweatshop labor; the rise of migrant workforces, including transnational elites, sex workers, and domestic workers; and the persistent importance of informalization (see Nagar et al., 2002 for an overview). In particular, feminist research examines how structural adjustment processes of privatization and (neo)liberalization have restructured social reproduction in this new phase of global accumulation, relying on the unpaid work of women to subsidize reallocations of capital and labor (Dalla Costa and Dalla Costa, 1993; Elson, 1995; Raynolds, 1998; Pearson, 2000). A broad range of case studies are demonstrating that informal work actively constitutes and subsidizes globalization as weak states and economies liberalize through feminization of workers, services, and productive activities, as in Chase's (2002) study of the privatization of a Brazilian mining town, and Rai's (1999) research on relations between Indian street vendors and the state. Other work examines land rights and gender, exploring how indigenous claims to communal land rights in the face of neoliberal privatization across Latin America can obscure women's land rights in the name of indigenous solidarity (Deere and Leon, 2001).

Global cities have received much attention yet much of that work overlooks gender. Feminist scholarship genders the flow of immigrants to global cities, looking at work experiences of low waged women (and men) who service the global economy (England and Stiell, 1997; Kofman, 2000; Yeoh et al., 2000). Some research focuses on transnational elites "on assignment," finding that they are usually men, many of whom are single, or whose partners are not supported in finding paid employment because firms lack formal policies on dual-career reloca-tion packages. Recently feminists have made visible these "invisible partners" and global "trailing spouses" (Yeoh et al., 2000; Hardill, 2002). Others focus on the functionally lone mother families left "at home" and "astronaut families" – a migra-tory pattern first associated with Asian families migrating to the Antipodes and Canada, from where the father "commutes" to Asia (Waters, 2002).

Feminists also examine transnational workers in the global food system, global-ized agroexport, and aquaculture industries. This work considers how local land-scapes have changed in response to agrarian transformation, shifting farming methods and labor practices, including farm labor migration patterns; see Raju in this volume and Jarosz and Qazi (2000) on the globalized apple industry in Washington state. Barndt (2002) follows the tomato trail from a field in Mexico to a Canadian supermarket, highlighting the experiences of women workers along a globalized food commodity chain. Ramamurty (2000) develops a gendered analy-sis of the globalized cotton commodity chain that constructs women as workers in an international division of labor from Japan to India. She demonstrates that cost-cutting pressures in the global cotton industry are differently expressed across places depending on indigenous patriarchies, specific workings of imperialism and nation-alism, the mix of state policies in each setting, and women workers' feminist acts in those places.

Conclusion

We demonstrate that feminist geographers' epistemological and methodological approaches to work contribute in vital ways to broader discussions within critical geography and feminist studies. Some feminists focus exclusively on women and their work, while others analyze gender relations and identities as these are trans-formed in dynamic relation with the restructuring of work. Intellectually feminist theorizations of work are an important arena through which researchers first inte-grated feminist and Marxist theories, presaging the move beyond an exclusive focus on women and gender to a broader focus on intersections among Marxist, feminist, anti-racist, postcolonial, and queer theory in many arenas of feminist scholarship. Recent research is marked by the exploration of the gendering of work through a focus on identity, culture, and representation. Along these lines, feminist researchers also examine the ways in which the gendering of work is linked to various refor-mulations of identity, subjectivity, and economic security.

Feminist geographers examine how the separation of work-production-public and home-reproduction-private obscures and devalues activities defined as "women's work." Through context-specific, grounded research, feminist analyses of work relations is opening up new ways of theorizing "work." For instance, femi-nists interrogate the complex formations of economic security and diverse subjec-tivities that exist in relation to capitalist systems. While exploding boundaries sophisticate our understandings of work relations, feminist researchers must be alert to the ways in which this emphasis might reinscribe and even create stereotypes about particular women based on their spatial location. For example, the exploita-tive character of domestic work in the West is contributing to a frequent conflation of domestic workers and women from the Global South. However, feminist research can usefully disrupt this pervasive representation, both in recovering white women in the West engaged in domestic work, as in Barbara Ehrenreich's (2001) *Nickel and Dimed*, and in making visible the diverse positionings of women in and from the Global South across all social locations and forms of work, including masculinized, elite, and leadership positions (see Raghuram and Hardill, 1998; Kofman, 2000; Yeoh et al., 2000).

Finally, we observe that English language research on work remains overly focused on concepts and problematics informed by Western industrial and market economies. Much feminist research frames work in terms of binaries of workplace/home, private/public as it conceptualizes the sites and activities of work. Even where research is in, and on, diverse settings, Western concepts often organize the analysis, partly because of the dominance of capitalist social relations through colonial relations and now under economic globalization. This means there is the risk of capital-centrism (Gibson-Graham, 1996), which excludes other ways of conceptualizing the sites, activities, and relations that could constitute an even more expanded and nuanced concept of work. Formations of work are not simply direct engagements with capitalist social relations, but also entail complex, contradictory, and differentiated forms across diverse places, cultures, and histories. Feminist geographers, with their emphasis on a relational world view and a grounded approach to analyzing experiences of work, continue to make important interventions into understandings of economic restructuring.

BIBLIOGRAPHY

Armstrong, P. (1997) The state and pay equity: juggling similarity and difference, meaning and structures. In P. M. Evans and G. R. Wekerle (eds), *Women and the Canadian Welfare State: Challenges and Change*. Toronto: University of Toronto Press.

Aspaas, H. R. (1998) Heading households and heading businesses: rural Kenyan women in the informal sector. *The Professional Geographer*, 50(2), 192–204.

Aufhauser, E. and Hafner, S. (2002) Regional and local labour market policy in the EU. *Geojournal*, 56(4), 253–60.

Barndt, D. (2002) *Tangled Routes: Women, Work and Globalization on the Tomato Trail*. Lanham, MD: Rowman and Littlefield.

Baylina, M. and Garcia-Ramon, M. D. (1998) Homeworking in rural Spain: a gender approach. *European Urban and Regional Studies*, 5(1), 55–64.

Beneria, L. (2001) Changing employment patterns and the informalization of jobs: general trends and gender dimensions. Working Paper for Women at Work Conference, Santiago, Chile.

Beneria, L. and Roldan, M. (1987) *Crossroads of Class and Gender: Industrial Homework, Subcontracting and Household Dynamics in Mexico City*. Chicago: Chicago University Press.

Bartham, R. and Shobrook, S. (1998) You have to be twice as good to be equal: placing women in Plymouth's Devonport Dockyard. *Area*, 30(1), 59–65.

Binnie J. (1997) Invisible Europeans: sexual citizenship in the New Europe. *Environment and Planning A*, 29(2), 237–48.

Blumen, O. (1998) The spatial distribution of occupational prestige in Metropolitan Tel Aviv. *Area*, 30(4), 343–57.

Boyer, K. (1998) Place and the politics of virtue: clerical work, corporate anxiety, and changing meanings of public womanhood in early 20th century Montreal. *Gender, Place and Culture*, 5(3), 261–76.

Bühler, E. (1998) Economy, state or culture: explanations for regional variations in gender inequality in Swiss employment. *European Journal of Urban and Regional Studies*, 5(1), 27–39.

Carlson, V. L. and Persky, J. J. (1999) Gender and suburban wages. *Economic Geography*, 75(3), 237–53.

Carney, J. (1992) Peasant women and economic transformation in The Gambia. *Development and Change*, 23(2), 67–90.

Chant, S. (1989) Gender and urban production. In S. Chant and L. Brydon (eds), *Women in the Third World: Gender Issues in Rural and Urban Areas*. New Brunswick, NJ: Rutgers University Press.

Chase, J. (2002) Privatization and private lives: gender, reproduction and neoliberal reforms in a Brazilian company town. In J. Chase (ed.), *The Spaces of Neoliberalism. Land, Place and Family in Latin America*. Bloomfield, CT: Kumarian Press.

Clark, W. A. V. and Withers, S. D. (2002) Disentangling the interaction of migration, mobility, and labor-force participation. *Environment and Planning A*, 34(5), 923–45.

Cooke, T. (1997) Geographic access to job opportunities and labor-force participation among women and African Americans in the greater Boston Metropolitan Area. *Urban Geography*, 18(3), 213–27.

Cravey, A. (1998) *Women and Work in Mexico's Maquiladoras*. Lanham, MD: Rowman and Littlefield.

Dalla Costa, M. and Dalla Costa, G. F. (1993) *Paying the Price: Women and the Politics of International Economic Strategy*. London: Zed Books.

Deere, C. D. and Leon, M. (2001) Individual versus collective land rights: tensions between women's and indigenous rights under neoliberalism. In J. Chase (ed.), *The Spaces of Neoliberalism: Land, Place and Family in Latin America*. Bloomfield, CT: Kumarian Press.

Domosh, M. and Seager, J. (2001) *Putting Women in Place*. New York: Guilford Press.

Duncan, S. and Pfau-Efflinger, B. (eds) (2001) *Gender, Economy and Culture in the European Union*. London: Routledge.

Dyck, I. (1999) Body troubles: women, the workplace and negotiations of a disabled identity. In R. Butler and H. Parr (eds), *Mind and Body Spaces: Geographies of Illness, Impairment and Disability*. London: Routledge.

Ehrenreich, B. (2001) *Nickle and Dimed: On (not) getting by in America*. New York: Metropolitan Books.

Elson, D. (ed.) (1995) *Male Bias in the Development Process*. Manchester: Manchester University Press.

England, K. (1993) Suburban pink collar ghettos: the spatial entrapment of women? *Annals of the Association of American Geographers*, 83(2), 225–42.

England, K. (1996) *Who Will Mind the Baby? Geographies of Childcare and Working Mothers*. London: Routledge.

England, K. and Stiell, B. (1997) They think you're as stupid as your English is: constructing foreign domestic workers in Toronto. *Environment and Planning A* 29, 195–215.

England, K. and Gad, G. (2004) Pink collar ghettos and glass ceilings in Canadian banking. Working manuscript.

Fagnani, J. (1993) Life course and space: dual careers and residential mobility among upper middle class families in the Ile de France region. In C. Katz and J. Monk (eds), *Full Circles: Geographies of Women over the Life Course*. New York: Routledge.

Fincher, R. and Panelli, R. (2001) Making space: women's urban and rural activism and the Australian state. *Gender, Place and Culture*, 8(2), 129–48.

Forsyth, A. (1997) "Out" in the valley. *International Journal of Urban and Regional Research*, 21(1), 38–62.

Freeman, C. (2001) Is local:global as feminine:masculine? Rethinking the gender of globalization. *Signs*, 26(4), 1007–37.

García-Ramon, M. D. and Monk, J. (1996) *Women of the European Union: The Politics of Work and Daily Life*. New York: Routledge.

George, S. (1997) Women's work and fulfillment in a changing world. In E. Sizoo (ed.), *Women's Lifeworlds*. London: Routledge.

Gibson-Graham, J. K. (1996) *The End of Capitalism (As We Knew It): A Feminist Critique of Political Economy*. Oxford: Blackwell.

Gilbert, M. R. (1998) "Race," space, and power: the survival strategies of working poor women. *Annals of the AAG*, 88(4), 595–21.

Gregson, N. and Lowe, M. (1994) *Servicing the Middle Classes*. London: Routledge.

Hahner, J. (1976) *Women in Latin American History, Their Lives and Views*. Los Angeles: UCLA Latin American Center Publications.

Halford, S. and Savage, M. (1997) Rethinking restructuring: embodiment, agency and identity in organizational change. In R. Lee and J. Wills (eds), *Geographies of Economies*. London: Arnold.

Hanson, S. and Pratt, G. (1995) *Gender, Work and Space*. London: Routledge.

Hapke, H. (2001) Petty traders, gender and development in a south Indian fishery. *Economic Geography*, 77(3), 225–49.

Hardill, I. (2002) *Gender, Migration and the Dual-Career Household*. London: Routledge.

Hayford, A. (1974) The geography of women: an historical introduction. *Antipode*, 6, 1–18.

Hays-Mitchell, M. (1993) The ties that bind: informal and formal sector linkages in streetvending: the case of Peru's ambulantes. *Environment and Planning A*, 25, 1085–102.

Holloway, S. L. (1998) Local childcare cultures: moral geographies of mothering and the social organisation of pre-school education. *Gender, Place and Culture*, 5(1), 29–53.

Hubbard, P. (1998) Sexuality, immorality and the city: red light districts and the marginalisation of female street prostitutes. *Gender, Place and Culture*, 5(1), 55–72.

Jarosz, L. and Qazi, J. (2000) The geography of Washington's world apple: global expressions in a local landscape. *Journal of Rural Studies*, 16(1), 1–11.

Jarvis, H. (1999) The tangled webs we weave: household strategies to co-ordinate home and work. *Work, Employment and Society*, 13(2), 225–47.

Johnson, L. (1990) New patriarchal economies in the Australian textile industry. *Antipode*, 22(1), 1–32.

Johnston-Anumonwo, I. (1997) Race, gender and constrained work trips in Buffalo. *Professional Geographer*, 49, 306–17.

Joseph, A. E. and Hallman, B. C. (1998) Over the hill and far away: distance as a barrier to the provision of assistance to elderly relatives. *Social Science and Medicine*, 46(6), 631–9.

Kempadoo, K. (2001) Freelancers, temporary wives, and beach-boys: researching sex work in the Caribbean. *Feminist Review*, 67(1), 39–52.

Kodras, J. E. and Padavic, I. (1993) Economic restructuring and women's sectoral employment in the 1970s: a spatial investigation across 380 United States labor market areas. *Social Science Quarterly*, 74(1), 1–27.

Kofman, E. (2000) Beyond a reductionist analysis of female migrants in global European cities: the unskilled, deskilled and professional. In M. Marchand and A. Runyan (eds), *Gender and Global Restructuring. Sitings, Sites and Resistances*. London: Routledge.

Kwan, M. P. (1999) Gender, the home–work link, and space–time patterns of non-employment activities. *Economic Geography*, 75(4), 370–94.

Laurie, N. (1997) Negotiating femininity: women and representation in emergency employment in Peru. *Gender, Place and Culture*, 4(2), 235–51.

Law, R. (1999) Beyond "women and transport": towards new geographies of gender and daily mobility. *Progress in Human Geography*, 23(4), 567–88.

Lawson, V. (1995) Beyond the firm: restructuring gender divisions of labor in Quito's garment industry under austerity. *Society and Space*, 13, 415–555.

Lawson, V. (1999) "Tailoring is a profession, seamstressing is work!" Resiting work and reworking gender identities among artisanal garment workers in Quito. *Environment and Planning A*, 31, 209–27.

Leslie, D. and Butz, D. (1998) "GM suicide": flexibility, space and the injured body. *Economic Geography*, 74, 360–78.

Liepens, R. (1998) The gendering of farming and agricultural politics: a matter of discourse and power. *Australian Geographer*, 29(3), 371–88.

Lind, A. (2000) Negotiating boundaries: women's organizing and the politics of restructuring in Ecuador. In M. Marchand and A. Runyan (eds), *Gender and Global Restructuring*. London: Routledge.

Longhurst, R. (2001) Managing managerial bodies. In *Bodies: Exploring Fluid Boundaries*. London: Routledge.

McCall, L. (1998) Spatial routes to gender wage (in)equality: regional restructuring and wage differentials by gender and education. *Economic Geography*, 74(4), 379–404.

McDowell, L. (1997) *Capital Culture: Gender at Work in the City*. Oxford: Blackwell.

McDowell, L. and Massey, D. (1984) A woman's place? In D. Massey and J. Allen with J. Anderson (eds), *Geography Matters! A Reader*. New York: Cambridge University Press.

Mackenzie, F. (1992) "The worse it got, the more we laughed": a discourse of resistance among farmers of Eastern Ontario. *Environment and Planning D: Society and Space*, 10(6), 691–713.

Mackenzie, S. and Rose, D. (1983) Industrial change, the domestic economy and home life. In J. Anderson, S. Duncan and R. Hudson (eds), *Redundant Spaces in Cities and Regions? Studies in Industrial Decline and Social Change*. London: Academic Press.

McLafferty, S. and Preston, V. (1997) Gender, race, and the determinants of commuting: New York in 1990. *Urban Geography*, 18(3), 192–212.

Massey, D. (1984) *Spatial Divisions of Labor*. New York: Methuen.

Massey, D. (1995) Masculinity, dualisms and high technology. *Transactions of the IBG*, n.s., 20, 487–99.

Mattingly, D. J. (2001) The home and the world: domestic service and international networks of caring labor. *Annals of the AAG*, 91(2), 370–86.

Meier, V. (1999) Cut-flower production in Colombia – a major development success story for women? *Environment and Planning A*, 31(2), 273–89.

Mez, J. and Bühler, E. (1998) Functional and spatial segregation in the Swiss financial sector: pink-collar ghetto and male bastion. *Environment and Planning A*, 30(9), 1643–60.

Mies, M. (1999) *Patriarchy and Accumulation on a World Scale: Women in the International Division of Labour*, 2nd edn. London: Zed Books.

Momsen, J. H. (ed.) (1999) *Gender, Migration, and Domestic Service*. London: Routledge.

Moore-Milroy, B. and Wisner, S. (1994) Communities, work and public/private sphere models. *Gender, Place and Culture*, 1(1), 71–90.

Moser, C. (1987) The experience of poor women in Guayaquil. In E. Archetti, P. Cammack and B. Roberts (eds), *Sociology of Developing Societies. Latin America*. London: Monthly Review Press.

Mullings, B. (1999) Sides of the same coin? Coping and resistance among Jamaican data-entry operators. *Annals of the AAG*, 89(2), 290–311.

Munk-Madsen, E. (2000) Wife the deckhand, husband the skipper: authority and dignity among fishing couples. *Women's Studies International Forum*, 23(3), 333–42.

Nadel-Klein, J. (2000) Granny baited the lines: perpetual crisis and the changing role of women in Scottish fishing communities. *Women's Studies International Forum*, 23(3), 363–72.

Nagar, R., Lawson, V., McDowell, L. and Hanson, S. (2002) Locating globalization: feminist (re)readings of the subjects and spaces of globalization. *Economic Geography*, 78(3), 257–84.

Oberhauser, A. M. (2002) Relocating gender and rural economic strategies. *Environment and Planning A*, 34(7), 1221–37.

Odland, J. and Ellis, M. (1998) Variations in the labor force experience of women across large metropolitan areas in the United States. *Regional Studies*, 32(4), 333–47.

Ong, A. (1987) *Spirits of Resistance and Capitalist Discipline: Factory Women in Malaysia*. Albany: State University of New York Press.

Peake, L. and Trotz, A. (1999) *Gender, Ethnicity and Place: Women and Identities in Guyana*. London: Routledge.

Pearson, R. (2000) Moving the goalposts: gender and globalization in the twenty-first century. *Gender and Development*, 8(1), 10–19.

Perrons, D. (1995) Gender inequalities in regional development. *Regional Studies*, 29(5), 465–76.

Pratt, G. (1997) Stereotypes and ambivalence: the construction of domestic workers in Vancouver, British Columbia. *Gender, Place and Culture*, 4, 159–78.

Preston, V., Rose, D., Norcliffe, G. and Holmes, J. (2000) Shift work, child care and domestic work: divisions of labour in Canadian paper mill communities. *Gender, Place and Culture*, 7(1), 5–29.

Pringle, R. (1989) *Secretaries Talk*. London: Verso.

Pulido, L. (1997) Community, place and identity. In J. P. Jones III, H. Nast and S. Roberts (eds), *Thresholds in Feminist Geography: Difference, Methodology, Representation*. New York: Rowman and Littlefield.

Radcliffe, S. (1990) Ethnicity, patriarchy and incorporation into the nation: female migrants as domestic servants in Peru. *Environment and Planning D: Society and Space*, 8, 379–93.

Raghuram, P. (1993) Invisible female agricultural labor in India. In J. Momsen and V. Kinnaird (eds), *Different Places, Different Voices. Gender and Development in Africa, Asia and Latin America*. London: Routledge.

Raghuram, P. and Hardill, I. (1998) Negotiating a market: a case study of an Asian woman in business. *Women's Studies International Forum*, 21(5), 475–84.

Rai, S. (1999) Fractioned states and negotiated boundaries: Gender and law in India. In H. Afshar and S. Barrientos (eds), *Globalization and Fragmentation in the Developing World*. London: Macmillan Press.

Ramamurty, P. (2000) The cotton commodity chain, women, works and agency in India and Japan: the case for feminist agro-food systems research. *World Development*, 28(3), 551–78.

Raynolds, L. (1998) Harnessing women's work: restructuring agricultural land and industrial labor forces in the Dominican Republic. *Economic Geography*, 74(2), 149–69.

Reskin, B. F. and Roos, P. A. (1990) *Job Queues, Gender Queues: Explaining Women's Inroads into Male Occupations*. Philadelphia: Temple University Press.

Rose, D. (1993) Local childcare strategies in Montréal, Québec: the mediations of state policies, class and ethnicity in the life course of families with young children. In C. Katz and J. Monk (eds), *Full Circles: Geographies of Women over the Life Course*. New York: Routledge.

Rose, D. and Villeneuve, P. (1998) Engendering class in the metropolitan city: occupational pairings and income disparities among two-earner couples. *Urban Geography*, 19(2), 123–59.

Rowbotham, S. (1998) Weapons of the weak: homeworkers' networking in Europe. *European Journal of Women's Studies*, 5(3/4), 453–67.

Salzinger, L. (2001) Making fantasies real: producing women and men on the Maquila shop floor. *NACLA Report on the Americas*, 34, 13–19.

Schroeder, R. (1999) *Shady Practices. Agroforestry and Gender Politics in the Gambia.* Berkeley: University of California Press.

Smith, F. M. (1999) Discourses of citizenship in transition: scale, politics and urban renewal. *Urban Studies*, 36(1), 167–87.

Smith, G. D. and Winchester, H. P. M. (1998) Negotiating space: alternative masculinities at the work/home boundary. *Australian Geographer*, 29(3), 327–40.

Staeheli, L. (2003) Women and the work of community. *Environment and Planning*, 35(5), 815–31.

Stiell, B. and England, K. (1999) Jamaican domestics, Filipina housekeepers and English nannies: representations of Toronto's foreign domestic workers. In J. H. Momsen (ed.), *Gender, Migration, and Domestic Service.* New York: Routledge.

Timar, J. (2002) Restructuring labour markets on the frontier of the European Union: gendered uneven development in Hungary. In A. Rainnie, A. Smith and A. Swain (eds), *Work, Employment and Transition: Restructuring Livelihoods in Post-Communism.* London: Routledge.

Tonkin, L. (2000) Women of steel: constructing and contesting new gendered geographies of work in the Australian steel industry. *Antipode*, 32(2), 115–34.

UNIFEM (2000) *Progress of the World's Women.* New York: UNIFEM.

Valentine, G. (1995) Out and about: geographies of lesbian landscapes. *International Journal of Urban and Regional Research*, 19(1), 96–111.

Valentine, G. (1996) An equal place to work? Anti-lesbian discrimination and sexual citizenship in the European Union. In M. D. García-Ramon and J. Monk (eds), *Women of the European Union: The Politics of Work and Daily Life.* London: Routledge.

Van Eyck, K. (2002) Neoliberalism and democracy? The gendered restructuring of work, unions and the Colombian public sphere. PhD dissertation, University of Washington, Seattle.

van Hoven, B. (2001) Women at work: experiences and identity in rural East Germany. *Area*, 33(1), 38–46.

Waters, J. L. (2002) Flexible families? Astronaut households and the experience of lone mothers in Vancouver, BC. *Social and Cultural Geographies*, 3(2), 117–34.

Whatmore, S. (1990) *Farming Women: Gender, Work and Family Enterprise.* Basingstoke: Macmillan.

Wolf, D. (1992) *Factory Daughters: Gender, Household Dynamics, and Rural Industrialization in Java.* Berkeley: University of California Press.

Wolfe, J. (1993) *Working Women, Working Men: São Paulo and the Rise of Brazil's Industrial Working Class, 1900–1955.* Durham, NC: Duke University Press.

Wright, M. (1997) Crossing the factory frontier: gender, place, and power in the Mexican Maquiladora. *Antipode*, 29(3), 278–302.

Wright, R. and Ellis, M. (2000) The ethnic and gender division of labor compared among immigrants to Los Angeles. *International Journal of Urban and Regional Research*, 24(3), 583–601.

Yeoh, B. and Huang, S. (1995) Childcare in Singapore: negotiating choices and constraints in a multicultural society. *Women's Studies International Forum*, 18(4), 445–63.

Yeoh, B., Huang, S. and Willis, K. (2000) Global cities, transnational flows and gender dimensions, the view from Singapore. *Tijdschrift voor Economische en Sociale Geografie*, 91(2), 147–58.

Chapter 7

Shea Butter, Globalization, and Women of Burkina Faso

Marlène Elias and Judith Carney

Shea butter, or *karité*, may not yet be a household word in the USA, but it is stealthily making its way into American homes in cosmetics and skin products. Prized for its superb healing and moisturizing properties, shea butter represents one of the growing lines of "natural" products popularized commercially in sunscreens, skin creams, lip pomades, and hair products. Shea nuts are collected from a tree indigenous to Africa's Sudano-Sahelian region and processed into butter by women.[1] Shea provides Sudano-Sahelian Africa with its prime contender in the global market, an export stimulated by the West's rediscovery of non-timber forest products (NTFPs) as cosmetic ingredients that had been replaced by chemicals in the nineteenth and twentieth centuries. The "green consumerism" promoted by companies such as The Body Shop is reawakening global markets for a handful of NTFPs that provide natural sources of vegetable butter or oil. In response to Western demand, advocates for gender equity, sustainable development, and fair trade are converging upon shea butter commercialization as a strategy to enhance the incomes of impoverished female producers.

Global demand for shea has stimulated production and export throughout the African Sahel. This is especially evident in Burkina Faso, the region's leading shea producer (Compaoré, 2000), which accounts for about 25 percent of world supply (Terpend, 1982; Booth and Wickens, 1988).[2] *Taaga* or *tanga*, as shea is termed in the Mossi language, ranks third in the country's agricultural export volume and remuneration (World Bank, 1989; ANDINES, 2002). In this landlocked former French colony, which boasts few resources and income opportunities, shea development projects sponsored by UNIFEM, the UNDP, bilateral aid agencies, and non-governmental organizations (NGOs) have proliferated over the past decade. By teaming up with women, the traditional shea nut collectors and butter producers, project boosters predict a sixfold increase in female earnings if the product is marketed through fair trade rather than conventional channels (ANDINES, 2002). In promoting women's economic development and improved producer prices, the projects also aim to protect the slow-growing shea tree from deforestation.

Yet the outcome of these projects regarding rural women's labor and property relations remains unevaluated. This chapter consequently examines the role of shea commercialization in contemporary Burkina Faso, with emphasis placed on women's access to shea nuts and the labor demands of butter processing.[3] We argue that while efforts to promote "fair trade" production of shea butter (primarily for cosmetic products) can increase women producers' incomes, there are many pitfalls to shea production as a development panacea. It is crucial to consider the global and the local conditions of shea production: from the shifting demands of a global shea market still dominated by the food industry – rather than the cosmetics trade – to local gendered property relations and geographies of poverty that prevent women from capturing the value of their product. As shea butter production is a development strategy premised upon a time- and labor-intensive activity, its sustainability is contingent upon its respect for the other labor demands necessary to the daily reproduction of Burkinabè producers – particularly women. Evaluation of the "success" of shea production rests on placing that production in the context of gendered lives and complex survival strategies, as well as in the context of globalized marketing constraints.

This chapter contributes to a geographic tradition of looking critically at gender, labor, and development (Whitehead, 1981; Momsen and Townsend, 1987; Carney and Watts, 1991; Hanson and Pratt, 1995). It particularly aims to expand debates concerning gender, scale, and place in the context of economic globalization (Escobar, 2001; Nagar et al., 2002). Gendered relations of shea production within Burkinabè communities is linked to regional and national output and global markets (see Kelly, 1999). The shea commodity chain is a female one and rests upon the importance of place for women at both ends of production and consumption. Rural women's access to nut trees, production of butter, and assistance from development projects are place-based, as is the local female knowledge that guides how and when to prepare the nuts. Western women consumers similarly access shea products in the specific places where they are sold, typically the cosmetic counter of a department store chain. This chapter's exploration of gender and labor relations in shea butter production highlights the need to situate the specificities of the local for adequately understanding contemporary globalization and alternatives to the mainstream neoliberal model. By taking a multiscalar approach, the chapter positions dynamics of gender and place in Burkina Faso within the context of international women's advocacy organizations and global demand for natural products.

The International Policy Context for Women's Shea Projects

Women's shea butter projects, currently supported by international aid agencies, emerged from three distinctive patterns in the global political economy during the 1980s. First, due in part to the declining value of traditional export commodities, the debt crisis swept Africa and other regions of the global South during the 1980s, subjecting many countries to the economic and state-oriented restructuring requirements of the International Monetary Fund (IMF). In short, IMF structural adjustment programs (SAPs) demanded a smaller role for the state and a reduction of expenditures believed to impede economic efficiency. Health and education budgets of poor countries became casualties of SAP reforms, as did state marketing boards that provided price guarantees and subsidies to agricultural producers (Jackson and Pearson, 1998). Emphasis on non-traditional exports such as shea butter, with auspicious market trends, received policy support for economic diversification (Haugerud et al., 2000). The door opened for the private sector and global corporate expansion, while NGOs stepped in to bridge the socioeconomic freefall that occurred with the contraction of governmental infrastructure.

Second, during the 1980s, development scholars and policy-makers – largely in response to Ester Boserup's landmark book *Women's Role in Economic Development* – began focusing on female poverty in poor countries. The fact that the United Nations proclaimed 1975–85 as the International Decade for Women reinforced this turn: global development agencies increasingly committed to addressing issues particularly pertinent to women in the South. This led to an array of Women in Development (WID) initiatives, which gained legitimacy in prominent development agencies such as the United Nations Development Program (UNDP), the World Bank, and NGOs.

Third, the development of women's shea projects in Africa can be understood in the context of "fair trade" consumerism that exploded in the 1990s. The 1989 collapse of the Soviet Bloc paved the way for the global expansion of neoliberal capitalism and an invigorated emphasis on free trade. Supplanting the withering hopes of commodity cartels to raise producer prices in poor countries, "fair trade" materialized as a counter discourse to contest the growing inequality between North and South that characterized capitalist expansion over previous decades. Efforts to improve producer prices shifted the focus to consumers in the North and capitalized on their growing awareness around consumption choices (Watts and Goodman, 1997; Dicum and Luttinger, 1999; Rice, 2001). Fair trade products are thus marketed in a new light to affluent Western consumers whose concerns with natural products, the environment, and social justice offer the opportunity to marshal support for greater producer remuneration.

Shea projects that encourage the female production and marketing of horticultural commodities and NTFPs (such as shea, babassu, and Brazil nuts) thus respond to three large-scale dynamics: non-traditional export promotion imposed by the IMF; WID initiatives that proliferated in the 1980s; and the current Western craze for "natural" cosmetics, skin, and hair products produced according to "socially just" standards. The long-term success of shea projects and their ability to satisfy such varied equity expectations will reveal a great deal about the prospects of neoliberal initiatives and NGO efforts to distribute benefits more equitably in the current period of global capitalism. This chapter represents a preliminary engagement with such questions as they pertain to female shea producers in Burkina Faso.

The Globalization of Shea Markets

International commerce in shea butter is longstanding in Burkina Faso. As early as the fourteenth century, references by Muslim scholars to the product's trade in the West African subregion confirm that sale of shea butter dates to at least that period (Lewicki, 1974). During the nineteenth century, shea nuts were collected and sold to merchant trading houses, which commercialized the ingredient as a component of margarine and as a substitute for cocoa butter in the manufacture of chocolate. While this phase of trade with Europe did not result in high prices for primary African NTFP exports, production was voluntary. The onset of colonialism in the 1880s aggravated the situation for Sudano-Sahelian producers. The imposition of forced labor, taxation, and crop requisitions forced Africans to expand commodity production. Shea butter was among the products peasants collected for colonial officials in Burkina Faso, and women bore the brunt of nut collection and butter processing (Massa and Madiéga, 1995; Freidberg, 1996). The emphasis on intensifying commodity production during colonialism forced Burkinabè producers into the international market in ways that increased economic vulnerability when climatic cycles resulted in poor yields, export prices declined, or NTFPs like shea were replaced by synthetic or other natural substitutes (Hopkins, 1973). Inequitable market share thus characterized relations between rural African producers and European buyers of shea nuts from the earliest phase of the commodity's global trade through independence of producer countries in the 1960s.[4]

Since independence, Burkina Faso has witnessed a growing global demand for shea. Over the 1970s and 1980s, despite a few years of poor production, Burkinabè shea nut exports steadily climbed, reaching more than 40,000 tons in 1984. Burkina's earnings from these exports came in second after its traditional leading export: cotton (Harsch, 2001, p. 6). At the time, demand derived largely from the food industry in France, Scandinavia, Great Britain, and Germany.[5] However, the world price for shea nuts plunged in 1986–7, reducing the price paid to producers from 70 to 15 f CFA/kg (CSPPA, 1988). The decade of the 1980s closed with fewer nut exports, although by the mid-1990s market demand for shea revived considerably.

Over this period of volatile nut prices, markets, and nut harvests, Burkina Faso's state marketing board, the *Caisse de Stabilisation des Prix et des Produits Agricoles* (CSPPA, the Agricultural Commodity Price Stabilization Board), guaranteed producers a base price for nuts, buffering them from international price fluctuations and deteriorating terms of trade. It additionally helped rural producers to get shea products to market. However, declining commodity prices had weakened the board's solvency by 1992, resulting in considerable market instability for female nut collectors and butter producers over the following years. In 1994, the CSPPA was disbanded, producer price-support ended, and women found themselves without a secure buyer outlet and with little remuneration for their product. Shea commercialization in Burkina Faso became increasingly informal and disorganized (Compaoré, 2000).

As economic liberalism and structural adjustment dismantled marketing boards, WID advocacy organizations stepped into the breach with shea projects. This latest phase of global market integration is unfolding amid challenges to the historic legacy of exploitative trade relations between Africa and the West. The hope is to increase the producer share of the shea market through fair trade mechanisms that improve rural women's incomes.

Trends in Burkina Faso's twentieth-century shea trade must be seen in the context of specific geographical and socioeconomic considerations. Burkina Faso is a small, landlocked country in Sudano-Sahelian West Africa. It is one of the poorest countries in the world, where the average life expectancy of its 12 million inhabitants is

46 years, 80 percent of the population is rural, and estimated average per capita income reaches just US$234 per annum. The national economy is dominated by agriculture and livestock rearing, in which 90 percent of the population is engaged. Sixty-five percent of export revenues derive from these activities (MEF, 1998), rendering the Burkinabè economy extremely vulnerable to global price fluctuations in primary commodities. Deteriorating international values for these products since the 1970s have resulted in worsening terms of trade and declining per capita incomes. In 1995, nearly half (45 percent) of the Burkinabè population survived on less than US$80 per year, and women comprised the majority of the poor (MEF, 1998). Burkina Faso suffers from a swollen debt burden and balance of payments deficit and has been undergoing successive IMF structural adjustment programs since 1991. Against this bleak economic backdrop, the country has experienced a rise in international aid for development projects (Compaoré, 2000). Those based on shea remain a donor favorite. No wonder hopes run high that shea producers have found an "anti-wrinkle cash cow" (Hale, 2003).

The global demand for shea remains centered on its ability to serve as a substitute for cocoa butter in the manufacture of chocolates, which requires a commodity price inferior to that of cocoa butter (Conti, 1979). In 1997, the food industry still absorbed 90 percent of international shea nut demand (UNIFEM, 1997). Shea's function as a "hidden" product ingredient will likely expand in the years to come, with the European Union's decision in 2000 to allow up to 5 percent of non-cocoa vegetable fats in chocolate manufacturing (Harsch, 2001).[6] Since the food industry aims for cheaper cocoa substitutes and can use either shea nuts or the derived butter, there is little prospect for West African shea butter producers to improve incomes through chocolate manufacturers.

Yet while nut exports for the manufacture of chocolate and margarine continue, the rising international demand for shea butter in recent years derives from its role as a "natural" substitute for chemicals in the multibillion dollar cosmetics industry. Several foreign cosmetics firms (The Body Shop, l'Oréal, l'Occitane) have begun using the butter in their lines of natural products. Convinced that women's income opportunities rest on the production and marketing of shea butter, rather than the raw nuts, UNIFEM and other WID proponents stepped into action. In 1997 one ton of unprocessed shea nuts sold domestically for f CFA 70,000 and externally for f CFA 100,000, while the same ton processed into shea butter obtained f CFA 148,000 (Harsch, 2001). WID projects thus focused on developing shea cooperatives to improve product quality, access to market information, and trading relationships with distant importers (Crélerot, 1995). While, since the late 1960s, the international community favored importing raw shea nuts rather than the processed butter (Conti 1979), WID interventions resulted in a revival of shea butter exports from Burkina Faso.

The contemporary export demand for shea butter reflects its growing popularity with cosmetics firms and WID interventions, which have improved the butter quality. UNIFEM recently helped to broker a deal with l'Occitane, the French cosmetics company, to purchase butter directly from a Burkinabè producer association, ensuring the profits go unswervingly to producers rather than middlemen. In 2001, the company contracted to purchase 60 tons of shea butter and planned to increase the amount to 90 tons in 2002 (Harsch, 2001).

The prospects for income gains with shea butter are thus far more promising in the cosmetics industry (Compaoré, 2000), especially as buyers like l'Occitane have shown a willingness to pay producers a higher-than-market value for the butter. The marketing of such commodities highlights the product's provenance, while humanizing its commercialization through alternative trade networks (EFTA, 1998). While much of the press attention on shea celebrates "making trade work for poor women" (Harsch, 2001), there is relatively little research on how its commercialization affects rural women's labor and access rights to nut trees. The next sections provide an overview of such issues for Burkinabè producers.

The Access Rights of Female Shea Nut Collectors

Shea occupies such a valued place in the environment and regional economy of Sudano-Sahelian Africa that when land is cleared for agriculture, the tree is generally spared. Thus, while the shea tree (*Vitellaria paradoxa*) is not deliberately planted, the species is actively selected for, managed, and protected by thousands of small-scale producers (Dalziel, 1937; Boffa, 1995; Lovett and Haq, 2000). For environmental organizations concerned with desertification and deforestation, protection of a NTFP like the shea tree, and promotion of its extraction, offers a sustainable livelihood option and the prospect for diversifying rural incomes. As the shea tree provides an excellent source of firewood, donor efforts have focused on preventing nut trees from being cut down for this purpose (Kessler, 1992). Providing reasonable producer prices to stimulate shea butter production for international markets thus furthers donor objectives to arrest deforestation in the Sahel (Elias, 2003).

The shea butter nut tree is found solely in Africa's Sudano-Sahelian region, where it thrives with just 500–1000 mm of rainfall (figure 7.1). It is long-lived: individuals can attain 200–300 years of age and are characterized by a slow growth cycle. Shea trees begin producing flowers as of their fifteenth year and attain full fruit production after approximately 45–50 years. The shea cycle is seasonal. In Burkina Faso, fruiting begins at the end of the dry season in May and continues throughout the rainy season until mid-September (Ruyssen, 1957; Terpend, 1982; Schreckenberg, 1996). An adult tree will produce on average 20 kilograms of fresh fruit annually. These typically yield about 4 kilograms of dried nuts and between 0.7 and 2.5 kg of butter, depending upon the extraction procedure (Terpend, 1982).

African women's longstanding involvement with shea is the tree's most significant feature. The product is one of the few economic commodities in the region under the control of women, who have carried out the nut's collection and processing into butter for centuries (Lewicki, 1974). Emerging global markets for shea butter thus offer a window of opportunity for Burkinabè women whose income opportunities are extremely limited (Compaoré, 2000). But shea commercialization also depends on female access rights to the trees and consideration of the labor involved to process the nuts into butter.

Access rights to shea trees and products correspond with land type. Shea trees found on household fields encompass two types of land: fields farmed by family members for collective food and income, and plots worked by individual household members. Women most commonly gather shea nuts from trees growing on land

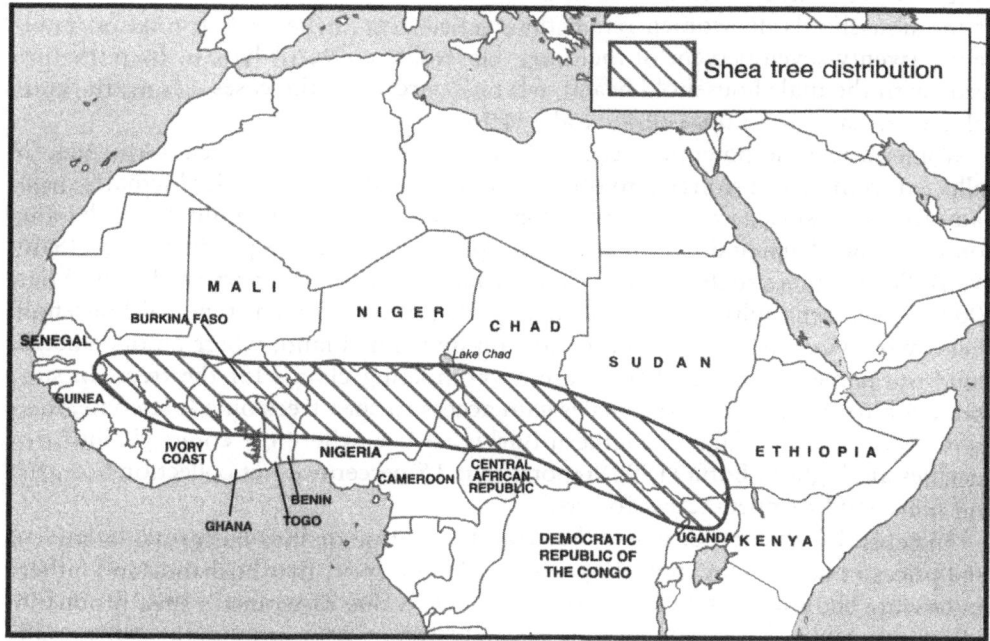

Figure 7.1 Shea tree distribution.

claimed communally by rural households (Ruyssen, 1957; Boffa et al., 1996). In a pattern typical of Sudano-Sahelian common property systems (Carney, 1988), rural Burkinabè households divide their landholding into individual and family fields. Each field type confers decision-making upon a specific person, who retains rights to the collected nuts (Saul, 1988; Boffa et al., 1996). In exchange for labor on household fields devoted to collective subsistence, females are granted individual plots that return the product of their labor directly to them. If a shea tree grows on her individual field, she alone is entitled to collect the fruits and the derived value of shea products (Terpend, 1982; Boffa et al., 1996).

However, the greater share of cultivated family land comes under the control of the male household head, who perpetuates family property relations through patrilineal lineages. On fields used for collective subsistence, he grants his wives and the spouses of his sons access to valued shea trees. As the key decision-maker and manager of household resources, the product of trees found on family fields ostensibly belongs to the male household head, who uses the derived benefits for the family's collective welfare. However, in practice, until the recent export demand for shea began, rural women collected the nuts for family consumption and sold surplus butter for petty cash without male household heads asserting claims to its value (Elias, 2003).

Recent research suggests this pattern is changing with the current phase of shea commercialization. There is evidence that male household heads are appropriating some of the value from nuts collected on family land. In Thiougou, located in the bountiful shea area of southern Burkina Faso (and near regional export markets in

Côte d'Ivoire), nut collection remains a female activity. However, in only two-thirds of the households do women retain the entire value of harvested nuts on family fields. Twenty-seven percent of the time, the female collector has to share the proceeds with the male household head, who in 7 percent of the cases claims the entire value of female shea sales (Boffa et al., 1996).

Women's income potential becomes even more vulnerable when shea nuts are collected from fields belonging to other households. When "outsider" females negotiate access to shea trees on land belonging to non-related families, the lending household head maintains rights to appropriate part of the product value (Saul, 1988). Such rights usually involve claims to a portion of the harvested nuts (Elias, 2003). Given the multiple claims to nuts gathered on different types of household fields, many women favor collecting nuts on unclaimed land, where an open access land-use regime prevails and there are no competing claims for the trees. In such areas, competition is keen, as females gain access to shea trees on a first-come, first-serve basis. In one village in southern Burkina Faso, where the country's shea tree densities are highest, Boffa (1995) reports that 15 percent of nut collection occurred on fallow land or open-access forests.

In central Burkina Faso, where the economic value of shea has grown relative to the prices of traditional export crops, researchers report that husbands and fathers-in-law are claiming an increasing portion of the value of women's shea production (Terpend, 1982). As a crop or processed product gains economic value, rights of access and control are often disputed and renegotiated between the resource decision-maker and its users. This process has frequently erupted into gender conflict between husbands and wives in areas of sub-Saharan Africa where development schemes target women as beneficiaries (Carney, 1988; Schroeder, 1999; Dolan, 2001; Hart, 2002). Whether the pattern under way in central Burkina Faso will prevail in WID shea projects cannot at this point be determined.

Shea Commercialization: Female Labor and Processing Demands

The rising value of shea butter in global cosmetics markets offers renewed hope to Sudano-Sahelian countries like Burkina Faso, where stagnating prices for traditional exports and deepening impoverishment have increased the need for economic diversification. However, the process is exerting ever-greater pressure on women to produce butter for export (Terpend, 1982). There are limits to the intensification of female labor for shea butter production, unless labor-saving technologies become available and men assist in the collection of nuts, water, firewood, and butter processing.

A key bottleneck for raising shea butter production and sales is the opportunity cost women face with collecting and processing the crop. The shea tree comes into fruit on the eve of the rainy season and continues fruiting throughout most of the agricultural cycle. This is the period of the year when females are busiest, as crops are planted, weeded, and harvested, and women must add field labor to their considerable household work demands. As a result, shea collection and processing directly compete with cultivated crops for female labor and time. This part of the agricultural calendar is also known as the hungry season (French: *soudure*); the period when cereal reserves from the previous year's harvest are at their lowest. Gra-

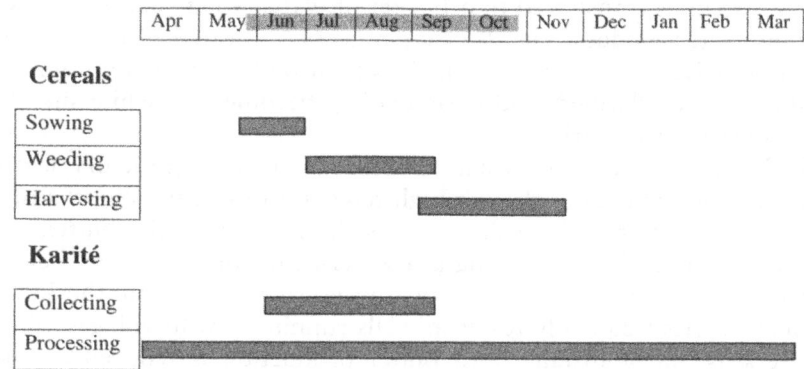

Figure 7.2 Seasonality of women's agricultural work in areas of shea production. *Source*: Elias fieldwork data, 2000.

naries are nearly empty, and families have difficulty securing basic nutritional needs when labor demands are greatest. Shea-associated activities are therefore added upon a full labor calendar when household food reserves are stressed. Figure 7.2 illustrates the overlap between periods of nut collection and the agricultural cycle.

Once collected, shea nuts undergo the initial processing phases in preparation for storage prior to transformation into butter. The fruits are first pulped, boiled, and dried. These preliminary steps are labor- and time-consuming, as women must collect firewood and water to carry out this initial phase of processing. As it is a female responsibility to ration the family's food reserves, shea represents one way to extend supplies by direct consumption of the butter produced for cooking oil as well as through the sale of a portion of the transformed product. While revenues from surplus butter sales seem meager, they frequently represent the only way women can purchase food to supplement the dwindling cereal reserves in household granaries. Both shea butter and shea nuts are sold during the wet season (Gosso, 1996). However, the frequency of rain during this period interferes with drying the nuts and processing them for commercial butter demand.

The making of shea butter is not a simple task. It involves intensive physical labor and considerable amounts of water and firewood.[7] Young unmarried daughters and little boys assist their mothers in nut collection and butter making. Preparation takes days and involves numerous stages. Once the fruit is collected, the pulp is eaten or deliberately removed. The nuts are then boiled and dried. As boiling arrests nut germination, this step must be performed immediately after collection, before sprouting or desiccation occurs.[8] Upon completion of these preliminary steps, nuts can be stored for several months until ready for transformation into butter. The remainder of the process can be carried out during the dry season, after the agricultural harvest, when women are less burdened with field tasks.

The processing of shea nuts into butter involves collective female labor. Nuts are first shelled by hand, then crushed, roasted, and pounded in a mortar with a pestle. Water is added to create a paste, which must be kneaded. Two or three women jointly reach into the thick shea batter to beat the paste so the foam floats to the surface. This foam is transferred to a bucket of water, where subsequent washings eliminate unwanted residues. The cleansing process – repeated as many as four times – yields progressively whiter foam, which is then boiled for many hours. The top layer is skimmed, or clarified, and upon cooling becomes the white shea butter so desired in international markets.

The final steps of shea transformation ideally occur during the dry season when women have more time, even though high temperatures in the months of January to March render butter making difficult (as the heat melts the butter, making it harder to manipulate). But processing is a water-demanding activity; it cannot take place without adequate water availability. This is crucial given that the Sahel's notorious drought cycles frequently result in wells running dry. In villages where water sources are seasonal or distant, shea butter production is reduced or suspended during the dry season months. Lobi women in southwestern Burkina Faso, for instance, spend as many as three to five hours a day collecting and transporting water to their households during this period. They consequently make shea butter during the rainy season (Crélerot, 1995). The seasonality of water availability thus remains another critical factor for assessing the labor burdens posed by donor efforts to commercialize rural women's shea production.

What are the rewards rural women currently derive from butter production? Local prices for shea butter vary annually, regionally, and seasonally according to fruit production. Within a productive cycle, the economic value of shea nuts and butter is lowest between June and September, when the fruits mature and the nuts abound in local markets. The cost of shea nearly doubles in the dry season when nuts are scarce. The average producer price of 500 f CFA per kilogram of butter, calculated from local Burkinabè markets, is based upon annual mean production for the decade 1990–2000. Revenues suggest that a woman who transforms her entire stock of nuts into butter from a typical yearly collection of 560–650 kg can earn between 50,000 and 58,500 f CFA, or between US$100 and 117 annually (ANDINES, 2002).

Such figures, however, overestimate the real value of women's earnings, since most nuts currently made into butter are retained for family consumption rather than commercialized. Boffa (1995) estimates that 60–90 percent of shea nuts are consumed as butter by Burkinabè families, while the remaining collected nuts are marketed raw or as butter. With an average household retention rate of 75 percent of collected nuts, a Burkinabè woman actually earns between 12,500 and 14,600 f CFA, or just $US25–29 per year from butter sales.[9]

On average, the conversion of 10 kg of shea nuts to butter demands eight to ten hours of an individual woman's labor, and 560–650 kg of nuts are annually made into butter. The associated labor process demands between 450 and 650 hours of female labor per year. This figure does not include the work involved in nut collection, the time spent in water and firewood procurement, or the initial pulping, boiling, and drying stages that precede butter processing. Thus, quite modest sums are typically returned to female producers for their labor in shea production.

In view of this staggering workload, WID shea projects aim to improve women's financial returns by organizing female producers into cooperatives with labor-saving shea presses that will increase butter output and quality. They additionally attempt to link shea producers directly with the global cosmetics market through trade fairs and workshops, and by brokering negotiations with firms. But project goals depend upon the continuing interest of cosmetics companies to negotiate contracts favorable to producers and to market butter through incipient "fair trade" networks. Such contracts have fetched two to six times the prevailing market value of the butter per kilogram and provided women with advanced payment for their product (ANDINES, 2002).

While successful in augmenting female producer remuneration per kilogram of butter sold, WID projects that focus on shea commercialization are frequently conceptualized without regard to the broader household work demands facing rural women or concern over the added labor burden the projects may incur. As in comparable WID projects, intra- and inter-household relations that bear directly on female access to and control over resources (in this case, shea nuts), decision-making, and income opportunities, are often ignored (Carney and Watts, 1991; Leach, 1991; Pearson and Jackson, 1998; Schroeder, 1999). Likewise, there is little focus on the consequences of changes in the customary use and economic value of shea on intra-household resource claims and benefits. Is an increase in female income, stemming from women's involvement in expanded shea markets, resulting in new claims on them financially by their husbands or within the household? Are women reallocating some of their labor from subsistence crop cultivation to shea butter production? It is crucial that WID projects, so prevalent in sub-Saharan development policy, engage in such fundamental issues as they pertain to shea in order to circumvent pitfalls that may increase women's work burdens or weaken their economic position.

Despite WID interventions that assist women in shea processing and marketing, shea producers and their advocates remain in a vulnerable market position for they must rely upon green consumers and cosmetics firms, in the words of the Body Shop's founder Anita Roddick (2001), "to make conscious choices to change the world." It is uncertain how long the West's infatuation with natural lines of products will last. It also remains unclear when African shea production will exceed the demand by firms like the Body Shop, l'Occitane, and l'Oreal. But one issue is certain: 90 percent of commercial demand for shea remains in the food industry, where the humanitarian commitment to Sudano-Sahelian shea producers is not being advanced. Given the new standards for EU chocolate manufacturers, surplus production of a quality butter product will easily find a market niche, but not at the favorable producer prices upon which WID and fair trade networks stake their hopes and efforts.

One remaining issue tempers the income-earning prospects of WID shea projects. Unless donor networks include both rural and urban producers, rural female nut collectors may not share equitably in shea development. Prior to donor involvement, nut gathering and processing were exclusively rural activities. However, the growth in the global shea market, as well as urbanization, makes large numbers of poor unemployed urban women available to work in the urban butter projects now being developed in closer proximity to donor offices in cities (Compaoré, 2000). Locating the processing facility in a city or a town offers access to donor agencies, paved roads, electricity, storage, and export firms. Urban women are thus better able to take advantage of market information, technology transfer, training, and credit opportunities than those in rural areas.

This tendency is evident in the largest women's shea cooperative in Burkina Faso, Songtaaba, located in the country's capital, Ouagadougou. Songtaaba is supported by UNIFEM and now involves over 2000 women, including hundreds of urban dwellers. Donors have introduced processing equipment, which reduces female labor demands, while ensuring a product of desired export quality. The Songtaaba cooperative purchases nuts from rural female collectors for processing into butter. While the cooperative guarantees higher nut prices to rural shea gatherers, the added value realized from export butter sales primarily benefits the urban cooperative's members.

Such cooperatives are also needed in rural areas, where butter production traditionally occurs, thus enabling the nut collectors themselves to capture the full market value added by processing. Efforts to develop rural cooperatives must overcome some problems. As most rural Burkinabè women are neither numerate nor literate, someone must be trained to calculate returns to labor so the butter is not otherwise sold at a loss (Ouédraogo, 1997). Rural women also face transport and marketing difficulties. Their frequent reliance on private entrepreneurs to purchase, transport, and market shea results in far lower producer prices than those attained in WID projects. Donor assistance with the formation of associations that include rural butter cooperatives would thus greatly extend the benefits of shea commercialization beyond urban areas.

Evidence already indicates that the benefits and profits of shea can be equitably distributed when donor efforts support a network of urban and rural butter cooperatives. In northern Ghana, WID advocates created an association of women's groups encompassing ten villages, providing them with equipment to facilitate butter extraction. In 1994, the Body Shop began purchasing shea butter from the association. By 1997, orders had reached eleven metric tons and were still growing (TBS, 1997).

Unless such rural–urban linkages are strengthened in Burkina Faso, Biquard (1992, p. 178) forecasts an eventual takeover of the export shea market by urban entrepreneurs at the expense of rural shea producers. She further notes that "if peasant communities do not achieve the means to somewhat develop shea processing techniques, others, such as urban businessmen, will do so by introducing the shea butter production chain."[10] Female and male entrepreneurs already involved in purchasing surplus butter from rural producers not organized into shea associations are well placed to capture profits from the product's growing international demand (Biquard, 1992).

Conclusions

There is no question that contemporary shea butter commercialization in Africa's Sudano-Sahelian region presently offers the prospect to improve female income opportunities. However, the sustainability of these projects depends on many factors outside the control of Burkinabè women: continuing donor support, green consumer markets, fair producer prices, sustained shea demand, and labor-saving technologies that are appropriate and do not make additional demands on scarce environmental resources such as water and firewood.

One potential problem with WID shea projects, and reported elsewhere with development schemes that single out women, is their propensity for gender conflict.

In correctly wishing to assist impoverished rural women, gender-equity advocates at times lose track of the generalized poverty that husbands also face. As the economic situation of rural African producers continues to deteriorate from low producer prices, policy-makers encourage economic diversification, especially the production of non-traditional exports such as horticulture and NTFPs, which depend disproportionately on female labor. When income opportunities emerge in an overall context of wrenching rural poverty, it is after all perhaps unsurprising that husbands find ways to appropriate part of women's earnings. It would be a misinterpretation of the difficulties facing poor households to attribute such efforts merely to patriarchy, greed, and individualism. Both women and their husbands face poverty together, even if development projects target one or the other.

While the neoliberal development model currently offered to Sudano-Sahelian producers offers ample opportunities to create WID projects, it is premised on intensifying the labor of already overworked rural women. A key question remains. Does a development strategy that intensifies rural women's work provide a sustainable approach for lifting impoverished households out of poverty?

The fair trade movement confronts contemporary neoliberalism with efforts to arrest the deepening divide between rich and poor. With shea increasingly a household word in the West, fair trade asks privileged female consumers in department stores from Los Angeles to Montréal and London to Tokyo to support the effort of African women producers to struggle out of poverty. Thinking about a commodity in this way recalls the crucial insight of Sidney Mintz (1985, p. 214): "We must struggle to understand . . . the relationship between producers and consumers, of the meaning of work, of the definition of self, of the nature of things. . . . In understanding the relationship between commodity and person, we unearth anew the history of ourselves."

ACKNOWLEDGMENTS

The authors would like to thank Dr Edmond Keller and the UCLA Globalization Research Center-Africa for critical funding support. Our appreciation also extends to Hamidou Savadogo for his assistance with data collection and to Lise Nelson for her invaluable comments on the manuscript.

NOTES

1 The Sahel, which means shore in Arabic, refers to the drought-prone climatic belt that extends south of the Sahara to the tropical humid zone across Africa. The region includes the countries of Senegal, Mali, Niger, Chad, and The Gambia, as well as northern Nigeria, Guinea-Bissau, Burkina Faso, Ghana, and Côte d'Ivoire (Franke and Chasin, 1980, pp. 21–5; Soto Flandez, 1995).

2 Other key shea exporting countries include Mali, Ghana, Côte d'Ivoire, and Nigeria (Compaoré, 2000).

3 This chapter is the result of two months of fieldwork in Burkina Faso, interviews with NGO personnel, Burkinabè academics, and government officials involved in shea projects, participation in NGO-led workshops with female butter producers, participant observation among the Mossi in the province of Boulgou in rural Burkina Faso, and a

review of relevant primary and secondary sources. The authors would like to thank the UCLA Globalization Research Center-Africa for funding this study of the outcome of new markets for economically marginalized female African producers.

4   From 1906 to 1929, shea nut exports varied from 0.6 to 21 tons, while post-war exports climbed to 9,109 tons in 1952 but tumbled seven years later, to 380 tons (Pehaut, 1976).

5   Japanese and North American demand emerged later.

6   Four large European importers – namely Aarhus, Karlshamns, Unilever, and Van Dermoortele – dominate the shea market. They are likely to increase butter purchases with improved product quality. With no commitment to fair trade, manufacturers are poised to buy surplus butter production at rates they establish. Operating in a highly secretive manner, each company presents vastly dissimilar figures for shea imports, making it difficult to analyze demand and prices.

7   On average, approximately 8.5–10 kg of firewood are required to produce a single kilogram of shea butter (Niess, 1988).

8   The shea processing methods described correspond to those of the Mossi, the largest ethnic group of Burkina Faso. However, there exist a variety of nut harvesting and processing techniques, according to ethnic groups and regions. A description of these different techniques lies beyond the scope of this chapter.

9   Figures are based upon an exchange rate of 500 f CFA to US$1 (2001).

10 Translation by authors.

BIBLIOGRAPHY

ANDINES (2002) Beurre de karité: Groupe LAAFI (Burkina Faso) et ANDINES (France). Ile Saint Denis: ANDINES SA.

Biquard, A. (1992) Femmes et innovations technologiques: Pertes sans profit. L'exemple du beurre de karité (Mali). In J. Bissiliat, F. Pinton and M. Lecarme (eds), *Relations de genre et développement: femmes et sociétés*. Paris : ORSTOM.

Body Shop (1997) Promotional literature in brochure form.

Boffa, J. M. (1995) Productivity and management of agroforestry parklands in the Sudan zone of Burkina Faso, West Africa. PhD thesis, Purdue University.

Boffa, J. M., Yaméogo, G., Nikiéma, P. and Knudson, D. M. (1996) Shea nut (*Vitellaria paradoxa*) production and collection in agroforestry parklands of Burkina Faso. In *Domestication and Commercialization of Non-timber Forest Products in Agroforestry Systems: Non-wood Forest Products*, 9. Rome: FAO.

Booth, F. and Wikens, G. (1988) Non-timber uses of selected arid zone trees and shrubs in Africa. *FAO Conservation Guide*, 19, 34–45.

Boserup, E. (1970) *Women's Role in Economic Development*. London: Allen & Unwin.

Carney, J. (1998) Struggles over crop rights and labour within contract farming households in a Gambian irrigated rice project. *Journal of Peasant Studies*, 50(3), 334–9.

Carney, J. and Watts, M. (1991) Disciplining women? Rice, mechanization and the evolution of Mandinka gender relations in Senegambia. *Signs*, 16(4), 651–81.

Compaoré, P. N. (2000) Femmes, développement et transfert de technologies. Le cas des presses à karité au Burkina Faso. PhD dissertation, University of Montréal.

Conti, A. (1979) The commercialization of shea butter in Upper Volta: its implication for women (unpublished).

Crélerot, F. (1995) Importance of shea nuts for women's activities and young child nutrition in Burkina Faso: the case of the Lobi. PhD dissertation, University of Wisconsin-Madison.

CSPPA (n.d.) *Shea Export Figures from 1965 to 1992*. Ouagadougou: Caisse de Stabilisation des Prix de Produits Agricole.

Dalziel, J. M. (1937) *The Useful Crops of West Tropical Africa.* London: Crown Agents.

Dicum, G. and Luttinger, N. (1999) *The Coffee Book: Anatomy of an Industry from Crop to the Last Drop.* New York: The New Press.

Dolan, C. S. (2001) "The good wife": struggles over resources in the Kenyan horticultural sector. *Journal of Development Studies,* 37(3), 39–70.

European Fair Trade Association (1998) *Fair Trade Yearbook: Towards 2000.* Brussels: EFTA.

Elias, M. (2003) Globalization and female production of African shea butter in rural Burkina Faso. Masters thesis, University of California, Los Angeles.

Escobar, A. (2001) "Culture sits in places": reflections on globalism and subaltern strategies of localization. *Political Geography,* 20, 139–74.

Faucon, M., Sauvageau, A. and Bahl, S. (2001) *Coût de production pour le beurre de karité du Groupement Laafi.* Ouagadougou: PACK, UNIFEM, CECI.

Franke, R. W. and Chasin, B. H. (1980) *Seeds of Famine: Ecological Destruction and the Development Dilemma in the West African Sahel.* New York: Allanheld, Osmun.

Freidberg, S. (1996) Making a living: a social history of market-garden work in the regional economy of Bobo-Dioulasso, Burkina Faso. PhD dissertation, University of California, Berkeley.

Gosso, D. (1996) *Étude de l'importance socio-économique du karité et des dangers qui menacent sa survie.* Thesis for Diplôme de Technicien Supérieur (DTS), Ouagadougou (unpublished).

Hale, B. (2003) Africa hopes for anti-wrinkle cash cow. *BBC News World Edition,* Business Section, online, 17 January.

Hanson, S. and Pratt, G. (1995) *Gender, Work and Space.* London: Routledge.

Harsch, E. (2001) Making trade work for poor women. *Africa Recovery,* 15(4), 6.

Hart, G. (2002) Geography and development: development/s beyond neoliberalism? Power, culture, political economy. *Progress in Human Geography,* 26(6), 812–22.

Haugerud, A., Stone, P. and Little, P. (eds) (2000) *Rethinking Commodities: Anthropological Views of the Global Marketplace.* Boulder, CO: Rowman and Littlefield.

Hopkins, A. G. (1973) *An Economic History of West Africa.* New York: Columbia University Press.

Jackson, C. and Pearson, R. (1998) Introduction. Interrogating development: feminism, gender and policy. In C. Jackson and R. Pearson (eds), *Feminist Visions of Development: Gender Analysis and Policy.* London: Routledge.

Kelly, P. (1999) The geographies and politics of globalization. *Progress in Human Geography,* 23, 379–400.

Kessler, J. J. (1992) The influence of karité (*Vitellaria paradoxa*) and néré (*Parkia biglobosa*) trees on sorghum production in Burkina Faso. *Agroforestry Systems,* 17, 97–118.

Leach, M. (1991) Engendered environments – understanding natural resource management in the West African forest zone. *IDS Bulletin,* 22(4), 17–24.

Lewicki, T. (1974) *West African Food in the Middle Ages.* New York: Cambridge University Press.

Lovett, P. N. and Haq, N. (2000) Evidence for anthropic selection of the Sheanut tree (*Vitellaria paradoxa*). *Agroforestry Systems,* 48, 273–88.

Massa, G. and Madiéga, Y. G. (1995) *La Haute Volta Coloniale: Témoignages, recherches, regards.* Paris : Karthala.

Ministère de l'Economie et des Finances (1998) *Repères socio-démographiques.* Ouagadougou: MEF.

Mintz, S. (1985) *Sweetness and Power.* New York: Penguin.

Momsen, J. H. and Townsend, J. G. (eds) (1987) *Geography of Gender in the Third World.* Albany: State University of New York Press.

Nagar, R., Lawson, V., McDowell, L. and Hanson, S. (2002) Locating globalization: feminist (re)readings of the subjects and spaces of globalization. *Economic Geography*, 78(3), 257–84.

Niess, T. (1988) *Technologie appropriée pour les femmes des villages. Développement de la presse a karité au Mali*. Eschborn: Gate.

Ouédraogo, S. (1997) *Analyse des contraintes à l'amélioration de la qualité des noix de karité, Projet Filière Karité*. Ouagadougou: CECI/CIDA.

Pearson, R. and Jackson, C. (eds) (1998) *Feminist Visions of Development: Gender Analysis and Policy*. London: Routledge.

Pehaut, Y. (1976) *Les Oléagineux dans les pays d'Afrique Occidentale associés au marché commun, Vol. 1 and 2*. Paris: Éditions Honoré Champion.

Rice, R. A. (2001) Noble goals and challenging terrain: organic and fair trade coffee movements in the global marketplace. *Journal of Agricultural and Environmental Ethics*, 14(1), 39–66.

Roddick, A. (2001) *Take It Personally: How to Make Conscious Choices to Change the World*. Oakland, CA: New Harbinger Publications.

Ruyssen, B. (1957) Le karité au Soudan. *Agronomie Tropicale*, 12, 143–72, 279–306, 415–40.

Saul, M. (1988) Money and land tenure as factors in farm size differentiation in Burkina Faso. In R. E. Downs and S. P. Reyna (eds), *Land and Society in Contemporary Africa*. London: University Press of New England.

Schreckenberg, K. (1996) Forests, fields, and markets: a study of indigenous tree productions in the woody savannahs of the Bassila region, Benin. PhD dissertation, University of London.

Schroeder, R. (1999) *Shady Practices*. Berkeley: University of California Press.

Soto Flandez, M. (1995) Dry forest silviculture in the Sudano-Sahelian region: Burkina-Faso's experience. *Unasylva*, 181(2), 13–17.

Terpend, M. N. (1982) La filière karité: produit de cueillette, produit de luxe. *Les dossiers faim-développement*, February.

UNIFEM (1997) *Le karité, l'or blanc des africaines*. Dakar: UNIFEM Bureau Régional.

Watts, M. and Goodman, D. (1997) Agrarian questions: global appetite, local metabolism: nature, culture, and industry in *fin de siècle* agro-food systems. In D. Goodman and M. Watts (eds), *Globalizing Food: Agrarian Questions and Global Restructuring*. London: Routledge.

Whitehead, A. (1981) "I'm hungry mum": the politics of domestic budgeting. In K. Young, C. Wolkowitz and R. McCullagh (eds), *Of Marriage and the Market*. London: Routledge & Kegan Paul.

World Bank (1989) Burkina Faso. In *Trends in Developing Economies*. Washington, DC: World Bank.

Chapter 8

Working on the Global Assembly Line

Altha J. Cravey

A place that was ever lived in is like a fire that never goes out. (Eudora Welty, 1998 [1944])

The global assembly line is a mundane reality for millions of the world's women. Getting a factory job in export production can provide empowerment, friendship, social connections, self-respect, and a regular wage. The exact same job can be a source of boredom, anxiety, illness, grueling routine, long-term health hazards, family disintegration, disappointment, insecurity, and intimidation. Workers in Mexican assembly plants confront these contradictions in their everyday lives. A focus on household and gender dynamics helps to explain some of the ways maquila[1] workers cope with these kinds of dilemmas. A close examination of their lives, in turn, provides scholars and activists with a more complete understanding of contemporary globalization processes as well as alternatives. Due to their particular situation, the place-based passions of these workers – the unquenchable fire that Welty describes above – are at once highly intimate and distinctly global.

My research seeks to understand the daily lives of workers who participate in globalized labor markets, whether they participate in extremely competitive export sectors or in industries that draw on transnational migrant labor. The former workplaces rely on global trade and investment flows, while the latter require transnational human movement (often at considerable risk). In this chapter, I examine the lives of workers on the "global assembly line" who have been swept up in highly volatile global production geographies. Often working in export-oriented assembly factories in distinctly regulated and conspicuously demarcated areas known as export processing zones, these women (and men) participate in intensely global and intimately local processes every day. Drawing on extensive field research in Mexico, I explore the ways in which contemporary forms of globalized production – such as the Mexican maquila – are influencing the nature of gender, work, place, and resistance.

A feminist approach to understanding global production geographies is advantageous in several regards. A feminist angle links structural transformation to embodiment and lived experience, and implies a commitment to social change and social justice, bringing together epistemology and experience in previously unimagined ways. From a feminist geography perspective, this includes exploring place-based passions that arise in one's household, neighborhood, workplace, or other well traveled circuits of daily life, passions that have a potential to simultaneously ignite new ideas and new ways of living.

A feminist perspective relies on relational thinking and sees gender itself as something that is contested, negotiated, and thus constructed in the course of social relationships. In my research, I seek to understand geographies of production in relation to geographies of social reproduction.[2] The geographies of export production, and women's involvement in these transnational production chains, can only be understood in relation to unpaid gendered labor. I thus engage closely with feminist debates about work that broaden the concept of work to include many unpaid activities such as childrearing, emotional work, community and household upkeep, cooking, cleaning, and caring for the sick and elderly (Harris, 1981; Mackintosh, 1981; Kessler-Harris, 1982; McDowell and Massey, 1984; Laslett, 1989; Stitchter, 1990). These debates and intellectual advances help us to understand the enormity (and necessity!) of feminist transformative projects. If, for instance, we see gender in the lives of workers on the global assembly line as a socially constructed category, we can then begin to grasp the ways in which specific gender norms, practices, and expectations are highly dynamic and open to transformation. I will demonstrate how women as active subjects reconfigure gender norms in a variety of arenas.

In the same way that most feminists consider gender to be a relational category, feminist geographers view space in a relational way, linking concrete material and symbolic processes as well as different scales of analysis – from the local to the global. My own research on women's lives in the maquila zones of northern Mexico connects women's lives at home and the production geographies of transnational capital. Moreover, by focusing on the place-based passions of maquila workers – passions for creating a better life, for organizing – I expand our understanding of globalization itself, how the process is gendered, and how ordinary people shape global processes in distinct contexts. Feminist geographers in particular have demonstrated that "place" is imbued with subjective meanings often revolving around notions of "home" and "belonging" (Massey, 1994; Jones et al., 1997; Women and Geography Study Group, 1997). As "a portion of geographic space" (Johnston et al., 2000, p. 582), "place" as an analytical category refers to a setting in which social relations and identities are constituted, one that is produced through social practices operating across larger geographical domains (Massey, 1994). Using the categories of space and place thus allows me to explore how women's lives change on the "global assembly line" and, in turn, how women (and men) influence Mexico's development trajectory as well as other processes operating at wider geographical scales (McDowell and Massey, 1984; Massey, 1984).

Using parallel case studies of Nogales and Ciudad Madero, the former a border town where maquiladoras predominate and the latter a town dominated by a previous model of industrialization, I document the simultaneous and differentiated reorganization of social reproduction and production in northern Mexico, where foreign investment fueled rapid export industrialization. I demonstrate that households (and makeshift households such as single-sex dormitories) are sites of gender struggle that play out differently in the two cities, differences linked to the fact that channels of public social provision are much less abundant in border cities than in the interior locations due to the legacy of previous development models historically aimed at interior industrial locations. In both places, ordinary women and men shaped local communities, households, and workplaces in ways that influenced regional and national trajectories.

Mexican Maquilas

The global assembly line in Mexico is known as the maquiladora industry, or simply maquila. The factories first emerged in 1965 as an *ad hoc* experiment at isolated sites on the USA–Mexico border and expanded rapidly into a sophisticated comprehensive federal program to attract export investment. Interestingly, before embracing maquila production, Mexico had a highly diversified industrial plant that grew from a long-term project of state-led import substitution industrialization (ISI), an approach that itself had been a key aspect in Mexico's high growth "miracle" years of the 1940s, 1950s, and 1960s.[3] In the 1980s, through the debt crisis, Mexican industrial strategy was reoriented to foreign investment for export production. Maquila production thus became a symbol and a means of wider philosophical and political transformations guided by neoliberal principles – principles that call for less state intervention and the expansion of the "free market." This shift in industrial policy was marked by a geographic shift from a centralized industrial core to dispersed northern sites, a change in the nature of the state, and a corresponding reorientation of the state's productive strategies. A range of inward-oriented industrialization policies was reversed to encourage internationalization, liberalization, and privatization of the economy. With the implementation of the North American Free Trade Agreement (NAFTA) in 1994, the neoliberal principles enshrined in maquila production were institutionalized and pushed further along the same trajectory.

These transformations did not occur in a vacuum: they influenced, and were influenced by, Mexican social structure. Consideration of gender and household relations helps to explain these wider transitions in Mexican history, geography, and development choices. Industrial workers in northern Mexican export production facilities tend to be young women rather than male breadwinners. They typically pool several incomes and form extended households that reorganize domestic tasks and incorporate new members as needed. As wage earners, women maquila workers gain certain improvements in status. Still, this advance entails considerable contradictions and social costs. We can explore these gendered tradeoffs by considering two locations in Mexico. The first, Ciudad Madero, Tamaulipas, was profoundly shaped by earlier state-led industrial programs, while the second, Nogales, Sonora, is deeply involved in maquila production (see figure 8.1).

Ciudad Madero: State-led Industrialization

Mexico enjoyed several decades of industrial expansion and diversification from about 1940 to about 1976. The country's import substitution industrialization (ISI) policies were so successful in terms of economic growth that Mexico was dubbed

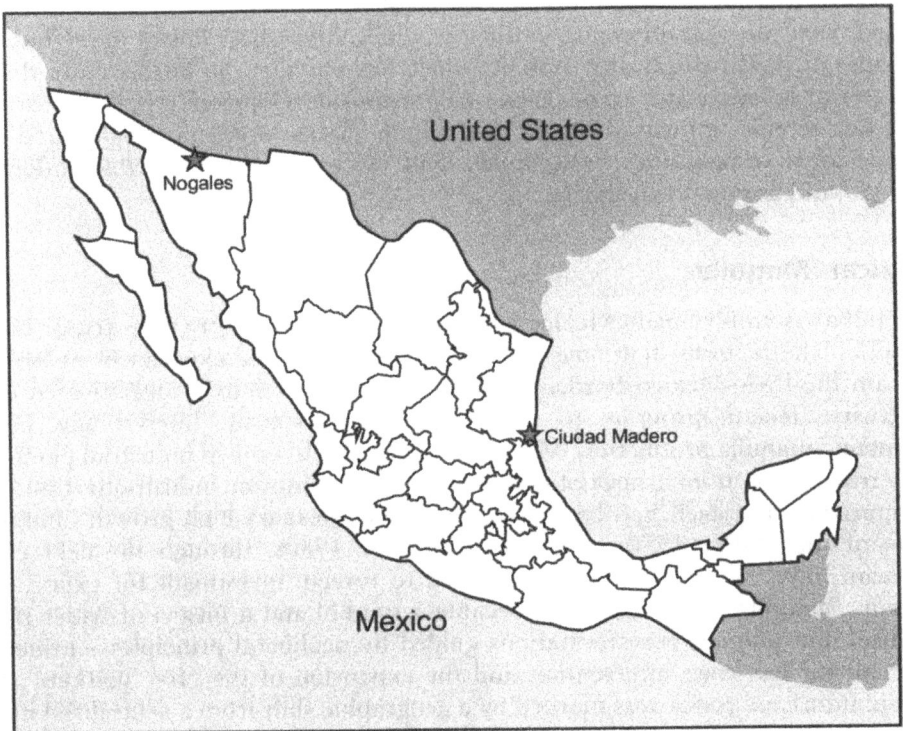

Figure 8.1 The two case study sites.

a "miracle" by many international observers (Hamilton, 1982; Gereffi and Wyman, 1990, Haggard, 1990; Maxfield, 1990; Middlebrook, 1995). The economic and social policies of the time shaped a specific set of gender and household norms that can still be observed in neighborhoods and communities of interior industrial districts within a highly concentrated and centralized geographical industrial pattern in the country. Ciudad Madero, Tamaulipas, on the Gulf Coast of Mexico (adjacent to Tampico) is one such place.

Nuclear households are the norm among industrial workers in Ciudad Madero. In 1992, I interviewed twenty industrial workers and found only two who lived in extended households. One young man explained his unusual situation (of living in an extended household) in this way: "My mother is a widow and enjoys having us all [himself, his sister, his brother, and his brother's wife] here in the house with her" (Sanchéz, personal interview, July 30, 1992). In the "miracle years," industrial workers such as those in Ciudad Madero were seen as the cutting edge of Mexico's development strategy and were rewarded with relatively high salaries and generous benefits understood as a family wage. A typical industrial household consisted of a male breadwinner, a female homemaker, and their children. In talking with people in Ciudad Madero, I found that women did most of the housework in these nuclear households, although there was some adult male contribution in 30 percent of households. This male input is considerably less than we will see in the gender-division-of-domestic labor in the Nogales example below. Feminists have documented and theorized the historical emergence of nuclear families, various ideologies of the family, and the "family wage." Feminist geographers are particularly attentive to the role of place and context in the emergence (or disintegration) of specific household forms.[4]

Some industrial workers in Ciudad Madero work for private industry, while others work for government-owned industries such as the local Pemex oil refinery (Pemex stands for the state owned oil company *Petroleos Mexcianos*). In both cases, unions, cooperatives, and other worker organizations gained government subsidies that funded many neighborhood services such as community schools, clinics, recreational facilities, and cooperatives. From 1940 to around 1976, high wages, comprehensive medical care, housing subsidies, educational, and childcare programs shaped a distinctive way of life in which many social goods seemed eternally abundant. These communities, in turn, shaped distinctive gender expectations that a male wage (in leading sectors of the economy) was sufficient for raising a family and for supporting a wife who would do most of the domestic tasks.

In contrast to what we will see in the Nogales case, the political struggles that shaped a particular way of life in Cuidad Madero were largely focused on economic concerns. That is, the "politics of place" in the formative years of the 1930s and 1940s was a politics of class and class formation. Male workers and their powerful unions (in Ciudad Madero and elsewhere) aligned themselves with national struggles and with nationalist development ideologies. A corporatist state, on the other hand, manipulated unions and male workers through patronage and generous social provision (Hamilton, 1982; Middlebrook, 1995). Gender thus shaped the myriad local struggles that created a distinctive way of life in Ciudad Madero; yet as the factory regime took shape, the gendered politics of place in the city was popularly understood and experienced as a politics that was primarily about class struggle and class formation.

The workers I interviewed in Ciudad Madero had access to a range of social services. All had state health insurance; workers at the state-owned oil conglomerate Pemex had company health plans and the others had federal health care provided through the *Instituto Mexicano de Seguro Social* (Mexican Institute of Social Security, IMSS), a comprehensive federal health care provider funded through employer and employee contributions. All those I spoke with said they were happy with their access to medical care. Federally funded childcare is abundantly available through both Pemex and IMSS agencies. While a small proportion of those I spoke with were renting their homes, most either own their houses or are in the process of buying them. Eight of the workers I talked with had benefited from a state housing program, two through *Instituto National del Fondo de la Vivienda para los Trabajadores* (National Institute for Worker Housing, INFONAVIT) and six through a Pemex program. Such public social programs helped to create and sustain the old factory regime as well as a particular form of gender relations in nuclear households. These normative ideas and practices about gender concern several social sites – household dynamics, industrial workplaces, and state policies and practices – in ways that are all closely aligned with each other. To distinguish between the gender norms that shaped earlier and later industrial strategies in Mexico, I refer to these comprehensive set of forces as a "factory regime."[5]

The old factory regime, represented by Ciudad Madero, involved various types of social provision which lent stability to state–worker relations and in turn encouraged a particular gendering of capitalist production. Feminist frameworks provide a way to include the subjective experience of such social programs and link these experiences with economic and cultural changes. What are the specific details of these transformed social policies? In Ciudad Madero and other centralized industrial communities the state provides benefits through two channels: direct provision from the state and provision through a state-owned industry and its labor union. As briefly sketched above, several categories of necessary goods (health care, housing, childcare, food, and recreation) are decommodified or partly decommodified in neighborhoods and communities dominated by the old factory regime. Households, individuals, neighborhoods, indeed whole industrial regions enjoyed long stable periods of growth and well-being during Mexico's miracle years.

If industrial policies had been overturned and dismantled in these very same places, the history of Mexico might have been very different. Specifically, we might have seen much more active resistance (e.g. strikes, protests, community actions) to neoliberal policies and to maquiladora-style production specifically than actually occurred. As Mexico's economic history unfolded, however, geographical displacement and dispersal of industry meant that new places in Mexico's vast and sparsely populated northern reaches became the sites for the erection of radically different production geographies (see figure 8.2). Struggles that might have emerged over economic policies were thus transferred to new regions of the country and to dispersed locations within that region. Furthermore, the focus of such struggles shifted from a primarily economic one (understood primarily through class) to one that linked social reproduction, community concerns, and economic issues.

Nogales: Maquila Industrialization

Nogales (population 300,000) sits across the border from Nogales, Arizona, and represents an entirely different production regime than the one described above in Ciudad Madero. It also supports quite distinct processes of social reproduction and gender dynamics. In comparing Ciudad Madero and Nogales, the starkest difference is the lack of state social provisions for worker, family, and community reproduction in Nogales. These political economic differences unsurprisingly shape contrasting household composition and organization in the two cities, as well as distinct channels of social access for their residents.

The two cities reflect the social and urban geographies of different productive and state-regulatory regimes. The regulatory role of the state between 1940 and 1976, which had nurtured domestic industrial capitalists and industrial workers in the old factory regime (Ciudad Madero), was completely revamped in favor of transnational capital accumulation (Nogales). Social policies in the new globalized spaces of Nogales support industrial investment rather than workers or communities, in contrast to institutions still evident in the interior industrial regions. Specifically, state policies and institutions in Nogales are geared toward providing transnational corporations with a dependent and relatively quiescent workforce by dismantling, deregulating, and privatizing social provision.

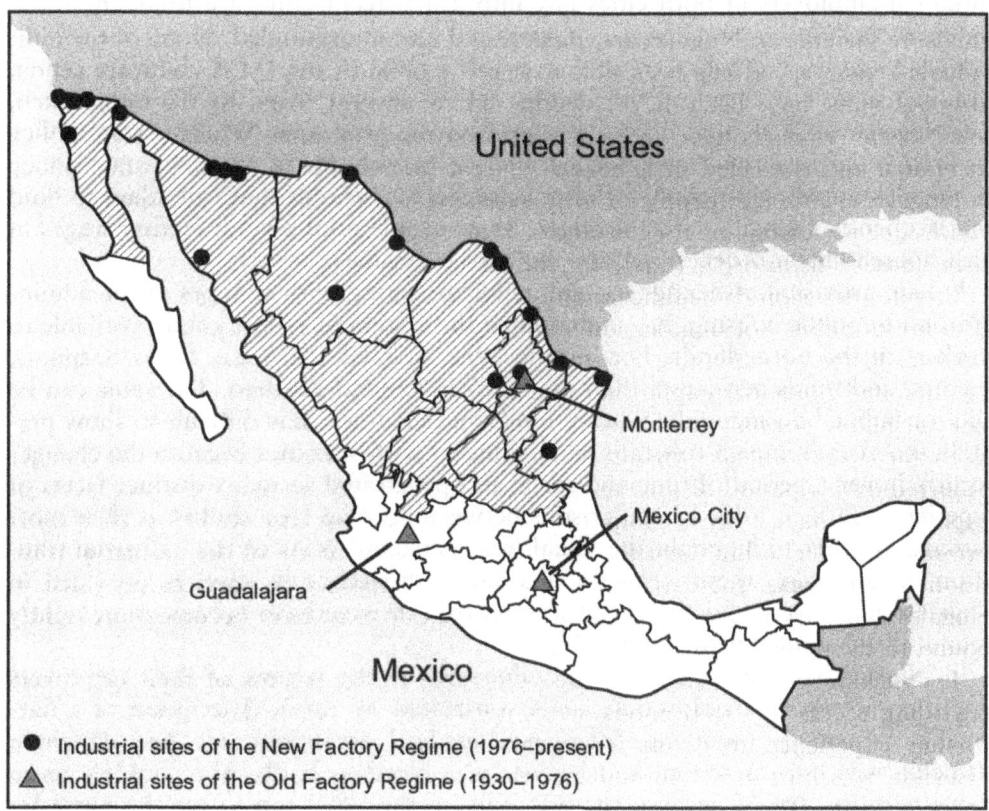

Figure 8.2 Major sites of the old and new factory regimes.

In short, the state emerged as a powerful ally of the employer by dismantling traditional networks of social provision in new transnational production sites, making workers more dependent on their employers. Some necessary goods, such as healthcare, childcare, and housing, were recommodified or became available only through the employment relationship. Goods considered social *rights* in Ciudad Madero became social *privileges* in Nogales and in the new factory regime. The ways in which workers resist, cope, negotiate, and reshape these processes are explored further below.

Industrial workers in the north have had very different experiences with federal programs of social provision. All Nogales squatter residents I talked with are insured through IMSS and many report that they have access to a doctor or nurse in the factory as well. In spite of having free benefits at IMSS, however, many visit private doctors because they cannot afford the time needed to see an IMSS doctor in understaffed northern clinics. The sentiment of a 29-year-old man from Veracruz is typical: "We never go to Social Security [IMSS]. We can't afford to wait all day to see the doctor" (Hurtado, personal interview, May 23, 1992). Unlike the workers in Ciudad Madero, few Nogales workers have access to federally funded childcare.

Although employers in both cities pay into a 5 percent childcare fund, the IMSS childcare facilities in Nogales are understaffed and underfunded. None of the individuals I interviewed has been able to enroll a child in the IMSS childcare center, although some have been on the waiting list for several years. By the same token, few Nogales workers have access to state housing programs. Whereas state policy facilitated and rewarded male-headed nuclear households in Madero, state policy in Nogales encourages people to form extended households that can adapt to fluid and frequently changing circumstances. Most people pool several factory wages in such households in order to pay for the bare necessities.

Private provision of healthcare and childcare, along with changes in the administration of public housing, has undermined the system of public goods available to workers in the borderlands. For instance, the Nogales IMSS has fewer hospitals, doctors, and funds per capita than the IMSS in Ciudad Madero. The same can be said for public housing and childcare funds and facilities. It is difficult to show precisely the overall impact (of state retrenchment) with statistics because the changes occurred over a period of time and space and penetrated so many distinct facets of experience. In fact, I draw comparisons between my two case studies *at their most extreme* in order to highlight the social (and personal) costs of the industrial transition. In any case, there is also a qualitative difference in services provided in Nogales: in a variety of ways, social services that do exist have become more tightly bound to the employment relationship.

In Nogales workers are far more vulnerable to the whims of their employers regarding access to social goods, as demonstrated by Ana's description of a harrowing experience involving emergency medical treatment for her daughter. "Ismelda was hurt at school and her ear was bleeding badly. She couldn't go to Social Security [IMSS] because she was born in the US. I ran to see the nurse [at the Avent maquiladora] and she called a private doctor who treated her for free" (Hernández Navarette, Personal interview, March 30, 1992). This mother's good record of employment at the factory enabled her to secure the medical care her daughter urgently needed. Even though help was unavailable from the state, her long and close relationship with her employer guaranteed the necessary treatment. The insecure nature of social goods and the way they are tied to the employment relationship is especially significant given the high turnover rates of maquiladora workers. On average, Nogales maquila workers lasted 15–16 months on a particular job in 1992.[6] By contrast, workers in Ciudad Madero in the old factory regime considered their jobs to be a lifelong arrangement. In turn, such changes in social provision weaken workers' positions when bargaining over conditions of employment. The geographically uneven nature of social provision reinforces differences in the development of the two industrial regions.

How does all this affect daily life? In what ways are gender and household dynamics caught up in larger globalization processes in northern Mexico? A feminist perspective is useful in illuminating the daily efforts to survive in Nogales and linking these daily struggles to wider sets of relationships. In contrast to workers in earlier high growth Mexican industrialization zones, maquila workers do not tend to form nuclear households and those who prefer nuclear households find one factory wage to be insufficient. Instead, maquila workers often pool multiple factory incomes in various types of extended household arrangements, sometimes incorporating unrelated individuals and expanding and contracting the household as needed over time.

Domestic chores, a key item in the work of social reproduction, are particularly interesting to consider in these households. I found that household work is a source of intense and ongoing negotiation and conflict in maquila households in Nogales. In contrast to earlier industrial workers who relied almost exclusively on female domestic labor, these tasks have been reorganized in Nogales. Men in these maquila households contribute twice as often and in more significant ways to household labor (Cravey, 1998a). In fact, some men contribute in regular and systematic fashion. For example, a man who has small children may take the night shift and a woman the day shift, or vice versa, so that one or the other can be home to care for children at any hour of the day or night. Such sequential scheduling is an unusual and creative response to the changing conditions in the new economic areas. As one young mother explained, "My husband takes care of the children. He has to take care of them because I work [in the factory] at night. We take turns with it every day" (Ortiz Mora, personal interview, May 17, 1992).

In these ways, both paid work and unpaid social reproductive work are reorganized in the context of rapidly globalizing labor markets. A relative dearth of public social programs (and a rapid retrenchment over time) has shifted a greater burden to individuals and households and in this way intensified conflicts over the division of domestic labor. Gendered negotiations over domestic work influence and are influenced by gendered negotiations in the (paid) workplace. As wage earners, women access a different source of power in northern households. Still, they are burdened by widespread shortages of social goods in their communities, as well as insufficient and fluctuating incomes.

A severe housing shortage exists in Nogales. For this reason, many industrial workers live in rudimentary makeshift housing in squatter areas that are "regularized" over time. Organizing themselves in large groups for orchestrated "land invasions" is one direct means of exercising place-based passions and providing shelter for one's self. In this way, a "politics of place" in Nogales is much more broadly focused than in the old factory regime, encompassing community and neighborhood concerns, issues of health and social reproduction, as well as identity, gender, and cultural politics. In the case of land invasions, women are frequently in the forefront of obtaining land and in the long process of gaining title and services (i.e. electricity, water, sewage disposal). To be sure, politics suffuse factory work as well, yet in the formative years of the new factory regime Nogales workers consistently lost ground (as workers), while finding their neighborhoods and households were potential sites for organizing and for making beneficial changes in their lives. In this way, a "politics of place" in Nogales has consistently involved questions of gender, of organizing "at home," and of social reproduction.

Some industrial workers in Nogales live in single-sex worker dormitories. These temporary households were a corporate solution to the housing shortage in Nogales, and they are not found in other border locations. Local government officials, labor leaders, and business people (who in some cases are the same individuals) use former factory buildings to create a system of highly regulated corporate dormitories in order to supply more than twenty local subcontractors with a flexible source of wage labor.

For workers who live in these large dormitories, the smallest details of life are regulated and the routines of daily life are individualized, a structure that reconfigures gender norms for many workers. Failure to comply with this system of total, twenty-four hour discipline may result in expulsion from the dormitory and from employment. For the most part, the individuals who reside in the dormitories are teenagers from interior rural areas. Many end up in Nogales after signing a contract with labor recruiters. Sixteen-year-old Ramon Delgado's experience is typical: "I heard about the jobs on a radio program in Carranza [his hometown]. I thought I could make more money than I make in the fields. My cousin and I signed up and got a ride to Nogales on a private bus" (Delgado, personal interview, May 3, 1992). The individualization of domestic labor within the dormitories has an immediate impact on residents' notions of masculinity and femininity because all individuals must provide domestic labor for themselves. Enrique Gutierrez confided that he was learning how to cook during his stay in the dormitory and that "getting together with a couple of other guys for meals is easier than fixing my own" (Gutiérrez, personal interview, May 17, 1992). On the other hand, some of the women in the dormitories viewed the individualization of tasks in the dormitories as liberatory. Three female cousins happily told me, "we work as a collective and only spend about an hour a day" on such tasks as laundry, cooking, and cleaning (López, personal interview, May 10, 1992). The flexibility of this dormitory arrangement has allowed local business leaders to maintain a tighter control on labor than at other border locations. The dormitory system also has allowed corporations to adjust the labor supply with little overt worker resistance. More workers can be recruited for peak moments of local economic cycles and subsequently dismissed during recessionary periods.

An examination of the Mexican transition to a new factory regime suggests that those very pressures that produced nuclear households in the old factory regime are causing further fracturing of the industrial household structure in the new deregulated northern industrial zones. Production in the new factory regime is far less protected from international competition than had been the case in the old factory regime. In the northern region, the effort by transnational employers to lower the costs of social reproduction relies on the expansion of the labor market to include younger female and male workers. These workers, in turn, find the wage insufficient for the nuclear family norm of industrial workers elsewhere and therefore develop other household forms. My research suggests that the same forces of fragmentation that created a nuclear family norm now threaten to further atomize the social fabric.

The case studies of Nogales and Ciudad Madero illustrate that state social programs were unevenly applied in the context of economic restructuring. Key state institutions supporting working families (such as IMSS clinics) did not keep pace with rapid industrial expansion in the north. Thus, social programs were defunded and dismantled in the northern sites of new transnational factory regimes in the north, while they continue to exist – albeit attenuated – in the interior regions (see Cravey, 1998a). In Ciudad Madero, state industrial and social policies regulated employment relationships and created distinctly gendered forms of capitalist production based on the male breadwinner and the nuclear family. In contrast, the gendered forms of capitalist production in the north are marked by a youthful, femi-nized workforce who more often live in extended households in which a variety of income sources may be pooled.

The differences between Nogales and Ciudad Madero, and the experiences of gendered workers in these two sites, demonstrate the importance of gender analy-sis of political economic transformation at a global scale. Gender analysis helps to explain how and why export production could extract more surplus labor and thus is more successful at attracting international capital investment. The dormitories of Nogales, which supply a steady stream of young, non-unionized labor, attest to the importance of social reproduction regimes to transnational capital. The Mexican state, in its pursuit of illusive "development" goals, is a key actor in the gender struggles that have shaped distinct factory regimes in different parts of the country.

A comparative analysis of gender and factory regimes in Cuidad Madero and Nogales illustrates the way feminist state theory might improve abstract conceptu-alizations of state power and control (Enloe, 1989). As we have seen, the Mexican government transferred social costs to new regions of the country and to fresh labor pools of youthful female workers in the 1980s when development planners embraced a market-led export-oriented development philosophy. A gender analysis of state strategy (and its reversal) helps to explain new sociospatial patterns of control associated with the maquila factory regime. Control is extreme in certain communities, as the existence of the Nogales dormitories demonstrates.

The country's embrace of neoliberal policy coupled a severe social policy retrenchment with the dismantling of a state-led industrial program based on a family (male) wage. The implications of this transformation in development policy are best understood by analyzing the gendered nature of the state, and, more impor-tantly, the gendered nature of state policy changes. Without knowledge of the complex gendered geographies of households, workplaces, and communities who were directly affected by the old and new factory regimes, we would not know how the state had been so effective in dismantling its former policy.

Conclusion

In the Mexican industrial transition to the maquiladora model, production geogra-phies became more and more global, while social reproduction geographies were increasingly localized, privatized, and individualized. The retrenchment of public social programs and the transfer of social reproduction responsibilities to smaller geographic scales facilitated the globalization of production in Mexico. The geo-graphic shift to northern factory sites helps to explain how such radical social trans-formations occurred without more active resistance in Mexican workplaces and communities.

What broader conclusions can be drawn from this example? Clearly the global assembly line is an intriguing place to examine the everyday impact of globaliza-tion on ordinary lives, and the way these same lives shape global realities. For this reason, the global assembly line is an excellent vantage point from which to see active resistance and imagine alternative forms of globalization. Shifting our per-spective back and forth from the micro-geographies of social reproduction to the global geographies of production gives us some preliminary tools to imagine alternative models of globalization that might spring from more egalitarian social relationships, or from social and economic justice ideals, or directly from creative new ways of living and self-expression. In this way, shifting the scale of analysis can reveal crucial dynamics that, when examined closely, suggest other versions of globalization, based less on governmental and economic understandings and more on personal, day-to-day lives of individuals swept up in globalization processes.

Thus, while details of the daily lives of maquila workers are highly specific to the situation in northern Mexico, the insights may be profitably linked to research and activism in East Asia, Guatemala, and elsewhere (Cravey, 1999; Traub-Werner and Cravey, 2002). Indeed, anti-sweatshop activists (such as the United Students Against Sweatshops) use such research to link activism and strategy in various places. Problematizing the space (and social reality) that lay beyond my case study sites allowed me to speculate more carefully about potential alliances and avenues of social change in northern Mexico.

My research on geographies of work suggests that we can have more impact in this regard if we recognize that geographical scale is actively produced and that ordinary people create global realms as well as local ones (Smith, 1992; Marston, 2000). Because their lives are profoundly global as well as local, the struggles of maquila workers shape global and local production geographies. Maquila research lends support to disciplinary suggestions that geographical scales are socially constructed (Marston, 2000). That is, various scales are interconnected in multiple complex ways for those who work on the global assembly line. Transformations at the scale of the body, the home, and the neighborhood are intertwined with transformations at the nation-state scale, the regional scale, and the global scale. Of these, the national scale is of particular analytic interest because of the way state policies and practices are gendered and mediate the global/local dialectic as described above (Cravey, 1998b).

If feminist geographers want to imagine (and construct) alternative forms of globalization that push beyond some of the dilemmas which maquila workers face on a daily basis, we need to closely examine our disciplinary conceptualizations of place, space, and geographic scale. Place, in particular, has recently captured the attention of many feminist geographers, anthropologists, and other social scientists (Prazniak and Dirlik, 2001; Escobar and Harcourt, 2002; Harcourt and Escobar, 2002; Hurtig et al., 2002). This enthusiasm for thinking about place and the "politics of place" is largely due to the fact that we may experience deep intimate bonds with a particular place, especially when this place has been home to us or shaped our own identity. As Eudora Welty eloquently put it, "A place that was ever lived in is like a fire that never goes out." The "politics of place" has a particular appeal for feminists and others who want to understand and intervene in social change. To recognize the full potential power of place and its connection to lived experience is to recognize the power of place as a tool in social change. My research in Mexico suggests that a feminist geographical perspective that situates a "politics of place" in a larger "politics of space" may be most effective in this regard because such conceptualizations avoid romanticizing place (Massey, 2002). In this way, place-based passions can be effectively harnessed to wider social transformations and social responsibilities at national, regional, and global scales (Massey 2002).

ACKNOWLEDGMENTS

I would like to thank Lise Nelson for constructive suggestions that substantially improved this chapter.

NOTES

1   The global assembly line in Mexico is known as the maquiladora industry, or simply maquila.
2   Social reproduction is defined as the tasks envolved in the daily and generational restoration of workers, encompassing mundane household chores (i.e. laundry, meal preparation, cleaning) as well as childcare and biological reproduction. It includes work that is not directly part of the wage labor system yet is essential to the long-term maintenance of the system.
3   During those decades, the average annual growth rates reached 8 percent.
4   See Cravey (1997) for more discussion of these debates.
5   Gender analysis of Mexico's economic strategies engages and extends Michael Burawoy's (1985) notion of "factory regimes."
6   These figures showing high turnover rates are consistent with other maquila research findings.

INTERVIEWS

Delgado, Ramón, Nogales dormitory resident, May 3, 1992.
Gutiérrez, Enrique, Nogales dormitory resident, May 17, 1992.
Hernández Navarette, Ana, Nogales squatter resident, March 30, 1992.
Hurtado, Raúl, Nogales dormitory resident, May 23, 1992.
López, Marisa, Nogales dormitory resident, May 10, 1992.
Ortiz Mora, Laura, Nogales squatter resident, May 17, 1992.
Sánchez, Gabriel, Ciudad Madero Quimica del Mar worker, July 30, 1992.

BIBLIOGRAPHY

Burawoy, M. (1985) *Politics of Production: Factory Regimes under Capitalism and Socialism*. London: Verso.
Cravey, A. J. (1997) The politics of reproduction: households in the Mexican industrial transition. *Economic Geography*, 73(2), 166–86.
Cravey, A. J. (1998a) *Women and Work in Mexico's Maquiladoras*. Lanham, MD: Rowman & Littlefield.
Cravey, A. J. (1998b) Engendering the Latin American state. *Progress in Human Geography*, 22(4), 523–42.
Cravey, A. J. (1999) Toothless tigers and mouldered miracles: geography and a global gender contract in the NICs. Unpublished manuscript.
Enloe, C. (1989) *Bananas, Beaches, and Bases: Making Feminist Sense of International Politics*. Berkeley: University of California Press.

Escobar, A. and Harcourt, W. (2002) Responses [to Harcourt and Escobar]. *Development*, 45(1), 25–7.

Gereffi, G. and Wyman, D. (eds) (1990) *Manufacturing Miracles: Paths of Industrialization in Latin America and East Asia*. Princeton, NJ: Princeton University Press.

Haggard, S. H. (1990) *Pathways from the Periphery: The Politics and Growth of the Newly Industrializing Countries*. Ithaca, NY: Cornell University Press.

Hamilton, N. (1982) *The Limits of State Autonomy: Postrevolutionary Mexico*. Princeton, NJ: Princeton University Press.

Harcourt, W. and Escobar, A. (2002) Women and the politics of place. *Development*, 45(1), 7–13.

Harris, O. (1981) Households as natural units. In K. Young, C. Wolkowitz and R. McCullagh (eds), *Of Marriage and the Market: Women's Subordination Internationally and Its Lessons*. London: Routledge.

Hurtig, J. D., Frazier, L. J. and Montoya del Solar A. R. (eds) (2002) *Gender's Place: Feminist Anthropologies of Latin America across the Americas*. New York: St Martin's Press.

Johnston, R., Gregory, D., Pratt, G. and Watts, M. (2000) *The Dictionary of Human Geography*, 4th edn. Oxford: Blackwell.

Jones, J. P. III, Nast, H. J. and Roberts, S. (eds) (1997) *Thresholds in Feminist Geography: Difference, Methodology, Representation*. Lanham, MD: Rowman and Littlefield.

Kessler-Harris, A. (1982) *Out to Work: A History of Wage-earning Women in the United States*. Oxford: Oxford University Press.

Laslett, B. and Brenner, J. (1989) Gender and social reproduction: historical perspectives. *Annual Review of Sociology*, 15, 381–404.

McDowell, L. and Massey, D. (1984) A woman's place? In D. Massey and J. Allen (eds), *Geography Matters!* Cambridge: Cambridge University Press.

Mackintosh, M. (1981) Gender and economics: the sexual division of labor and the subordination of women. In K. Young, C. Wolkowitz and R. McCullagh (eds), *Of Marriage and the Market: Women's Subordination Internationally and Its Lessons*. London: Routledge.

Marston, S. (2000) The social construction of scale. *Progress in Human Geography*, 24(2), 219–42.

Massey, D. (1984) *Spatial Divisions of Labor: Social Processes and the Geography of Production*. London: Macmillan.

Massey, D. (1994) A global sense of place. In *Space, Place and Gender*. Cambridge: Polity Press.

Massey, D. (2002) Don't let's counterpose place and space. *Development*, 45(1), 24–5.

Maxfield, S. (1990) *Governing Capital: International Finance and Mexican Politics*. Ithaca, NY: Cornell University Press.

Middlebrook, K. (ed.) (1995) *The Paradox of Revolution: Labor, the State, and Authoritarianism in Mexico*. Baltimore: Johns Hopkins University Press.

Prazniak, R. and Dirlik, A. (eds) (2001) *Places and Politics in an Age of Globalization*. Lanham, MD : Rowman and Littlefield.

Smith, N. (1992) Contours of a spatialized politics: homeless vehicles and the production of geographical scale. *Social Text*, 33, 54–81.

Stichter, S. (1990) Women, employment and the family: current debates. In S. Stichter and J. Parpart (eds), *Women, Employment and the Family in the International Division of Labor*. Philadelphia: Temple University Press.

Traub-Werner, M. and Cravey, A. J. (2002) Spatiality, sweatshops, and solidarity in Guatemala. *Social and Cultural Geography*, 3(4), 383–400.

Welty, E. (1998) Some notes on river country. In *Eudora Welty: Stories, Essays, and Memoir*. New York: The Library of America (originally published 1944).

Women and Geography Study Group (1997) *Feminist Geographies: Explorations in Diversity and Difference*. Harlow: Longman.

Chapter 9

From Migrant to Immigrant: Domestic Workers Settle in Vancouver, Canada

Geraldine Pratt in Collaboration with the
Philippine Women Centre

Filipineza

In the modern Greek dictionary, the word "Filipeniza" means "maid."

If I became the brown woman mistaken
for a shadow, please tell your people I'm a tree.
Or its curling root above ground, like fingers without a rag,

without the buckets of thirst to wipe clean your mirrorlike floors.
My mother warned me about the disappearance of Elena.
But I left her and told her it won't happen to me.

The better to work here in a house full of faces I don't recognize.
Shame is less a burden if spoken in the language of soap and stain.
My whole country cleans houses for food, so that

the cleaning ends with the mothers, and the daughters
will have someone clean for them, and never leave
my country to spend years of conversations with dirt.

When I get up, I stand like a tree, feet steady, back firm.
From here, I can see Elena's island, where she bore a child
by a married man whose floor she washed for years,

whose body stained her memory until she left in the thick
of rain, unseen yet now surviving in the uncertain tongues
of the newly-arrived. Like the silence in the circling motions

of our hands, she becomes part myth, part mortal, part soap.
(Bino A. Realuyo, 2002[1])

Bino Realuyo writes of a bargain struck by a Filipina domestic worker, to clean
houses for food "so that the cleaning ends with the mothers, and the daughters will
have someone clean for them, and never leave [their] country." Such bargains are

struck by Filipino domestic workers working in more than 130 countries, in what Parrenas (2001, p. 361) identifies as a "female labor diaspora." This vast labor diaspora is a striking example of the high costs of global restructuring, borne disproportionately by particular groups of people. The Filipino diaspora is a product of global uneven development, the labor export policy of the Philippines' government (conceived within the context of a long history of US colonialism in the Philippines, massive foreign debt, and the discipline of the International Monetary Fund), and the demand in industrial nations for migrant women to provide low-wage service work, in particular to carry out the tasks of social reproduction that many resident women no longer wish or are able to perform. Yet conditions for this labor diaspora differ from country to country, and a convincing feminist analysis requires a nuanced appreciation of this geographic specificity. It is this insistence on concrete geographic particularity that is a defining characteristic of feminist geographers' contributions to discussions of transnationalism, diaspora, and globalization (Mitchell, 1997; Katz, 2001; Pratt and Yeoh, 2003).

A peculiarity of the Canadian context is that there is a federal work visa program, called the Live-in Caregiver Program (LCP), that allows Canadian families to sponsor and employ non-Canadians as live-in caregivers. By international standards, the LCP is unusual because registrants have an opportunity to apply for permanent resident status in Canada after 24 months in the LCP. This promise of (eventual) citizenship weighs heavily in assessments of the ethics of the LCP and claims to Canada's relative superiority in protecting rights of migrant domestic workers. Few would argue that the program is more than exploitative in the short term, and the federal government is explicit that the LCP exists because few Canadians are willing to perform this live-in job. The "wrong" of Canadians taking such obvious advantage of economic misery elsewhere is set "right" by bestowing permanent residency on those who function responsibly within the LCP. When nanny agents were interviewed in Vancouver in 1994, one made this argument succinctly: "You're not getting 19 bucks an hour [working as a domestic worker], but . . . you're getting Canadian citizenship. . . . [Activist groups will] wreck it for these women, who come in and 99.9 percent of them are happy to be here" (Pratt, 1999, p. 222).

In a more scholarly vein, an early and astute critic of the LCP, Audrey Macklin, ventures that "the entitlement to permanent membership [can function] as a measure of recognition and compensation," and "the more exploitative the consequences of temporary status, the greater the marginal value extracted from the worker, . . . the greater the entitlement" (Macklin, 2003, p. 487).[2] She argues that a polity cannot relegate migrants to permanent "temporary" status without "committing a basic injustice" (ibid.). The compensatory entitlement to citizenship is a distinctive characteristic of the LCP and distinguishes Canada from other industrialized nations, such as the United States, where large numbers of domestic workers live illegally and hence in a perpetual state of impermanence (Mattingly, 2001). But what are the implications of the notion of compensatory citizenship, and how are we to think through such concepts toward a transnational feminist ethic of global justice? This is a pressing question, especially because so many women in the global north – including feminists – are compromised as beneficiaries of these global circuits of labor.

In this chapter, we approach this issue by returning to the participants of an earlier research project after a period of seven to eight years, during which time all have left the LCP and become permanent residents.[3] In 1995, fifteen migrant workers, all registered in the LCP, participated in an extended research project at the Philippine Women Centre (PWC). This was a participatory action project, a collaboration between a white academic and a Filipino activist group, which emerged out of and attempted to address concerns about the exploitative practices of feminists of the global north studying women of the global south (Pratt, 2000). Alternating between expressions of deep sadness, frustration, anger, and remarkable humor, domestic workers told rich stories of being propelled from the Philippines by economic necessity and the difficulties of their hard and stigmatized lives working and living in middle-class Canadian homes (Pratt in collaboration with the PWC, 1998, 1999). Eight years on, we listen to them speak of their experiences settling in Vancouver, to assess whether and how permanent residency compensates for the earlier period of intense exploitation. How are we to measure relations of justice if former domestic workers continue to find themselves, now as permanent residents of Canada, enmeshed within deeply exploitative labor practices?

Commenting on the Swiss guest worker program, Frisch (1986) remarked: "We asked for workers. We got people instead" (quoted in Macklin, 2003). The same is true of the LCP. This insight – that workers are also people – has been fundamental to feminist geographers, who have insisted on the many ways that social relations beyond the workplace tend to ghettoize women in certain occupations and cheapen their wages (Hanson and Pratt, 1995; McDowell, 1999). A focus on migrants extends our thinking about the geographical extent of such social relations, well beyond the boundaries of the nation-state. Former domestic workers are operating within a global labor market and their settlement in Canada is conditioned by extensive and continuing obligations in other spaces, notably the Philippines. This is to say what is now commonplace in transnational studies: immigrant settlement and labor market experiences cannot be understood within the boundaries of the "receiving" country (Pratt and Yeoh, 2003). But the point goes beyond this. An understanding of immigrant settlement and "integration" demands more than a narrowly economistic and individualistic framework. It requires that we look beyond individuals' wages and occupational statuses. We attempt to understand domestic workers' settlement experiences as embodied, transnational ones, paying special attention to how social networks and familial obligations shape their immigrant experiences.

Labor Market Experiences in Vancouver

Our earlier research was a fully collective enterprise. We met many times as a group and the research itself was a mechanism for these domestic workers to develop deep friendships and a common interpretation and critique of their situations. Eight years on, their lives and experiences are more dispersed and, although we attempt to develop generalizations about their settlement experiences, we begin with short descriptions of the lives of three domestic workers who have worked very hard to settle successfully, to demonstrate how they have done this in different ways.[4]

April

It was April who coined the phrase "from RN to RN" to describe her trajectory from Registered Nurse in the Philippines to Registered Nanny in Canada, a phrase that we have made much use of in analyses of the LCP (e.g. Pratt, 1999). April's contract through the LCP ended in 1995 and she became a landed immigrant in 1996, a process that was delayed by its considerable expense (C$1600 for various processing fees). From the moment of entering Canada, April was intent on returning to nursing, but she soon found that restrictions associated with the LCP and the reluctance of professional bodies in British Columbia to accredit her nursing training dictated that she take a one-year, full-time "refresher" course. The first four years after the LCP were spent in an assortment of full-time and part-time jobs (live-out nanny, cashier, cleaner, waitress, elder care, babysitting) in an effort to save enough to go back to school. In 1997 and 1998 she was working three and then four jobs. Finally, accepted into the refresher course at a community college, and at the discretion of a low-level bureaucrat, April was able to put together a mixture of EI (Employment Insurance) benefits, student loans, and savings to finance the one-year course.[5] She took her final exams in June 2000 and was hired even before she received her (positive) exam result. April, now thirty-four years old, has been employed full time in the same suburban hospital for the past three years but continues to augment her earnings with extra hours at another hospital. April has thus come full circle: from Registered Nurse to Registered Nanny, and back again.

Anna

Anna left the Philippines when she was eighteen. She had graduated from a midwifery course but could not be licensed until twenty-one years of age. She worked first as a domestic worker in Singapore for five and a half years and then came to Canada under the LCP. She finished her LCP contract in 1996 after twenty-four months in Canada but for reasons unclear to her it took one and half years to obtain a landed visa and she continued to work as a live-in domestic worker with the same employer during this time. After finishing her LCP contract, she inquired about updating her midwifery skills but judged it to be too expensive and took a course in long-term care. Her jobs began to multiply. When she first finished her course, she worked from Monday to Friday as a nanny, and then Saturday and Sunday at a nursing home. By 1999 she was still living with her employer, and working three jobs: nanny full-time, cleaning at Future Shop from 6 p.m. until 1 a.m., and then Saturday and Sunday at a nursing home. In 2000 she moved into housing adjacent to the Kalayaan Centre, and began working on a casual basis as a cashier at Army Navy and at the Superstore. The jobs began "to rain down" but all were on a casual or contract basis and she was holding five jobs simultaneously. She had been active at the PWC for many years and contract jobs began to emerge from the contacts that she developed there: at ELP (End Legislated Poverty), at FNSG (Filipino Nurses' Support Group), and eventually at Bridge Transition (a women's transition shelter), where she now works full-time.

Delia

Delia, now forty-two, is the second eldest of thirteen children. Her family owns a prosperous farm in Nueva Ecija: "my father was making good money in vegetable production. I did not even have to go abroad at that time." Though she has a BEd in elementary education, her movement to and through Canadian society has been shaped by different systems of patronage. A wealthy family, which owns almost all of the land in her village but had moved to Canada, paid for her six-month training to qualify for the LCP and then all of the expenses involved in applying and coming to Canada. They brought her to Canada in 1994 through the LCP and she worked for various members of the family in Vancouver and Toronto for the next three years, for less than $300 a month in recognition of her debts to the sponsoring family. After completing her LCP contract, Delia continued to live in her patron's home, and the employer's daughter, who worked as an accountant at a Catholic church, told Delia of a vacancy there for a housekeeper. In 1997 Delia started working at the Church part-time (three days a week) and supplemented this with cleaning jobs (in private homes at $10 per hour), which she found through her friends. Within the year, the job as housekeeper went full-time and she continues to work there (earning $12 per hour with benefits), cleaning the residences of the priests. Delia's life now revolves almost exclusively around the church and her residence is effectively tied to her employment. She lives with a former domestic worker, another participant in this research project, in an apartment adjacent to the church. Hearing of her daily routine, Cecilia Diocson (interviewer) observed, "You [and your roommate] are really like nuns. You are even more than the nuns," and asked, "Are you feeling happy?" Delia responded, "Yes. I am happy and contented."

We begin with these stories because they simultaneously introduce experiences common to a good number of the fifteen research participants and embed them within some of the complexity and specificity of individuals' lives. Filipinas often have the experience of being stereotyped, simplified, and overgeneralized (e.g. all Filipinas are nannies or sex workers), and we have no desire to repeat this tendency. The individual stories indicate a range of experiences – from recovering previous occupational standing to significant deskilling. As a generalization, the latter has been the more common experience and most who participated in this study continue to do domestic work in other forms (table 9.1). We want to examine variations in individual life experiences to understand why this has been the case.

Understanding LCP Registrants' Downward Occupational Mobility

A nanny agent who was interviewed in Vancouver in 1994 made a stark and disturbing contrast between Europeans, who use the LCP as a "jumping board" to a career ("they have their own plan, their own career, their own training back [home]"), and Filipinas, who remain as nannies and housekeepers. He reasoned that "Some of them [Filipinas] will really have a little bit more plans for their life than being that way, but the majority: nannies. That's what they're going to do: housekeeping" (Pratt, 1997, p. 172). While his description of the general pattern of Filipinas' occupational achievement in Vancouver is to some extent true, he places it in a potent narrative of cultural limitation and underachievement. To counter such cultural stereotyping, it is important to recognize how Filipinas' experiences of downward occupational mobility are the outcome of both rational planning and situational constraints. In particular, we examine the social networks, extending from Vancouver to the Philippines, which are central to this downward trajectory.

Table 9.1 Summary of occupations for 15 research participants

Occupation in the Philippines	2003 Occupation
Social work (1)*	Housekeeping in hotel (1)
Midwife (5)	Home care (2) Factory assembly (1) Housekeeping (1) Women's transitional shelter (1)
Teacher (7)	Housekeeping (2) Early childcare education (casual jobs at 3 different daycare centers) (1) Babysitting and low-level retail service jobs (1) Waitress and nurses' aid (1) Community settlement worker (1) Janitorial (1)
Registered nurse (1)	Registered nurse (1)
Telephone operator and nurses' aid (1)	Stay-at-home mother and part-time job in floral shop (1)

*Numbers in parentheses indicate the number of women who fall into this occupational category.

Remittances

It is commonplace to note the importance of remittances within transnational families more generally and for the economy of the Philippines in particular. Monetary remittances of overseas workers are the largest source of foreign exchange earnings in the Philippines. It was estimated in 1994 that overseas labor migrant workers sent over US$2.6 billion to the Philippines through the formal banking system. Money sent though private finance companies, letters, and return migrants was thought to bring the total closer to US$6 billion (Karp, 1995, cited in Parrenas, 2001).

In the Jamaican context, Mullings (1999) argues that receipt of overseas remittances has the effect of loosening Jamaican women workers' immediate dependency on their jobs and hence employers' control over them. But the labor market outcomes are very different for those who are sending remittances. With few exceptions the women in our study continued to send large sums of money home on a regular basis. Anna describes her intense and continuing financial commitments to her family.

> But even if I have no child of my own, I have the whole family to support back home . . . basically I am running the whole family on my own. And I am sending five nieces and nephews to school at one time. It's like you with the kids, having joy when you come home. Now I am happy that I was able to help five of them. . . . I have a seaman. I have an architect. I have a midwife and one, next year, will finish civil engineering. I also have a computer secretary. . . . When classes started I was getting crazy to think where I would get the money. That meant thousands [of pesos]. Like I think for me, I was only lucky that I could find enough jobs, or if there is not enough money, I have to tighten my belt.

Anna was admired (and teased) by her friends for her capacity to "tighten her belt."

Mira: She is not spending for things at all. She likes keeping money with her. She is a bank.
Mhay: And she is not buying any nice clothes. I think her jeans are still from Singapore.
Anna: Oh now, come on. They wouldn't fit me anymore.
Mhay: That's her bag, since a few years ago.
Anna: Yes, this is my bag from when I was still in Singapore in 1992.

Indeed, Anna's thrift and her hard work at multiple jobs have allowed her to purchase many things for her family in the Philippines, beyond the educations of her nieces and nephews. She has also built a house in which her brother lives. But the remittances have had a decisive effect on her career choices in Vancouver. Early on, she enquired into the possibilities of practicing the job for which she is trained (midwife), and quickly determined that she could not pursue this option: "I can't. I need the money. For me, being the breadwinner of the family, I cannot jeopardize my nieces' chances of going to school. So I decided that I would rather not spend that money for my own education." Instead, she took a six-month course in long-term care aid, which cost $1700 but met only on Saturdays and Sundays. This course prepared her to work in a nursing home, where (as it turned out) she was able to get only irregular hours.

Like Anna, most of the research participants have continued to send home remittances, some more and some less than when they were in the LCP, depending on their family's needs in the Philippines and their own circumstances in Vancouver. Mira, for example, remitted $500 of the $650 earned monthly while registered in the LCP but stopped sending money when she was pregnant (as a single mother) and unemployed, and she has only recently resumed now that she is employed again.

One of the exceptions to this rule is well worth noting. This is April, the one woman in the group who has returned to school for sustained retraining (one year, full-time) and regained her professional standing in Vancouver. April is one of only two women who told us that they now send little money home. April sends money "only when it is needed" but there are no monthly remittances: "Not monthly except for the medication. My sister [who also came to Vancouver through the LCP and with whom she now lives] and I share in the medical expenses for our family." She notes only one exceptional time – a "life and death situation" – when half of her monthly salary was sent home to her family in the Philippines.

Assessing the "rights" and "wrongs" of sending large or small remittances home to the Philippines is a difficult matter. Anna understands her commitment to her family as something that distinguishes her from an individualistic white society: "That's really our difference from the whites. They ask, 'How come you're still supporting your family? You have your own life.' Because for them, once your leave their house, they don't care anymore whatever might happen to their parents." There is little doubt, however, that remittances adversely affect the occupational mobility of former LCP registrants in Vancouver.

Social networks

The research participants are enmeshed in social relations beyond their families, and these too affect the types of jobs they have found in Vancouver. The importance of social networks for finding and landing jobs is well known. The fact that networks are often gendered (i.e. women tend to find jobs through women, and men through men) is one important explanation for why the sex-based occupational segregation persists (Hanson and Pratt, 1995), and a similar observation can be made about ethnic-based networks.

The women who participated in this research have drawn on different types of networks that span from Vancouver to the Philippines. Ruby, for instance, found her job in a private nursing home on Vancouver Island through a friend, now also living on Vancouver Island, who was a classmate in her midwifery course in the Philippines. Delia found her job at the church through the daughter of her former LCP employer, who came from her home village in the Philippines. Anna came to her job at the women's transitional shelter through her work in the activist, non-profit community in Vancouver. Though the sample is small, we want to distinguish between two types of networks – those that are closed and relatively static and those that are dynamic and open – and argue that, although the latter are critically important for regaining the type of occupational status that every single woman in the LCP has lost, in fact the former are more common.

Yolly's case exemplifies the limits of some types of social networks. Forward looking and trained as a midwife in the Philippines, Yolly began to retrain even before she finished the LCP. She took a long-term care aid course at a private school called Gateway. The course cost $2000 and met on Saturdays from 9 to 5.30 for six months. Once she finished her LCP contract in 1997, she found a job at Classic Caregiver Services through a newspaper advertisement. She now works at Greater Vancouver Community Services (a job that she also found through the newspaper) providing home care. Her job is unionized and she earns C$18 an hour. Her hours have been cut back, however, and what the agency calls "regular" hours is in fact a 25–30-hour week. This is arranged as a split shift, the details of which Yolly is informed of every two weeks. "Sometimes I work from 8.00 a.m. to 1 p.m. . . . and then there's a few hours in between. Then I go back to work at 5 p.m.: 5 p.m. to 7, 7 to 8, or 8 to 9. Sometimes I work overnight. Like I start at 12 a.m. until 6 a.m."

> *Cecilia*: But you have no social life.
> *Yolly*: Yeah.
> *Cecilia*: You don't see your friends.
> *Yolly*: I miss half my life.

Yolly has applied to a hospital and other care facilities (where hours and shifts could be better and wage premiums are paid for night time or weekend hours) but they do not recognize the school from which she graduated and she has been told that she must attend a six-month full-time course, which costs $4000. She does not see herself doing this. Asked whether she knew about the limitations of the Gateway course, she replied "No, because I just followed my friends." Clearly, not all of Yolly's networks have served her well.

In contrast, the networks of April and Anna have supported their efforts to regain their lost occupational status. There was a point when April got very discouraged in her efforts to regain her credentials as a registered nurse, and she began to train as a medical office assistant. She found the support to persevere in her efforts to retrain as a registered nurse from two sources: the Filipino Nurses' Support Group (FNSG) and her former employer (from the LCP): "my former employer was pushing me to go back to nursing. So it felt that I was being pulled back into nursing."[6] April had the skills to maneuver through a vast bureaucracy and what appears to be the "blind luck" of convincing a low-level bureaucrat to reconsider her decision not to process April's application for EI benefit (see note 4). It is the rich, overlapping mixture of state assistance, self-determination, and the push and pull of support from multiple sources that eventually sustained April's efforts to regain her credentials as registered nurse.

Anna has developed different but similarly rich and diverse networks that have led her out of the frantic pace of holding five low paying casual (or short-term contract) jobs simultaneously. Beginning her community work at the Kalayaan Centre through the PWC and FNSG, her contacts have spiraled out into other anti-poverty and women's organizations. As her web of networks multiplies, both within and beyond the Filipino community, Anna's job stability appears to increase. In Anna's words:

> being involved in the Centre gives me opportunities to learn other avenues, which we can benefit from or access. Like with ELP [End Legislated Poverty], when I applied there. Because they have a high respect for the Centre. . . . That's how I was able to get the job. And at Bridge [her current job], my involvement in the community helped too . . . during the interview, there was a question in there about being an activist, like: "what do you think of colonialism?" So I could easily answer that.

Preston (2003) calls attention to the importance of the type of contact for landing stable jobs, arguing that business associates lead recent immigrant women to more secure jobs than friends and relatives. Our sample is small and so our observations are speculative, but we would like to expand the definition of business associate and embed it within a fuller institutional and social context. Both April and Anna have had critically important institutional support (e.g. the FNSG, the PWC, and state assistance – in April's case directly, and in Anna's case indirectly through state support of the community-based social service sector). Beyond this, the density and expansive nature of their networks also seems noteworthy. April was encouraged not only by the FNSG but by her former employer when she lost faith in her capacity to retrain as a registered nurse. Unfortunately, the richness and expansive nature of Anna's and April's networks of support was relatively uncommon among the women in this study. Yolly's case seems closer to the norm.

Transnational Accounting of the LCP

If the opportunity to settle permanently in Canada is the "pay off" of the LCP experience, how do we assess the justice of this exchange? Because LCP registrants are people and not just workers, whose lives exist transnationally, this is no simple

Table 9.2 Housing in Vancouver

Rents shared basement suite with roommates/ extended family	5
Rents self-contained apartment (market housing)	1
Rents BC Housing (state subsidized) apartment	1
Rents apartment adjacent to church	2
Lives in shared accommodation at Kalayaan Centre	1
Rents a portion of house	1
Owns trailer with extended family	1
Owns single family house	1
Has moved to USA	2

matter. In an effort to expand the discussion of the ethics of the LCP, we consider material benefits that are invisible in Canada, as well as the long-term stigmatization and profound family dislocation experienced by LCP registrants.

Building a house in the Philippines

The most striking asset that has been purchased by three women in our study, made possible through their LCP experience, is a house in the Philippines. This is a significant asset that is invisible in Canada (table 9.2). Ruby provides a striking example of the discrepancy between housing purchased in the Philippines and in Canada. She rents a portion of a house in Richmond, a Vancouver suburb, for $650 a month, where she lives with her husband, their four-year-old child and his two older children (aged seventeen and nineteen) from a previous marriage. Previously they rented an entire house in Richmond for $1200 a month but they lost this when the owner returned to live in it himself. Ruby misses this arrangement – "That pushes me to really have my own house" – and these concerns have led her to plan a move to a town on Vancouver Island, where housing is more affordable. But these local arrangements must be considered in relation to Ruby's significant accomplishments in securing substantial and high status housing for her family in the Philippines. Since the LCP, Ruby has worked as an assembler in an electrical components factory, as a nurses' aid (in a private household), babysitting, and selling Avon products. Ruby has large responsibilities for her family in the Philippines, and she is able to manage them because her husband has agreed that the income that she earns in Canada can be sent to them: "Of the money I make, my husband does not touch it. But the money he makes, that's the one we use for our family here." Ruby sends home about 25,000 pesos (C$625) monthly. She supports one sister who has three children, whose husband was gunned down in 2002. This sister lives in Ruby's house in the Philippines, along with her own three children and Ruby's mother. Ruby has almost finished paying the seven-year mortgage on this house. The house is located in an exclusive compound, where many of the houses have been purchased by overseas workers and are inhabited by relatives of these workers. Ruby describes the meaning of her house for herself and her family:

R: I made plans a long time ago. That house in the Philippines is for my old age and retirement.

C: Do you still want to go back, Ruby?

R: I want to retire there. That house is ready for me. When I start receiving my pension, that's where I will be. . . . I would like to stay there, say, for a year or two, and then come back here.

C: You've missed the Philippines?

R: I missed it. And my house is my project that I worked hard for since I was single and young. . . . It's not a big house but my parents could not afford to have a house like that because we are poor. I bought it without my parents knowing about it. It was a surprise to them.

C: Really?

R: I brought them to that house after it was finished. They did not know about it. I bought a house and lot. It was already completed when my sister brought them over to that house. When they saw it and when my sister said that they should move in that house, my father cried because he said that he could not afford that house. I really felt proud about that – knowing that your parents were living a difficult life and now they can live in what people called "millionaire's row" in that part of the town.

C: That house that you bought, it's through your work here in Canada.

R: Yes. I worked hard for that in Canada. I worked three jobs at that time. I used to make $3000 a month and I would send most of that to them. Also, I would share that money with my siblings who also do not make enough for themselves. For me, that house is not only for me but for the whole family.

C: How many years did you work for that house?

R: This started in 1997. It was only a lot when I bought it. Then the real estate agency built the house.

While not all former LCP registrants have been able to build houses on a "millionaire's row," in all cases the purchase of a house is a significant achievement that has been accomplished through their employment in Canada.

Personal costs of stigmatization and family dislocation

For many, such accomplishments come at a very high personal cost. One cost is extremely difficult to measure: the intangible loss of self-esteem through the stigmatization of the LCP within the Filipino community in Vancouver. This was expressed most explicitly and fully by April. April tells of the stigmatization that she felt, first as a domestic worker and then as a waitress.

C: Did six years of being a domestic worker, a nanny and a waitress, have any impact on you personally?

A: [Crying] . . . actually, there is. It lowered one's self-esteem because the way people looked at and treated you. I don't mind working as a nanny or a waitress, but then, something hits you after a while of doing these types of jobs. . . . Especially in our community, people looked down on nannies and because you are working as a waitress, they think you are no good. You're nothing and your reputation is tarnished. . . . Even now, some people who knew me when I was a waitress would still derogatorily say that "Oh, she was a waitress." And then they thought that you are down there among the lowlies. Even last year, someone told me that I was a "sugar mommy" when I was working as a waitress.

Along with the difficulties of measuring the loss of self-esteem, it is virtually impossible to quantify another cost, and this is the cost of family dislocation. Only a minority have been joined in Canada by another family member,[7] and in two of these cases, this has been their own child from whom they were separated, for five years in one case and nine in the other. The latter family reunions have not been easy and in the case of Bing, her daughter has returned to the Philippines (and then to Vancouver) in an effort to find her own place within her transnational family.

A further detail makes this family dislocation more poignant: six (almost half) of the women from the study are single, and most are between their early thirties and early forties. When we first met in 1995 and introduced ourselves to each other, April said by way of introduction, "I'm looking for Mr Right." Eight years on, she lives with her sister. At the same workshop, Delia spoke of a troubled relationship: "My boyfriend calls me up [from the Philippines], but we just quarrel a lot over the phone." By summer 2003, she has come to prefer a quiet, controllable life:

> I think I am going to be single in my whole life. But I'm happy with my life right now. [Laughing] I want a quiet life. And I am not sure what would happen if I got married. I don't want that uncertainty. When I see my married brothers and sisters with their problems, I said that I did not want that kind of life. It's full of headaches.

Yolly is less resigned. She continues to go to singles dances organized within the Filipino community in Vancouver a few times a year and states: "I am looking for my a double [partner]. [Laughing] Being single, I feel like my life is empty."

Toward Transnational Feminist Solidarity

Assessing the LCP requires a nuanced analysis, one that moves between honoring these women's struggles and accomplishments, and a critique of government policies and social practices that entrap them. On the one hand, it is too simple to argue that the LCP is purely and simply exploitative. Ruby's house on "millionaire's row" is too significant and concrete an achievement to ignore, and to do so would seem an insult to her agency and self-determination. But, on the other hand, it is far too simplistic to see Canadian citizenship as a fair trade for two years of labor as an indentured servant. Leaving aside the ethics of this "modern day" slavery,[8] it is clear that the effects of the LCP are not contained within the time of the program, and thus the notion of compensatory citizenship requires recalibration. The LCP reverberates through registrants' lives for years – likely a lifetime. The majority of women who participated in our earlier research have never really escaped domestic work. They continue to perform it as home care workers and as housekeepers. And to fully understand the long-term repercussions of the LCP, we must also look beyond this downward occupational mobility and other labor market outcomes, to assess these women's lives more fully – especially their experiences of profound family dislocation. In our view, on balance these personal costs are too great, and it is unjust for Canadian families to benefit and thrive precisely because of and through this severe family disruption.

Such dislocation is no doubt felt by Filipina migrant domestic workers throughout the world (e.g. Yeoh and Huang, 1995; Constable, 1997; Momsen, 1999), but it remains important to develop analyses within specific national contexts. Stasiulis and Bakan (2002) rightly note that the LCP only appears progressive against the backdrop of the taken-for-granted oppression of female migrant workers world-wide. It is important to critique the long-term effects of the LCP precisely to disrupt Canadians' complacency about their relative virtue.

But it is also important simultaneously to make use of the possibilities opened by the LCP. Large numbers of former LCP registrants are now landed immigrants or full citizens of Canada: in 1998, 46 percent of "landings" of Filipinos (and in that year Filipinos ranked fifth in terms of number of immigrant arrivals to Vancouver) came through the LCP program (McKay and PWC, 2002). We have documented experiences of hard work, unrelenting work schedules, the exercise of civic responsibility through intense community involvement, and either the waste of previous educational training or great efforts expended to retrain in order to utilize existing credentials. Canadian feminists can join in a struggle to transform conditions in the Philippines so that so many Filipinas need not seek labor op-portunities overseas, but there is immediate work to be done closer to home. This involves a careful assessment of specific strategic opportunities. At this moment, one such opportunity is to join in solidarity to reform immigration laws so that Filipinas can enter Canada as immigrants (rather than migrants). Filipinas ought not to suffer the multiple injuries of a prolonged period of temporary status. Working toward such an immediate project, we have attempted to show that such temporary status is a "basic injustice" that reverberates long after the LCP.

NOTES

1   We thank Melissa Wright for drawing our attention to this poem.
2   Macklin is developing this argument in relation to "exotic dancers," many of whom come to Canada from Eastern Europe. Macklin recognizes the deeply problematic nature of this exchange of citizenship for time-limited slavery, and the extent to which self-help advocacy groups must lobby for and support domestic workers to maintain levels of exploitation within tolerable limits.
3   All but two have settled in Canada and the two exceptions married US citizens and settled in the USA. One of the latter two had attained an open visa in Canada but her sister was keen for her to join her in the USA and introduced her to an older white man, whom she eventually married. The other married an older African-American man and moved with him to his home in the Southern USA.
4   There are some methodological lessons to be learnt here. We began the present research project with focus groups but eventually switched to in-depth interviews. This was neces-sitated by the difficulty of assembling women who, much more so than when registered in the LCP, did not share common work schedules. Their paths have also moved in dif-ferent directions and in-depth interviews were required to appreciate the complexity of their individual lives.
5   She tells the story thus: "EI directed me to the Training Assistance Centre [TAC]. TAC interviewed me. By then I was already accepted at Kwantlen [college] and had applied for a student loan. The waiting period at TAC required two weeks. They told me that I may not be able to make it because there was no time for processing my paper [before the start of the school term]. I told her that I was already admitted and if I cannot get in during this session, I will have to go back all over again from the beginning because my

RNABC application will have become due . . . [and] my TOEFL and other requirements will have expired. That would mean that I would have to retake the [TOEFL] exam . . . reapply to RNABC and you know how long it will take to gather another set of documents. It took me two to three years at least to do this the first time around. . . . I tried to argue and explain to her. . . . Finally she relented, and considered my application. Instead of a six-month EI benefit, she extended it to almost one year. . . . That's how I survived for one year."

6   It is noteworthy that this is the only instance where the support of a non-Filipino employer is noted by any of the women in this study.

7   The relatively low level of family sponsorship is evident in the data reported in McKay and PWC (2002, p. 8). In 1996, 824 Filipinos "landed" in Vancouver through the LCP and only 198 LCP dependents. The equivalent numbers are 607 and 87 in 1997, and 875 and 139 in 1998.

8   It is worth settling on this term for a moment, which may – at first glance – seem overblown. For employees tied to one particular employer and residence, without basic freedoms of mobility, this is indeed slavery.

BIBLIOGRAPHY

Constable, N. (1997) *Maid to Order in Hong Kong: Stories of Filipina Workers*. Ithaca, NY: Cornell University Press.

Hanson, S. and Pratt, G. (1995) *Gender, Work and Space*. London: Routledge.

Katz, C. (2001) On the grounds of globalization: a topography for feminist political engagement. *Signs: Journal of Women in Culture and Society*, 26, 1213–34.

Karp, J. (1995) A new kind of hero. *Far Eastern Economic Review*, 158, 42–5.

McDowell, L. (1999) *Gender, Identity and Place: Understanding Feminist Geographies*. Cambridge: Polity Press.

McKay, D. and the PWC (2002) Filipina identities: geographies of social integration/exclusion in the Canadian metropolis. Research on Immigration and Integration in the Metropolis, Vancouver Centre of Excellence, Working Paper Series, No. 02–18.

Macklin, A. (2003) Dancing across borders: exotic dancers, trafficking and immigration policy. *International Migration Review*, 37, 464–500.

Mattingly, D. (2001) The home and the world: domestic service and international networks of caring labor. *Annals of the Association of American Geographers*, 91, 370–86.

Mitchell, K. (1997) Transnational discourse: bringing geography back in. *Antipode*, 29, 101–14.

Momsen, J. (ed.) (1999) *Gender, Migration and Domestic Service*, London: Routedge.

Mullings, B. (1999) Sides of the same coin? Coping and resistance among Jamaican data-entry operators. *Annals of the Association of American Geographers*, 82, 290–311.

Parrenas, R. S. (2001) Transgressing the nation-state: the partial citizenship and "imagined (global) community" of migrant Filipina domestic workers. *Signs: Journal of Women in Culture and Society*, 26, 1129–54.

Pratt, G. (1997) Stereotypes and ambivalence: the social construction of domestic workers in Vancouver, BC, Canada. *Gender, Place and Culture*, 4, 159–77.

Pratt, G. (1999) From registered nurse to registered nanny: discursive geographies of Filipina domestic workers in Vancouver, BC. *Economic Geography*, 75, 215–36.

Pratt, G. (2000) Research performances. *Environment and Planning D: Society and Space*, 18, 639–51.

Pratt, G. in collaboration with the PWC (1998) Inscribing domestic work on Filipina bodies. In H. Nast and S. Pile (eds), *Places through the Body*. London: Routledge.

Pratt, G. in collaboration with the PWC (1999) Is this really Canada? Domestic workers' experiences in Vancouver, BC. In J. Momsen (ed.), *Gender, Migration and Domestic Service*. London: Routledge.

Pratt, G. and Yeoh, B. (2003) Transnational (counter)topographies. *Gender, Place and Culture*, 10, 159–66.

Preston, V. (2003) Does neighbourhood matter? A case study of immigrant women's employment in Toronto. Paper presented at the annual meetings of the Canadian Association of Geographers, Victoria, BC, May.

Realuyo, B. A. (2002) Filipineza. *The Nation*, February 18, 37.

Stasiulis, D. and Bakan, A. (2002) Negotiating the citizenship divide: foreign domestic worker policy and legal jurisprudence. In R. Jhappan (ed.), *Women's Legal Strategies in Canada: A Friendly Assessment*. Toronto: University of Toronto Press.

Yeoh, B. and Huang, S. (1995) Childcare in Singapore: negotiating choices and constraints in a multicultural society. *Women Studies International Forum*, 18, 445–61.

Chapter 10

Borders, Embodiment, and Mobility: Feminist Migration Studies in Geography

Rachel Silvey

Feminist migration research has provided critical interventions into conceptualizations of the spaces and subjects at the center of migration studies. The growing feminist literature on migration examines the construction, persistence, and reorganization of relations of gender and difference as they shape unequal geographies of mobility and displacement. This chapter traces these core concerns in feminist migration studies through examples drawn from recent work in the subfield.

Migration is a socially embedded process, such that it reflects and reinforces social organization along the lines of gender, race, class, nation, sexuality, caste, and religion, among other differences (for reviews, see Kofman et al., 2000; Willis and Yeoh, 2000). As such, spatial mobility is more adequately conceived not as an epiphenomenon, nor primarily in terms of structural causes or consequences, but rather as interconnected in its meaning and operation to changes in the economic and cultural landscapes of which it is a constitutive part. For analyses of gender and difference, the social embeddedness of migration means that mobility is organized and ascribed with meanings in and through existing hierarchies and spatialities of power, rather than as a result of them (Lawson, 1999).

Beginning in the 1970s, as England and Lawson's introductory chapter to this section discusses, both internal and international migration flows grew increasingly feminized along with global labor, and growing numbers and proportions of women of color began participating in migration flows and waged employment. This chapter examines the meanings, divisions, and social hierarchies associated with these grounded changes in migration patterns as they are intertwined with the social differentiation of mobility patterns, regulations, and experiences.

The chapter focuses on the gender politics animating migration and the production of scale and space as these are addressed in the context of feminist migration research. The discussion is divided into two thematic sections – borders and embodiment – around which I organize the discussion of geographers' interventions into feminist migration studies. I also draw on examples from my research on

Indonesian migration to Saudi Arabia to further illustrate how critical theoretical attention to borders and embodiment transforms scholarship on migration.[1]

Borders and Boundaries

Feminist migration research approaches geographic boundaries as socially constructed, laden with power, and inflected by gender and difference (Marston, 2000; Boyle, 2002; Hyndman, 2004). Historically, most migration scholarship understood boundaries as empirically identifiable delimitations across which to examine the causes and consequences of migration (for a review, see Brown and Lawson, 1985). By contrast, feminist research makes boundaries themselves the focus of inquiry and addresses questions about the political and gender-specific processes tied to the making and remaking of them (Nagar, 2002). As Hyndman (2004) writes, examining the social and political constructedness of boundaries, borders, and scales allows researchers to "interrogat[e] the taken-for-granted meanings of dominant . . . discourse . . . to reveal the assumptions, constructions, and power relations that are foundational to such apprehensions." Such research queries the ways that boundaries define the limits and contours of, for example, the nation, the household, the region, the body, and the supranational, and how these political processes refract geohistorically specific social cleavages (Creswell, 1996; Nagel, 2002).

For instance, feminists ask how the boundaries of the nation are developed in conjunction with particular social hierarchies. As Yeoh and Huang (1999) demonstrate in their research on migrant domestic workers in Singapore, the boundaries of the nation are produced in part through the privileging of particular identities and the marginalization of others. Their work shows that the nation is founded on conceptions of citizenship that exclude specific women, specifically in this case international women migrants who work as domestic servants. They reveal both the ways in which women's migration across national borders is tied to the construction of social boundaries between groups and the ways in which this particular view of the nation contributes to the exploitation of migrant women who have crossed national borders to work as domestics in Singapore. It underscores the politically produced and exclusionary operation of national boundaries, as these play out in migration research and in the lives of migrants (see also Radcliffe, 1990; Huang and Yeoh, 1996). In addition, recent research has examined the gender politics of the state and the boundaries defining states' jurisdiction in relation to migrants (Fincher, 1997; Desbiens, 2004; Fan, 2004; Mountz, 2004; Walton-Roberts, 2004).

In addition to gendering the nation and the state, feminist migration research has also reworked conceptualizations of the boundaries of the household. As in the example of the nation, attention to constructions of gender and difference within and through the household reveal that the household's boundaries intersect with those of labor markets (Chant, 1991, 1998; Lawson, 1998) and international networks of social reproduction (Wright et al., 2000; Mattingly, 2001; Bailey et al., 2002). Mattingly, in research centered on the racial–ethnic organization of immigrant domestic labor, writes that for "immigrant domestic workers and their employers, networks of caring labor interlace the home and the world" (Mattingly, 2001, p. 384). Her research demonstrates that where the household begins and ends, and how household membership is defined in relation to domestic space, are processes of boundary-making that are forged in dynamic relation to the organization of labor markets, divisions of reproductive labor, and gendered international migration streams.

Analyzing social boundaries as they are cross-cut by gender, race, religion, and class, Nagar (1998), Dwyer (1999, 2000), Mohammad (1999), Elmhirst (2000), and Secor (2002) explore the processes through which community boundaries are forged. The categories of difference that organize place-based belonging and exclusion, and the ways that immigrants are affected by these processes, are key questions in this work. Such research rejects sociospatial fixity, yet continues to examine specific empirical geographic processes. Like Doreen Massey, these authors' approaches to the production of boundaries serve to "re-imagine place . . . in a way [that is] i) not bounded ii) not defined in terms of exclusivity iii) not defined in terms of an inside and an outside, and iv) not dependent on false notions of an internally-generated authenticity" (Massey, 1999, p. 40). In each of these studies, gender and difference are understood as crucial to defining the boundaries that delimit the identities of migrant groups, and migrants are understood to participate in producing the boundaries in the context of the power relations and "community" politics that shape the possibilities of migrants as subjects.

Borderline Cases: Indonesian Migrants in Saudi Arabia

Such interventions into the analysis of boundary construction are also illustrated in research on Indonesian transnational migrants in Saudi Arabia (Robinson, 1991, 2000; Spaan, 1999; Tirtosudarmo, 2000; Hugo, 2002; Silvey, 2004a, b). Beginning in the 1980s, the Indonesian New Order (1965–98) state aggressively promoted the out-migration of women to work as domestic laborers in Saudi Arabia. In the eyes of Indonesia's ambassador to Saudi Arabia at the time, overseas employment meant jobs for unemployed Indonesian nationals as well as crucial foreign exchange that would come in the form of remittances (Robinson, 1991). In 1984, the Indonesian state developed a unit within the Department of Manpower in order to monitor, regulate, promote, and train overseas labor.[2] In that the training programs targeted women, the Indonesian state's efforts to increase employment and income promoted women's labor migration in particular.

Since the state's inception of the labor export programs, the numbers of Indonesian women migrating abroad have risen rapidly and dramatically. By the early 1990s, close to 70,000 Indonesian women were migrating to Saudi Arabia each year (Amjad, 1996), and until very recently, the annual rates of out-migration continued to increase (Hugo, 2002). The majority (59 percent) of all documented overseas workers from Indonesia between 1989 and 1994 chose to migrate to Saudi Arabia (Hugo, 1995, p. 280); two-thirds of the migrants were women; and more than 80 percent of these women were estimated to work as domestic servants (Amjad, 1996, p. 346; Chin, 1998, p. 103).[3]

Despite the Saudi state's enforcement of strict immigration controls and the frequent repatriations of undocumented workers over the past two decades, there remain persistently high levels of in-migration of Indonesian domestic workers (Pujiastuti, 2000).[4] The Saudi state's inclusion of Saudi women in higher education, its selective and partial incorporation of women in the labor force, and its formal restrictions on women's spatial mobility have together fueled the demand for domestic workers and contributed to the isolation of these migrants from one another in the homes of their employers. For the most part, the Saudi state has turned a blind eye to the large numbers of undocumented migrant female domestic workers employed within its territory.

Both the Indonesian and Saudi states have contributed to constructing domestic work as a woman's job that garners low wages, grants little security and few benefits, involves high rates of multiple forms of abuse, and provides only slim chances of occupational mobility (e.g. Radcliffe, 1990; Huang and Yeoh, 1996; Pratt, 1997; Yeoh and Huang, 1998; Momsen, 1999). Interestingly, however, once many Indonesian female migrants began working overseas, they began to demand state protection for themselves as "heroes of national development" (Robinson, 2000). In the late 1990s Indonesian social activists concerned with international women migrants' rights sought to extend the territorial and scalar boundaries of the state's jurisdictional scope. Specifically, they pressured the state to expand its concern for Indonesian citizens into the Saudi Arabian national context.

The large-scale migration of Indonesian women to Saudi Arabia has thus coincided with challenges to the geography of the Indonesian state's regulatory and protective capacities. Women migrants' NGOs have pressured the Indonesian state in particular to reorganize the traditional legal–spatial ties linking the sovereign nation-state to the household scale. The NGOs have argued that the Indonesian state must internationalize and attend to labor carried out in the domestic sphere in order to better protect overseas migrant domestic workers. In so doing, these NGOs have expanded the boundaries of the nation and the household to include overseas women migrants as citizens and laborers.[5] In this case, the limits of gendered state regulation and jurisdiction are both reflected and transformed in association with women's migration. Specifically, Indonesian women citizens migrate abroad to work as domestic servants in homes that both the Indonesian and Saudi states construct as "private," "women's" spaces. However, as the activists' efforts mentioned above highlight, the boundaries of the sites, spaces, and scales through which migrants are controlled and empowered are struggled over, entered into by activists, and affected by the political agency of migrants and non-state actors (see also Katz, 2001; Silvey, 2003).

Gender relations and norms are constitutive of – and not just coincidental to – social and spatial boundaries, and they can raise different stakes for the state and the nation when they travel transnationally (Nagar et al., 2002; Walton-Roberts, 2004). For instance, whereas domestic workers' rights were not a widely contested political issue within Indonesia's national boundaries, in the Saudi context they have emerged as a political flashpoint. Activists, some government officials, and international NGOs have portrayed the widespread abuse, rape, and high rates of suicide and disappearance among Indonesian women in Saudi Arabia as representative of the Indonesian state's spatial–juridical limitations. It is through the transnational migration of women that gendered sites and spaces have become a critical arena of political action, and the Indonesian state has been pressured to expand the gendered boundaries of the protections it accords to citizens. The struggles to promote women migrants' rights are played out in contestations over borders at various scales, and as the following section explores, the migrant body is crucial for understanding the gender dimensions of migration politics.

Embodiment: Mobile Bodies and the Limits to Transgression

In the mid-1990s, feminist migration studies began to examine the body as a scale of analysis (for overviews of feminist work on the body, see Duncan, 1996; McDowell, 1999, especially chapter 2; Tyner and Houston, 2000; Longhurst, 2001; Mountz, 2004). This work argues that the body is a theoretically powerful starting point from which to examine migration, and it focuses analytical attention on embodied subjectivities and the roles of migrant bodies in producing space and place. In revising gender-neutral approaches to the subject, concern with embodiment reworks unmarked masculinist assumptions about the migrant. The socially differentiated migration process itself is not just understood as an outcome of gendered bodies, but is viewed as part and parcel of the various gender politics constructing migrant bodies and processes of embodiment in particular places.

To examine embodiment as a political process is to question the *a priori* ontological status of bodies, to ask what interests are served when certain bodies are viewed as "illegal" or "undocumented," and to uncover the corporeal experiences that underpin the migration of specific social groups. For instance, Pratt (1997, 2004) and Tyner (1994, 1997) analyze the ways that Filipina migrants' bodies are constructed through the categories of "immigrant," "woman," and "Filipina." In different ways, their research highlights the political nature of migrant bodies themselves and of the social dynamics tied to them. In a similar vein, Walton-Roberts's (2004) research addresses migration strategies and migration behavior linking the Punjab, India and Vancouver, Canada to show how state regulations of immigration and marriage operate across space through migration to affect the construction of migrants' bodies in gender-specific ways.

Concern with the embodiment of migration has also led to research that works against understanding spatial mobility exclusively in terms of transgressive, agency-driven, potentially empowering moves (Pratt and Hanson, 1994). Hyndman's work (2000) argues against mobility as opportunity, focusing instead on displacement. She examines the politics of humanitarian discourse surrounding refugee resettlement, and the ways in which forced migration, as well as efforts to ameliorate its consequences, lead to limits on refugees' bodies and agency. Sharing Hyndman's concern with violence directed at migrant bodies, Wright (1997, 1999, 2001) examines the effects of the construction of women workers' bodies in Mexico's maquiladoras as "cheap," "docile," and disposable (see also Cravey, 1998). She traces the ways in which these stories contribute to invisibilizing the high rates of murder and rape of factory women in the region. For feminist migration studies, this work puts forth a complex reading of power in and through bodies that refuses dualistic, structure/agency polarizations, and insists that migration be viewed through embodied cultural struggles of both migrants themselves and the forces that control their mobility.

Attention to migrants' bodies parallels the broader feminist concern with identity and its political refractions (Yeoh and Huang, 2000). From its beginnings, feminist migration research has dealt with women's subjugation to patriarchal limitations on the self, but more recently, it has also examined the ways in which discourses are circulated and transposed into gendered migrant notions of the sexed self (Domosh, 1999; Cravey, 2004). An example is the work by Hubbard (1998) that examines the discourses and practices that serve to marginalize street prostitutes in Birmingham (UK). In examining the "notions of appropriate sexual, gender, and racial behavior [that are invoked] in [the] identification of prostitutes as immoral" (Hubbard, 2000, p. 55), he shows how prostitutes are othered. In his work, mobile subjects are policed by discourses defining the boundaries of appropriate bodies performing acceptable physical acts (see also Cresswell, 1996, 1999; Yeoh and Huang, 1998).

Sexed bodies are imbricated in spatialized power relations.[6] In transnational contexts, the most financially and politically powerful migrants tend to be either explicitly or implicitly masculinized, such that in the literature on transnationalism, women are:

> alternatively taken to be truants from globalised economic webs, stereotyped as subservient or victimised, or relegated to playing supporting roles, usually in the domestic sphere. . . . [In this literature,] there tends to be both a deep utopic hope that transnationalism may offer opportunities to realign and equalise gender relations, and a knowing scepticism that patriarchal relations return in different guises at different times. (Pratt and Yeoh, 2003, p. 160–1)

Such contradictory representations of gender and global migration are also apparent in Indonesian women's migration to Saudi Arabia (Silvey, 2004a), as the following section explores.

Embodying Iconic Tensions: Indonesian Migrants in Saudi Arabia

Indonesian transnational migrants are involved in opening up new labor opportunities and political spaces, while simultaneously seeing their bodies entrapped across greater distances. Indonesian migrant domestic workers face problems that result from the ways their bodies and their employment niches are constructed and managed. Most importantly, women migrants who are undocumented are particularly vulnerable to all forms of abuse, including sexual and physical assault and harassment. Yet, as mentioned above, both the Saudi and the Indonesian states construct domestic workers' bodies as beyond their respective jurisdictional scopes (Silvey, 2004b). Undergirding the erasure of immigrant women's suffering and the widespread violation of their human rights are popular and state-supported understandings of women's bodies as "private," making abuses more difficult to expose than if they were conceived as a "public" issue (Staeheli, 1996; Nagar, 2002).

Women's NGOs in Indonesia, such as Kalyanamitra and the Center for Indonesian Migrant Women, have focused on the scale of the body in their efforts to protect women migrants. These organizations have expressed particular concern with the material dangers and sexual violence that individual women migrants face (Robinson, 2000; Hugo, 2002).[7] One public demonstration in October 1998 provided an example symbolizing this change: women whose Saudi employers had raped them and left them pregnant rallied in front of a labor supply agency, demanding that the agency and the government help to provide support for the children born of these crimes (Robinson, 2000, p. 277). In effect, activists called upon the migrant female body as a vessel and emblem of the nation itself.

Activism that brings the "private" spaces of the domestic sphere and the body to the streets can challenge masculinist conceptions of women's spaces and the relationships between scales (Staeheli, 1996). Attention to the individual gendered body resonates with women migrant activists' concerns, and it points to women's rights on intrinsic and individual grounds, rather than resiting their value within the household or as primarily instrumental for family welfare and national order, or as a source of national revenue (Nagar, 2002). At the same time, as activists take tales of migrants' embodied experiences into the transnational NGO world, they can be seen to be "scaling up" their issues. Women's NGOs active in West Java, Indonesia, while diverse in their approaches, tend overall to understand violence against women as produced at multiple scales (see *Yayasan Jurnal Perempuan*, 1999). As Mitra, a leader at Solidaritas Perempuan, said, "The transformation of domestic labour into an international labour issue will have repercussions for the way we think about women's rights, whether they are within families, in our country, or in the world." She thus suggests that her organization's advocacy work does not so much "jump scale" as interlink processes within and across scales.

These women migrant rights activists understand embodiment as constituted through a range of intersecting, sometimes competing, forces and processes, and they see migrants' physicality as shaping their agentic roles in these processes. Such an approach parallels feminist migration researchers' emerging understanding of the body. As Mountz (2004, p. 328) puts it, "the body is a crucial element to understand the operation of power in relations between states and migrants. . . . Embodiment locates power relations and contextualizes decision-making within workplace settings and life histories." By examining mobility as embodied in this way, migration research moves toward a deeper appreciation of the everyday, material, sexed, and socially differentiated dimensions of mobility.

Migration research focuses on the mobility of bodies across borders, and feminist geography emphasizes thinking beyond essentialist conceptions of such boundaries and processes of embodiment.[8] Feminist migration research also aims to identify and unpack the power relations embedded in, shaped through, and reinforced by migrants' bodies in particular places and across space. Through persistent feminist engagement with these themes, through reclaiming bodies as analytically central and as lived sites of power, and through understanding borders as liminal arenas of control and possibility in the lives of migrants, ultimately critical migration research aims to contribute to more progressive future geographies of mobility.

ACKNOWLEDGMENTS

Many thanks are due to Lise Nelson for her astute editorial work, extraordinary patience, and boundless energy. She made this chapter possible. I'd also like to thank

Shirlena Huang, Paul Boyle, Vicky Lawson, and Altha Cravey for helpful feedback on an earlier version of the chapter.

NOTES

1 Thorough reviews of the gender and migration literature in geography can be found elsewhere (Chant, 1992; Kofman and England, 1997; Boyle and Halfacree, 1999; Momsen, 1999; Kofman et al., 2000; Willis and Yeoh, 2000; Silvey, 2004c). While some of the research discussed in this chapter may not be explicitly informed by feminist geography, its foci overlap with those of feminist geography (Graham, 2000; Graham and Boyle, 2001; Longhurst, 2002).

2 In 1994, the Pusat AKAN's name was changed to Direktorat Jasa Tenaga Kerja Luar Negeri (Directorate of Overseas Manpower Services).

3 As early as the late 1980s, migrants were also destined for sites throughout the Middle East, Southeast Asia, and East Asia. It is likely that the numbers of undocumented emigrants increased for all destinations after the beginning of the Indonesian financial crisis, as growing numbers of people sought income abroad (Jones, 2000). But the numbers of documented emigrants fell during the same period, in part because the cost of documentation and travel became prohibitive for many prospective migrants (Ananta et al., 1998, p. 321; see also Spaan, 1999, p. 157). Further, many overseas contract workers, particularly in Malaysia, have been forcefully repatriated and housed in refugee camps, as the economies of the receiving countries have not provided the surplus necessary to continue to pay overseas workers (Ananta et al., 1998). During the writing of this chapter, newspaper reports indicated that the Saudi government has stopped providing visas to Indonesian workers (*Pikiran Rakyat*, April 16, 2003).

4 In addition, the decisions in 1983 by Bangladesh and Pakistan to ban domestic workers' out-migration (Chin, 1998, p. 102) limited the supply of female domestics and increased the demand for Indonesian women workers.

5 Except where otherwise noted, research reported in this chapter is based on the author's interviews with NGO activists and field research carried out in a migrant sending community in West Java in 1998, 2000, and 2002.

6 In March 2004, the Social Science Research Council's Working Group on Gender and Migration formally recognized geography as a discipline through the inclusion of a geographer in the group. According to the SSRC website, "The [SSRC's] International Migration Program established the Working Group on Gender and Migration in 2002 . . . to assess the contributions of current scholarship on gender to the study of international migration and promote scholarly attention to gender both as a topic of research and as an analytical concept within the field of migration studies." For more information, see: http://www.ssrc.org/programs/intmigration/working_groups/gender

7 In part, the growing attention to sexual violence came about in response to the widespread rape of ethnically Chinese women during the riots of May 1998. Importantly, the attention to sexual violence and the rights of women to control their bodies has emerged in force in the post-1998 political context in Indonesia, and it reflects part of the popular rejection of New Order methods of repression.

8 While an extended discussion of methodology in feminist migration studies is outside the scope of this chapter, recent discussions have examined the promise of multimethod research (e.g. Graham, 1999; McKendrick, 1999). These are important methodological developments for feminist migration studies, in that methodological plurality and critical interrogation of methodologies are key feminist concerns with far-reaching implications for knowledge about gender and migration.

BIBLIOGRAPHY

Amjad, R. (1996) Philippines and Indonesia: on the way to a migration transition. *Asian and Pacific Migration Journal*, 5(2/3), 339–66.

Ananta, A., Kartowibowo, D., Wiyono, N. and Chotib, H. (1998) The impact of the economic crisis on international migration: the case of Indonesia. *Asian Pacific Migration Journal*, 7(2/3), 313–38.

Bailey, A., Wright, R., Mountz, A. and Miyares, I. (2002) (Re)producing Salvadoran transnational geographies. *Annals of the Association of American Geographers*, 92(1), 125–44.

Bondi, L., Avis, H., Bankey, R., Bingley, A., Davidson, J., Duffy, R., Einagel, V.I., Green, A.-M., Johnston, L., Lilley, S., Listerborn, C., McEwan, S., Marshy, M., O'Connor, N., Rose, G., Vivat, B. and Wood, N. (2002) *Subjectivities, Knowledges, and Feminist Geographies: The Subjects and Ethics of Social Research*. Lanham, MD: Rowman & Littlefield.

Boyle, P. (2002) Population geography: transnational women on the move. *Progress in Human Geography*, 26(4), 531–43.

Boyle, P. and Halfacree, K. (eds) (1999) *Migration and Gender in the Developed World*. London: Routledge.

Brown, L. and Lawson, V. (1985) Migration in third world settings, uneven development, and conventional modelling: a case study of Costa Rica. *Annals of the Association of American Geographers*, 75, 29–47.

Chant, S. (1991) Gender, migration and urban development in Costa Rica: the case of Guanacaste. *GeoForum*, 22(3), 237–53.

Chant, S. (ed.) (1992) *Gender and Migration in Developing Countries*. London: Belhaven Press.

Chant, S. (1998) Households, gender, and rural–urban migration: reflections on linkages and considerations for policy. *Environment and Urbanization*, 10(1), 5–21.

Chin, C. (1998) *In Service and Servitude: Foreign Female Domestic Workers and the Malaysian "Modernity" Project*. New York: Columbia University Press.

Cravey, A. (1998) *Women and Work in Mexico's Maquiladoras*. Lanham, MD: Rowman and Littlefield.

Cravey, A. (2004) Desire, work and transnational identities. *Ethnographies* (in the press).

Cresswell, T. (1996) *In Place/Out of Place: Geography, Ideology, and Transgression*. Minneapolis: University of Minnesota Press.

Cresswell, T. (1999) Embodiment, power and the politics of mobility: the case of female tramps and hobos. *Transactions, Institute of British Geographers*, 24, 175–92.

Desbiens, C. (2004) "Women with no femininity": gender, race and nation-building in the James Bay. *Political Geography*, 23(3), 347–66.

Domosh, M. (1999) Sexing feminist geography. *Progress in Human Geography*, 23(3), 429–36.

Duncan, N. (ed.) (1996) *Bodyspace: Destabilizing Geographies of Gender and Sexuality*. London: Routledge.

Dwyer, C. (1999) Contradictions of community: questions of identity for young British Muslim women. *Environment and Planning A*, 31(1), 53–68.

Dwyer, C. (2000) Negotiating diasporic identities: young British South Asian Muslim women. *Women's Studies International Forum*, 23(4), 475–86.

Elmhirst, R. (2000) A Javanese diaspora? Gender and identity politics in Indonesia's transmigration resettlement program. *Women's Studies International Forum*, 23(4), 487–500.

Fan, C. (2004) The state, the migrant labor regime, and maiden workers in China. *Political Geography*, 23(3), 283–305.

Fincher, R. (1997) Gender, age and ethnicity in immigration for an Australian nation. *Environment and Planning A*, 29(2), 217–36.

Gidwani, V. and Sivaramakrishnan, K. (2003) Circular migration and the spaces of cultural assertion. *Annals of the Association of American Geographers*, 93(1), 186–213.

Graham, E. 1999. Breaking out: the opportunities and challenges of multi-method research in geography. *Professional Geographer*, 50(11), 76–89.

Graham, E. 2000. What kind of theory for what kind of population geography? *International Journal of Population Geography*, 6, 257–72.

Graham, E. and Boyle, P. (2001) Editorial introduction: (re)theorising population geography: mapping the unfamiliar. *International Journal of Population Geography*, 7, 389–94.

Halfacree, K. and Boyle, P. (1999) *Migration and Gender in the Developed World*. London: Routledge.

Huang, S., Teo, P. and Yeoh, B. (2000) Diasporic subjects and identity negotiations: women in and from Asia. *Women's Studies International Forum*, 23(4), 391–8.

Huang, S. and Yeoh, B. (1996) Ties that bind: state policy and migrant female domestic helpers in Singapore. *Geoforum*, 27, 479–93.

Hubbard, P. (1998) Sexuality, immorality and the city: red-light districts and the marginalisation of street prostitutes. *Gender, Place and Culture*, 5(1), 55–72.

Hugo, G. (1995) Labour export from Indonesia. *ASEAN Economic Bulletin*, 12(2), 275–98.

Hugo, G. (2002) Women's international labor migration. In K. Robinson and S. Bessell (eds), *Women in Indonesia: Gender, Equity, and Development*. Singapore: Institute of Southeast Asian Studies.

Hyndman, J. (2000) *Managing Displacement: Refugees and the Politics Humanitarianism*. Minneapolis: University of Minnesota Press.

Hyndman, J. (2004) Mind the gap: bridging feminist and political geography through geopolitics. *Political Geography*, 23(3), 241–366.

Jones, S. (2000) *Making Money off Migrants: The Indonesian Exodus to Malaysia*. Hong Kong: Asia 2000; Wollongong: Centre for Asia Pacific Social Transformation Studies, University of Wollongong.

Katz, C. (2001) On the grounds of globalization: a topography for feminist political engagement. *Signs: Journal of Women in Culture and Society*, 26(4), 1213–34.

Kofman, E. and England, K. (1997). Editorial introduction. Citizenship and international migration: taking account of gender, sexuality, and race. *Environment and Planning A*, 29(2), 191–4.

Kofman, E., Sales, R., Phizucklea, A. and Raghuran, P. (eds) (2000) *Gender and International Migration in Europe*. London: Routledge.

Lawson, V. (1998) Hierarchical households and gendered migration: a research agenda. *Progress in Human Geography*, 22(1), 32–53.

Lawson, V. (1999) Questions of migration and belonging: understandings of migration under neoliberalism in Ecuador. *International Journal of Population Geography*, 5, 261–76.

Longhurt, R. (2000) Geography and gender: masculinities, male identity and men. *Progress in Human Geography*, 24(3), 439–44.

Longhurst, R. (2001) Geography and gender: looking back, looking forward. *Progress in Human Geography*, 25(4), 641–8.

Longhurst, R. (2002). Geography and gender: a "critical" time? *Progress in Human Geography*, 26(4), 544–52.

McDowell, L. (1999) *Gender, Identity and Place: Understanding Feminist Geographies*. Oxford: Blackwell.

McKendrick, J. (1999) Multi-method research: an introduction to its application in population geography. *Professional Geographer*, 50(11), 40–50.

Marston, S. (2000) The social construction of scale. *Progress in Human Geography*, 24(2), 219–42.

Massey, D. (1999) Power-geometries and the politics of space-time, Hettner-Lecture. Heidelberg: Department of Geography, University of Heidelberg.

Mattingly, D. (2001) The home and the world: domestic service and international networks of caring labor. *Annals of the Association of American Geographers*, 91(2), 370–86.

Mohammad, R. (1999) Marginalisation, Islamism and the production of the "other's" "other." *Gender, Place and Culture*, 6(3), 221–40.

Momsen, J. H. (ed.) (1999) *Gender, Migration, and Domestic Service*. London: Routledge.

Mountz, A. (2004) Embodying the nation-state: Canada's response to human smuggling. *Political Geography*, 23(3), 323–45.

Nagar, R. (1998) Communal discourses, marriage and the politics of gendered social boundaries among South Asian immigrants in Tanzania. *Gender, Place and Culture*, 5(2), 117–39.

Nagar, R. (2002) Women's theater and the redefinitions of public, private, and politics in North India. *ACME – An International E-Journal for Critical Geographies*, 1(1), 55–72.

Nagar, R., Lawson, V., McDowell, L. and Hanson, S. (2002) Locating globalization: feminist (re)readings of the subjects and spaces of globalization. *Economic Geography*, 78(3), 257–84.

Nagel, C. (2002). Geopolitics by another name: immigration and the politics of assimilation. *Political Geography*, 21, 971–87.

Nelson, L. (1999) Bodies (and spaces) do matter: the limits of performativity. *Gender, Place and Culture*, 6(4), 331–54.

Pratt, G. (1997) From registered nurse to registered nanny: discursive geographies of Filipina domestic workers in Vancouver, BC. *Economic Geography*, 75(3), 215–36.

Pratt, G. (2004) *Working Feminism*. Philadelphia: Temple University Press.

Pratt, G. and Hanson, S. (1994) Geography and the construction of difference. *Gender, Place and Culture*, 1(1), 5–29.

Pratt, G. and Yeoh, B. (2003) Transnational (counter) topographies. *Gender, Place and Culture*, 10(2), 159–66.

Pujiastuti, T. N. (2000) The experience of overseas workers from Indonesia. Unpublished MA thesis, Department of Geographical and Environmental Studies, University of Adelaide.

Radcliffe, S. (1990) Ethnicity, patriarchy and incorporation into the nation: female migrants as domestic servants in Peru. *Environment and Planning D: Society and Space*, 8, 379–93.

Robinson, K. (1991) Housemaids: the effects of gender and culture in the internal and international migration of Indonesian women. In G. Bottomley, M. de Lepervanche and J. Martin (eds), *Intersexions: Gender/Class/Culture/Ethnicity*. Sydney: Allen and Unwin.

Robinson, K. (2000) Gender, Islam, and nationality; Indonesian domestic servants in the Middle East. In K. Adams and S. Dickey (eds), *Home and Hegemony: Domestic Service and Identity Politics in South and Southeast Asia*. Ann Arbor: University of Michigan Press.

Secor, A. (2002) The veil and urban space in Istanbul: women's dress, mobility and Islamic knowledge. *Gender, Place and Culture*, 9(1), 5–22.

Silvey, R. (2003) Spaces of protest: gendered migration, social networks, and labor protest in West Java, Indonesia. *Political Geography*, 22(2), 129–57.

Silvey, R. (2004a) Transnational domestication: Indonesian domestic workers in Saudi Arabia. *Political Geography*, 23(3), 245–64.

Silvey, R. (2004b) Transnational migration and the gender politics of scale: Indonesian domestic workers in Saudi Arabia. Special issue on gender and transnational migration in Southeast Asia. *Singapore Journal of Tropical Geography*, 25(2), 141–55.

Silvey, R. (2004c) Power, difference, and mobility: feminist advances in migration studies. *Progress in Human Geography*, 28(4), 1–17.

Silvey, R. and Lawson, V. (1999) Placing the migrant. *Annals of the Association of American Geographers*, 89(1), 121–32.

Spaan, E. (1999) *Labour Circulation and Socioeconomic Transformation: The Case of East Java, Indonesia*. Groningen: Rijksuniversiteit Groningen.

Staeheli, L. (1996) Publicity, privacy, and women's political action. *Environment and Planning D: Society and Space*, 14, 601–19.

Staeheli, L. (1999) Globalization and the scales of citizenship. *Geography Research Forum*, 19, 60–77.

Tirtosudarmo, R. (2000) Indonesian domestic workers in Saudi Arabia. Mimeo, International Institute of Asian Studies, Leiden.

Tyner, J. (1994) The social construction of gendered migration for the Philippines. *Asian and Pacific Migration Journal*, 3, 589–617.

Tyner, J. (1996) Constructions of Filipina migrant entertainers. *Gender, Place and Culture*, 3(1), 77–93.

Tyner, J. (1997) Constructing images, constructing policy: the case of Filipina migrant performing artists. *Gender, Place and Culture*, 4(1), 19–35.

Tyner, J. A. and Houston, D. (2000) Controlling bodies: the punishment of multi-racialized sexual relations. *Antipode*, 32(4), 387–409.

Walton-Roberts, M. (2004) Rescaling citizenship: gendering Canadian immigration policy. *Political Geography*, 23(3), 265–81.

Willis, K. and Yeoh, B. (eds) (2000) *Gender and Migration*. Cheltenham: Edward Elgar.

Wright, M. (1997) Crossing the factory frontier: gender, place, and power in the Mexican Maquiladoras. *Antipode*, 29(3), 278–302.

Wright, M. (1999) The dialectics of still life: murder, women, and Maquiladoras. *Public Culture*, 11(3), 453–74.

Wright, M. (2001) A manifesto against femicide. *Antipode*, 33(3), 550–66.

Wright, R., Bailey, A., Miyares, I. and Mountz, A. (2000) Legal status, gender and employment among Salvadorans in the US. *International Journal of Population Geography*, 6, 273–86.

Yayasan Jurnal Perempuan (Women's Journal Organization) (1999) http://www.yjp.or.id/yjp-upload/index.htm (accessed March 10, 2004).

Yeoh, B. and Huang, S. (1998) Negotiating public space: strategies and styles of migrant female domestic workers in Singapore. *Urban Studies*, 35(3), 583–602.

Yeoh, B. and Huang, S. (1999) Spaces at the margins: migrant domestic workers and the development of civil society in Singapore. *Environment and Planning A*, 31, 1149–67.

Yeoh, B. and Huang, S. (2000) "Home" and "away": foreign domestic workers and negotiations of diasporic identity in Singapore. *Women's Studies International Forum*, 23, 413–30.

Chapter 11

The Changing Roles of Female Labor in Economic Expansion and Decline: The Case of the Istanbul Clothing Industry

Ayda Eraydın and Asuman Turkun-Erendil

Introduction

Since the 1970s, there have been substantial changes in the organization of production throughout the world. Under waves of deregulation, flexible ways of organizing production have been implemented in order to open markets, to achieve sustained economic growth, and to adapt to changing conditions (Schoenberger, 1988; Hirst and Zeitlin, 1991; Amin, 1994). Increasing integration into world markets through export-oriented production undoubtedly creates new employment opportunities; however, for many developing countries, the only opportunity to take part in the world economic system has been through the export of labor-intensive products. Women have been recruited into these globalizing production systems on a massive scale as low paid, and often temporary, workers (Peck, 1996). Women's position in export-oriented relations of production includes not only their widespread participation in marginal *formal* sector jobs, but their extensive participation within informal work relations, including but not limited to homework (Beneria and Roldan, 1987; Benton, 1990; Lawson, 1992; Peck, 1992; Phizacklea and Wolkowitz, 1995). Although working conditions are not satisfactory, gaining access to waged employment has been an important step for many women.

Decreasing barriers between nation-states and increasing flows of capital and goods have increased the interdependence between different parts of the world (Dicken, 1988; Leyshon and Thrift, 1997). While liberalization provides new opportunities, many argue that nation states have increasing difficulty regulating national economies, which has meant increasing economic volatility and shorter economic cycles (Ohmae, 1990, 1995; Drache and Gertler, 1991; Gerny, 1993, 1996; Jessop, 1994). The crisis periods in these cycles negatively impact employment opportunities and the working conditions of disadvantaged segments of labor markets, especially female labor.

In this chapter, we use our research on female labor in Istanbul's clothing industry to examine the effects of industrial boom and bust cycles on women's lives.[1]
First, we trace how women gained entry into new globally oriented production
systems during the clothing industry boom period (1980–95), exploring how entry
into factory production shifted women's identities and roles both in the family and
in society. We argue that the restructuring of production not only generates new
labor processes, but also creates new relations between home and work (see also
Nippert-Eng, 1996; Castells, 1997; Weyland, 1997; Felstead and Jewson, 2000).
Second, we examine how this segment of labor has been affected during the periods
of vulnerability and economic downturn after 1995. Our analysis demonstrates that
as the state loses capacity to intervene during cyclical economic downturns, women
workers suffer most directly because of their more marginal position in the labor
market.

The article is divided into four main sections. The first section briefly discusses
theoretical debates that shape our inquiry, while the second section examines the
structural characteristics of a rapidly expanding clothing industry during the late
1980s and early 1990s in Turkey. The third section turns to the changing work patterns and identities of women workers during those years of rapid growth in the
clothing industry. We argue that the incorporation of women into the clothing industry, usually second-generation migrants from rural Turkey, had a significant impact
on gender identities and roles within migrant families. The fourth section traces the
ripple effect of economic crisis, and the contraction of the clothing industry
(2000–1), on women's identities and family survival strategies. Our conclusion
reflects upon the challenges of analyzing the dynamics of gender and work on global
assembly lines prone to cyclical downturns such as those that have occurred in the
Turkish textile industry.

Female Labor in Industrial Restructuring: Theoretical Aspects

Theories of economic restructuring and the emergence of new systems of production have focused mainly on forms of flexibility that enable producers to adapt to
volatile conditions in globalized markets. Several studies have explored the impact
of restructuring and flexible work organization on labor processes and labor
markets (Storper and Walker, 1983; Massey, 1984). However, while these studies
consider gender as a characteristic of labor, they do not view gender divisions of
labor as an integral part of the restructuring processes (Christopherson, 1989;
Jenson, 1994).

The role of gender in industrial restructuring has been theorized in two key ways.
One approach, which can be defined as the marginalization thesis, emphasizes the
role of women in the family and domestic divisions of labor, arguing that gender
divisions in the home put women in a disadvantageous position in labor markets
(Kessler-Harris and Sacks, 1987; Walby, 1990). From this perspective, gender
inequalities in labor markets arise from cultural factors, such as patriarchal relations within families (Mies, 1986; Pessar, 1994; Stratigaki and Vaiou, 1994; White,
1994). The second approach explains the disadvantaged position of women at work
in terms of the structure of the labor market itself. According to this perspective,
the labor market is divided between a primary sector required to meet the

technological needs of producers, and a secondary labor market consisting of other workers whose particular skills are expendable and who must therefore accept lower wages and insecure conditions (Doeringer and Piore, 1985). Women are generally confined to the latter (Benton, 1990). According to this approach, female labor is exploited when necessary and used as a reserve army.

When thinking through these two different approaches in the context of globalization it is possible to analyze the disadvantaged position of female labor, especially in export-oriented production. According to these paradigms, industrial restructuring, especially in developing countries, increases women's share of employment due to the absolute and relative growth in the informal sector and in the share of temporary, low paid jobs with no social security (Beneria and Roldan, 1987; Pessar, 1994). Domestic and international capital draws on female labor because it is the cheapest and most exploitable.

While these theoretical debates increase our understanding of the ways in which industrial restructuring, gender divisions and labor markets interact (Peck, 1992) they often overlook considering workers as "active agents" in the processes in which they participate. Consequently, the individual strategies and aspirations of employees responding to changing characteristics of labor markets are not taken into account. While it is undoubtedly true that labor processes and gender divisions are also shaped by labor markets (Peck, 1992), women's attitudes to employment opportunities and their economic strategies require further attention if the role of female labor in industrial restructuring is to be understood.

In the context of women's lives and families, the transition to formal factory employment is usually a life changing experience. How do these industrial processes transform gender relations, at home, at work, and in the larger society? As importantly, as these same industries experience contraction, how does this process affect these same women and their newly constructed lives and identities? What are their strategies of survival?

According to the literature, during periods of prosperity women search for external job opportunities, especially in the formal sector, whereas during crisis periods, they try to find solutions with the collaboration of family members. Some of these "internal" strategies include increasing domestic food production and cooperation among families (Gilbert, 1994; Delarrocha, 1995), changing consumption norms, buying cheaper products and services, decreasing expenditures for education, housing, and health (Kanji, 1995), returning back to home-villages (Drakakis-Smith, 1996), and enforcing the participation of all members for family income (Eraydın and Erendil, 1999a, b; Turkun-Erendil, 2002). Our analysis of women in the clothing industry shows that under the new survival conditions, the burden is not only on women, but also on the other members of the family, including children. This view is supported by findings of several studies in Turkey (Kumbetoglu, 1996; Kandiyoti, 1988, 1998; Hattatoglu, 2000, 2001; Dedeoglu, 2002).

These perspectives inform our inquiry into women's involvement in Istanbul's clothing industry, from the early 1990s through 2002, by showing us the new demand and supply conditions in this era. As demand is mainly shaped by new types of flexible production labor requirements, supply is related to the changing roles of women within their families and the society as well as the conditions of the economy in transition. Therefore, the new roles of women in the production process have to be evaluated not only as an economic issue, but also as a process of social transformation.

Industrial Restructuring of the Clothing Industry and Female Labor

The 1980s witnessed a turning point in economic policy in Turkey, from the protectionist strategies that dominated in the 1960s and the 1970s, to liberalization and privatization policies designed to promote the country's integration into the world economy. As in many developing countries, the export-oriented growth of the Turkish economy began with labor-intensive export production that nominally made best use of Turkey's "comparative advantages" in the international market. The textile industries were among the first to expand as a result of these liberalization policies, stimulated by direct government support, the falling costs of labor, and increasing productivity levels (Yeldan, 2001).[2]

New economic development policies in the 1980s encouraged many domestic firms to become involved in export production. The Turkish state began to offer direct financial incentives for export industries, and officials created a myriad of institutions designed to facilitate export production, such as Eximbank. Export promotion policies and the deregulation of trade enabled the rapid growth of textile exports, growth initially dependent on existing industrial capacity. The value of clothing exports rose dramatically from US$130 million in 1980 (9 percent of total exports) to US$2898 million in 1990 (22.4 percent of total exports) and to US$6100 million in 1995 (28 percent of total exports) (ITKIB, 2001). Despite this phenomenal growth, however, the income generated from exports was not enough to overcome the lack of domestic investment capital, a structural constraint on the further expansion of textile production after the mid-1990s.[3]

The growing internationalization of the Turkish economy in the 1980s created some economic growth, but it also made Turkey vulnerable in new, unexpected ways. After the 1990s, Turkey experienced a shortening cycle of successive economic contractions tied to events in global financial markets. The 1994 crisis, caused by local financial policies, was followed by a severe downturn in 1997 and 1998 related to instability and contraction in the Asian and Russian markets. Finally, Turkey experienced successive financial upheavals in November 2000 and February 2001, ones tied to financial instabilities caused by overvalued national currency. These economic downturns all had important negative impacts on employment and income at all levels of society. Unemployment levels reached up to 7.7 percent in 1998 and rose further to 8.5 percent in 2001 (DİE, Statistics on Turkey, 2001). Many firms simply closed. In the year 2000, the share of newly established firms in the total was reduced to 20.1 percent, while closing companies reached up to 27.8 percent of the total (DİE, Statistics of Companies, Cooperatives and Firms, 2000, 2001).

Cyclical downturns in the economy and rising global competition negatively affected some clothing firms. The 1994 crisis led to shrinking export levels, especially in the textile and clothing sectors in Turkey. This crisis was particularly serious for producers exporting standardized lower quality goods and the large numbers of subcontractors that proliferated during the boom years. The crises experienced in 1998 and 2001 led to similar contractions but also demonstrated the differences between firms more clearly. Firms with stable export relations with EU countries or the USA, particularly those producing high quality products with fewer competitors, even profited from crisis conditions due to substantial devaluations and depressed wage levels.[4] In contrast, firms with high debt loads or ones that depended on failing export markets, such as Russia, experienced dramatic economic reversals. Lower quality standardized textile producers experienced a contraction of demand during this cyclical downturn and quite a large number of firms experienced bankruptcies, while others tried to survive by decreasing production, firing workers, lowering wages, and halting investments. Some firms, on the other hand, migrated to other production sites, namely Bulgaria and Romania, due to more advantageous profit-making conditions in those locations.[5]

Parallel to the rapid growth in production and exports, official statistics show that between 1985 and 2000 registered employment in the clothing industry increased by about 50 percent, reaching 7 percent of total manufacturing employment in 1990 and 9.6 percent in 2000. Women make up almost 60 percent of these workers. If undocumented and temporary employees are considered, these figures about the clothing industry and the share of women in total employment are likely to account for a substantially higher percentage of the total (DİSK/Labor Union of Textile Workers, 2002). Women thus form the laboring "backbone" of the textile industry and fueled its expansion during the boom years.

These working conditions changed dramatically after the years of rapid expansion (1990–5) due to successive crises in the second half of 1990s. Unsurprisingly, vicious economic cycles and losses in a significant number of Turkish clothing firms generated difficult conditions for workers, especially female labor employed in the small informal firms that suffered the most during the downturns.[6] Workers we interviewed in 1995 felt they could choose among different available jobs, both formal and informal, whereas by 2002 workers felt compelled to hold on tightly to jobs previously considered undesirable (see also Bora, 2002; Erdoğan, 2002; Işık and Pınarcıoğlu, 2001). The ways in which these economic cycles impacted women's work strategies, identities, and family survival strategies are explored in the next two sections.

Women, work and the expansion of Istanbul's clothing industry during the period 1990–5

The Turkish experience indicates that competitive conditions in the manufacture of clothing depend upon flexibility in gendered labor supplies as well as flexible structures of production. Labor flexibility is achieved in turbulent market conditions through high rates of labor turnover and the use of temporary employment to meet increased labor demand during short-term economic cycles (Braverman, 1974; Jenson, 1994). Studies of the extensive subcontracting arrangements indicate that the subcontracting of sewing to small firms has created intricate networks among firms, through which costs are minimized and swift response to changing demand is enabled (Cinar et al., 1988; Eraydın and Erendil, 1999). This system has become prevalent particularly in labor-intensive stages of textile production (Eraydın, 1994) in which mainly women are employed, including homeworkers (Lordoglu, 1990; White, 1994; Kumbetoglu, 1996).

In order to understand the role of female labor in the restructuring of Istanbul clothing industry, we set out to interview 500 women working in the industry. Sampling proceeded as follows. First, we selected at random 240 clothing firms from the 2397 listed by the Union of the Chambers of Industry and Trade in May 1995. In August and September 1995, we conducted interviews with the managers or owners of these firms and gathered information about the organization of production, labor processes, and employees.

Following these interviews, we selected 150 of the firms for the second stage of data collection, ensuring that we included firms involved in all the main stages of production.[7] In February 1996, we interviewed 428 women working for these firms, together with 24 female entrepreneurs. The sample included 35 family workers and 48 homeworkers and the rest were wage-earners in workplaces. These proportions were based on the data collected in the first stage. A standard schedule was used, with additional sections included for particular groups. As a third stage, in order to identify the new trends and the effects of the crises in 1998 and 2001, a short survey was conducted in Istanbul in July 2002. This survey included interviews with the key actors in the clothing sector and women in different segments of production.

The data collected in 1995–6 reveal that the main source of female labor in Istanbul's textile industry is drawn from migrant families. In Turkey, a high rate of natural population increase has combined with considerable rural-to-urban migration to produce very large increases in urban population. Metropolitan areas have received the largest migratory flows: between 1960 and 1990, the population of Istanbul increased by almost 5 million, with migrants accounting for 50 per cent of this increase.

Our surveys and interviews indicate that the first-generation migrant women were often reluctant to work outside the home, except to do paid housework, because neither they nor the family members were ready to change their traditional gender roles. Attitudes often began to change among second-generation migrants. Daughters of rural migrants, either born in metropolitan areas or having migrated at early ages, did not usually secure educational or vocational qualifications, a situation also related to gender roles and expectations. Instead, second generation women migrants became incorporated into the labor market mainly via the informal sector and, during the 1980s and 1990s, they became the main source of labor for the clothing industry.

According to the 1995–6 survey, on average, women employed in the Istanbul clothing industry were young (54 percent were below the age of 25) and poorly educated (64 percent did not proceed beyond primary school). Just over half were unmarried. The survey shows that women came from relatively large families and many had other family members also working in clothing production.[8] For many immigrant families it became common in the early 1990s for female children to immediately begin working in the clothing industry once they reached working age. This situation indicates the importance of the clothing industry as a means of survival within the city. Wages, however, were low: 11 percent of the paid workers and 75 percent of the homeworkers earned less than the official minimum wage. The average figure was reported to be US$196 per month, which was considerably lower than the average wages of registered employees in either private manufacturing industries (US$278 per month) or the state-owned manufacturing industries (US$384 per month). Four major groups involved in the clothing industry were identified through the research, namely paid workers (77 percent of those interviewed), factory owners (5 percent of those interviewed), homeworkers (10 percent of those interviewed), and family workers (8 percent of those interviewed). We consider each of these groups in turn.

Among paid workers there were significant variations among those in different occupations. Large numbers were employed directly in production processes. They were usually drawn from the poorest families and have the most limited education (65 percent had only primary schooling and 5 percent did not even finish primary school). At the opposite extreme, some women (4 percent of those interviewed) worked as intermediaries between local and foreign firms. These women were university graduates from middle-income families. Women involved in design stages (designing, pattern making, and grading), marketing, public relations, and management occupy an intermediate position in terms of their educational and social backgrounds.

In 1995–6, only half of the paid workers were registered employees protected by employment laws, and half of them worked in unregistered firms. The majority of those who worked in unregistered firms were employed in small workshops located in squatter housing areas. In fact the primary reason these small workshops arose in squatter settlements was to facilitate their access to young girls as employees. These firms did not have a long life: some of them closed down after a production season, while others that were more successful went on to register officially despite continuing to use informal labor. Employers generally deduct social security costs directly from wages; therefore, many workers readily accepted informal conditions in order to secure the higher take-home pay offered in the informal sector. For employers this is advantageous since they avoid the employment regulations for working conditions and also additional costs associated with legal rights related to marriage, birth, and children. For this reason, employers in this sector were not concerned about the marital status of their workers.

The second group of women included in the study was homeworkers. Under the pressure of highly competitive market conditions, employers often use homeworkers for several reasons: to save on labor costs, to achieve flexibility in the volume of production, and/or as strategy aimed at deunionization (see also Peck, 1992). According to many scholars, homeworking is also attractive to some women in that it provides an income-earning opportunity without creating a conflict within the family (Ozbay, 1993; Kumbetoglu, 1996; Eşim, 2002; Dedeoglu, 2002). It conforms to dominant gender ideologies that limit women to "domestic" work, which means that the demands of flexible production and the conditions of the female labor supply come to support each other to result in the continuation of this system (Turkun-Erendil, 2002). Some other researchers claim that this is not always a necessary situation but depends on various contingent factors, such as the preference of some women to organize their time more freely and creatively according to the needs of their families (Hattatoglu, 2000, 2001).

Evidence from several countries indicates that growth in export-oriented garment production often coincides with a rise in the number of homeworkers (Beneria and Roldan, 1987; Pineda-Ofreneo, 1988; Singh, 1990). The same pattern had been expected in Istanbul. However, data made available by firms in this study, together with anecdotal evidence from several participants, led us to believe that home-working in 1995–6 was decreasing in the Istanbul clothing industry. There are two broad reasons for this trend. The first concerns the labor process. Young girls pro-vided an alternative to homeworkers: they too could be employed on a very flexi-ble and low-wage basis. Further, various costs associated with the employment of homeworkers did not apply: materials did not need to be transported; workers could be shifted to different stages of production very quickly; and quality control could be undertaken more easily. Second, as we elaborate further in due course, signifi-cant changes were detected in women's attitudes to work: more and more women wanted to work on the shopfloor instead of at home.

Interviews and surveys point to the complex transformation of gender identity and gendered distinctions of public and private domains due to the increasing number of migrant women who started to become involved in the clothing indus-try. In traditional Turkish families, women are usually *permitted* to work outside only if there is no other way that the family can survive. In other words, it does not affect the dominant view that the main role of women is at home (White, 1994). The families and social groups to which these women belong may provide assis-tance to "enable" women to work outside the home, but they do not want women to become dedicated to such work. Married women carry very substantial domes-tic responsibilities, which are even greater for those with children.[9] Among married women working in the Istanbul clothing industry, 89 percent have children, although there are signs that family size is decreasing. While women are able to draw upon local networks to organize child care, these are generally family-based and serve to strengthen women's obligations to their families (for a comparison see Chant and McIlwaine, 1995).

Although women began to be incorporated into the labor market for the pur-poses of increasing family income, the boom period led to important changes in their attitudes to work. Two key changes were demonstrated by our research, and demonstrate the changing identities and family structures among rural–urban migrants over time. First, women no longer defined their participation in waged work exclusively in terms of their family roles. One consequence of this is that they rarely left work for reasons such as marriage or motherhood. Moreover, although marriage was still a major means of achieving social status and respectability, women seemed to be less willing to get married as they started to earn their own wages. Second, there was unambiguous evidence that women now made their own decisions about whether or not to work outside the home themselves: 80 percent stated that it was solely their decision to take on such work; only 16 percent reported making the decisions with their husbands or with other family members. The great majority spoke of waged work as a way of life and did not contemplate quitting.

According to the findings of our research in 1995–6, the female labor force of the clothing industry clearly did not fit the image of being passive and dependent. Moreover, while some resistance to women working outside the home persisted, the majority of families participated in this transformation in attitudes (Ecevit, 1991; Senyapili, 1992). That is why the decreasing employment opportunities in the crisis years of the 1990s are very important. Once they have become a part of the labor market, it is now more difficult to accept the earlier role restricted by being a house-wife with lower family incomes.

While the great majority of women wanted to work outside the home (81 percent of those interviewed), this did not mean that they were satisfied with their work or the conditions at work and at home. During the rapid growth period, when job opportunities were increasing, they followed several strategies in order to enhance their situation. Three such strategies could be identified. The first strategy was related to searching for better working conditions through job mobility. Women tried to find easier types of work with better conditions when they had alternatives. Second, female employees tried to increase their wages. Women took advantage of the intense demand for their labor associated with the rapid expansion of clothing production in the 1980s. For subcontractors in the informal sector in particular, experienced female workers acquired significant bargaining power. These women were needed by new firms keen to establish themselves very swiftly within production networks. In these conditions workers could move from one firm to another in pursuit of higher wages. This mobility and the employment in the informal sector did not create a problem in the access to healthcare services, since in the Turkish system it is possible to use these services if one person in the family is registered.

The third strategy entailed attempting to escape from patriarchal relations, both at home and in the workplace. Patriarchal workplace norms and practices, such as demanding extra "domestic service work" and controlling their relationships with other workers, remained prevalent despite some changes over the years. In the context of a rapidly expanding textile industry, women could challenge patriarchal workplace relationships by leaving and finding work in a different firm. In addition, a substantial number of paid workers (40 percent) indicated that they wanted to work a considerable distance from home for similar reasons. Most of these women were single and they clearly wanted to escape from the control exerted by their immediate families, other relatives, and neighbors. They said that "if we work in a place far from the house we can be able to go out for a walk and meet with our [boy]friends at the lunch breaks and have free time of our own, which is not possible in our neighborhood."

As for female factory owners, the interviews showed that almost half of them had entered the clothing industry as paid workers, subsequently establishing firms with friends or relatives. During the rapid growth of industry, many small female-headed subcontracting workshops were established by raising the necessary capital via family circles. Their family origins were broadly similar to those of other women in the study. A second group of female factory owners came from a very different social background. They were the daughters or wives of men who owned large businesses and who wished to invest in this sector. While some of these firms were essentially subcontractors, others were firms specialized in finished fashion products.

In this study very few typical family enterprises were encountered. Only newly established small firms fit the profile of family enterprises. In most such businesses, the owners' sons ran the workshops while their fathers organized outwork. But as workshops grew and the number of paid workers increased, capitalist labor relations became dominant even in family enterprises. Thus, most of the young girls working in family businesses received weekly wages in the same way as the other workers. Nevertheless, female members of these families, especially daughters and sisters, tended to express dissatisfaction about work because of the extension of patriarchal familial relations into the workplaces. A young girl working in a family workplace complained about the extra work she did for days and nights during the busy production seasons. "I really do not want to work in this place, since being a worker and the sister of the owner is disadvantageous. I am getting some amount of wage which will be used as my trousseau, but the extra work I do is not paid at all."

This brief review of the four different categories of workers and entrepreneurs in the clothing industry provides some general understanding of the characteristics of women and their working conditions in Istanbul's garment industry. Although women in all the categories are more willing to earn their own money and work outside home, their access to different niches of the labor market differentiates according to their education, skills, and family income and background. Our analytical findings show that crude, structural analyses of labor markets and economic restructuring ignore the high degree of variability among women in this sector, their work experiences and different survival strategies. Therefore, a deeper understanding of the variable positions and conditions of women in the labor market is necessary to define their efforts and abilities to stay in the labor market even during periods of contraction.

The impact of crisis conditions on female workers

Obviously, the new conditions faced since the second half of the 1990s affected all these groups negatively. According to the findings of the Labor Union of Textile Workers, in 2002 the average wage in the clothing sector declined from US$144 to $132 in formal establishments, from $110 to $97 in informal workshops (60 percent of the total number of establishments), and from $83 to $69 in workshops where foreign and child labor is used. The fear of unemployment, especially after the crisis of 2001, has put a higher burden on the shoulders of women, limiting their options of changing their position at work and home, and leaving many without hope of preparing a better future for their children.

The impact of crisis conditions on wage earners is different according to their place of work. Many of the firms with stable export relations have passed through the crises with substantial gains and almost no job losses. Labor in those firms accepted a reduction or stagnation in wages due to the threat of unemployment. This threat has become a widespread discourse, which employers use to pull down wages. On the other hand, quite a large number of firms have difficulties during these cyclical crises and slide into bankruptcy, leaving many workers unemployed. Therefore, there has been a general loss of money on the part of wage-earners even if they do not lose their jobs completely.

Severe unemployment conditions for both male and female labor in the formal sector since 2001 seem to have been reversing the decline in homeworking apparent in our 1995–6 surveys. After the crises of 2001, many homeworkers acknowledge, in ways few did in 1995, that homeworking is an important family survival strategy (Hattatoğlu, 2000; Ozbay, 1993; Kumbetoglu, 1996). As formal sector job losses increase for both men and women, the female homeworker has become the breadwinner in a large number of families.

Government officials and civil organizations are also beginning to recognize the importance of homework, bringing new attention to working conditions inside the home. For instance, some local governments organize bazaar areas for women who want to sell their homemade products. In addition, in one of the districts of Istanbul, a cooperative has been established by a group of women homeworkers in order to negotiate with employers and intermediaries who try to decrease the payments. This organization, the Home-Based Workers Cooperative, is supported by the Working Group for Women Home-Based Workers, and represents an unprecedented political act on the part of homeworkers. Nevertheless, these organizational efforts have coincided with a very difficult time. In the context of the economic crisis, most women workers are afraid of losing their very modest incomes and shun taking part in such organizations because textile firms and intermediaries have many options for finding cheaper labor elsewhere in the city. In addition, in these low-wage conditions, many women choose to leave textile production entirely, searching for other work options such as cleaning houses and offices. One of the women stated that "many women in our neighborhood stopped doing homework because it brings very little income and they prefer doing cleaning jobs for two or three hours after the closing time of offices or shops." Therefore, we can claim that for some women who cannot find work in the textile industry, there is a shift from production to services and housework, but it is difficult to estimate the extent of this shift.

Obviously, the mobility strategies women used during the years of expansion to escape oppression do not work during crisis periods. While the negotiation power of this segment of the labor market is very limited, they now face the threat of losing their jobs completely; that is why they are eager to accept almost all the conditions of the employers. There are many workers who have accepted half of their wages for a certain time (from six months to a year) in order to help the workplace stay in production, since the ones who have lost their jobs have limited alternatives. Surveys conducted in 2002 indicate similar crisis conditions for female factory owners. At all levels, they have been prone to the difficulties faced by many firms in the sector. Many have slid into bankruptcy or have tried to survive by decreasing production costs and by accepting lower profit levels. Clearly, crisis economic conditions have negative effects on women entrepreneurship.

As expressed in the interviews with various family enterprises, hard times are usually endured by firing workers and drawing family members into work without any payment. In order to survive, family enterprises draw on female family members, who are expected to stay at home, or invite some relatives living in other cities or home-villages to help the family business on a temporary basis. Therefore, especially in small enterprises, during expansionary periods capitalist labor relations become established, but these relationship break down during crisis and contraction conditions. During downturns, family enterprises draw on family labor (especially female labor) and extended kin networks.

Concluding Remarks

As the findings indicate, the flexible production system within the Istanbul clothing industry, which realized a high growth rate until the late 1990s, depended upon a pool of second generation female migrants in order to compete within global markets. During this period of growth, women, more than 80 percent of whom came from families migrating to Istanbul from rural areas, became an integral part of the labor market. In doing so, their own attitudes to work and behavior as workers changed. During an era of rapid industrial expansion they became active agents in the labor market and used their bargaining power to earn more under better working conditions, a situation that undermines the label of "marginal" or "reserve labor" common in structural analysis. However, under current conditions, they are more unprotected and more open to exploitation. Still, many women do not want to be drawn in and out of the labor market according to demand conditions: they want to incorporate work into their lives permanently, in one way or another adapting to new situations and finding creative solutions.

Women want to earn more and work in more advantageous positions. However, the severe competition in international markets and fluctuations in domestic demand bring about difficulties in retaining their existing employment. We know that in order to sustain their competitive position in difficult times, factory owners try to decrease labor costs. Some of the factory owners interviewed in 1995–6 indicated their intention to expand production in new areas outside Istanbul to search for lower wages and realized this during the late 1990s. As discussed earlier, some of them relocated in Eastern European countries, where wages are lower and subsidies are higher. Only a small portion of these firms could sustain their competitive positions by changing their product combinations into high value-added products. In 1999, we claimed that due to the risks associated with retaining competitive advantage based on low wages in the clothing industry, the possibility of retaining the enhanced bargaining power for women workers may be limited and temporary (Eraydın and Erendil, 1999a, b). While the rapid rise in export-oriented clothing production brought advantages for women on the margins of the metropolitan area, it is difficult to achieve sustained growth in this type of production where external competition is very high and the domestic market is very volatile.

In this chapter we have incorporated a feminist perspective to the transformation of the division of labor under the pressure of increasing exports during one period and economic crisis a few years later. We hope our analysis contributes to feminist approaches to gender and work in the context of globalization, emphasizing the importance of treating gender and global assembly lines as a dynamic phenomenon, not a static one. The repeated invocation of women as "subordinated" in global manufacturing processes elides the complex changes in women's lives generated in part by their participation in global production systems. It also obscures the effects of the expansion and contraction of global markets. In the case of Turkey, we demonstrate that the transformation of divisions of labor in the past few decades cannot be understood fully if the gender differences are not taken into account. It is certain that transformation in production systems causes radical changes for all working people, but a disproportionate burden falls on female workers.

NOTES

1   This chapter is based on two research projects (1995–6 and 2001–2), each consisting of extensive surveys and interviews with workers and employers in Istanbul's clothing industry. The second stage of the project is based on selective interviews with prominent actors in the sector, such as labor unions, exporters' associations, exporting firms, and women

in different categories. The information on homeworkers flows from workshops prepared by the Working Group for Women Home-Based Workers (member of HomeNet organization), of which Asuman Turkun-Erendil is a member. More detailed methodological procedures are provided in the following sections of the chapter.

2   For example, in 1998, labor costs (US$/hour) in the clothing sector were 18.04 in Germany, 13.60 in Italy, 13.03 in France, 3.70 in Portugal, and 2.69 in S. Korea, compared to 1.33 in Turkey. However, there are countries with lower labor costs in this sector, such as Thailand, 0.78, China, 0.43, India, 0.39, and Pakistan, 0.24 (Werner International, DİSK/Labor Union of Textile Workers, Research Department, 2002).

3   Although the government had invited foreign capital into Turkey since the beginning of 1960s, this bottleneck in the textile industry remained acute because the state continued to limit the amount of foreign direct investment relative to total private investment.

4   In the first and second quarters of 2000, wages decreased by 5.5 and 3.5 percent (Yeldan, 2001).

5   These include high tax exemptions, lower labor costs, lower energy prices, and the opportunity of using the unused quotas (Kaya, 2001; DİSK/Labor Union of Textile Workers, Research Department, 2002).

6   The economic problems experienced since the early 1990s not only reflected low rates of national economic growth and even a decline in GDP per capita, but also shaped an entirely new era of vulnerability for a variety of income groups. For example, economic crisis negatively impacted higher income groups engaged in formal sector employment and ownership. This situation is well presented by the figures on the declining income inequalities between the years 1994 and 2001. The share of the highest quintile declined from 54.9 to 43.8 percent, while the share of poorest quintile rose from 4.90 to 6.3 percent. The decreasing income inequalities do not mean, however, that the lower income groups are in a better situation. The poverty threshold decreased in Turkey during that same period: people defined as poor in 2001 subsisted on a smaller income then the poor of 1994. Ironically, the Marmara region (where Istanbul is located) has the highest per capita income in Turkey, but it is the region that experienced the most drastic increase in poverty (Özcan, 2002).

7   The production stages are grouped into three: (a) those involving pattern making, grading, cutting, and public relations, which tend to involve the most highly skilled tasks; (b) sewing; and (c) quality control and packaging.

8   The average number of people in the families of the women in our sample was 4.6 and the average number of people working in those families was 2.5.

9   According to the research in Istanbul, 51 percent of the married women took on full responsibility for all housework within their own homes, and 24 percent got help from other female members of their families. Only 13 percent reported sharing the housework with their husbands, while 12 percent reported paying others to work in their homes.

BIBLIOGRAPHY

Allen, S. and Wolkowitz, C. (1987) *Homeworking: Myths and Realities*. London: Macmillan.

Amin, A. (ed.) (1994) *Post-Fordism: A Reader*. Oxford: Blackwell.

Beneria, L. and Roldan, M. (1987) *The Crossroads of Class and Gender: Industrial Homework, Subcontracting and Household Dynamics in Mexico City*. Chicago: University of Chicago Press.

Benton, L. (1990) *Invisible Factories: the Informal Economy and Industrial Development in Spain*. Albany: State University of New York Press.

Bora, A. (2002) Olmayanin nesini idare edeceksin? Yoksulluk, kadinlar ve hane. In N. Erdogan (ed.), *Yoksulluk Halleri: Turkiye'de kent yoksullugunun toplumsal goruntuleri*. Istanbul: Demokrasi Kitapliği.

Braverman, H. (1974) *Labor and Monopoly Capital*. New York: Monthly Review Press.

Castells, M. (1997) *The Power of Identity*. Oxford: Blackwell.

Chant, S. and McIlwaine, C. (1995) Gender and export manufacturing in the Philippines: continuity and change in female employment? The case of the Mactan Export Processing Zone. *Gender, Place and Culture*, 2, 147–76.

Christopherson, S. (1989) Flexibility in the US service economy and the emerging spatial division of labour. *Transactions of the Institute of British Geographers*, 14, 131–43.

Cinar, M., Evcimen, G. and Kaytaz, M. (1988) The present day status of small-scale industries (sanatkar) in Bursa, Turkey. *International Journal of Middle East Studies*, 20, 287–301.

Dedeoglu, S. (2000) Toplumsal cinsiyet rolleri acisindan Turkiye'de ail eve kadin emegi. *Toplum ve Bilim*, Fall, 139–70.

Dicken, P. (1988) *Global Shift: Industrial Change in a Turbulent World*. London: Paul Chapman.

DİE (2000, 2001) *Statistics of Companies, Cooperatives and Firms*. Instanbul: State Statistics Institute.

DİE (2001) *Statistics on Turkey*. Instanbul: State Statistics Institute.

DİSK/Labor Union of Textile Workers: Research Department (2002) *Turk Tekstil ve Hazir Giyim Sektorleri: Mevcut Sorunlar ve Cozum Onerilerimiz*. Research Report.

Delarrocha, M. G. (1995) The urban family and poverty in Latin America. *Latin American Perspectives*, 22(2), 12–31.

Doeringer, P. B. and Piore, M. J. (1985) *Internal Labor Markets and Manpower Analysis*, 2nd edn. New York: M. E. Sharp.

Drache, D. and Gertler, M. S. (1991) The world economy and the nation-state: the international order. In D. Drache and M. S. Gertler (eds), *The New Era of Global Competition*. Quebec: McGill-Queens University Press.

Drakakis-Smith, D. (1996) Third World cities: sustainable urban development, population, labour and poverty. *Urban Studies*, 33(4/5), 673–99.

Ecevit, Y. (1991) Shop floor control: the ideological construction of Turkish women factory workers. In N. Redclift and M. T. Sinclair (eds), *Working Women: International Perspectives on Labour and Gender and Ideology*. New York: Routledge.

Eraydın, A. (1994) Changing spatial distribution and structural characteristics of Turkish manufacturing industry. In F. Senses (ed.), *Recent Industrialization Experience of Turkey in a Global Context*. London: Greenwood Press.

Eraydın, A. and Erendil, A. (1999a) *Yeni Uretim Surecleri ve Kadin Emegi*. Ankara: T. C. Basbakanlik Kadinin Statusu ve Sorunlari Genel Müdürlügğü.

Eraydın, A. and Erendil, A. (1999b) The role of female labour in industrial restructuring: new production processes and labour market relations in the Istanbul clothing industry. *Gender, Place and Culture*, 6(3), 259–72.

Erdoğan, N. (ed.) (2002) *Yoksulluk Halleri*. Istanbul: Demokrasi Kitapligi Yayincvi.

Eşim, S. (2002) Women's informal employment in Central and Asian Europe. Post Conference Report, NEWW Gender Policy Conference, July 10–15, Krakow, Poland.

Felstead, A. and Jewson, N. (2000) *In Work and at Home: An Understanding of Homeworking*. New York: Routledge.

Gerny, P. H. (1993) The deregulation and reregulation of financial markets in a more open world. In P. G. Gerny (ed.), *Finance and World Politics*. Aldershot: Edward Elgar.

Gerny, P. H. (1996) International finance and the erosion of state policy. In P. G. Gerny (ed.), *Globalization and Public Policy*. Aldershot: Edward Elgar.

Gilbert, A. (1994) Third World cities, poverty, employment, gender-roles and environment during a time for restructuring. *Urban Studies*, 31(4/5), 605–33.

Hattatoglu, Z. D. (2000) Evde çalısan kadinlar arasi dayanisma/orgutlenme. Unpublished PhD thesis, Mimar Sinan University, Istanbul.

Hattatoglu, D. (2001) Ev eksenli çalısma ve stratejileri: kadin ozgurlesmesi acisindan bir tartisma. In A. Ilyasoglu and N. Akgokce (eds), *Yerli bir Feminizme Dogru*. Istanbul: Sel Yayincilik.

Hattatoglu, D. (2002) Ev eksenli calismada cocuk emegi ve kadin emegi iliskileri. *Iktisat Dergisi*, 430, 54–7.

Hirst, P. and Zeitlin, J. (1991) Flexible specialization versus post-Fordism: theory, evidence and policy implications. *Economy and Society*, 20, 1–54.

Işik, O. and Pınarcıoğlu, M. (2001) *Nobetlese Yoksulluk*. Istanbul: Iletisim Yayinlari.

ITKIB (General Secretariat of Istanbul Textile and Apparel Exporters' Association) (1997–2002) *Export Performance Evaluation and the Factors Affecting Performance in Textile, Apparel, Leather and Rug Sectors*. Annual reports.

Jenson, J. (1994) The talents of women, the skills of men: flexible specialization and women. In S. Wood (ed.), *The Transformation of Work*. London: Unwin Hyman.

Jessop, B. (1994) Post-Fordism and the state. In A. Amin (ed.), *Post-Fordism: A Reader*. Oxford: Blackwell.

Kandiyoti, D. (1988) Bargaining with patriarchy. *Gender and Society*, 2(3), 274–90.

Kandiyoti, D. (1998) Gender, power and contestation: Rethinking bargaining with patriarchy. In C. Jackson and R. Pearson (eds), *Feminist Visions of Development: Gender Analysis and Policy*, London: Routledge.

Kanji, N. (1995) Gender, poverty and economic adjustment in Harare, Zimbabwe. *Environment and Urbanisation*, 7(1), 37–55.

Kaya, E. S. (2001) Romanya ve Bulgaristan'daki Turk tekstil ve hazirgiyim yatirimlari uzerine bir degerlendirme: yatirim yarislari icindeki isverenlerimiz, algilamalar ve gercekler, *Finansal Forum*, August.

Kessler-Harris, A. and Sacks, K. B. (1987) The demise of domesticity in America. In L. Beneria and C. Stimpson (eds), *Women, Households and The Economy*. New Brunswick, NJ: Rutgers University Press.

Kumbetoglu, B. (1996) Gizli isciler: kadinlar ve bir alan arastirmasi. In S. Cakir and N. Akgökce (eds), *Kadin Arastirmalarinda Yontem*. Ankara: Sel Yayincilik.

Lawson, V. A. (1992) Industrial subcontracting and employment in Latin America: a framework for contextual analysis. *Progress in Human Geography*, 16, 1–23.

Leyshon, A. and Thrift, N. (1997) *Money Space: Geographies of Monetary Transformation*. London: Routledge.

Lordoglu, K. (1990) *Eve Is Verme Sistemi Icinde Kadin Isgucu Uzerine bir Alan Arastirmasi*. Istanbul: Friedrich Ebert Vakfi Publications.

Massey, D. (1984) *Spatial Divisions of Labour: Social Structures and the Geography of Production*. London: Macmillan.

Mies, M. (1986) *Patriarchy and Accumulation on a World Scale*. London: Zed Books.

Nippert-Eng, C. E. (1996) *Home and Work: Negotiating Boundaries through Everyday Life*. Chicago: University of Chicago Press.

Ohmae, K. (1990) *The Borderless World*. New York: Harper Business.

Ohmae, K. (1995) *The End of Nation State: The Rise of Regional Economies*. New York: Free Press.

Ozbay, F. (1993). Kadinlarin evici ve evdisi ugraslarinda degisme. In S. Tekeli (ed.), *1980'ler Turkiye'sinde Kadin Bakis Acisindan Kadinlar*. Istanbul: Iletisim.

Özcan, Y. Z. (2002) Measuring poverty and inequality in Turkey, 2001: preliminary results. Unpublished report.

Peck, J. (1992) Invisible threads: homeworking labour market relations and industrial restructuring in the Australian clothing trade. *Environment and Planning D*, 10, 671–89.

Peck, J. (1996) *Work-Place: The Social Regulation of Labor Markets*. New York: Guilford Press.

Pessar, P. R. (1994) Sweatshop workers and domestic ideologies: Dominican women in New York's apparel industry. *International Journal of Urban and Regional Research*, 18, 127–42.

Phizacklea, A. and Wolkowitz, C. (1995) *Homeworking Women*. London: Sage.

Pineda-Ofreneo, R. (1988) Philippines domestic outwork: subcontracting for export oriented industrialization. In J. G. Taylor and A. Turton (eds), *Sociology of Developing Societies: South East Asia*. Basingstoke: Macmillan.

Schoenberger, E. (1988) From Fordism to flexible accumulation: technology, competetive strategies, and international location. *Environment and Planning D: Society and Space*, 6, 252–62.

Senyapili, T. (1992) A new stage of gecekondu housing in Istanbul. In I. Tekeli (ed.), *Development of Istanbul Metropolitan Area and Low Cost Housing*. Istanbul: Turkish Social Science Association.

Singh, A. M. (1990) *The Political Economy of Unorganised Industry: A Study of Labour Process*. New Delhi: Sage.

Storper, M. and Walker, R. (1983) The theory of labour and the theory of location. *International Journal of Urban and Regional Research*, 7, 1–43.

Stratigaki, M. and Vaiou, D. (1994) Women's work and informal activities in Southern Europe. *Environment and Planning A*, 26, 1221–34.

Turkun-Erendil, A. (2002) Turkiye'de ev eksenli calisma uzerine yapilmis arastirmalar ve calismalar. *Iktisat Dergisi*, 430, 36–47.

Walby, S. (1990) *Theorizing Patriarchy*. Oxford: Blackwell.

Weyland, P. (1997) Gendered lives in global places. In A. Öncün and P. Weyland (eds), Space, Culture and Power. London: Zed Books.

White, J. B. (1994) *Money Makes Us Relatives: Women's Labor in Urban Turkey*. Austin: University of Texas Press.

Yeldan, E. (2001) *Kuresellesme Surecinde Turkiye Ekonomisi*. Istanbul: Iletişim Yayinlari.

Chapter 12

Female Labor in Sex Trafficking: A Darker Side of Globalization

Vidyamali Samarasinghe

Introduction

The concept "sex trafficking" refers to the coercive dimensions of recruiting and transporting persons to work in the global sex industry. "Trafficking" of human cargo in the contemporary world is not limited to women and children who are pushed into sex work. However, the highly successful economic base of the commercial sex industry rests primarily on the labor of the female sex worker, and increasingly depends on trafficking to provide its labor supply. While the global sex industry currently flourishes as a lucrative economic enterprise that employs a significant number of workers and attracts many customers, public opinion generally regards it as a deviant enterprise that should be more strictly controlled or indeed suppressed altogether (Weitzer, 2000, p. 2).

A range of feminist scholarship demonstrates that the current forces of economic globalization have had far-reaching effects on the volume and structure of female sex trafficking (Hughes, 2002; Samarasinghe, 2003; Shelley, 2003). Saskia Sassen (1988), for example, contends that two overarching characteristics closely associated with globalization, i.e. the "feminization" of global labor and of migration, have contributed significantly to the increasing incidence of trafficking of females to provide the labor for the fast growing global sex industry. I expand on this literature by examining how sex trafficking operates in particular historical and geographical contexts, place-specific dynamics that complicate any singular understanding of "the global sex trade."

The literature on the interaction of place and space, a core theme in the study of human geography, has introduced important theoretical perspectives, methodological tools, and grounded empirical reviews on understanding the diversity of human experiences (Tuan, 1974; Soja, 1980, 1989; Harvey, 1990). Feminist geography has demonstrated that interaction of place and space is gender-specific and produces different experiences for women (Rose, 1993; Samarasinghe, 1997). In this chapter, based on specific examples drawn from the Philippines, Nepal, and Cambodia, I demonstrate that while the current forces of globalization have had a decisive effect on the general form and structure of both the supply side and the demand side of female sex trafficking, country-specific sociocultural histories and economic conditions create a diversity in the form and nature of female sex trafficking and women's experiences. If scholars do not examine the interactions of place and space they risk creating a monolithic category of "global sex trafficking" that obscures how such trafficking operates within historically specific contexts and how it affects women's and girls' lives differently across space.

Women's Labor: Uncounted under the Stigma of Prostitution

The social stigma associated with sexual activity outside marriage for women and girls in many societies of the South, and the illegality of the activity, renders voluntary recruitment for the sex trade extremely difficult. Trafficking has become the strategy of recruitment for commercial sex work, since the demand quite clearly outpaces the voluntary supply. Given the clandestine nature of the sex industry in general, data on most aspects of the trade are only estimates. Women who are trafficked into this sector may be required to sell sex directly as street prostitutes or as employees of brothels. They may be employed as bar girls, in massage parlors, in strip clubs, escort services, or nightclubs, and would be expected to provide different types of sexual services to clients. Women who are trafficked into the sex trade are enslaved by their owners, often subjected to long periods of debt bondage imposed by the employers/recruiters, and usually lack knowledge of the terms of their debts (Truong, 1990). What has made prostitution based on trafficking an immensely lucrative industry is the amount of profit generated from this work through minimally paid or unpaid labor, largely of women and children (Nakashima Brock and Thistlewaite, 1996).

Illegally acquired gain for a trafficker operating in Europe is on average 150,000 euros per suspect and the police estimated that from each trafficked victim the criminals could earn an average of 45,000 euros per year (van Dijk, 2003). One sex trafficking broker arrested in Japan in 2002 had earned 10 million yen a month for placing 400 trafficked Columbian women in strip clubs in Japan (*Financial Times*, February 6, 2003). As Shelley (2003) observes, the illegal trade of trafficking fits into several business models ranging from small networks to highly organized groups, all of which are in the business because of high profits.

Not only do private individuals profit from sex trafficking, but many states have a vested interest in directly or indirectly promoting the trade. In an era of globalization and debt restructuring, the "development" policies of many countries of South and Southeast Asia are focused on increasing foreign exchange earnings. While not as visible in policy documents, the suite of policies designed to foster foreign currency earnings and debt repayment undoubtedly encourages the growth of prostitution and the consequent trafficking of females because it generates foreign exchange and revenue from the remittances of exported female sex workers (Truong, 1990; Lim, 1998). Indeed, the illegality of the sex trade does not obviously preclude it from being a sizable revenue earner to the governments of these countries. Policymakers have to deal with an industry that is highly organized and increasingly sophisticated and diversified, as well as having close interlinkages with the rest of the national and international economy. Although accurate numbers for the revenue generated are also hard to come by, one estimate calculates a US$1.2–3.3 billion per year contribution to Indonesia's GDP and US$22.5–27 billion a year for Thailand's economy (Lim, 1998).

The female sex worker earns the least in this profitable global sex industry. The client, procurer/recruiter, entertainment/tourist industry entrepreneurs, brothel owners, pimps, military establishments, travel agency owners, and crime syndicates, who are overwhelmingly male, capture most of the profits of this industry. The industry, for example, does not have to bear any cost for training its workers. In female sex trafficking, women's bodies become a commodity, separate from the person. "Training" for the job is deemed to be unnecessary, because the work expected of the trafficked sex worker is based on her natural physical sexual makeup of "being" female. Access to health and medical facilities is virtually non-existent for trafficked victims, and physical and sexual abuses are the norms that govern their work.

Globalization and the Socialization of Reproductive Labor

Female sex trafficking has been a feature of human behavior for millennia. However, the current forces of globalization add new dimensions to the pattern and structure of trafficking. These forces of globalization compress social space, which David Harvey describes as "annihilation of space by time," and reduce the power of the nation-state (see Harvey, 1990, 2000). Globalization has indeed entrenched the power of capital over labor, leading to a clear division between the capital-rich North and the capital-poor and labor-surplus South. Translated into a spatial form, as Sassen (1998) argues, globalization has thus created a global grid of new economic geography of centrality and marginality.

The "feminization" of flexible and poorly remunerated labor in assembly line jobs has become a hallmark of the processes of globalization. In the current globalized economy, women from the South have provided a key source of labor for export manufacturing industries such as garments and electronics, as extensively documented (Standing, 1989, 1999; Beneria and Roldan, 1992; Elson, 1996, 1999; Kabeer, 2000). Similarly, the increasing numbers and visibility of women in the "international maid trade" are also a part of globalization dynamics. Deregulation and economic restructuring not only displaced peasant and other local economies, they also eased travel regulation across borders for women from poorer countries seeking work as domestic servants in relatively wealthier countries, in both the South and the North (Eelens and Speckman, 1990; Tyner, 1996; Chin, 1998; Samarasinghe, 1998; Pyle, 2001).

Undergirding these globalization processes are new transnational constructions of gender that have expanded the attractiveness of women workers. Women's work in the reproductive sphere, which had been hitherto largely uncounted and invisible, has been socially reconstructed to support the emergence of a gendered globalized labor force. Women are expected to "know" naturally as women how to perform reproductive activities and the resultant "deskilling" of such activities in the labor market has translated into lower wages for women. Most importantly, these transformations have resulted in the creation of new structures of employment for women on a global scale, whether they are in the factory or in the kitchens of wealthy households. It has opened up new spaces of work for women, some legitimate spaces and other less legitimate and more exploitative spaces.

As women have been drawn into global assembly line production, they have also formed the largest pool of potential transnational migrants. While transnational migration has a long history among men in Asia, the feminization of overseas migration in Asia has been overwhelming during the past two decades (D'Cunha, 2002). With prospects for survival in the village getting bleaker, and with "work" potentially available in the cities within or across borders, and mobility no longer discouraged, it is not surprising that women are willing to take a risk to ensure a better life for them and their families. As Todaro (1976) argues, the phenomenon of labor migration is driven not necessarily by actual opportunities, but by the perceived work opportunities that the vistas of globalization appear to create. The only information that they may have is that job opportunities are more available outside their restrictive, usually rural, communities. This perception is actively cultivated and nurtured by smooth talking recruiters who dangle the prospects of such legitimate employment to lure women into the lucrative commercial sex industry.

Together these two processes, "feminization" of labor and "feminization" of migration, are perhaps the most significant unintended consequences of the forces unleashed by globalization. They create, as Saskia Sassen (1988) argues, "counter-geographies" of globalization. According to Sassen the past two decades have seen a growing presence of women in a variety of cross-border circuits that have become a source for livelihood, profit-making, and accrual of foreign currency. While these circuits are diverse, they have one common feature, i.e. they are profit- and revenue-making ventures developed "on the backs of the truly disadvantaged" (Sassen, 2000, p. 503). Sassen identifies women who venture in search of work in newly globalized system as the key actors in the creation of "counter-geographies" of globalization, clearly the darker side of globalization.

A fundamental difference between women workers in transnational manufacturing firms or overseas female domestic workers and female prostitution based on trafficking is that the latter is widely considered illegal and nurtures an underground economy based on stigma and humiliation. The impact of globalization on trafficking of females for prostitution is just as significant as for the other two forms of globalized female labor activity. The "time–space compression," the hallmark of contemporary globalization, refers in part to the phenomenal upsurge in efficient and accessible information technology. This phenomenon has a significant effect on the volume and structure of trafficking of females for commercial sex work. The internet has been highly effective in transcending borders to bring clients and traffickers/pimps closer and forge more efficient interconnections of crime syndicates that mastermind female sex trafficking operations (Hughes, 1999). It allows the demand and the supply components of commercial sex work to move more freely to create a more spatially flexible market. Furthermore, globalization has profoundly impacted the perception of waged work opportunities on the part of poor women in the Global South, particularly in rural areas where such waged work for women is rare. Success stories of a few female returnees, who have obviously been successful in making money by traveling outside the community, seem to mitigate the risks of trusting the recruiters or travel to new destinations (Newar, 1998; Evans and Battarai, 2000).

Labor in Female Sex Trafficking: Labor Exploitation in Nepal, the Philippines and Cambodia

Among the three countries, the Philippines is a source, transit, and destination for internationally trafficked persons. Nepal is a source of women and girls trafficked primarily across the border to India for purposes of sexual exploitation and bonded labor. Cambodia is a source, destination, and transit country, with a significant component of internal trafficking of women and children into the commercial sex industry (US Department of State, 2003).

In highly urbanized centers such as Metro Manila, rural to urban migration of women in search of employment becomes a viable source for trafficking, to service the domestic as well as the overseas sex trade (Santos, 2002). Government encouragement of the tourist industry and associated "entertainment" tourism often involves, implicitly, state promotion of the commercial sex industry (Jeffreys, 1999). In the Philippines, the internet in particular plays an important role in helping the traffickers/brokers to advertise the sex industry, flash the pictures of women across the globe to clients in the West and Australia, ease payment with a click of a computer key, and set up chat rooms so that the client may get the "best deals" from trafficked sex workers (Hughes, 2002). When the Philippine government banned "mail order bride" agencies from the Philippines because it was reported that women were subject to abuse and sexual and labor exploitation, internet sites sprang up to fill the void (Hughes, 1999). In short, the internet is a tool that boosts the demand side, while increasing the exploitation of the supply side.

Trafficking of Filipino women overseas for sexual purposes has been greatly facilitated by the state policy encouraging women to seek employment overseas. Export of labor has been an acknowledged policy strategy of successive Filipino governments (Sassen, 2000). Hailed by the policy-makers as "new heroes," the overseas contract workers (OCWs) contribute billions of dollars to boost the economy of the Philippines (see chapter 9, this volume). Santos (1999) notes that international migration has been the anchor of trafficking, not only for the labor but for sex as well. According to the Philippine Overseas Employment Administration (POEA), more Filipino women than men migrate overseas in search of employment. There has been an steady increase of female OCWs from about 12 percent of the total in 1975, to 47 percent in 1987 and 58 percent in 1995 (POEA, 1998).

Female OCWs are mostly in the service sector, of which housemaids constitute the largest number. While those who migrate as housemaids and as "mail order" brides also get trafficked into the commercial sex sector, women who run the greater risk of being trafficked into prostitution are those who are recruited as "entertainers," which is widely known as a euphemism for the sex trade (Santos, 1999). In 1998, nearly 50,000 Filipinos migrated overseas as "entertainers" and 95 percent of them went to Japan (POEA, 1998).

It is helpful to understand the growth of this migration, and of sex trafficking, through analysis of both push and pull factors that shape women's choices. While rural poverty and the lack of suitable employment opportunities are important in pushing women into migration streams, this structural push factor is exacerbated by the cultural norms inculcated among young women that they are expected to help their families. Olfreno and Olfreno (1998, p. 105) note that many female sex workers cite the need to repay their "debt of gratitude" to parents and support their natal and/or conjugal families as a major reason why they enter and stay in the sex trade. The "pull" factors consist not only in women's perceptions of a better economic opportunities overseas, and expectations of making a higher income in the richer countries, but also in the ease of travel and active encouragement for such travel by dollar-hungry policy-makers. It seems clear to many women that selling their labor in a foreign country is a promising way of supporting their families and moving ahead.

The overwhelming majority of those who are recruited to travel to Japan as entertainers are young women, between the ages of sixteen and twenty-eight. Their education ranges from middle school to university level. The internet is an important tool of information for overseas job opportunities for potential victims of trafficking as bar girls to Japan, who are often misled by slick recruitment campaigns for sex trafficking. While deprivation and lack of appropriate work opportunities are push factors that propel the women into trafficking streams, most often they are not the poorest of the community. Despite efforts to screen and document migrant workers in the Philippines, an estimated one to two million women OCWs remain undocumented. Those most likely to circumvent the law are those who are aggressively recruited as entertainers/bar girls. As Barry (1995, p. 13) estimates, 65 percent of Filipino entertainers migrating overseas had been trafficked to Japan. Fraud and deceit is very much a part of recruitment and it is also noted that money they earned was withheld from them or they received such small payments that they were forced to continue in sex work, beholden to the entertainment industry entrepreneurs (Raymond, 2002).

The impact of globalization on sex trafficking is somewhat different in Nepal than what is observed in the Philippines. Nepal ranks as one of the poorest countries, with a per capita gross national product of US$230 a year in 2002 (World Bank, 2003). Nepal is a landlocked country, with a 1500-kilometer porous border with India, which is larger and better off economically, and also has legalized red light districts. There are no reliable data on how many women and children are trafficked each year from Nepal. Some scholars estimate that 100,000–200,000 Nepali girls are currently employed in the brothels of India, mainly in Mumbai (Bombay), Kalkot (Calcutta), Siliguri, Kanpuir, Gorakkhpur, Lucknow, and New Delhi (Sangroula, 2001). Other scholars, while acknowledging the fact that Nepalese women and children are indeed trafficked to brothels in India, observe that the estimated numbers are exaggerated (Evans and Bhattarai, 2000).

Trafficking in Nepalese girls and women to Indian brothels emerged in the 1960s. By the 1970s, criminal links between sex traders and the Nepalese pimps were firmly established (Sangroula, 2001). The emergence of trafficking historically was facilitated by a 1950 Indo-Nepal Friendship treaty that ensured no immigration control or record keeping between Nepal and India (Asian Development Bank, 2002). Growing concern about trafficking generated a new regulation in 1999, issued by the Ministry of Labor Nepal, that limits the overseas travel of single women (under thirty-five years of age) and minors unless they are accompanied by a relative and can show proof of consent from a guardian (Sanghera and Kapur, 2000). Despite such legal provisions, the long land border between India and Nepal ensures the clandestine structure of trafficking of females from Nepal to brothels in India.

Evidence from Nepal attests to the fact that as income from agriculture is decreasing, migration plays an important part in the livelihood strategies of rural households (Seddon et al., 1999). Female trafficking in Nepal is a complex, multicausal phenomenon (Acharya, 1998). Poverty and sexism together create unequal bargaining power between men and women, which results in female vulnerability that nurtures female sex trafficking (Mahendra et al., 2001). The desperate need to sell female labor in order to survive pushes families to sell the girls to traffickers who approach the families with false promises. "If somebody comes to our house and says that he will find employment for our daughters in the circus and also pays first month's salary, we wouldn't call it trafficking. We would do so out of compulsion for livelihood" (Participant in a focus group discussion, quoted in Mahendra et al., 2001, p. 35).

One of the fundamental issues that emerges with regard to the labor in female sex trafficking, especially in Nepal, is a disjuncture in the accepted definition of a "minor" that exists between the international law makers and the local society. There is an international agreement that child prostitution and child trafficking of any kind is a violation of the rights of the child, as such activities are designated as criminal. However, as many scholars, activists, and development organizations have noted, children's labor contribution is an accepted norm in household survival strategies among the poor in Nepal (Save the Children/UNICEF, 2000). The legal age for majority in Nepal is sixteen years, and not eighteen as put forth by the UN. Many children, mainly girls, are married and sexually active once they reach puberty, which is before they reach the age of sixteen years. Hence, a migration of a young person below the age of eighteen in search of employment is not an unusual occurrence and their sexual activity as commercial sex workers may not be construed as a violation of a children's rights in Nepali society.

The analysis of the Nepali situation with regard to female sex trafficking shows that while the incidence of exploitation of women's time and labor for sexual purposes is no different from the Philippine case, younger females are more vulnerable to trafficking in Nepal than in the Philippines. This is in part due to the fact that Nepal is a poorer country, and many girls have a much lower level of education and get married at a younger age than in the Philippines. At the local level in Nepal, trafficking of women and girls involves low levels of female education, deep-rooted gender discrimination, poverty, and a lack of economic opportunities in the rural areas, particularly the marginalization of certain social groups (Evans and Battarai, 2000). The physical geography of the two countries also brings in a marked difference in the nature of trafficking. The long, unmonitored, porous border with India facilitates the clandestine nature of trafficking from Nepal straight into the brothels of India. In the case of the island nation of the Philippines, the exit points are mainly the airports or the seaports, and most of the women who are trafficked into various forms of overseas sex trades leave the country ostensibly for legitimate employment as housemaids or dancers/entertainers, or on fiancée visas as "mail order brides." They are relatively more educated than the Nepali recruits, and are looking for job opportunities overseas that usually provide foreign currency earnings.

In contrast to either the Philippines or Nepal, female sex trafficking in Cambodia is shaped significantly by the fact that the country has been subjected to more than twenty-five years of civil war. It is one of the poorest countries of the region, with a gross national product per capita of US$280 per year in 2002 (World Bank, 2003). Cambodia shares porous land borders with Vietnam, Laos, and Thailand. Two factors stand out with regard to trafficking activities in Cambodia. First, while Cambodian women and girls are trafficked to neighboring countries for sexual exploitation, it also remains a main destination point for female trafficking, especially from Vietnam. Second, there is a significant internal trafficking of Cambodian women and girls to brothels in Phnom Peng, Sihanoukville, and Kampang Cham (National Assembly of Cambodia, 1997).

Swain notes that prior to 1975, Cambodian women were used as prostitutes, especially for foreign men who were brought there by the wars in South East Asia (Swain, 1995). While during the Khmer rouge regimes and subsequent Vietnam occupation prostitution was suppressed, the arrival of 10,000 civilian administrators of the United Nations Transitional Authority (UNTAC) in 1991 created a demand for women and girls in prostitution and numbers increased. Sex tourism in Cambodia has developed alongside the mainstream tourism industry since the country regained peace after nearly three decades of civil war. Child sex trafficking to serve both local and foreign clients is significant in Cambodia (Agence France, 2002). Especially in the case of foreign clients, internet services provides information on the price for services, prime locations, and access to brokers and traffickers (Hughes, 2000).

The disruption of families and livelihoods resultant from the civil war has been further accentuated by globalization and its impact on Cambodian society. Globalization's urban bias is usually manifested in terms of market-led industrial locations, where the infrastructure and potential for skilled employment are better than in the impoverished rural areas. Cambodia has not attracted global capital for industries and is not considered a large source area for overseas migrant workers. Nonetheless, the comparatively small number of globally oriented economic ventures, and associated service sectors, are focused on the capital, Phnom Peng. Hence, in the absence of work opportunities in the rural areas, Phnom Peng has become an attractive magnet for internal migrants (Human Rights Task Force on Cambodia, 1996). The growth in migrants to Phnom Peng has seen a concomitant rise in sex workers. Parts of the sprawling shanties of Phnom Peng have virtually become red light districts, a known destination for young females from rural areas trafficked into the sex industry.

Marjorie Mueke notes that the historical practice of selling women, which can be found in Southeast Asia, whereby men could sell their daughters for labor to get relief from poverty, has provided an important precedent for the current practice (Mueke, 1992). On the part of the girls, the prevalent cultural norm that daughters are expected to provide financial support for the families makes it convenient to entice young women into sex trafficking (Derks, 1997). Yet the cultural conditions in Cambodia virtually make *coerced* trafficking a necessity to feed the demand for the sex trade. While virginity before marriage for young girls is celebrated in

Cambodia, prostitution is looked at with contempt. Marriage and family are expected norms for Cambodian women. Furthermore, prostitution is illegal in Cambodia. Hence, the supply for the commercial sex sector in Cambodia resorts to abduction, kidnapping of girls, and giving false promises of legal employment. Many of them are forced to work to pay off debts, money the recruiters have allegedly paid their families (Caouette, 1998). The girls are in effect sex slaves: they receive no money, only food and armed guards stop them from running away, and the trade operates with virtual impunity due to the ineffectiveness of the law and the corruption of law enforcement officers (Marks, 2003).

The cross-border trafficking in Cambodia has two streams. One stream of migrants is trafficked from Vietnam across the border to Cambodia to supply the demand for commercial sex along the Thai/Cambodian border, frequented by military personnel and Thai men who cross the border to visit the brothels. The other stream is from Cambodia across to Thailand to bars, massage parlors, and brothels, especially in Bangkok. In most cases, while the trafficked girls were expected to work off their debts, it has been commonly observed that the debts kept increasing. Furthermore, the trafficked commercial sex workers on the Thailand side are fearful of leaving the brothel because of the risk of being apprehended by the authorities for illegal immigration (GAATW, IOM, CWDA, 1997).

The increased demand for female sex trafficking during the presence of the peacekeeping forces in Cambodia in the early 1990s shows no signs of abating. It has created a momentum that has sustained an industry, which is based on the supply of females mainly from Cambodia and neighboring Vietnam. The central location of Cambodia has made it a virtual hub for traffickers in the region.

Conclusions

Globalization produces the feminization of work and the feminization of migration in search of work. It has opened new spaces for women's employment, albeit employment largely stemming from the transformation of reproductive labor into wage labor. While globalization has no doubt opened legitimate workspaces for poor women in the South, they are clearly not sufficient to meet the demands for work for all female job aspirants. This has created a bonanza for the traffickers, who have stealthily grasped the opportunity to recruit women and girls into the sex industry with false promises of legitimate job offers. For traffickers, it is a well organized economic venture based on the movement of people solely to profit from the trafficked person's sexual labor. Hidden behind the stigma associated with trafficked female sex workers is the relentless skimming of profits from the exploitation of female bodies and their time. It builds a circuit that creates "counter-geographies" of globalization, highlighting the exploitative side of a global economic activity (Sassen, 2000).

As noted by Soja (1980, 1989), "space" in itself may be primordially given, but the organization and meaning of space is a product of social translation, transformation, and experience. "Place" is one of geography's most fundamental concepts and places differ from one another due to the complex interactions of environmental, socioeconomic, and political dimensions (Rose, 1993). In analysis of human experiences, the interaction of space and place produces different values, structures, and organizations among different societies. The female sex trafficking situations in the Philippines, Nepal, and Cambodia highlight the differential impacts of globalization on women's pursuit of "work" and represent an important site for feminist geographers to contribute to debates about sex trafficking.

Primarily due to the clandestine nature of the trafficked driven female sex work, accurate and reliable data are hard to come by. Consequently, there is a widespread tendency to generalize on the form and structure of trafficked women's involvement and victim hood in the sex industry. Such generalizations often lead to the implementation of flawed policies to combat female sex trafficking. What is required for the implementation of realistic anti-trafficking programs is a multilayered analysis combining issues of global relevance with an exploration of country- or region-specific sociocultural, environmental, and political dynamics that trigger the particular form of sex trafficking at the ground level.

BIBLIOGRAPHY

Agence France (2002) Foreign pedophiles: the tip of Cambodian epidemic of child abuse. *Agence France Presse*, August 28.

Asian Development Bank (2002) *Combating Trafficking of Women and Children in South Asia*. Country Paper: Kingdom of Nepal. Manila: ADB.

Acharya, U. D. (1998) *Trafficking in Children and Their Exploitation in Prostitution and Other Intolerable Forms of Child Labor in Nepal*. Country Report. Support from ILO/IPEC. Katmandu.

Barry, K. (1995) *The Prostitution of Sexuality*. New York: New York University Press.

Beneria, L. and Roldan, M. (1987) *Crossroads of Class and Gender: Industrial Homework, Sub-contracting and Household Dynamics in Mexico City*. Chicago: Chicago University Press.

Coalition Against Trafficking in Women (CATW) (1999) Declaration of rights for women in conditions of sex trafficking and prostitution. *Re/reproduction*, 2, April.

Caouette, T. M. (1998) *Needs Assessment on Cross-border Trafficking in Women and Children – Mekong Sub-region*. Bangkok: United Nations.

Chin, C. (1998) *In Service and Servitude: Foreign Female Domestic Workers and the Malaysian "Modernity Project."* New York: Columbia University Press.

D'Cunha, J. (2002) Thailand: migration and trafficking in women. In J. Raymond (international coordinator), *A Comparative Study of Women Trafficked in the Migration Process*. New York: Ford Foundation and CATW.

Derks, A. (1997) *Trafficking of Cambodian Women and Children to Thailand*. Geneva: International Organization on Migration.

Deutsche Presse Agentur (2002) Trafficking in kids a flourishing trade in Southern Asia. *Deutsche Presse Agentur*, May 6.

Doezema, J. (1998) Forced to choose: beyond the voluntary v. forced prostitution dichotomy. In K. Kempadoo and J. Doezema (eds), *Global Sex Workers: Rights, Resistance and Redefinition*. London: Routledge.

Eelens, F. and Speckman, J. D. (1990) Recruitment of labor migrants for the Middle East: the Sri Lankan case. *International Migration Review*, 24, 297–22.

Elson, D. (1996) Appraising recent developments in the world development for nimble fingers. In A. Chachhi and R. Pittin (eds), *Confronting State, Capital and Patriarchy: Women Organizing in the Process of Industrialization*. London: Macmillan.

Elson, D. (1999) Labor markets as gendered institutions: equality, efficiency and empowerment issues. *World Development*, 27(3), 611–27.

Evans, C. and Battarai, P. (2000) *A Comparative Analysis of Anti-trafficking Intervention
Approaches in Nepal*. Kathmandu: The Asia foundation and Population Council,
Horizons.

Global Alliance Against Trafficking in Women (GAATW), International Organization on
Migration (IOM), Cambodian Women's Development Association (CWDA) (1997) *Two
Reports on the Situation of Women and Children Trafficked from Cambodia and Vietnam
to Thailand*. Bangkok: GAATW, IOM, CWDA.

Harvey, D. (2000) *Spaces of Hope*. Berkeley: University of California Press.

Harvey, D. (1990) *The Conditions of Postmodernity: An Enquiry into the Origins of Cultural Change*. Oxford: Blackwell

Hughes, M. D. (2002) The use of new communications and information technologies for
sexual exploitation of women and children. *Hastings Women's Law Journal*, 13(1),
127–46.

Hughes, M. D. (2000) Welcome to rape camp: sexual exploitation and the internet in Cambodia (http://www.Uri.edu/artsci/wms/Hughes/rapecamp.htm).

Hughes, M. D. (1999) Introduction. In D. M. Hughes and C. Roche (eds), *Making the Harm
Visible: Global Sexual Exploitation of Women and Girls*. Kingston, RI: Coalition Against
Trafficking in Women.

Human Rights Task Force (1995) *Prostitution and Sex Trafficking: A Growing Threat to
Women and Children in Cambodia*. Phnom Peng: Human Rights Task Force.

Jeffreys, S. (1999) Globalizing sexual exploitation: sex tourism and traffic in women. *Leisure
Studies*, 18, 179–96.

Kabeer, N. (2000) *Power to Choose: Bangladeshi Women and Labor Market Decisions in
London and Dhaka*. London: Verso.

Kempadoo, K. (1998) Introduction: globalizing sex worker's rights. In K. Kempadoo and
J. Doezema (eds), *Global Sex Workers: Rights, Resistance and Redefinition*. London:
Routledge.

Lerner, G. (1987) *The Creation of Patriarchy*. Oxford: Oxford University Press.

Lim, L. L. (1998) The economic and social bases of prostitution in South East Asia. In L. L.
Lim (ed.), *The Sex Sector: The Economic and Social Bases of Prostitution in South East
Asia*. Geneva: International Labor Office.

Mahendra, V. S., Battarai, P., Dahal, D. R. and Crowley, S. (2001) *Community Perceptions
of Trafficking and Its Determinants in Nepal*. New Delhi: The Asia Foundation and Population Council Horizons.

Marks, K. (2003) The skin trade. *Independent* (London), January 22.

Muecke, M. M. (1992) Mother sold food. Daughter sells body. The cultural continuity of
prostitution. *Social Science Medicine*, 7, 891–901.

National Assembly of Cambodia (1997) *Report on the Problem of Sexual Exploitation and
Trafficking in Cambodia*. Phnom Peng: Commission on Human Rights and Reception of
Complaints. mimeograph

Natashima Brock, R. and Brooke Thistlewaite, S. (1996) *Casting Stones: Prostitution and
Liberation in Asia and the US*. Minneapolis: Fortress Press.

Newar, N. (1998) My sister next? *Himal Magazine*, 11(10) (retrieved from http://www.himal-
mag.com/98oct/sister.htm).

Olfreno, R. and Olfreno, R. P. (1998) Prostitution in the Philippines. In L. L. Lim (ed.), *The
Sex Sector: The Economic and Social Bases of Prostitution in South East Asia*. Geneva:
International Labor Office.

Philippine Daily Inquirer (2003) More Pinays leave for Japan. Luzon Bureau, February 11.

Philippine Overseas Employment Administration (POEA) (1998) *Information Primer for the Deployment of Overseas performing Artists*. Manila: POEA Manpower Development, Employment Branch.

Physicians for Human Rights (1997) *Commercial Sexual Exploitation of Women and Children in Cambodia. Personal Narratives: A Psychological Perspective*. Boston: Physicians for Human Rights.

Pyle, J. L. (2001) Sex, maids, and export processing: risks and reasons for gendered global production networks. *International Journal of Politics, Culture and Society*, 15(1), 55–76.

Raymond, J. (2002) Intersection between migration and trafficking. In J. Raymond (international coordinator), *A Comparative Study of Women Trafficked in the Migration Process*. New York: Ford Foundation and CATW.

Rose, G. (1993) *Feminism and Geography: Limits of Geographical Knowledge*. Minneapolis: University of Minnesota Press.

Samarasinghe, V. (1997) Counting women's work: intersection of time and space. In J. P. Jones III, H. J. Nast and S. M. Roberts (eds), *Thresholds in Feminist Geography: Difference, Methodology, Representation*. Lanham, MD: Rowman and Littlefield.

Samarasinghe, V. (1998) Feminization of foreign currency earnings: women's labor in Sri Lanka. *Journal of Developing Areas*, 32(2), 302–26.

Samarasinghe, V. (2003) Confronting globalization in anti-trafficking strategies in Asia. *Brown Journal of Foreign Affairs*, 10(1), 91–104.

Sanghera, J. and Kapur, R. (2000) *An Assessment of Laws and Policies for the Prevention and Control of Trafficking in Nepal*. Kathmandu: The Asia Foundation and Population Council, Horizons.

Sangroula, Y. (2001) *Trafficking in Girls and Women in Nepal-Building a Community Surveillance System for Prevention*. Kathmandu: Kathmandu School of Law.

Santos, A. (2002) The Philippines: migration and trafficking in women. In J. Raymond (international coordinator), *A Comparative Study of Women Trafficked in the Migration Process*. New York: Ford Foundation and CATW.

Santos, A. (1999) Globalization, human rights and sexual exploitation. In D. M. Hughes and C. Roche (eds), *Making the Harm Visible: Global Sexual Exploitation of Women and Girls*. Kingston, RI: Coalition Against Trafficking in Women.

Sassen, S. (1988) *The Mobility of Labour and Capital: A Study in International Investment and Labour Flows*. Cambridge: Cambridge University Press.

Sassen, S. (1998) *Globalization and Its Discontent*. New York: New York University Press.

Sassen, S. (2000) Women's burden: counter-geographies of globalization and the feminization of survival. *Journal of International Affairs*, 53(2), 503–24.

Save the Children/UNICEF (2000) *Bringing up Children in a Changing World: Whose Rights? Conversations with Families in Nepal*. Katmandu: Save the Children/UNICEF.

Seddon, S., Jagganath, J. and Gurung, G. (1999) *Foreign Labor Migration and the Remittance Economy of Nepal*. Report Prepared for the DFID, Nepal Country Office. Katmandu.

Shelley, L. (2003) Trafficking in women: the business model approach. *Brown Journal of World Affairs*, 10(1), 119–32.

Soja, E. W. (1989) *Postmodern Geographies: The Reassertion of Space in Critical Social Theory*. New York: Verso.

Soja, E. W. (1980) The socio-spatial dialectic. *Annals of the Association of American Geographers*, 70, 207–25.

Standing, G. (1999) Global feminization through flexible labor: a theme revisited. *World Development*, 27(3), 538–602.

Standing, G. (1989) Globalization through flexible labor. *World Development*, 17(7), 1077–95.

Swain, J. (1995) *River of Time: A Memoir of Vietnam and Cambodia*. New York: St Martin's Press.

Todaro, M. (1976) *International Migration in Developing Countries*. Geneva: International Labor Office.

Truong, T.-D. (1990) *Money and Morality: Prostitution and Tourism in South East Asia*. London: Zed Books.

Tuan, Y.-F. (1974) *Topophilia: A Study of Environmental Perception, Attitude and Value*. Englewood Cliffs, NJ: Prentice Hall.

Tyner, J. A. (1996) The gendering of Philippine international labor migration. *Professional Geographer*, 48(November), 405–16.

UNIFEM (1998) *Trafficking in Women and Children: Mekong Sub-region*. New York: United Nations.

United States Department of State (2003) *Trafficking in Persons Report*. Washington, DC: United States Department of State.

van Dijk, E. (2003) The role of organized crime in trafficking in human beings. Paper presented at the Trafficking in Persons Conference, Human Rights Law Centre, University of Nottingham, June 26–27.

Weitzer, R. (2000) *Sex for Sale: Prostitution, Pornography and the Sex Industry*. London: Routledge.

Wijers, M. and Lap-Chew, L. (1997) *Trafficking in Women, Forced Labor and Slavery-like Practices in Marriage, Domestic Labor and Prostitution*. Utrecht: Foundation Against Trafficking (STV).

World Bank (2003) *Making Services Work for Poor People: The World Development Report 2004*. Washington, DC: Co-publication of the World Bank and Oxford University Press.

Chapter 13

Changing the Gender of Entrepreneurship

Susan Hanson and Megan Blake

My name's Josephine, but I'm known as Jo. People will call up [my engineering company] and ask to speak with the owner. So, my secretary will say, OK, I'll put you through to Jo. I get on the phone, and you know, it's like – yeah, it was difficult in the very beginning. (Josephine, founder of a land surveying and civil engineering company)

A customer will come in to the store for the first time, and . . . the reaction when they see me behind the counter is phenomenal. Some will turn around and walk right out, others will sort of reluctantly come over and ask me about an item or whatever, but it's very obvious to me that they're surprised to see a black woman as the proprietor, if you will. Very often they'll say to me, may I speak to the owner? Or, sometimes I have a white young man who works for me handling the inventory, the heavy furniture and stuff, and invariably, if he and I are in the shop at the same time, a customer will come in and go to him immediately. I'll be sitting behind the counter at the cash register, but they assume he's the owner. And he's very young. I'm 56 and he's 26, OK? (Alice, founder of a used furniture and second-hand shop)

You always get the guys that look at you and go, oh my god, she's driving a truck. I always have to laugh because they'll drive by me and do a double take. I have a bumper sticker on the back of it now that reads "Real Women Drive Trucks." (Pam, owner of a trucking company)

What words and images come to mind when *you* hear the word "entrepreneur"? Would you be as surprised as the customers of Jo, Alice, and Pam were by the identity of these business owners? Words that are often associated with "entrepreneur" are "risk-taking," "autonomous," "powerful," "knowledgeable," "independent" – all words that are more widely associated with being male than with being female in many societies today. It's not by chance, then, that the terms "entrepreneur" and "self-made man" are practically synonymous for many people: entrepreneurship in the USA, the UK, and many other places is coded as "male," as an activity that is appropriate for men but not unquestionably so appropriate for women. Precisely because "woman entrepreneur" is practically an oxymoron, we wish to explore how gender and geography are related to entrepreneurship. We do so in this chapter by focusing on women who run businesses in male-dominated sectors of the economy like trucking, engineering, or construction as a way to understand how women business owners are changing the gender of entrepreneurship.

Associations between entrepreneurship and masculinity persist despite dramatic increases in women's business ownership in recent decades. In 1972 women-owned businesses (WOBs) accounted for less than 5 percent of privately held businesses in the US, a figure that had jumped to 35 percent by 1992, and women are currently starting businesses at twice the rate that men are (U S Census, 1972, 1992; Fagenson and Marcus, 1991; SBA, 1996). In the UK, according to Carter (1993), the rates have been similar. Between 1981 and 1987 British women's rates of self-employment grew by 70 percent, such that by the early 1990s women accounted for 25 percent of all the self-employed in the UK (Carter, 1993). More recent UK data show that while men's rates of self-employment decreased by 2.5 percent between 1992 and 2002, women's rates of self-employment increased by 5.8 percent (Office of National Statistics, 2002).

Because almost everyone who starts a business has previously been employed in the paid labor force, one reason for the rise in women's business ownership in the USA and the UK is the increase in women's labor force participation in the past 30 years. That is, as women gain labor market experience, they are more likely to start businesses. In the USA, 60 percent of working-aged women (ages 16–64) were in the paid labor force in 2000, up from 43 percent in 1970. The comparable figures for the UK are 69 percent of working aged (16–60) women in the paid labor force in 2000, up from 59 percent in 1984 (Office of National Statistics, 2000). Moreover, most people who launch a business do so in a field in which they have had prior experience working as a paid employee. So, for example, a seamstress is unlikely to launch a construction company, and a college English professor is unlikely to start an accountancy firm. In addition, we know that the labor market in which business owners have gained their work experience is sharply segregated along gender lines, with women and men working in different *industries* (e.g. healthcare, manufacturing, retail), different *occupations* (e.g. nursing, truck driver, cashier), and different *jobs* (e.g. head psychiatric nurse, long-distance driver, entry-level cashier) (Reskin and Hartmann, 1986).

As a result, the kinds of businesses that women usually start differ substantially from the ones that men start, and the profiles of women's and men's businesses differ markedly. In general, women's businesses are disproportionately in the retail and service sectors, and they are smaller and less profitable than men's businesses. The kind of business most often associated with women – and the kind that is therefore seen as most appropriate for women if they are to own a business – is a small, home-based, part-time, craft-oriented retail business. In our own empirical studies we have found that women are still not welcome – and are often considered anomalous mavericks – as owners of businesses in male-dominated industries such as excavating, auto repair, or commercial real estate.

In this chapter we explore how some women entrepreneurs are challenging and changing the gender of entrepreneurship by running businesses in male-dominated arenas. We are particularly interested in how geography shapes processes of gendered entrepreneurship. Just as gender and geography shape wage and salary labor markets differently in different places (Hanson and Pratt, 1995), so too are gender and geography likely to shape entrepreneurship differently in different places, even different places within a single metropolitan area. The ideas and the evidence we present in this chapter come from our own studies of entrepreneurship in Worcester, Massachusetts. In the next section we briefly describe these studies so that you can understand the nature of the data we draw upon. Following the description of the data, we describe: (a) the nature of gender segmentation in entrepreneurship in the USA and in the Worcester area; (b) the role of gender and geography in shaping entrepreneurship; and (c) the ways in which woman-owned businesses in male-dominated industries are changing the meaning of gender in the places in which they are located.

Description of Our Studies

We have carried out two separate studies of entrepreneurship in Worcester, Massachusetts, an urban area about 50 miles west of Boston with a population of more than 510,000 in the year 2000. Megan Blake's study sought to understand the role of context in shaping women's participation in business ownership. To achieve this aim, she conducted in-depth personal interviews in 1996–7 with 30 agents likely to assist people in business ownership (e.g. bankers, lawyers, accountants, business counselors) and 16 women business owners. Questions posed to those who assist in business start-up focused on the nature of assistance provided, opinions regarding attributes of a successful entrepreneur, and impressions regarding differences or similarities between male and female business owners. Questions posed to business owners focused on the nature of the business; the start-up process; links to previous employment; any help or advice that the business owner received, sought, or did not seek; the perceived contributions of the business to the business owner's community; and the business owner's plans for the future of her business. The interviews were then followed by a mailed questionnaire sent to a larger sample of business owners in the Worcester metropolitan area.[1]

Susan Hanson's study likewise entailed in-depth interviews and mailed surveys. In 1998–9 some 200 men and women business owners in the Worcester metropolitan area were interviewed; questions focused on the start-up process, networks of support (both personal and professional), relationship of the business to the owner's previous job history, location decisions, problems encountered, and relationship of the business to the community. The mailed survey, which yielded 350 responses, covered similar topics but in much less depth. Both the interview sample and the survey sample were stratified by gender of owner and gender type of the industry based on proportion of the labor force that was female;[2] within each category (e.g. woman-owned businesses in male-dominated industries) owners were selected at random from a purchased list of business owners.

Gender Segmentation of Entrepreneurship

Before we consider gender segmentation in entrepreneurship, we need to reflect on what we mean by entrepreneurship. Almost all studies of entrepreneurship and self-employment discuss the difficulties of defining these terms. Aronson (1991,

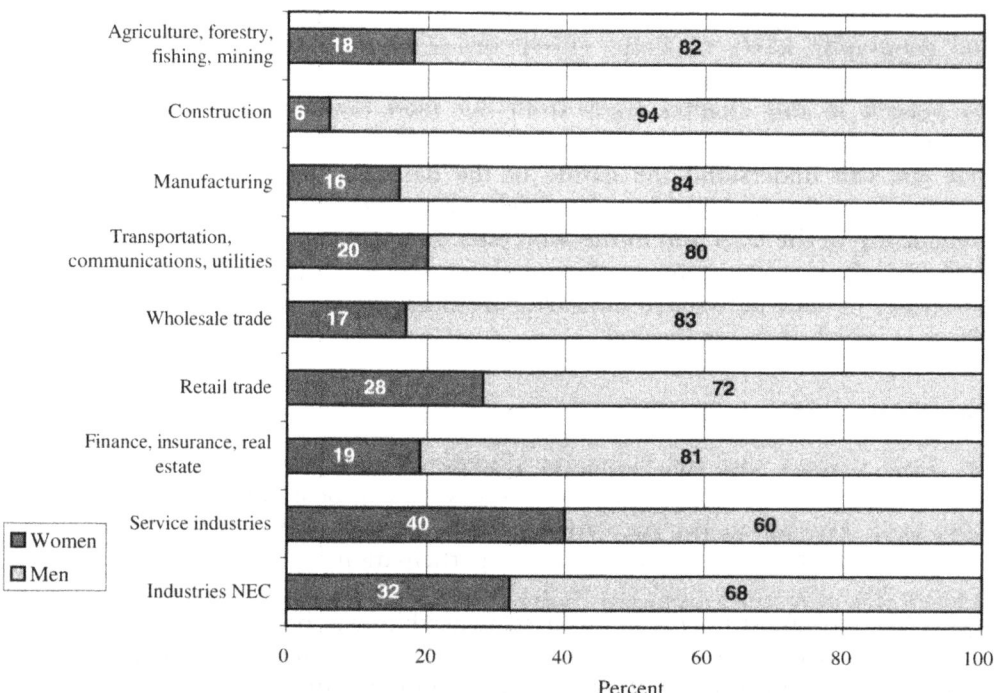

Figure 13.1 Business owners by gender, Worcester metropolitan area, 1997 (percentages).

p. xi) and others distinguish between entrepreneurship and self-employment, reserving the term "entrepreneur" for someone who innovates rather than someone who simply works for herself or who forms a new firm. Light and Rosenstein (1995) point to the difficulty in measuring innovation and call for a broadly inclusive conceptualization of entrepreneurship. We follow Lavoie, cited in Moore (1990, p. 276), by defining an entrepreneur as someone "who has taken the initiative of launching a new venture, who is accepting the associated risks and the financial, administrative, and social responsibilities, and who is effectively in charge of its day-to-day management."

In the USA and the UK the national government collects data on woman-owned businesses, but defining a woman-owned business is not as straightforward as it might seem. In fact, in the USA, the definition used in the gathering of government statistics changed between the 1992 survey of businesses and the 1997 survey, which is the most recent one available. In 1997, the US Bureau of the Census began distinguishing woman-owned businesses as those where at least 51 percent of the "interests, claims, or rights in the business" are held by women (US Bureau of the Census, 1997, p. 6).

The data collected by the US Department of Commerce are useful for comparing gender differences in patterns of business ownership. Using the 1997 data (US Bureau of the Census, 1997), figure 13.1 shows, for the Worcester metropolitan

Figure 13.2 Comparison of women's business ownership and employment, Worcester metropolitan area, 1997 (percentages).

area, the proportion of businesses in each industry type that are owned by women and men. You can see that men own well over half of all businesses in every industry, but considerable differences exist between, say, construction, where men own 95 percent of the businesses, and service industries, where they own "only" 60 percent. These patterns of business ownership in Worcester closely resemble those for the USA as a whole; that is, women are most likely to own businesses in the industry sectors (retail trade and services) that account for the most businesses overall, but they are noticeably underrepresented in the ownership of businesses in manufacturing, transportation, wholesale trade, and FIRE (finance, insurance, and real estate).

To what extent can these patterns be explained by women's employment histories? That is, does women's underrepresentation as business owners in certain industries match their underrepresentation as workers in those industries? Figure 13.2 compares the proportion of businesses in each industry that are owned by women (also shown in figure 13.1) with the proportion of persons employed in each industry who are women. The graph clearly shows that in no industry are women business owners represented in the same proportion as their proportion of the labor force in that industry. For example, women comprise more than half (51 percent) of the labor force in retail trade, and nearly two-thirds of the labor force in FIRE and in services (65 and 64 percent, respectively), but they own only 28 percent of retail businesses, 19 percent of FIRE businesses, and 40 percent of businesses in the service industry. Although the proportion of the labor force that is female in an industry does affect the proportion of WOBs in that industry ($r = 0.78$), notice that FIRE and services, with almost equal proportions of female employees, have strikingly different proportions of WOBs (19 versus 40 percent). What do you think accounts for the difference between these two proportions?

Our analysis in this chapter counts as a "male-dominated industry" any specific industry category, such as auto repair or plastics manufacturing, in which, according to the people we interviewed, at least 70 percent of the businesses are *owned* by men. Owning a business in a male-dominated arena provides a woman business owner with an income premium: just as working in a male-dominated occupation yields a higher income on average than does working in any other occupation (Hanson and Pratt, 1995), owning a business in a male-dominated field leads to higher gross sales and more personal income.

If women lack experience working in male-dominated fields, how do they become business owners in such areas? One important avenue is via taking a female-dominated job in a male-dominated industry. Sylvia worked as the office manager in a plastics manufacturing firm for many years; this position was an excellent vantage point from which to learn the business and to make valuable contacts with suppliers and customers, all of which enabled Sylvia to launch her own plastics manufacturing firm. Carol used her job as a file clerk in a web-based advertising firm to learn enough about that type of business to start her marketing consulting firm. Another way in which women become owners of businesses in male-dominated industries is to join with, or take over from, a male relative when he retires or dies. When her husband died, Gloria took over her husband's funeral home business after attending mortician school to become a certified funeral director (his uncle, also a funeral director, kept the business going during the two years Gloria was in training). Others, like Jo the civil engineer, had worked in male-dominated occupations in the industries in which they eventually launched a business.

A fourth avenue involves forming alliances with those men or women who already have the necessary skills in the male-dominated industry. For example, Pam purchased an established trucking business from a man who was retiring because of illness. What she got for her money was his truck, all his existing customers, and advice on how to run her business. Kate, who owns a construction firm, initially partnered with established male builders and thereby apprenticed while she started her business.

How Gender and Geography Shape Entrepreneurship

Traditional approaches to entrepreneurship suggest that an individual spots an economic opportunity that is not currently being met within the market and then creates an organization to pursue the opportunity (Bygrave and Hofer, 1991; Herron et al., 1991). Recent research on entrepreneurship argues, however, that geographic context is likely to affect the birth rates and longevity of firms. Every entrepreneur starts a business somewhere, and the structures and institutions that constitute the places where individuals begin their businesses encompass the local availability of resources (e.g. money and information), the local state, industrial structure, and mediated social relations (Brush, 1992). In this section we focus on two interconnecting gendered dimensions of context that affect women's participation in male-dominated entrepreneurship activities: (a) resource systems; and (b) local social structures and gender cultures.

Resource systems

Resources, such as access to capital and information, circulate through social networks that are often group-specific (Portes and Zhou, 1992; Malecki, 1994; Granovetter, 1995; Light and Rosenstein, 1995; Zhou, 1996). Social networks, therefore, are important to entrepreneurship, and the makeup of these networks can influence gender-based segmentation in business ownership (Barrett, 1995; Blake, 2001). A key part of the entrepreneur's network in the start-up phase, according to those who aid entrepreneurs in this process, is what one lawyer in Blake's study called a "community of resources" (COR). A COR can include a small business development counselor, a banker, a lawyer, and an accountant. Because the COR has the information and resources that a person needs to start a business, COR members act as gatekeepers to entrepreneurship and help to shape the gender-based segmentation in entrepreneurship that we described in the previous section.

The business owner needs to maintain the support of the COR to establish legitimacy with it (Arnold et al., 1996). Establishing legitimacy involves conforming to a set of norms, rules, and requirements (Aldrich and Fiol, 1994; Arnold et al., 1996). At start-up, COR agents measure the would-be entrepreneur against legitimizing standards, which are often grounded in gender stereotypes. One loan officer at a bank, for example, told the story of a woman who wanted to start a business selling automobile antennae flags to car dealers. He said he was the only banker of five to support the loan and indicated that the real reason the others had denied her a loan was that they could not envision her doing the type of male-dominated business she was proposing. He described her as "a woman in a man's world." Another banker, who had underwritten a loan to a woman to launch a cleaning business, said that his bank would have had "a greater degree of difficulty granting her a loan if she had wanted to start a business fixing cars" because customers are not likely to patronize female auto mechanics and therefore such a business would be doomed to failure. In fact, this woman's previous work experience had been as an auto mechanic! Construction was another sector that COR agents identified as being reserved for men, while "services" and "beauty shops" were perceived as suitable for women.

These perceptions of COR agents as to what kinds of businesses are, and are not, appropriate for women shape women's access to entrepreneurial resources and thereby shape the gender of entrepreneurship in places. Paula (who owns an auto repair business with Sharon) was on the receiving end of gender-based discrimination by COR agents. She discusses how difficult it was for her to find a building and buy equipment:

> People did not take us seriously. For example, it took us eight months to find a location; six of these months were finding a realtor that would work for us. Equipment vendors would not give us the time we needed at the beginning in order to assess the equipment and decide which equipment was best for us. The one that did we buy from, we continue to do so. We only do business with people who took us seriously before we were a big account. . . . I would say it directly related to their assessment that we were not going to make it and they made that assessment because we were women.

The quotes from the first banker and from Paula also illustrate two of the ways that resources may be withheld. The first is through outright denial or refusal to provide financing or information. While it is illegal in the United States for bankers to deny loans on the basis of gender, they can focus on a credit blemish that they would otherwise overlook as a justification for denial of the loan. The second way that resources may be withheld is through discouragement. Evidence from other research shows that women initiate the process of applying for loans as often as men do (Aldrich et al., 1995), but Blake's study indicates that a large proportion of women do not complete the application process. Only 25 percent of the survey respondents used a bank loan to start their business, but more than half of the women business owners (59 percent) had contacted a bank manager to discuss business credit. Of those who had contacted a bank officer, a staggering 46 percent did not, in the end, complete the application. Pam, who owns a trucking firm, provides an explanation for why women do not complete the application:

> [The loan process] was discouraging. . . . Often I never got as far as filling out the loan papers at the banks that said no. I went in and sat down and told them what I was looking for, and they told me "you can fill them out if you want too, but . . ."

This procedure for denying loans is widely used by bank officers in the United States largely because they must report loan acceptance rates by gender and race to the government. If a bank is found to be turning down a higher proportion of women and minority applicants, it is fined for committing racial or gender discrimination. Banks do not have to report the number of loan inquiries, however. By discouraging people from making an application, the bank's rate of approval increases. By discouraging women from applying for loans, the bank cannot be accused of discriminating on the basis of differential approval rates of completed applications for men and women.

Local social structures and gender cultures

Because the norms, rules, and requirements that define legitimacy are established and interpreted locally (Lomi, 1995), the ease with which women are able to establish themselves and gain access to resources varies from place to place, sometimes at a very fine spatial scale. For instance, Pam discusses the way she eventually was able to get the loan she needed to acquire her trucking firm.

> I went to the Small Business Development Center and spoke with the counselor. . . . She was very helpful. . . . When I was going for a loan she helped me to work on a business plan. I was able to get enough to keep me going. . . . She told me to stick with a small-town bank like the one in Spencer [a semi-rural suburb of Worcester], which is what I did.

Responses to Blake's mail survey reveal similar urban/suburban differences concerning the ability of women to gain access to financing. Women in Worcester's suburbs were less likely than women in the City of Worcester to be discouraged from applying for a bank loan. Among loan seekers, 22 percent of the urban women (versus 17 percent of suburban women) succeeded in getting a loan only after they experienced a failure or discouragement. Moreover, one-third of urban women who sought a bank loan, compared to 27 percent of suburban women, failed to secure any loan at all.

This spatial variation in access to resources is related to the way that business communities define themselves through their location and through the customers they serve. Location is a symbolic indicator of the market orientation of the firm (McDowell, 1997). COR agents located in the central business district of a city are often targeting customers that they (the COR agents) believe will enhance their status and position within a key market as identified by the city's economic development agents. In the case of Worcester City, these customers include the high-technology and manufacturing sectors, sectors that are owned primarily by men and employ mainly men. One lawyer summed up the relationship between a firm's customers and the identity of accountancy and legal firms: "I will spend more time with those clients that will benefit [how others perceive] my business." Lawyers indicated that they would give "less important" work to the paralegal to handle, while focusing their own time on those customers with more "sophisticated needs."

Joanne, the owner of an IT firm, had been treated this way until she was able to find a lawyer who matched her in terms of market identity:

> It is noteworthy that initially [my lawyer in Worcester] wondered whether or not she had the background, and she referred me to a firm in Boston that specializes in intellectual property. This firm was a big law firm, very prestigious, and I never got a response from them. I kept e-mailing my attorney [in that Boston firm]. She said "your contracts are in the mail. Your contracts are in the mail." Meanwhile, [my lawyer in Worcester] and her firm were doing personal work for me. Finally, I just said to her, "you are so responsible, please work for me." . . . I get more attention from her [than from the Boston firm]. The others were unhelpful because I was a small fish.

The business community in many of Worcester's suburbs (unlike that in Worcester City) supports locally oriented firms. Paula (auto repair) illustrates this:

> We are members of the Chamber of Commerce and as Ann [president of the local Chamber] mentioned to you, [this town] is very proactive in encouraging people to do business locally, through the Welcome Wagon and the Chamber. A lot of people in [our town] will come to us, they heard about us opening and want to patronize a local business. . . . So our success is a combination of being women and being local.

This local orientation is, moreover, often more accepting of women holding important roles in the business community. One female Chamber of Commerce president told me that it is much more likely that a woman will run her local Chamber than is the case in larger urban areas. The orientation of the business community helps to create an environment that is more conducive not only to women's overall participation in entrepreneurship, but also to women's ownership of businesses in male-dominated sectors. Compare this woman's comments about the Worcester Chamber of Commerce to those expressed by Paula, above, about a suburban Chamber: "Worcester is a large politicized organization, unfriendly to women. Everyone in power positions is male. If you read their monthly newsletter, you will see that men are noted for their business accomplishments and women are noted for their social accomplishments." This quote illustrates that in addition to market orientation, local gender relationships and associations affect the degree to which women are accepted into a business community. For instance, Paula discussed how she and Sharon chose the location for their auto repair business.

> What I found was in the area in which I lived, which is south of Worcester bordering on Rhode Island, there were many people that thought a business like this would not work. It is a very Italian area, so I looked for an area where there were more women who took responsibility for their own cars, as well as have the money to pay for repairs. That is how I targeted [this] area. Neither one of us live here. We live between 20 and 30 miles away. We found this area by looking for independent women. I was looking for a certain population density and an area that would have a lot of businesses that employed women in other than your traditional pink collar occupations. With the high technology and I knew that there were a lot of women in this area who worked in the area or lived in the area or both.

Changing the Gender of Entrepreneurship in Place

Entrepreneurship helps to promote local economic development by supplying places with diversity, flexibility, and innovative energy that are difficult to sustain within larger, more structured organizations (Malecki, 1994; Peck, 1996). A dynamic and diverse small-firm sector generates, in turn, greater local entrepreneurial vitality. Entrepreneurship also contributes to local employment growth and the local tax base (Light and Rosenstein, 1995). Women's entrepreneurship supports local development in all of these ways, but, in addition, through entrepreneurship, women gain autonomy and control over their working lives (Goffee and Scase, 1983; Stevenson, 1988; Carter and Cannon, 1992). Moreover, as we have shown, through entrepreneurship in male-dominated sectors, women gain even more financial independence than they would have by owning a business in a different sector. In addition, our research suggests that the *kind* of economic development that entrepreneurship nourishes in a place depends upon the ways in which entrepreneurship is gendered there.

The contributions of women entrepreneurs to the places in which their businesses are located are not limited to providing jobs, paying taxes, and creating income for themselves and their families. By choosing entrepreneurship over wage and salary employment, women are helping to change gender relations in the local labor markets in which their businesses are located. Women who own and run successful businesses in male-dominated industries play an especially important role in reshaping the gender contours of entrepreneurship and the gender contours of their local labor markets; such women do this in several ways. One is by changing opportunities for women in the places where their businesses are located by hiring women as employees and adopting labor practices that differ from those of male business owners. Another is by serving as mentors for would-be female entrepreneurs. A third way is by confronting, challenging, and eroding gender stereotypes in their everyday lives on the job. We take up the first two of these ways briefly and then explore the third in some detail by listening to the experiences of some women entrepreneurs who work in a "man's world."

According to Brush (1990), women business owners improve employment opportunities for women because they are more likely than their male counterparts to employ women. When women who own businesses in male-dominated sectors hire women, they open opportunities for those women to become familiar with a segment of the labor market that, by definition, has been relatively closed to women and that generally commands higher pay. Hiring women to work in male-dominated industries is one way in which business owners increase the likelihood that more women will follow in their footsteps. Compared to their male counterparts, women entrepreneurs also tend to follow different labor practices and have different relationships with their employees (Brush, 1990); women, for example, are more likely to adopt non-hierarchical management strategies and incubator relationships with associates (Carter and Cannon, 1992). Self-employment itself is a different form of labor practice for these women because they are taking control of their own employment (Aronson, 1991).

Moreover, whether they are aware of it or not, women business owners are role models, opening up opportunities for other women to become business owners (Portes and Zhou, 1992). Hanson's interviews in Worcester revealed that women business owners serve an important role as mentors for other would-be or nascent women entrepreneurs; fully two-thirds of women business owners in the interview sample had consciously mentored other entrepreneurs, and 35 percent of those who had been a mentor said that they had mentored only women. Here is another way in which women entrepreneurs working in a male-dominated arena are changing the gender of entrepreneurship.

In our interviews, women who had launched businesses in male-dominated sectors recounted again and again the myriad ways in which they were, consciously or not, challenging, confronting, and eroding gender stereotypes every day on the job. Several noted with some irony, and usually with a smile, that through such confrontations they had changed the attitudes of their (mostly male) customers. The civil engineer, Jo, whose story opens this chapter and who started her own engineering firm 25 years ago, is one such example. One of the main reasons she left the large firm she had been working for to launch her own business was the "attitude toward women" she had encountered in that firm: "I really didn't like their attitude toward women, so [when that firm merged with another company] I said, 'That's it; I'm out of here'." She continued:

> A lot of people are very surprised when they find out an engineering business is owned by a woman, especially 25 years ago. . . . There weren't very many women civil engineers then. I have some great stories about the chauvinist pigs. It was very difficult in the beginning. But it has gotten much better. You know, you kind of have to prove yourself. I've walked on sites and guys have said to me, "I'm not working with a broad and I'm not doing this and I'm not doing that." I had one particular client that did that to me. Now, ten years later I'm still doing all his engineering work. We had quite a session the first day we met, but you know, it's getting better over the years. You do have to prove that you can do it.

Donna, who runs an appliance repair business, made a similar point about the attitudes of her male customers:

> Because I have to ask technical questions over the phone, I sell parts, they say to me "boy you are pretty knowledgeable, kind of smart for a woman." You know I get this question at the other end of the line, "how come you know so much?" Gee whiz, at the same time I'm getting accolades because "Gee you've been so good with me . . ."

> This is coming from a man. They are still very unsure about women having a brain. It's true. These are the male customers not the service representatives. Female customers probably relate to me and I know I'm helpful, extremely helpful.

Ruth runs a diner, which, as she points out, is traditionally considered a man's form of entrepreneurship. She explains that some men feel free to harass women in diners and that she will not tolerate such harassment in her establishment.

> One thing I find is being a woman owner, a diner is typical of men. I'd say 90 percent of diner businesses are owned and operated by men. You find a lot of, um, you get a little bit of harassment, and that part I don't like. Even though you own it, a lot of men associate diners with less respect to women in the food business; in a diner especially, and that's where you have to keep them in line, keep yourself above their remarks. I find there's a lot of discrimination as far as a woman in the diner business being the owner. Because most of them are owned by men. And half of my customers are women now, and only because I'm a woman owner. Prior to that, it was mostly men.

We asked, "So, women don't want to come into a diner that's man-owned?" and she replied, "Right. Now that it's woman-owned, and I own it, and I don't tolerate any abuse. I'm highly respected for that." By changing the behaviors (and perhaps the attitudes) of her male customers, Ruth has created a woman-friendly diner, a safe place for her women customers, a place that is changing the meaning of gender in her working-class community.

Other women with whom we spoke pointed out that their businesses often benefited from their position of being a woman running a business in a traditionally male field. It is interesting that this point often came out in response to the question "What do you see as innovative about your business?" Sandy, for example, who sells and services vacuum cleaners, said in answer to this question:

> Most of it is the fact that I am a woman doing this. Every other vacuum shop in the area is – you know, it's just something you think of an old man standing in the door . . . fixing vacuum cleaners. And the other thing is that, because I owned a cleaning service [before starting this business] and because I am a woman, people actually trust what I tell them as far as what would work and what they need. I try to sell them only what they need, and I'm not trying to sell them a particular product. Like the lady that was just here, she had a Kirby salesman [try to sell her a] $1200 to $1800 dollar vacuum cleaner. She doesn't need that.

Viola, who runs a contracting firm, also spoke of the advantages of being a woman in a male-dominated industry:

> I deal with mostly office people. I deal with a lot of women that are either brokers or office personnel that run the relocation divisions. Not just women, both men and women, I guess. But my ability to talk with them and get along real well [is important]. They feel comfortable. . . . Contractors in general have a very bad reputation of being real rough around the edges, and where I came from, not construction, I didn't grow up with a hammer in my hand like most of the contractors out there did. They can be a little rough around the edges and I'm not, and that's what puts a lot of people at ease.

The owners of both of the woman-owned auto repair businesses in our studies emphasized that many of their customers choose to patronize their shops because they believe that women are more trustworthy than men. When we asked "What do you see as innovative about your business?" one replied, "mainly honesty, and comfort level for women coming in for car repairs." Paula and Sharon almost flaunt the female ownership of their auto repair business: "Yeah, it's like if the name doesn't scare [them] off, then the pink door definitely will give them a clue that if they don't want a woman working on their car, they need to go next door."

Conclusion

The quotes with which we opened this chapter reveal the gendered assumptions that people make about entrepreneurship: women don't run engineering firms, own furniture stores, drive trucks. The data in figure 13.1 show just how unusual it still is for women to own businesses in certain industries. Our interviews with entrepreneurs in Worcester, Massachusetts, make clear some of the reasons for the dearth of women entrepreneurs in certain (typically male) arenas. COR agents are key gatekeepers, determining which would-be and nascent entrepreneurs will gain access to the resources they need to launch a successful business. Many COR agents make resource allocations on the basis of gendered stereotypes that rule out the possibility of successful woman-owned businesses in trucking, auto repair, and the like. These COR agents simply cannot believe that such businesses will be able to attract enough customers to survive, so their denial of resources like a bank loan is based in assumptions about gender norms in local places. We found that local social structures and gender cultures, as they affect women's entrepreneurship, vary between city and suburb, with suburban areas being more amenable to women's business ownership.

Our Worcester interviews also demonstrate that, although they may not be numerous, women who own businesses in male-dominated fields are changing the gender of entrepreneurship in significant ways. Simply through their daily interactions with their employees, customers, suppliers, and other business owners, these women are subtly – and often not so subtly – challenging gender stereotypes and forcing others to accept their competencies, recognize their skills, and understand that "you don't have to be a certain gender to do a certain job." In doing so, they are opening opportunities for other women, who, like Pam, the owner of a trucking firm, want the chance to do what they enjoy doing. Perhaps in your lifetime, "typically male" and "typically female" lines of work and types of businesses will become obsolete, and gender will not shape access to opportunity. The next time you hear about Mary who owns a construction company or Lydia who runs an automobile towing service, think of that!

ACKNOWLEDGMENTS

Susan Hanson's study was funded by the National Science Foundation (award number SBR 9730661), the Sloan Foundation, and The William and Flora Hewlett Foundation (grant no. 2000-5633), which supported her fellowship at the Center for Advanced Study in the Social and Behavioral Sciences, Stanford, CA, 2001-02. Megan Blake's study was funded by the National Science Foundation (award number 9811231), and the Piper and Pruser-Holzhauer Fellowships, both from the Clark University School of Geography.

NOTES

1   The response rate for the survey was 28 percent. Of those who did not respond to the initial survey, 40 were asked an abbreviated set of questions over the telephone. Postal and telephone questionnaires covered similar topics to the interviews but in much less depth.

2   Because data on gender of ownership are not available by industry, it is not possible to identify those industries in which ownership is predominantly male or female. Therefore we used employment data (percentage of workforce in industry that was female) to define male-dominated (employment less than 30 percent female) and female-dominated (employment more than 70 percent female) industries.

BIBLIOGRAPHY

Aldrich, H., Elam, A. and Reese, P. (1995) Strong ties, weak ties, and strangers: do women owners differ from men in their use of networking to obtain assistance? Working paper 4, Small Business Foundation of America, Washington, DC.

Aldrich, H. E. and Fiol, C. M. (1994) Fools rush in? The institutional context of industry creation. *Academy of Management Review*, 19, 645–70.

Arnold, S. J., Handelman, J. and Tigert, D. J. (1996) Organizational legitimacy and retail store patronage. *Journal of Business Research*, 35, 229–39.

Aronson, R. L. (1991) *Self-employment: A Labor Market Perspective*. Ithaca, NY: ILR Press.

Barrett, M. (1995) Feminist perspectives on learning for entrepreneurship: the view from the small business. In *Frontiers in Entrepreneurship Research*. Wellesley, MA: Babson College.

Blake, M. (2001) It takes a village: women's entrepreneurship in Worcester, Massachusetts. PhD dissertation thesis, Clark University, Worcester, MA.

Brush, C. (1990) Women and enterprise creation. In *Enterprising Women: Local Initiatives for Job Creation*. Paris: OECD.

Brush, C. (1992) Research on women business owners: past trends, a new perspective and future directions. *Entrepreneurship Theory and Practice*, 16, 7–30.

Bygrave, W. D. and Hofer, C. W. (1991) Theorizing about entrepreneurship. *Entrepreneurship Theory and Practice*, 15, 13–22.

Carter, S. (1993) Female business ownership: Current research and possibilities for the future. In S. Allen and C. Truman (eds), *Women in Business: Perspectives on Women Entrepreneurs*. London: International Thompson.

Carter, S. and Cannon, T. (1992) *Women as Entrepreneurs*. London: Academic Press.

Fagenson, E. A. and Marcus, E. C. (1991) Perceptions of the sex-role stereotypic characteristics of entrepreneurs: women's evaluations. *Entrepreneurship Theory and Practice*, 14, 33–47.

Goffee, R. and Scase, R. (1983) Business ownership and women's subordination: a preliminary study of female proprietors. *Sociological Review*, 31, 625–48.

Granovetter, M. (1995) The economic sociology of firms and entrepreneurs. In A. Portez (ed.), *The Economic Sociology of Immigration: Essays on Networks, Ethnicity, and Entrepreneurship*. New York: Russell Sage.

Hanson, S. and Pratt, G. (1995) *Gender, Work, and Space*. London: Routledge.

Herron, L., Sapienza, H. J. and Smith-Cook, D. (1991) Entrepreneurship theory from an interdisciplinary perspective: 1. *Entrepreneurship Theory and Practice*, 15, 7–12.

Light, I. and Rosenstein, C. (1995) Expanding the interaction theory of entrepreneurship. In A. Portes (ed.), *The Economic Sociology of Immigration: Essays on Networks, Ethnicity, and Entrepreneurship*. New York: Russell Sage Foundation.

Lipman-Blumen, J. (1984) *Gender Roles and Power*. Englewood Cliffs, NJ: Prentice Hall.

Lomi, A. (1995) The population ecology of organizational founding: location dependence and unobserved heterogeneity. *Administrative Science Quarterly*, 40, 111–14.

McDowell, L. (1997) *Capital Culture: Gender at Work in the City*. Oxford: Blackwell.

Malecki, E. J. (1994) Entrepreneurship in regional and local development. *International Regional Science Review*, 16, 119–53.

Moore, D. P. (1990) An examination of present research on the female entrepreneur – suggested research strategies for the 1990s. *Journal of Business Ethics*, 275–81.

NFWBO (1996) *Giving Something Back: Volunteerism among Women Business Owners in the US*. Silver Spring, MD: National Foundation of Women Business Owners.

Office of National Statistics (2000) *Labour Market Statistics* (http://www.statistics.gov.uk). Accessed March 14, 2003.

Office of National Statistics (2002) *Labour Market Statistics* (http://www.statistics.gov.uk). Accessed March 14, 2003.

Peck, J. (1996) *Work Place*. New York: Guilford Press.

Portes, A. and Zhou, M. (1992) Gaining the upper hand: economic mobility among immigrant and domestic minorities. *Ethnic and Racial Studies*, 15, 491–522.

Reskin, B. and Hartmann, H. (1986) *Women's Work, Men's Work: Sex Segregation on the Job*. Washington, DC: National Academy Press.

SBA (1996) *The Third Millennium: Small Business and Enterprise in the 21st Century*. Washington, DC: Small Business Association (www.sba.gov).

Stevenson, L. (1988) Women and economic development: a focus on entrepreneurship. *Entrepreneurship and Economic Development*, 18, 113–26.

US Bureau of the Census (1972) *Economic Census: Women Owned Businesses*. Washington, DC: US Bureau of the Census.

US Bureau of the Census (1992) *Economic Census: Women Owned Businesses*. Washington, DC: US Bureau of the Census.

US Bureau of the Census (1997) *Economic Census: Survey of Women-owned Business Enterprises*. Washington, DC: US Bureau of the Census.

Zhou, Y. (1996) Inter-firm linkages, ethnic networks and territorial agglomeration: Chinese computer firms in Los Angeles. *Papers in Regional Science*, 75, 265–91.

Chapter 14

Gender and Empowerment: Creating "Thus Far and No Further" Supportive Structures. A Case from India

Saraswati Raju

Introduction

In feminist movements across the world, discussions around women's "empowerment" started in the latter half of the 1970s when Third World feminists and women's organizations first invoked the concept. "Empowerment" through a transformation of economic, social, and political structures at national and international levels was explicitly used to frame the struggle for social justice and for women's equality as feminist scholars and activists started to formulate a comprehensive framework within which developmental processes could be placed and reworked.

Empowerment has been defined as a process of "undoing internalized oppression," and in the case of women's empowerment it is also about changing social and cultural forms of patriarchy that remain the sites of women's domination and oppression. In feminist theoretical frameworks, along with recognizing women's oppression and subordinated location in specific family, community, market, and state structures, women's agency as a catalyst for change is also recognized. Moreover, the *process* of empowerment has to operate both intrinsically and extrinsically, whereby women become self-confident and aware about their own capabilities, transforming and deconditioning age-old images of powerlessness to gain access to resources of all kinds. Empowering processes should also enable them to challenge gender and social power relations.[1] That is to say, an empowerment process that simply facilitates women to gain access to resources, but does not aim at redefining existing social and political power structures, can indeed be quite counterproductive or disabling (Mohanty, 2001). Feminist notions of empowerment thus see women as proactive agents and not as passive beneficiaries, clients, or participants in social change.

In terms of international development, in recent years the feminist vocabulary has been co-opted by the mainstream (GOI, 1995), and state-led developmental interventions – the result of bureaucratic initiatives – now typically have a marked gender focus often at the behest of foreign funding that makes it mandatory. Of late, such projects have been conceptualized well in identifying various stages towards women's empowering processes.[2] However, when it comes to implementing – "doing" – empowerment on the ground, the majority of the programs and projects have facilitated women's access to productive resources and encouraged their active participation in development processes by promoting an instrumentalist logic (i.e. women's development as an input in development *per se*), but has failed to realize that women have everyday lives in families, which are predominantly controlled by men, with women having little power to change that (Jandhyala, 2001; Chopra, 2003). Thus, at times the outcomes have been such that empowerment in one sphere may often mean disempowerment in another (Nagar and Raju, 2003).

This chapter is an outcome of the author's recent engagement with gender and empowerment issues and their embeddedness in the Indian context, more specifically in northern India, which is known to have a relatively more rigid gendered social space (Karve, 1965; Sopher, 1980; Dyson and Moore, 1983; Kishore, 1993; Raju and Bagchi, 1993; Agarwal, 1994; Uberoi, 1994; Murthi et al., 1995; Chen, 1995; Raju et al., 1999; Menon-Sen and Shiva Kumar, 2001; Dreze and Sen, 2002; Rustogi, 2003).

Drawing on an impact assessment of an integrated women's empowerment project in northern India, this chapter addresses several contested issues. It is argued that despite creating "supportive influences" (Dreze and Sen, 1997) and extending the "outer boundary of the existing power structure" (SinghaRoy, 2001) in empowering women, several strategic issues have remained unresolved. This failure, in the author's opinion, stems from a misplaced notion about women's empowerment that is not inclusive of broader economical, social, and political structures and participation by stakeholders other than women alone. It concludes that gender issues are complex and require spatially contextualized strategic interventions, as there cannot be a metanarrative or a single blueprint for gender-inclusive public policy.

In any evaluation, particularly those done for governmental organizations, conventional hardcore data and validity issues become important. At the same time, the author is aware of postmodernist critiques of quantitative data production for their *a priori* assumptions. Thus the challenge has been to adopt a combination of quantitative and qualitative methodologies in such a way that the inadequacies of both could be minimized. That this reflects the complications of a researcher located at the crossroad of intellectual traditions with engagements across disciplinary boundaries is an issue that needs discussion, but that task is outside the purview of this chapter.

Impact Assessment of Interventions: Search for the Appropriate Methodology

We are studying a women's development project in northern India that ran between 1995 and 2002; the project included 174 villages in two phases (1995–8 and 1999–2002), and was funded by a bilateral foreign donor and the state government. Given the consistently poor record of the project area in terms of women's position, the project aimed at improving the status of women by generating awareness about health, civic amenities, legal and livelihood matters, and rights, with the formation of women's collectives to act as initiators and leaders of change.

My role in this was to conduct an impact assessment just as the project was about to be over and advise the respective agencies whether or not there should be an extension of the project. Along with a research team consisting of my postgraduate students and a colleague from a local college in the project area, I spent a few months in the project area in 2001, assessing the nature, degree, and extent of the efforts and impact of this empowerment project on women.

Development projects have often banked on creating a collective subjectivity or what might be called a new social identity through the formation of women's collectives to enhance their bargaining power in public space. The project that I examined was organized in this fashion: the project area had small groups of women with one leader-motivator in each village – creating a slot in women's daily routine and a place to assemble (Mahila Samakhya, 1996) through provisioning for regular meetings among the members. The leader-motivators were put through extensive training and awareness-generation camps and went through periodic refresher courses. The idea was that this cohort of leaders, along with a select group of women who volunteered to form groups in each village, would work as change agents to create an enabling environment in which women develop collective bargaining power to deal with practical and strategic needs and thus improve lives for women in general through what can be termed a "ripple effect."

For assessing the impact, conceptually the stylized model of intervention can be thought of as forming concentric circles:

1 The innermost circle, i.e. an information/awareness-rich *core group* consisting of the leader-motivator and the select volunteer women members.
2 An information/awareness-rich *extended group*, i.e. women who had continuous interaction with the members of the innermost circle.
3 An information/awareness-poor *excluded group*, i.e. the remaining women in the village and thus outside the information/awareness loop.

The idea was to see how empowering information moved from innermost to extended to outermost circles.

In the absence of comparable pre-intervention and post-intervention temporal data, such a tripartite classification helped us to attempt a cross-sectional comparative analysis. For this, the study obtained quantitative and qualitative data on all three groups of women through several channels. We conducted: a semi-structured questionnaire-based survey of 380 households, with the questions largely drawn on the basis of project statements;[3] focus group discussions; and in-depth interviews with key stakeholders including village elders, panchayat (decentralized village-level institutional arrangement for governance) members, state and district officials, political representatives, police personnel, and grassroots functionaries from the Departments of Health, Education, and Women and Children. Additionally, we conducted a content analysis of 120 "activity registers" of the leader-motivators in which they had kept a mandatory account of issues discussed and activities undertaken in their monthly meetings; these provided a fairly good account of how frequently specific issues were being taken up for discussion.[4]

However, since we were charged with conducting an impact assessment – which connotes a "before" and "after" – and the requirement of the study was such that observations had to be "validated," the key question remained as to how to sift through changes that might have occurred due to other social processes. Our biggest methodological challenge was how we might be able to "attribute" social changes to the project's interventions *per se*. If, for example, the sampled households within the three subgroups differed in their basic social attributes (one group was better-off than the others in its initial endowment in terms of caste location, education, income, and so on), it could be argued that it was the built-in economic and social differences at the base between the groups and not the project's beneficial intervention that had created the change or the illusion of change. In order to overcome this, we first had to conduct a comprehensive demographic assessment of the sample population in the project area. We developed a social matrix sample, including information on caste, education levels, and household income, and compared these results for women and households across the three "circles" of project involvement. From this study, we concluded that there were no significant prior differences among the households in the three circles or groups.[5]

Having established that there were no prior categorical sociodemographic differences among the groups, we then moved on to the specific study of project impact: if the observed differences in the level of information, awareness, and practice between the subsets (i.e. those in the core group and the extended group on one hand and the excluded group on the other) are significant, the project could claim that the enhanced levels of information, awareness, and practice in the former groups were due to the project interventions. We tried to assess the degree of "significant" differences quantitatively *and* qualitatively – for example, we counted outcomes (such as number of childhood immunizations or number of women with bank accounts across the three groups) and we also interviewed women in all three circles of relationship to the project. The result is an assessment that includes both "hard data" and a qualitative assessment of intervention success.

Group Mobilization and Women's Agency

One of the significant steps in empowerment is a shift from "I cannot" (a state of powerlessness) to a state of "we can" (a state of collective self-confidence); empowerment increases women's sense of agency, or what has been termed as "the power within" by feminists (Kabeer, 1999). In moving toward a state of empowerment, the formation of collectives is emphasized because collectives provide a platform to women where they can speak up and discuss matters that concern them individually as well as collectively and in the process relate to others in their struggle.

Weekly meetings of women in the core group had indeed provided a space for women to meet and to access and share information on wide-ranging issues – from sex-selective abortion to alcoholism – that require interventions. These women had also been able to manifest the collective courage externally in their interaction with the outside world and were quick to point out the changes they were able to facilitate as a result – from accessing and managing drinking water, to arranging for getting village streets cleaned to arranging loans, getting money for weddings, and getting old-age pensions to widows. The core group women, through interaction with and encouragement from leader-motivators, also took advantage of existing public goods in the area of health, finances, education, etc. For example, a leader-motivator quotes from her daily diary:

> I helped a widow in getting [financial] help for marriage of her daughter by first approaching the Block Development Officer and on his suggestion the departments of Social Welfare and then of Rural Development. I was told that only those "below the poverty line" are eligible for any such help. I was directed to other officials but with no avail. This continued and I had to run from pillar to post, but I was adamant and instead of giving up, I tried every avenue as directed. Ultimately it was the Red Cross office where I managed to get some help. Although it took quite a bit of time and efforts in this whole process, it was a learning experience for me that I would use in dealing with similar cases. Depending upon problems women tell me I also direct them to Anganwadi and help the elderly in their pension cases.

That women have changed is a perception shared by men. The senior state officers see village women who were earlier shy in dealing with the outside world flash their identities of motivator-leaders as "passports" to facilitate their access to almost all government departments and a much higher visibility of village women in their daily public grievances meeting. Village men see them as better "home-makers and managers." The interpersonal familial perceptions are, however, fraught with tensions, a point we take up later in the discussion.

Whether increased mobility and contacts with the outside are synonymous with an empowering process for women can be a contested issue. In the Bangladesh context, Naila Kabeer (1999) has questioned equating the mobility of Bangladeshi Muslim women with an empowering process, as moving in public spaces may be contrary to their religious sensibilities. While this question has to remain open, in our project area women reported that one of the most important changes in their lives was their enhanced ability to move out of their homes and attend to a variety of work on their own. Crossing over to the public domain may seem inconsequential at the outset, but what it means to women themselves in terms of self-esteem and dignity and confidence is important (Dave and Krishnamurty, 2000).

Many group processes have helped to change the lives of individuals; however, empowerment as a group cannot be delinked with empowerment at the individual level and vice versa. This interlinked process is quite visible in individual women's realization of personhood. Many women (in the core and extended groups) took pride that they can now write their names instead of putting their thumbprints. When we asked them during interviews, 'How does this matter, since you can still not read or write?' it was important to them, they said, that they are now recognized as individuals and that their identities are no longer masked under anonymous thumbprints. They also narrated how before the project some of them would not even talk face-to-face with strangers – and how that has changed (Nagar and Raju, 2003). At the existential level, this shift is a giant leap forward for women, as one leader-motivator said:

> When I organized a meeting with members to talk about the ills of dowry and urged them to pledge against accepting or giving dowry in our children's wedding, a few members said, "it is simpler said than done." I decided I would do it myself in order to set an example. I married my daughter without dowry, exchanging symbolically only

Rs1 and a coconut. I also fixed a few such marriages within my immediate family. That did the trick and several others followed.

The confidence and enhanced self-esteem that individual women experienced had impacts in many areas of their lives. A woman in the project area remarked, "yes, he is my husband and I listen to him, but how long can I go on listening? I now tell him, 'I am going to do this or that.' It is better if he says yes because in any case I would go ahead and do what I think is right. By saying yes, he can avoid being openly defied – good for him. He would not admit it, but let me tell you he does not really object." This is an example of how power relations are expressed not only through direct agency and choice, but also through strategic moves whereby women "allow" their husbands to retain their supremacy in a non-confrontational move and still get things done as per their own wishes (Agarwal, 1997; Kabeer, 1999; Dave and Krishnamurty, 2000).

Issues of Access: Maternal and Reproductive Health

A key focus of the project was on women's health. The project provided information and also regular training in an interactive framework to leader-motivators about matters such as reproductive health, including issues related to pregnancy, delivery, post-delivery care, child health, and family planning. They were then supposed to bring this knowledge back to women volunteers during the monthly meetings.

It can be conceptualized that if through individual and collective empowering processes women were gradually attaining a stronger sense of independent person-hood, they should be able to use their awareness to access information and use that information for better outcomes for themselves and their children. In terms of health access, the argument is that a heightened sense of self-worth, among other positive changes, should result in informed health-seeking behavior. For example, one effort in this project was to give women information about maternal health, starting with knowledge about complications during pregnancy, precautions to be observed during pregnancy and childbirth, and the availability of health personnel.

It was clear to us in our assessment that in the project area, women within the information loop (the core group and the extended group) were not only aware, but their information on various issues was also more specific than that held by the women belonging to the excluded group. For example, the former groups identified nutrition and proper medical care as crucial elements in care during pregnancy, whereas the women in the excluded group did not have this level of detailed infor-mation. In common-sense matters such as proper rest and exercise, women belong-ing to different groups were not different from each other. In order to capture how awareness and knowledge about care during pregnancy translate into actual prac-tice we looked at measures taken during pregnancy and childcare – for example, we looked at whether women went for regular medical check-up, followed the pre-scribed regime of iron and folic acid supplement, ate nutritional meals, and took general care. For childcare, the questions asked were about breast-feeding and immunization. Once again, the performance of women in groups A and B within the information loop was significantly better than that of women in group C outside the information loop.

In this context, it may be pointed out that one of the important components of reproductive and child health has been to encourage institutional deliveries under proper hygienic conditions. However, women in all the three groups had over-whelmingly large percentages of births still taking place at home. The provision of adequate infrastructural support for women in childbirth does not simply follow from women's knowledge: whether women knew about a preference for these facil-ities does not change the fact that in most parts of the project area such facilities simply were not available. Under such circumstances, instead of institutional deliv-eries, the presence of trained birth personnel becomes more crucial; in this respect women within the information loop had an edge over the women outside it in terms of the proportion of deliveries attended by trained personnel.

The same can be said of the extent to which children were immunized. Women in the inner two circles of the information loop expressed a higher degree of pref-erence to have their children immunized; immunizations, however, were not always available.

Issues of Access: Authoritative Resources and Autonomy

Another key focus of the project was on women's economic autonomy. The project both provisioned for institutional support for market interventions and educated women about personal finances.

In the project area, we found that male members in the core and the extended groups were more willing to entrust women with the management of finances. Women in the inner two groups also had bank accounts in their name at a much higher rate than in the excluded group. Handling the account and/or having the accounts in their names is crucial for women because it brings a sense of control as well as freedom to spend the money without seeking explicit permission from others. As one woman puts it:

> Earlier whenever I needed money, I had to ask my husband. He would ask a million questions and also where I spent the money. Not that I squander the money now that I operate the account myself, but I do not have to depend upon my husband every time I need the money. Also, he never asks where I spent it. This feels good in other ways too. I feel I am worth it and in control – no longer that sense of dependency is there. Once in a while I buy something for myself without much hassle.

This financial freedom appears to be a fallout of women's increased self-esteem to assert themselves, combined with the perception among men about women's increased ability to handle finances. In several cases, husbands of these women take covert pride in what their wives were able to achieve and reported a change in their own behavior. A male teacher in the project area, for example, reports:

> I remember, almost a decade ago, I had bought a tractor and until after the tractor reached home, my wife did not know about it. Now I cannot buy even a pair of bul-locks without first conferring with my wife as she knows exactly what the earnings were and where and how they are to be spent. She would object vehemently if I go out and spend the money as I wish and to tell you the truth I secretly admire my wife for what she has become now.

With a share of their own earnings going into the bank, accompanied by freedom to spend as per their wish, women have been able to divert their expenditure toward children's education and securing the future with savings, a definite signal of relatively better control over their lives and aspirations for children. One woman says:

> I save from selling milk. My husband also gives me some money that he earns from selling agricultural produce. I have an independent bank account since two years. I spend money on children's clothing, their education. They live away from home. Sometimes I pay for their railway tickets. I feel I can spend without many restrictions. I am respected in my home because I have this money.

Issues of Access: Participation in Political Space

It has been often argued that sustained and critical changes are not possible in women's lives unless the "personal" becomes "political" and they are able to articulate their subordination as an outcome of wider social and political structures. Political awareness and participation in village affairs can prove to be a crucial first step in claiming public spaces and gaining legitimacy in women's struggle for gender equality. The project had attempted to create political awareness by exposing women to electoral processes and grassroots activism.

Overall, women in the project area across all three levels of project involvement showed little awareness about various schemes that run in the villages, especially for women. Awareness of the already existing village-level committees – the *Gram Sabha* – was significantly higher among the women who belonged to the information loop. Yet in terms of attending the meeting of these committees, very few women did so. Significantly, on being asked why they did not attend the meetings, the general refrain was that "women do not go there." Also, *Gram Sabha* and *Panchayat* were seen as domains belonging to men even among the more enlightened women belonging to the information loop.

This observation acts as shorthand for understanding the relationship between structure and individual (Kabeer, 1999). Although enhanced awareness has encouraged women to cross over from the domestic sphere in many respects, certain spaces are still seen as outside their reach. In this northern part of the country, if women follow the deeply entrenched rules, norms, and practices, they were likely to gain more respect. Also, village-level committees had elderly male members from women's own households, with whom some of them observed purdah. This brings forth the point that empowering processes cannot simply be conceptualized at the level of individual and collective actions, but must be situated in values embedded in the wider societal context (Niraula and Morgan, 1995; Kabeer, 1999).

Gender Relations: Negotiations and Contestations

The impact analysis makes it amply clear that social change is a complex issue and interventions such as those envisaged in the project could change certain aspects of women's lives, whereas deeply entrenched and traditionally coded ideas about women's and men's primary responsibilities were resistant to change. Identification of specific gendered domains starts early in life and is a part of socialization from which even the "empowered" women could not escape. Such "duality" of accepted behavioral codes results in contradictions. For example, in the project area, boys were usually sent for higher education, but more often girls were not allowed to do the same. Higher education for girls after a certain level was not encouraged because the respondents felt that the girls would then not get proper bridegrooms – yet the importance of girls' education was unequivocally accepted by many.

On the one hand, as already indicated, in the aftermath of the project husbands and other male villagers clearly took pride in what "their" women were now capable of doing and there was a certain amount of crossing over of the domains in terms of husbands/male members sharing household responsibilities. In fact, as already mentioned, government functionaries at the highest level of hierarchy acknowledged the visibility and proactive roles of women in the public domain during the focus group discussions. On the other hand, men typically expressed presumptuous stands toward women. During one of the interviews, for example, the man of the house tried to undermine the initiative the woman was taking in responding to our queries by saying "purely paper main zero laga dijiye, usko kuch nahin ata" ("put zeros against all queries, she does not know anything").

Equality in gender relations is a contested domain and there were visible tensions. Despite expanding spaces for women, visible and invisible limits were clearly drawn. Women tried to redefine and expand their spaces through individual and collective empowering processes wherever and in whatever way they could without being too radical in their approach. And at the risk of being controversial, one would like to suggest that they used the existing social constructs of male behavior, even if it meant following in many instances the existing gender codes. Earlier examples also showed that women avoid a confrontational approach and try to carve out extended spaces for themselves within the constructs already available, a strategy that can be termed "incremental pragmatism" (Raju, 2001).

A Bridge too Far?

The state-initiated project interventions had achieved positive results in certain spheres – in women's enhanced self-esteem and dignity, sense of agency, organization of collectives, and peer support for negotiating the day-to-day challenges in an unequal world. These changes in individual and contextual/collective domains resulted in several positive changes.

However, the interventions had paid very little – if any – attention to enabling women to negotiate and assert their rights in critical areas of interpersonal relationships. And it is important to note that, by and large, the behavioral changes induced in both men and women were those that reinforced women's role in the family, i.e. maternal and childcare, cleanliness, and better domestic, water, and food management.

At this juncture it is important to delve into yet another issue, which refers to the role of the leader-motivators who were the key players in translating intervention goals into practice. Despite the thorough gender training and sensitization, these leader-motivators came from the same socioeconomic and cultural milieu as the rest of the women. They had carried their own baggage of patriarchal values and ideas about coded gendered roles. The project seemed to leave most of these intact.

In its conceptualization, the project had visualized a broad canvas for social change aimed at sensitizing women in legal literacy, community participation in developmental activities, and livelihood issues. However, the content analysis of the activity registers of these leader-motivators showed that the discussions remained largely confined to reproductive health issues and women's role as mothers. Since women undertook these activities within the confines of domesticity, they had usually been equated with "women's needs" in the government discourses and also in implementing the project. It was easier too because attending to these needs did not specifically call for any changes in the subordinate position of women and the status quo in the region.

It was true that in meeting practical challenges, women increasingly developed a deeper awareness of and sensitivity to the necessity for more fundamental changes in power structures between sexes. Yet the project did not fully explore its latent potential to address, question, and interrogate embedded stereotypical gender constructs that were the hallmark of asymmetrical and oppressive power structures in the region.

Without positing women in a combative manner, the project could have made it possible for women to go beyond familial domains and start questioning male dominance as an underlying construct of patriarchy. The project might have given women the tools and the confidence, for example, to start questioning double-speak around sexuality that treated extramarital relationships for men as signs of masculinity and virility, but the same as a matter of family *izzat* (honor) for women.

Similarly, questions could have been asked about whether women were in a position to negotiate with their men for condom use for safe sex. The project assessment we undertook was not able to probe these questions in depth, but informal discussions with leader-motivators and other women in the villages clearly indicated this to be a largely untrodden path. And in the empowering processes throughout the project there existed very little discussion on this. At the same time, however, there has to be an element of dealing with reality as it exists that cannot be ignored. For example, members in the core and extended groups did manage to deal with alcoholism and violence when it was in excess and carried out in public, or if concerned women wanted to bring it out in the open – but not as an instrument of domestic violence on a wider scale.

This raises the question of whether the changes brought about by the project were mostly cosmetic. Perhaps not, but the kinds of issues the project was committed to on paper – empowering women in an integrated manner not only to enhance their capabilities and give them access information and resources, but also to enable them to lead a life wherein they could have "an alternative life role" – would have entailed systemic and structural changes to be effective. These changes ought to occur both at the level of the household (i.e. gender relations at home) and at the level of the wider community, and such a vision has to enfold within its orbit stakeholders other than women – in this case husbands, community leaders, opinion builders, and even older decision-making women.[6] Such a process of social change needs to be institutionalized so that what happens hinges not upon individuals, but upon collective concern articulated through a cohesive group – a broad-based ownership – that lends ideals of gender equity a sense of legitimacy.

Is this possible to achieve within the confines of a state-driven project? The point of how to attain gender equity as a legitimate claim is an extremely complicated question and focusing on women's empowerment *per se* may not provide all the answers. One of the reasons why the project under scrutiny has not been able to address issues such as dowry and sex-selective elimination of girls more effectively seems to be due to inadequate attention paid to locating the process of women's empowerment in broader structural, social, and political dynamics of power relations between women and men that in this specific context assign men greater affectivity than women to engineer social changes in institutional rules.[7] What this means in operational terms is that although women may be seen as a primary agency for change,

> to reclaim the transformative aspects of the concept of empowerment [as originally conceptualized] we need to move [beyond the issues of] "access to," "participation in," and "self-help" to ideas of contestation, and struggle, in which alliances can be formed with different groups, *including men* in specific times and places, for purposes of transforming institutions of the state, the economy, the community and the family. (Bisnath and Elson, 2002, emphasis added)

It is important, however, to note that this perspective was not totally missing in the project. There had been a proviso for male partnership and the activities required for such changes, such as programs for adolescent boys and meetings for men, at the conceptual level of the project, but these activities had received at best token attention in the project at the implementation stage. In fact, most of the success stories, whether at the level of individuals or the group, are those where there has been a well coordinated effort between women and men. In focus group discussions and general meetings this point was often discussed and reiterated at various levels.

Admittedly, addressing men is not a simple task. As pointed out by Kabeer (1994) and others, as long as the existing power structure privileges men, they would be strategically interested in negating inequities that may in fact be socially constructed and therefore may be subjected to challenge and transformation. In contrast, women may not relate to their long-term interests embedded in such transformations in a way they can relate to day-to-day domestic routine, and in turn may be socially conditioned to subscribe to prevailing gender norms as "divinely ordained, biologically given or economically rational" (Kabeer, 1994, p. 299). The project has taken women beyond such social conditioning to a certain extent – and we have to consciously think about how to move the agenda forward in a more holistic way.

Taking along the constituency of men in our opinion is especially important in the project area where ground realities are starkly patriarchal. Undoubtedly, there is an inherent danger in this proposition: if men were to be brought in at the initial stage itself, given the nature of patriarchy in the project area, it could have curtailed women's participation altogether and would have proved counterproductive. Having acknowledged this, however, there are ways and means and non-invasive tools available to address men without jeopardizing women's interests (Raju and Leonard, 2001; Nagar and Raju, 2003).

For long lasting processive changes, the transformative aspects of the concept of empowerment need reclaiming. In operational terms, this entails broader structural changes in power relations between the sexes in social, economic, and political realms so that empowered individuals, in this case women, find the extrinsic environment enabling for them to realize their dreams without eroding the mutually interdependent nature of gender relations *sans* exploitation. That would mean that other actors, including older women as well as men, would have to be brought in more effectively within the intended interventions. These dimensions require the rethinking and the repositioning of certain basic attributes of the "empowering processes" as circumscribed by this project.

NOTES

1   Greater self-confidence and overcoming of the internalized sense of being deprived, marginal, and oppressed can enable one to access external resources without feeling barred, while control over the external world or resources can lend one's personality a boost to self-expression in variety of ways (Sen, 1997, cited in Savitri and Elson, 2002).

2   The survey team consisted of the project director, an informal advisor from the local area, and four postgraduate researchers – all trained in conducting socioeconomic field surveys and qualitative methods. The team was briefed about the purpose of the survey, what each question meant, and what kind of probing would be required without being invasive or overbearing. The team did a pretesting of the questionnaire. Throughout the pretesting as well as the entire survey, the team was accompanied by the project director and occasionally by the local advisor. We made it mandatory that the teams share their field experience at the end of the day, exchange notes, and go through their filled questionnaire and "clean" the information. Doing this in the field itself proved useful in case some information was missing or required a recheck.

3   Although questionnaires are often seen as problematic because of preconceived assumptions and categories, we still adopted the method because we wanted to have a few lead questions to organize our own thinking in terms of parameters to be taken and also to help us have guided conversation and dialogue. However, we kept the questions – some of them open-ended – simply worded with ample room for modification in the field itself in terms of both content and vocabulary. Also, there was no precoding to avoid limiting the thought process and responses.

4   For content analysis, we read through each activity register and noted various issues that were raised and discussed in weekly meetings under different heads. These heads were not precoded, but were free-listed as they appeared in the registers. Later, the issues were grouped together under broad heads. For example, feeding, vaccination, etc. were classified as "childcare." The outcome is indicative of how frequently particular issues were discussed, irrespective of the time that might have been devoted to each issue.

5   In order to test the significance of the differences between sample proportions p_1 and p_2, we use the properties of the sampling distribution of $|p_1 - p_2|$ which is found to be normal with mean zero and standard error of $\sqrt{p_1q_1 + p_2q_2}$ where $q_1 = 1 - p_1$ and $q_2 = 1 - p_2$ and n_1 and n_2 are the size of the two samples. Using the properties of the normal distribution the statistical significance of the absolute difference $p_1 - p_2$ can be tested for 1 and 5 percent levels of significance. Thus if $p_1 - p_2$ lies between two and three standard errors, it is significant only at a 5 percent level of significance. Alternatively, if it is greater than three standard errors, it is significant at a 1 percent level of significance.

6   We have, for example, a plethora of studies clearly indicating that mothers rarely make fertility decisions alone, and this is only one example.

7   This can be seen most clearly when, concerning questions of dowry or feticide in the
    project area, a common refrain even from responsible key persons at both the village and
    district levels is *it is women who do it*. There is no questioning of or attempt to decon-
    struct the underlying processes of decision-making and the submissive location of women
    therein. For example, at one site, we were told that they just had a meeting on "female
    feticide" that was attended by more than eighty people. When we asked how many men
    attended the response was "it is women's problem – why men?" In another instance, a
    news report downloaded from the internet (hindustantimes.com) has this to say: the
    women folk in the study subdivision have a strong representation in village *panchayat*.
    Yet, female feticide is assuming alarming proportions here.

BIBLIOGRAPHY

Agarwal, B. (1994) *A Field of One's Own: Gender and Land Rights in South Asia.*
    Cambridge: Cambridge University Press.
Agarwal, B. (1997) Bargaining and gender relations: within and beyond the household.
    Feminist Economics, 2(1), 1–50.
Batliwala, S. (1997) What is female empowerment? Lecture delivered in Stockholm, April
    25.
Bisnath, S. and Elson, D. (2002) Women's empowerment revisited. Progress of the World's
    Women: A New Biennial Report (UNIFEM).
Chen, M. (1995) A matter of survival: women's right to employment in India and Bangladesh.
    In M. Nussbaum and J. Glover (eds), *Women, Culture and Development*. Oxford: Oxford
    University Press.
Chopra, R. (2003) From violence to supportive practice: family, gender and masculinities.
    Economic and Political Weekly, 38(17), 1650–7.
Dave, A. and Krishnamurty, L. (2000) Home and the world: Sangha women's perceptions of
    "empowerment": some reflection, mimeo, November.
Dreze, J. and Sen, A. (1997) *Economic Development and Social Change*. Delhi: Oxford Uni-
    versity Press.
Dreze, J. and Sen, A. (2002) *India Development and Participation*. Oxford: Oxford Univer-
    sity Press
Dyson, T. and Moore, A. (1983) On kinship structure, female autonomy and demographic
    behaviour in India. *Population and Development Review*, 9, 35–60.
GOI (1995) *Fourth World Conference on Women, Beijing 1995: Country Report*. New
    Delhi: Department of Women and Child Development, Ministry of Human Resource
    Development.
Jandhyala Kameshwari (2001) State initiatives. Paper presented at the seminar Towards
    Equality: A Symposium on Women, Feminism and Women's Movement.
Kabeer, N. (1994) *Reversed Realities: Gender Hierarchies in Development Thought*. London:
    Verso; New Delhi: Kali for Women.
Kabeer, N. (1999) *The Conditions and Consequences of Choice: Reflections on the
    Measurement of Women's Empowerment*. New Delhi: UNRISD.
Karve, I. (1965) *Kinship Organization in India*. Mumbai: Asia Publishing House.
Kirkwood, B. R., Cousens, S. N., Victoria, C. G. and Zoysa, I. D. (1997) Issues in the design
    and interpretation to evaluate the impact of community-based interventions. *Tropical
    Medicine and International Health*, 2 (11), 1022–9.
Kishore, S. (1993) May God give sons to all: gender and child mortality in India. *American
    Sociological Review*, 58, 247–65.

Mahila Samakhya (1996) *Beacons in the Dark: A Profile of Mahila Samakhya*. Bangalore: Karnataka.

Menon-Sen, K. and Shiva Kumar, A. K. (2001) *Women in India: How Free? How Equal?* New Delhi: UNDP.

Mohanty, M. (2001) On the concept of empowerment. In D. K. SinghaRoy (ed.), *Social Development and the Empowerment of Marginalised Groups*. New Delhi: Sage.

Murthi, M., Dreze, J. and Guio, A. (1995) Mortality, fertility and gender bias in India: a district level analysis. London School of Economics, Development Economics Research Programme, STICERD, Discussion Paper 61.

Nagar, R. and Raju, S. (2003) Women, NGOs and the contradictions of empowerment and disempowerment: a conversation. *Antipode*, 35(1), 1–13.

Niraula, B. B. and Morgan, P. (1995) Marriage formation, post-marital contact with natal kin and autonomy: evidence from two Nepali settings. *Population Studies*, 50(1), 35–50.

Raju, S. (2001) Negotiating with patriarchy: addressing men in reproductive and child health. *Economic and Political Weekly*, 36, December 8–14, 4589–92.

Raju, S. and Baghchi, D. (ed.) (1994) *Women and Work in South Asia: Regional Patterns and Perspectives*. London: Routledge.

Raju, S., Atkins, P. J., Kumar, N. and Townsend, J. G. (1999) *Atlas of Women and Men in India*. New Delhi: Kali for Women.

Raju, S. and Leonard, A. (2000) *Men as Supportive Partners in Reproductive Health: Moving from Rhetoric to Reality*. New Delhi: Population Council.

Rowlands, J. (1997) *Questioning Empowerment: Working with Women in Honduras*. Oxford: Oxford University Press.

Rustogi, P. (2003) *Gender Biases and Discrimination against Women: What Do Different Indicators Say?* New Delhi: UNIFEM.

Savitri, B. and Elson, D. (2002) *Women's Empowerment Revisited*. Progress of the World's Women: A New Biennial Report. UNIFEM.

SinghaRoy, D. K. (2001) Critical issues in grassroots mobilisation and collective action. In D. K. SinghaRoy (ed.), *Social Development and the Empowerment of Marginalised Groups*. New Delhi: Sage Publications.

Sopher, D. E. (1980) *An Exploration of India: Geographical Perspectives on Society and Culture*. New York: Cornell University Press.

Uberoi, P. (ed.) (1994) *Family Kinship and Marriage in India*. New Delhi: Oxford University Press.

Part III City

Chapter 15

Feminist Geographies of the "City": Multiple Voices, Multiple Meanings

Valerie Preston and Ebru Ustundag

Women's daily lives are shaped by the cities in which they live, the cities portrayed in the media, and the cities of our imaginations. Simultaneously, women construct the material, social, and symbolic circumstances of daily life in cities worldwide.[1] The interrelations between the social, economic, political, and cultural processes that constitute gender and the gender relations and identities that mold the urban environment are readily apparent in the media. Privileged urban settings in popular television shows glamorize the lives of affluent and white young women. The glamorous urban backdrops in situation comedies contrast with the images of destruction that dominate the daily news. Daily, stories about women living in bulldozed and bombed buildings in Jerusalem, Gaza, and Baghdad, the impact of HIV/AIDS on women in Johannesburg, Los Angeles, and Harare, and violence in Washington, DC, Toronto, Ontario, and other North American cities recur. In each urban place, women actively reconstruct their everyday environments, often struggling against long odds to improve their own lives and those of their households and communities. Women's activities alter popular images of urban places and, in the process of remaking these images, women reconstitute their own identities.

To capture the dynamic relationships between urban environments and gender, feminist geographers began by analyzing the material conditions of everyday life that contribute to gender inequality, concentrating on economic circumstances. Current research articulates the relations of production for all types of capital: social, economic, cultural, and symbolic (Bourdieu, 1984). This theoretical stance underscores women's agency in all spheres of their lives, not just their material circumstances. Moreover, it encourages the eclectic mix of theories and methods that characterizes contemporary feminist research about women's lives in urban places. They include quantitative studies of the journey to work, policy studies of homeless women, and discursive analyses of stereotypes regarding foreign domestic workers (Johnston-Anumonwo, 1997; Pratt, 1997; Yeoh and Huang, 1999; Klodawsky Farrell and D'Aubry, 2002).

Feminist geographers have a dynamic view of contemporary cities which recognizes that urban places are not simply containers within which women organize their daily lives. Cities are the locations from which global movements of capital and information emanate, the locations of everyday lives that are buffeted, restructured, and terrorized by national and international forces, and the places where women act to mold social, economic, political, and cultural processes of change at all spatial scales. Although women's agency is a theme of longstanding importance in feminist geography, contemporary research emphasizes the ways that women alter representations of places, events, and people, particularly women's own identities, by changing their material circumstances in different places.

While a materialist reading of gender in contemporary urban life opens the door to greater understanding of the fragmentation of cultural identity and daily experience (Bondi and Christie, 2003, p. 320), we acknowledge that gender is only one of many bases of inequality (Anderson and Jacobs, 1999). The effects of gender depend upon its interrelations with other aspects of identity, such as class, ethnicity and race, disability, and sexual orientation. In our discussion, we aim to show how recent feminist geography has examined the complex ways that gender is articulated in relation to other identities through spatial practices in specific urban places.

The review concentrates on literature published in English since 1990. These articles and books often emphasize the experiences of women living in Western Europe, Oceania, selected cities in Asia, and Anglo-North America. The geographical specificity of this literature affects its theoretical, empirical, and political scopes (Nagar et al., 2002). We have noted some omissions, but we can only underline the urgency of Saraswati Raju's recent call for dialogue among scholars from various locations (Raju, 2002).

In this chapter, we outline recent transitions in urban research by feminist geographers. From its origins, feminist geography has examined the changing spatial relations between paid and unpaid work in urban places, so we begin by commenting on current spatial analyses of women's work in contemporary cities. The restructuring of urban economies worldwide has transformed women's lives. Some women have benefited, while many others have suffered. Examining the links between work and home in contemporary cities, feminist geographers have tracked the impact of each wave of restructuring on gender inequality. The uneven development of contemporary cities in which rich and poor live cheek by jowl in spaces that are increasingly segregated from each other also heightens gender inequality for women at certain locations, while reducing inequality for others at different locations (Bondi, 1991; Smith, 1996; Bunting and Filion, 2000). Feminist geographers are now exploring how the internet and other social and technological changes in the organization of work are reconstituting the geographic links between paid and unpaid work in many cities, paying particular attention to the uneven impacts of the benefits and disadvantages (see the chapters by Hanson and Blake, and Boyer in this book).

Recognizing that movement links women's various activities in different urban spaces, the next section reviews the growing literature about women's mobility, physically and figuratively. Economic restructuring, armed conflict, and environmental degradation are encouraging increasing numbers of women to migrate across regional and national borders, settling in cities worldwide. Striving to maintain social connections at multiple locations, migrants live transnational lives that are situated simultaneously in their cities of destination and places of origin (Mahler, 1999). Suburbs derided as homogeneous enclaves of the privileged are now home to growing numbers of recent immigrants, the elderly, and empty nesters from all class backgrounds (Wyly, 1999). Responding to the increasingly diverse character of urban places and their populations and the dictates of contemporary social theory, feminist geographers now emphasize that women are active agents of change laying claim to various spaces in urban areas. This dynamic view of women is readily apparent in recent literature examining how telecommunications and access to the internet are transforming women's lives and the urban environments in which they live (see the chapter by Gilbert and Masucci in this book).

We end with an exploration of the body as a new site for geographic inquiry by reviewing recent research about gendered landscapes of violence and fear. The links between representations of urban places and the construction of gendered identities are explored in recent studies that connect women's fear in urban spaces to representations of men's and women's bodies (Mehta and Bondi, 1999). A discursive analysis of the representations of urban spaces and women has replaced the earlier emphasis on the material conditions of specific places. The impact of this theoretical shift is evident in studies (Koskela, 1997; Nagar, 2000a) of Finnish women whose "bold walk" allows them to reclaim public space and women's engagement in political and social movements to reduce domestic violence in Dar es Salaam. In both cases, the discussion concentrates on competing views of femininity that originate in taken-for-granted notions about the demarcations between home, body, and community and the gendered relations of power that underlie them (see the chapters by Koskela, Kamiya, and Nagar and Swarr in this book).

The City as the Place Where the Work Is Done

The city was a critical object of inquiry for early feminist geographers who explored the effects of segregated land uses in contemporary cities. The spatial separation of home and work was associated with a highly gendered division of labor between masculine paid work and feminine unpaid work (Mackenzie, 1999). Men were associated with public, productive spheres, including paid work outside the home, while women were associated with private, reproductive spheres that confined them within the home. Drawing on critical social theory, feminist geographers examined the spatiality of social relations in the household and the workplace. These dualisms have been challenged by changes in the geography of contemporary cities and the evolving gender division of labor.

New spatial divisions of labor have emerged in which women constitute a large and growing proportion of the labor force in urban centers of industrialized countries (Massey, 1995; Sassen, 2000; McDowell, 2001). The feminization of employment has redrawn the urban map of employment. Some firms relocated to the suburbs to attract a female workforce that preferred to work near home (Wekerle, 1984, p. 12), while others employing equally large numbers of female workers located at central locations that were the hubs of public transit services, a mode of transportation used more by women than by men (Hanson and Pratt, 1992). Firms' diverse locational strategies and varied demand for labor have resulted in a complex and largely undocumented geography of women's paid employment among cities and within them (Bagchi-Sen, 1995; Hanson and Pratt, 1995). Research has concentrated on "global cities," the command and control centers for international capital (Sassen, 2000). We know much less about the feminization of paid work in regional urban centers and, even more seriously, the processes by which women enter waged labor in cities of all sizes outside the wealthy, industrialized countries.

Since initial feminist interest in the geography of women's paid work, women's experiences of the workplace have diversified (McDowell, 2001). The differentiation of women's experiences in urban labor markets has coincided with growing attention to context in feminist theory. Feminism's concern for "context" is rooted in the view that there is no universal perspective, since all knowledge is necessarily subjective (Nast, 1998). While sharing a general concern for context with all feminist scholars, feminist geographers have concentrated on its geographic aspects. For example, McDowell (2001) examined regional variations in women's employment in the United Kingdom. In the 1980s, a distinct regional geography intersected with class differences in women's paid work. A cadre of well educated professional women, concentrated in London and other urban areas in the Southeast, made important gains in earnings and occupational status, while many women in other cities, particularly those in northern England, suffered job losses, increasingly precarious employment, and declining real wages.

Within urban areas, the geography of women's employment is also finely textured. For some women, local expectations about paid work, the women's qualifications, and their family circumstances interact with local employers' views about hiring women to create highly local labor market practices that vary from one neighborhood to another (Hanson and Pratt, 1995). Class intersects with the effects of gender. While some working-class women are spatially entrapped and immigrant women in the same city arrange rides to central workplaces with co-workers, well educated professional women commute almost as long and far as their male counterparts.

Women sometimes coordinate their dual responsibilities by altering the geographic locations of home and work. Gentrified neighborhoods are attractive to well paid professional women and the single mothers who can afford them, but little research has examined women's lives in the transformed landscapes (Rose, 1988). A fascinating exception is Liz Bondi's (1998) case study of two gentrified neighborhoods in Edinburgh where newcomers feel the activities of female prostitutes limit the mobility of newly arrived middle-class women within the neighborhood. The lack of attention to women's lives outside central workplaces is a stark contrast to the emphasis on these women in contemporary popular culture: witness *Sex and the City*.

The transformation of the suburbs is as profound as that of gentrified neighborhoods. A growing literature underscores the social diversity of suburbs (Chambers, 1997; Baxandall and Ewen, 2000; Preston, 2002). Few North American suburbs ever fit the popular stereotypes associated with *Leave It to Beaver* and unrelieved domesticity (Baxandall and Ewen, 2000). In the United States and Canada, the social diversity of the suburbs is being magnified by the uneven geography of economic restructuring that benefits some suburban locations while harming others, the direct settlement of immigrants from diverse social backgrounds, population aging that closes schools and other community facilities just as needs for specialized services for the aged increase, and exclusionary municipal politics that are implemented to benefit the affluent (Murdie and Teixeira, 2000; Pulido, 2000; Bourne and Rose, 2000). The impacts of these changing geographies on the links between home and work have not been explored much with the exception of a few studies that have examined women's views of working mothers and the social practices by which suburban mothers successfully combine home and work (Dyck, 1989; Dowling, 1998).

Technology has the power to dissolve the walls of the workplace, relocating it to the home or liberating workers to roam independent of a geographically fixed work-place. Telecommuting, working from home through cyberspace, was once hailed as an emancipatory strategy by which women could combine paid and unpaid work at one location (Christensen, 1993). Recent analyses are more sophisticated, empha-sizing the benefits of cyberspace as a source of information for women and a means of overcoming distance that will allow women to participate in dispersed commu-nities of like-minded individuals linked in cyberspace, while recognizing the constraints of telecommuting (Light, 1995). There is no denying the benefits of communication over the internet or its importance as a source of information, par-ticularly for promoting women's activism. By harnessing cyberspace, women may have the opportunity to overcome the isolation from facilities and each other that current gender norms and urban land use patterns promote. However, the benefits of telecommunications and the internet are not equally available to all women. Pro-gressive policies and programs are needed to insure that the benefits of cyberspace are widely distributed and its potentially adverse impacts are mitigated (Phizacklea and Wolkowitz, 1995).

Research still emphasizes urban workplaces, a surprising omission according to Domosh (1998), who observes "We have not moved much beyond the front door in our analysis of the home."[2] When we consider our understanding of women's work in urban environments, the omission is surprising for three reasons. The res-idential environment is a crucial influence on women's decisions to enter the paid workforce and the types and locations of paid employment that they will consider (Hanson and Pratt, 1988). For example, in one Vancouver suburb, women held diverse and contradictory views about motherhood, some deeply disapproving of working mothers and others tolerating working mothers (Dowling, 1998). There is also growing evidence that parental attitudes and experiences rooted in the home shape young people's beliefs about the labor market. In Cambridge and Sheffield, a booming high-tech center and a declining industrial center, young men's aspirations in the labor market and related views of masculinity were related to their fathers' attitudes and experiences despite the transformations in the local labor markets (McDowell, 2002a). Nor should we assume that well educated and successful pro-fessional women are immune to unequal power relations in the household. In Singapore, regulations restricting the employment of foreign workers, combined with the demands of relocation and local social practices, encourage many part-ners of relocated executives and professionals to withdraw from paid work and participate in voluntary organizations despite the women's prior professional success in their countries of origin (Yeoh et al., 2000).

Career success for many professional women depends on turning the home into a workplace for paid domestic workers (Gregson and Lowe, 1994; Moss, 1997; Stiell and England, 1997). The contradictions inherent in the home being a workplace for one occupant and a place of private refuge for another exaggerate inequalities due to race, ethnicity, and place of birth. The privacy of the dwelling trumps the rights of the worker in most cases, putting poorly paid, immigrant minority women in their places (Pratt, 1997; Stiell and England, 1997; Westwood and Phizacklea, 2000). The skills and work experience of domestic foreign workers are devalued, as we might expect from geographic theory that emphasizes that the home is a space for reproduction outside the public sphere. However, place matters. There are important differences in the experiences of foreign domestics and the impact of paid domestic work on the women's future life chances related to institutions, informal social practices, and economic circumstances at the local urban level. At the very local level within individual dwellings, the power relations between the female employer and the female domestic are shaped by land use, employment, and immigration policies at the local, regional, and national levels (Mattingly, 1999; Yeoh and Huang, 1999).

Urban poverty is increasingly the norm for female-headed, single-parent households in the industrialized world, who are often dismissed as members of a stigmatized underclass (Morris, 1994, p. 114). A growing literature notes that women are poor because they experience disadvantages in the labor market and in the home (Thomas, 1994). Local circumstances, particularly the extent and nature of local social networks, may ameliorate or exacerbate poverty. In Worcester, Massachusetts, poor black women benefited from strong local social contacts that helped them to juggle the competing demands of paid and unpaid work (Gilbert, 1998). In contrast, minority women living on welfare in Grand Rapids, Michigan, had limited social networks that were impoverished in terms of resources. Consequently, they relied mainly on formal services for assistance (Peake, 1999). Research that specifies the factors that contribute to these local variations in social capital is an essential prerequisite for developing public policies that will ameliorate the circumstances of poor women in different cities. Among the most important considerations may be the stigmatized representations of poor minority women that are often associated with residence in marginalized and undesirable locations (Fincher, 1998).

At the most intimate level, researchers are now examining the ways that gender is inscribed on working bodies and the ways that working bodies shape gender identities (McDowell, 1997; Moss, 1997). By concentrating on the body, attention also turns to constructions of masculinity and femininity and the multiple identities associated with each. We have more complex and dynamic readings of gender that recognize how men's and women's multiple identities intersect in place-specific ways (McDowell, 1997). McDowell (2002a, b) addresses the reproduction of masculinities among white working-class youth in contemporary Britain. Comparing the experiences of poorly educated young men in the booming high-tech economy of Cambridge with those of their counterparts in declining Sheffield, she finds that in both cities, young men are increasingly disadvantaged in the local labor market. Surprisingly, their views of masculinities and their identities as men are still very traditional despite their difficulties in finding and keeping the steady and fairly well paid jobs to which they aspire.

Moving among the Façades

Feminists have always recognized that women's access to resources depends on the ability to move among locations (Massey, 1995). Drawing on time–space geography, geographers documented the spatial constraints that limited women's movements across urban space (Tivers, 1985). Although spatial constraints were often associated with suburbanization, subsequent research underscored how women actively overcame the tyranny of distance. In suburban neighborhoods, women drew on local social resources to fulfill the competing responsibilities of home and workplace (Dyck, 1989). The strategies are very similar to those employed by poor minority women in central neighborhoods (Gilbert, 1998).

Persistent inequities in access to transportation still constrain the life chances of certain groups of women. Dependence on public transit lengthens the work trips of minority women in American cities (McLafferty and Preston, 1991; Johnston-Anumonwo, 1995). Their lives are made even more difficult by steady increases in commuting time caused by congestion and sprawl in many urban places. For women, who still undertake most unpaid domestic work, long work trips undermine their quality of life. Although access to the automobile is sometimes promoted as a solution to women's limited mobility, its effects are likely to be less effective than we might imagine. A case study of middle-class suburban residents of Sydney, Australia, confirms that in a low-density environment with infrequent public transit, women are spatially entrapped without regular access to a car. However, in some cases, the sense of constraint derives from social norms related to the definition of a good mother (Dowling and Gollner, 1997). In neighborhoods where "good mothers" enrol children in after-school activities chosen on the basis of reputation rather than convenience, delivering children to after-school activities is very difficult without a car. The task is highly feminized. Women are largely responsible for scheduling and organizing after-school activities. The growing geographic literature about men's and women's roles as parents underscores the continuities and stabilities in our identities as mothers and fathers that are reinforced by a contemporary geography that still confines children and their caregivers to homes that are separated sharply from other locations (Aitken, 1999).

The findings from the Sydney suburbs point to the need for new directions in feminist research about mobility (Law, 1999). The journey to work is only one aspect of mobility and mobility, both actual and symbolic, is critical to all aspects of women's lives. As cities have been transformed from cities of production to cities of consumption, more attention is being paid again to trips made for purposes other than work,[3] particularly trips required for shopping and other consumer activities. The expanded definition of women's mobility complements a growing geographical literature about gender and spaces of consumption (Thrift and Glennie, 1996). In contemporary cities of consumption, shopping is highly gendered, yet feminist analyses of contemporary consumption are still relatively rare. Investigations of second-hand shopping by Gregson et al. (2002), a historical analysis of women's presence in consumer and leisure spaces of nineteenth-century New York (Domosh, 2001), and an examination of social constructions of femininity inherent in a Canadian department store (Dowling, 1993) are welcome exceptions.

The constraints on mobility uncovered in the Sydney suburbs also underscore the ways that mobility is socially constructed. Movements across space reflect the interrelations between available resources and social identities. In Sydney, limited access to an automobile constrains the mobility of some middle-class mothers. Other women experience mobility limitations by virtue of their identities as women of color, immigrant women, lesbians, and women with disabilities. Despite its persistence as a social marker, little geographic research documents how gender and color jointly and independently influence even the most straightforward aspects of mobility such as the spatial range of activities. In the United States, the omission may reflect the unexamined assumption that the end of racist laws barring people of color from specific locations erased racial differences in mobility. In Canada, studies in Montréal have established the importance of mobility to recently arrived immigrant women who consider learning to use public transit second in importance only to learning French (Ray and Rose, 2000). The ability to move among locations in the city that many women take for granted is prized by women living on the margins.

Sexual orientation and disability also influence men's and women's mobility but their effects are profoundly gendered (Valentine, 1995). Lesbians and gays have very different impacts on urban landscapes, with concomitant implications for their mobility. Comparison of gay communities in American, British, and Australian cities with the lesbian communities in New York's Park Slope neighborhood confirms that lesbians often have fewer resources to shape their local environments than gay men (Rothenberg, 1995; Knopp, 1998). Lacking the financial resources to remake urban landscapes where they are the dominant group, lesbians move through spaces in which they are almost always outsiders. Unlike gays, who have been able to use territorial concentration as a strategy to promote broader political agenda, lesbians rarely attain the same territorial base. Lesbians' experiences of mobility are akin to those of women with disabilities who also lack all types of capital (Chouinard, 1999). Sexual orientation and disability set women apart as the other, women on the outside and out of place, a status that is reinforced by restrictions on their abilities to move among urban spaces.

Women's experiences of limited mobility begin in childhood, contributing significantly to environmental influences on gender identities. In many industrialized societies, the spatial extent of children's activity spaces is narrower for girls than for boys (Valentine, 1997). Through complex negotiations and deception, children test and subvert the boundaries set by parents. Responding to local events and social norms, their own views of each child's spatial competencies and the demands of daily life, parents continuously revise their rules about the spatial limits of children's activities. The practices that result are highly differentiated by gender, age, and other dimensions of social identity. Spatially limited activity spaces and fewer opportunities for independent movement may occur more often for girls than for boys, but gender differences are fluid and dynamic. Valentine's research implies that the links between gender identities and mobility are malleable. Research is needed to identify how social, cultural, and political circumstances in specific places interact to promote greater mobility for girls and women.

Women Remaking Urban Spaces

Responding to mobility and resource limitations, women have contested the actions of the state in urban areas. In the same way that studies of mobility documented the constraints on women's movement, research about urban activism began by documenting women's access to public services and facilities. The provision of public services is often predicated on highly gendered, racialized, and agist assumptions about the populations living in suburban and central locations and women's ability and willingness to participate in voluntary organizations (Fincher, 1991). The needs of women who are "out of place" – for example, immigrant women and young single mothers in suburban areas – are considered rarely by policy-makers (Rose and Chicoine, 1991; Rose, 1993; Truelove, 2000; Permezel, 2002). The importance of the local state is highlighted in an analysis of the provision of preschool child-care in Sheffield (Holloway, 1998). Despite a local government policy to enhance the provision of preschool care in deprived areas, children from a more privileged area enjoyed greater access to the service. Local geographies of mothering that are strongly related to class combined with more financial resources mean that preschool services are provided earlier through a mix of public and private facilities to children in more privileged areas.

Research has also documented spatial differences in women's urban activism (Staeheli, 1996; Fincher, 1998; Fincher and Panelli, 2002). In the Australian case, the current cohort of professional, working women living in the inner suburbs[4] has campaigned successfully for state-regulated and publicly funded childcare, while the cohort of women who care for elderly relatives in aging postwar suburbs has neither sought nor benefited from state assistance. Again, the geography of women's engagement with the state is underpinned by distinct moral geographies concerning women's roles as caregivers and mothers. Widely held views about gender identities that are inherently geographical buttress spatial variations in the provision of services and women's actions to claim services.

Where dominant gender relations often encouraged women to "stay home," women have resisted confinement to the private sphere of the home and its neighborhood by participation in voluntary organizations. Women's participation reshaped the city and its social relations, often by creating liminal spaces in the community where women had more power and authority than in either the home or the workplace (Moore-Milroy and Wismer, 1994). Literature (Sibley, 1995; Spain, 2001; Hayden, 2002) has highlighted women's historical contributions to urban planning in North America. Building settlement houses, shelters, and women's clubs and associations in large cities throughout the United States, women in the nineteenth century provided services for immigrants, African-Americans who had migrated from rural areas, and the urban poor. Although many services were available for men and women, the volunteers sought to assist women who were highly stigmatized in the Victorian era, particularly single mothers (Spain, 2001). They also challenged dominant ideologies about women's place and women's space (Hayden, 2002). According to Sibley (1995), masculine control of the production of knowledge has hidden women's active participation in the planning and construction of nineteenth-century cities, leaving us with an incomplete historical record that reinforced traditional and unequal gender identities and gender roles.

The "Take Back the Night" campaigns in many European and North American cities in which women have contested their rights to urban spaces also aim to transform gender identities and gender roles (Wekerle and Peake, 1996). By participating in the campaigns, women lay claim to identities as active urban citizens enacting new forms of substantive citizenship (Lister, 1997). Despite the vigor of recent campaigns, women still face barriers to their presence in many urban spaces. Women are twice as likely as men to report feeling unsafe even though men are far more likely to be the victims of crime in public spaces (Wekerle, 2000, p. 47). Women's fear in public spaces has many sources, but it relates first and foremost to social constructions of femininity and masculinity and men's and women's bodies.

Although the body has been central to the feminist agenda in geography, "the body has acted as geography's Other; it has been denied and desired depending on the particular school of geographical though under consideration" (Longhurst, 1995, p. 102). Feminist geographers have used Foucauldian analysis to articulate the body as a site of resistance (Longhurst, 1995; Nast, 1998). Women's bodies are constantly shaped by the steady demands and pressures of capitalism as it responds to technological change (Mitchell, 2000, p. 220). Haraway (1991) has interrogated the impact of technological change, calling attention to the changing ways that humans work with machines to produce new bodies and subjectivities. Different media are also implicated in the construction of female bodies. The "androcentric" vision which imposes itself as neutral and has no need to spell itself out as legitimate (Bourdieu, 2001, p. 9) imposes certain ways of being in the urban scene.

The regulation of bodies in space was accentuated in the wake of the destruction on September 11, 2001. The media emphasized a few images to represent the destruction, its aftermath, and the victims and perpetrators. Here we emphasize the representations of Afghanistan in which repeated portrayals of Afghan women as oppressed bodies behind the veil were common. There is no denying the misogyny of the Taliban government that damaged the lives of many Afghan women, but the singular media interpretation of veiling and the veiled woman as a symbol of underdevelopment and patriarchy neglects the multiple meanings of veiling. As a site of resistance, the veil can establish a mobile honor zone from within which young women may interact with men without fearing the loss of their reputations (White, 2002, p. 220). Some Muslim women have adopted the veil as a brave act of defiance against the social corruption of a Western-oriented market economy (Moghissi, 1991; Afsaruddin, 1999). These multiple interpretations were heard rarely in the Western media following the events of September 11, 2001. Instead, the veil emerged as a potent symbol for the "otherness" of Islam to the West (Gole, 1996).

The interpretation of the veil is not surprising in light of dominant discourses that have constructed women as "an irruption in the city, a symptom of disorder, and a problem" (Wilson, 1991). The veil simultaneously hides women and marks them as different in contemporary industrialized cities. It reinforces women as out of place in public spaces. In contemporary cities, women rarely enjoy the same social and political rights as men. Even where equity is the official stance, the development, implementation, and administration of public policies is highly gendered, and highly racialized (Kobayashi, 1993). According to Yuval-Davis (1997), women carry the "burden of representation" for the collectivity's identity and future destiny. Women are associated in the collective imagination with children and therefore with the collective as well as the familial future. As a result, the ways in which women are represented in political discourse, the degree of formal emancipation they have achieved, their forms of participation in economic life, and the nature of the social movements through which they express their demands are linked to state-building processes (Kandiyoti, 1997) that are inherently spatial. Until women enjoy the same choices as men to act as the *flâneur* in contemporary cities, gender inequality in rights to the city will persist – a condition that women will contest by claiming spaces to provide services for women and their families, by campaigning to insure women's rights to public space, and by pressuring governments to fund equitably public services essential to women's well-being. The historical record indicates that as women remake cities, they will also reconstitute gender relations and gender identities.

Rethinking Gender Inequality in Urban Places

In current feminist research, the city is no longer the sole object of study. Instead, the urban environment is the context within which gender identities and gender relations are negotiated and this context is reformed by changing gender identities and relations. Thinking about the relations between gender and the urban environment is more differentiated and more dynamic than before. By paying attention to the interrelations between gender and other dimensions of social identity, feminist geographers have developed a richer understanding of how the urban environment constitutes gender and how gender relations and identities affect urban inequality. The analysis of transnationals in Singapore (Yeoh et al., 2000) highlights the many ways that gender intersects with other aspects of identity in a specific urban setting.

Current research recognizes that the urban context is a changing set of situated social relationships that are influenced by economic, social, political, and cultural changes operating at various spatial scales and interacting with each other. Change in urban places continuously alters women's lives and the gender inequalities associated with them. The fluidity of current urban environments may also create opportunities for women to make progressive changes through individual and collective action. The city in flux may be more amenable to women's influence. Capturing the dynamic nature of urban places and the impacts of changing economic, social, and political relations among urban residents is an important task that remains.

After the "cultural turn," feminist geographers have drawn on psychoanalytical theories, postcolonial theories, and poststructural theories that emphasize discourses. How women represent themselves and others and how women are represented play a critical role in the production of gender inequalities (Fraser, 1989). In examining these representations, more attention needs to be paid to the intersections between class and race, sexual orientation, and disability, and their interactions with space and place. We also need to move beyond the Anglo-American world. We need to enter into dialogue with feminist geographers from other parts of the world to examine feminist issues outside North American, European, and Australian cities. If we take seriously the argument that urban spaces and urban places shape gender relations and gender identities, then we need to examine the interrelations between geography and gender in many more cities. More research outside North America and Western Europe would also open new research questions. For example, armed conflict is part of daily life in many cities outside North America and Western Europe, but few feminist geographers have addressed its implications for women's lives (Hyndman, 2001).

In their analyses of urban processes, feminist geographers have already begun to specify the spatiality of the social processes that form gender relations and gender identities. We still need to know more about how place intersects with gender roles and gender identities in specific contexts. An example will illustrate this argument. A recent evaluation of welfare-to-work programs in San Francisco found that improved access to transportation was likely to help only about half the women, those who already had the education and skills to qualify for jobs (Chapple, 2001). For the other half, access to transportation was a minor issue. They needed help acquiring the training and job experience desired by employers and developing social contacts that might eventually provide information about possible jobs. Some were unlikely to ever have a job for a variety of reasons, including health problems that could only be managed with medical insurance, a benefit that is rarely available from the entry-level jobs for which most of these women qualified. The example illustrates the necessity of critical, grounded research that takes place seriously so that feminist geography can contribute to contemporary policy debates that daily alter the lives of the most vulnerable women.

Claims to rights as citizens are one perspective by which we can advance recent research concerning gender relations and gender identities in urban places. The literature about rights to the city recognizes the complex interrelations between gender and other dimensions of social identity that arise in specific geographical locations. In their research about cities, feminist geographers have already contributed to this emerging literature. Their research on women's paid work in contemporary cities has shown that spatial inequality contributes to persistent gender differences in occupations and earnings. Research about mobility confirms that many women still suffer material and symbolic limitations on movement among urban locations, restrictions that reduce many women's social mobility and progress toward gender equality. Finally, women have had more success in the informal sector as active citizens participating in voluntary organizations than in the formal political sphere where they are still underrepresented in most urban governments. The outcome of these trends is a persistent geography of gender inequality within and among cities.

We want to end by emphasizing the potential for feminist geographers studying the city to contribute to current debates about rematerializing contemporary human geography (Lees, 2002). As this brief review has indicated, geographical analyses of gender issues in cities already explore geographic variations in women's material circumstances and the ways that they constitute gender identities. The eclectic mix of theories and methods upon which feminist geographers draw means they are ideally situated to undertake research that links discussions of gender identities with analyses of the material circumstances that underpin these representations. Recent explorations of the material processes that give rise to different views of masculinity and femininity are a good example of this approach (McDowell, 2002a, b). Although the research is in its very early stages, it promises to help us to identify the strategies by which women have engaged in the production of various forms of social, economic, symbolic, and cultural capital that will advance their rights to the city (Lefebvre, 1996).

NOTES

1 For more discussion of the relations between gender and environment that are the basis of feminist geography, see Mackenzie (1999) and Domosh and Seager (2001).
2 The same opinion is expressed by Holloway (1998).
3 The recent attention to consumption represents a return to the broad scope of early studies of women and transport that took account of many types of travel (Hanson and Hanson, 1981; Preston and Takahashi, 1988).
4 In Australian cities, suburb refers to the vast majority of the built-up area. Sydney is a typical Australian city, with an inner city, the municipality of Sydney, that is approximately 2.5 km^2 in area.

BIBLIOGRAPHY

Afsaruddin, A. (ed.) (1999) *Hermeneutics and Honor: Negotiating Female "Public" Space in Islamic/ate Societies*. Cambridge, MA: Distributed for the Center for Middle Eastern Studies of Harvard University by Harvard University Press.

Aitken, S. C. (1999) Putting parents in their place: child rearing rites and gender politics. In E. K. Teather (ed.), *Geographies of Personal Discovery: Places, Bodies and Rites of Passage*. London: Routledge.

Anderson, K. and Jacobs, J. M. (1999) Geographies of publicity and privacy: residential activism in Sydney in the 1970s. *Environment and Planning A*, 31(6), 1017–30.

Bagchi-Sen, S. (1995) Structural determinants of occupational shifts for males and females in the US labor market. *The Professional Geographer*, 47(3), 268–79.

Baxandall, R. F. and Ewen, E. (2000) *Picture Windows: How the Suburbs Happened*. New York: Basic Books.

Bondi, L. (1991) Gender divisions and gentrification. *Transactions of the Institute of British Geographers*, 16(2), 190–8.

Bondi, L. (1998) Gender, class and urban space: public and private space in contemporary urban landscapes. *Urban Geography*, 19(2), 160–85.

Bondi, L. and Christie, H. (2003) Working out the urban: Gender relations and the city. In G. Bridge and S. Watson (eds), *A Companion to the City*. Oxford: Blackwell.

Bourdieu, P. (1984) *Distinction: A Social Critique of the Judgement of the Taste*. Cambridge, MA: Harvard University Press.

Bourdieu, P. (2001) *Masculine Domination*. Stanford, CA: Stanford University Press.

Bourne, L. S. and Rose, D. (2000) The changing face of Canada: the uneven geographies of population and social change. *The Canadian Geographer*, 45(1), 105–19.

Bunting, T. and Filion, P. (2000) Uneven cities: addressing rising inequality in the 21st century. *The Canadian Geographer*, 45(1), 126–31.

Chambers, D. (1997) A stake in the country: women's experiences of suburban development. In R. Silverstone (ed.), *Visions of Suburbia*. London: Routledge.

Chapple, K. (2001) Time to work: job search strategies and commute time for women on welfare in San Francisco. *Journal of Urban Affairs*, 23, 155–73.

Chouinard, V. (1999) Life at the margins: disabled women's explorations of ableist spaces. In E. K. Teather (ed.), *Embodied Geographies, Space, Bodies and Rites of Passage*. London: Routledge.

Christensen, K. (1993) Eliminating the journey to work: home-based work across the life course of women in the United States. In C. Katz and J. Monk (eds), *Full Circle: Geographies of Women over the Life Course*. New York: Routledge.

Daphne, S. (2001) *How Women Saved the City*. Minneapolis: University of Minnesota Press.

Domosh, M. (1998) Geography and gender: home, again? *Progress in Human Geography*, 22(2), 276–82.

Domosh, M. (2001) The "Women on New York": a fashionable moral geography. *Environment and Planning D*, 19(5), 573–92.

Domosh, M. and Seager, J. (2001) *Putting Women in Place: Feminist Geographers Make Sense of the World*. New York: Guilford Press.

Dowling, R. (1993) Femininity, place, and commodities: a retail case study. *Antipode*, 25(4), 295–319.

Dowling, R. (1998) Suburban stories, gendered lives: thinking through difference. In R. Fincher and J. M. Jacobs (eds), *Cities of Difference*. New York: Guilford Press.

Dowling, R. and Gollner, A. (1997) *Understanding Women's Travel: From Transport Disadvantage to Mobility*. Sydney: NRMA.

Dyck, I. (1989) Integrating home and wage workplace: women's daily lives in a Canadian suburb. *The Canadian Geographer*, 33, 329–41.

Fincher, R. (1998) In the right place at the right time? Life stages and urban spaces. In R. Fincher and J. M. Jacobs (eds), *Cities of Difference*. New York: Guilford Press.

Fincher, R. (1991) Caring for workers' dependents: gender, class and local state practices in Melbourne. *Political Geography Quarterly*, 10, 356–81.

Fincher, R. and Panelli, R. (2002) Making space: women's urban and rural activism and the Australian state. *Gender, Place and Culture*, 8(2), 129–48.

Fraser, N. (1989) *Unruly Practices: Power, Discourse and Gender in Contemporary Social Theory*. Minneapolis: University of Minnesota Press.

Gilbert, M. (1998) Race, space and power: the survival strategies of working poor women. *Annals of the Association of American Geographers*, 88(4), 595–621.

Gole, N. (1996) *Forbidden Modern: Civilization and Veiling*. Ann Arbor: University of Michigan Press.

Gregson, N., Crewe, L. and Brooks, K. (2002) Discourse, displacement, and retail practice: some points from the Charity Retail Project. *Environment and Planning A*, 39, 1661–83.

Gregson, N. and Lowe, M. (1994) *Servicing the Middle Classes: Class, Gender and Waged Domestic Work in Contemporary Britain*. New York: Routledge.

Hanson, S. and Hanson, P. (1981) The impact of married women's employment on household travel patterns: a Swedish example. *Transportation*, 10(12), 165–83.

Hanson, S. and Pratt, G. (1988) Reconceptualizing the links between home and work in urban geography. *Economic Geography*, 64(4), 299–321.

Hanson, S. and Pratt, G. (1992) Dynamic dependencies: a geographic investigation of local labor markets. *Economic Geography*, 68(4), 373–402.

Hanson, S. and Pratt, G. (1995) *Gender, Work and Space*. London: Routledge.

Haraway, D. (1991) *Simians, Cyborgs, and Women: The Reinvention of Nature*. New York: Routledge.

Hayden, D. (2002) *Redesigning the American Dream, Gender, Housing, and Family Life*. New York: W. W. Norton and Company.

Holloway, S. L. (1998) Local childcare cultures: moral geographies of mothering and the social organization of pre-school education. *Gender, Place and Culture*, 5(1), 29–53.

Hyndman, J. (2001) Towards a feminist geopolitics. *Canadian Geographer*, 45(2), 210–22.

Johnston-Anumonwo, I. (1995) Racial differences in the commuting behavior of women in Buffalo, 1980–1990. *Urban Geography*, 16(1), 23–45.

Johnston-Anumonwo, I. (1997) Race, gender, and constrained work trips in Buffalo, NY, 1990. *The Professional Geographer*, 49(3), 306–17.

Kandiyoti, D. (1997) Women, Islam and the state. In J. Benin and J. Stork (eds), *Political Islam: Essays from the Middle East*. Berkeley: University of California Press.

Klodawsky, F., Farrell, S. and D'Aubry, T. (2002) Images of homelessness in Ottawa: implications for local politics. *The Canadian Geographer*, 46(2), 126–43.

Knopp, L. (1998) Sexuality and urban space: gay male identity politics in the United States,
the United Kingdom, and Australia. In R. Fincher and J. M. Jacobs (eds), *Cities of Difference*. New York: Guilford Press.

Kobayashi, A. (1993) Multiculturalism: representing a Canadian institution. In J. Duncan
and D. Ley (eds), *Place/Culture/Representation*. London: Routledge.

Koskela, H. (1997) Bold walk and breakings: women's spatial confidence versus fear of violence. *Gender, Place and Culture*, 4(3), 301–19.

Law, R. (1999) Beyond "women and transport": towards new geographies of gender and
daily mobility. *Progress in Human Geography*, 23(4), 567–88.

Lees, L. (2002) Rematerializing geography: the "new" urban geography. *Progress in Human
Geography*, 26(1), 101–12.

Lefebvre, H. (1996) *Writing on Cities*. Oxford: Blackwell.

Light, J. (1995) The digital landscape: new space for women. *Gender, Place and Culture*,
2(2), 133–46.

Lister, R. (1997) *Citizenship: Feminist Perspectives*. New York: New York University Press.

Longhurst, R. (1995) The body and geography. *Gender, Place and Culture*, 2, 97–105.

McDowell, L. (1997) *Capital Culture: Gender and Work in the City*. Oxford: Blackwell.

McDowell, L. (2001) Father and Ford revisted: gender, class and employment change in the
new millennium. *Transactions of the Institute of British Geographers*, 26(4), 448–64.

McDowell, L. (2002a) Transitions to work: masculine identities, youth inequality and labour
market change. *Gender, Place and Culture*, 9(1), 35–59.

McDowell, L. (2002b) Masculine discourses and dissonances: strutting "lads," protest masculinity and domestic respectability. *Environment and Planning D: Society and Space*,
20(1), 97–120.

Mackenzie, S. (1999) Restructuring the relations of work and life: women as environmental
actors, feminism as geographic analysis. *Gender, Place and Culture*, 6, 417–30.

McLafferty, S. and Preston, V. (1991) Gender, race, and commuting among service sector
workers. *The Professional Geographer*, 43(1), 1–15.

McLafferty, S. and Preston, V. (2003) Poverty and geographical access to employment: minority women in America's inner cities. In D. Janelle, B. Warf and K. Hansen (eds), *World-
minds: Geographical Perspectives on 100 Problems*. Dordrecht: Kluwer.

Mahler, S. J. (1999) Engendering transnational migration: a case study of Salvadorans.
American Behavioral Scientist, 42(4), 690–719.

Massey, D. J. (1995) *Spatial Divisions of Labor: Social Structures and the Geography of
Production*. New York: Routledge.

Mattingly, D. J. (1999) Making maids: US immigration policy and immigrant domestic
workers. In J. Momson (ed.), *Gender, Migration, and Domestic Service*. New York:
Routledge.

Mehta, A. and Bondi, L. (1999) Embodied discourse: on gender and fear of violence. *Gender,
Place and Culture*, 5, 67–84.

Mitchell, D. (2000) *Cultural Geography: A Critical Introduction*. Oxford: Blackwell.

Moghissi, H. (1991) *Feminism and Islamic Fundamentalism: The Limits of Postmodern
Analysis*. London: Zed Books.

Moore-Milroy, B. and Wismer, S. (1994) Communities, work and public/private sphere
models. *Gender, Place and Culture*, 1(1), 71–90.

Morris, L. D. (1994) *Dangerous Classes: The Underclass and Social Citizenship*. New York:
Routledge.

Moss, P. (1997) Spaces of resistance, spaces of respite: franchise housekeepers keeping house
in the workplace and at home. *Gender, Place and Culture*, 4(2), 179–96.

Murdie, R. and Teixeira, C. (2000) Towards a comfortable neighbourhood and appropriate housing: immigrant experiences in Toronto. CERIS Working Paper 10, Toronto.

Nagar, R. (2000a) I'd rather be rude than ruled: gender, place and communal politics among South Asian communities in Dar Es Salaam. *Women's Studies International Forum*, 23(5), 571–85.

Nagar, R. (2000b) Mukhe jawab do! (answer me!): women's grass-roots activism and social spaces in Chitrakoot (India). *Gender, Place and Culture*, 7(4), 341–62.

Nagar, R., Lawson, V., McDowell, L. and Hanson, S. (2002) Locating globalization: feminist (re)readings of the subjects and objects of globalization. *Economic Geography*, 78(3), 257–83.

Nast, H. J. (1998) Unsexy geographies. *Gender, Place and Culture*, 5(2), 191–206.

Peake, L. (1999) Toward a social geography of the city: race and dimensions of urban poverty in women's lives. *Journal of Urban Affairs*, 19(3), 335–61.

Peck, J. (2001) *Workfare States*. New York: Guilford Press.

Permezel, M. (2002) The practice of citizenship: place, identity and the politics of participation in neighborhood houses. Unpublished PhD dissertation, Faculty of Architecture, Building and Planning, University of Melbourne.

Phizacklea, A. and Wolkowitz, C. (1995) *Homeworking Women: Gender, Racism and Class at Work*. Thousand Oaks, CA: Sage.

Pratt, G. (1997) Grids of difference: place and identity formation. In R. Fincher and J. M. Jacobs (eds), *Cities of Difference*. New York: Guilford Press.

Preston, V. (2002) Moving to Mississauga: immigration in our midst. In A. von Gernet and F. Dieterman (eds), *Mississauga: The First Ten Thousand Years*. Mississauga: The Mississauga Heritage Foundation.

Preston, V. and Takahashi, S. (1988) An analysis of multistop grocery shopping trips. *The Professional Geographer*, 32, 339–46.

Pulido, L. (2000) Rethinking environmental racism: white privilege and urban development in Southern California. *Annals*, 90(1), 12–40.

Raju, S. (2002) We are different, but can we talk? *Gender, Place and Culture*, 9(2), 173–7.

Ray, B. and Rose, D. (2000) Cities of the everyday: socio-spatial perspectives on gender, difference, and diversity. In T. Bunting and P. Filion (eds), *Canadian Cities in Transition: The Twenty-First Century*, 2nd edn. Don Mills, ON: Oxford University Press.

Rose, D. (1988) A feminist perspective on employment restructuring and gentrification: the case of Montreal. In J. Wolch and M. J. Dear (eds), *The Power of Geography*. Winchester, MA: Unwin Hyman.

Rose, D. (1993) Local childcare strategies in Montreal, Quebec: mediations of state policies, class and ethnicity in the life courses of families with young children. In C. Katz and J. Monk (eds), *Full Circles: Geographies of Women over the Life Course*. London: Routledge.

Rose, D. and Chicoine, N. (1991) Access to school daycare services: class, family, ethnicity and space in Montreal's old and new "inner city." *Geoforum*, 22(2), 185–201.

Rothenberg, T. (1995) "And she told two friends": lesbians creating urban social space. In D. Bell and G. Valentine (eds), *Mapping Desire: Geographies of Sexualities*. London: Routledge.

Sassen, S. (2000) *The Global City: New York, London, Tokyo*, 2nd edn. Princeton, NJ: Princeton University Press.

Sibley, D. (1995) Gender, science, politics and geographies of the city. *Gender, Place and Culture*, 2(1), 37–49.

Smith, N. (1996) *The New Urban Frontier: Gentrification and the Revanchist City*. London: Routledge.

Sorkin, M. and Zukin, S. (2002) *After the World Trade Center, Rethinking New York City.* New York: Routledge.

Spain, D. (2001) *How Women Saved the City.* Minneapolis: University of Minnesota Press.

Staeheli, L. (1996) Publicity, privacy and women's political action. *Environment and Planning D: Society and Space,* 14, 601–19.

Stiell, B. and England, K. (1997) Domestic distinctions: constructing difference among paid domestic work in Toronto. *Gender, Place, and Culture,* 4(3), 339–59.

Thomas, S. (1994) *Gender and Poverty.* New York: Garland Publishing.

Thrift, N. and Glennie, P. (1996) Consumers, identities and consumption spaces in early modern England. *Environment and Planning A,* 28, 25–43.

Tivers, J. (1985) *Women Attached: The Daily Lives of Women with Young Children.* London: Croom Helm.

Truelove, M. (2000) Services for immigrant women: an evaluation of locations. *The Canadian Geographer,* 44(2), 135–51.

Valentine, G. (1995) "Out and about": a geography of lesbian communities. *International Journal of Urban and Regional Research,* 19(3), 96–111.

Valentine, G. (1997) Making space: separatism and difference. In J. P. Jones III, H. J. Nast, and S. M. Roberts (eds), *Thresholds in Feminist Geography: Difference, Methodology, Representation.* Lanham, MD: Rowman and Littlefield.

Wekerle, G. (1984) A woman's place is in the city. *Antipode,* 6(3), 11–20.

Wekerle, G. (2000) From eyes on the street to safe cities. *Places,* Winter, 44–9.

Wekerle, G. and Peake, L. (1996) New social movements and women's urban activism. In J. Caulfield and L. Peake (eds), *City Lives and City Forms, Critical Research and Canadian Urbanism.* Toronto: University of Toronto Press.

White, J. B. (2002) *Islamist Mobilization in Turkey: A Study in Vernacular Politics.* Seattle: University of Washington Press.

Westwood, S. and Phizacklea, A. (2000) *Transnationalism and the Politics of Belonging.* London: Routledge.

Wilson, E. (1991) *The Sphinx in the City: Urban Life, the Control of Disorder, and Women.* Berkeley: University of California Press.

Wyly, E. K. (1999) Continuity and change in the restless urban landscape. *Economic Geography,* 75(4), 309–38.

Yeoh, B. and Huang, S. (1999) Spaces at the margins: migrant domestic workers and the development of civil society in Singapore. *Environment and Planning A,* 31(7), 1149–67.

Yeoh, B., Huang, S. and Willis, K. (2000) Global cities, transnational flows, and gender dimensions, the view from Singapore. *Tijdschrift voor Economische en Sociale Geografie,* 91(2), 147–58.

Yuval-Davis, N. (1997) *Gender and Nation.* Thousand Oaks, CA: Sage.

Chapter 16

Spaces of Change: Gender, Information Technology, and New Geographies of Mobility and Fixity in the Early Twentieth-century Information Economy

Kate Boyer

Introduction

Feminist geographers have had a longstanding interest in the way power and identity are expressed in the workplace (Mackenzie, 1988; England, 1993, 1995; McDowell, 1994, 1997, among others). At the same time, a growing body of scholarship within geography has begun to look at the ways in which the information and communications technologies are remapping relations of access and opportunity at the level of the workplace, city, and region (Mitchell, 1996; Wheeler et al., 2000; Graham and Marvin, 2001). Neither of these processes is new, nor are they separate.

The financial services sector was one of the first North American industries to direct itself toward the organization and flow of information rather than material goods. In the late nineteenth and early twentieth centuries the financial services sector greeted with enthusiasm the invention of devices that would facilitate the ordering of information, and adopted new information and communications technologies such as the telephone, typewriter, dictaphone, and mimeograph machine on a grand scale. These technologies changed the way office work was done, the way businesses operated, and even the way people thought about cities. By the early twentieth century new information technologies allowed companies within the financial services sector to dramatically expand their reach into cities, towns, and more remote parts of the continent, as well as overseas. Within individual companies, relations between branches and head offices came to function as an increasingly complex network, keeping flows of information and capital in motion across a greater and greater territory.[1] At the scale of individual offices, work was becoming more and more rationalized and segmented by task. At the same time, demographics within the financial services sector were also changing dramatically, so that by the end of the 1920s most of the work in the white-collar office was being done by women (Fine, 1991; Kwolek-Folland, 1994).

Through an analysis of the financial services sector in early twentieth-century Montreal, Canada, the central hub of the Canadian financial services industry until mid-century, this chapter considers early twentieth-century networked capitalism as a *gendered* system. After tracing the scope and scale of women's entrance into the white-collar workplace in early twentieth-century Montreal by way of background, I examine how new information technologies influenced social and spatial relationships at the level of the body, the city, and within broader networks of branch-banking. I argue that two processes – the feminization of clerical work and the emergence of the modern "wired" workplace – created new geographies of fixity and mobility, and that these differed both by sex and by scale.

This chapter is based on research conducted for a dissertation on white-collar workplace culture and changing meanings about women's use of public space in early twentieth-century Montreal (Boyer, 2001). This larger project was based on an analysis of employee files, intradepartmental memoranda and employee journals from the five largest financial institutions in Canada in the early twentieth century;[2] archival materials from clerical schools; English and French newspapers and women's magazines; and census records, including a 10 percent sample of the 1901 nominal census for the island of Montreal ($N = 5500$). The only existing scholarship on the financial services sector in Quebec at the time of the project focused exclusively on the smaller French banking sector (Rudin, 1986; Dagenais, 1987, 1989). I thus decided to focus primarily on the larger English sector. However, I brought in French sources where they were available in order to analyze the cultural construction of religious and linguistic identity – and difference – within the sector as a whole. Based on this research I argued that the early white-collar workplace constituted a new kind of work culture, and that changes in workplace practices influenced broader discourse and debate about women's "place" in society, helping to create an image of modern womanhood that was consummately urban (Boyer, 1998, 2001).

The Feminization of Clerical Work

The revolution in office technologies which began in the late nineteenth century was marked by the invention of machines with broad applicability like the telephone and typewriter and specialized office equipment such as the dictaphone, mimeograph, and hollerith machine. These new technologies were mobolized to create an explosion of paperwork that required keeping track of. In turn, for companies in the financial services sector, this created the need for a large number of routinized office jobs at the bottom of the organizational structure (Lowe, 1981, 1984). Over the first two decades of the twentieth century, clerical work expanded as an employment market in cities throughout North America. For class-conscious companies attuned to their public image and operating in the context of sexist culture, it made sense to hire educated, middle-class women who could be paid wages equivalent to a male day-laborer to perform office work. For young women whose parents could afford even a short course in professional training beyond high school, clerical work was an appealing option. Though not well paid in comparison to men's wages, clerical work paid better than nearly any other type of employment open to women. With shorter hours, and workplaces which were clean and modern, clerical work had become the third most important source of employment for women in Quebec by 1911, after manufacturing and domestic service.[3]

The demographics of who performed clerical work in early twentieth-century Montreal was conditioned in part by the cultural dynamics of the city. As today, Montreal was a rich mix of language, religion, and ethnicity, with English-Protestant, French-Catholic, and Irish-Catholic constituting the three major cultural groups. Montreal was also home to significant Jewish, Italian, and other smaller immigrant communities (Olson, 1991; Linteau, 1992; McNicholl, 1993). Most Montreallers spoke French as their first language and claimed Catholicism as their religion. Of those claiming British origins, about two-thirds were Protestant, and one-third Irish-Catholic. While the Catholic Church controlled an impressive array of social institutions, economic power was concentrated in a bourgeoisie comprised largely of Anglo-Scottish Protestants, with Irish- and French-Catholics composing the majority of the middle and working classes (Dickinson and Young, 1988; Germain and Rose, 2000). Montreal served as the center-point for the national banking and financial services industry, as well as for the smaller French banking system. The financial services sector overall was disproportionately English-Protestant in composition, in terms of both language spoken and religious-linguistic background of the workforce, though among clerical employees Irish-Catholics were overrepresented relative to the city as a whole (Boyer, 2004).

Not unlike the North American economy of the late 1990s, the 1910s saw the emergence of an employment market that sucked workers in. Montreal served as the major Canadian destination point for overseas immigration throughout this period, as well as a key destination for rural Canadians seeking better employment prospects than were available in the countryside. Demographics within the clerical workforce echo this trend. For example, as early as 1901 a third of the women claiming employment as clerical workers, stenographers, or typists on the Montreal census were born outside the city itself (Boyer, 2001). Of the 588 women employed at the Head Office of the Bank of Montreal between 1902 and 1923 for whom place of birth is known, more than half were born outside of Montreal, with one-quarter hailing from the United Kingdom and smaller portions from the United States and other countries (figure 16.1).

As a group, clerical workers in the early twentieth-century city were overwhelmingly young and single. Reinforcing the findings of Dagenais and Rudin, my research found that the average age at hiring for all women employed at the Bank of Montreal between 1902 and 1923 was twenty years old. The average length of tenure was two years, and over 98 percent were single (Boyer, 2001). Though for a small number of women clerical work served as a means to remain financially independent throughout adult life, for most it represented a time of independence and relative freedom between young womanhood and married life.[4]

Employed in modern workplaces and associated with technology's cutting edge, clerical workers became associated with an emergent landscape of leisure and consumption expressed through the cinema, telephone, and "street-car city."[5] A 1931 illustration in *MacLean's Magazine* for an article singing the praises of office work offers a vivid example of how the freedom enjoyed by clerical workers was con-

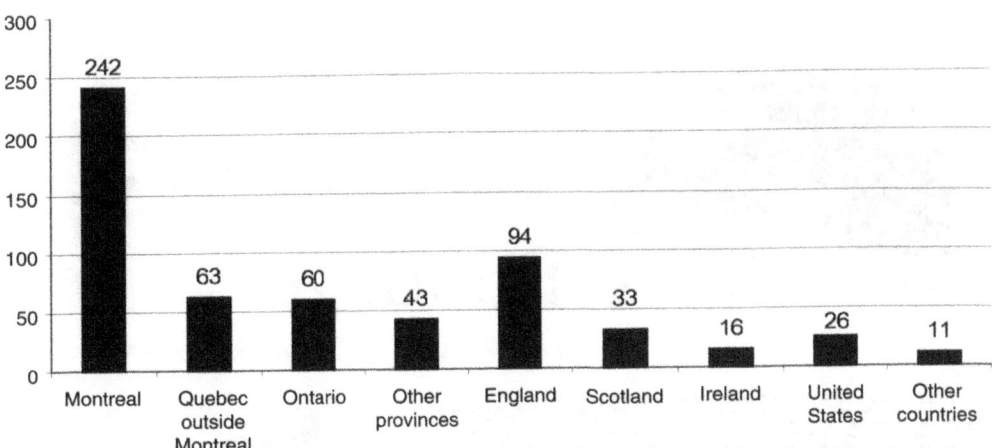

Figure 16.1 Birthplaces of women employed by the Bank of Montreal, 1902–1923. *Source*: Employee, files, Bank of Montreal Archives.

trasted in popular media to the constraints of life for the young mother at home (figure 16.2). In it, a young woman shackled to the "ball and chain" of home and family gazes longingly toward an image of life in a busy office. In turn, if life at home was represented by the ball and chain, life as a clerical worker was represented by the bright lights of urban nightlife, as suggested in this excerpt from a 1926 article in the *Montreal Daily Herald*:

> Eight o'clock and the entrance to each of the many theatres that open their jaws along St Catherine Street holds its group of waiting ones, a stenographer waiting for her girl-friend, a saleslady keeping a tryst with her fellow. . . . [A] young sheik waits for a dance-hungry jazz baby inside the portal of the hall of the theatre just down the street. (*Montreal Daily Herald*, January 7, 1926, p. 3)

Through their advice columns, articles with such titles as "Saturday Talks to Business Girls," and contributions to discussions on the need for working women's clubs in the city, newspapers of the day cast clerical workers as a recognizable urban type, characterized by their youth, freedom, and (importantly) unmarried status, and set in contrast to the confined lives of their "married sisters" (Boyer, 2004).

Inside the Early Wired Workplace

The twin processes of the administrative revolution and the feminization of clerical work also coincided with efforts to rationalize work processes within the financial services sector by employing principles of scientific management or Taylorism (Lowe, 1984). In the white-collar office, scientific management took the form of differential pay-scales, job-classification schedules, human resources departments, and the quantification of labor in terms of words per minute and/or units of work.[6] In offices, the drive for efficiency was manifested in spatial practice through various

Figure 16.2 Office work as freedom?
Source: MacLean's Magazine, October 1, 1931, p. 15.

forms of bodily, or corporeal, discipline designed to control workers' movements through time and space. Employees clocked in and out, ate their lunches at designated times, and punched time cards documenting attendance and punctuality. At Sun Life Insurance Company, one of the largest employers in the financial services sector in early twentieth-century Montreal, employees were not allowed to leave their departments unless on company business, and department heads were instructed to report offenders.[7] Loitering in locker rooms, washrooms, and corridors was specifically forbidden. Employees wishing to leave the building during business hours required a pass from the department head, to be presented to the hall porter in exchange for passage through the building.[8]

In addition to this monitoring of employees' movements within the building, workstations were arranged to be efficient and cost-effective. At most of the corporations under review by the 1920s, stenography and clerical work were centralized within company headquarters, and performed at desks in large, open-floored workspaces. The Bank of Nova Scotia had a typing pool by 1911 (Lowe, 1982), and in that same year the Sun Life employee journal (then *Sunshine*) described the "army of typewriters," and typing pools filled with women.[9]

The degree of workplace rationalization and spatial constraint a worker experienced differed by his or her position within the occupational hierarchy. Farther up the employment ladder were those whose labor was less mechanized and more highly paid (such as supervisors and managers). These jobs allowed employees freedom to circulate, patrolling workstations and monitoring other employees' output levels. At the lower end of the ladder were those whose labor was more mechanized, and it is at this level that we find the most standardization and monitoring of the work process. For employees whose work was identified most closely with a certain piece of machinery (such as a typewriter), being anywhere but at one's workstation meant not "doing one's job." Because women were barred from advancing up the managerial hierarchy in the time period under review, they were subject to higher levels of workplace rationalization and surveillance in the white-collar workplace than were men.

With each ascending level on the hierarchy, employees in the institutions under review had larger workspaces with more autonomy and more privacy.[10] Employees at the lower end of the administrative ladder toiled in densely packed workstations designed for efficient work output, spatially bound to their desks in order to do their jobs. In addition to saving on building costs, the open floor plan denied privacy, curtailed space for individual expression or taste, and allowed for easy surveillance of a large workforce (figure 16.3). In contrast, upper management (at Sun Life and elsewhere) typically worked in large offices decorated with middle-class accoutrements such as paintings and lounge chairs. Freed from the typewriter or dictaphone, these men worked at desks designed for conversing, reading, and thinking, in offices equipped with doors and secretaries to buffer and regulate contact with others.

As Foucault has shown, institutional space has long been characterized by the surveillance of bodies and the ordering of their movements in time and space.[11] Though to a lesser degree, the rules governing spatial practice in the early twentieth-century white-collar workplace have their origins in the mechanisms of bodily control exercised in factories, prisons, schools, hospitals, and mental institutions.[12]

Figure 16.3 Policy Department, Sun Life, 1920s.
Source: Sun Life Archives.

In the case of the white-collar workplace in the financial services sector in which women were barred from moving up the occupational ladder, disciplinary practices took on a distinctly gendered character. Although clerical work offered women a new kind of employment opportunity as compared to other choices available to them, women and men were positioned very differently in the white-collar work-place. Whereas women were "tied" or tethered to spatially fixed pieces of equip-ment such as a typewriter or dictaphone in workspaces that more easily lent themselves to visual and auditory surveillance, those higher up the employment ladder typically worked in private offices with a higher degree of freedom, privacy, and mobility. In contrast to both the highly mobile nature of so many clerical workers' lives – migrating from farms and even from overseas – and media repre-sentations which highlighted this group's freedom, within the workplace itself cler-ical workers were spatially fixed: "tied down" to their machines and workstations.

Gendered Networks: Branch Banking in the City and Beyond

In addition to the ways new technologies changed spatial patterns in the white-collar office, technological innovation also changed the role financial institutions played within cities, and the networks of capital and information they supported. The expansion of branch banking served as an engine for urban growth more generally, spurring the expansion of infrastructural networks to serve transportation, com-

Figure 16.4 Sun Life head office, Montreal, *c.*1950.
Source: Sun Life Archives.

munications, and energy needs. Taking advantage of interior steel-frame construction developed in the late nineteenth century, head offices served to concentrate power in the form of information, people, and capital in "command and control" centers, creating new geographies of centrality in the larger cities. Imposing head office buildings gave companies a new symbolic significance on the urban landscape, and gave ideas about rationality, modernity, efficiency (and in the Canadian case, imperialism) physical form. They represented both the company itself (as on letterhead and in advertising), and the cities in which companies were headquartered: contributing to a process of urban branding which continued throughout the century.

Though situated in urban areas rich in services and amenities, the trend in the case of the companies under review was for head offices to function as much as possible as total worlds, in which employees could engage in both work and non-work activities. A vivid example of this is to be found in the Sun Life Insurance Company headquarters, completed at the end of the 1920s (figure 16.4). The structure was heralded as the largest building in the British Empire at the time of completion. Architects and engineers reveled in describing its monumental size and, even more, its innovative use of technology. Take, for example, the following description of the company kitchen and employee dining area, described in a 1929 issue of *Construction: A Journal for the Architectural Engineering and Contracting Interests of Canada*:

> nutritional needs are provided for generously and scientifically by a cafeteria . . .
> its array of modernly equipped steel-topped services and conveniences, smart and

comfortable furniture, and skilful planning for light, air and space . . . is a commend-
able model of its kind. The equipment of the adjoining kitchen is to the last degree
interesting to those who study modern methods of swiftly feeding hungry people in
numbers to be reckoned in hundreds.[13]

The author reinforced his point by describing various pieces of particularly impressive kitchen equipment, including a 70-gallon coffee urn, mixing machines capable
of handling 45 pounds of bread dough, and a dishwasher that could clean and san-
itize 10,000 plates an hour.

To E. F. Chackfield, an engineer reporting in 1931 on the structural marvels of
this building in the *Municipal Review of Canada*, the 24-story head office amounted
to no less than "a city within a city."[14] Assuring readers that the 35 acres of interior space was every bit as impressive as its exterior, the author highlighted the 38
elevators, 5½ miles of telephone cables, and three basements housing entertainments
such as archery ranges, shooting galleries, and a bowling alley, all to amuse the
10,000–12,000 employees in off-hours.

By drawing on what was then technology's cutting-edge, head offices could keep
employees linked to each other and the outside world, while at the same time sat-
isfying as many employee needs as possible (including dining, sports, and other
leisure activities) in-house, under watchful eyes. Descriptions such as these highlight
the fact that not only were twentieth-century head offices impressive, they were
impressive in a way that older examples of corporate architecture were not. They
suggest a system for measuring the value and importance of commercial space based
on magnitude of scale and level of technological sophistication which would endure
throughout the century and beyond (Boyer, 2004).

While head offices served as control points through which the most information
flowed and where the most far-reaching decisions were made, the early twentieth-
century information economy was also marked by tremendous spatial dispersion to
hinterlands and overseas. For Sun Life, with branches throughout the British
Empire, the goal was clearly to be global (figure 16.5). The same can be said of the
Bank of Montreal, which in 1912 had 167 branch offices nationwide, as well as
branches in Great Britain, the USA, and Mexico, and affiliate branches in major
cities in Asia, Europe, Australia, New Zealand, Argentina, Bolivia, Brazil, Chile,
Peru, and British Guiana.[15] Expansion on this scale required that capital and information flow smoothly from one part of the system to another. Toward this end,
technology functioned not on its own or in isolation from social relations, but in
and through the tight management of labor and the spaces and bodies in which it
was performed. As we saw above, this happened in a highly gendered way on the
information shop-floor.

Expansion of the financial services sector in the late nineteenth and early twentieth centuries created massive, geographically dispersed networks within individual corporations. In the companies under review, men were transferred to and from
head offices and between branches as part of managerial training. This practice
served to familiarize management trainees with different positions within the
company, and to standardize practices across the network. Transferring also helped
to build corporate affinity. As Benedict Anderson has observed, spatially mobile
employees can be thought of as human "nodes" which strengthen a sense of shared
corporate identity (Anderson, 1983). As they moved from one place to another

Figure 16.5 "Sun Life on Top of the World."
Source: Sun Life Archives.

within the corporate network, everywhere meeting people with whom they built and shared common cultural references, these mobile (male) corporate emissaries (and their wives and children) built webs of community nationally and internationally.

Whereas men were expected to be mobile within this network, and were routinely transferred as part of their rise up the managerial hierarchy (or occasionally for breaches of discipline), women were barred from transferring, mirroring the constraint on women's movement at the level of the workstation discussed in the previous section. Employee files from the Bank of Nova Scotia suggest that requests from women seeking to transfer within the same job category during this period, even when they offered to cover their own travel and relocation expenses, were systematically denied.[16] The denial of transfer requests from female employees both suggests the lower value placed on their labor and reinforces the argument that women functioned as fixed points within networks of the twentieth-century financial services sector. As a contributor to the Royal Bank employee magazine put it in 1927:

> As regards promotion, it should be remembered that in one respect the female employees are at a disadvantage. A young man entering banking with the intention of making it his life-work is prepared to undertake service at any branch, however remote, whenever directed to do so by Head Office . . . the circumstances and inclinations of female employees, as a rule do not permit them to serve in any but the branch or branches located in their hometown or city.[17]

As we saw in the first section, this testament to women's supposed immobility stands in contrast to the highly mobile nature of women's lives as suggested by the 1901 census and places of birth of Bank of Montreal employees. In spite of evidence to the contrary, however, within the middle-class ideology that informed gender relations in the early twentieth-century white-collar workplace, women were imagined as more immobile and "home-based" than men, even when they were at work.

Conclusions/Trajectories

As I have argued, the adoption of new information technologies within the early twentieth-century financial services sector was not a neutral process. Instead, it regulated, defined, and positioned bodies and spaces in ways that privileged some bodies over others, and reinforced certain relationships of power. On the one hand, clerical work opened up new opportunities for women in the context of limited employment options. Yet regimes of office rationalization intended to ensure the efficient flow of capital and information at the level of the workplace had gendered effects. Women in the early wired workplace were more likely to be "tethered" to specific machines in the office, and, owing to rules barring their advancement, were subject to higher levels of surveillance and corporeal constraint than their male counterparts.

I have further argued that these effects were echoed at the broader scale of networked branch banking. In addition to advancing goals of profit-making and augmenting imperial glory, networked branch banking and insurance companies constituted a system for career advancement that worked to further constrain women, both physically and metaphorically. While men moved through the system, making human networks and moving up the career ladder, women were conceived of as fixed points, tethered to their branch and hometown as they were to their typewriter or dictaphone. In these ways, new technologies reinforced power relations of class, sex, and ethnicity, even as they opened new opportunities for both sexes.

I have told a story about a technological regime that pulled women into the center of the information economy, and also into the heart of the city. In today's world, "stenographer" is being phased out of civil service job classification schedules nationwide. Instead, this work is increasingly talking place in data-entry warehouses on the ex-urban fringe (or even offshore), far out of public view. But as the information economy has expanded and flourished, so have the same forms of discipline, surveillance, and spatial restraint I have talked about here. As the spaces of information processing withdraw farther from public view, I would suggest that it is that much more vital to use examples such as this from the past to shine light on the *contemporary* world of information processing and ask questions along the lines that I have asked here – interrogating the ongoing relationship between changing information systems and the politics of who is served and who is silenced.

NOTES

I would like to thank Bob Sparrow at the Sun Life Insurance Company Archives.

1. For more about the nature of twentieth-century networked capitalism see Castells (1991).
2. Sun Life Insurance Company, the Bank of Nova Scotia, the Bank of Montreal, Royal Bank, and the Canadian Imperial Bank of Commerce.
3. 1911 Census of Canada, Vol. IV, Table XI.
4. Though high relative to other kinds of work open to women, clerical wages capped out after a few years, making it difficult to support oneself for a sustained period of time.
5. For an analysis of how certain nineteenth-century urban sites came to feminized, see Domosh (1996).
6. For more about scientific management as an expression of gender relations in North America, see Banta (1993).
7. *Company Rules and Regulations*, circa 1920, Box 406, Sun Life Archives.
8. Ibid.
9. *Sunshine*, November 1911, p. 142. See also Florence M. Richards, "Half a century of girls," *The Sun Life Review*, January 1945, p. 23. Both: Sun Life Archives.
10. For more on gender, power, and the allocation and arrangement of floor space in the white-collar workplace, see van Slyck (1992) and Spain (1992).
11. For more on corporeality in the white-collar workplace, see Pringle (1989).
12. For more on the importance of ordering workers' movements in systems of advanced capitalism, see Pred (1990).
13. Author Unknown, *Construction: A Journal for the Architectural Engineering and Contracting Interests of Canada*, 22(2), February 1929, pp. 40–8, quoted passage, p. 42.
14. E. F. Chacksfield, "A city within a city," *The Municipal Review of Canada*, 27(3), March 1931, pp. 12–13.
15. Report of Annual General Meeting 1912, Bank of Montreal Archives.
16. E. J. McClaren, Case 31, pp. 242–3; Case 50; M. L. McDougal, Case 28. All Bank of Nova Scotia Archives.
17. W. L. G. Cumming, "The duties of country junior," *The Royal Bank Magazine*, August 1927, p. 8. Royal Bank Archives.

BIBLIOGRAPHY

Anderson, B. (1983) *Imagined Communities: Reflections on the Origin and Spread of Nationalism*. London: Verso.

Banta, M. (1993) *Taylored Lives: Narrative Productions in the Age of Taylor, Veblen, and Ford*. Chicago: University of Chicago Press.

Boyer, K. (1998) Place and the politics of virtue: clerical work, corporate anxiety, and changing meanings of public womanhood in early twentieth-century Montreal, *Gender, Place and Culture*, 5(3), 261–76.

Boyer, K. (2001) The feminization of clerical work in early twentieth-century Montreal. PhD dissertation, McGill University, Montreal.

Boyer, K. (2004) "Miss Remington" goes to work: gender, space and technology at the dawn of the information age. *The Professional Geographer*, 56(2), 201–12.

Castells, M. (1991) *The Informational City: A New Framework for Social Change*. Toronto: Centre for Urban and Community Studies.

Coombs, D. (1978) The emergence of a white-collar workforce in Toronto, 1895–1911. PhD dissertation, York University, Toronto.

Dagenais, M. (1987) Division sexuelle du travail en milieu bancaire: Montréal, 1900–1930. Memoir présenté a l'Université du Quebec à Montréal comme exigence partielle de la matrise en histoire Université du Québec à Montréal.

Dagenais, M. (1989) Itinéraires professionnels masculins et féminins en milieu bancaire: le cas de la Banque d'Hochelaga, 1900–1929. *Labour/Le Travail*, 24, 45–68.

Dickinson, J. and Young, B. (1988) *A Short History of Quebec: A Socio-Economic Perspective*. Toronto: Copp Clark Pitman.

Domosh, M. (1992) Corporate cultures and the modern landscape of New York City. In K. Anderson and F. Gale (eds), *Inventing Places: Studies in Cultural Geography*. Melbourne: Belhaven Press.

Domosh, M. (1996) The feminized retail landscape: gender ideology and consumer culture in ninteenth-century New York City. In N. Wrigley and M. Love (eds), *Retailing, Consumption and Capital*. New Haven, CT: Yale University Press.

England, K. (1993) Suburban pink collar ghettos: the spatial entrapment of women? *Annals of the Association of American Geographers*, 83(2), 225–42.

England, K. (1995) Girls in the office: recruiting and job search in a local clerical labor market. *Environment and Planning A*, 27, 1995–2018.

Fine, L. (1991) *Souls of the Skyscraper: Female Clerical Workers in Chicago 1870–1930*. Philadelphia: Temple University Press.

Foucault, M. (1979) *Discipline and Punish: The Birth of the Prison*. New York: Random House.

Germain, A. and Rose, D. (2000) *Montréal: The Quest for a Metropolis*. New York: Wiley.

Graham, S. and Marvin, S. (2001) *Splintering Urbanism: Networked Infrastructures, Technological Mobilities and the Urban Condition*. London: Routledge.

Kwolek-Folland, A. (1994) *Engendering Business: Men and Women in the Corporate Office, 1870–1930*. Baltimore: Johns Hopkins University Press.

Linteau, P. (1992) *Histoire de Montréal depuis la confederation*. Montréal: Boréal Press.

Lowe, G. (1981) The administrative revolution in the Canadian office: an overview. In K. Lundy and B. Warme (eds), *Work in the Canadian Context: Continuity Despite Change*. Toronto: Butterworths.

Lowe, G. (1982) Class, job and gender in the Canadian office. *Labour/La Travail*, 10 (Autumn), 11–37.

Lowe, G. (1984) Mechanization, feminization, and managerial control in the early twentieth century Canadian office. In C. Heron and R. Storey (eds), *On the Job: The Labour Process in Canada*. Toronto: McGill-Queen's University Press.

McDowell, L. (1994) Social justice, organizational culture and workplace democracy: cultural imperialism in the City of London. *Urban Geography*, 15(7), 661–80.

McDowell, L. (1997) *Capital Culture: Gender at Work in the City*. Oxford: Blackwell.

Mackenzie, S. (1988) Building women, building cities: toward gender sensitive theory in the environmental disciplines. In C. Andrew and B. Milroy (eds), *Life Spaces: Gender and Household Employment*. Vancouver: University of British Columbia Press.

McNicholl, C. (1993) *Montréal: Une société multiculturelle*. Paris: Editions Belin.

Mitchell, W. (1996) *City of Bits: Space, Place and the Infobahn*. Boston: MIT Press.

Olson, S. (1991) Ethnic strategies in the urban economy. *Canadian Ethnic Studies*, 33(2), 39–64.

Pred, A. (1990) *Lost Words and Lost Worlds: Modernity and the Language of Everyday Life in Late Nineteenth-Century Stockholm*. Cambridge: Cambridge University Press.

Pringle, R. (1989) *Secretaries Talk: Sexuality, Power, and Work*. London: Verso.

Rudin, R. (1986) Banker's hours: life behind the wicket at the Banque d'Hochelaga, 1901–1921. *Labour/Le Travail*, 18 (Fall), 63–76.

Spain, D. (1992) *Gendered Spaces*. Chapel Hill: University of North Carolina Press.

van Slyck, A. (1992) Gender and space in American public libraries, 1880–1920. Working Paper No. 27, Southwest Institute for Research on Women, University of Arizona.

Wheeler, J., Aoyama, Y. and Warf, B. (eds) (2000) *Cities in the Telecommunications Age: The Fracturing of Geographies*. New York: Routledge.

Chapter 17

Gender and the City: The Different Formations of Belonging

Tovi Fenster

This chapter highlights the different formations of gendered belonging as they are expressed in women's and men's daily practices in the city. It emphasizes not only the formal expressions of belonging built in to the different definitions of citizenship or the sacred dimensions of belonging expressed in individuals' and communities' religious and national attachment to territories but also the "everyday" nature of this sentiment that men and women develop in their daily practices in cities today.

This analysis is based on research carried out between 1999 and 2002 in which residents of London and Jerusalem were interviewed regarding their everyday experiences as related to three notions – comfort, belonging, and commitment – with regard to the various categories of their environment, home, building, street, neighbourhood, city centre, city, urban parks (Fenster, 2004). The research is based on a qualitative–content analysis methodology of peoples' narratives about their perceptions of sense of comfort, belonging, and commitment. People told their stories about their lives in the city as related to these categories and from their daily experiences we drew out our understanding on the gendered aspects of belonging in the city. The people interviewed represent both the "majority" hegemonic – that is, the Jewish secular in Jerusalem and the white middle-class English in London – and also the "minority," the "other," whether Bangladeshi immigrants in London or Palestinians in Jerusalem. This wide range of cultural expressions and ethnicities enabled exploration of the multilayered expressions of belonging both in their formal structures as citizenship definitions and in their personal, intimate, private expressions in daily practices in the city.

Why look at these two cities? Because they reflect contrasting images and symbolism. Jerusalem is a home for people of diverse identities, especially in the light of its image as one of the holiest cities in the world, a place of symbolism for Muslims, Christians, and Jews. And it is also a city that is associated with rigidity, perhaps fanaticism, strict rules, and boundaries, which sometimes find their expressions in spaces of *sacred belonging* that sometimes exclude women (see Romann,

1991; Be'tselem, 1997; Bollens, 2000; Cheshin et al., 2000; Fenster, forthcoming). London is a city famous for its globalization impacts and its images of cosmopolitanism, openness, and tolerance, but also for its negative and depressing connotations, especially for non-English immigrants and other types of newcomers (see Raban, 1974; Forman, 1989; Thornley, 1992; Fainstein, 1994; Jacobs, 1996; Pile, 1996). Comparing the narratives of women and men living in these two cities helps to expose the multilayered nature of gendered belonging which is constructed in urban daily practices.

What is a sense of belonging? Probyn (1966, cited in Yuval Davis, 2003) has emphasized the affective dimensions of belonging – not just of be-ing, but of longing or yearning. The Oxford Dictionary defines "belonging" through three meanings: first, to be a member (of a club, household, grade, society, state, etc.); second, to be resident or connected with; third, to be rightly placed or classified to or fit in a specific environment. These dimensions emphasize the membership component of belonging and its multilayered dimensionality (Yuval Davis, 2003). As we see later, in many cases belonging is also associated with past and present experiences and memories and future ties connected to a place, which grow with time (Fullilove, 1996; Crang, 1998).

Side by side with these definitions, there is a large body of literature that deals with the everyday practices of belonging. De Certeau's book *The Practice of Everyday Life* (1984) constructs the notion of belonging as a sentiment, which is built up and grows out of everyday life activities. De Certeau terms this "a theory of territorialization" through spatial tactics. In his work he draws the distinction between "place" and "space" as being, somewhat confusingly, that space is place made meaningful (Leach, 2002), or in de Certeau's words: "*space is a practical place.* Thus, the street geometrically defined by urban planning is transformed into a 'space' by walkers" (de Certeau, 1984, p. 117). For de Certeau, corporal everyday activities in the city are part of a process of appropriation and territorialization: "The ordinary practitioners of the city live 'down below,' below the thresholds at which visibility begins. They walk – an elementary form of this experience of the city; they are walkers, *Wandersmanner,* whose bodies follow the thicks and thins of an urban 'text' they write without being able to read it" (ibid., p. 93). This everyday act of walking in the city is what marks territorialization and appropriation and the meanings given to a space. What de Certeau constructs is a model of how "we make a sense of space through walking practices, and repeat those practices as a way of overcoming alienation" (Leach, 2002, p. 284). De Certeau actually defines the process in which a sense of belonging is established, a process of transformation of a place, which becomes a space of accumulated attachment and sentiments by means of everyday practices. Belonging and attachment are built here on the base of accumulated knowledge, memory, and intimate corporal experiences of everyday walking. A sense of belonging changes with time as these everyday experiences grow and their effects accumulate.

A significant aspect of "everyday belonging" develops through men's and women's spatial knowledge of environments. In both London and Jerusalem women's and men's narratives reveal the connection between daily walking practices and a sense of belonging. The knowledge of the area reinforces a sense of belonging: "I know the street, I live here, I know the building, every stone of it.

... I know it more and more. A very intimate knowledge" (Susana, thirties, married with one child, Jewish-Israeli, Jerusalem, July 13, 2000). People living in London also mention this intimate knowledge of the area, of its little alleys and shortcuts: "knowing the streets makes me feel I belong," "knowing shortcuts – it shows I know the neighborhood" (Robert, thirties, single, British-White, London, September 1, 1999). Also, people who walk their dogs several times a day mentioned this daily ritual as contributing to their sense of belonging to the area.

This dimension of belonging, which is based on everyday ritualized use of space, has a clear gendered dimension. Several women interviewed said that they felt much more attached to their environment after they became mothers. As a result of their gendered divisions of roles in the household these young mothers began to use the environment near their home more intensively than before, especially for shopping or taking the children to school or walking with their baby strollers. Their role as mothers is one of the significant aspects of their embodied knowledge as related to the notion of belonging. Their daily household duties made their attachment to the environment stronger because of their gendered division of roles more than for their partners. Men, on the other hand, didn't mention their fatherhood as a significant indicator in their experiences of a sense of belonging to their environment.

In what follows we elaborate on these themes of belonging, looking in particular at the notion of exclusion, the formal one expressed in the different definitions of citizenship, and the personal one expressed in the life-accumulated experience and memory.

Formal Structures of Belonging: The Discourse around Citizenship

Popular definitions of citizenship mention equality, communality, and homogeneity as part of what citizenship means – almost in implicit contrast to notions of difference and cultural, ethnic, and gender diversity. Citizenship is interpreted by Marshall (1950, 1975, 1981) as "full membership in a community," encompassing civil, political, and social rights. The discussion on citizenship during the past decade among academics in the field of political science, sociology, and geography is viewed by many as the result of political and social crises, wherein the exercise of power is challenged and thus the widely used definition of citizenship has shifted to a more complex, sophisticated, less optimistic interpretation of exclusions (Kofman, 1995). The idea of citizenship as a formation of national belonging is now used analytically, to expose differences in the *de jure* and *de facto* rights of different groups within and between nation-states (Smith, 1994). The concept is also used normatively to determine how a society, sensitive to human rights, should incorporate its individuals and communities in normative frameworks of belonging.

Does belonging in the context of citizenship definition have a spatial dimension? Expressions of citizenship in space have been coined as "spaces of citizenship" (Painter and Philo, 1995). "Spaces of citizenship" refer to the expression in space of the relationship between the state and its citizens from its social and political aspects of rights, and the ways in which spaces of inclusion and exclusion are defined. Another interpretation of "spaces of citizenship" is the quest for equality; that is, whether all citizens get equal treatment from the state in matters which involve equal access to resources (i.e. the provision of equal access to natural resources such as land, water, or minerals), as well as equal access to infrastructure, welfare services, education, employment, and knowledge. This relates both to cases of discrimination against those who are defined legally as citizens but do not receive equal levels of state services (such as the Bedouin and the Ethiopian in Israel) and to those whose citizenship is denied (see Fenster, 1998, 1999a). This connects notions of citizenship, (i.e. the formal expressions of belonging) to the discourse around exclusion.

Citizenship, Belonging, and Exclusion

Legitimized forms of exclusion or lack of belonging are expressed in many different definitions of citizenship. Many critics from both the political left and right recognize that citizenship by definition is about exclusion rather than inclusion for many people (McDowell, 1999). Thus, citizenship definitions are identity-related in that they dictate which identities are included within the hegemonic community and which are excluded. Definitions of "full citizenship" – either formally or informally expressed – have negative effects on women, children, immigrants, people of ethnic and racial minorities, gay men and lesbians, and sometimes on elderly people too. Citizenship definitions are also spatial. They dictate in which *representations of space* (Lefebvre, 1992) the rights and duties of citizen are relevant and in which spaces they are not. One such spatial distinction is the separation between private and public spaces, which usually affects women's exclusions in most cultures (Fenster, 1999b). As global urban spaces become more and more diversified in terms of their citizens' identities, these notions of citizenship are becoming more crucial and relevant to the discussion on belonging in global cities.

The various definitions of citizenship can be viewed as one of the legitimate ways to exclude "strangers" by way of clarifying the boundaries between "us" and "them." A conflict is inherent here between the recognition that "strangers" are in fact a socially diverse group of people with different abilities and needs, and the strong belief that in a democratic society we should all have equal access to all goods and resources that society offers. This conflict between equality and difference has its expressions in the gendered dimensions of citizenship. Women's efforts to achieve full "political citizenship," for example, tackle issues of equality and difference, especially in women's participation in formal and informal politics. Women's "social citizenship" also refers to equality in earning and in access to welfare services, while paying attention to the gendered differences in access to these resources (Lister, 1997).

The Spatialities of Gendered Belonging and Exclusions

Space is where we can see most tangibly that cultural citizenship values exclude women – literally. This is because cultural construction of space has inherent in its symbolism the legitimacy to exclude women from power and influence. This section highlights some of these symbolic constructions of space that are formulated by the patriarchy.

The most common are private/public devices, which for many women in different cultural contexts mean the construction of permitted/forbidden spaces (Fenster, 1999a). The "home" is the "private" – the women's space, the space of stability, reliability, and authenticity – the nostalgia for something lost which is female. "Home is where the heart is and where the woman (mother, lover) is also" (Massey, 1994). The "public" is perceived as the white, middle- or upper-class, heterosexual male domain. This sometimes means that women in both Western and non-Western cities simply cannot wander around the streets, parks, and urban spaces alone, and in some cultures cannot wander around at all (Massey, 1994; Fenster, 1999a). In many social constructions, women belong to the "private" only. A lot has been written about the different definitions and perspectives of the public/private (gendered) divide: its cultural orientation (Charlesworth, 1995; Fenster, 1999b); the associations of public spaces with the political sphere (Cook, 1995; Yuval Davis, 1997); its roots in Western liberal thought and different forms of patriarchy (Pateman, 1988, 1989); and feminist perspectives. Most feminist interpretations critique the public/private divide and argue that this distinction is actually a false one, but one that is invoked largely to justify female subordination and exclusion and to separate the abuse of human rights at home from the public (Bunch, 1995; Hyndman, 2003). The arbitrary nature of private/public divides becomes even more apparent in global economies as services, spaces, and activities such as health, childcare, and elderly care that used to be "public" in industrialized welfare states are now privatized (Eisenstein, 1996). Another point raised by feminist critiques regarding this division is its separative connotation between the state, the market sector, and the provision of welfare services and the patriarchal separation (Lister, 1997).

Such symbolic spaces of private and public are mostly relevant with regard to the practicalities of gendered belonging, as they play major roles in the construction and defense of cultural and ethnic collectivities. They are often the symbol of a particular national collectivity, its roots and spirit (Yuval Davis, 1997). Therefore, women's spatial mobility is very much dictated if not controlled by these cultural symbolic meanings of space. In this way, cultural and ethnic norms create "spaces of belonging and dis-belonging" which then become, for example, forbidden and permitted spaces for women in certain cultures (Fenster, 1998, 1999b). The boundaries of these spaces are usually dictated by the *male* "cultural guards" of society (Fenster, 1999b).

How is space culturally constructed as an entity in which women and men feel a sense of belonging or dis-belonging? First and foremost, it is the very intimate space – our bodies – that is culturally constructed (Sibley, 1995). This means that norms of cleanliness, dirt, odor, and modes of dressing which reflect values of shame and disgrace with the body covering actually become norms of inclusion and exclusion. Bourdieu (1984) analyzes the importance of cultural notions of an "appropriate" performance of the self. He argues that "the body is the most indisputable materialization of class taste" (Bourdieu, 1984). He developed the concept "habitus" to describe the distinguishing aspects of behavior, taste, and consumption which are combined to create flexible, rather than rigid, categories of "class." These categories reproduce themselves by display of certain tastes and the foundations of the body, its size, volume, demeanor, ways of eating and drinking, walking, sitting, speaking, etc. It is through the bodyspace that certain cultural habits are transformed depending on one's own identity and class – and gender. Thus, through body activities the boundaries between private and public and between who belongs and who does not are underlined. "Eating, like the other sin of the flesh, sex, has been constructed as a notoriously privatized activity" (Valentine, 1998). It belongs to the home or any other "privatized spaces" within the public domain but not to the street. Valentine argues that the custom of eating serves to regulate the boundary between the "private" and the "public" and what is "forbidden" or "permitted" in those spaces.

In what follows we analyze how these binary categorizations effect daily practices of women in men in London and Jerusalem.

Gendered Belonging and Memory

My home, memory, ownership, family, friends . . . I don't belong anywhere else.

This was Aziza's response to the question "What makes you feel you belong to your home?" Aziza defines herself as a Palestinian. She lives in West Jerusalem. She has Israeli citizenship although she doesn't identify herself as an Israeli. What are her reflections on her sense of belonging?

In spite of the associations of "belonging" in "the holy city" of Jerusalem with notions of sacredness, rituality, religious territoriality, and conflicts, we can identify in Aziza's narrative the meanings of "everyday belonging," the private secular sense of belonging. Aziza actually related to the multidimensional meanings of this sentiment. As a Palestinian citizen of Israel she talks about her home and associates it with memory. It could be her personal home but perhaps she also related to a national communal "home," a place of belonging and memory for the Palestinians in their homes before 1948 when the state of Israel was established.[1] Her personal sense of belonging here is associated with notions of nationalism and citizenship. We have elaborated earlier on the connections between belonging and citizenship. Here we can mention another dimension of belonging which is associated with nationalism, a link that has gained attention in the literature especially with regard to the discourse around citizenship and human rights of indigenous people (Fenster, 1999a; Read, 2000; Yuval Davis, 2003). Belonging in this respect is linked to notions of participation and inclusion in the construction of citizenship identity and membership in one's own nation. The notions "politics of belonging" and "politics of recognition" are related to such a rationalization.

Aziza also mentions belonging as associated with ownership. In this context, a sense of belonging to the physical environment can also be associated with an emotional attachment created between a person and a physical place which is based on the person's subjective meanings of that place. Belonging here is linked to its material and physical aspect. In addition, Aziza mentions a sense of belonging as connected to family and friends. Belonging here is linked to the people living at home more than to the physical home. Aziza finally says: "I don't belong anywhere else." Indeed, the notion of belonging is associated with the discussion on inclusion and exclusion, issues that are developed below.

For Suna, an Egyptian living in London, belonging is also multilayered. She says: "Belonging is very complex, subtle, nothing concrete, often it is familiarity with, memory from childhood with palm trees, I feel more belong [to the childhood landscape] as it is more familiar to me" (Suna, forties, single, Egyptian (British), London, July 29, 1999). Here Suna connects belonging to the notion of memory. It is perhaps one of the most explicit expressions of a sense of belonging and a part of one's own identity. Memory is either "real" – that is, a personal memory of childhood's reminiscences – or it is a "symbolic memory" – which in the narrative of another woman, Eleonore, consists of memories of ancestral graves:

> No sense of belonging to the house I live in . . . I have a deep sense of belonging to the town of my family where "the graves are" – there I feel rooted, sense of belonging to where my mom lives. . . . I ask myself "where do I belong?" and I don't know! (Eleonore, fifties, single, British-White, London, September 1, 1999)

Sandercock (1998) connects belonging and life in the cosmopolitan city. She mentions three elements in the city which are crucial for creating a feeling of belonging: city of memory, city of desire, and city of spirit:

> Memory, both individual and collective, is deeply important to us. It locates us as part of something bigger than our individual existences, perhaps makes us seem less insignificant, sometimes gives us at least partial answers to questions like: "Who I am?" and "Why am I like I am?" Memory locates us, as part of family history, as part of a tribe or community, as a part of city building and nation making. Loss of memory is, basically a loss of identity. (Sandercock, 1998, p. 207)

Memory in fact creates and consists of a sense of belonging. It could be a short-term memory based on intimate knowledge, which builds upon everyday life practices, of daily use of the streets, the paths, the pedestrians, and the city center. This is a corporal memory because it engages bodily experiences in using these spaces: walking, driving, cycling. It is also an identity-related memory as it engages experiences affected by one's own identities as a woman, gay, black, disabled, etc. Memory is also long term. It goes back to the past and consists of an accumulation of little events from the past, our childhood experiences, our personal readings and reflections on specific spaces, which are associated with significant events in our personal history. Such memories build up a sense of belonging to those places where these events took place. Thus Suna, who is an immigrant living in London, develops a sense of belonging and attachment to her childhood places, where her past memories took place, as a major part of her identity. This role of memory as part of one's own identity or as part of a collective identity of the community is becoming more and more evident in people's narratives:

> What makes me belong to my home is the people I love and they love me, the objects I love, colors, things which are part of my past – experiences, part of my home in me. (Suzana, thirties, married with one child, Israeli-Jewish, Jerusalem, July 13, 2000)

> I feel I belong to the Old city of Jerusalem as it brings memories from school days and boarding school, we used to go there every week. It makes me feel connected, it brings memories of my school days, in front of the Orient house. I used this area a lot in my life. (Saida, thirties, single, Palestinian-Muslim-Arab, Jerusalem, December 30, 2000)

For both Suzana and Saida memory becomes part of their own identity but also part of their collective identity. For Saida in particular, who as a Palestinian currently lives in what she experiences as daily oppressions and humiliations, memory to places she "territorialized" in her everyday practices in childhood is almost the only possibility to feel attached and connected to the city. Because of the political situation in Jerusalem her past memories become the essence of belonging to the city more than her current rather limited everyday practices.

Claim, Belonging, and Exclusion in Public Spaces

Can claim to a space be perceived as a form of belonging? A claim over "public" space is one of the expressions of belonging in everyday life. Such a claim is usually "informal," taking place as part of casual daily encounters between people or groups. It usually takes place when individuals wish to appropriate sections of public settings for various reasons, sometimes to achieve intimacy or anonymity or for social gatherings, which are mostly temporary. Claim and appropriation of space are a construct of everyday walking practices that de Certeau notes. These practices, which are repetitive, connect belonging with what is labeled by Viki Bell (1999) as performativity. Performativity means a replication and repetition of certain practices and it is associated with ritualistic repetitions with which communities colonize various territories. These performances are acted in certain spaces and through them a certain attachment and belonging to place is developed (Leach, 2002).

Space-claiming is also a class issue. It can take various forms of appropriation and territorialization. These are the implicit but sometimes explicit rules of inclusion and exclusion that play a role in the structuring of society and space in a way which some find oppressive and others appealing. One such example is the big shopping malls in many cities around the world, which are actually meant to serve the needs of certain groups in society while making efforts to exclude others (youth, poor, lower class, blacks, immigrants, etc.), and achieve that by introducing a lesser or a greater degree of surveillance. Public parks in big cities are another example of public spaces which are "appropriated" by middle-class people, excluding the poor and the homeless (Zukin, 1997). These spaces are usually guarded and watched with the intention of excluding those who are not following the norms. Thus, as Yuval Davis (2003, p. 3) mentions, "the politics of belonging form norms of inclusion and/or exclusion that result out of boundary formation which differentiates between those who belong and those who do not, determine and color the meaning of the particular belonging." The "boundaries of belonging" are usually symbolic and they may change according to the needs and goals of the hegemony. The power to exclude, which is based on "the boundaries of belonging," becomes in many cities the power of urban planning, of monopolizing space through zoning, and the relegation of weaker groups in society to less desirable and attractive spaces (see Fenster, 2004).

Do the narratives of belonging as related to public spaces refer to any of the above notions of "everyday walking practices," "performativity and belonging," "senses of exclusion and inclusion"? Let us look at how people in the two cities narrate their sense of belonging, especially as related to their neighborhoods. In both cities people associated a strong sense of belonging with a desire for homogeneity. This is the case in both secular and ultra orthodox neighborhoods in Jerusalem, and in Banglatown in London, a neighborhood that mirrors the cultural sense of belonging of its Bangladeshi residents.

How do the Bangladeshi people, for example, express their sense of belonging to their London neighborhood? In spite of the fact that they usually express ambivalent feelings toward a sense of belonging to their home in London (see elaboration in the next section), they feel a strong sense of belonging to their neighborhood sometimes more than to their home. This is because the Bangladeshi community in the East End of London have formulated a community or a "network of belonging" similar to what they had back home in the Sylhet district in Bangladesh (Forman, 1989). The social composition of the residents of the Brick Lane–Spitalfields area is made out of people from nearby villages from the same district in Bangladesh. This is probably why Ahmed, who has lived in the area for the past 30 years, says: "Yes, I feel belong to the Brick Lane area but sometimes I feel that this is actually part of Bangladesh. I live in this area and the family who live below us is from my village, the other family is from the next village and so on" (Ahmed, forties, married with six children, Bangladeshi-British, London, August 7, 2001).

The Bangladeshi immigrants have formulated their own "networks of belonging" in the global city – London. It is a neighborhood-oriented network and it keeps their old community ties and their imagined sense of belonging to their homeland. Elsewhere (Fenster, 2004) I have elaborated on how "physical spaces of belonging" have been created by the Bengali people in "Banglatown" constructing spaces and services that serve the cultural and ethnic needs of the community, such as food stores, traditional cloth shops, street names in Bengali, music stores, mosques, and travel agencies for their frequent travels to Bangladesh. Banglatown has become in many ways a representation of spatial belonging and difference, mostly dictated and shaped by the males in the Bangladeshi society, leaving the women to their private domains. This is expressed in the dominant presence of men in the streets of Banglatown. Women also walk in the streets but it is primarily male space. The fact that most Bangladeshi women are dressed traditionally also makes it a restricted space for women; that is, for most of them the traditional dress, which is dictated and guarded by the males, is a prerequisite for their walking in public spaces.

Similarly, the ultra orthodox neighborhoods in Jerusalem can also be seen as a reflection of belonging but, as seen later, these formations of spatial belonging exclude women especially because of their clothing; in contrast, the Bangladeshi in Banglatown don't seclude people of other cultures and don't enforce their norms and traditions as much as the ultra orthodox people do in their neighborhoods in Jerusalem. Walking in the Brick Lane area one can notice the mixture of women's clothing. Some Bangladeshi women are dressed traditionally but other women are dressed in Western style. It is the same as in other public spaces such as Hyde Park in London, which is used by a mixture of people of different identities, such as the Arab Muslim population who live nearby and people of other identities who equally share the use of these public spaces.

In Jerusalem, the ultra orthodox sense of belonging is more gender-exclusive. In these neighborhoods, it is women's clothing that becomes the most explicit and visible element of exclusion. It reflects cultural norms and determines where it is forbidden for women to go if they are not dressed according to the specific norms dominant at these spaces. This situation is echoed in women's narratives. All the women we interviewed living in Jerusalem, Jewish and Palestinians alike, mentioned the *Mea Shearim* (Hebrew for one hundred gates) neighborhood of the ultra ortho-dox Jews as the place with extreme expressions of spatial *exclusion* and a space most forbidden for them. The "guards of honor" do not trust women's own judg-ment as to what "modest clothing" means and therefore they elaborate: "long sleeved shirt," "no trousers," "no tight skirts," etc. In some of the shops there is a clear sign saying: "It is forbidden for women dressed immodestly to enter this shop" (see elaboration in Fenster, 2004, forthcoming).

Immigrants and Indigenous Gendered Belonging

The articulation of "belonging at home" for the Bangladeshi immigrants living in London sheds another light on the different meanings of a sense of belonging for immigrants and a sense of belonging for indigenous people. *The meaning of belong-ing at home* is a complicated issue for them to define. As immigrants they perceive "home" not as a house or a flat but more as homeland; home is perceived as an emotional place, not as a physical one. Those who were born in Bangladesh and came to London during the 1970s as children find it hard to relate to the notion of belonging at home in its narrow meaning.

"Home is home there because we are here for economic reasons," says Harun, who is in his forties, married with five children, and who has lived in London for the past 38 years (London, August 7, 2001). He defines his identity as British-Bengali-Muslim. He didn't choose to live in London, he came as a child and stayed because of what he termed "economic reasons," and also because his children were born and live in London and will not go back to Bangladesh. He represents a gen-eration of Bangladeshi who feel trapped between worlds. They didn't choose to move to London and now they can't go back: "We are trapped in a time zone. We can't go forward and not backward" (London, August 7, 2001).

Both Harun's emotional "imagined" home back in Bangladesh and his "real" home in London are places where he doesn't feel he belongs. He can't live in Bangladesh anymore but he doesn't feel "at home" in London. He feels neither British nor Bangladeshi. He is "in-between-homes." A sense of belonging for an immigrant is therefore an ambivalent issue and as such it is different from a sense of belonging of indigenous people such as Hassan. He is a Palestinian who lives in the Old City of Jerusalem in a house that has belonged to his family for the last 600 years. Hassan is in his forties, married with four children. He says: "the house is part of me – the house is me" (Jerusalem, May 3, 2001). For Hassan a sense of belonging to the house is deeply rooted in the chain-history of his family. He says he will never sell the house although it is too small for the whole family because it is part of him, of his identity. Two identities, immigrant identity and indigenous identity, construct two forms of belonging: a dialectic sense of belonging of an immi-grant who is also a member of cultural minority, and in contrast a strong sense of belonging of an indigenous person who expresses a strong bonding to his home and city. The two men feel the same sense of belonging to their homeland and home; the difference is that the Bangladeshi left his home and the Palestinian stayed.

As we notice, these are male expressions of belonging, while the voices of Bangaldeshi women are somewhat left out in these narratives. When I asked my

Bangladeshi male interviewees if I could interview their wives they said it is a bit complicated because their wives are not used to talking to strangers and their English is not good enough. From conversations with them I understood that for the Bengali men to allow their wives to meet an "outsider" is probably something that threatens the patriarchal norms in their society. Thus the most clear expression of patriarchy in this research is probably in the absent voices of the Bangladeshi women. Their absence is actually a reflection of their non-appearance in public activities and spaces in the global city of London.

But patriarchy doesn't remain private. It has its explicit expressions in the shaping of public spaces as well. As mentioned elsewhere (Fenster, 2004), patriarchy raises its head when the city's planning, management, and network decline. Several Palestinian women experience a much more rigid and fierce patriarchal atmosphere in the streets of East Jerusalem now than before the Intifada. It is not that the atmosphere in East Jerusalem before this last Intifada (the Palestinian uprising which began in 2000) was totally inviting for women, as patriarchy existed then as well, but lately it became more explicit and rude because of the city's decline, which is very much connected to discriminatory politics of planning and development.[2]

As mentioned above, patriarchy is also visible in the streets of Banglatown in London, and there it is reflected in Muslim women's traditional clothing and head cover and their absence from the public sphere. This is, however, a different type of patriarchy, connected to intercommunity cultural values more than to the ways the "politics of planning" and city management perpetuate these norms in public spaces. Thus the two groups of women, the Bangladeshi and the Palestinian, live in cities which are famous for their flows of commodities, communications, capital, and corporations, yet their lives and the intersection between their cultural and social duties at home and the possibilities of exploiting the choices that the city provides are still very limited.

Finally, a different scale of a sense of belonging is associated with food and spices that are identity-related, especially among immigrants:

> And the spices that I usually eat and cook with ... Indian spices that you can get everywhere ... I learnt the names of the spices in English only five years ago when I arrived in London because at home [Canada] I would shop in Indian shops – food makes me feel I belong. (Mandy, 28, single, Canadian-Indian, Jerusalem, June 16, 2000)[3]

One of the first things that immigrants establish when they arrive in a new country is the ethnic food stores that sell specific food and spices. Sandercock (2000) calls this "market mechanisms," which also include services such as lawyers, tax accountants, and shoe repair. London has been famous for its large variety of food stores and restaurants that provide food from around the world. This is one of the consequences of the "Ethno-towns" such as Chinatown and Banglatown, and the broad variety of restaurants and supermarkets established in these areas, which sell food and spices from their places of origin. Jerusalem too has a large selection of ethnic restaurants and places to find peculiar spices. The Old City Markets are known for their diversity of spices, especially those associated with Oriental cuisine. When Mandy was interviewed she lived in Jerusalem and she enjoyed the wide range of spices suitable for her own cultural food, Indian cuisine.

Conclusions

This chapter exposes the multilayered nature of gendered belonging in the city. As analyzed above, a sense of belonging can be a personal, intimate, and private sentiment as well as a formal, official, public-oriented recognition of belonging. This chapter thus conceptualizes the different "formations of belonging," both the collective and the personal. Let us conclude the various meanings of belonging as they were highlighted in people's narratives:

- *Belonging as a form of citizenship* is one of the more common interpretations of this term. Official belonging is usually formalized in patterns of citizenship. In Jerusalem, forms of belonging and citizenship are connected to the abuse of human and citizen rights of the Palestinians living in the city, usually by means of politics of planning and development that promote Jewish interests.
- *Belonging and walking practices.* Repetitive daily walking practices are one of the mechanisms of creating an "everyday" sense of belonging. We all belong because we all have repetitive daily uses of city spaces by foot or car or public transportation. Our daily practices help us to draw our "private city" and to underline the intimate allies and paths that we use in our daily practices in the city. Walking practices are usually gendered, as women's daily walking routine is usually dictated by their household gendered divisions of roles, e.g. taking care of the children, doing the shopping, working in the vicinity to the home. This is usually more explicit for young mothers who walk with baby strollers, and for dog owners, whose daily repetitive practices create their sense of belonging to the environment.
- *A sense of belonging is associated with memory.* The place one was born in, the family one belongs to, deeply shape a sense of belonging. Other aspects of belonging are changing: places of living, homes, and neighborhoods. Some components of belonging are short term, while others are based on long-term memories, such as childhood memories. Belonging has its personal aspects, belonging to places and people that are connected to personal experiences, personal memories. A sense of belonging is also collective. It is based on collective memories and shared symbolism of a community. Its significance in one's own life is a result of one's own affiliations, beliefs, and ideology.
- *A gendered sense of belonging is about power relations and control,* even in intimate and private spaces such as home. The larger the category of space the more significant is the role of power relations on one's own feelings of belonging. In public spaces power relations are identified as "claim," "appropriation," "exclusion," "discrimination." Power relations also dictate "the boundaries of belonging." They are formed by the hegemony and exclude the "other," those that are not considered by the hegemony to be part of it, such as Palestinians in Jerusalem and to a lesser extent Bangladeshis in London. The latter feel excluded from the boundaries of Englishness but they do feel included within the "boundaries of Britishness." A sense of belonging and power relations are associated with the "private" – the power to exclude – and the "public" – the power to gain access.
- *"The right to belong"* can be identified as the right of people of different identities to be recognized and the right to take part in civil society in spite of one's

own identity differences, what Sandercock (2000) terms "the right to differ-ence." The right to belong in contested spaces can be perceived as a deeper expression of "citizenship in the global city." The right to belong relates to the situations where one's own rights to equality and one's own rights to maintain identity difference are fulfilled. These rights are connected to communities' priv-ileges to maintain their sites of memory and commemoration and these sites are acknowledged and preserved by politics of planning and development.

- *Belonging and urban planning.* An important connection between a sense of belonging at home and urban planning is the association of order and belong-ing. Deciding upon the order of things is actually one of the basic activities of physical planning on each level of space and the more people are involved in the decision-making about "the order of functions" in their own street, neighbor-hood, or even city center the deeper the sense of belonging they develop to these environments. Urban planning is the field where expressions of spatial citizen-ships and belonging are made. As elaborated in the chapter, Jerusalem's "poli-tics of planning and development" is a tragic example of abuse of formations of citizenship and belonging in the field of urban planning.

NOTES

1 The Palestinians in Jerusalem make up 32 percent of Jerusalem's total population, which was in the year 2000 657,500 inhabitants, 11 per cent of Israel's total population. Jerusalem is the capital and the largest city in Israel, extending over 1.26 million hectares. It consists of the West (predominantly Jewish) and the East (predominantly Palestinian). East Jerusalem was occupied and annexed after the 1967 war by Israeli law, bringing it under its sovereignty and taking the whole and unified Jerusalem as the capital of Israel. The Palestinians reject the city's unification by force and see East Jerusalem only as their capital, although using West Jerusalem services. Because of its uncertain status as the capital of Israel in the eyes of international law, the municipality and the Israeli govern-ment's "politics of planning and development" in the city has been targeted to maintain a Jewish majority there in what is termed "the battle over demography." The demographic balance in the year 2000 shows the success of these policies: 68 percent of the city's pop-ulation are Jews, 32 percent are Palestinians.

2 As mentioned before, policies of discrimination against the Palestinians in Jerusalem derive from one of the main targets of the policies of planning and development in Jerusalem, which is to maintain the Jewish majority in the city. The two main channels to realize this goal are a vast Jewish development and expansion and a lack of Palestin-ian development and improvement. The result is that although Israeli governments declare Jerusalem as united, the city is managed from only one side (Yiftachel and Yacobi, 2002). This means that urban governance, urban economies, and services are targeted on the Jewish inhabitants in spite of the fact that the Palestinians consist of nearly a third of its population. Practically speaking, most economic and planning efforts are targeted on expanding and modernizing Jewish areas as part of the policy of Judaization of large parts of East Jerusalem and its surrounding hills. The means to achieve this goal are many: expropriating Palestinian lands, construction of Jewish neighborhoods or settlements on these lands, restrictions on Palestinian building and land use through the adoption of planning policies, residency regulations and other measures, determining restricted housing capacity for the Palestinian population while encouraging the Jewish population by means of financial subsidies to move to Jerusalem, declaring "open landscapes areas"

near existing Palestinian villages, thus preventing their natural expansion, declaring houses already built in these areas as illegal, and at the same time building Jewish neighborhoods on areas previously declared as green areas.

3 Mandy was living at the time of interview in Jerusalem but has studied before in London.

BIBLIOGRAPHY

Be'tselem (1997) *A Policy of Discrimination: Land Expropriation, Planning and Building in East Jerusalem*. Jerusalem: Be'tselem.

Bell, V. (1999) Performativity and belonging: an introduction. *Theory, Culture and Society*, 16(2), 1–10.

Bollens, S. (2000) *On Narrow Ground: Urban Policy and Ethnic Conflict in Jerusalem and Belfast*. Albany: State University of New York Press.

Bourdieu, P. (1984) *Distinction: A Social Critique of the Judgement of Taste*. London: Routledge.

Bunch, C. (1995) Transforming human rights from a feminist perspective. In J. Peters and A. Wolper (eds), *Women's Rights, Human Rights*. New York: Routledge.

Cheshin, A., Hutman, B. and Melamed, A. (2000) *Separated and Unequal: The Inside Story of Israeli Rule in East Jerusalem*. Cambridge, MA: Harvard University Press.

Charlesworth, H. (1995) What are women's international human rights? In R. Cook (ed.), *Human Rights of Women*. Philadelphia: University of Pennsylvania Press.

Cook, R. (1995) Women's International human rights: the way. In R. Cook (ed.), *Human Rights of Women*. Philadelphia: University of Pennsylvania Press.

Crang, M. (1998) *Cultural Geography*. London: Routledge.

de Certeau, M. (1984) *The Practice of Everyday Life*. Berkeley: University of California Press.

Eisenstein, Z. (1996) Women's publics and the search for new democracies. Paper presented at the conference: Women, Citizenship and Difference, London, July.

Fainstein, S. (1994) *The City Builders*. Oxford: Blackwell.

Fenster, T. (1998) Ethnicity, citizenship and gender: the case of Ethiopian immigrant women in Israel. *Gender, Place and Culture*, 5(2), 177–89.

Fenster, T. (1999a) Culture, human rights and planning (as control) for minority women in Israel. In T. Fenster (ed.), *Gender, Planning and Human Rights*. London: Routledge.

Fenster, T. (1999b) Space for gender: cultural roles of the forbidden and the permitted. *Environment and Planning D: Society and Space*, 17, 227–46.

Fenster, T. (2001) Planning, culture, knowledge and control: minority women in Israel. In O. Yiftachel, J. Little, D. Hedgcock, I. Alexander (eds), *The Power of Planning*. Dordrecht: Kluwer.

Fenster, T. (2004) *The Global City and the Holy City: Narratives on Planning, Knowledge and Diversity*. London: Pearson.

Fenster, T. (forthcoming) Globalization, gendered exclusions and city planning and management: beyond tolerance in Jerusalem and London. *Hagar – International Social Science Review*, in the press.

Forman, C. (1989) *Spitalfields: A Battle for Land*. London: Hilary Shipman.

Fullilove, M. T. (1996) Psychiatric implications of displacement: contributions from the psychology of place. *American Journal of Psychiatry*, 153(12), 1516–22.

Hyndman, J. (2003) Beyond either/or: a feminist analysis of September 11th. *ACME: An International E-Journal for Critical Geographies*, 2(1), 1–13.

Jacobs, J. M. (1996) *Edge of Empire: Postcolonialism and the City*. London: Routledge.

Kofman, E. (1995) Citizenship for some but not for others: spaces of citizenship in contemporary Europe. *Political Geography*, 14, 121–37.

Leach, N. (2002) Belonging: towards a theory of identification with space. In J. Hillier and E. Rooksby (eds), *Habitus: A Sense of Place*. Aldershot: Ashgate.

Lefebvre, H. (1992) *The Production of Space*. Oxford: Blackwell.

Lister, R. (1997) *Citizenship: Feminist Perspectives*. New York: New York University Press.

MacDowell, L. (1999) City life and difference: negotiating diveristy. In J. Allen, D. Massey and M. Pryke (eds), *Unsettling Cities*. London: Routledge.

Marshall, T. H. (1950) *Citizenship and Social Class*. Cambridge: Cambridge University Press.

Marshall, T. H. (1975) *Social Policy in the Twentieth Century*. London: Hutchinson.

Marshall, T. H. (1981) *The Right to Welfare and Other Essays*. London: Heinemann.

Massey, D. (1994) *Space, Place and Gender*. Cambridge: Polity Press.

Painter, J. and Philo, C. (1995) Spaces of citizenship: an introduction. *Political Geography*, 14, 107–20.

Pateman, C. (1988) *The Sexual Contract*. Cambridge: Polity Press.

Pateman, C. (1989) *The Disorder of Women*. Cambridge: Polity Press.

Pile, S. (1996) *The Body and the City*. London: Routledge.

Raban, J. (1974) *Soft City*. London: Harvill Press.

Read, P. (2000) *Belonging: Australians, Place and Aboriginal Ownership*. Cambridge: Cambridge University Press.

Romann, M. and Weingrod, A. (1991) *Living Together Separately*. Princeton, NJ: Princeton University Press.

Sandercock, L. (1998) *Towards Cosmopolis*. London: Wiley.

Sandercock, L. (2000) When strangers become neighbours: managing cities of difference. *Planning Theory and Practice*, 1(1), 13–30.

Sibley, D. (1995) *Geographies of Exclusion*. London: Routledge.

Simith, N. (1994) Marxist geography. In D. Gregory, R. Johnston and D. Smith (eds), *The Dictionary of Human Geography*. Oxford: Blackwell.

Thornley, A. (ed.) (1992) *The Crisis of London*. London: Routledge.

Valentine, G. (1998) Food and the production of the civilized street. In N. R. Fyfe (ed.), *Images of the Street*. London: Routledge.

Yiftachel, O. and Yacobi, H. (2002) Planning a bi-national capital: should Jerusalem remain united? *Geoform*, 33, 137–45.

Yuval Davis, N. (1997) *Gender and Nation*. London: Sage.

Yuval Davis, N. (2003) Belongings: in between the indegene and the diasporic. In U. Ozkirimli (ed.), *Nationalism in the 21st Century*. Basingstoke: Macmillan.

Zukin, S. (1997) *The Cultures of Cities*. Oxford: Blackwell.

Chapter 18

Urban Space in Plural: Elastic, Tamed, Suppressed

Hille Koskela

Introduction

Space is produced in everyday encounters on the street. Space is, at present, commonly understood to be a social category, produced not only in political and economic processes but in the practices and power relationships of daily life – including gender relations. Space shapes the way in which gender identities are formed and, reciprocally, gender identities and gendered social relations shape space (see Massey, 1994, p. 186; Domosh, 1998). The ostensibly trivial gendered practices and power structures of the everyday confine women's space, and thus produce and (re)produce space that is "gendered" (see Rose, 1993). Among other key features of this, it has been amply demonstrated that "warnings about the potential for sexual victimization are a central feature of women's socialization" (Macmillan et al., 2000, p. 308); an essential part of this socialization turns out to be spatial. In this and myriad other ways, the interplay between public space and (gendered) social relations is crucial for understanding both how space is produced and how gender is constructed.

In this chapter, my aim is to examine some of the practices that contribute to gendering of urban space. I will provide different perspectives on this issue by pointing at three different spaces of the everyday life. The first section, "Elastic space," examines how gendered space is produced in a red light district where women are constantly gazed at and harassed. By exploring the places where the "usual" types of harassment are aggravated, it is possible to provide exceptionally glaring examples of the processes that go on also elsewhere and that seem otherwise unremarkable. In the next section, "Tamed space," I discuss how women negotiate space for themselves by conscious practices of boldness to counteract their fear of violence. Women appear to be "experts" in interpreting different elements of urban behaviour. The third section, "Suppressed space," examines the relationships between security, surveillance technology, and the gendered practices of visual control. New urban surveillance technologies have the double effect of, on the one hand, possibly reducing the level of street harassment and violence women experience, but, on the other hand, of magnifying and extending the nature of such harassment. Two central theoretical points of view shape all these sections: the production of urban space and the politics of looking.

Examining the "politics of looking" is a key theme in feminist urban geography. Power relationships intertwine with the field of vision, including acts of seeing and being seen, as well as the cultural meanings of the visual and its representations. Often even the seemingly innocent questions of visuality are gendered (see Rose, 1993; Nast and Kobayashi, 1996; Robinson, 2000, among others). What is characteristic of all lived urban space is "the semiotics of the street." In encounters on the street we read each other – as well as the built environment – as *signs*. The plays of looking and being looked at form essential components in this process. Space is simultaneously *read* and *produced*. In urban space women are more likely than men to be the ones who are looked at, the objects of the gaze (Massey, 1994). This, as I shall argue, applies both to face-to-face encounters and to "mediated gazes"; for example, the "electronic gaze" of a surveillance camera. Further, women are constantly reminded that an (invisible) observer is a threat, and this potential observer is presented as male. Indeed, one of the main reasons why women feel unsafe in public space is their "exaggerated visibility" (Brown, 1998, p. 218). The cultural codes and politics of seeing and being seen are deeply gendered.

"Elastic Space": Street Prostitution and a Woman as a "Sign of a Whore"

The question of safety is an issue which acutely illustrates the gendered nature of space. There is a long tradition of feminist research on "the geography of fear" (Smith, 1987; Valentine, 1989; Pain, 1991, among others). Women's insecurity has both a structural and a contextual dimension. It is now widely understood that there is no single female experience, but the patterns of structural vulnerability are of great importance (e.g. McDowell, 1993). Feelings of insecurity are bound into the structurally subordinate positions of some women; for example, ethnic or sexual minorities.

However, gender relations in space vary also according to each particular context. A glaring example of this is how (gendered) social relations are constructed *in a red light district*. Street prostitution and its side effects – kerb-crawling, scrutiny, and other forms of sexual harassment – define women's spatial experience in these zones. Sexual harassment in the public sphere can be understood as a form of "non-criminal" street violence: women face a range of behavior that is perceived as offensive but would not necessarily be considered criminal. The contribution of such incidents to women's everyday lives is easily trivialized. However, harassment has a remarkable impact on how women perceive particular neighborhoods, and how freely they feel they are able to move around. While such harassment is certainly experienced in all parts of the city, kerb-crawling tends to heighten the level of intimidation, and hence in prostitution quarters the situation is aggravated – both for prostitutes and for women who are just pedestrians, passing through or living in the district.[1]

Red-light districts exist in virtually all cities throughout the world. Research on these areas has generally focused on four topics: on legal questions of the regulation of prostitution and the "toleration zones" (Høigård and Finstad, 1992; Kilvington et al., 2001); on local activism and efforts to displace prostitution (Hubbard 1999; Weitzer, 2000; Tani, 2002); on the stigmatization and the place identities of established red light districts (Hart, 1995; Tani, 2001); and on the marginalization of the prostitutes and the conceptualization of them as "the Other" (Duncan, 1996; Hubbard, 1998; Koskela and Tani, 2005). Although many studies on prostitution areas touch on the issue of sexual harassment, it has rarely been the primary object of study. However, the existence of prostitutes and their customers in a particular area has the effect of making it more difficult for *all* women to use that space, especially when prostitution occurs in a residential area. When women actually live in a prostitution district, it is impossible for them to avoid walking around where prostitution goes on and, hence, also impossible to avoid (uncomfortable) interactions on the street. It doesn't matter whether an individual woman is involved with prostitution or not: the mere visibility of women in or near a red light district is assumed by men to imply their sexual availability (Rendell, 2001, p. 114). This process is contextual but not exceptional.

The production of social relations is defined by temporality and spatiality. Temporality is realized in individual experiences. Women feel that their freedom to use urban spaces varies over the course of the day. Prostitution is, mostly, part of the nocturnal side of urban life. In daytime urban space is not assumed by most women to be as contested, unpleasant, and frightening as nighttime. Spatiality makes a difference both to the images of place and to individual experiences. Eventually, temporality and spatiality are intertwined, creating a particular urban experience: space becomes *elastic* – it is different according to the circumstances, to the time of the day, to who is passing by, and to how you feel at that moment (Koskela and Tani, 2005).

The "male gaze" has a crucial role in street harassment of women, and even more so in red light districts. Women walking on the streets have little room to escape the male gaze. "Looking" is a central element in the process by which men evaluate who might be a prostitute, and which women they find desirable. For kerb-crawlers, the streets of red light districts become *spaces of power* (Koskela and Tani, 2005) in which men assume roles as active subjects: they can gaze from the distance (often from a car), make judgments about the women they survey, and make sexualized decisions about those women, but also they can escape from difficult situations if necessary. They appear as *flâneurs* – men who take "visual possession of the city" (Wilson, 1995, p. 65). For women, the streets turn into almost exclusively *sexualized spaces* in which they are objects, not subjects. Wherever prostitution is connected to heterosexuality, to straight women as prostitutes and straight men as their clients, the streets become also strongly *heterosexual* spaces. This creates a vicious circle, reinforcing the gender differences again and again.

For women, being dressed up in a known street prostitution district means taking an increased risk of being taken for a prostitute. Dressing can be seen as a means of reproducing power relations; in Foucauldian terms, it is a way of being one's own overseer. Women are thus often in a position of reproducing oppressive gendered power relations by policing their own dress. If they dress "down," to be anonymous, then they participate in their own erasure (Rose, 1993, 143); if they are harassed, they are told it is their fault for dressing "provocatively" (regardless of what their actual state of dress is). Women are constantly advised to manipulate their dress for the sake of preventing crime – as well as in the name of conforming to "desirable" heterosexual images (Gardner, 1990; McDowell, 1995; Munt, 1995).

Despite the efforts of women to protect themselves from harassment by adjusting their own appearance, there is no evidence that harassment does indeed depend on women's outfits. A case study in the prostitution quarters of Helsinki (see Koskela et al., 2000; Koskela and Tani, 2005) proves this point: all women can be and are harassed, no matter how they are dressed, whether they are carrying shopping bags or walking a dog. Men simply seem to be unable – or unwilling – to "read the signs of prostitution." There is a gap between women's interpretation of what they look like and men's interpretations of this. The atmosphere in a red light district enables – even encourages – men to assert power through harassment. Women are transformed into representations of sexuality, availability and, commodity (see Rendell, 2001). Gender alone is enough to be read as a sign, and any woman becomes a sign of a "whore."

Two forms of resistance can be seen rising as a response to street harassment: collective civic activism and individual strategies. In Helsinki, a movement called "Prostitution Off the Streets" mobilized civic activism in the red light district, starting in summer 1995. Activists monitored the streets, making notes about harassment, kerb-crawling, and prostitution activity, mobilized residents to sign petitions against street prostitution and the fact that it was not illegal in Finland, and delivered information to authorities about activities in the district. In addition, the activists organized different street-happenings to mediate their message: for example, they sold apples – "forbidden fruit" – on the street to emphasize the point that in Finland street vending is subject to license, while prostitution is not (Tani, 2002, p. 352). This movement, however, was controversial: it can be interpreted as an anti-prostitute moral crusade, and further, it can be seen to promote "a form of 'cleansing' to maintain purity of residential space" (Hubbard, 2000, p. 255).

In terms of individual strategies (beyond "preventative" dressing strategies, which were entirely ineffective) women residents of the Helsinki red light district described particular responses they had developed to cope with the sudden, unexpected – but everyday – harassment encounters on the streets. They employed various verbal and non-verbal practices of resistance, ranging from ignoring the unpleasant encounters and using "mental distancing," to more proactive strategies such as mentioning "a price" for their presumed sexual services that was beyond all expectations. Some women were so annoyed about being constantly harassed in their own neighborhood that they started behaving aggressively. Kicking cars was a typical form of physical resistance. Also gathering information on the cars – or pretending to do so – worked as a form of resistance, as did the use of mobile phones. To dial with a mobile phone was usually enough to drive away the kerb-crawler. Women used their mobiles to talk to someone they knew while they kept on walking towards their destination. Especially in threatening cases of harassment this clearly made a difference to how the situation was experienced. Hence, mobile phones blur the difference between public and private, presence and absence, being alone and being accompanied. The outcome of (gendered) power relations "cannot be determined" (Robinson, 2000, p. 81). Who appears to be "a sign of a whore" turns out to be a subject of her own being: capable of resisting, technologically equipped, and spatially competent. Elastic space is temporally and socially contextual. Sexual harassment contributes to the "gendering" of urban space but so do the strategies of resistance. In this respect, individual strategies and deeds contribute to the collective process of "the production of space": to defining, controlling, and rewriting the social space of the street.

"Tamed Space": The Dynamics of Fear and Boldness

Gender relations in space vary not only according to social circumstances but also according to *personal* experiences and feelings, and, furthermore, often these two overlap. Judging whether a path across a dark park is safe to take is a practical question of everyday life, and it is a matter of personal feelings. Yet it also reveals the power structures that produce social space. The question of fear and courage – women's ability to use public space – is a question of (re)defining and (re)producing space as well as managing the self. In everyday life, the dynamics of fear and boldness form a constant, internal negotiation in three respects: a spatial dimension – i.e. *where* to go – a temporal dimension – *when* to go – and a social dimension – *with whom* to go.

Spatial relations, including restricted access to public space and limited mobility because of fear of violence, can be seen as a test for equality – *a parameter of empowerment* (Koskela, 1997, p. 302). Therefore, fear of violence is to be interpreted not only as a result of crime but also as a sensitive indicator of gendered power relations that constitute society and space. It is important to challenge the view (sometimes held unconsciously) that fearfulness is an essentially female quality. If women's fear is generally regarded as "normal" and their boldness thought to be risky, then the notion of women as victims is unintentionally reproduced (Koskela, 1997). It is *not an inborn quality* of women to be fearful. Women are not "readily victims," but also respond to and work against oppressive behavior and have agency over their own lives (Alcoff, 1996, p. 26).

Many women are quite confident in relation to their daily environments. Women have various practices of boldness.[2] For most individual women, the feeling of boldness relates to their confidence in their social competence – in their ability to interpret who is dangerous and who is not. If women have the feeling of being capable of interpreting the environment they walk in, they can feel spatially confident. Since they have grown up with constant alertness, women appear to be acting as *experts in urban semiotics* – perhaps more qualified than men at reading the urban scene.

When women are asked how they are able to read the urban scene, it becomes apparent that part of this expertise is intuitive. Women are not able to tell exactly how they know that something or someone is dangerous. Women read signals from the eyes and looks of other people, from their gestures, from the movements of their bodies, and from their fashion and style. Likewise, safe situations are interpreted by indescribable intuition. Boldness and spatial confidence – in other words, the semiotic expertise – can be as "feminine" as fearfulness. This other side of the relationship between gender, emotions, and space has been examined much less than fear; most research on this topic restricts itself narrowly to issues of safety and protection, thus perpetuating "the cultural script" of male violence and female victimisation (Wilson, 1991, p. 10; Walkowitz, 1992, p. 244).

Women have multiple ways of being spatially confident and taking possession of space. "The ontology of emotions" would help to understand this. "Feelings" or perceptions of safety and risk are not a mathematical function of *actual* risks but highly complex products of each individual's experiences, memories, and relations to space. Intuition and learned knowledge can be contradictory, and often women's feelings are based on both. The origin of feelings about safety and space sometimes seems to be irrational, and their essence internally contradictory, but even then resulting habits and reactions can be logical. In their everyday lives many women know perfectly well when and where to be careful or confident, although there is no measurable way to explain this. It is clear, however, that the capacity to distinguish "safe" from "unsafe" men is loaded with interpretations of race, class, and visual appearance, and the men who "look safe" in that context may actually be the most dangerous ones. Relying on "senses" can never entirely guarantee one's safety (e.g. Stanko, 1996). Yet still, instead of trying to connect fear or boldness to the statistical risks of being a victim of known violence – which is, from a perspective of daily life, actually an absurd idea – these feelings should be respected *as such* (Koskela, 1997, p. 304).

Women report that when they have had the courage to do something that is normally seen as spatially daring (such as walking in the "wrong" neighborhood), this is often followed by a certain remorse. In light of the social production of fear this is not surprising. Women's fear is socially produced in parental instructions, in daily warnings issued in discussions with friends and relatives, in crime and "danger" news reporting in the media, in violence in films and other fiction, and, finally, in the cultural reproduction of ideologies about women and the family (Valentine, 1992; Gardner, 1995). In addition, the tradition of victim-blaming, reproduced by media images and discourse on violence against women, supports the notion that women should blame themselves if something happens when they enter a situation regarded as "not suitable for women" (Gardner, 1990). Further, this all is reproduced in crime prevention advice explaining to women how to avoid risks. Women are told not to walk alone in potentially dangerous areas, to avoid being accompanied by strangers, or even to keep their curtains tightly closed whenever it might be possible for someone outside to see inside (Gardner, 1995, explaining the US context). It seems to be so strongly expected that women be afraid all the time, that their boldness and social and spatial confidence is a *taboo*. This is maintained by the social production of carefulness and restraints.

One of the boldness strategies that women use is an internal dialogue about "reason." In frightening situations, women use "reasoning" to convince themselves that they should not be afraid but should keep up their courage (Koskela, 1997, p. 306). They try to draw on the facts they have heard about statistically calculated risks, or persuade themselves that they ought to walk toward the potential attacker and not to avert their eyes or change their direction. By reasoning fear of violence out of their minds, women can gain more confidence and, again, reclaim space for themselves.

Courage can also be gained through an awareness of the cultural relativity of danger, where experiences in different cultures make women feel confident in a more familiar environment. Sense of danger is a cultural construction, and strange environments are commonly perceived to be more dangerous than familiar ones (Merry,

1981; Valentine, 1989). It is easier to interpret the signs of danger, both the verbal and non-verbal ones, in familiar cultural contexts, and this ease is often associated with a sense of confidence. Further, danger is culturally constructed also in that there are differences in terms of women's freedom in physical and social mobility – in how much freedom of mobility girls have in their childhood and youth. Within the Western world there are differences; for example, in Scandinavia girls are allowed to move around freely compared to the USA. Independent mobility at an early age, although it might be perceived as a risk, can be claimed to support gender equality and women's spatial confidence.

Courage could be interpreted as emancipatory. Crucial to a strong feeling of belonging is often the fact that one uses the space a great deal. Using space often can be a way of *demystifying* it. As I argued, fear is socially constructed in numerous ways. These indirect sources produce *the rhetoric of danger and threat*. The "map" of everyday experiences is in sharp contrast to the maps of one's parents, the media, or other indirect sources. Making the use of space a part of one's daily routine erases the myth of danger from it. By routinizing space, women are *taming* it for themselves.

The urban has often been described as hostile and dangerous for women, but it can also be seen as a precondition of female emancipation, fostering a positive womanhood with fascinating freedoms and possibilities (Wilson, 1991, p. 25). Part of the feeling of taking possession of space is the cultivation of an "urban mentality." Risky urban environments can be interpreted as exciting when contrasted to quiet countryside. Positive personal relations to the city are expressed in spatial terms. Part of the urban mentality is to enjoy the spatial experience: the pleasure of gliding fluently and confidently across the city. A city can be perceived as a labyrinth, in which every path is likely to bring new experiences. Urban space is unrestrained, "wild" space – *in succession with* it being "tamed." When not being afraid, it becomes possible for women to turn the gaze around. Women can be active "voyeurs": to glance, to flirt, to invite with their eyes (Wilson, 1991; Munt, 1995). Women, too, can be *flâneurs* (Wilson, 1995, p. 68). They can also use a (hostile, evaluating, or humiliating) gaze as a strategy of resistance.

The subjective dilemma of being bold or being "wise" (which essentially means submitting to the fearful role) is a reflection of gendered power. The space shaped by fear and boldness is the space of female embodied knowledge; an internally contradictory space, beyond the mind–body split. It is simultaneously subjective and intersubjective. The women who feel confident reclaim space for themselves through everyday practices and routinized uses. By daring to go out – by their very presence in urban sphere – women produce space that is more available for other women. Hence, spatial confidence is a "manifestation of power" (Koskela, 1997, p. 316). Walking on the street can be seen as a political act: women "write themselves onto the street."

"Suppressed Space": The Politics of Surveillance

The third theme I discuss here – namely, video surveillance – is a response to fear. In most Western cities, *visual control* is increasingly being seen as a solution to the problem of insecurity. In city after city, surveillance cameras are now part of the ordinary "architecture of fortification" which "braces the structures from the threat of a supposedly violent other" (Epstein, 1997, p. 139), and visual control has become a permanent condition. Surveillance can be used for reinforcing the "purification" and "homogenization" processes of urban space (Davis, 1990; Mitchell, 1995; Ellin, 1997). It is also feared to have a negative "chilling effect" on urban life – to *suppress* urban space.

Until recently, not much attention was paid to the emotional aspects of surveillance nor to the complex relationship between security, surveillance technology, and gender (see, however, Ainley, 1998; Brown, 1998; Koskela, 2000, 2002). Gender relations are by no means the only dimensions of power within this practice. Instead, gender is one example of the many forms of power associated with surveillance. Indeed, surveillance is used to monitor and exclude "suspicious youths," political activists, people of color, sexual minorities, or people generally who look as though they cannot afford to "consume" (e.g. Judd, 1995; Lees, 1998). Visual appearance forms the basis for prejudice: urban visual control is "ridden with racism and sexism" (Graham, 1998, p. 491), and concern for its misuse cuts across age, class, race, and gender.

Where gender comes into this picture is not simple or straightforward. On the one hand, since the purpose of surveillance is to increase safety, and since women are known to be the ones who are most often afraid, surveillance might be especially beneficial for women. In this way, surveillance might facilitate women's access to urban space and to the resources and benefits of urbanity. If video surveillance is able to deter violent crime – even a minor part of it – it can be considered as worth it.

But, on the other hand, surveillance includes multiple power relationships intertwined with "looking" and the field of vision. To put it simply: for most women, urban harassment is a common reason for fear; scrutiny is a common form of harassment; and (remote) surveillance is an increasingly common form of scrutiny. Surveillance will increase rather than diminish the "exaggerated visibility" of women – the visibility of women to an unseen and remote voyeur. The politics of looking inevitably extends the issue of surveillance beyond the crime prevention debate.

The variety of women's attitudes to video surveillance ranges from favorable and trusting to strongly opposed.[3] When being monitored, women can feel embarrassed, uneasy, guilty without a reason, irritated or angry – and even fearful. But they may also think that if the cameras are exploited properly, they can prevent crime and, thus, make the city safer. The emotional experience of being under surveillance is often ambivalent or mutable. A surveillance camera can make one feel safe but then, all of a sudden, change to be "a sign of danger." Surveillance appears to evoke collateral positive and negative feelings. Under urban surveillance, women can *simultaneously* feel more secure and more fearful (Koskela, 2002). The "emotional space" that surveillance creates is unstable, nebulous and unpredictable – "like a liquid" (Koskela, 2000, p. 259).

The act of surveillance is loaded with gendered social practices. Surveillance could be interpreted as a part of "male policing in the broadest sense" (Brown, 1998, p. 217). The "masculine culture" of urban video surveillance is causing mistrust: often women do not rely on those behind the camera because of the reproduction of patriarchal power by the guards and the police responsible for and in control of surveillance (e.g. Wajcman, 1991; Fyfe, 1995; Herbert, 1996). The surveillance systems "tend to inflate stereotypes" in the minds of the observers (Lyon, 2001, p. 63). The overseers bring along not only their professional skills and experience but also their emotions, preferences, prejudices, and social habits. Their observations about what the monitor shows are the result of their interpretation of the cultural codes they are aware of. They overlook things for which they do not have a ready knowledge or understanding. The interpretation of what the camera shows depends on the viewer.

Surveillance is insensitive to issues that are of particular importance to women – it is largely unable to identify situations where a sensitive interpretation of social interaction is needed. Under surveillance, social contact is *reduced to the visual*. Surveillance embodies and magnifies the power of the visual; its meaning *overpowers* other senses. Therefore, the ordinary in-passing verbal street harassment women routinely face is unobserved (Koskela, 2002). Alcohol-related disturbances are also often considered – from the perspective of the overseers – not to be serious enough to be interrupted. Surveillance is not able to "change the general intimidation, verbal harassment, staring, and drunken rowdiness amongst groups of men which constrains women's movement most strongly" (Brown, 1998, p. 218).

This insensitivity can be understood as a "passive" relation between surveillance and harassment. However, there is also an "active" relation (see Koskela, 2000). It has been shown that there is considerable concern among women about the "potential 'Peeping Tom' element" of surveillance cameras (Honess and Charman, 1992; Koskela, 2002). Women are worried that surveillance extends the reach of male "voyeurism" (Trench, 1997; Brown, 1998), and that cameras are placed in spaces of any intimate nature (Koskela, 1999).

Indeed, it is possible to use surveillance cameras as an actual means of harassment. There is some voyeuristic fascination in looking, in being able to see – especially without being seen (Ainley, 1998). Moreover, voyeurism itself is generally identified as a masculinist characteristic. Arguably, in Western – and other – cultures, the female body is placed to be an object of a gaze in a far different sense than is the male body. An anecdote, which illustrates this well, is an advertisement of a major Finnish department store, Stockmann, that lately announced above some pictures of the latest women's fashion: "You perform for the surveillance cameras every day. Are you dressed for it?" Now, urban women are gazed on not only by men on the street but also by the hidden (male) gazes behind the camera.

Used by an abuser, a "look" can be a weapon. Surveillance can be a way of reproducing and reinforcing male power. The gaze of a surveillance camera is "calculated to exclude" (Munt, 1995). A camera represents total one-way-ness of the gaze by making it impossible to look back. One may see the camera but eye contact with it is impossible. There is no "mutual" gaze. It would be ridiculous to *flirt* with a surveillance camera. Its objects are constantly seen but with no possibility of "responding to" or "opposing" the gaze. One can only be the observed, not the observer. Surveillance is "opening up new possibilities for harassment and stalking" (Ainley, 1998, p. 92). Furthermore, it does not replace or erase other forms of embodied objectification: women still encounter harassment and objectifying attitudes in face-to-face contacts. Surveillance can be understood as the "re-embodiment" of women, as "an extension of male gaze"; surveillance reproduces the sexualisation of women, and contributes to the process of masculinisation of space.

There is accumulating evidence of exactly the kind of voyeuristic use of surveillance cameras that women are most worried about (in greater detail see, for example, Hillier, 1996; Koskela, 2002). There are countless stories of male operators who have been observed spying on women, or of cameras placed in women's changing rooms. Real and manipulated images from surveillance systems have been edited onto tapes for commercial purposes. Individual sequences from intimate surveillance has been edited onto (pornographic) tapes to be shown at house parties.

Surveillance is about the "regulation of bodily and other visible activities" (Hannah, 1997, p. 171). Being conscious of being controlled can lead to the internalization of control (Foucault, 1977, 1980). What ensures discipline simultaneously erodes confidence. Also, unverifiability is a crucial dimension for maintaining power. While everything and everybody under vigilance is becoming more visible, the forces behind this are becoming less so. It is impossible to infer whether there is somebody looking at you at a particular moment or not. The public is "under surveillance," but by invisible unknown persons in an unknown place. The camera is reduced – as per the idea of the Panopticon – to be just "the gaze," a depersonalized and distant overseer (Foucault, 1980). Consequently, the camera renders the monitored persons as passive, potential victims (Koskela, 2002, p. 268). It treats them as objects and does not increase their feeling of being "in control," but places them "under control." Although Foucault in his interpretations of the Panopticon prison was not focusing on gender issues, in contemporary urban space this element is pervasive and prominent. Surveillance contributes to perpetuating the existing imbalance in gender relations rather than challenging it.

What must not be forgotten, however, is that control is never completely hegemonic: "power is always contaminated with resistance" (Robinson, 2000, p. 68). Surveillance can be turned to "counter-surveillance," to a weapon for those who are oppressed, and "used by women for their own ends" (Ainley, 1998, p. 92). An example of this was recently given in Finland by a woman who had experienced violence and used surveillance to protect herself. She had to prove that the offender had violated the restraining order given by the court, and installed a surveillance camera on her own front door to catch him on tape. Indeed, gender relations can be turned upside down.

By examining the issues of gender and surveillance it is possible to increase understanding of the interacting emotional and power-related processes that play a role in producing space. Space is produced both "behind" and "in front of" a surveillance camera, as well as in the virtual environments of cyberspace. Ever-increasing surveillance will challenge us to reconstruct our notions of space. Surveillance is likely to change some conventional concepts: re-evaluation of the concepts of presence and absence, hiding and revealing, reality and truth, the vagueness of the images, the significance of the visual, etc. While visual images are loaded "with the promise of reality" (Groombridge, 2002, p. 38), it is clear that the "reality" of a surveillance camera tape is a social product: "surveillance does not find knowledge, but creates it" (Allen, 1994, p. 144). In urban space the ostensible aim of surveillance is to make space more accessible and available to different groups. However, if surveillance is used to exclude "the Other," as it arguably has been, it can also be seen as making space "less public." It is not "opening up public spaces" (Lyon, 2001, p. 135) but creates predetermined, calculated, and rigid urban space – *a suppressed space.*

Conclusions

On a street level, where economic and political processes are not the (only) dimensions that produce space, "gaze" has a crucial role. It matters who is seen and who is not, who is able to look and who is not, whose gaze is producing power and who will avert their eyes. Paradoxically, "being seen" can mean threat or protection, repression or promise, depending on the circumstances. The production of urban space is "necessarily a mutual process, actively involving both the gazing subject and the subject of the gaze" (Robinson, 2000, p. 81).

In the three examples I have discussed in this chapter, the gaze has a crucial role. There is always an external real or virtual gaze which modifies urban experiences. In a red light district the gaze is an essential part of the social processes that produce space. In the dynamics of fear and boldness the personal emotional aspects of being seen or being able to see make the difference. And in the practices of urban control, the gaze is technologically mediated.

All the examples have also shown that women are not merely objects in space where they experience restrictions and obligations; they also actively produce, define, and reclaim space. Meanings about gender and space are modified in social practices and "are not natural or fixed but continually contested and subverted" (Boys, 1999, p. 193). Despite the often reproduced image of women as "spatially restricted," they are capable of working as active agents, occupying practices of resistance, and using power.

Urban space is gendered in many ways and it is, indeed, produced in everyday encounters on the street. These encounters, however, are quite surprising. The "semiotics of the street" is more complex than it seems at first sight, the "signs" we read being both contextual and mutable. Women *do* face threats and embodiment but they can still be unsubmissive. Resistance takes many forms. Urban space should be discussed in plural: it is elastic, tamed, and suppressed, sometimes alternately, sometimes simultaneously.

NOTES

1 This section is based on a survey conducted in 2000 in Helsinki by the research department of the City of Helsinki in collaboration with the author, the aim of which was to find out how prostitution influenced female residents' lives in a well known red light district. The study was made in the form of a postal questionnaire, which included discussion-style open questions, to enable an interpretative approach. The results of this project are available elsewhere (see Koskela et al., 2000; Tani, 2001, 2002; Koskela and Tani, 2005).

2 My arguments on boldness and women's spatial confidence are based on a qualitative study conducted in 1996–7 in Helsinki, Finland. It consisted of individual and group interviews with women who had very different relationships to urban space: some were outgoing, whereas others lived more privatized and spatially restricted lives. Women were

recruited with the help of the National Consumer Research Center, which had collected "a panel," a group of citizens willing to take part in research projects. Hence, women did not participate because of their levels of fear. The results of the project have been reported in other publications in greater detail (see Koskela, 1997, 1999; Koskela and Pain, 2000).

3 Women's emotional responses to surveillance have been studied in two series of interviews in Helsinki and Edinburgh. In Helsinki the interview sessions covered many subjects related to fear of violence (see note 2), whereas the ones in Edinburgh were more specifically focused on video surveillance. From these data I have produced a typology of women's attitudes to surveillance as well as more general reasoning on surveillance and the changing nature of urban space (see Koskela, 1999, 2000, 2002).

BIBLIOGRAPHY

Ainley, R. (1998) Watching the detectors. Control and the Panopticon. In R. Ainley (ed.), *New Frontiers of Space, Bodies and Gender*. London: Routledge.

Alcoff, L. M. (1996) Feminist theory and social science. New knowledge, new epistemologies. In N. Duncan (ed.), *BodySpace. Destabilizing Geographies of Gender and Sexuality*. London: Routledge.

Allen, M. (1994) "See you in the city!" Perth's citiplace and the space of surveillance. In K. Gibson and S. Watson (eds), *Metropolis Now. Planning and the Urban in Contemporary Australia*. Sydney: Pluto Press.

Boys, J. (1999) Positions in the landscape? Gender, space and the "nature" of virtual reality. In Cutting Edge – The Women's Research Group (eds), *Desire by Design. Body, Territories and New Technologies*. London, I. B. Tauris.

Brown, S. (1998) What's problem girls? CCTV and the gendering of public safety. In C. Norris, J. Moran and G. Armstrong (eds), *Surveillance, Closed Circuit Television and Social Control*. Aldershot: Ashgate.

Davis, M. (1990) *The City of Quartz. Excavating the Future in Los Angeles*. New York: Vintage.

Domosh, M. (1998) Those "gorgeous incongruities." Polite politics and public space on the streets of nineteenth-century New York City. *Annals of the Association of American Geographers*, 88, 209–26.

Duncan, N. (1996) Renegotiating gender and sexuality in public and private spaces. In N. Duncan (ed.), *BodySpace. Destabilizing Geographies of Gender and Sexuality*. London: Routledge.

Ellin, N. (ed.) (1997) *The Architecture of Fear*. New York: Princeton Architectural Press.

Epstein, D. (1997) Abject terror: a story of fear, sex and architecture. In N. Ellin (ed.), *The Architecture of Fear*. New York, Princeton Architectural Press.

Foucault, M. (1977) *Discipline and Punish. The Birth of a Prison*. London: Penguin Books.

Foucault, M. (1980) The eye of power. In C. Gordon (ed.), *Power/Knowledge. Selected Interviews and Other Writings 1972–1977 by Michel Foucault*. Lewes: Harvester Press.

Fyfe, N. R. (1995) Policing the city. *Urban Studies*, 32, 759–78.

Gardner, C. B. (1990) Safe conduct. Women, crime and self in public places. *Social Problems*, 37, 311–28.

Gardner, C. B. (1995) *Passing By. Gender and Public Harassment*. Berkeley: University of California Press.

Graham, S. (1998) Spaces of surveillant simulation. New technologies, digital representations, and material geographies. *Environment and Planning D: Society and Space*, 16, 483–504.

Groombridge, N. (2002) Crime control or crime culture TV? *Surveillance and Society*, 1, 30–6 (www.surveillance-and-society.org).

Hannah, M. (1997) Space and the structuring of disciplinary power. An interpretive review. *Geografiska Annaler*, 79B, 171–80.

Hart, A. (1995) (Re)constructing a Spanish red-light district. Prostitution, space and power. In D. Bell and G. Valentine (eds), *Mapping Desire. Geographies of Sexualities*. London: Routledge.

Herbert, S. (1996) The geopolitics of the police. Foucault, disciplinary power and the tactics of the Los Angeles Police Department. *Political Geography*, 15, 47–57.

Hillier, J. (1996) The gaze in the city. Video surveillance in Perth. *Australian Geographical Studies*, 34, 95–105.

Høigård, C. and Finstad, L. (1992) *Backstreets. Prostitution, Money and Love*. Cambridge: Polity Press.

Honess, T. and Charman, E. (1992) *Closed Circuit Television in Public Places*. Crime Prevention Unit Series, paper 35. London: Home Office Police Research Group.

Hubbard, P. (1998) Sexuality, immorality and the city. Red-light districts and the marginalization of female street prostitutes. *Gender, Place and Culture*, 5, 55–72.

Hubbard, P. (1999) *Sex and the City. Geographies of Prostitution in the Urban West*. Aldershot: Ashgate.

Hubbard, P. (2000) Policing the public realm. Community action and the exclusion of street prostitution. In J. R. Gold and G. Revill (eds), *Landscapes of Defence*. Harlow: Pearson.

Judd, D. R. (1995) The rise of the new walled cities. In H. Liggett and D. C. Perry (eds), *Spatial Practices. Critical Explorations in Social/Spatial Theory*. Thousand Oaks, CA: Sage.

Kilvington, J., Day, S. and Ward, H. (2001) Prostitution policy in Europe. A time of change? *Feminist Review*, 67, 78–93.

Koskela, H. (1997) "Bold walk and breakings." Women's spatial confidence versus fear of violence. *Gender, Place and Culture*, 4, 301–19.

Koskela, H. (1999) *Fear, Control and Space. Geographies of Gender, Fear of Violence and Video Surveillance*. Helsinki: Publications of the Department of Geography, University of Helsinki.

Koskela, H. (2000) "The gaze without eyes." Video surveillance and the changing nature of urban space. *Progress in Human Geography*, 24, 243–65.

Koskela, H. (2002) Video surveillance, gender and the safety of public urban space. "Peeping Tom" goes high tech? *Urban Geography*, 23, 257–78.

Koskela, H. and Pain, R. (2000) Revisiting fear and place. Women's fear of attack and the built environment. *Geoforum*, 31, 269–80.

Koskela, H. and Tani, S. (2005) "Elastic spaces": street prostitution and the (hetero)sexualization of urban space. In A. Klinik and G. Stahl (eds), *Night and the City. Reflections on the Nocturnal Side of Urban Life*. London: Verso.

Koskela, H., Tani, S. and Tuominen, M. (2000) *"Sen näkönen tyttö." Tutkimus katuprostituution vaikutuksista helsinkiläisten naisten arkielämään*. Helsinki: City of Helsinki Urban Facts Research Series, no. 5.

Lees, L. (1998) Urban renaissance and the street. Spaces of control and contestation. In N. R. Fyfe (ed.), *Images of the Street. Representation, Experience and Control in Public Space*. London: Routledge.

Lyon, D. (2001) *Surveillance Society. Monitoring Everyday Life*. Buckingham: Open University Press.

McDowell, L. (1993) Space, place and gender relations. Part II. Identity, difference, feminist geometries and geographies. *Progress in Human Geography*, 17, 305–18.

McDowell, L. (1995) Bodywork. Heterosexual gender performances in city work places. In D. Bell and G. Valentine (eds), *Mapping Desire. Geographies of Sexualities*. London: Routledge.

Macmillan, R., Nierobisz, A. and Welsh, S. (2000) Experiencing the streets. Harassment and perceptions of safety among women. *Journal of Research in Crime and Delinquency*, 37, 306–22.

Massey, D. (1994) *Space, Place and Gender*. Cambridge: Polity Press.

Merry, S. E. (1981) *Urban Danger. Life in a Neighborhood of Strangers*. Philadelphia: Temple University Press.

Mitchell, D. (1995) The end of public space? People's Park, definitions of the public, and democracy. *Annals of the Association of American Geographers*, 85, 108–33.

Munt, S. (1995) The lesbian *flâneur*. In D. Bell and G. Valentine (eds), *Mapping Desire. Geographies of Sexualities*. London: Routledge.

Nast, H. J. and Kobayashi, A. (1996) Re-corporealizing vision. In N. Duncan (ed.), *Body-Space. Destabilizing Geographies of Gender and Sexuality*. London: Routledge.

Pain, R. (1991) Space, sexual violence and social control. Integrating geographical and feminist analyses of women's fear of crime. *Progress in Human Geography*, 15, 415–31.

Rendell, J. (2001) "Bazaar beauties" or "pleasure is our pursuit." A spatial story of exchange. In I. Borden, J. Kerr, J. Rendell and A. Pivaro (eds), *The Unknown City: Contesting Architecture and Social Space*. Cambridge, MA: The MIT Press.

Robinson, J. (2000) Power as friendship. Spatiality, femininity and "noisy" surveillance. In J. P. Sharp, P. Routledge, C. Philo and R. Paddison (eds), *Entanglements of Power. Geographies of Domination/Resistance*. London: Routledge.

Rose, G. (1993) *Feminism and Geography. The Limits of Geographical Knowledge*. Minneapolis: University of Minnesota Press.

Smith, S. J. (1987) Fear of crime. Beyond a geography of deviance. *Progress in Human Geography*, 11, 1–23.

Stanko, E. (1996) Reading danger. Sexual harassment, anticipation and self-protection. In M. Hester, L. Kelly and J. Radford (eds), *Women, Violence and Male Power*. Buckingham: Open University Press.

Tani, S. (2001) Bad reputation – bad reality? The intertwining and contested images of a place. *Fennia*, 179, 143–57.

Tani, S. (2002) Whose place is this space? Life in the street prostitution area in Helsinki, Finland. *International Journal of Urban and Regional Research*, 26, 343–59.

Trench, S. (1997) Safer transport and parking. In T. Oc and S. Tiesdell (eds), *Safer City Centres. Reviving the Public Realm*. London: Paul Chapman.

Valentine, G. (1989) The geography of women's fear. *Area*, 21, 385–90.

Valentine, G. (1992) Images of danger. Women's sources of information about the spatial distribution of male violence. *Area*, 24, 22–9.

Wajcman, J. (1991) *Feminism Confronts Technology*. Cambridge: Polity Press.

Walkowitz, J. R. (1992) *The City of Dreadfull Delight. Narratives of Sexual Danger in Late-Victorian London*. London: Virago.

Weitzer, R. (2000) The politics of prostitution in America. In R. Weitzer (ed.), *Sex for Sale. Prostitution, Pornography, and the Sex Industry*. London: Routledge.

Wilson, E. (1991) *The Sphinx in the City. Urban Life, the Control of Disorder and Women*. London: Virago Press.

Wilson, E. (1995) The invisible *flâneur*. In S. Watson and K. Gibson (eds), *Postmodern Cities and Spaces*. Oxford: Blackwell.

Chapter 19

Daycare Services Provision for Working Women in Japan

Kamiya Hiroo

Introduction

This chapter examines the daily activity patterns of Japanese working women in metropolitan areas,[1] focusing on their increasing participation in the labor market. This is seen in the context of a National Census report in 1990 that brought about a so-called "1.57 shock." This was the sensational reaction to the remarkable decline in the overall fertility rate among Japanese women (which had dropped to 1.57, well below the national population replacement level). The news prompted public concern that such a sharp decline in the fertility rate of Japanese women would lead to severe labor shortages in the near future.

As geographers, we can see the increasing number of women who have no children or only one child as a protest by women who are juggling multiple roles as wage earners and mothers. In the past 20 years, feminist geographers' interest in child rearing has focused on this "dual role" issue (Tivers, 1985). Gradually, research concern has also been shifting to examining the broader mothering and parenting culture (Dyke, 1990; Fincher, 1991; Valentine 1997). In the Japanese context, a sociologist recently suggested that Japanese traditional patriarchy places priority on the mothering role more than on the wife role (Sechiyama, 1996). This priority induces married women to work only on a part-time basis – because the traditional gender role of women in Japan is strongly oriented to child rearing. This strong responsibility of Japanese women for child rearing is reinforced by the "Japan-focused East Asia social welfare regime," which is characterized as little intervention and relatively low social welfare expenditure by the government (Peng, 1996).

The strong emphasis on mothering means that women with young children tend to quit their jobs when they become pregnant. On the other hand, the demand for female labor is expected to rise in the future: the rapidly aging population, combined with low fertility rates, will create a shortage of labor. Under such conditions, a social support system that helps working women to bring up their children is

urgently needed. Public and business entities have begun to appreciate the importance of such a support system and have taken action to improve the situation. For
example, the Ministry of Health and Welfare has recently initiated a comprehensive project to help working women by providing public support for child rearing.
At the same time, some private companies have attempted to provide or expand
daycare services for their workers. Other companies have started providing daycare
services for working mothers who commute from the suburbs to the city core in
metropolitan areas.

This study depicts the daily lives of households based on activity diary data. It
attempts to identify the constraints of married women's childcare role on their daily
activities when they choose to work as wage earners. In particular, by analyzing the
daily activity patterns of urbanites in Japan, the public provision of daycare is evaluated from the standpoint of accessibility and service hours. Daycare service is considered from a time–geographic perspective, treating the study population not as an
aggregated unit but as indivisible entities. The time–geographic approach presumes
that people have their own specific social, environmental, and family backgrounds.
By taking this perspective, we attempt to identify various alternatives for improving working women's welfare.

Daycare service is evaluated mainly from the mother's perspective. Though childcare service involves other family members, such as the father and the children themselves, women with young children are most strikingly affected by the child rearing
activity. In addition, many Japanese women still quit their job when they become
pregnant. This study's focus on the mother's activities in Japanese urban settings
will widen our understanding of how women cope with their dual role as paid
workers and unpaid family caretakers, and deepen our understanding of women's
daily activities.

Before we discuss the empirical study, a general outline of women's labor in Japan
is presented. This overview covers the "M-shaped" age profile of women's labor
activity rate, which many scholars depict as a typical feature of female labor in
Japan. Then, the delivery of daycare service in three municipalities is scrutinized as
well as the delivery of daycare service at the national level. After that, based on the
survey conducted in 1988 and 1990, the daily lives of women with young children
in these areas are considered in detail. That involves sketching out the activity path
of every household member using time–geographic representation methodology. In
decoding such a diagram, particular attention is drawn to when, who, how, and in
what travel mode parents or other family members make trips to escort young children to and from daycare facilities. In the final section, after considering a recent
policy shift in daycare service provision, some policy implications for future service
delivery are discussed.

Female Labor in Japan

As shown in figure 19.1, the ratio of women who work for wages (RWP: number
of female workers in the female population between age 15 and 64) has remained
virtually unchanged. There was only a small fluctuation from 1955 to 2000. But
when we consider the RWP for metropolitan areas and for non-metropolitan areas
separately, we see contrasting trends for the two areas. For instance, in 1955 the

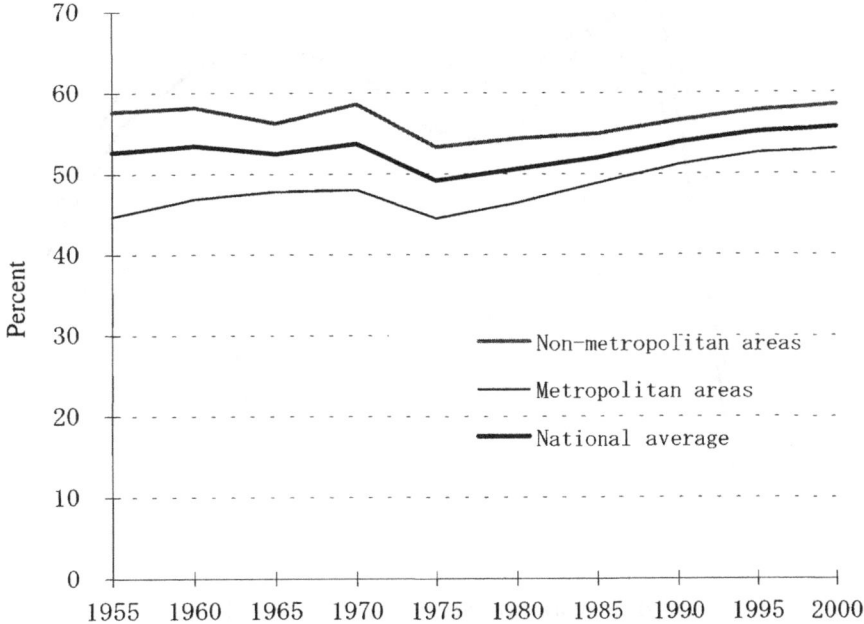

Figure 19.1 The ratio of women who work for wages (RWP: number of female workers in the female population between age 15 and 64) from 1955 to 2000.

RWP for non-metropolitan areas was much higher than that for metropolitan areas: many women had jobs in agriculture. While the RWP for non-metropolitan areas declined gradually from 1955 to 2000, the RWP for metropolitan areas showed steady growth. Consequently, the two trends had converged to almost the same level by 1985, though the former was still a little higher than the latter.

For the same period, there was a substantial shift in the age profile of female labor (figure 19.2). At first glance, the M-shaped age profile is the same for metropolitan areas and non-metropolitan areas, in both 1955 and 2000. But on closer examination, in 1955 the RWP for the age group 20–24 is much higher than that for middle-aged women, while in 2000 the ratio was almost the same for the two age groups.

An M-shaped age profile of the female labor force is unusual in developed countries. Sweden, the United States, and France, for instance, used to show an age profile similar to that of contemporary Japan. They now have a reversed U-shaped profile, with fewer women of child rearing age quitting their jobs. The United Kingdom resembled Japan to some extent, in that the female labor force had a similar age profile, though it has been approaching a reversed U-shape in recent years. In sum, the frequent exit of women from the labor market is a phenomenon that is now distinctive to Japan among the developed countries.

Turning to the regional level, figure 19.3 depicts the female labor force participation rate (LFPR) by municipality in the Nagoya metropolitan area. The outer areas have a higher ratio in general, as the share of women engaged in agriculture

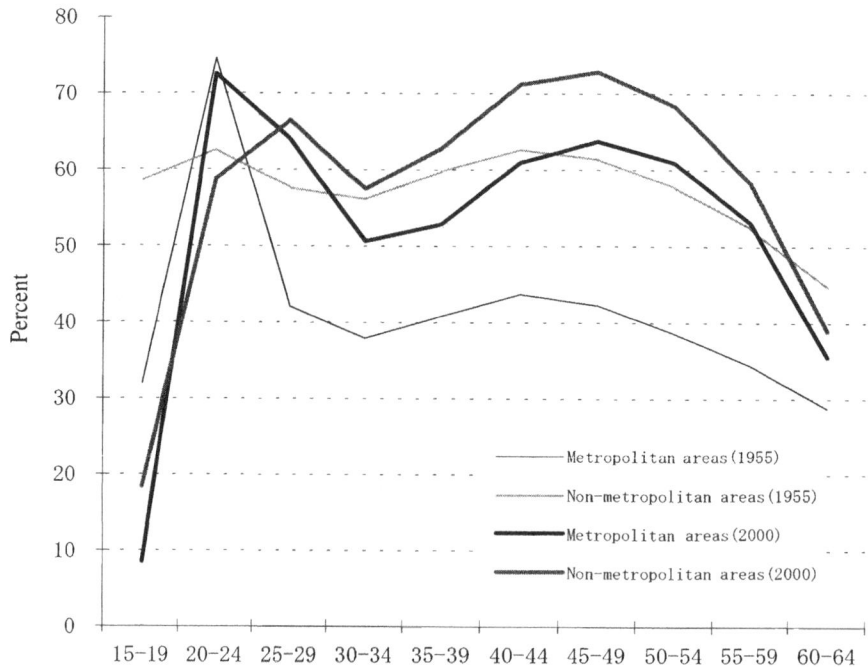

Figure 19.2 Female ratio of working persons by region and age.
Source: Statistics Bureau, Population Census of Japan, 1955, 2000.

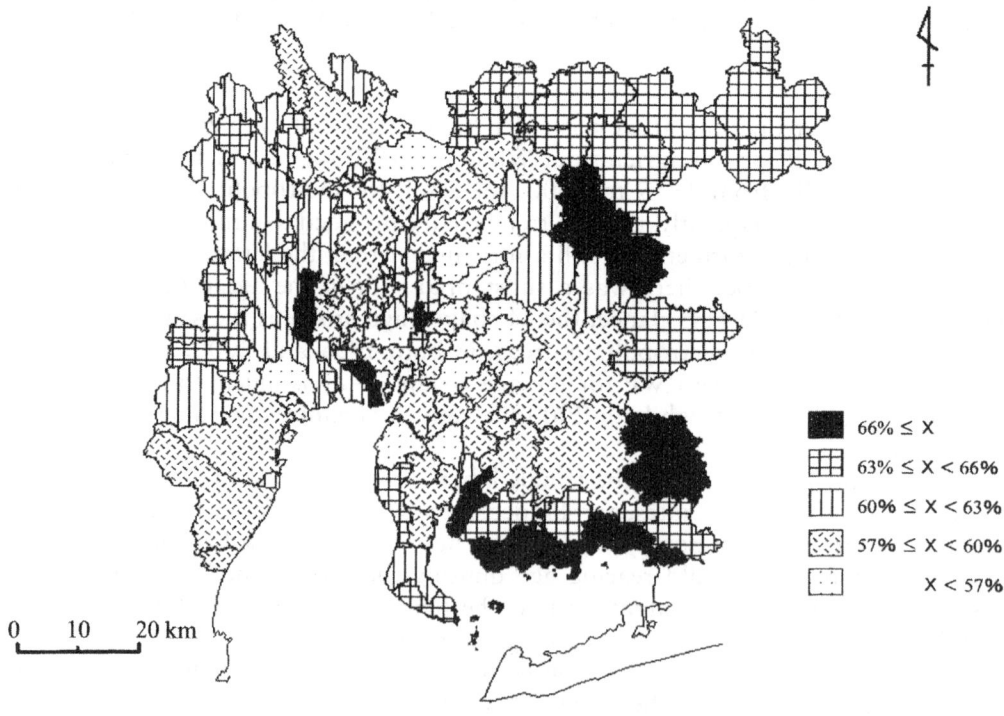

Figure 19.3 The female labor force participation rate by municipality in the Nagoya metropolitan area.

is higher in this area. This type of regional pattern is rather similar to that observed in figure 19.1. Meanwhile, typical suburban municipalities, like the eastern areas in and around Nagoya City, tend to show a lower LFPR than average. This may be due to the large share of white-collar workers, higher household incomes, and a higher rate of residential moves.

Two of the three study areas are situated within the Nagoya metropolitan area and Tokyo metropolitan area: Nissin Town and Kawagoe City. The third study area is Shimosuwa Town, located in Nagano Prefecture, 140 km to the northeast of Nagoya. The LFPR was 64.9 percent in Shimosuwa Town, 49.2 percent in Nissin Town, and 52.1 percent in Kawagoe City. There was a rather large discrepancy between Shimosuwa Town and the latter two, which reflects the fact that suburban areas have a lower labor activity rate for women than other areas. Thus, Shimosuwa Town can be regarded as typical of a small town, and Nissin Town and Kawagoe City as representative of suburban municipalities.

Daycare Service Provision in Japan

Trends in daycare service in Japan

Most preschool children in Japan experience life in a group by attending a day nursery or a kindergarten. From the viewpoint of child welfare, local governments have a legal responsibility to provide day nurseries for children who are not otherwise being cared for in the daytime. In other words, children are supposed to be raised at home, unless their mother is working. Kindergartens are considered a discretionary option for parents, as they are provided for educational purposes. This principle is clear, in that daycare service in nurseries is under the authority of the Ministry of Health and Welfare, while kindergartens are under the auspices of the Ministry of Education. In short, day nurseries are for children whose mother is working, while kindergartens are for children whose mother is not a wage earner. Hereafter, the focus is on the day nursery services, for we are mainly interested in the daily activity pattern of women who are working.

Daycare service delivery by local government must conform to some guidelines. Among these, three important ones concern: (a) screening procedures by which local governments select children who may receive care; (b) service hours of daycare; (c) fees for care to be charged to the parents. But the selection criteria for children who are in "want of care in the daytime" are becoming blurred, as more women work on a part-time basis. In most daycare centers, service is provided for eight hours. Only a few centers offer extended service hours, usually from 7.00 a.m. to 7.00 p.m. The reasons for not being open longer are insufficient subsidies from the central government, expected cost increases for hiring more staff, and complicated working schedules. The fees charged depend upon the parents' income, with the amount of subsidy determined according to a formula set by the central government. Half of the subsidy comes from the central government, a quarter from the prefectural government (Ken), and the rest from the municipality.

The total number of daycare centers in Japan is shown in table 19.1. As this table indicates, the number of centers has declined in recent years; this is due to the declining fertility rate. On the other hand, various attempts are in progress to upgrade

Table 19.1 Number of day care centers and accepted children

Year	No. of centers	No. of children
1965	11,248	799,622
1970	13,818	1,109,862
1975	18,009	1,561,397
1980	21,960	1,940,793
1985	22,899	1,770,466
1990	22,703	1,723,775
1995	22,438	1,678,866
2000	22,199	1,904,067

Source: Ministry of Health and Welfare, Social Welfare Statistics.

the quality of service. One approach is to expand the service provided by local government. The Ministry of Health and Welfare has promoted various forms of daycare to cope with the increased number of working women, the shrinking family size, variable employment status, and diversified life styles. In addition to the standard eight hours of service per day, subsidies for infant care, extended hours, and night care started around 1970.

Infant care (for babies less than 12 months old) was provided at 4340 centers (19.1 percent of the total) in 1990. In practice, the screening criteria applied by daycare centers accepting infants differ from one municipality to another. Some municipalities take in infants when the maternity leave from work ends, while other municipalities only accept infants over six months old.

Extended hours service was provided at 826 centers (3.6 percent) in 1990. This number is lagging far behind the target of 2000 set by the minister. Nighttime service is scarce, particularly in metropolitan areas where demand for such care is great. In 1990, nighttime care was only available at 33 centers.

Alternative forms of care include baby hotels in the private sector, day nurseries attached to large office buildings or hospitals, nursery facilities run by volunteer groups, and home care by neighbors. The central government does not subsidize such care, but some local governments do offer financial support.

The increased women's labor force participation has generated a larger demand for daycare services in metropolitan areas, particularly in suburban areas. The municipalities are forced to meet this demand. Below, I describe the measures taken by local government to cope with these increased demands, based on a mail survey of 27 suburban municipalities in the Tokyo Prefecture. Valid answers were obtained from 16 municipalities. This overview will shed light on the measures that suburban municipalities employ in response to the demand for daycare service.

Some municipalities give subsidies to women whose children are already grown up. These women can spare time for child rearing and accept children from neighbors, caring for them in their homes. In addition, the Tokyo Prefectural Government, an upper-tier political body, also provides additional financial support for this system. The only women who are allowed to take in children are those qualified as a nurse, midwife, teacher, etc.

Table 19.2 Number of attached care facilities

Year	Municipal	Private	Total
1979	349	1336	1685
1981	419	1559	1978
1983	430	1637	2067
1985	413	1612	2025
1987	412	1767	2179
1989	423	1831	2254
1991	426	2117	2543
1993	449	2763	3212
1995	480	3249	3729
1997	472	3389	3861
1999	443	3388	3831
2001	421	3372	3793

Source: Japan Child Beneft Association, list of attached care facilities.

The Tokyo Prefectural Government sets a higher standard for workers in day nurseries to improve the level of care compared with the national standard. Some municipalities also allocate more nurses to the daycare service. For instance, in the case of three-year-old children, the central government provides subsidies based on a standard of one nurse for eight children; the prefectural government calculates on the basis of one nurse to six children; and Higashi-kurume City calculates on the basis of one nurse to five children.

Levels of extended hours of care vary by municipality, but many centers are open between 7.30 a.m. and 7.00 p.m. This is a little longer than the standard daycare hours, from 8.30 a.m. to 5.00p.m., at national level. Daycare service for infants is provided in many suburban municipalities. For instance, in Kokubunji City all municipal and private day nurseries accept infants of less than 12 months. Municipal centers are founded and run by local governments, while many private centers are founded and run by non-profit organizations. In general, private centers are more likely than municipal centers to accept infants for extended hours care.

There is still a large unsatisfied demand for daycare as a result of the continuing growth of female labor. It is often argued that the number of day nurseries attached to large offices and nursery facilities operated by volunteer groups has been increasing in recent years in response to this unsatisfied demand. Since few of these facilities receive financial support from the government, it is not easy to estimate the number of such facilities accurately. According to the available data on the facilities attached to large office buildings and hospitals that receive subsidies from the public sector, these facilities have been growing slightly in number due to the relative labor shortage in present-day Japan (table 19.2). In 1989, there were 2254 attached nursery facilities in total, of which 81.2 percent were private and 18.8 percent were public. Children under two years old constituted 54.5 percent of the total number of accepted children in these facilities. The proportion of this group in attached facilities is considerably higher than that in public daycare centers. This is because private companies recommend that parents transfer children of over three years old to public daycare centers in order to cut the cost of providing daycare services.

Looking at the type of industry that provides care facilities, over half have been set up in medical institutions for hospital nurses working on shifts. The medical sector is followed by retail and wholesale industries and service. The number of attached facilities at manufacturing companies, both offices and plants, is declining. In the retail sector, like stores selling cosmetics and soft drinks, and in the service industries, such as those employing female golf caddies, the number is increasing.

For instance, Seibu Department Store in Tokyo is pushing forward a project to stabilize the spinning off of female labor and the development of workers' capabilities. The reason is that female labor in retailing is more essential than in other industries. Since child rearing is the main obstacle for women who want to continue their work, Seibu Department Store had set up an attached facility only a few minutes' walk away from the store in 1982. The facility accepts children between seven months and three years old. Service hours are limited to the eleven hours between 8.30 a.m. and 7.30 p.m. Extended service is available if the mother is working overtime.

Another example is drawn from Yakult, a soft drink retailing company. An attached facility was opened in an office building of the company on the outskirts of Tokyo. The company aimed to recruit women with young children. Children aged one to three years are accepted and cared for from 9.00 a.m. to 2.00 p.m. As the company bus takes children to and from the workers' homes, women do not have to escort their children to the facility.

Taking into account the recent movements to develop a higher level of care by both public and private bodies, we will proceed to describe the daily lives of women in the three areas covered in this study.

Daycare service in three municipalities

As noted above, the female labor force participation rate in the small town of Shimosuwa is much higher than in the metropolitan suburbs of Nissin Town and Kawagoe City. Parallel to this difference in the activity rate, there is also quite a significant difference in provision of daycare services (table 19.3).

In 1990, daycare services were provided at seven centers run by Shimosuwa Town. Children were cared for from 8.30 a.m. to 4.00 p.m.; this amounts to slightly fewer hours compared with the other two areas. Only one center accepted children under three years old. The three-year age rule is rather strict compared with eligibility in the other two municipalities. The availability of extended services from 8.00 a.m. to 6.00 p.m. is limited to two centers.

Ten centers in Nissin Town provide daycare service; nine of them are municipal and the other one is private. The standard opening hours are between 8.00 a.m. and 4.00 p.m., while extended care is available between 7.30 a.m. and 6.00 p.m. Infants over six months old are accepted at four centers.

Public daycare in Kawagoe City is provided at 20 municipal centers and eight private centers. Care hours vary from center to center, but usually they are from 8.30 a.m. to 5.00 p.m. at municipal centers. Extended care at seven municipal

Table 19.3 Child care provision in three municipalities

	No. of centers[a]	Service hours		Age of acceptance
		Standard[b]	Extended[c]	
Shimosuwa Town	7	8.30 a.m.–4.00 p.m. (8.30 a.m.–3.30 p.m.)	8.30 a.m.–5.45 p.m. (2)	3 years (6) below 3 years (1)
Nissin Town	10 (1)	8.00 a.m.–4.00 p.m.	7.30 a.m.–6.00 p.m. (5)	12 months (6) 6 months (4)
Kawagoe City	28 (8)	8.30 a.m.–5.00 p.m.	7.30 a.m.–6.00 p.m. (12) 7.30 a.m.–6.30 p.m. (3)	2 years (6)[d] 12 months (3) 8 months (15)[e] 3 months (4)

[a] Figures in parentheses indicate number of private centers.
[b] Hours in parentheses indicate service in winter.
[c] Figures in parentheses indicate number of centers.
[d] Including one center with a one-year and six-month minimum age.
[e] Including one center with a six-month minimum age.

centers is available from 7.30 a.m. to 6.00 p.m. All private centers provide longer extended service hours than municipal centers. The age qualification for acceptance also varies, but centers in Kawagoe City tend to accept younger children than centers in the other two areas. The municipality also subsidizes women who accept and care for children of neighbors.

This brief overview covers the somewhat divergent levels of daycare service in three municipalities. One might expect to find a higher standard of service in Shimosuwa Town because of the higher economic activity rate of women in this area, but the facts are quite different. This is partly because women in non-metropolitan areas are more likely than women in cities to be members of extended families and to get help from their parents.

Daily Lives of Women with Young Children

This section considers how much of the child rearing activity that hinders women's working is taken over by other family members or by public childcare. It is based upon data drawn from an activity diary collected in Shimosuwa Town, Nissin Town, and Kawagoe City. Detailed information on the data set is provided elsewhere (Kamiya et al., 1990; Kamiya, 1993). The final section briefly discusses the policy implications of this investigation. The methodology for this analysis is rather descriptive. A representative sample household was drawn from the data set, and the daily activities of all household members were recorded. The activity diary covers two days, Sunday and Monday, for all three areas. Here, information from the data on Monday is analyzed because this chapter is mainly concerned with the activities of working women.

Sample households are extracted according to the following procedures. First, households with preschool children are selected from all the households surveyed for each area. As a result, 21 households for Shimosuwa Town, 48 for Nissin Town,

Table 19.4 Childcare provision in three municipalities

Sample no.	Employment status of wife[a]	Workplace of wife[b]	Family type[c]	Form of child care	Escort	No. of cars
SSW-7	P	N	N	Day center	Wife	1
SSW-17	F	N	E	Grandparents	–	2
NSN-117	F	C	N	Day center (ext.)	Wife	2
KWG-179	F	C	N	Day center (ext.) and friend	Wife (and husband)	1
NSN-22	F	S	N	Voluntary facility	Wife	1

[a] N, no job; P, part-time job; F, full-time job.
[b] N, neighborhood; S, suburb; C, city center.
[c] N, nuclear family; E, extended family.

and 46 for Kawagoe City remain in this category. Among these households, those which seem to exemplify the prototype activity patterns are selected for closer examination. In table 19.4 and the following description, the abbreviation SSW stands for the Shimosuwa data set, NSN for Nissin, and KWG for Kawagoe. For instance, SSW-7 means household number seven in the Shimosuwa data set.

First, a household with a woman working on a part-time basis and leaving her child in a daycare center is chosen (SSW-7). Women with young children can work on a full-time basis when someone or some institution takes over her responsibility for childcare. Therefore, an extended family with grandparents caring for the children is taken as a representative case (SSW-17). When assistance from kin is not available to women in nuclear families, they have to find an alternative strategy for care in order to work on a full-time basis. To show this, one household which uses the extended service hours of a daycare center is selected (NSN-117). If the opening hours of the extended day center are not sufficient, those women must find additional means for childcare. One strategy is to take their children to some person other than family members (KWG-179). Another strategy is to use a nursery facility run by a voluntary group instead of a municipal daycare service (NSN-158).

Table 19.4 summarizes household attributes and forms of childcare in the households sampled. The daily activities of households are described in line with the listed attributes and forms of care.

First, we examine the nuclear family, where the wife is working on a part-time basis (figure 19.4). The household consists of a husband and wife, a son (ten years old), and two daughters (five and eight years old) (SSW-7). In this household, the wife takes her youngest daughter to and from the daycare center every day. In Shimosuwa Town, provision of daycare service with extended hours is rather limited. The standard opening times for daycare in Shimosuwa Town are from 8.30 a.m. to 4.00 p.m. Within her "window" of free time, it is not easy to work on a full-time basis. Therefore, she is forced to work part-time. Like this woman, many married women in Japan are working on a part-time basis.

Second, the case of an extended family is examined (figure 19.5). In this household, the grandparents take care of the children in the daytime and the wife works

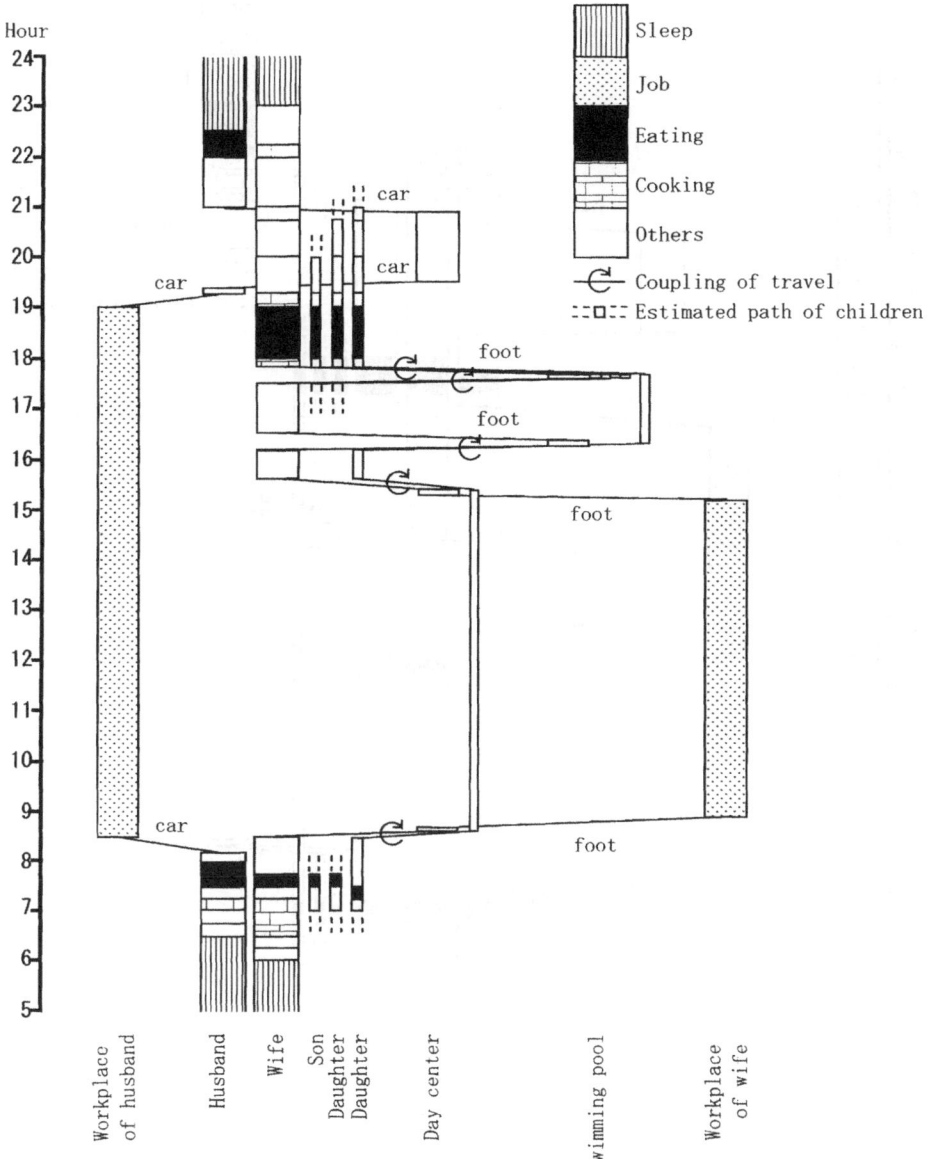

Figure 19.4 Daily lives of a nuclear family with a wife working on a part-time basis (SSW-7).

on a full-time basis (SSW-17). Grandparents, husband and wife, and two sons (a one-year-old and a seven-year-old) and a daughter (of six years old) constitute the household. The wife works from 8.00 a.m. to 5.00 p.m. every day; during her absence, the grandparents stay at home to provide care for the younger son. The grandfather also takes the role of escorting the daughter to the day center by car. On the day surveyed, the wife left home earlier and came home much later than the

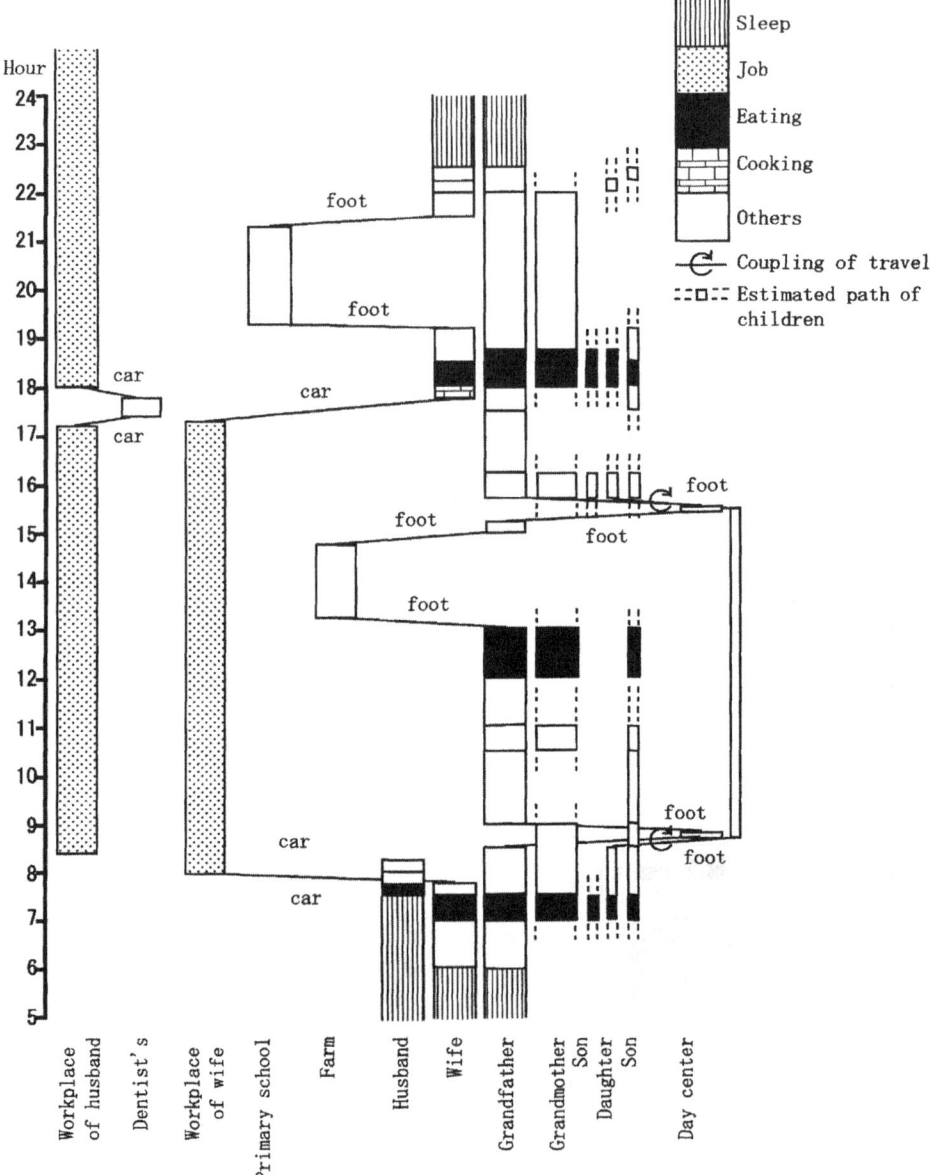

Figure 19.5 Daily lives of an extended family with a wife working on a full-time basis (SSW-17).

daughter. In addition, late in the evening the wife was able to participate in leisure activities with the help of childcare by the grandparents. Though this family typifies the substitute role played by grandparents, it is unnecessary to say that not all wives in extended families can expect such assistance.

In nuclear families, parents usually expect no help from the grandparents. This is the third type of household shown here (NSN-117). In this household, the wife

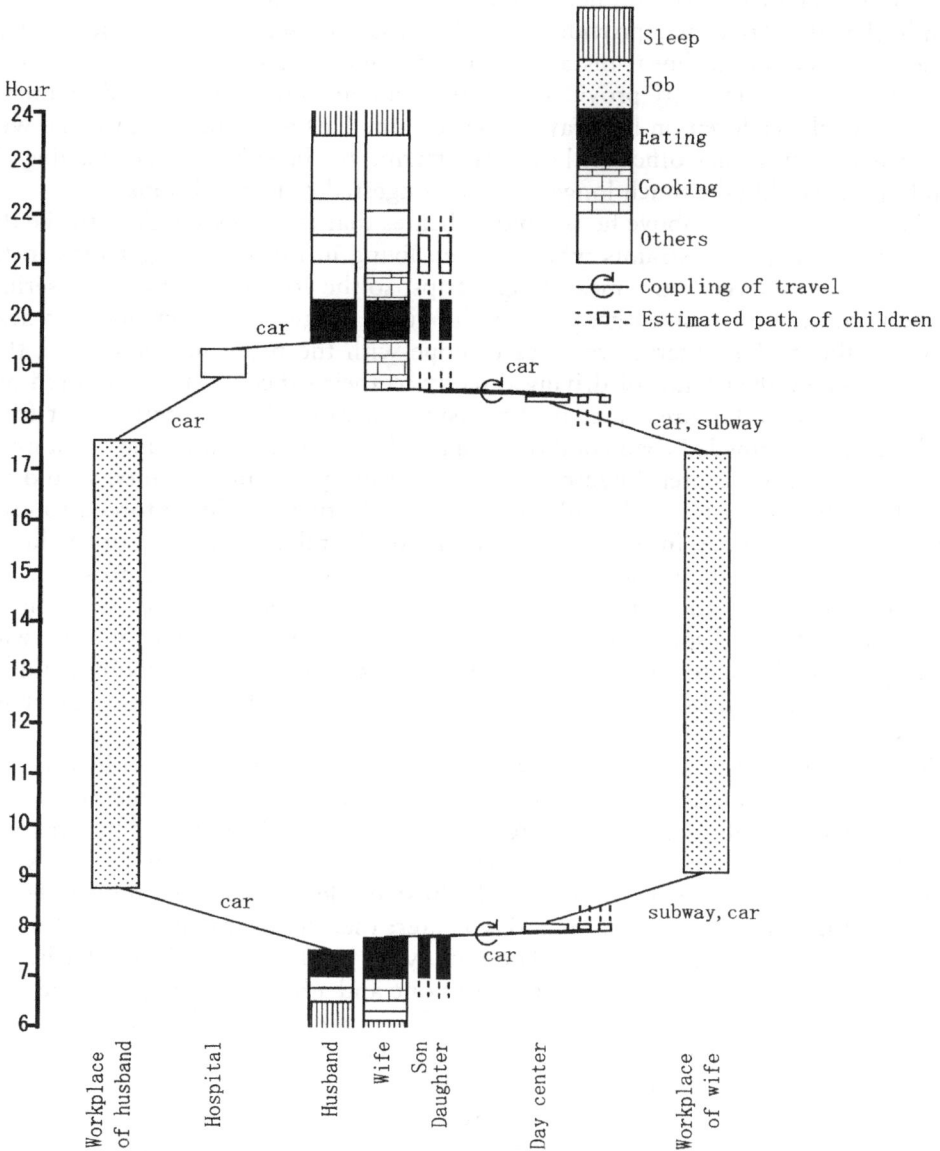

Figure 19.6 Daily lives of a nuclear family using the extended hours service of a daycare center (NSN-117).

held a full-time job using the extended service hours of care (figure 19.6). This household consists of a husband and wife, a daughter (of four years old), and a son (of three years old). The household has two cars, and the husband and the wife each use one of them for everyday commuting. The workplace of the husband is located on the outer fringe of the metropolitan area, so he drives for about 45 minutes from home to his office. Every day, the wife drives to the subway station and transfers to the subway to reach her office in the city center. Both children are taken to the daycare center by the wife on her way to the station. It is very convenient for the family that the day center provides extended hours of service and is located very close to the station (in linear distance, it is only 400 m from the day center to the station). If the service were provided for only standard hours, she would be unable to drop off the children on her way to the station. This is because even if she were to drive straight to her office without transferring to the subway after leaving the children, it would take much longer due to congested road conditions.

The situation in the above household indicates that escorting children to the day center poses major constraints when women living in the suburbs get jobs in the city center, since larger cities mean longer trips, so the constraints are more serious in the large cities. If women get jobs on the outer fringe of the metropolitan area, they are able to find alternative ways to cope with the problem. Otherwise, they have to change their habit of driving directly to their office with the children and leaving them in a daycare center in the business district. That pattern is not feasible because the crowded road conditions make the commute more time consuming. Furthermore, a day center downtown, which is run by the municipality, would not accept children who live in the suburbs. As a result, they would be forced to use a center near their home. In the sample household, the relative location of the home, the day center, and the station put the wife in a very advantageous position, and the extended service hours at the center are also very favorable to her. In addition, having a car at her disposal is important; even though it is just a 15-minute walk from home to the center, it is very difficult to walk there with two young children. Apparently, working full-time makes it necessary for a wife to have a car at her disposal.

The fourth sample household shows that when service provision is insufficient, wives are forced to obtain additional care (KWG-179). This household consists of a husband and a wife and a son (figure 19.7). The husband commutes by car to his workplace in the outer suburbs, and the wife commutes by railway to her workplace in the city center. Every morning, the husband leaves home with his wife and son; first they drop off the son at a day center; then he drops the wife off at the station and drives on to his office. The wife, who works on a full-time basis, gets off work at 6.00 p.m. Usually, it takes about one and a half hours to get home from her office. The husband also comes home late, arriving just before 11.00 p.m. because of his overtime. As both parents come home late, they ask a friend to pick up their son from the day center and keep him at her house. Thus, the wife stops at the friend's house on her way home from work to collect her son. As mentioned before, extended hours of service are offered at some centers in Kawagoe City, but they close at 6.30 p.m. So this household must rely upon such strategies.

The fifth household has a situation similar to that described above. In this household, the woman works full-time and takes the child to a nursery facility operated by a volunteer group (NSN-22). A husband and wife and a daughter (of eight months old) form this household (figure 19.8). The husband works in the city center, while the wife works in a suburban area. So the husband commutes by bus and the wife by car. Every day, the wife leaves her daughter at the daycare facility on the way to work. The reason why the parents use this kind of facility for childcare may be severe time constraints. The extended service hours provided at the municipal

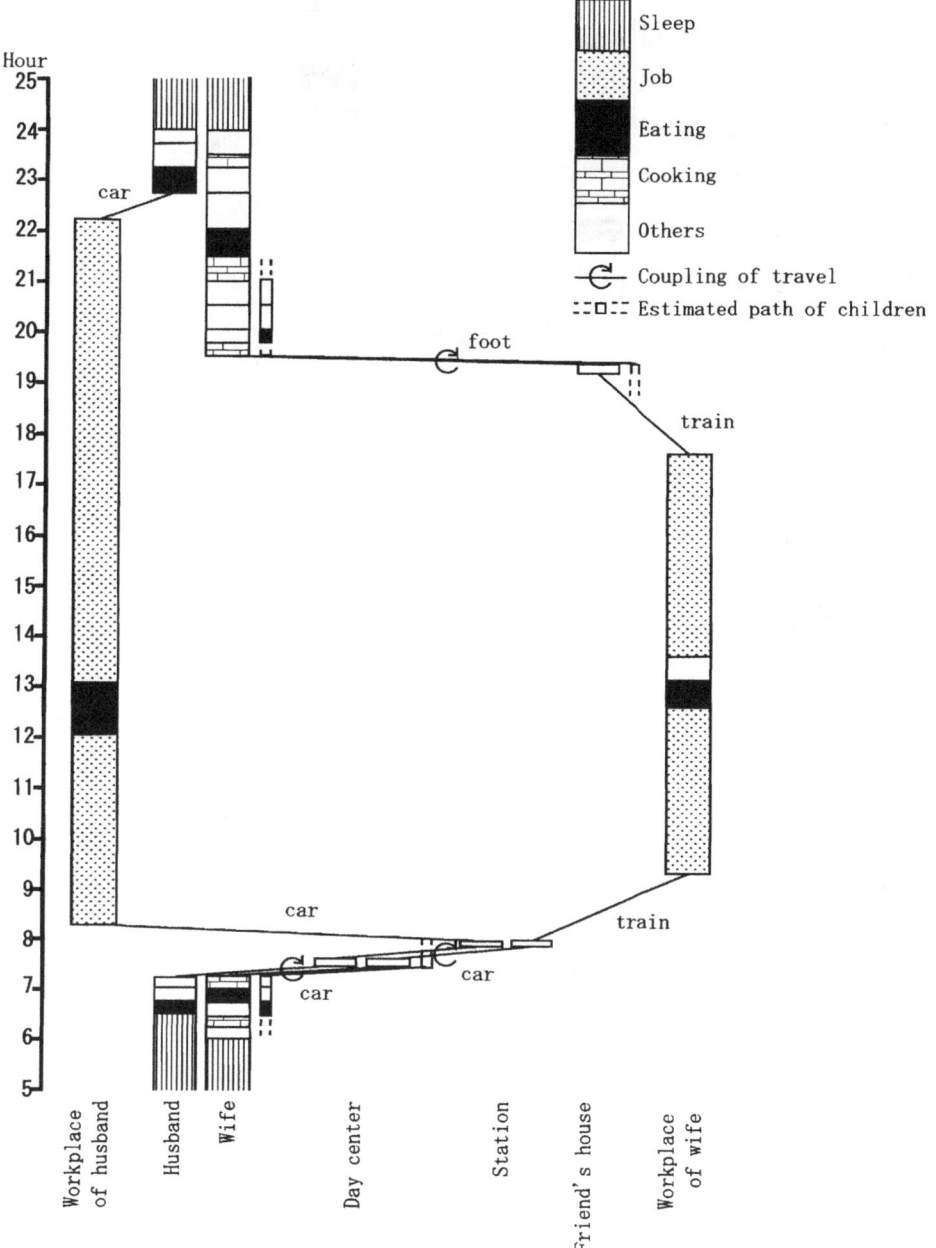

Figure 19.7 Daily lives of a nuclear family depositing their child with a friend (KWG-179).

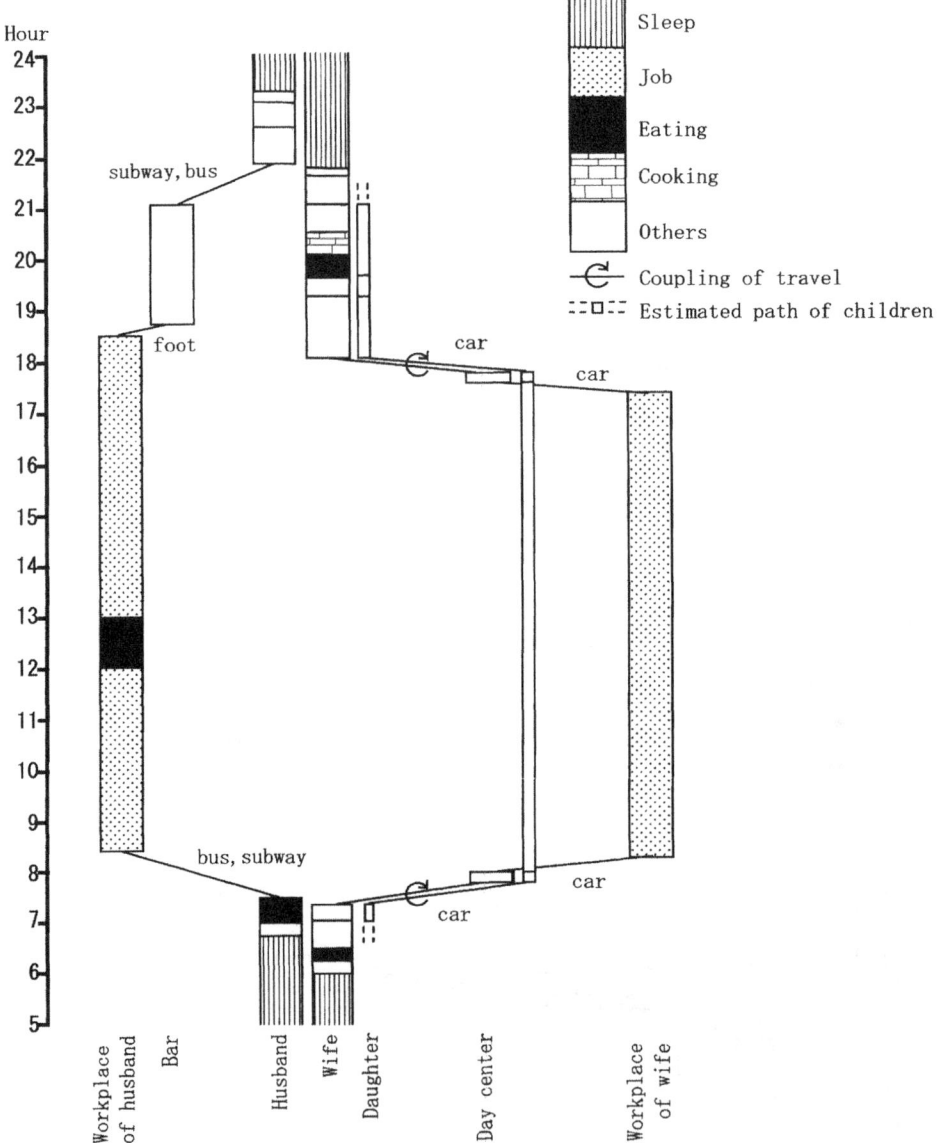

Figure 19.8 Daily lives of a nuclear family depositing their child at a voluntary facility (NSN-22).

center are from 7.30 a.m. to 6.00 p.m., and her working hours are from 8.15 a.m. to 5.15 p.m. The service just fits with her working hours, even adding 20 minutes for driving time from the center to her workplace.

Another reason to choose a facility near the wife's workplace may be easy access. This is particularly important when the infant is so young and likely to get sick. If the infant were left in a center near home, it would be impossible for the wife to see her daughter during her work break. On the other hand, there is an economic disadvantage stemming from their choice. Because this non-authorized facility receives less subsidy from the local government, the parents pay more than they would at the municipal day center. Perhaps the household would have made the decision to leave the infant in this non-authorized facility after taking all these factors into account and evaluating easy access as the most important one. Such a decision gives some indications about application of the US linkage policy on urban development in the Japanese context. Linkage policy, as applied in San Francisco, obliges developers to dedicate part of their capital investment to infrastructure, including daycare facilities. But when the same policy is applied in Japan's large cities, the benefit a mother derives from the proximity between her office and day centers would be negated by traffic conditions and fragmented jurisdictions of local government. But this linkage policy in US cities can serve as a benchmark to measure the effectiveness of the central government policy in Japan of subsidizing the attached facilities for daycare that are installed in office buildings. Further discussion of whether to place care facilities near the home or near the workplace is deferred to the final section, as it is inextricably related to city planning practices.

So far, the daily activities have been described by citing at length sample households with young children. To conclude this section, these observations are combined with a discussion of the level of daycare service as sketched above.

As noted in the daycare service provision in Japan section, there are marked differences in the service levels provided in the three areas studied. Urban settings also differ in some respects, notably in the job opportunities open to the wives, the time required for commuting, the availability of cars for women in the daytime, and the incidence of extended families.

The sample households in Shimosuwa Town, where the service level of childcare in terms of extended hours is relatively low, indicate that it is hard for women with young children to work full-time. Abundant job opportunities and widespread extended families lessen the problems of women confronting such difficulties. Yet we must conclude that women living in Shimosuwa Town and working full-time are in a less favorable situation due to the poor level of daycare service there.

Living with grandparents is an important way for women with young children to lessen the childcare load and continue to work. However, such assistance is not easily available to residents of metropolitan areas. Even so, grandparents living apart but not too far from the home may help working women to care for the children. But as assistance from kin is not commonly available to women, one should not conclude that such measures are effective for women in metropolitan areas. Therefore, extended service hours and infant care in public day centers would provide women with more effective measures. Women who make use of the extended service hours are engaged in full-time jobs. In metropolitan areas, unless current working hours are changed dramatically, women must have easy access to public help, such as extended service hours.

Two of the women utilizing the extended service hours commute to their offices in the city center; such people also need sufficient attention. As indicated elsewhere, most women change their employment status around age 30 to 35. Before marriage or childbirth, the commuting distance of women is almost the same as that of men; after marriage, the former becomes much shorter than the latter. Moreover, frequent entrance into and exit from the labor market is typical of middle-aged workers. In this light, it would seen that many female workers with young children commuting to the city center might not have quit their jobs when they got married or had children, though our data set does not contain information about this. These women are regarded as having the most urgent need for extended hours of care. Yet the above discussion suggests that they may not comprise a large segment of the population. This view may need to be revised because it is derived from the end result we observed in the women's daily activities. If extended service hours or flexible service hours would be commonly provided, the number of women not quitting their jobs and continuing to commute to the city center would be likely to increase.

It is debatable whether the provision of extended service hours could really satisfy the needs of all the working mothers. As the sample household in Kawagoe City indicated (KWG-179), the existing schedule of extended service hours is insufficient for commuters to the city center. Even a slight adjustment of working hours makes it difficult for women to leave their children in public care. Nevertheless, any changes in the existing service other than extended hours may bring women better service. For example, locating the daycare center in a convenient place would improve their access to this service (NSN-117). Commuters to the city center can enjoy the benefit of a good location if the daycare center is between home and the workplace. For women in metropolitan areas, the most attractive location may be in the proximity of a railway station, since many of them commute by train and the suburban station is the focal point of such commuters. A movement to establish care facilities near railway stations is now in progress. For instance, the Ministry of Health and Welfare is currently investigating the plan to subsidize such facilities, and a private company started up such a business for profit in 1993.

Along with the expansion of existing forms of public care, one should look for alternative measures at home to provide daycare in the future. Though we have not presented this as an example, insufficient care provision by municipal day centers can be supplemented by the husband even if the wife works on a full-time basis. In order to make it possible, the working hours of the husband must be shortened. The same kind of effect can be expected if the husband's workplace shifts from the city center to a nearby suburb or home. Promotion of flexible working hours and office relocation from the city center to the suburbs, aided by telecommunications, may be welcomed by households with young children.

Some consideration must be given to the effects of city planning practices upon childcare. As mentioned above, the linkage policy in San Francisco includes the obligation to set up care facilities. In Japan, this practice would not be feasible in a metropolitan environment because the congested public transportation would prevent parents from bringing young children to facilities in the city center. This idea is worth considering in Japan's medium-sized cities, however. For the large metropolitan areas, planning policy should be directed toward a home-based location of daycare facilities near stations or toward a relocation of places of work from the city center to the suburbs. Both possibilities would lessen the constraints of escorting children from home to the workplace. To proceed in the latter direction, relocation of the workplace, it is necessary to reorient current office and housing development to provide for more flexible land use.

Changing Daycare Service and Female Labor

Finally, I will describe changes in the past decade in the government's attitude to daycare provision.

The continued low fertility rate throughout the 1990s prompted the Ministry of Health and Welfare to launch the comprehensive project called the "Angel Plan" for supporting childcare. This plan was intended to lessen the burden of child rearing by providing various kinds of service in day centers, such as extended hours of care, weekend care, and infant care. In addition, government control of day centers was abolished in 1998, and private companies were permitted to enter the childcare market. Such deregulation was closely related with welfare pluralism (Johnson, 1987), and aimed at increasing the service level of daycare through competition. At the same time, the Equal Opportunity Law was revised in 1997 to ban discrimination in hiring, promotion, and job training against women. Maternity leave and care-for-elderly leave were enacted for workers in the same year. Although these arrangements aimed at helping women to work, the longstanding stagnation of Japanese economy has shrunk the labor market. Flexible employment strategies are now widely adopted by management, and more women are now employed as temporary and contingent workers.

It seems too early now to judge the net effect of these shifting policy directions. However, if the M-shaped age profile of Japanese women's labor activity rate does not change substantially in the near future, alternative causes and measures should be sought to enable less constrained women's labor force participation.

NOTES

This chapter draws in part on my article "Daycare services and activity patterns of women in Japan," *Geojournal*, 48, 207–15 (1999).

1 Metropolitan areas include the three largest conurbations of Tokyo, Osaka, and Nagoya.

BIBLIOGRAPHY

Dyke, I. (1990) Space, time and renegotiating motherhood: an exploration of the domestic workplace. *Environment and Planning D: Society and Space*, 8, 459–83.

Fincher, R. (1991) Caring for worker's dependants: gender, class and local state practice in Melbourne. *Political Geography Quarterly*, 10, 365–81.

Johnson, N. (1987) *The Welfare State in Transition: The Theory and Practice of Welfare Pluralism*. Lewes: Wheatsheaf.

Kamiya, H. (1993) Day care service provision and daily lives of women in Nissin town. *Geographical Report of Aichi Kyoiku University*, 76, 18–35 (in Japanese).

Kamiya, H., Okamoto, K., Kawaguchi, T. and Arai, Y. (1990) A time-geographic analysis of married women's participation in the labor market in Shimosuwa Town, Nagano Prefecture. *Geographical Review of Japan*, 63, 766–83 (in Japanese).

Peng, I. (1996) The East Asian welfare states: peripatetic learning, adaptive change, and nation building. In G. Esping-Andersen (ed.), *Welfare States in Transition: National Adaptations in Global Economies*. London: Sage.

Sechiyama, K. (1996) *Patriarchies in East Asia: Comparative Sociology on Gender*. Tokyo: Keiso-shobo (in Japanese).

Tivers, J. (1985) *Women Attached: The Daily Lives of Women with Young Children*. London: Croom Helm.

Valentine, G. (1997) "My son's a bit dizzy." "My wife's a bit soft." Children and cultures of parenting. *Gender, Place and Culture*, 4, 37–62.

Chapter 20

Organizing from the Margins: Grappling with "Empowerment" in India and South Africa

Richa Nagar and Amanda Lock Swarr

From a tiny NGO staffed by rural and neoliterate women in Bangladesh to the "gender planning cells" of the international financial institutions in Washington, DC, the idea of *empowerment* has captured the imagination of those operating in a wide variety of sociopolitical, geographical, and institutional locations. Although defined and interpreted in diverse ways by these multiple actors and institutions, the so-called "empowerment approach" is often associated with "women and development" discourses where the subordination of the poorest women in the South is seen as interwoven with all aspects of their lives, including family, work, legislation, and state structures. Activists operating within an empowerment framework often recognize that gender interlocks in crucial ways with power relations based on other structures of difference, such as race, class, caste, ethnicity, and sexual orientation. As a political intervention, the empowerment approach seeks to mobilize marginalized actors and raise their consciousness, not with an objective of reversing the existing power hierarchies, but to enable them to make their voices heard and to exercise greater control on the processes that shape their everyday lives (Weiringa, 1994, pp. 832–3).

While the language of enabling choice and "giving voice" to the marginalized sounds powerful in theory, in reality the terrain of empowerment is uneven and highly contested. This problematic emerges partly from a lack of shared political vision(s) or agenda(s) among those who advocate empowerment; partly from the gaps between feminist theory and praxis; and partly from the narrow frameworks within which development planners frequently attempt to "simplify and fix" long-term and deep rooted social inequities and injustices (Moser, 1993; Weiringa, 1994; Kabeer, 1999). At the same time, critics also recognize that empowerment discourse – like the NGOs and donor institutions with which it has become thoroughly intertwined – is here to stay, and therefore needs to be engaged with, problematized, and claimed in ways that are sensitive to contextual specificities and political imperatives. Given the centrality of "empowerment" in development and NGO politics, it is surprising that very few feminist geographers have elaborated on how to productively engage with empowerment while also recognizing its limits. In fact, feminist literature in the interdisciplinary social sciences as a whole has largely registered a silence when it comes to exploring the multiple and contextual meanings of empowerment, or the processes that co-produce empowerment and disempowerment in different spaces and scales.

In this chapter, we illustrate how discourses of empowerment and freedom from oppression have been deployed to advance different kinds of grassroots struggles in two countries of the so-called South where NGO-based activism has become increasingly prominent and refined: India and South Africa. The Indian case study looks at the deployment of empowerment by an NGO working with rural women from the "scheduled castes" and "scheduled tribes" in their efforts to gain critical literacy.[1] The South African example, by contrast, focuses on a religious body working with black lesbians and gay men living in periurban township areas on issues of sexuality, poverty, and race. Through these examples, our collaboration highlights the connectivities and divergences in the ways that a dominant discourse of empowerment is interpreted, critiqued, and/or reappropriated by grassroots activists in line with their own political agendas and context-specific realities and idioms.

Empowerment, Transnationalism, and Feminist Praxis

> Any effective politics challenging a capital inspired globalization must have similar global sensitivities, even as its grounds are necessarily local. This is different from a "place-based" politics. It is not merely about one locale or another, nor is it a matter of just building coalitions between such diverse places, vital as that is. Precisely because globalization is such an abstraction, albeit with varying forms, struggles against global capital have to mobilize equivalent, alternative abstractions. Built on critical triangulation of local topographies, countertopographies provide exactly these kinds of abstractions interwoven with local specificities and the impulse for insurgent change. (Katz, 2001, p. 1232)

Anthropologist Arjun Appadurai (2002, p. 22) argues that the alliances and divisions of the contemporary global political economy are marked by two broad types of grassroots political movements. The first comprises those who have chosen to address their problems of inclusion, recognition, and participation through armed, militarized means, while the second is constituted by those opting for a "politics of partnership between traditionally opposed groups such as states, corporations, and workers." As a concept and discursive strategy, empowerment has emerged prominently in the latter type of organizing. Not only has it allowed organizers to define their place-based strategies, it has also become a vehicle for local organizations working in different parts of the world to link up and build alliances with each other.

It is important to recognize that building linkages with local organizations located within different national boundaries of the global South is one critical piece of creating a globally sensitive understanding of grassroots politics. The other essential part, as Katz reminds us, is to understand how global sensitivities operate in local topographies,[2] and how these topographies can be critically linked to produce new countertopographies capable of imagining and enacting transformative politics.

Countertopographies offer "a multifaceted way of theorizing the connectedness of vastly different places made artifactually discrete by virtue of history and geography but which also reproduce themselves differently amidst the common political-economic and socio-cultural processes they experience" (Katz, 2001, p. 1229).

Although one can think of a variety of ways to generate countertopographies, here we examine how the notion of empowerment – with all its global institutional affiliations – is deployed on the ground by activists organizing for sociopolitical change in North India and South Africa. We focus on the complexities and contradictions associated with empowerment by considering how it is interpreted, interrogated, and (re)conceptualized by actors in extremely marginalized socioeconomic locations. Our analysis suggests the need for scholars to move beyond an easy categorization of political interventions as feminist versus non-feminist or rural versus urban, and to engage with the complex ways in which multiple scales, as well as the politics of intimacy, sexuality, and access to resources, intersect in the articulation of grassroots struggles for social and redistributive justice (Swarr and Nagar, 2004). We believe that collaborative analyses that embrace transnational comparison as a method and theoretical framework can play a crucial role in producing countertopographies and in providing useful models for undertaking and extending current formulations of transnational feminist praxis (Grewal and Kaplan, 1994; Nagar, 2002).

Limited Empowerment? Feminist Pedagogy in a Modernization Framework

> There is no doubt that Mahila Samakhya transformed many of us from overworked domestic creatures into fiery feminists. . . . It lit a fire in our hearts by showing us how we were oppressed as women. . . . But when we started searching and addressing the most profound and hidden causes of our oppressions, it removed the support from under us. . . . What was the use of lighting the fire in our hearts, then? Just to cause more burnout and bitterness? . . . Call me hypercritical, but . . . what do we do with a band-aid feminism that is happy when we label our feudal men as our enemies, but terribly unhappy when we critique the government, take issues with water privatization policies, or challenge the community forestry schemes of the World Bank? (Shivani, interviewed by Richa in Tehri Garhwal, May 1999)

This quote from Shivani, a former worker of the Mahila Samakhya Programme (henceforth MS) in the Indian state of Uttar Pradesh, is not intended to dismiss or undermine the tremendous accomplishments of what has been recognized as one of the world's most innovative and radical government-supported schemes for women's empowerment. Instead, Shivani's commentary is striking because of the way it brilliantly encapsulates the limits of "empowerment" as a discourse and praxis in many donor-funded feminist interventions in the South. To develop a better understanding of how women are at once enabled and confined by the specific ways in which empowerment is conceptualized and implemented at the grassroots level, this example focuses on the curriculum for rural women designed in the Banda district (later divided into Banda and Chitrakoot) where MS Uttar Pradesh worked for over a decade. But first some background on MS is in order.

MS Banda: a backdrop

The adoption of the National Policy on Education by the government of India in 1986 is seen by many as a major accomplishment of the Indian feminist movement. This policy emphasized the need for the government to play a "positive interventionist role" in the empowerment of the poorest rural women and led to the launch of the MS Programme in Uttar Pradesh, Karnataka, and Gujarat in 1989, with joint funding from the Dutch government. Envisioned by dynamic feminist activists working at the national level, this innovative scheme for women's education operated in collaboration with gender-progressive NGOs at the district level. Headquartered at the state level, MS programs were implemented through district-level units. In this multitier structure, feminist thinkers and activists from metropolitan centers such as Delhi and Bangalore played a prominent role as "consultants" and "trainers," while urban women with formal education held the official positions in each district-level organization. The pivotal forces, however, were the rural women who worked as coordinators and mobilizers in the villages, organized meetings with village women, and helped them to form action groups to collectively reflect on their conditions, constraints, and needs, as well as to determine concrete strategies and goals for their empowerment. Rather than being imposed from above, the literacy component was introduced only when the women themselves demanded it. In this way, MS created a space for the most disadvantaged rural women to collectively seek empowerment on their own terms according to their varying place-specific realities (Agarwal, 1994; Nagar, 2000).

In the drought-prone, impoverished, and violence-ridden district of Banda in Uttar Pradesh, it did not take long for MS women to make their mark. As acute water scarcity and governmental apathy pushed rural and illiterate women from the scheduled castes and Kol "tribe" to master the technology of fixing handpumps, the mainstream media complimented them on their "incredible feat" and compared them with national heroines such as Bachendari Pal, the Everest climber, and Kalpana Chawla of NASA. The experiences of women handpump mechanics, in turn, triggered among women critical analyses of caste, class, and gender relations, and an intense desire to attain formal literacy. Beginning with women's literacy camps, these developments soon gave birth to a residential school called Mahila Shikshan Kendra (MSK), where rural women came to acquire critical literacy in six-month long training sessions. The emergence of *Mahila Dakiya*, a broadsheet published by the neoliterate women of MSK in their local dialect, *Bundeli*, is regarded as a direct result of this successful literacy campaign. A rich combination of narratives, pictures, poems, songs, and commentary, *Mahila Dakiya* became a vehicle for women to discuss, articulate, and circulate their views on everything ranging from caste conflicts and gender violence in their villages to parliamentary discussions in New Delhi.[3]

The entanglements of empowerment and disempowerment

So what role did the residential school/MSK play in empowering the women of MS Banda? Here, we reflect briefly on the themes that gained prominence in the MSK curriculum as a way to understand how empowerment was conceptualized in/for

MS Banda and how these conceptualizations allow us to understand the possibilities, contradictions, and limitations embedded in MS Banda's approach to empowerment.

With a commitment to build on local knowledges, feminist trainers from a Delhi-based organization, Nirantar, collaborated with the teachers of MSK to develop a curriculum around water, land, forests, society, and health. Falling under these five content areas were a number of themes and subthemes that reflected the everyday materialities of women's lives (see table 20.1).

Table 20.1 reveals that intersectionality was at the core of the curriculum developed in MSK Banda. Each content area allowed women to reflect deeply on the ways in which their gender, caste, sexuality, and positions as landless laborers enmeshed to define their access to resources, and the violence they suffered in their homes, villages, and forests. The curriculum also allowed women to develop a critical awareness of how systemic practices of gendered discrimination and violence were sanctioned by religion, myth, and tradition and how these, in turn, were reinforced by family and caste structures. This collective and critical analysis among women was facilitated by MSK's pedagogy, which encouraged women to share their first hand experiences, combined with group activities such as singing, dancing, role-plays, and learning of skills such as biking, which were hitherto "forbidden" or inaccessible to women.

At the same time, the curriculum suffered from three major drawbacks. First, there was no effort to educate women about corporate globalization and its effects on rural communities. In addition, any critical analysis of the state and its projects of development remained superficial and largely disconnected from the ways in which caste- and family-based systems of gender discrimination intersected with the violence unleashed by state policies against the rural poor. Due to these limitations, the MSK curriculum remained preoccupied with critiquing patriarchy and feudal "cultural" practices that constructed women's interests as always oppositional to men's. Second, by dismissing local beliefs as myth, tradition, and superstition and contrasting these to a more modern and superior "scientific knowledge," the curriculum defeated its own goal of building on local knowledges and instead marginalized them. Finally, despite its use of collective learning as pedagogy, MSK's curriculum focused on educating *individual* women for self-empowerment. In so doing, it failed to organize a support network that could be extended or used by women for their collective empowerment.

These limitations of the literacy program emanated from several sources. To begin with, MS Banda was a government-funded program and, therefore, not in a position to critique the state and its projects. Another problem arose from the fact that despite its focus on local knowledges and women's self-empowerment, the knowledge was disseminated by formally educated trainers from the cities who came as "experts" and whose approach to empowerment was necessarily shaped by their own histories and social locations. "Training," moreover, was always seen as a one-way process – flowing from the experts/consultants to the village women, never vice versa. Also, coinciding in the mid-1990s with the involvement of the World Bank in MS, organizational officials at state and national levels increasingly felt the need to evaluate and measure the levels of empowerment MS women in each village had achieved. The issue of measuring empowerment to ensure continued funding from

Table 20.1 Summary of themes and subthemes covered in the five content areas of the MSK curriculum

Content area	Central themes	Subthemes
Water	Who controls water?	Ownership of village wells; caste, class, and constructions of purity and pollution The availability of irrigation and drinking water; government's role in making water available
	Gender division of labor around water	
	Access to water	How caste and landlessness shape access to water
	Gender, power and technology	Bundeli women as handpump mechanics
	Water and health	Water-borne diseases; people's right to potable water
	The science of water	Water in the ecosystems; natural and human-made water sources; properties of water; water resource mapping
Land	Agriculture	Land, topography, and agriculture; gender and agricultural labor; minimum wages; equal wages for men and women
	Access to land	Gendered and caste dimensions of land distribution, government land reform measures and social movements, gender and inheritance
	Planet Earth	Rotation and revolution; Earth's exterior and interior; soil and natural resources; pollution
Forests	Forests as an ecosystem	Local flora and fauna; reproductive cycle of plants
	Social, economic, and administrative aspects	"Tribal" culture and forests Government management of forests History of changing rights over forests: forest policies, displacement, resettlement The forest economy: minor forest produce; gendered labor and violence against women engaged in the forest economy
	Commercialization and deforestation, and afforestation in the local and national context	Struggles over forests; women as protectors of the forest, e.g. Chipko Movement in the Himalayas
Society	Concepts: historical time and lived memory; history versus myth	
	Changing sociopolitical, cultural, and technological formations	"We make history"
	India before colonial rule	Precolonial India; British colonial rule; struggle for freedom; partition of India and Pakistan
	Post-independence India	Governance and citizenship; gendered legal rights (marriage, divorce, maintenance); labor laws; rural communities and police

Table 20.1 *Continued*

Content area	Central themes	Subthemes
		Decentralized village administration: Panchayats and women
		Social movements of peasants and women
Health	Critical understanding of health	Reasons for poor health: health infrastructure, drinking water, environmental degradation
		Understanding the government health care system
	Gendered politics of health	Work and violence
		Early marriages and multiple pregnancies
		Taboos, myths, and restrictions
	Understanding our body	Gender differentials in health status
		Food and nutrition; functioning of the body and related diseases; changes in the body; dealing with common ailments
	Reproductive health	Male and female reproductive systems
		Menstruation and associated taboos and practices
		Conception, reproduction, and local birthing practices
		Fertility, contraception, and the politics of population control
		Women's health movements and campaigns against coercive population control policies, hormonal injectibles and implants, and sex preselection

This table is based on the descriptions of the curriculum provided by Nirantar (1997) and Nagar's analysis of course materials used by MSK teachers in 1998–9.

the donors was vigorously debated and recognized by many within (and outside of) MS as fraught with tensions. In practice, however, it translated into a labor-intensive production of sophisticated bi-monthly activity reports complete with statistics on how many women had achieved "low," "medium," or "high" levels of empowerment in various villages covered by each district where MS operated. In this scenario, the education that adult rural women received at MSK easily became a measurable indicator of empowerment where the competencies of learners came to be evaluated mainly on the basis of their success at securing admission in Grade Five in the formal school system.

This growing need that teachers and students felt for "transition from an alternative educational space into the mainstream" (Nirantar 1997, p. 129) essentially made it impossible for an institution such as MSK to build a revolutionary agenda. What it meant, unfortunately, was that despite its radical vision, MS could only incorporate the poorest rural women – now literate instead of illiterate – at the margins of the system: as newly trained feminists, they had to unlearn what they had learned in their villages and homes, absorb the new knowledge in a "modern" language (e.g. Hindi instead of Bundeli in the case of Banda), be examined on what they had learned according to the rules of the mainstream educational system, and, at the end of their six-month training, return to their villages where their material options in terms of how they were going to live their lives remained the same as they were six months before.

For example, let us consider the case of Bala (16 years old) and Rukmini (15 years old), two bright sisters who graduated from MSK, but could not continue their education due to grinding poverty. Instead, within a few months of their graduation, their parents arranged their marriages with two brothers in a Dalit ("untouchable") family of landless agricultural laborers. The sisters' parents thought it was a perfect match because Sitara, the mother of the grooms, had been one of the first handpump mechanics of MS Banda and subsequently studied at MSK. One year after the wedding, however, Richa met Sitara, Bala, and Rukmini in a household where Sitara regularly starved and overworked her daughters-in-law. Sitara forced Bala (who had given birth to a four-pound baby six days before Richa's visit) to work in the fields until her waters broke, and expressed intense displeasure when she discovered that the newborn was a girl. The baby died within two weeks of her birth. Reflecting on the situation, Rukmini (interview, April 6, 1999) said:

> My eyes used to shine with excitement and hope when we were at MSK. But this past year has broken me. Everything we learned at MSK seems unreal. . . . The only thing that is different in my life because of MSK is that I get a lot more depressed than I would have been if I hadn't gotten my hopes up at MSK.

Rukmini's testimony is reminiscent of the bitterness that Shivani expressed in the opening quote of this story. MSK gave women the tools to reflect critically on patriarchal subjugation at home, but it did so in a limited way. In the absence of a fuller understanding of the interlocking structures within which their daily oppressions were located and constituted, MSK failed to equip women with concepts and strategies that could enable them to advance their political analyses or actions. Organization of external support networks, or any broader work to create community change, fell outside the purview of MSK. Thus, even as women received "feminist training" at MSK, they were deprived of an opportunity to collectively build their own countertopographies for insurgent change. If workers such as Shivani developed revolutionary understandings within this framework and started mobilizing village women to become publicly critical of the government, they soon found out that their politics could no longer be accommodated by the very organization that gave them the tools to analyze the injustices in their lives.

Freedom within Material Constraints: Gay and Lesbian South African Theology

For our second case study, we move beyond an understanding of empowerment as confined to traditional development projects and explore some of the ways in which this concept is employed in other organizational spaces dedicated to changing the lives and circumstances of those who have been marginalized in the global South.

For black gay and lesbian Christians in Johannesburg, South Africa, the Hope and Unity Metropolitan Community Church (HUMCC) is a site of inspiration, community, and empowerment for its congregation. From its inception, the HUMCC has been a space for members to engage in a concerted attempt to reconcile and integrate their sexual, religious, and African identifications.[4] One of the most important objectives of the HUMCC has been to reverse the fragmentation in the lives of its members and to promote spiritual healing and self-love through a process that can be likened to development notions of "empowerment." In articulating empowerment as enabled by the HUMCC, we build on the Indian case study above to further complicate the simultaneity of empowerment and disempowerment, while recognizing the creative strategies of organizational leaders that are responsive to place-specific struggles and, here, encourage members to find a sense of inclusion in the Kingdom of God.

The HUMCC: a rhetoric of freedom

Decades of colonialism and apartheid have led to immense contradictions of wealth and poverty, repression and resistance in South Africa and spawned a political current of intense and effective activism. South Africa's anti-apartheid activism has been widely recognized since the Soweto uprisings of 1976, and the end of apartheid brought a similarly grassroots-based emerging gay and lesbian rights movement that coalesced in 1994 when South Africans drafted the first constitution in the world to protect citizens from discrimination on the basis of sexual orientation. In this same year, and with many of the same motivations, black gays and lesbians formed the HUMCC, which, until 2000, operated out of the Harrison Reef Hotel in Hillbrow, Johannesburg.

The pastors and congregants of the HUMCC combat widely held ideas that homosexuality is "un-African"[5] and that Christianity is incompatible with same-sex orientations by creating community through religious services and events including picnics, discussion and Bible study groups, parents' luncheons, and same-sex union ceremonies. Most congregants in the HUMCC, while devoutly Christian, have had painful experiences in churches as a result of their sexual orientations. A survey conducted by South African scholar Graeme Reid within the HUMCC found that respondents felt "alienation, embarrassment, discomfort, depression and fear" (Reid, 1997, p. 102) during their experiences at other churches. Pastor Nokuthula Dhladhla, a clergywoman of the HUMCC, describes how when she was raped her family and members of her former congregation saw it as "a punishment sent from God because I was still possessed with a lesbian demon" (Dhladhla, 2001). It was then that she joined the congregation of the HUMCC. Many members similarly seek out the support and theologies of the church in the face of familial discrimination. Responses to HUMCC members "coming out" to their families included: "initial shock and subsequent acceptance, religious intolerance, and an expression of loss akin to a death in the family; in one particular case, the respondent's life was endangered when a relative attempted to shoot her" (Reid, 1997, p. 106).

The sense of home and empowerment of the HUMCC offers a welcome contrast to the unpredictability of family and community responses to members' homosexuality. For instance, Cora describes how comfort and security were important factors in her decision to change churches. Speaking of the HUMCC, she explains,

"I just felt comfortable there. Welcome. So, I left my church, said: 'Bye guys, see you.' Then I joined HUMCC. . . . I mean, I felt as if I belonged there" (Mahlaba, 2000). The HUMCC serves as a source of support and offers affirming arguments to counter mainstream perceptions of Christianity and homosexuality. For many members a combination of desires for spiritual fulfillment, new information about homosexuality, acceptance and comfort, and increased possibilities for social networking with other gays and lesbians led them to the HUMCC.

In the HUMCC, empowering ideologies of self-acceptance and deep spiritual faith counter internalized and institutional discrimination. Sermons are one way these ideologies are imparted. Empowerment in this context is sometimes conceptualized as "freedom" – freedom from individual feelings of self-doubt and fragmentation, from community-based discrimination, and from institutional oppressions based on race and sexual orientation. One sermon by the late Pastor Tsietsi Thandekiso perhaps characterized this sentiment most clearly:

> Those people who say: "You know, Bra Tsi, I have been afraid. I am a coward, I am gay, but I look down, I'm oppressed . . . I fear people. I am not living my life. I am going to get rid of that spirit. I want to be free. I want to be free." . . . Come over and stand next to me. . . . If you want to pray this morning for the spirit of freedom. If you are forgiven you start walking in freedom. Come over quickly. Come and stand next to me so that I can pray with you. . . . If you want to pray for spirit of freedom, just freedom. I know there are people, they pretend at work, they pretend at home. They are living a lie. It's not good for you. It's not good for God. It's not good for anybody. . . . I want to be free . . . I'm going to be free . . . I am going to receive my freedom. (Thandekiso, in Reid, 1997, p. 115)

Through their involvement with the HUMCC, many congregants seek and find this empowerment, or freedom. However, it is not simply material freedom but spiritual freedom that allows them to put their trust in God and makes the impact of their often difficult daily lives less significant. Their difficulties are addressed not only through sermons, but through the testimonies in which members share their experiences that are routinely part of services. Through testimonies about suicidal feelings, job losses, and relationships, for example, members articulate their gratitude for God and his plans for them and their trust in Him, even when their paths are unclear. These testimonies and sermons demonstrate the ways that HUMCC members have overcome discrimination and affirm their faith in God and Jesus.

Revisiting entanglements of empowerment and disempowerment

While these discourses of freedom and empowerment have positive spiritual and psychological effects on the lives of HUMCC congregants, most members experience contradiction between the empowerment promoted by the church and their material circumstances. For instance, sermons have often suggested that gay and lesbian congregants should "come out" to their families to avoid the fragmentation that so many of them find psychologically disconcerting. However, the HUMCC is unable to offer much material support for those who lose the financial support of their families or, in more extreme cases, experience violence. Even the church itself has been plagued by financial concerns, such as the congregation's inability to pay rent in 2000, which led to its eviction from the Harrison Reef Hotel and left the church itself homeless. However, leaders and members of the church still found opportunity in these circumstances, drawing parallels between their own unpredictable lives and the situation facing the church. Further, they look for both material and spiritual solutions to these difficulties and reaffirm the constancy of their faith in the face of material uncertainty.

Another factor that illustrates the contradictions in the support of the church is the Hope and Unity Church's failed affiliation with the United States-based and internationally oriented United Fellowship of Metropolitan Community Churches (UFMCC). Despite the similarity of their names and their commitments to anti-homophobic Christian theologies, initially the affiliation between the UFMCC and the HUMCC was merely symbolic. When the HUMCC attempted to enact formal association in the mid-1990s,

> A representative from the UFMCC visited Johannesburg to evaluate the HUMCC's application to affiliate, but this was not successful. According to the Reverend Thandekiso the UFMCC was not satisfied with the administrative procedures of the church or the fact that the congregation was unable to support the salary of a full-time pastor, apparently one of the UFMCC's criteria for affiliates. (Reid, 1999, p. 8)

Reid further quotes from a 1995 letter from the UFMCC Administrator for Global Outreach addressed to Reverend Thandekiso, which poses questions and makes requests such as "Tell us about your plans for buying a church facility in Johannesburg" and "What financial goals has your church made to salary the pastoral leadership?" (Reid, 1999, p. 63). Such questions illustrate a disjuncture between the needs and resources of the HUMCC, whose members are predominantly unemployed, and the financial and ideological assumptions made by Northern organizations such as the UFMCC. Like the Indian case study above, this example shows how empowerment and material limitations are interwoven in complex and contradictory ways. However, despite their lack of formal association and financial support the HUMCC displayed the UFMCC flag on the altar of the church and used its affiliation to gain social legitimacy. This is another instance in which an organization advances its own interests creatively and with potentially radical effects.[6]

While the discourses and material circumstances of empowerment and disempowerment that shape the experiences of members of the HUMCC exist simultaneously, they do so within a valuable critical framework that can provide a model for successful activism and spiritual empowerment. The ideology of the HUMCC clearly identifies the difficulties of the multiple and intersecting struggles of church members and tries to reconcile them. Many of the sermons and discussions facilitated by church leaders in the mid-1990s, for instance, suggested ways that members' multiple identifications might be less personally trying if they were honest about their sexuality in all areas of their lives. Members confront the discrimination facing homosexuals within most denominations of Christianity, the belief that homosexuality is "un-African" and the ways this belief renders them non-existent, the social ostracism of friends and family, and the poverty that is a legacy of colonialism and apartheid. They also face disconnections between their intersecting yet contradictory identifications – as black, gay, African, mostly poor, Christians – and affiliations, such as feeling excluded from most Christian communities, from some white and Northern gay discourses, and from the financial stability that they seek.

Leaders of the HUMCC are cognizant of these disconnections. According to these leaders, the solution to the problems their congregants face is personal: individuals should be empowered to make changes in their lives and accept themselves. But discrimination faced by members, especially homophobic discrimination, is also rooted in the sociopolitical context of South Africa. According to the ideology put forth by the HUMCC, much of this discrimination is based in the social and legal condemnation of homosexuality. Thus the solution to this problem, in contrast to the values of the churches from which most HUMCC members came, is not simply personal and individual, but also political. Congregants are encouraged to resolve their fragmented identities and to work toward a society that will be more accepting of these intersecting identifications through both personal empowerment and political activism. When the sexual orientation clause was first proposed, for instance, congregants participated in the political project of writing letters to the new government about their views and lives as gay and lesbian Christians.

The transition to democracy in the mid-1990s in South Africa has been accompanied by dramatic political and economic shifts in the face of globalization. Both the anti-apartheid movement and the gay and lesbian movement in South Africa have had an enormous impact on the political, legal, and social landscape of the country. Black gay and lesbian South Africans, including those in the HUMCC, have personally benefited from these transformations and played a significant role in attaining constitutional and legal protection from discrimination (lobbying, for instance, for inclusion of the sexual orientation clause in the new constitution). Although the HUMCC has not presented an overt analysis of globalization's role in shaping the struggles facing congregants, the seeds for such a critique are emerging from church members themselves. Increasingly, members share ideas about how to make political change in South Africa and build countertopographies connected to global struggles for social justice.

Conclusions

The stories of MSK and the HUMCC exemplify how empowerment is a partial and contradictory terrain, where empowerment in one realm is often inextricably enmeshed with disempowerment in another. Despite these entanglements, however, discourses of empowerment are actively interpreted, critiqued, and reappropriated by activists at multiple geographic scales according to their own political agendas and context-specific realities. The quest to reconsider the possibilities of locally, nationally, and globally sensitive strategies and connections between political movements is essential to those struggling to organize from the margins. Clearly, these questions of strategy and political frameworks are attentive not only to the connections among grassroots organizations but also to how activists locate themselves.

In the first case study, we see how MSK, with its theoretical and political claims about respecting local knowledges and its commitment to empowering women, facilitated women's coming together in critical literacy efforts. However, like other similar well intentioned initiatives, these literacy programs became a means by which modern culture is diffused, and whereby the identities of individuals and groups become aligned with the state, as an anchoring point for the universal subject of modernity (Kamat, 2002, p. 146). Though not based in a traditional "development" framework, the discourses of the HUMCC similarly worked to "empower" its members by confronting their internalized oppression as well as the social inequities they faced. At the same time, the organization's ability to conceptualize empowerment in more effective and far reaching ways was constrained by its marginalized socioeconomic status, and the initial lack of support from the Northern-based UFMCC.

By considering these case studies in a comparative framework, we highlight the complex interplay of multiple scales and political understandings in the articulation of grassroots struggles for social justice. In an era when non-governmental organizations have become important vehicles to enact social change, imagining and producing countertopographies necessarily entails a sensitive engagement with the creative and often contradictory place-based strategies of local and national organizations, while also recognizing the impact of global institutional structures and power hierarchies. By systematically analyzing the interventions, strategies, and critiques emerging from the margins of the global "South," we can extend the often partial and isolated academic theorizations of development and queer politics and productively contribute to the project of creating feminist countertopographies.

ACKNOWLEDGMENTS

This collaboration was supported by a fellowship from the University of Minnesota's Graduate School Partnership Program and both authors have contributed equally to this chapter's conceptualization and production. Our field research in South Africa (conducted by Amanda with Susan Bullington from September 1999 to November 2000) and in India (conducted by Richa from December 1998 to May 1999) was supported by the University of Minnesota Graduate School, Center for Advanced Feminist Studies, the MacArthur Program, and a Grant-in-Aid of Research, Artistry and Scholarship. We are indebted to members of the HUMCC and MSK for sharing their ideas and resources with us. In order to retain individuals' anonymity, we have used pseudonyms and changed identifying details throughout this chapter. We also thank Susan Bullington and David Faust for their valuable critical feedback.

NOTES

1 We use the term "critical literacy" to refer to literacy that is simultaneously committed to developing critical understandings of social structures and processes.
2 Katz uses the term "topography" to delineate a research method; in her view, "To do a topography is to carry out a detailed examination of the material world, defined at any scale from the body to the global, in order to understand its salient features and their mutual and broader relationships . . . producing a critical topography makes it possible to excavate the layers of process that produce particular places and see their intersections with material social practices at other scales of analysis. Revealing the embeddedness

of these practices in place and space in turn invites the vivid revelation of social and political difference and inequality" (Katz, 2001, p. 1228).

3 *Mahila Dakiya* was so unusual and powerful that it earned MS Banda the prestigious Chameli Devi Award of India's Media Foundation for "outstanding woman journalist," an honor usually reserved for an individual.

4 In this chapter we focus only on the Hillbrow congregation, though the HUMCC has a sister church in Durban.

5 One of the most often-cited advocates of the view of homosexuality as "un-African" is Zimbabwean President Robert Mugabe, who in 1995 famously referred to homosexuals as "dogs and pigs."

6 While attempts to affiliate with the UFMCC were initially unsuccessful, the HUMCC is now officially connected to the UFMCC.

BIBLIOGRAPHY

Appadurai, A. (2002) Deep democracy: urban governmentality and the horizon of politics. *Public Culture*, 14(1), 21–47.

Agarwal, B. (1994) *A Field of One's Own*. Cambridge: Cambridge University Press.

Dhladhla, N. (2000) *In the Same Boat* (www.behindthemask.org.za).

Grewal, I. and Kaplan, C. (eds) (1994) *Scattered Hegemonies: Postmodernity and Transnational Feminist Practices*. Minneapolis: University of Minnesota Press.

Kabeer, N. (1999) Resources, agency, achievements: reflections on the measurement of women's empowerment. *Development and Change*, 30, 435–64.

Kamat, S. (2002) *Development Hegemony: NGOs and the State in India*. New Delhi: Oxford University Press.

Katz, C. (2001) On the grounds of globalization: a topography for feminist political engagement. *Signs: Journal of Women in Culture and Society*, 26(4), 1213–34.

Mahlaba, C. (2000) Series of personal interviews with Amanda Swarr and Susan Bullington. Gauteng, South Africa.

Moser, C. O. N. (1993) *Gender Planning and Development: Theory, Practice and Training*. London: Routledge.

Nagar, R. (2000) *Mujhe jawab do*: feminist grassroots activism and social spaces in Chitrakoot (India). *Gender, Place and Culture*, 7(4), 341–62.

Nagar, R. (2002) Footloose researchers, traveling theories and the politics of transnational feminist praxis. *Gender, Place and Culture*, 9(2), 179–86.

Nirantar (1997) *Windows to the World: Developing a Curriculum for Rural Women*. New Delhi: Nirantar.

Reid, G. (1997) Coming home: visions of healing in a Gauteng church. In P. Germond and S. de Gruchy (eds), *Aliens in the Household of God: Homosexuality and Christian Faith in South Africa*. Cape Town and Johannesburg: David Phillip.

Reid, G. (1999) Above the skyline: integrating African, Christian and gay or lesbian identities in a South African church community. Master's thesis, University of the Witwatersrand.

Swarr, A. L. and Nagar, R. (2004) Dismantling assumptions: interrogating "lesbian" struggles for identity and survival in India and South Africa. *Signs: Journal of Women in Culture and Society*, 29(2), 491–516.

Weiringa, S. (1994) Women's interests and empowerment: gender planning reconsidered. *Development and Change*, 25, 829–48.

Chapter 21

Moving Beyond "Gender and GIS" to a Feminist Perspective on Information Technologies: The Impact of Welfare Reform on Women's IT Needs

Melissa R. Gilbert and Michele Masucci

Introduction

In headline-grabbing reports called "Falling through the net," the United States Department of Commerce began reporting in 1999 an increasing gap between those who have access to information technology and those who do not, with poor people, minority populations, elderly people, and women disproportionately losing ground (Department of Commerce, 1999; NTIA, 1999). The digital divide can be seen as part of a broader polarization of resources that benefits those who already possess technological skills, access to information (and information technologies), and a command of multiple literacies. The gender and race aspects of the digital divide relate to other disparities in that women and people of color are disproportionately likely to hold low-skill, low-wage jobs, with few benefits and little opportunity for advancement (Merrifield et al., 1997). These are often the types of jobs that provide the least access to learning about and using information technologies in workplace settings, which further exacerbates the digital divide across all segments of technology use contexts.

Digital divide concerns are also interrelated with other policy issues, including health, education, and welfare disparities among people of different social, economic, and racial backgrounds. We focus on the effects of "welfare reform" in relationship to disparities in accessing information technology for women. Despite the disadvantaged position of many women and people of color in the labor market, welfare policy was dramatically changed in 1996 to include work requirements in exchange for time-limited assistance. As a result, poor women have been involved in either taking jobs that do not provide wages high enough to support their families or facing the loss of their welfare benefits. And the five-year time limit for many poor women to receive public assistance has run out. There seems little doubt that the key to economic stability for the average American family is increasingly linked to education and job skills, which in today's market include knowledge of information technology. Those who do not have access to information technologies are going to continue to fall behind.

We sought to examine both the possibilities and limitations of information technologies as a factor in changing the circumstances of poor women and understanding the barriers to accessing technology and related educational programs that prevent them from being able to participate on an equal basis with others in the economic system. Moreover, we began to address the ethical implications of programs that seek to overcome the digital divide because of the inherent inequalities among partners with vastly different resource capacities.

We have been involved in a project that assesses the experiences of poor women through developing a demonstration community technology center (CTC) and associated educational programs in collaboration with community partners at Harrison Plaza Public Housing Development in North Philadelphia (Pennsylvania). The partners involved in establishing the demonstration CTC included the Harrison Plaza Residents' Association, Temple University, the Department of Housing and Urban Development, and the Philadelphia Housing Authority. At the time the CTC was implemented (1999–2001) it was an unprecedented approach to providing access to information and communication technologies because it involved providing extensive opportunities for training that linked technology skills and computer access to basic literacy, basic skills, and job training. The implementation of the CTC was based on a community need for accessing information technology that the partners translated into a request for new computers, improved access to the internet, and the provision of computer-based literacy and job training on location in the public housing development's community center. The community center provided services not only to residents of Harrison Plaza public housing development but also to the 7500 residents of the 11th Street Corridor where Harrison Plaza is located.

We undertook the project as a pathway to understanding the relationship between the circumstances of poor women receiving public assistance and barriers they faced in accessing technology. We viewed this interrelationship as intrinsic to a broader conceptualization of the digital divide, and fundamental to identifying appropriate strategies for addressing barriers to accessing information.

Our objective in synthesizing the lessons learned from the demonstration CTC is to address the outcomes of the project and to argue that the perspectives on information resources of women marginalized by race, class, and poverty suggest that feminist geographic research needs to move beyond a "gender and geographic information system (GIS)" framework to more broadly encompass information technologies. Drawing on broad cross-disciplinary discussions of "GIS and society" and the emerging discourse on feminism and GIS, we will argue that the contemporary discourse fails to acknowledge the social realities of poor women and the social, economic, and political context in which they operate. If poor women's empowerment is to be the goal of information technology research, then more attention needs to be paid to the factors that shape poor women's daily lives, most notably to the institutional constraints imposed through welfare reform, which essentially determine that poor women have neither the time nor the energy to engage with information technology, at least in the manner in which the CTC was originally conceived.

To make our argument, we will first elaborate on the nature of the digital divide and draw on the discourses of "GIS and society," public participation GIS (PPGIS), and feminist critiques of GIS to establish a framework for examining the digital divide in terms of poor women's lack of access to IT. Second, we will describe the initial goals and strategies of the Harrison Plaza Demonstration CTC and discuss why the reality of life under welfare reform for the women required those of us providing IT services and training to poor women to shift strategies. Finally, we will conclude by suggesting what this experience may indicate about a feminist research agenda for IT.

Overview of Digital Divide Issues in the Contemporary USA

The concept of the digital divide, while frequently used in US public policy debates, is ill defined (DiMaggio et al., 2001). The US Department of Commerce (1999) report does provide a beginning point for defining the digital divide, in that it identifies differential access to computers and the internet faced by low-income, minority populations and women as compared with more affluent white and male populations at home and at work. The report found that households with incomes of $75,000 and above are 20 times more likely to have access to the internet than those at the lowest income level. Even as the percentage of all families with computers and internet access has grown, the gap between whites and non-whites has widened during the past 15 years. More significantly in terms of accessing information resources of the internet, Hispanics and blacks are less likely to access the internet from home (9 percent) than from work (10 and 12 percent, respectively), whereas whites are more likely to access the internet from home (27 percent) than from work (19 percent). Furthermore, in inner-city neighborhoods, only 21 percent of Hispanic and black families have household computers, in contrast to 47 percent of all white (non-Hispanic) families. Furthermore, the study found that female heads of households with dependent children are the population most impacted by the digital divide: their access to computers and the internet is 18 times lower than that of married families with children (the group with the greatest access to computers and the internet).

One of the solutions to close the digital divide that has recently gained prominence is the creation of partnerships (of varying kinds) between technology-rich organizations and communities and community organizations that have limited technology access, with the goal of transferring computers, software, and training across the divide (Mark et al., 1997). Another widespread approach to overcoming differential access has been the placement of computers connected to the internet in public libraries. And many private corporations and large institutions donate old computers to public schools and to community organizations serving the needs of low-income families. Non-governmental organizations, such as NetDay and CTCNet, have also worked to provide technical support to public schools and community centers across the country to overcome infrastructure barriers that prevent internet access within public facilities.

In general, these approaches reflect two means of increasing internet accessibility and thereby reducing the digital divide. One is to find mechanisms to make computers with internet access available to as large a number of people as possible by upgrading facilities in public institutional settings such as libraries. The second is to find mechanisms for providing facilities, needed equipment, technical assistance, and educational training directly to communities negatively impacted by the digital divide. Yet different concerns drive each approach: the effort to improve capabilities of public institutions is related to a broad societal objective of democratizing information – one of the central tenets that drove the origins of the world wide web; in comparison, the effort to increase internet accessibility within communities negatively impacted by the digital divide is often linked to other interrelated "social service" or community empowerment needs and goals, such as improving job skills and employment opportunities.

Despite the proliferation of models attempting to overcome differential internet and computer access, little research has explored what the gap represents to people with the least access. There is an assumption that the digital divide has profound implications for poor women as a sort of information poverty that exacerbates and reinforces their social and material conditions. But this assumption does not take into account an understanding of how the condition of poverty translates into practical issues that poor women must face. Issues such as balancing work and family responsibilities, accessing health and welfare resources, and interacting with social networks to mitigate circumstances could be reflected in alternative approaches to improving access to information technologies. For instance, community technology center access that does not provide daycare or evening hours means that women with children have less access to resources provided than others that the CTC might serve. And if the facility is located outside of the immediate community location, families may not be served effectively.

Perspectives on how to examine inequality within GIS and society and feminist geography discourses

While the current work on "GIS and society" and the critiques that feminist geographers are making of GIS do not adequately address digital divide issues, the two areas of work are interrelated. We think that by examining empowerment and ethics in relationship to partnerships that arise to develop community GIS, some of the elements of a broader examination of digital divide barriers faced by poor women can be outlined. And we submit that in establishing an approach for describing information technology resources and how they may be accessed, we can address issues raised in relationship to both GIS and IT.

Discussions around "GIS and society," "PPGIS," and feminist critiques of GIS collectively examine the relationships and dynamics of the interfaces between society, information technologies, information use, empowerment, and representation. In particular, feminist geographers have underscored themes that have been prevalent within the GIS and society realm, though not explicitly defined as feminist concerns. The points of commonality among these discourses include: the focus on scale (Kwan, 2002a, b; Niles and Hanson, 2003); the implications of examining the social construction of GIS (Schroeder, 1999; Harvey, 2000, 2001; Sieber, 2000; Craig et al., 2002); and the epistemological underpinnings of GIS and how that relates to the projects that GIS is applied to (Mark, 2000; Kwan, 2002c; Pavlovskaya, 2002; Schuurmann and Pratt, 2002). The feminist critique that diverges the most sharply from the thematic direction of other GIS and society concerns is a focus on the lack of inclusion of women and feminist technical experts in the development of GIS projects (Kwan, 2002b; McLafferty, 2002; Schuurman and Pratt, 2002) and how this creates, as McLafferty argues, a sort of technocracy that seemingly inevitably returns to a mainstream agenda that mitigates empowerment objectives that women may have.

Marking the formulation of an organized critique of GIS with the publication of *Ground Truth*, Pickles (1995a) provides a foundation for examining the disjuncture between technology use and achieving socially relevant objectives. *Ground Truth* draws on geographic conceptualizations of social problems and the discipline's long-standing tradition of linking technological advances, technological knowledge, and geographic knowledge to the analysis of social problems. One of the key themes throughout *Ground Truth* is a concern with the persistent disparities between those who have access to technology and those who do not. Yet the hope forwarded in *Ground Truth* is that simply placing information technology in the hands of marginalized communities, such as the urban poor, would produce such social benefits as more democratized decision-making, improved opportunities for economic advancement, and improved dialogue among the members of marginalized communities (Pickles, 1995a). This view holds that there is an inherent relationship between technology and power, noting that many information technologies were developed purposefully as "instruments of policy making" (Curry, 1995). Harris et al. (1995) assert a transformative relationship between technology and accessibility more directly, stating that since access to information forms a critical element of social power relations, access to information technology represents empowerment. Moreover, they describe the conditions in which a lack of access to technology in fact signifies social and political marginalization.

Examining the issue of access to technology has also been the focus of two research initiatives of the National Center for Geographic Information Analysis (NCGIA), including I-19 (The GIS and Society Initiative) and Project Varenius (Geographies of the Information Society). The central theme of these research initiatives is that society faces mounting problems because of the paradox of widespread dissemination of computers and information technologies alongside significant disparities in the use of and access to these technologies. One of the impacts of these two formal initiatives of NCGIA has been to provide a basis for more nuanced assessments of the implications to society of the rapid advancement of information technologies. Because of this work, more scholars are considering both negative and positive outcomes for both society as a whole and different communities at multiple geographic scales and in varying social contexts (Harris et al., 1995; Onsrud, 1996; Curry, 1998; Wellman et al., 2002).

The idea of participatory and community information technology programs is to develop and use information technology on behalf of and in conjunction with the public, community or marginalized groups (Kellogg, 1999; Nedovic-Budic and Pinto, 1999; Sieber, 2000; Carver, 2003). An important critique that feminist geographers have raised is that, despite the focus on social disparity in regard to geographic information technology (GIT) that these initiatives provoked, and despite the work of scholars across a wide array of partnership projects aimed at disseminating GIT (see Craig et al., 2002, for examples of PPGIS), simply placing information technologies at the nexus of community has failed to produce empowerment for marginalized peoples and women (McLafferty, 2002). Feminists point out that unless resources are deployed with an understanding of the realities of gender-differentiated power structures within communities, simply giving resources to "the community" will not necessarily benefit women.

As currently constructed, participatory information technology development and use is mostly interinstitutional, following an evolution of the idea that technology and information is transferred from more to less privileged organizational and community settings. McLafferty (2002) points out the limitations that this can imply for women: they may not be perceived to be technically capable of engaging community or mainstream efforts to develop information resources; specific resources developed may not adequately depict or meet women's information needs; and women may not be positioned powerfully enough to affect the decision-making processes that ensue as a result of enhancing a community's information resources.

In practice, the focus of participatory models is typically on transference of knowledge about how to develop technology capacity, as opposed to how communities might best represent themselves and benefit from technology. The concern raised by feminist geographers related to power dynamics in developing GIS is further exacerbated when poor women's circumstances are taken into account. Herein lies an important area of research concern that our project addresses – that is, the relative differences in the empowerment of partners in specific programs and the relationship of power dynamics within partnerships to the goals that marginalized people, such as poor women, have for achieving empowerment through such practical advances as gaining higher skilled, higher waged jobs and educational improvement (Martin, 2002; Monk et al., 2003). Since the basis of most interinstitutional partnerships is that one partner is more privileged in terms of knowledge about the technology even as the ideal nature of these partnerships might be participatory and reflexive, we need to understand the limitations and advantages of collaborative partnerships among groups with vastly unequal resources, as in the case of the Harrison Plaza Demonstration CTC.

One critical aspect of looking at the power dynamics among collaborating institutions in technology-related partnerships is a need to critically examine the roles of all partners, not just the outcomes to those identified as the population to be served. This means that alongside the need to understand poor women's perspectives is the need to better understand the dynamics of all partners involved in shaping the information technology use context of poor women. While depicting the framework of information technology use from the perspective of poor women transitioning from welfare-to-work marks a strong contrast with how issues about technology use have traditionally been addressed, such concepts as technology transfer, organizational capacity building, and training are still relevant to a discussion of how to improve the use of technology within marginalized communities. Further, while there is a growing discourse that seeks to evaluate technology use partnerships beyond merely looking at the outcomes of the "transference of technology" (Sheppard, 1995) the idea of a *non*-top down information technology partnership is illusive. We argue that by shifting the attention to the actual technology use setting and partners we can better address the persistent concerns of accessibility, technology impacts on organizations, and empowerment (or lack of empowerment) as related to technology use.

Analyzing scenarios of the use and development of information technology from the standpoint of different organization objectives, institutional resources, and human resources capabilities represents an important new area of research in technology adoption. For example, in the field of GIT there have been many studies that identify and assess on a case-by-case basis the organizational approaches to adoption and use of technology in planning and government institutional settings from mainstream and feminist perspectives (Calkins, 1991; Obermeyer and Pinto, 1994; Campbell and Masser, 1995; Masucci, 1995, 1996; Obermeyer, 1995; Lopez, 1998; Masser, 1998; Kellogg, 1999; Schroeder, 1999; Carver, 2003). However, information about community contexts as settings for the use of information technologies is all too often embedded within case study descriptions, even though there is important theoretical work being done to depict how an analysis of community actors is critical to gaining an understanding of how marginalized communities fight to improve their circumstances through participation in mainstream establishments (Escobar and Alvarez, 1992; Perritt, 1998; Pulido, 1998).

Since participatory development and the use of information technologies rely on alliances between mainstream organizations and alternative organizational structures – that is, ones that lie beyond the realm of the public sector and government support – we also think that an understanding about the roles that partner organizations and institutions play in the technology use of marginalized communities is fundamental to understanding the technology use context of the poor. One rationale for this focus is provided by Perritt and Masucci (1997), who argue that community actions often stem from a desire to produce a public awareness of interests not represented in mainstream decision-making structures. This idea is important to the analysis of technology use among marginalized people, such as the urban poor, because these communities are often the only ones that can address the social concerns that are neglected by government and private sector resources, often providing the leadership and impetus for mainstream organizations to create new strategies and approaches that address advocated concerns. As we support democratization of decision-making, it will be critical to democratize access to information technology. In order to accomplish this task, we need to improve our understanding of the context in which poor women, as one important marginalized community, are working to improve their lives (Martin, 2002).

Shifting the focus of "GIS and society" and "feminist critiques of GIS" narratives toward community concerns underscores that if we begin an examination of social inequality from the vantage point of the community, GIS may or may not be how information resource needs are framed. Our interactions around the Harrison Plaza Demonstration CTC support the idea that there is a broader concern with digital divide disparities in terms of access to information technologies and opportunities to learn about them as compared with participating in development of community information resources, including GIS resources, *per se*. In this way, we connect the concerns of geographers with the impact that GIS has on addressing social inequality because if we begin with GIS as the nexus for partnership, we have already implicitly generated a power dynamic between technical expert and community around organizing geographic information. The same could be said of organizing to implement the CTC, yet we note that its development originated within the community.

As we also take into account the particular dynamics of the community with which we collaborated, a more specific policy directive, welfare reform, permeated the entire implementation discourse, and the actual implementation of the CTC and associated educational programs dramatically reflected the realities confronted by women managing their lives within the context of TANF requirements.

Welfare reform in Pennsylvania and limiting effects on accessing information technologies

A review of the mechanics of welfare reform policy within Pennsylvania will begin to show the complexity faced by women served by the demonstration CTC. The Personal Responsibility and Work Opportunity Reconciliation Act of 1996, more commonly referred to as "welfare reform," eliminated the federal guarantee of cash assistance to poor people and replaced it with a system operated by the states that contains stringent work requirements in exchange for time-limited welfare assistance. The basic program provisions of Temporary Assistance to Needy Families (TANF), established by the federal legislation with states having the option to make requirements more stringent, are designed to promote rapid movement of people from welfare into the labor force, with less emphasis on long-term education and skills attainment.

Pennsylvania's TANF plan, in effect since March 3, 1997, is designed to promote the rapid movement into private sector employment of women receiving welfare benefits. It established a program known as the Road to Economic Self-Sufficiency through Employment and Training (RESET) in which all welfare recipients are enrolled. Despite the title, RESET is based on the unsubstantiated argument that emphasizing a work-first approach (rather than, for example, long-term education) will result in women establishing a work history with increasing wages and benefits that will ultimately lead to economic self-sufficiency. Therefore, when a woman joins the RESET program she is required to develop with her caseworker a personal plan, called an Agreement of Mutual Responsibility (AMR), which specifies the woman's plans to achieve self-sufficiency and how she will conduct the initial job search.

Upon enrolling in RESET, welfare recipients are required to immediately undertake an eight-week job search either independently or through a state-sponsored job search program. If the recipient does not find a job in that eight weeks, she is required to participate in work activities outlined in her AMR. During this period, which ends when the recipient finds a job or after two years' receipt of cash assistance, the woman must participate in work activities such as: independent job search; self-directed community service; adult education, literacy, or English as a second language programs; and welfare-to-work programs operated by independent service providers under contract to the state. No more than 12 months of full-time education, however, can be counted as work activity.

After 24 months, if the woman has not yet found a job, she is required to participate in a more restricted set of work activities for at least 20 hours per week, including: an unsubsidized job; subsidized employment; work programs operated by independent service providers that will arrange 20 hours per week of work activities; on the job training; or community service jobs. In other words, training and education can only be undertaken *in addition* to the mandatory 20 hours per week of work activity. If a recipient does not find a job within 60 months of joining RESET, she will receive no more cash assistance. Furthermore, if a woman willfully fails to participate in any aspect of the welfare program, she will be sanctioned, resulting eventually in loss of benefits. Finally, there is a lifetime limit of 60 months' assistance, after which welfare recipients are cut off from any further cash assistance at any time in their life. However, this time limit is not being enforced.

One of the underlying assumptions of this welfare "reform" is that poor women with children will become economically self-sufficient through employment. Yet against this optimistic projection are the facts that most women are in sex-segregated and race-segregated occupations with the attendant low wages and lack of opportunity for advancement, that there is still a significant gender/race wage gap, and that many women's wages are less than adequate to support a family (Reskin and Hartmann, 1986; Hanson and Pratt, 1991, 1995; Kodras and Jones, 1991; Institute for Women's Policy Research, 1995; Blank, 1997; Gilbert, 1997, 1998). Indeed, the contemporary US labor market is undergoing dramatic changes that include a polarization of wages, an increase in low-wage service sector jobs, and increases in part-time and temporary employment, all of which have further disadvantaged many women and minorities (Kodras and Jones, 1991; Jezierski, 1995; Blank, 1997; Dresser and Rogers, 1998; Harrison and Weiss, 1998).

The result of these trends is that many women who work for wages are little or no more financially secure than they would be if they had received payments from "welfare" (Spalter-Roth et al., 1993). Additionally, research has demonstrated that 43 percent of former welfare-recipient mothers worked substantial hours but still could not find enough work to lift their families out of poverty (Institute for Women's Policy Research, 1995). Furthermore, research has demonstrated that welfare recipients are disproportionately low-skilled and many lack the basic education skills necessary to acquire and retain employment – particularly for jobs that pay living wages (Burtless, 1995; Olson and Pavetti, 1996; Heinrich, 1999). These data, in the context of a welfare reform that has increased the numbers of working poor women and the competition for low-wage jobs, suggest that we urgently need to examine whether increasing access to technology and related educational programs will help poor women to attain employment at wages that will lift their families out of poverty.

Harrison Plaza Demonstration CTC

Because of strict guidelines related to RESET in terms of obtaining paid work, one of the key themes expressed by the Harrison Plaza residents' association was that in developing the demonstration CTC, educational programs that would emphasize building information technology skills were to be prioritized above other concerns. A further community requirement was that the demonstration CTC was to be housed within the public housing development. A final constraint was that it could not be networked to the housing authority because of concerns related to privacy.

After a year of planning, the Harrison Plaza Demonstration Community Technology Center demonstration project opened in 2000–1, centrally located in North Philadelphia's 11th Street Corridor, serving one of the most economically distressed communities of the city. This geographic area is home to over 7500 residents, including 3318 residents living in four public housing developments. The inner-city area served by the Harrison Plaza CTC is characterized by extreme concentrated poverty, with the associated social characteristics (i.e. illiteracy, school drop-out, and lack of job networks) and spatial isolation due to a history of deindustrialization, disinvestment, and racial discrimination in housing and labor markets (Wilson, 1996; Jargowsky, 1997). A report based on 1990 census data and other sources describing the community and school characteristics of William Harrison Middle School, located adjacent to the Harrison Plaza CTC, provides a profile of a community in need of resources (Yancey et al., 1995). The residents of the area are predominantly minorities (82.6 percent African-American and 7.6 percent Latino). Educational attainment levels are low: 45 percent of adults have less than a high school education and 29 percent have a high school degree or equivalent. The residents experience the highest rates of unemployment and poverty in the city, resulting in significant numbers of households requiring public assistance. Consequently, the mean household income in 1990 was $10,035; nearly 55 percent of the population and 74 percent of youth live below the poverty line. Furthermore, 95 percent of students attending the middle school receive free or reduced cost lunches.

If we focus solely on residents of Harrison Plaza, the picture is even more devastating. According to Housing and Urban Development information, as of April 1999, the average annual income for residents, all of whom are African-American, is $6906. Almost all residents are living in extreme poverty, defined as below 30 percent of median household income. Only 22 percent of households receive any wages and 65 percent are receiving welfare benefits.

The large numbers of female headed households in this area who rely on public assistance and public transit, combined with the metropolitan labor market characteristics of an advanced service economy (i.e. the suburbanization of employment, polarization of jobs in terms of skill levels, wages, and benefits, and occupational segregation by sex and race), have presented significant challenges to the implementation of Pennsylvania's welfare reform plan. This means that one of the challenges addressed in implementing the demonstration CTC was how best to address the needs of a group of women who are among the most marginalized in terms of access to technology and in terms of access to jobs and services.

The demonstration CTC was designed to be used by residents of the four public housing developments located nearby the community center, as well as other residents in the surrounding area. The community perceived the CTC to be a critical resource for its ability to balance welfare reform impacts for women and their families. Prior to the establishment of the CTC, there was only one functioning computer at the center available for use by residents and one continuing after-school program that served approximately 25 elementary and middle school students.

Through the collaborative approach of this center, partners worked together to obtain funding from HUD for a technical assistance grant to the Residents' Council of Harrison Plaza to support upgrading the equipment and software capabilities of the CTC. Partners also collaborated on identifying, developing, implementing, and managing new educational programs. Program expansion was made possible by drawing on the interests and strengths of various partners, providing new technical and educational resources for populations that had been previously not served by the CTC, with the major emphasis on serving adult women involved in the transition from welfare to work. CTC partners supported the administration of the center and its educational programs, and recycled computers to distribute to residents for home use. The CTC functioned as a location for basic and technology literacy training, basic skills training, and technology-related job training during the year the demonstration project was implemented.

Project outcomes

The demonstration CTC eventually consisted of seven new personal computers with Pentium III processors, modem connections for three of these computers, and a printer available for one computer. Specialized software included vocabulary building, literacy training and reading applications that were age-appropriate, math and science games for children, and typing and resume-building applications for adults. Through two years of discussion about the particulars of the CTC, no consensus on the composition of computers was ever achieved, and despite intense efforts to address such issues as how workstations should be configured, the appropriate balance of Macintosh and Windows personal computers for specific populations and purposes, and developing a community approach for identification of appropriate software, again no consensus among the collaboration partners was ever achieved. The default approach for implementing specific choices was that the on-the-ground service providers worked behind the scenes of the formal planning and implementation process to purchase computers and begin a patchwork network that ultimately comprised the CTC resource.

Temple University's role within the collaboration eventually focused on developing and implementing pilot educational programs made available during the year (2000–1) of the operation of the demonstration CTC. These programs emphasized the development of technological literacies alongside basic skills and literacy among the individuals served. The programs included an adult basic computer education course, an after-school program, open computer access hours throughout the day and evening, and two internet access courses. The center was open from 9 a.m. to 9 p.m. in the fall and from 9 a.m. to 6 p.m. in the spring. The CTC and associated educational programs were maintained through an informal coalition of community residents, the Harrison Plaza Residents' Council, and Temple University faculty, students, and researchers.

The original objective of these programs was to provide opportunities for the adult residents of Harrison Plaza (mostly women) to gain access to computers and basic computer education as an effort to overcome the digital divide issues that challenge the Harrison community. As the programs evolved it became apparent that time constraints – due to balancing the needs of accessing services, participating in welfare reform mandated activities (such as paid employment, job search, and job training), and raising children – presented a significant barrier to the direct participation of women in programs of the CTC. Within the bounds of the demonstration project, there was no way to compensate for these demands on the women; thus, their direct use of CTC ended up being very limited.

However, the participation of children at the CTC was extensive – nearly 50 children accessed the CTC on a daily basis through participation in the After School Program (the After School Program consisted of homework assistance, computer training, and life development skills and recreation). And through participation of children in the After School Program, the CTC was involved in an outreach program to address needs of adults interested in gaining access to resources of the CTC. Several community events were organized to welcome adult family members into the CTC through supporting children in the After School Program. These included assemblies, open house events, and celebrations of achievements of children. For many adult family members, engaging with the activities of the CTC through children's activities was the first time they had accessed a computer lab.

Even though the demonstration year is over, as a result of the experience of the CTC program, Temple University faculty (such as ourselves) have developed a long-term approach to addressing digital divide issues in the Harrison community. Through an extensive process of interacting with families, literally door-to-door, we have worked to establish a relationship in which there is a commitment to working with families to enhance access to computer resources and associated educational programs, and to strengthen linkages between university students and families in the Harrison community. Our commitment to families has resulted in collaboration to implement ongoing education programs in association with the Harrison Family Center (adjacent to the property of the Harrison Plaza Community Center). The CTC and associated educational programs have been integrated within the programs of the Harrison Family Center. The Harrison Family Center continues to involve community members and Temple University faculty and students in the development and delivery of educational programs. And the Harrison Family Center facilities include a computer lab, as well as classroom space for the After School Program, which provide continuity for families served during the demonstration CTC program.

Finally, in assessing the challenges associated with implementing the CTC and associated educational programs, we identified the following barriers that continue to create a digital divide for this community, above and beyond the actual limitations of computers and software available for instruction and access to computers and the internet:

- limited access to high speed internet service providers and limited networking capability in community and school computer lab facilities;
- lack of facilities that can be allocated to computer labs and associated programs;
- limited availability of qualified staff and investment in supplies, hardware and software upgrades, and network administration;
- extreme geographic, social, and economic isolation of the Harrison Community, constraining the development of an adequate infrastructure to support a networked lab with universal access to the internet;
- limited access to trained community service providers in the area of technology use;
- limited overall computer access in the community.

Ultimately, however, the most significant barriers to poor women's participation in the CTC and related programs were the result of women attempting to fulfill their multiple roles of family providers, mothers, employees, students, etc. within the constraints imposed by welfare policy, and more broadly the changing political economy, institutionalized racism, and gender relations.

We suggest that if the beginning point for collaboration had focused on developing specific information resources as opposed to developing a context within which information resource discussions might take place, the specific information that is relevant within this community might never have emerged. Moreover, the issues we identify above give a more nuanced perspective on why developing community technology resources is such a complex process.

An anecdote often cited during planning meetings related to implementing the demonstration CTC was that another housing development had implemented a $1 million state-of-the-art computer center linked to the internet via a T-1 connection. According to the Harrison Plaza Residents' Council, the lab failed because it had been located in what had previously been an apartment building against the wishes of residents who wanted it to remain as housing. Residents refused to use the facility, which sat empty, to the dismay of the architects, housing development planners, and consultants. In developing the Harrison CTC, the Residents' Council steadfastly resisted such a top-down approach, and while the digital divide barriers identified above persisted, the small steps in implementing the developed resource did support organizing and educational goals of the community.

Relevance of the Demonstration CTC Project to Feminist Research on GIS

One of the most important issues that our project addresses is how to better conceptualize the digital divide. In our view, a better conceptualization involves examining the interrelationships among the technology use context, social networks, and the social policies and institutions regulating IT use. We conclude that this more nuanced conceptualization will better allow us to understand the relationship between access to IT, the information sought and obtained by individuals, and its impact on personal decision-making. And we find that if feminist geographers broaden the focus from GIS as an approach to empowerment toward these interrelated elements of technology access, we may be better positioned to support empowerment objectives of poor women.

An improved conceptualization of the digital divide also has strong implications for how feminist geographers relate to information technology issues, including addressing concerns about how to use GIS as part of feminist research projects. Improving technological literacy and ultimately gaining equal access to internet information resources seems to be inherent to this objective. However, an expanded notion of "technology accessibility," one which anticipates different interrelated facets of technology access that go beyond mere availability of computers and communication process, addresses how people interface with computer technology. This would involve an examination of the intersection between social science and information technology constructs on technology accessibility.

We suggest that when poor women's experiences are taken into account, it becomes clear that GIS research is inadequate without continued revisioning and reconceptualization as a tool for organizing and analyzing community information.

Examining digital divide barriers to accessing information technology underscores the difficulties encountered in the development of and participation in processes to develop community information resources for women who participated in the Harrison Plaza Demonstration CTC. We conclude by suggesting that GIT projects need to include a focus on how women differently perceive what information is relevant to their daily circumstances – and that this depends on race and class experiences. Further, we argue that organizing around information technologies, including developing and implementing geographic information systems, takes pre-eminence over the authentication of specific information technology implementation or the development of information systems from the standpoint of poor women.

BIBLIOGRAPHY

Blank, R. M. (1997) *It Takes a Nation: A New Agenda for Fighting Poverty.* New York: Russell Sage Foundation.

Burtless, G. (1995) Employment prospects of welfare recipients. In D. S. Nightengale and R. Havemen (eds), *The Work Alternative: Welfare Reform and the Realities of the Job Market.* Washington, DC: Urban Institute Press.

Calkins, H. W. (1991) GIS and public policy. In D. J. Maguire, M. F. Goodchild and D. W. Rhind (eds), *Geographical Information Systems: Principle and Applications, Volume 1: Principles.* Harlow: Longman.

Campbell, H. and Masser, I. (1995) *GIS and Organizations: How Effective are GIS in Practice?* London: Taylor and Francis.

Carver, S. (2003) The future of participatory approaches using geographic information: developing a research agenda for the 21st century. *URISA Journal,* 15, 61–71.

Craig, W., Harris, T. and Weiner, D. (2002) *Community Participation and Geographic Information Systems.* London: Taylor and Francis.

Curry, M. (1995) GIS and the inevitability of ethical inconsistency. In J. Pickles (ed.), *Ground Truth: The Social Implications of Geographic Information Systems.* New York: Guilford Press.

Curry, M. (1998) *Digital Places: Living with Geographic Information Technologies.* New York: Routledge.

Department of Commerce (1999) *Falling through the Net: Defining the Digital Divide.* Washington, DC: Department of Commerce.

DiMaggio, P., Hargittai, E., Neuman, R. W. and Robinson, J. P. (2001) Social implications of the internet. *Annual Review of Sociology,* 27, 307–36.

Dresser, L. and Rogers, J. (1998) Networks, sectors, and workforce learning. In R. P. Giloth (ed.), *Jobs and Economic Development.* Thousand Oaks, CA: Sage.

Escobar, A. and Alvarez, S. (1992) *The Making of Social Movements in Latin America: Identity, Strategy and Democracy.* Boulder, CO: Westview Press.

Gilbert, M. R. (1997) Identity, space, and politics: a critique of the poverty debates. In J. P. Jones III, H. J. Nast and S. M. Roberts (eds), *Thresholds in Feminist Geography: Difference, Methodology, and Representation.* Lanham, MD: Rowman and Littlefield.

Gilbert, M. R. (1998) "Race," space, and power: the survival strategies of working poor women. *Annals of the Association of American Geographer,* 88(4), 595–621.

Hanson, S. and Pratt, G. (1991) Job search and the occupational segregation of women. *Annals of the Association of American Geographers,* 81, 229–53.

Hanson, S. and Pratt, G. (1995) *Gender, Work, and Space.* London: Routledge.

Harvey, F. (2000) The social construction of geographical information systems. *International Journal of Geographical Information Science,* 14(8), 711–13.

Harvey, F. (2001) Constructing GIS: actor networks of collaboration. *URISA Journal*, 13(1), 29–37.

Harris, T. and Weiner, D. (1998a) Empowerment, marginalization, and "community-integrated" GIS. *Cartography and Geographic Information Systems*, 25(2), 67–76.

Harris, T. and Weiner, D. (1998b) Community-integrated GIS for land reform in Mpumalanga Province, South Africa. Paper presented at PPGIS meeting in Santa Barbara, October.

Harris, T. M., Weiner, D., Warner, T. and Levin, R. (1995) Pursuing social goals through participatory GIS: redressing South Africa's historical political ecology. In J. Pickles (ed.), *Ground Truth: The Social Implications of Geographic Information Systems*. New York: Guilford Press.

Harrison, B. and Weiss, M. (1998) Labor market restructuring and workforce development: the changing dynamics of earnings, job security, and training opportunities in the United States. In R. P. Giloth (ed.), *Jobs and Economic Development*. Thousand Oaks, CA: Sage.

Heinrich, C. J. (1999) Aiding welfare-to-work transitions: lessons from JTPA on the cost-effectiveness of education and training services. *Poverty Research News*, 3, 9–12.

hooks, b. (1990) *Yearning: Race, Gender, and Cultural Politics*. Boston: South End Press.

Institute for Women's Policy Research (1995) *Research in Brief. Welfare to Work: The Job Opportunities of AFDC Recipients*. Washington, DC: IWPR.

Jargowsky, P. (1997) *Poverty and Place: Ghettos, Barrios and the American City*. New York: Russell Sage Foundation.

Jezierkski, L. (1995) Women organizing their place in restructuring economies. In J. Garber and R. Turner (eds), *Gender in Urban Research*. Thousand Oaks, CA: Sage.

Kellogg, W. (1999) From the field: observations on using GIS to develop a neighborhood environmental information system for community-based organizations. *URISA Journal*, 11(1), 15–32.

Kodras, J. and Jones, J. P. III (1991) A contextual examination of the feminization of poverty. *Geoforum*, 22, 159–71.

Kwan, M. (2002a) Introduction: feminist geography and GIS. *Gender, Place and Culture*, 9(3), 261–2.

Kwan, M. (2002b) Is GIS for women? Reflections on the critical discourse in the 1990s. *Gender, Place and Culture*, 9(3), 271–9.

Kwan, M. (2002c) Feminist visualization: re-envisioning GIS as a method in feminist geographic research. *Annals of the Association of American Geographers*, 92(4), 645–61.

Leitner, H., Elwood, S., Sheppard, E., McMaster, S. and McMaster, R. (2000) Modes of GIS provision and their appropriateness for neighborhood organizations: examples from Minneapolis and St Paul, Minnesota. *URISA Journal*, 12(4), 43–56.

Lopez, X. (1998) *The Dissemination of Spatial Data: A North American–European Comparative Study on the Impact of Government Information Policy*. Greenwich, CN: Ablex.

McLafferty, S. (2002) Mapping women's worlds: knowledge, power and the bounds of GIS. *Gender, Place and Culture*, 9(3), 263–9.

Mark, D. (2000) Geographic information science: critical issues in an emerging cross-disciplinary research domain. *URISA Journal*, 12(1), 45–54.

Mark, J., Cornebise, J. and Wahl, E. (1997) *Community Technology Centers: Impact on Individual Participants and Their Communities*. Newton, MA: Interim Report to Informal Science Division/ESIE, National Science Foundation – Education and Human Resources Division, April.

Martin, D. (2002) Constructing the "neighborhood sphere": gender and community organizing. *Gender, Place and Culture*, 9(4), 333–50.

Masser, I. (1998) *Governments and Geographic Information*. London: Taylor and Francis.

Masucci, M. (1995) Developing GIS for low resource settings. In *GIS/LIS '95 Proceedings*. Bethesda, MD: ACSM/ASPRS, AAG, URISA, AM/FM.

Masucci, M. (1996) Developing GIS for citizen environmental monitoring and hazards mitigation in Alabama and São Paulo, Brazil. In *GIS/LIS '96 Proceedings*. Bethesda, MD: ACSM/ASPRS, AAG, URISA, AM/FM.

Merrifield, J., Bingman, M. B., Hemphill, D. and Bennett deMarrais, K. P. (1997) *Life at the Margins: Literacy, Language and Technology in Everyday Life*. New York: Teachers College Press.

Monk, J., Manning, P. and Denman, C. (2003) Working together: feminist perspectives on collaborative research and action. *ACME: An International E-Journal for Critical Geographies*, 2(1).

Nedovic-Budic, Z. and Pinto, J. (1999) Understanding interorganizational GIS activities: a conceptual framework. *URISA Journal*, 11(1), 53–64.

Niles, S. and Hanson, S. (2003) A new era of accessibility? *URISA Journal*, 15, 35–41.

NTIA (1999) *Falling through the Net II: New Data on the Digital Divide*. Washington, DC: National Telecommunications and Information Administration.

Obermeyer, N. (1995) The hidden GIS technocracy. *Cartography and Geographic Information Systems*, 22(1), 78–83.

Obermeyer, N. and Pinto, J. (1994) *Managing Geographic Information Systems*. New York: Guilford Press.

Olson, K. and Pavetti, L. (1996) *Personal and Family Challenges to the Successful Transition from Welfare to Work*. Prepared for the office of the Assistant Secretary for Planning and Evaluation and the Administration for Children and Families, The Urban Institute.

Onsrud, H. J. (1995) Identifying unethical conduct in the use of GIS. *Cartography and Geographic Information Systems*, 22(1), 90–7.

Pavlovskaya, M. (2002) Mapping urban change and changing GIS: other views of economic restructuring. *Gender, Place and Culture*, 9(3), 281–9.

Perritt, R. (1998) Sustaining community environmental movements in Brazil and Alabama. Paper Presented at the Annual Meeting of the Association of American Geographers, Boston.

Perritt, R. and Masucci, M. (1997) *Human Environmental Interchange: Managing the Effects of Recent Droughts in the Southeastern US*. Philadelphia: Harcourt Brace/Saunders.

Pickles, J. (ed.) (1995a) *Ground Truth: The Social Implications of Geographic Information Systems*. New York: Guilford Press.

Pickles, J. (1995b) Representations in an electronic age: geography, GIS and democracy. In J. Pickles (ed.), *Ground Truth: The Social Implications of Geographic Information Systems*. New York: Guilford Press.

Pulido, L. (1998) *Environmentalism and Economic Justice: Two Chicano Struggles in the Southwest*. Tuscon: University of Arizona Press.

Reskin, B. and Hartmann, H. (eds) (1986) *Women's Work, Men's Work: Sex Segregation on the Job*. Washington, DC: National Academy Press.

Schroeder, P. (1997a) A public participation approach to charting information spaces. In *1997 ACSM/ASPRS Technical Papers, Volume 5: Autocarto 13*. Bethesda, MD: ACSM/ASPRS, AAG, URISA, AM/FM.

Schroeder, P. (1997b) GIS in public participation settings. Paper presented at UCGIS Annual Assembly and Summer Retreat, Bar Harbor, ME, June 15–21.

Schroeder, P. (1999) Changing expectations of inclusion, toward community self-discovery. *URISA Journal*, 11(2), 43–51.

Schuurman, N. and Pratt, G. (2002) Care of the subject: feminism and critiques of GIS. *Gender, Place and Culture*, 9(3), 291–9.

Sheppard, E. (1995) GIS and society: towards a research agenda. *Cartography and Geographic Information Systems*, 22(1), 5–16.

Sieber, R. E. (2000) GIS implementation in the grassroots. *URISA Journal*, 12(1), 15–29.

Spalter-Roth, R., Hartmann, H. and Andrews, L. (1993) Mothers, children, and low-wage work: the ability to earn a family wage. In W. J. Wilson (ed.), *Sociology and the Public Agenda*. Newbury Park, CA: Sage.

Wellman, B., Boase, J. and Chen, W. (2002) The networked nature of community online and offline. *IT and Society*, 1(1), 151–65.

Wilson, W. J. (1996) *When Work Disappears: The World of the New Urban Poor*. New York: Vintage Books.

Yancy, W., Saporito, S. and Thadani, R. (1995) *Neighborhoods, Troubles, and Schooling: The Ecology of Philadelphia's Public Schools*. Philadelphia: The National Center on Education in the Inner Cities Center for Research on Human Development and Education, Temple University.

Chapter 22

Women Outdoors: Destabilizing the Public/Private Dichotomy

Phil Hubbard

Men – white middle-class men at least, and in particular – own the street without thinking about it. Women must always make a conscious claim, must each time assert anew their right to be "streetwalkers." (Wilson, 2001, p. 139)

Despite the rise of poststructural ways of thinking that disturb many of the dualisms that have characterized geographic thinking, much feminist scholarship continues to explore the consequences of deeply embedded binary categorizations. One of the most discussed of all dualisms is that between public and private. While the meaning of these terms remains hotly debated, and has changed markedly over time, exploring the gendered nature of public and private constitutes an especially important tradition in feminist geography. It is not difficult to understand why: the public/private dichotomy is an inherently spatial one. Hence, exploring the gendered nature of public space, and the ideology that associates men with the public realm and women with the private, remains a priority for feminist geographers (Duncan, 1996; Namaste, 1996).

Foremost among the public spaces that have attracted the attention of feminist theorists are "the streets" – spaces often romanticized as the realm of public mixing, democratic encounter, and spontaneous action (Sennett, 1993). Streets are of course both a social and spatial construct, being a set of architectural forms and layouts as well as a collection of rituals and expectations. These expectations coalesce to bequeath the streets a distinctive social etiquette; one of the central props of this etiquette is the idea that there are certain activities deemed acceptable in private but wholly inappropriate and ill mannered when performed on the streets. In the urban West proscribed public activities typically include bathing, defecating, spitting, and, in some circumstances, even eating, smoking, or drinking (see Valentine, 1998). The common denominator in these proscriptions is that these activities involve a transgression of the boundaries of the body, with liquids, gases, or solids crossing the threshold of the body. As Jervis (1999) relates, modern citizens are only supposed to allow the boundaries of their body to be crossed in private; to allow this to happen in view of strangers is to fail to observe the complex systems of manners that have been developed since the beginning of the Enlightenment. For example, while allowing the boundaries of the body to be penetrated by a partner of the opposite gender – and to a lesser extent, partner(s) of the same sex – in private space is generally regarded as benign, giving in to such sexual urges on the street is regarded as less than civilized – a grotesque and even animalistic mode of behavior deemed quite unacceptable in modern society (Jervis, 1999).

In this chapter, I want to offer a review and critique of the ways in which the distinction between public and private conduct is policed and regulated, focusing specifically on the regulation of sexuality in contemporary Western cities. It is evident that there is a complex, yet widely understood, series of expectations about public sexual conduct, and that these are often different and contradictory for men and for women. For instance, despite the premium placed on appearing "desirable" and "sexy" in contemporary Western societies, to appear in a state of undress – to expose one's self or to adopt particularly revealing modes of dress – in public is also to attract disapproval. Indeed, one need only reflect on the befuddled pronouncements of (British) High Court judges who with worrying frequency condemn rape victims for dressing "provocatively" to begin to grasp the complex and often contradictory codes of sexual comportment that structure streetlife – and that structure it differently for men and for women.

Examining the double standards and contradictions that shape the sexuality of the streets therefore allows us to examine how space is implicated in the making of sexual subjectivities. Indeed, many feminist writers have stressed that sexual politics of the street are hugely important in shaping women's place in society, with the visibility of particular sexual roles on the streets informing broader notions of women's ability to use and shape both public and private spaces (Wilson, 2001). Hence, while the equation between gender and sexual identities is far from straightforward, examining which of women's public performances provoke anxiety, and which are subject to regulation by the state and law, sheds much light on the way women are constructed as a feminine Other. At the same time, it also tells us about Western masculine identities, which Kirby (1996) suggests are typically scripted in terms of sexual conquest.

Given this, it seems that scrutinizing the sexual life of the city streets should constitute a central plank of feminist geographic enquiry. In the remainder of this chapter, I thus alight on some geographies of the street to demonstrate why the public/private dichotomy is so important in shaping gender relations, and why it needs to be destabilized in the interests of creating more democratic, empowering gender identities.

Streets of Freedom and Constraint

When discussing the sexual life of the streets, it is customary to begin by delimiting two paradigmatic figures: the flâneur and the prostitute. Now a trope among urban theorists, the flâneur first appears in Charles Baudelaire's Paris poems of the 1850s, resurfacing in the work of Walter Benjamin in the early twentieth century (see Buck-Morss, 1989). Broadly defined, the flâneur is a male pedestrian who finds delight and pleasure in ambling contentedly and unhurriedly through the city. Indeed, to promenade without purpose is the highest ambition of the flâneur; as Benjamin (1999, p. 417) observes, "an intoxication comes over the man who walks long and aimlessly through the streets . . . with each step, the walk takes on greater moment." In effect, the flâneur was a "painter of modern life," a spectator of contemporary manners and urban scenes who was at one and the same time part of the ebb and flow of the street but refused to surrender his individuality to the crowd. Inevitably, the flâneur was a member of the bourgeois elite, someone who had time on his hands and who could afford to dawdle. In Elizabeth Wilson's (2001) description of the flâneur, we thus learn of the lives of these bourgeois men of leisure, who, free from the need to work, spent their day looking at the urban spectacle. In Paris, for example, the flâneur was free to explore the spaces of urban leisure and pleasure that became synonymous with the nineteenth-century metropolis, including cafes, theatres, arcades, brothels, clubs, restaurants, panoramas, and pleasure grounds.

Though the flâneur has often been treated as an allegorical male urban type – a figure created at the interstices of urban journalism, literature, and social science – it should not be forgotten that he was also a real person. While no one knows just how many flâneurs there were in Paris and other cities of nineteenth-century Europe, their existence alerts us to the possibilities and pleasures that the streets offered for some. In this sense, Rendell (2002) stresses that the flâneur should be regarded as a specific incarnation of a more general type – the urban rambler – who in the eighteenth century was personified by the rakes and rovers who roamed the streets in their search for specifically sexual pleasure. Hence, as urban improvement and modernization created more hygienic, commodious, and ordered public spaces (see Ogborn, 1998), street life began to take new forms, with young, heterosexual, upper-class men forging a new masculine identity based on rambling through a diversity of public spaces where women could be encountered, evaluated, and consumed.

Foremost among these "women of the streets" was the prostitute: "The prostitute was the quintessential figure of the urban scene . . . for men as well as women, the prostitute was the central spectacle in a set of urban encounters and fantasies. Repudiated and desired, degraded and threatening, the prostitute attracted the attention of a range of urban male explorers" (Walkowitz, 1992, p. 233). The relationship between prostitutes and flâneurs was thus ambivalent: prostitutes were objects of visual pleasure consumed by the man about town, but their very existence threatened the patriarchal ideologies that empowered the flâneur. In nineteenth-century Western cities, the public presence of prostitutes was thus regarded as sexualizing and feminizing the public realm, thereby demonstrating to male authority that its control of the city was not complete. Cut loose from the bounds of monogamy, productive labor, and religious asceticism, prostitutes disturbed the masculine assumption that the public realm was a place for men and that women should flaunt their sexuality only in private. Janet Wolff (1985) argues that it was considered disreputable for women to be unaccompanied in public spaces, with "proper" bourgeois women associated with interior spaces, particularly the private sphere of the home and family. Thus, in the same moment that prostitutes reaffirmed the privilege and power of the flâneur (who treated the prostitute as a sexual commodity, to be bought and sold), the mobility of prostitutes provided cause for concern. Hence, one finds the prostitute attracting the attention of a variety of urban reformers, social philanthropists and "men-about-town" in the nineteenth century, bequeathing a detailed archival record of a group that was an object of both desire and disgust (see Corbin, 1990).

Examining the histories of Western urban female prostitution, one finds that male anxieties fueled a range of regulations designed primarily to restrict the public visibility of sex workers. For example, in Second Empire Paris, inspired by obsessive moral hygienist Parent-Duchâtlet, the result was the construction of strictly monitored *maisons de tolerance*; in Britain, a system of *magdalene* asylums was established to "cure" prostitutes, while the police invoked a variety of new powers such as the 1847 Town Police Clauses Act and the 1860s Contagious Diseases Acts to contain prostitution – or at least to render it invisible in all but the least prosperous areas (Corbin, 1990; Ogborn, 1998). The latter euphemistically titled Acts (of 1864, 1866, and 1869), for example, applied to specific garrison and port towns, forcing any woman suspected of prostitution to be subjected to humiliating medical inspections, with any woman showing signs of venereal disease subject to confinement. Notably, the Contagious Diseases Act did not apply to any men suspected of being the clients of prostitutes.

Similarly, today there are complex regulations designed to limit the visibility of sex work in Western cities. The location of female prostitution in contemporary British cities is a reflection of the social and cultural status of prostitutes, with their "placement" in specific sites mirroring male anxieties about the visibility of feminine sexuality. As I have argued elsewhere, this results in a complex and locally specific geography, as although popular understandings of public/private space inform structures of regulation and containment, the geography of prostitution is more fragmented and fluid than might be anticipated, cutting across public and private in sometimes unexpected ways (Hubbard, 1999). Specifically, in some cities prostitutes are "placed" out of sight in "private spaces" – brothels, massage parlors, or private flats – where sex work can be performed largely unfettered by the state and law, but where the ability of prostitutes to leave these spaces of confinement and enter the public realm as sex workers remains restricted. For example, in Edinburgh the authorities have sought to introduce a *de facto* licensing system whereby sex work in off-street massage parlors is tolerated, although the same tolerance has not been evident when prostitutes have sought to work in gentrifying public spaces (such as the former docks of Leith). Yet in other cities, sex work can be encountered in public view in "red light districts" where it is subject to the disciplining gaze of the state and law. In such circumstances, the sex worker's occupation of public space may still be highly circumscribed, subject to monitoring and regulation by the police, who – often erratically – will invoke standing statutes to prosecute prostitutes and, sometimes, their male clients. In either case, it appears that it is the sight of the sexed body of the prostitute *on the street* that is regulated, not the fact of prostitution itself. Prostitution in public view continues to disturb male assumptions that sexuality should be domesticised, with the state and law conspiring to prevent prostitution from "leaking out" into the wider public realm.

Whether or not the contemporary sex worker and kerb-crawler (client) are the modern-day equivalents of the nineteenth-century prostitute and flâneur is a moot point. Nonetheless, the continuing existence of legislation designed to limit the mobility of sex workers – and to a lesser extent, their clients – emphasizes that the regulation of sexuality is profoundly gendered. Indeed, Wolff (1985) suggests there cannot be a female flâneuse, only the prostitute, underscoring the idea that the freedom to roam is very much a male freedom. Arguing that the flâneur's license to take in the city's sights is the walking embodiment of the male gaze, Wolff concludes that women alone in public space are inevitably women "out of place," subject to sanctions and negative connotations. While others have argued that there is a risk of overstating this argument, pointing to the ways that women have been able to access a range of public spaces on their own terms (e.g. Wilson, 2001, on women's consumption spaces; Munt, 1999, on the lesbian flâneur), the fact that the term "women of the street" remains synonymous with sex workers provides evidence of the restricted sexual opportunities for women in urban public spaces. The history of prostitution underlines this: repeated efforts to restrict the public visibility of sex work demonstrate that uncontained feminine sexuality arouses concern. Yet it is also apparent that there has been little concerted effort to destroy prostitution, merely enclose it. Closeted away from the public gaze, it becomes part of a "restricted economy" which channels desire to capitalist ends. Indeed, as Brown (2001) notes in his exploration of "closet spaces," the net result of this process of concealing female sexuality is that sexual relations themselves become commodified: it is through this manipulation of space that a market for sex is created. As Brown shows, restricting the availability of sexual services through related strategies of spatial segregation and social stigmatization serves to valorize these very services, bequeathing a market that thrives on *scarcity*. Unfortunately, as we will see in our next section, this is a market based on profoundly unequal (and gendered) relations of exchange.

Streets of Concealment and Display

To highlight the differential sexual freedoms afforded to men and women on the streets is not to insist that the public realm is a solely male realm, nor the private realm exclusively female. Instead, as a multitude of feminist geographers have demonstrated, it is to assert that men and women occupy space on a profoundly different basis (Namaste, 1996). At the heart of this unequal occupation of space are iniquitous relations of exchange, whereby women move, or are moved, between men as commodities. Drawing on the influential work of Luce Irigaray, Rendell summarizes this exchange as follows:

> Women are exchanged, both socially and symbolically, as commodities with use values (sex and/or child-bearing and rearing) and exchange values (signifiers of male worth in terms of property and commodities) . . . men organise and display their activities of exchange and consumption, including the desiring, choosing, purchasing and consuming of female commodities, for others to look at in public space. (Rendell, 2002, p. 19)

This process of exchange is, as Rendell's own fascinating account of Regency London suggests, one in which visual consumption is paramount. For the man of the streets, women are to be looked at, part of the urban spectacle, while women are unable to return the gaze. Being looked at is passive and female, whereas looking is deemed active and male. This is certainly the case for the flâneur, who "enjoyed the freedom to look, appraise and possess" (Pollock, 1988, p. 79). Against this, the women who they looked at remain fixed, frozen in the male gaze – reinforcing the idea that there was no possibility of a female flâneur in the nineteenth century, only the prostitute.

Yet this take on the visual consumption of bodies is perhaps too simplistic, and perhaps fails to communicate the erotic topographies of the contemporary city. Indeed, Wilson (2001) insists that the turmoil and constant movement of the city creates multiple opportunities for women to return the male gaze, and in so doing resisting men who would seek to conquer and subdue women. Fighting the stereotypes of female passivity in public space, she argues that women can occupy a variety of viewing positions on the street as they enjoy a variety of sexual roles and pleasures. The lesbian flâneur, for example, has become a key figure in debates around the erotic possibilities of the city, highlighting that it is not always men who treat the streets as their sexual playgrounds (Munt, 1999).

Yet for the anxious and fragile male, the spectacle of being gazed on by women provokes anxiety. While men may take narcissistic pleasure in being gazed upon by other men, and in displaying their body to both men and women, if they are looked on by women as a sexual object, this triggers attempts to reassert sexual order, and put women in "their place." Likewise, the existence of the lesbian flâneur also prompts unease, not least among gay male populations who are sometimes antagonistic to the lesbian population (Nast, 2002). Generalizing male anxiety to a universal principle, attempts to reassert order are often justified with reference to the "public good"; legislation designed to curtail the visibility of feminine sexuality on the streets is often claimed to be necessary to protect "public morals." For example, contemporary British Obscenity Acts make it a criminal offence to publish or display material likely to "corrupt or deprave those likely to read, see or hear it" (Obscene Publications Act, 1959); with the 1959 Act couched in language that makes it clear the intention is to protect "public morals and decency." Yet these acts define the "decent citizen" in both gendered and aged terms, suggesting on the one hand that the obscene words and images against which the public needs protection are those that cater to male sexual urges, and on the other hand that it is only women and children who need protecting from their corrupting influence.

Such gendered censorship of public space is particularly apparent in a variety of regulations designed to restrict the public placement of obscene words and images. In Britain, this spatial regulation can largely trace its origins to the Victorian obscenity laws that were prompted by the increasing visibility of obscene prints and books in London. A by-product of the new commodity forms that accompanied Modern urbanism, sexually explicit images started to be commonplace in the streets of Victorian London; to social regulators, these threatened to draw "respectable" citizens into a state of sexual depravity (Nead, 1997). It was this threat, especially the idea that pornography compromised middle-class codes of female sexual comportment, that prompted the state to introduce forms of surveillance designed to control obscenity. Accordingly, the forces of law and order began to map the visual dangers of commercial sexuality on to, and out of, London's thoroughfares; vice was seen to originate in the streets, and was deemed to corrupt those on the streets. However, as Mort (1998, p. 889) outlines, it was not London as a whole that was implicated in this geography of immorality, but "specific zones and quarters" of the metropolis. Nead (1997) echoes this when she traces the origins of the 1857 Obscene Publications Act to the dense warren of streets around the Strand which produced and displayed indecent publications. Focusing on Holywell Street, "a precarious monument to Elizabethan London," her account describes concern about the visibility of "obscene" prints and "penny dreadfuls" in a street that was regarded as both aesthetically and morally lacking (Nead, 1997, pp. 664–5). This was mirrored in descriptions that contrasted the area with the bright, wide, and straight thoroughfares of respectable London, emphasizing the dangers associated with the dark side of desire. For example, parliamentary debates at the time suggested this was an uncontrolled urban haunt trading in "a poison more deadly than prussic acid" (Hunter et al., 1993, p. 61). On passing the Obscene Publications Act, its author Lord Campbell thus looked forward to a time when Holywell Street would be "the abode of honest, industrious handicraftsmen and a thoroughfare through which any modest woman might pass" (Hunter et al., 1993, p. 63). Such male anxieties are mirrored in more recent debates about the suitability of public advertising hoardings (billboards). One recent advert – for a new women's perfume, "Opium" – was a source of particular controversy. Featuring a naked British model, Sophie Dahl, reclining in high heels and diamonds, clutching one breast "rapturously," this image was designed, according to an executive of Yves Saint Laurent, the fashion house that manufactures Opium, to portray a woman who "looks like she's had too much of everything: too much food, too much sex, too much love" (Tom Ford, cited in *The Observer*, December 24, 2000). While the advert provoked little reaction when featured in British fashion magazines, when placed in the public realm it prompted 730 complaints – the most ever received by the Advertising Standards Authority. Further, many complaints emanated from women, with the French feminist group Chiennes de Garde condemning the image as "porno chic." The media commentator Libby Brooks suggests the reason why the advert was shocking for both women and men:

> Female sexuality, like the female body, is generally constructed as passive . . . the image of Dahl is threatening because she knows what she wants, and for all the progressive talk of bedroom liberation, that continues to alarm us. . . . The erotic landscape is male. Our public culture is suffused with the homogeneous image of What Turns Men On. . . . Despite the advances of feminism, female power continues to be experienced as dangerous or exotic, and not something that can be directly expressed. (Brooks, 2000, p. 24)

Emphasizing the homogeneity of the images of desire that are allowed to be seen in the public realm, Brooks alludes to the threat posed to established sexual identities by the sight of a sexually desiring and assertive female.

Winship (2000) has made a similar argument with respect to the advertising hoardings for Wonderbra (featuring Eva Herzigova modelling a bra next to the slogan "Hello boys") and French Connection clothes (i.e. the notorious "fcuk" campaign) that appeared in the 1990s. In her view, these posters provoked disquiet (and numerous complaints) because they were essentially ambivalent: on the one hand, the provocative poses adopted by the women in these adverts implied that the streets were a place where women were available to predatory males; on the other, they could be seen as threatening the masculinity of the male viewer by presenting an *image* of an assertive woman who was, in effect, returning his gaze. As in pornography, the posters' subtext was that the (male) observer would need to surrender (his masculine) control to satisfy the desire of the fantasized object/subject. This may be a strategy that ultimately succeeds in selling to (young) women – or it may not – but it can be no surprise that these "raunchy, assertive and intentionally shocking" adverts have been subject to censorship when placed in public view (Winship, 2000, p. 42).

In relation to British public morals legislation, these poster campaigns featured images that were not sexually explicit or obscene in legal terms. Yet they were widely condemned as though they did flout public morality laws, and were subject to sanctions from regulatory bodies (e.g. Yves Saint Laurent is now obliged to clear all its advertising campaigns with the ASA). In the light of the preceding discussion, we can best understand this condemnation by considering the highly gendered relations of exchange and display that inform what should, and should not, be seen on the street.

Such explanations also apply when we consider the recent moral panics concerning another form of "sex advertising" – namely, the placement of prostitutes' calling cards in public telephone boxes. Although data on this subject are conspicuously lacking, it is estimated that thirteen million prostitutes' cards are placed in telephone boxes in central London every year (the vast majority for female prostitution). Colloquially referred to as "tart" or "tom" cards, these are professionally produced, featuring a stylized photograph of a provocatively clad man or woman together with a contact number and a thinly veiled allusion to the type of sexual services on offer (e.g. "Spanking new teacher offers corrective lessons," "Asian babes offer double the pleasure," "New to London – needs to be taken in hand"). Most are suggestive rather than explicit in their content, though an emphasis on abject sexual practices (e.g. sado-masochism, bondage, domination) is often apparent. While these cards feature images that are largely innocuous (and legal by the standards of existing British obscenity laws), their increasing visibility has provoked considerable reaction. Recently, the Director of Westminster Planning Department, Carl Powell, claimed that the cards "project an unsavoury image of the city to visitors and are offensive to women and children." Working with the owners of most payphones, British Telecom, the Westminster City Council sought to remove these cards, collecting over a million in one eight-week operation, at the same time that they pressed for new legislation to deal with carding. Following the publication of a Home Office consultation paper, Sections 46 and 47 of the Criminal Justice and Police Act 2001 were ultimately introduced, making the display of some forms of advertising (specifically relating to the services provided by a prostitute), in some spaces (specifying public telephone boxes), a criminal offense. In the tradition of British obscenity laws, these powers represent a classic example of "moral panic" legislation, being designed to diffuse male anxieties about the visibility of obscene materials in public space, but justified with reference to the sensibilities of women and children (Hubbard, 2002).

Streets of Empowerment and Possibility

Viewed through the lens of different feminist analyses, it is possible to come to two different conclusions as to the desirability of the regulations that restrict the public visibility of sex workers and sex advertising. The first is that the commodification of women's sexuality is a core project of patriarchy, and that it is necessary to resist such commodification, including its removal from the public realm. Radical feminists condemn pornography and prostitution, arguing that they are degrading, dangerous, and immoral. For example, in her powerful and widely cited analysis of women's exploitation, Kathleen Barry (1995) insists on the necessity of making interconnections between sex work and other forms of sexual exploitation such as rape, sexual harassment, arranged marriage, trafficking, sexual violence, and child sex abuse. Viewed in this way, the visibility of women "for sale" on the streets reinforces the idea that women are sexually subordinate to men, available to be actively gazed upon, appraised, and consumed by male viewers.

Against this, there is a second discourse that claims that the sex industry is one of the few forms of labor where women can state their claims to full sexual citizenship. Emanating from many sex workers and feminist activists, this analysis suggests that sex work can be a form of sexual emancipation for women that resists male domination (Nagle, 1997). From this perspective, circumscribing the public visibility of sex workers and censoring assertive expressions of feminine sexuality naturalizes an oppressive heterosexual norm and perpetuates unequal gender relations in the public *and* private spheres. Accordingly, this analysis suggests that if women take to the streets – whether as "streetwalkers" or not – more democratic, empowering gender roles will emerge and the ideology that associates men with the public realm and women with the private would collapse (Squires, 1994).

Developing this argument, Nancy Duncan (1996) argues that sex workers need to use public space as a space of presence, forcing their existence to be recognized and demanding a reconceptualization of feminine and masculine sexuality. This argument draws sustenance from the fact that many sexual minority groups (e.g. gay, lesbian, bisexual, transsexual) try to promote their claims to full sexual citizenship by reterritorializing public space as sites of sexual diversity and respect between strangers (Bell, 1995). In this view, carnivalesque events like Sydney's Mardi Gras and London's Pride in the Park make dissident sexualities visible, and the queering of public space reminds society at large that sexual Others have claims to citizenship. After all, if a group does not exist in public, it is effectively invisible in the eyes of the state and the heterosexual majority, apparently having neither rights nor needs. Accordingly, proponents of sexual rights for minority groups have emphasized the potentiality of reclaiming the street, using public space to fight homophobia, prejudice, and misogyny. In his seminal analysis of gay rights, Castells (1983) argued that it was crucial that gay groups were "out" – that is to say, able to move between sequestered spaces of private intimacy and the public sphere of rights, property ownership, and political representation. In a somewhat similar sense, Jeffrey Weeks (1998) claims that new sexual movements have always had two characteristic elements: a moment of transgression followed by a moment of citizenship. He contends that moments of carnivalesque transgression in the public gaze challenge the status quo, conveying a claim to inclusion which ultimately allows equal access. In Weeks's view, these moments therefore go hand-in-hand; without the transgressive moment there can be no inclusion, no notion of equality. For Weeks (1998, p. 37), all sexual citizens need to "transcend the limits of the personal sphere by going public."

The idea that sexual dissidents can define themselves as sexual citizens by occupying public space on their own terms thus offers a tantalizing vision of a situation where a wide range of individuals are granted rights, recognition and respect irrespective of their sexuality or gender. In effect, they would have no need to hide their sexuality and to confine its expression to sites that are out of the public gaze (and the prying eyes of the state). This "ideal" geography would be one of "radical openness" (Duncan, 1996, p. 143), a geography that discourages the commodification, marginalization, or privitization of sexual Otherness. As sexual citizens, all urban dwellers would thus enjoy the pleasures traditionally only enjoyed by heterosexual men – the freedom to roam and to treat the city as a site of multiple sexual opportunities. Adopting terms originating in the work of Deleuze and Guattari (1987), Duncan suggests that this presents a very different vision of the sexual city; one characterized by a *smooth* geography of desire rather than a *striated* space of repression and containment.

Such visions of a sexually inclusive and democratic city thus drive many women to resist the patriarchal nature of public space by using the streets as spaces from which to make claims for sexual and gender rights through protests, parades, and vigils (Squires, 1994). Likewise, Brooks's (2000) call for more publicly visible and less homogeneous representations of feminine sexuality has been taken up not just by advertising agencies seeking publicity, but also by women artists and activists (e.g. see Mitchell, 1991, on the billboard art of Barbara Kruger).

Yet such efforts to assert women's right to the streets tread a difficult tightrope. In the first instance, this strategy demands that women draw attention to their sexuality, potentially reinforcing the trap of commodification and exploitation. Moreover, women's challenges to male dominance on the streets may simply provoke male attempts – often with violence – to re-establish the boundary between public and private. For such reasons, we should perhaps remain wary of claims that women need to seize the streets to claim their rights as sexual citizens. For example, those prostitute women who organized a May Day parade against Westminster City Council's policy on carding certainly drew attention to the issue, but in doing so sacrificed their anonymity and right to privacy. This emphasizes that the problem is not simply that women lack publicity, but that they need to fight for both publicity and privacy (where publicity is defined as *power to access*, privacy as *power to exclude*). Equating privacy with political inactivity, and publicity with empowerment, many activist women have fallen into a trap where they are left with neither (Squires, 1994).

Beverley Skeggs has likewise written of public visibility constituting a "trap" for sexual dissidents:

> It summons surveillance and the law, it provokes voyeurism, fetishism, the colonist/imperial appetite for possession . . . it reduces the body to the sign of identity. . . . Only some groups can positively and resourcefully spatialise the claim for recognition via visibility . . . and only some groups can legitimate and/or symbolically convert their visible claims. (Skeggs, 1999, p. 228)

Skeggs thus argues that women can only assert their sexuality in public through the adoption of certain sexual demeanors and modes of dress. For many women, this type of visibility is exactly what they would wish to avoid at present, given that it relies on attracting (and returning) the male gaze, and thus provokes unwelcome attention from those who would seek to maintain the boundaries of public/private. Rejecting the reification of the streets as the ultimate site for seeking sexual citizenship, I thus argue that women will only be able to have control over their own bodies, feelings, and relationships when they reject male distinctions of what belongs in public and private, and instead search for new modes of sexual citizenship that exclude those who seek to conquer and suppress, rather than celebrate, feminine sexuality.

Conclusion

This chapter has suggested that while the contemporary regulation of the streetscape banishes from sight performances and representations that are apparently offensive to all "decent" citizens, these regulations actually serve to reinforce grossly unequal gender roles. Imposing greater restrictions on the mobility of women on the streets than for men, the net result of contemporary legislation (and the moral assumptions that underpin it) is that women outdoors are situated in a distinctive relationship with men, so that they become objects for male contemplation (and consumption) whose visibility is fiercely monitored and policed by state fathers. To counter this, it is common for feminist commentators to implore women to seize the streets, making gender politics visible in public space by returning the male gaze. Yet, as I have argued, this strategy is problematic, and may well simply reinforce existing hierarchies of public/private by instigating new moral codes and legislation. Instead, there is a need for a more fundamental disruption of public/private, with women fighting against those who seek to control and repress their sexuality by destroying the distinction between public and private space. Documenting the way that the sexuality of the streets creates profoundly unequal relations of exchange is an important part of this struggle, but perhaps more important is exposing the gendered conceits on which the distinction of public and private is based. One thing is clear, however: feminist geographers will remain at the forefront of this struggle, and will continue to fight gender inequality wherever it is encountered.

BIBLIOGRAPHY

Barry, K. (1995) *The Prostitution of Sexuality: The Global Exploitation of Women.* New York, New York University Press.

Bell, D. (1995) Pleasure and danger: the paradoxical spaces of sexual citizenship. *Political Geography*, 14(2), 139–53.

Benjamin, W. (1999) *Charles Baudelaire: A Lyric Poet in the Era of High Capitalism.* London: Verso.

Brooks, L. (2000) Anatomy of desire. *Guardian* (London), December 12, 17.

Brown, M. (2001) *Closet Spaces.* London: Routledge.

Buck-Morss, S. (1989) *The Dialectics of Seeing: Walter Benjamin and the Arcades Project.* New York: Albany Press.

Califia, P. (1994) *Public Sex Act: The Culture of Radical Sex*. Pittsburgh: Cleis Press.

Castells, M. (1983) *The City and the Grassroots*. Berkeley: University of California Press.

Corbin, A. (1990) *Women for Hire: Prostitution and Sexuality in France after 1850*. Cambridge, MA: Harvard University Press.

Deleuze, G. and Guattari, F. (1987) *A Thousand Plateaus: Capitalism and Schizophrenia*. London: Athlone.

Duncan, N. (1996) Renegotiating gender and sexuality in public and private places. In N. Duncan (ed.), *Bodyspace: Destabilizing Geographies of Gender and Sexuality*. London: Routledge.

Hubbard, P. (1999) *Sex and the City: Geographies of Prostitution in the Urban West*. London: Ashgate.

Hubbard, P. (2002) Maintaining family values? Cleansing the streets of sex advertising. *Area*, 34(4), 353–60.

Hunter, I., Saunders, D. and Williamson, D. (1993) *On Pornography: Literature, Sexuality and Obscenity Law*. London: St Martin's Press.

Jackson, P., Stevenson, N. and Brooks, K. (1999) Making sense of men's lifestyle magazines. *Environment and Planning D – Society and Space*, 17(3), 353–68.

Jervis, J. (1999) *Transgressing the Modern: Explorations in the Western Experience of Otherness*. Oxford: Blackwell.

Kirby, K. (1996) *Indifferent Boundaries*. New York: Guilford Press.

Mitchell, W. J. T. (1991) An interview with Barbara Kruger. *Critical Inquiry*, Winter, 434–48.

Mort, F. (1998) Cityscapes: consumption, masculinities and the mapping of London since 1850. *Urban Studies*, 35(6), 889–907.

Munt, S. (1999) *The Lesbian Flâneur*. London: Cassell.

Nagle, J. (1997) *Whores and Other Feminists*. New York: Routledge.

Namaste, K. (1996) Genderbashing: perceived transgressions of normative sex-gender relations in public spaces. *Environment and Planning D – Society and Space*, 14(2), 221–40.

Nast, H. (2002) Queer patriarchies, queer racisms, international. *Antipode*, 34, 874–904.

Nead, L. (1997) Mapping the self: gender space and modernity in mid-Victorian London. *Environment and Planning A*, 29(4), 659–72.

Ogborn, M. (1998) *Geographies of Modernities*. New York: Guilford Press.

Pollock, G. (1988) *Vision and Difference: Feminity, Feminism and the Histories of Art*. London: Routledge.

Rendell, J. (2002) *The Pursuit of Pleasure: Gender, Space and Architecture in Regency London*. London: Athlone.

Sennett, R. (1993) *The Conscience of the Eye*. London: Faber and Faber.

Skeggs, B. (1999) Matter out of place: visibility and sexualities in leisure spaces. *Leisure Studies*, 18(2), 213–32.

Squires, J. (1994) Private lives, secluded spaces: privacy as political possibility. *Environment and Planning D – Society and Space*, 12(3), 387–401.

Valentine, G. (1998) Food and the production of the civilised street. In N. Fyfe (ed.), *Images of the Street*. London: Routledge.

Walkowitz, J. (1992) *The City of Dreadful Delight: Narratives of Sexual Danger in Late Victorian London*. London: Virago.

Weeks, J. (1998) The sexual citizen. *Theory, Culture, Society*, 15(3), 35–52.

Wilson, E. (2001) *The Contradictions of Culture: Cities, Culture, Women*. London: Sage.

Winship, J. (2000) Women outdoors: advertising, controversy and disputing feminism in the 1990s. *International Journal of Cultural Studies*, 3(1), 27–55.

Wolff, J. (1985) The invisible flâneuse: women and the literature of modernity. *Theory, Culture and Society*, 2(3), 37–47.

Part IV Body

Chapter 23

Situating Bodies

Robyn Longhurst

Bodies are conundrums, paradoxes, riddles that are impossible to solve. They are deeply embedded in psychoanalytic, symbolic, and social processes yet at the same time they are undoubtedly biological, material, and "real." Bodies are an effect of discourse but they are also foundational. They are referential and material, natural and cultural, universal and unique (Nast and Pile, 1998). Everyone has a body (indeed, *is* a body) but bodies are differentiated through age, ethnicity, sex, sexuality, gender, size, health, and so on. Bodies exist *in* places; at the same time they *are* places (McDowell, 1999).

Perhaps it is this paradoxical nature of bodies, and that there is no consensus on what bodies are and what bodies can do, that has led to the recent explosion of work on bodies in the English-speaking academy, especially in the social sciences and humanities. A vast array of terms such as *the* body, bod*ies*, bodily, gendered and sexed bodies, lived bodies, embodiment, body inscription, body schema, corpus, corporeal(ity), and Body without Organs (BwO) (Deluze and Guattari, 1987) have become far more commonplace over the past decade.

Geographers have not been immune from this "body fixation" (Simonsen, 2000, p. 7). Feminist geographers and geographers interested in sexuality have been especially influential in pushing bodies to the forefront of theoretical and empirical agendas. By the mid-1990s, by virtue of seeking new ways of understanding gendered/sex subjects, bodies were firmly on the geographic agenda.

In this chapter I look briefly at how feminists over the past few decades have thought about bodies and map out a number of key areas of interest, including constructionism, and the mind and body dualism. Second, I examine some of the reasons for the growing popularity of research on bodies. Third, I focus on the relationship between bodies, subjectivities, and spatialities and chart changes in feminist geographers' and others' work on bodies over the past decade. Finally, I close by signaling new and prospective directions in feminist geographers' and others' research agendas on bodies. Throughout the chapter I refer to my own longstanding research on pregnant bodies in public places.

Bodies and Feminisms

Bodies have long been a focal point for feminist theorizing and politics but the ways in which they have been conceptualized have varied greatly. Feminists have traversed numerous different intellectual and political routes in their attempt to understand bodies.

> Following de Beauvoir, many feminists tended to see women as prisoners of their bodies and to view the body *per se* as problematic and as something to be transcended. Some, like Shulamith Firestone (1971), looked hopefully to medical science as a means of reducing the reproductive differences between men and women. Others drew on social construction theory to establish a firm distinction between sex as a biological given, and gender which was a social creation through which the differences between the sexes were exaggerated and hierarchised. If sexual inequalities were based on socialisation processes rather than on biology, it was thought that they were more open to change. (McDowell and Sharp, 1999, p. 17)

Some feminists denied the relevance of women's bodily "difference" (men's bodies were constructed as the norm, women's bodies were constructed as "different"); others, such as Adrienne Rich (1976), celebrated it, arguing that women's bodies are powerful sites of reproduction, nurturance, and female sexuality.

Constructionist feminists have often criticized authors such as Rich on the grounds that the position they adopt is "essentialist," meaning that discourses which make reference to the physical, biological body serve to naturalize what is actually social difference. Constructionist feminists argue that bodies are discursively produced. They tend to be concerned with the processes by which bodies are written upon, marked, scarred, transformed, or constructed by various social and political regimes. For constructionist feminists, references to the biological body are seen to reinforce claims that women are naturally incapable of certain kinds of action.

Some theorists, such as Diana Fuss (1989), Moira Gatens (1991), and Vicki Kirby (1992, 1997), argue persuasively that the distinction between essentialism and constructionism (between biology and social inscription) does not hold. In adopting *either* an essentialist *or* a constructionist position, a binary distinction between nature and nurture, between sex and gender, is further entrenched. Kirby (1997) proposes collapsing the sex/gender distinction by examining the ways in which these categories share a complicitous relationship that produces material effects. Judith Butler (1990, p. 7) makes the point that "Sex itself is a gendered category." She argues: "The distinction between sex and gender turns out to be no distinction at all" (Butler, 1990, p. 7).

In my own research on pregnancy I have examined the biology of pregnant bodies. Weight gain, enlarged breasts, the need to urinate more frequently, and the growing fetus are undeniable biological changes during pregnancy but these biological changes are always historically and spatially located. Pregnant bodies are not simply sets of biomedical facts waiting to be uncovered by science (see Morgan and Scott, 1993, pp. 7–10). What it means to be pregnant shifts across time and space. The biology of pregnant bodies is discursively produced (Longhurst, 2001a, pp. 2–10). "Real," material pregnant bodies cannot escape their political, economic, cultural, and social settings. "There is no body outside of its context" (Cream, 1995,

p. 33). Context constructs the way the materiality of pregnant bodies is understood and experienced. Thinking about pregnant bodies in this way challenges the binary division between essentialism and constructionism, between biology and social inscription.

Feminists have also challenged the binary division between body and mind. The notion that the body is separate from the mind has dominated Western thinking since René Descartes published what has perhaps become the best known quotation in all philosophy: *cogito, ergo sum*; I think, therefore I am.

> Challenges to this particular dualism of valuing mind over body, as well as other manifestations of binary thinking such as binaries set up to value males over females, masculinity over femininity, and culture over nature, are now well ensconced in academic literature on the body. (Moss and Dyck, 2003, p. 59)

Kirby (1992) points out that contesting the phallocentricism inherent in binary thinking does not mean simply reasserting that side of the binary that has previously been devalued. She explains: "a binary division, contrary to its apparent meaning, is the double articulation of one term, not two" (Kirby, 1992, p. 13). It is not enough, therefore, to simply focus attention on the body. The division itself, *between* mind and body, must be disrupted. Minds cannot float freely in space and time – they are always embodied. Bodies cannot exist without minds. In an attempt to theorize the mind/body in non-dualistic terms some feminists such as Genevieve Lloyd (1993) have turned to the work of seventeenth-century (pre Descartes) philosopher Benedictus Spinoza, who argues that the mind is an idea of the body rather than separate from it.

Growing Popularity of Bodies

It would be misleading to give the impression that over the past two or three decades it has only been feminists who have paid attention to bodies. Feminists have had a longstanding interest in bodies but over the past ten or fifteen years others, including sociologists, cultural studies scholars, political theorists, medical and health researchers, artists, and art critics, have also paid considerable attention to bodies. There has been an upsurge of interest both inside and outside the academy.

Various commentators have put forward different reasons to explain this interest. Kathy Davis (1997) argues that feminism is largely responsible. Since the 1970s feminists have generated an enormous amount of research on bodies, especially the female body (see Davis, 1997, pp. 4–15). (*Our Bodies, Ourselves* by the Boston Women's Health Book Collective (1973) is an early example of the attention feminists have paid to bodies as sites of political contestation.)

Others argue that the current interest in bodies has been prompted by contradictions in intellectual discourse. Postmodern scholars, often inspired by Foucault, take bodies as sites *par excellence* for exploring the construction of different subjectivities (Davis, 1997, p. 4). While modernists contend that the body represents "the hard 'facts' of empirical reality, the ultimate justification for positivism and the Enlightenment quest for transcendental reason" (Davis, 1997, pp. 3–4), postmodernists contend that the huge variation in bodily appearance and cultural practices indicates they are socially constructed. Frank (1990, 1991) argues that it is this use of bodies for contradictory theoretical agendas that accounts for the current status of the body in the social sciences and humanities.

Still other writers, such as Mike Featherstone (1983) and Bryan Turner (1984), argue that the current popularity of bodies is due to radical changes in the social and cultural landscape of late modernity. Gill Valentine (2001, p. 33) explains that in late modernity affluent Western societies have experienced a "growth in mass consumption, the democratization of culture, a decline in religious morality and a postindustrial emphasis on hedonism and pleasure." Valentine (2001, p. 33) continues that it is in this context that "the body has emerged in consumer culture as an important bearer of symbolic value and as constitutive of our self-identities."

Interest in bodies has been promoted not only by feminism, intellectual movements, and changes in the social and cultural landscape, but also by changes in biomedicine. Davis (1997, p. 2, citing Crawford 1984, p. 80) points to recent medical advances as a reason for an upsurge in interest in bodies. Moss and Dyck (2003) discuss the huge increased presence of, and financial investments in, biotechnology. They note the emergence of new diseases, "scientific breakthroughs" in biomedicine, social implications of the human genome project, and the recent relaxation of human cloning laws in Britain for medical research as contributing to an increase in the attention paid to bodies.

Regardless of the reasons, there is little doubt that over the past ten to fifteen years bodies have come to be seen to be a trendy area of inquiry both inside and outside the academy and geography has not been immune to this body craze. Feminist geographers and geographers interested in sexuality have been especially influential in putting bodies "on the map." In 1986 Rich called for an examination of "the geography closest in – the body" (Rich, 1986, p. 212). Since then a great deal of exciting and ground-breaking work on bodies, subjectivities, and spatialities has been produced.

Bodies, Subjectivities, and Spatialities

It is possible to review contributions on bodies, subjectivities, and spatialities in a number of ways, such as grouping different types of bodies/subjectivities depicted in the literature (e.g. disabled bodies, men's bodies, young people's bodies, "racialized" bodies), exploring bodies as they are constituted in specific environments (e.g. bodies in paid workplaces, bodies at home, bodies in clubs and pubs), or looking at specific activities (e.g. recreating bodies, sporting bodies, working bodies). For the purposes of this chapter, however, I have chosen to chart the course of bodies in feminist geography chronologically because it illustrates some interesting changes that have occurred in understandings of "the subject" and subjectivity in the 1980s and 1990s.

In the late 1980s feminist geographers and others began to examine the potential of focusing on relationships between bodies and spaces.[1] This interest was fueled, in part, by a desire to critique the transparent, rational, masterful, and masculinist subject of much humanism and structuralism. Feminist political theorist Iris Marion Young, in her wellknown essay "Throwing like a girl," turned to existential phenomenology to trace "some of the basic modalities of feminine body comportment, manner of moving, in relation to space" (Young, 1989, p. 53). Young approached the body (focusing on bodily movement) as part of a prediscursive realm. She relied heavily on Maurice Merleau-Ponty's (1962) *Phenomenology of Perception*, in which he examines the relationship between consciousness and the world. A number of humanist geographers, such as David Seamon (1979, 1980) and Anne Buttimer (1976), also adopted this prediscursive approach to the body in an attempt to understand people's lifeworlds, but they did not pay explicit attention to gender.[2]

Feminist geographer Louise Johnson also saw some potential in phenomenology (and in psychoanalysis and the historical archeology of Foucault) for offering "a more elaborate framework for investigating the sexed body in space which challenges existing conceptions of space and time" (Johnson, 1989, p. 137). Feminist interpretations of various approaches to the body – phenomenological, psychoanalytic, and Foucauldian – argued Johnson, could potentially refigure the gendered/sexed subject and reconceptualize the discipline of geography.

Johnson was correct. In the early 1990s feminist geographers and others interested in spatiality did start to draw on these theories, although over the decade phenomenology was to prove less popular than psychoanalysis and especially poststructuralism. Influenced by the work of theorists such as Michel Foucault, Friedrich Nietzsche, Franz Kafka, and Gilles Deleuze, feminist geographers and others began to reconfigure the gendered/sexed subject. Bodies were increasingly understood as surfaces on which values, morality, social laws, and even cities are etched. For example, in 1992 feminist philosopher Elizabeth Grosz examined some of the ways in which bodies are inscribed by urban life. Grosz (1992) argued that there is a mutually constitutive relationship between bodies and cities – a two-way linkage or *interface* between bodies and cities.

I recall the early 1990s as an exciting time for feminist geographers interested in bodies. In 1992 I began doctoral research on pregnant women's experiences of public places in Hamilton, New Zealand (Longhurst, 1996). I explored the "lived" geographies of 31 women who were pregnant for the first time. Many of these women tended to withdraw from public places such as night clubs, bars, pubs, restaurants, and cafes, and from public activities such as sport and paid employment, during pregnancy. Two possible reasons for these pregnant women's withdrawal from public places during pregnancy were put forward: first, that pregnant women are often represented as being "seeping," "ugly," abject bodies who are not to be trusted in the public realms; second, that pregnant women are often represented as being emotional, irrational, and forgetful (read "hysterical"), and, therefore, not to be trusted in public space.

I wanted to create a "sexually embodied geography" that contested hegemonic, disembodied, masculinist geographies. I was inspired by the work of feminist geographers such as Gillian Rose who in the early 1990s was examining the epistemology and politics of knowledge production. In Rose's influential book *Feminism and Geography* (1993) she argues convincingly that there is something in the very claim to knowing in geography that tends to exclude women as producers of knowledge. There is a specific notion of knowing, and of knowledge, as masculine, exhaustive, rational, and associated with the mind rather than the body (as though the two can be separated), which marginalizes women in the production of geographical knowledge (also see Longhurst, 1995).

It was not only feminists at this time who were putting bodies on the geographical agenda. So too were geographers interested in sexuality.[3] In 1992 David Bell organized a "Sexuality and Space Network." The Network's principal purpose was to provide support, contact, and advice for researchers in what was a marginalized part of geography. They organized workshops, paper sessions at conferences, and seminars, including a one-day seminar in London entitled "Lesbian and Gay Geographies?" Also in 1992 Beatriz Colomina edited a collection of essays published in a volume entitled *Sexuality and Space*.

In the late 1980s and early 1990s it is likely that much of the research that was carried out on gendered and sexed bodies was not published or was published in conference proceedings rather than in "mainstream" publications (read: refereed material that reaches academic readers via a body of accepted journals and "respected" publishing houses). It was not until the mid-1990s that feminist geographic work on bodies really began to enter the "mainstream." Rose noted in 1995 that there is "a growing concern with the 'bodily' in geography" (Rose, 1995, p. 545).

The flurry of work in the mid-1990s was aided in part by the publication of a new journal, *Gender, Place and Culture*, in 1994, edited by Liz Bondi and Mona Domosh. Feminist geographers finally had a specific forum for debating issues of gender, sex, sexuality, and the body. In the first issue David Bell, Jon Binnie, Julia Cream, and Gill Valentine drew on cultural theorist Judith Butler's (1990) notion of performativity in order to think about the performance of sexual identities (such as the hypermasculine "gay skinhead" and the hyperfeminine "lipstick lesbian") in space.

The concept of performativity has been very influential among feminist geographers because it moves away from essentialist and static understandings of identity, instead theorizing identity as constantly re-enacted through bodily performance. In Butler's work the body does not exist as a "prediscursive" site; instead, the body is constituted through the compelled enactment and repetition of hegemonic discourses (of class, race, gender, and sexuality among others) (Butler, 1990). Drawing on Butler's notion of performativity, Linda McDowell and Gill Court (1994) interrogated the gendered bodily performance of merchant bankers in the City of London (also see McDowell, 1995). Similarly, many of the contributors to the collection *BodySpace* (Duncan, 1996) took the concept of performativity as central to their explorations of bodies and spaces (see Nelson, 1999).

In the later half of the 1990s feminist geographers and others began engaging increasingly with psychoanalytic approaches to bodies in an attempt to understand more about the acquisition of gendered/sexed identities and relationships with others. Rose (1993, p. 103) argued that "feminist psychoanalytic commentaries offer an eloquent critique of geography's white, heterosexual, masculine gaze, a gaze torn between pleasure and its repression." David Sibley (1995) drew on Julia Kristeva's notion of "abjection" (the feeling of disgust that the subject has in encountering certain images or matter – the horrible – to which it can only respond with aversion and nausea) to examine a tendency in Western culture for powerful groups to attempt to "purify" and dominate space – to exclude those who are considered abject and Other. Steve Pile (1996), inspired by the work of Henri Lefebvre, Sigmund Freud, and Jacques Lacan, used psychoanalytic concepts and methods, such as the unconscious and dream analysis, to sketch new cartographies of people and places. More recently, Nast (2000, p. 215, italics in original) developed: "ways for thinking through how the psyche can be understood as a structured and libidinized *spatial effect*, a repository of colonial violences of body and place, unspoken and hence repressed ('unconscious')." Nast put forward the term "racist-oedipalization" to signify the ways in which racism is fundamental to white oedipal family structures and norms (also see Bondi, 1999; Robinson, 2000).

By the late 1990s more and more research on bodies was being carried out not only by feminist geographers and geographers interested in sexuality but also by cultural geographers, health geographers, geographers interested in disability, postcolonial geographers, and urban geographers. For example, in 1998 Heidi Nast and Steve Pile published a collection of 21 essays entitled *Places through the Body*. A quick flick through this volume reveals geographers' broadening interest in bodies and a growing recognition that bodies are "entangled" (see Sharp et al., 2000) in multiple power relations (not just in gendered/sexed power relations) that are simultaneously real, imagined, and symbolic.

In the late 1990s research was being published on ill, impaired, and disabled bodies (Moss and Dyck, 1996; Butler and Parr, 1999), pregnant, birthing, and lactating bodies (Longhurst, 1998, 2000; Mahon-Daly and Andrews, 2002), racialized bodies (Skelton, 1995; Waller, 1998; Nast, 2000), working bodies (McDowell and Court, 1994; Massey, 1996; Pratt with the Philippine Women Centre, 1998), "classed" bodies (Dowling, 1999), tourist bodies (Veijola and Jokenin, 1994; Johnston, 2001), mobile bodies (Cresswell, 1999), virtual bodies (Fuchs, 1995; Wakefield, 1998; Parr, 2002), the bodies of geographic information system users (Lilley, 2002), and even dead bodies (Pointon, 1999). The research I have cited represents only a small sample of that which has been published but it is enough to illustrate that there has been growing recognition that for many geographers bodies are now "at the heart" of geography and of understanding people–place relationships.

Where to from Here?

In this concluding section I turn attention to the question: what exciting new work on bodies might lie around the corner for feminist geographers and others interested in spatiality? This question is difficult to answer. To begin, though, I suspect that feminist geographers and others will continue paying attention to embodied subjectivities over the next few years. For example, interestingly, in this volume (and not just in this section) there are a number of contributions on bodies. Pamela Moss examines "bodily notions of research." Matthew Hannah investigates "virility and the violation of bodies in the US 'war on terrorism'." I think over the next few years bodies are likely to remain a key concern for feminist geographers and others interested in theorizing lived experience, subjectivities, and power relations.

More specifically, I think over the next few years feminist geographers and others will increasingly attempt to examine and convey the complexity that exists in the interlocking of race, sexuality, gender, class, and other facets of embodied subjectivity. "Bodies are always irreducibly sexually specific, necessarily interlocked with racial, cultural, and class particularities" (Grosz, 1994, p. 19). Gender and sex are not the only variables that position bodies in webs of power.

In the late 1990s feminist geographers began to increasingly decenter gender as the primary category of analysis and I think this trend will continue (see Longhurst, 2002). While it is not always politically strategic to examine simultaneously multiple axes of embodied subjectivity, it can prove useful. For example, in the chapters that follow Kawango Agot examines HIV/AIDS but through the lens of African women's embodied experiences. Robina Mohammad focuses on women's bodies and the performance of community identity through the lens of race, culture, religion, and nation. These authors recognize that embodied subjects are continually repositioned in, and by, a complex array of discursive and material practices. Examining one variable – for example, gender or sex or disability or illness – may not convey adequately the complexity of body–space relationships under scrutiny. It is important to determine which aspects of subjectivity and space matter when, and how particular combinations can be examined to create more emancipatory social relations.

By the late 1990s geographers had also come to realize that not only are different dimensions of embodied experience inextricably intertwined but also bodies and places are inextricably intertwined. This trend looks set to continue. Bondi and Davidson (2003, p. 338, emphasis in original) explain: "To *be* is to be some*where*, and our changing relations and interactions with this placing are integral to understandings of human geographies." Similarly, Michael Brown and Larry Knopp (2003, p. 322, italics in original), in discussing queer geographies, argue: "From the closet to the body, to the city, to the nation and to the globe, new queer cultural geographies show us that a variety of subjectivities are performed, resisted, disciplined, and oppressed not simply in but *through* space." Some earlier work tended to map bodies on to places as though they were fixed or simply a backdrop to people's everyday lives. For example, in Longhurst (1994) I signal the possibility of examining the ways in which Centre Place shopping mall in Hamilton, New Zealand, inscribes the bodies of pregnant women, but I do not actually address this in my research. In contrast, today embodied subjectivities are more likely to be seen as possessing a "spatial imperative" (Probyn, 2003).

Consider, for example, Ruth Bankey's (2001) and Joyce Davidson's (2000, 2001) work on agoraphobia, in which they seamlessly blend bodies and spaces, blurring any boundaries between the two. During panic attacks agoraphobics often feel as though their bodies collude with place to become one. Susan Bordo (in Bordo et al., 1998, p. 80) also blurs the boundaries between bodies and spaces. Relaying her own experience of agoraphobia she explains that during a panic attack she feels like "a drowning person," her only thought to find air.

Another trend evident in feminist geographers' research on bodies, and one that is likely to continue in the future, is a desire to move beyond "an abstracted subject" embedded in many textual approaches to the body. Lise Nelson (1999, p. 332) examines the limitations of Butler's concept of performativity, particularly Butler's treatment of the body and the subject as the site of "compelled" discursive repetition. By privileging the moment of interpolation into discursive structures, Nelson contends that Butler's performative subject is "abstracted from personal, lived history as well as from its historical and geographical embeddedness." She argues that Butler problematically "jettisons agency altogether" (Nelson, 1999, p. 332) and urges geographers to critically deploy the concept of performativity.

Others, such as Michael Dorn (1998), have also argued in favor of including lived experience, our own and our research subjects', in our geographies. The linguistic turn dominant in poststructuralist theorizing heightened awareness that bodies are constituted by language – they are texts. However, they are texts that bleed, eat, defecate, make love, and so on. They are more than linguistic terrains. Bodies have a weighty and often "messy" materiality, viscosity, and fluidity but this is rarely acknowledged by geographers.

> The reason this [omission] is significant is that the "messiness" of bodies is often conceptualised as "feminised" and "Othered." Bracketing out questions about the borders of body/space relationships functions as an attempt to position geographical knowledge as that which can be separated out from corporeality, both of its subjects and producers. Ignoring the messy body is not a harmless omission, rather, it contains a political imperative that helps keep masculinism intact. (Longhurst, 2001b, p. 23)

In my own research on pregnant bodies in public places, managers' bodies in central business districts, and men's bodies in domestic bathrooms (Longhurst, 2001a), I have attempted to invoke the messiness of bodies. What bodies have in common is a fluidity, volatility, and abject materiality (although some bodies are commonly represented as more abject than others). This corporeality cannot be plucked from the spaces it constitutes and is constituted by.

Conclusion

Linda McDowell (1999, p. 68) explains:

> The placing of the body right at the center of social theory has perhaps been one of the most exciting moves in contemporary theoretical endeavours . . . questions of the sexed body – its differential construction, regulation and representation – are absolutely central to an understanding of gender relations at every spatial scale.

I concur fully with McDowell. Given the centrality of bodies to understanding gender and space relations, I think it is likely they will remain at the top of many feminist geographers' and others' agendas for some time to come. Feminist geographers have long recognized the importance of power and that "the ways in which we live out body/place relationships are political" (Nast and Pile, 1998, p. 2). Making bodies explicit in our theorizing and empirical investigations has provided new ways of understanding subjectivities, power, and politics. It has allowed us to develop a deeper understanding of the ways in which individuals, communities, and nations are simultaneously engaged in various mutually constituted power relations in different places and at different times.

In this chapter I have mapped the contours of some of the debates surrounding feminists' ideas about bodies over the past few decades and identified some of the major contributions made by feminist geographers and others to these debates. I have also attempted to signal new and prospective directions in feminist and critical research on bodies. My hope is that by "situating bodies" in this way I have provided at least a partial foundation for the chapters that follow, chapters that I am sure will provoke in readers further questions and debate about bodies, subjectivities, and spatialities.

NOTES

1   Clearly it is not just feminist geographers who are interested in issues of bodies, subjectivity, spatiality, and power. A great deal of invaluable work has been carried out, and continues to be carried out, by feminists and others in a range of disciplines, including architectural theory, politics, literary criticism, art history, and philosophy.

2   It was not just sexed and gendered bodies that were a focus of attention. Several years before Young published her essay "Throwing like a girl" Minnie Bruce Pratt explored issues of racism and anti-semitism focusing on her embodied experiences of being a white lesbian "living in a part of Washington DC that white suburbanites called 'the jungle'" (Pratt, 1984, p. 11). In short, during the mid/late 1980s authors began to articulate what many people (especially those who have experienced marginalization) had known for a long time – experiences of places are deeply embodied.

3   It is not always useful to conflate feminism and sexuality (see Bell and Valentine, 1995, pp. 11–12) but a number of geographers interested in sexuality do draw on feminist theorizing (see Binnie and Valentine, 1999, for a review of "geographies of sexualities" and Domosh, 1999, on "sexing feminist geography").

BIBLIOGRAPHY

Bankey, R. (2001) *La donna é mobile*: constructing the irrational woman. *Gender, Place and Culture*, 8(1), 37–54.

Bell, D., Binnie, J., Cream, J., and Valentine, G. (1994) All hyped up and no place to go. *Gender, Place and Culture*, 1(1), 31–47.

Bell, D. and Valentine, G. (eds) (1995) *Mapping Desire: Geographies of Sexualities*. London: Routledge.

Binnie, J. and Valentine, G. (1999) Geographies of sexualities – a review of progress. *Progress in Human Geography*, 23(2), 175–87.

Bondi, L. (1999) Stages on journeys: some remarks about human geography and psychotherapeutic practice. *The Professional Geographer*, 51(1), 11–24.

Bondi, L. and Davidson, J. (2003) Troubling the place of gender. In K. Anderson, M. Domosh, S. Pile and N. Thrift (eds), *Handbook of Cultural Geography*. London: Sage.

Bordo, S., Klein, B. and Silverman, M. K. (1998) Missing kitchens. In H. Nast and S. Pile (eds), *Places through the Body*. London: Routledge.

Boston Women's Health Book Collective (1973) *Our Bodies Our Selves*. New York: Simon and Schuster.

Brown, M. and Knopp, L. (2003) Queer cultural geographies – We're here! We're queer! We're over there, too! In K. Anderson, M. Domosh, S. Pile and N. Thrift (eds), *Handbook of Cultural Geography*. London: Sage.

Butler, J. (1990) *Gender Trouble: Feminism and the Subversion of Identity*. New York: Routledge.

Butler, R. and Parr, H. (eds) (1999) *Mind and Body Spaces. Geographies of Illness, Impairment and Disability*. London: Routledge.

Buttimer, A. (1976) Grasping the dynamism of the lifeworld. *Annals of the Association of American Geographers*, 66, 277–92.

Colomina, B. (1992) *Sexuality and Space*. New York: Princeton Architectural Press.

Crawford, R. (1984) A cultural account of "health": control, release and the social body. In J. McKinlay (ed.), *Issues in the Political Economy of Health Care*. New York: Tavistock.

Cream, J. (1995) Resolving riddles: the sexed body. In D. Bell and G. Valentine (eds), *Geographies of Sexualities: Landscapes of Desire*. London: Routledge.

Cresswell, T. (1999) Embodiment, power and the politics of mobility: the case of female tramps and hobos. *Transactions of the Institute of British Geographers*, n.s., 24, 175–92.

Davidson, J. (2000) ". . . the world was getting smaller": women, agoraphobia and bodily boundaries. *Area*, 32(1), 31–40.

Davidson, J. (2001) Fear and trembling in the mall: women, agoraphobia and body boundaries. In I. Dyck, N. Lewis and S. McLafferty (eds), *Geographies of Women's Health*. London: Routledge.

Davis, K. (ed.) (1997) *Embodied Practices: Feminist Perspectives on the Body*. London: Sage.

Deleuze, G. and Guattari, F. (1987) *A Thousand Plateaus. Capitalism and Schizophrenia, Volume 2*, trans. B. Massumi. Minneapolis: University of Minnesota Press.

Domosh, M. (1999) Sexing feminist geography. *Progress in Human Geography*, 23(3), 429–36.

Dorn, M. L. (1998) Beyond nomadism: the travel narratives of a "cripple." In H. Nast and S. Pile (eds), *Places through the Body*. London: Routledge.

Dowling, R. (1999) Classing the body. *Environment and Planning D: Society and Space*, 17(5), 511–14.

Duncan, N. (ed.) (1996) *BodySpace. Destabilizing Geographies of Gender and Sexuality*. London: Routledge.

Featherstone, M. (1983) The body in consumer culture. *Theory, Culture and Society*, 1(2), 18–33.

Frank, A. W. (1990) Bringing bodies back in: a decade review. *Theory, Culture and Society*, 7, 131–62.

Frank, A. W. (1991) For a sociology of the body: an analytical review. In M. Featherstone, M. Hepworth and B. S. Turner (eds), *The Body. Social Processes and Cultural Theory*. London: Sage.

Fuchs, C. (1995) Death is irrelevant: cyborgs, reproduction, and the future of male hysteria. In C. Gray (ed.), *The Cyborg Handbook*. London: Routledge.

Fuss, D. (ed.) (1989) *Essentially Speaking: Feminism, Nature and Difference*. London: Routledge.

Gatens, M. (1991) A critique of the sex/gender distinction. In S. Gunew (ed.), *A Reader in Feminist Knowledges*. New York: Routledge.

Grosz, E. (1992) Bodies-cities. In B. Colomina (ed.), *Sexuality and Space*. New York: Princeton Architectural Press.

Grosz, E. (1994) *Volatile Bodies: Toward a Corporeal Feminism*. St Leonards: Allen and Unwin.

Johnson, L. C. (1989) Embodying geography – some implications of considering the sexed body in space. In *New Zealand Geographical Society Proceedings of the 15th New Zealand Geography Conference*, Dunedin, August.

Johnston, L. (2001) (Other) bodies and tourism studies. *Annals of Tourism Research: A Social Science Journal*, 28(1), 180–201.

Kirby, V. (1991) *Corpus delicti*: the body at the scene of writing. In R. Diprose and R. Ferrell (eds), *Cartographies. Poststructuralism and the Mapping of Bodies and Spaces*. Sydney: Allen and Unwin.

Kirby, V. (1992) *Addressing Essentialism Differently . . . Some Thoughts on the Corporeal.* Occasional Paper Series, No. 4, University of Waikato, Department of Women's Studies.

Kirby, V. (1997) *Telling Flesh: The Substance of the Corporeal.* London: Routledge.

Lilley, S. (2002) Talking with the magician's apprentice: fleshing out GIS users. In L. Bondi, H. Avis, R. Bankey, A. Bingley, J. Davidson, R. Duffy, V. I. Einagel, A.-M. Green, L. Johnston, S. Lilley, C. Listerborn, M. Marshy, S. McEwan, N. O'Connor, G. Rose, B. Vivat and N. Wood, *Subjectivities, Knowledges, and Feminist Geographies: The Subjects and Ethics of Social Research.* Lanham, MD: Rowman and Littlefield.

Lloyd, G. (1993) *The Man of Reason: "Male" and "Female" in Western Philosophy.* London: Routledge.

Longhurst, R. (1994) The geography closest in – the body . . . the politics of pregnability. *Australian Geographical Studies,* 32(2), 214–23.

Longhurst, R. (1995) The body and geography. *Gender, Place and Culture,* 2(1), 97–105.

Longhurst, R. (1996) Geographies that matter: pregnant bodies in public places. DPhil thesis, Department of Geography, University of Waikato.

Longhurst, R. (1998) (Re)presenting shopping centres and bodies: questions of pregnancy. In R. Ainley (ed.), *New Frontiers of Space, Bodies and Gender.* London: Routledge.

Longhurst, R. (2000) "Corporeographies" of pregnancy: "bikini babes." *Environment and Planning D: Society and Space,* 18, 453–72.

Longhurst, R. (2001a) *Bodies. Exploring Fluid Boundaries.* London: Routledge.

Longhurst, R. (2001b) Trim, taut, terrific, and pregnant. In D. Bell, J. Binnie, R. Holliday, R. Longhurst and R. Peace, *Pleasure Zones. Bodies, Cities, Spaces.* Syracuse, NY: Syracuse University Press.

Longhurst, R. (2002) Geography and gender: a critical time? *Progress in Human Geography,* 26(4), 544–52.

McDowell, L. (1995) Body work: heterosexual gender performances in city workplaces. In D. Bell and G. Valentine (eds), *Mapping Desire: Geographies of Sexualities.* London: Routledge.

McDowell, L. (1999) *Gender, Identity and Place: Understanding Feminist Geographies.* Oxford: Polity Press.

McDowell, L. and Court, G. (1994) Performing work: bodily representations in merchant banks. *Environment and Planning D: Society and Space,* 12, 727–50.

McDowell, L. and Sharp, J. P. (eds) (1999) *A Feminist Glossary of Human Geography.* London: Arnold.

Mahon-Daly, P. and Andrews, G. J. (2002) Liminality and breastfeeding: women negotiating space and two bodies. *Health and Place,* 8, 61–76.

Massey, D. (1996) Masculinity, dualisms and high technology. In N. Duncan (ed.), *Body Space.* London: Routledge.

Merleau-Ponty, M. (1992) *Phenomenology of Perception,* trans. C. Smith. London: Routledge & Kegan Paul.

Morgan, D. H. J. and Scott, S. (1993) Bodies in a social landscape. In S. Scott and D. H. J. Morgan (eds), *Body Matters.* London: Falmer Press.

Moss, P. and Dyck, I. (1996) Inquiry into environment and body: women, work, and chronic illness. *Environment and Planning D: Society and Space,* 14, 737–53.

Moss, P. and Dyck, I. (2003) Embodying social geography. In K. Anderson, M. Domosh, S. Pile and N. Thrift (eds), *Handbook of Cultural Geography.* London: Sage.

Nast, H. (2000) Mapping the "unconscious": racism and the Oedipal family. *Annals of the Association of American Geographers,* 90(2), 215–55.

Nast, H. and Pile, S. (eds) (1998) *Places through the Body.* London and New York: Routledge.

Nelson, L. (1999) Bodies (and spaces) do matter: the limits of performativity. *Gender, Place and Culture*, 6(4), 331–53.

Parr, H. (2002) New body-geographies: the embodied spaces of health and medical information on the Internet. *Environment and Planning D: Society and Space*, 20, 73–95.

Pile, S. (1996) *The Body and the City. Psychoanalysis, Space and Subjectivity*. London: Routledge.

Pointon, M. (1999) Funerary and sexual topographies: the death and commemoration of Diana, Princess of Wales. *New Formations, Sexual Geographies, A Journal of Culture/Theory/Politics*, 37, 114–29.

Pratt, G. with the Philippine Women Centre (1998) Inscribing domestic work on Filipina bodies. In H. Nast and S. Pile (eds), *Places through the Body*. London: Routledge.

Pratt, M. B. (1984) Identity: skin blood heart. In E. Bulkin, M. B. Pratt and B. Smith (eds), *Yours in Struggle: Three Feminist Perspectives on Anti-semitism and Racism*. New York: Long Haul Press.

Probyn, E. (2003) The spatial imperative of subjectivity. In K. Anderson, M. Domosh, S. Pile and N. Thrift (eds), *Handbook of Cultural Geography*. London: Sage.

Rich, A. (1976) *Of Woman Born: Motherhood as Experience and Institution*. New York: W. W. Norton.

Rich, A. (1986) Notes towards a politics of location. In A. Rich, *Blood, Bread and Poetry: Selected Prose 1979–1985*. New York: W. W. Norton.

Robinson, J. (2000) Feminism and the spaces of transformation. *Transactions of the Institute of British Geographers*, 25(3), 285–301.

Rose, G. (1993) *Feminism and Geography. The Limits of Geographical Knowledge*. Minneapolis: University of Minnesota Press.

Rose, G. (1995) Geography and gender, cartographies and corporealities. *Progress in Human Geography*, 19(4), 544–8.

Seamon, D. (1979) *The Geography of the Lifeworld*. New York: St Martins Press.

Seamon, D. (1980) Body subject, time space routines and place ballets. In A. Buttimer and D. Seamon (eds), *The Human Experience of Space and Place*. London: Croom Helm.

Sharp, J. P., Routledge, P., Philo, C. and Paddison, R. (eds) (2000) *Entanglements of Power. Geographies of Domination/Resistance*. London: Routledge.

Sibley, D. (1995) *Geographies of Exclusion. Society and Difference in the West*. London: Routledge.

Simonsen, K. (2000) Editorial: the body as battlefield. *Transactions of the Institute of British Geographers*, 25(1), 7–9.

Skelton, T. (1995) "Boom, bye, bye": Jamaican ragga and gay resistance. In D. Bell and G. Valentine (eds), *Mapping Desire: Geographies of Sexualities*. London: Routledge.

Turner, B. (1984) *The Body and Society*. Oxford: Blackwell.

Valentine, G. (2001) *Social Geographies: Space and Society*. Englewood Cliffs, NJ: Prentice Hall.

Veijola, S. and Jokenin, E. (1994) The body in tourism. *Theory, Culture and Society*, 11, 125–51.

Wakefield, N. (1998) Urban culture for virtual bodies: comments on lesbian "identity" and "community" in San Francisco Bay Area cyberspace. In R. Ainley (ed.), *New Frontiers of Space, Bodies and Gender*. London: Routledge.

Waller, G. A. (1998) Embodying the urban Maori warrior. In H. Nast and S. Pile (eds), *Places through the Body*. London: Routledge.

Young, I. M. (1989) Throwing like a girl: a phenomenology of feminine body comportment, motility, and spatiality. In J. Allen and I. M. Young (eds), *The Thinking Muse: Feminism and Modern French Philosophy*. Bloomington: Indiana University Press.

Chapter 24

Bodies, State Discipline, and the Performance of Gender in a South African Women's Prison

Teresa Dirsuweit

Introduction

One of the key functions of the space of the prison is the normalization of aberrant criminal behaviors. Foucault (1991) argues that this occurs through a number of mechanisms which act on the body. He terms this the correct training of the self. In this formulation the body is acted upon; it is the site of the production of disciplinary power emanating from institutions in the form of inscribing norms. He goes on to argue that the self is produced in the conformance to these norms. The body is key to the operation of the capillary power, which Foucault (1991, pp. 140–1) terms bio-power: an "explosion of numerous and diverse techniques for achieving the *subjugation of bodies* [emphasis added] . . . the rudiments of anatomo- and bio-politics, created in the eighteenth century as *techniques* [original emphasis] of power."

Drawing on research in a South African women's prison, in this chapter I argue that the normalization of the prisoner through the correct training of the body represents a site of both subjection *and* resistance. Women criminals, particularly those who are involved in violent crime, are transgressive of both the law and gendered social norms. The correct training of women prisoners includes their "feminization" in accordance with accepted gendered norms. The first part of this chapter is devoted to an examination of how the institution defines and normalizes gendered identity through work and recreation. The second part explores how the normalization of gender identity is undermined by the prisoners.

Several geographers have interrogated Foucault's formulation of state discipline within the institution. Notably, Hannah (1997, p. 173) allows for a more fluid understanding of spatialized disciplinary power by acknowledging ruptures within disciplinary power.

Bribery, smuggling, differential treatment, abuse, sexual trafficking and deal-making have plagued practically every attempt to rationalize and regularize the functioning of institutions of control since they arose in their modern forms. In the invisible spaces [the interstices of invisibility and anonymity], and through the innumerable complicities of modern institutions, individual and collective resistance to control has become and remained a constant feature of confined life.

In a similar vein, I argue that there is a great deal of resistance to the correct training of the body which takes the form of transgressing the gendered norms associated with work and recreational activities in the prison. This resistance is associated with a counter-economy comprised of smuggling and trading food, drugs, and other goods. This counter-economy in turn shapes the adoption of more masculine gendered identities on the part of prisoners marked as "female" by the state. Furthermore, while the prison is in a state of constant transformation, I argue that a stabilized moral economy of gender and sexuality has developed between prisoners. Drawing on understandings of the body as socially constructed (see Butler, 1990, 1993; Grosz, 1994, 1995) and on interviews completed in a South African women's prison, my analysis demonstrates that gender and sexuality in the prisons develop at the intersections between state disciplinary practices, prison economies, and prisoner resistance strategies.[1]

Work, Recreation, and the Inscription of Feminized Normality

Prisoners are not required to work, but boredom and economic need mean that a large proportion of prisoners spend mornings in workrooms. There are several workspaces in the prison: the sewing room; the laundry; the kitchen; the hairdressing salon; and the dollmaking room. Women are to be "reformed" not only in terms of their crimes, but also in terms of their status as women. In the men's prison, prisoners are often taught skills such as locksmithing, carpentry, and metalworking. The only artisan work that is offered to women is hairdressing and dollmaking, and the workspaces associated with these activities are showcased to visitors as rehabilitative spaces. The dollmaking studio and the hairdressing salon are situated among the offices of staff members at the main gate (hoofhek). The proximity of these workspaces to hoofhek implies privilege: women who work in these activities are disciplined enough to be allowed to work there unsupervised.

The kitchen, laundry, and sewing workspaces are panoptic devices where direct surveillance by warders (referred to as "members" by both prisoners and correctional services staff) from a raised position ensures that prisoners perform their tasks in accordance with prison regulations. The machinery is arranged to face the raised dais or clear glassed office of the supervising member. The workspace connects the rhythm of the timetable to the rhythms of visibility and surveillance. It is here that the prison as a factory is exemplified. A work ethic is considered to be one of the primary ways in which the rehabilitation takes place and so the body is acted upon through the repetitive tasks of institutional food preparation, the sewing of prison pyjamas and endless washing, ironing, and folding. In several visits to these spaces, I observed that work was constant, with little by way of talking or singing to break the activity. Here Foucault's (1991) docile body is clearly in evidence with a twist; the rehabilitating body is the working body performing suitably feminized activities.

Whereas prisoners may reject the normalizing function of work, recreational activities are far more popular. These too reflect gendered norms of work entrenched in the state discipline of women prisoners and generally take the form of crafts such as macramé. Similarly, sporting activities include those that are deemed suitable for women, such as netball. However, perhaps the most popular recreational activity in the prison is the yearly beauty pageant. This event is an interesting example of collaboration between the state and broader society for the disciplining of gendered prison identities. The beauty pageants held at the different women's prisons are regularly reported in national television news. I managed to attend one of these events while I was doing my research. Beauty pageants in general are performances that function to uphold hegemonic notions of femininity. They also serve the normalizing function of the prison. It seems that women prisoners are to aspire to notions of feminine beauty as opposed to criminality. These pageants are not taken lightly; a number of prominent television and beauty industry personalities are invited to judge the event. The preparation beforehand takes weeks and the event is interspersed with entertainment acts from inmates of the women's and the men's prison. These range from solos to choral arrangements.

The presence and reporting of the event in the media is at first glance puzzling. Why would the media be interested in an event which, for all its celebrity judges, is not far from some high school beauty contest or at a stretch a small town's beauty festival? What about the event is so newsworthy that it would warrant coverage on national television news? The only parallel events are transvestite beauty pageants, which are also regularly reported in the media. These pageants have an ambiguous status, as they are both a celebration of queer life and a freak show. Similarly, coverage of the prisoner beauty pageant event plays into discourses of female criminality as being profoundly aberrant. Consider the following quotes from a magazine article:

> Suzanne's ethereal beauty belies the viciousness of her crimes, which include robbery, kidnapping and what the judge described as the "cruel and barbaric" murder of a wealthy 50-year-old Johannesburg business man ... whose cheque book and bank cards she also stole ... his bloodied trousers were found 13 metres from his half-naked corpse. He had been stabbed 32 times and his face had been crushed to pulp. (*Marie Claire*, April 2000, p. 58)

The prison houses a range of women and the crimes are equally varied. Because there are fewer women's prisons, the prison holds those who have shoplifted, convicted sex workers, hijackers, and those who have murdered (in many cases these are women who have suffered years of domestic violence and killed their husbands). The journalist, however, emphasizes the worst crimes because the abnormality of women who commit violent crime titillates. The journalist further plays into hegemonic discourses of female criminality in the following quote, where prisoners are associated with uncontrolled animals. "In the beginning the bitchiness was intense, as contestants stared at one another *like dogs on a street corner*, and tensions erupted into bitter fights (ibid., p. 59, emphasis added). The media coverage is fed back to prisoners; the beauty queens are powerful exemplars of hegemonic norms of femininity. The journalist reflects the institutional and broader social intention of the pageant: "The rest followed proudly, for this is their chance to hold their heads up high. . . . But for almost five hours, 730 women were transported to a world which seemed to be a million miles away from the bars and the cells. They felt so close to freedom it hurt" (ibid.). While beauty pageants may lend prisoners a sense of self-esteem and escape, they also reinforce institutional efforts to reinscribe hegemonic norms of femininity on the prisoner body. In this instance prison walls are permeable to external social influence.

Both Butler (1990, 1993) and Grosz (1994, 1995) offer a powerful means of understanding the feminization of the prisoner body in the space of the institution. Both argue that the body is brought into a gendered identity through socially constructed and power-laden discursive structures. Butler (1993) follows Foucault's argument that the subject is brought into being through medico-juridical discourse. She draws this notion into feminist theory by arguing that the body is further marked out through juridical, biological, and medical discourse which upholds a regulatory heteronormative gendered regime.

The feminine body is constructed, brought into being, through biological discourse distinguishing a male/female dualism which in itself serves the project of heterosexuality. These categories are normalized through naturalization and stabilization, which, Butler argues, occurs through the reiteration of gendered performances. In many respects, and indeed as the article from *Marie Claire* demonstrates, woman prisoners present a category of womanhood that is uncomfortably blurred. Women prisoners destabilize the hegemonic norms not only because they are criminals who have broken with social morality, the law. Their criminality also flouts gendered norms held by broader society because women are not expected to murder, hijack, commit armed robbery, and run drug cartels. There is a conflation of the act of breaking the law and the subversion of less overt gendered norms or "laws." Work and recreation, the rehabilitative activities of the prison, are intended to address both transgressions.

While Butler offers a powerful means of understanding the abnormalized woman prisoner body, Grosz provides a means of understanding the mechanism of work and recreation within the institution. For Grosz (1994, pp. 33 and 213) the body is a surface of inscription which is "marked, transformed, and written upon or constructed by the various regimes of institutional, discursive and nondiscursive power . . . both sex and sexuality are marked, lived, and function according to whether it is a male or female body that is being discussed." While Butler is often criticized for dismissing the materiality of the female body, Grosz argues that the inscribed surface of the body is not neutral and it affects the range of "texts" which can be ascribed to it. Her project is to desconstruct these texts within their spatiotemporal contexts: "exploring how the subject's exterior is psychically constructed; and conversely, how the processes of social inscription of the body's surface construct a psychical interior" (ibid., p. 105).

Grosz then applies her understanding of the body to space, a move that helps to shed light on gender dynamics in South African prisons. In her work on the relationship between the body and the city, she argues for the concept of interface: a fluid engagement of "assemblages or collections of parts, capable of crossing thresholds between substances to form linkages, machines" (ibid., p. 108). In prison, work provides an interface between the body and the institution; it engages the body politic in specific activities that signify hegemonic notions womanhood. And so the biotechnology (Foucault, 1991) of the institution is intended to reinscribe, to rescript the body within a chain of feminized signification. In this way the state normalizes a particular gendered identity within the institution.

Resistance, the Counter-economy and the Performance of Masculinity

The diverse ways that women prisoners engaged these "abnormalizing" discourses and practices reveals the gaps and fissures within state disciplinary practices. During interviews with prisoners, most avoided the theme of work in the prison. No one wanted to speak about it despite my increasingly less subtle prompting. The only comments that were made pertained to the unfairness of the wages:

> I've been working, working hard with the machine, what do I get.

> I have been working in the workshop for three years. I started earning fourteen rand and then the pay increased to thirty-six rand. . . . I took a draft [meaning: she asked to be transferred] to [another prison]. I started earning forty-one rand.

> I mean they're trying to make the workshop. They're trying to give something, but it doesn't pay. How can you, if you get the ten rand to buy the cream and it's finished. How it means you've got to work the whole month for that ten rand . . . everybody doesn't like to sit for nothing.

This lack of commentary on work was telling. Prisoners often rejected the work tasks of the prison and obtained cash through smuggling or networked connections. Work was considered as an act of compliance to the institution. To refuse work is an act of defiance against the institution and its feminized norms and an assertion of independence from the institution.

> But me I am so proud of myself, I will try in my own ways to get money. . . . You see in prison they want you to use yourself so that you tell yourself that you have no future, your future is crime.

> You know why I stopped working, let me tell you I am one person . . . I am an independent person.

Corruption in prison is an economic reality with particular codes of conduct. Prisoners use the narrow windows of their cells as an extensive smuggling system. The basis of prisoner economies of power lies in access to resources. Any resource, including food, drugs, money, toiletries, protection from violence, and sex, was extensively traded. "Panda[2] . . . I mean if I need something then I'm going to go out and panda or zula for this and zula for that." Sex and food and drug smuggling trangress the discourses of incarceration as a site of alienation and rehabilitation, and many of the prisoners involved in these activities also strongly reject state gender disciplinary practices by assuming overtly masculine identities. Prisoners speak of themselves as "men" and "women," using the two sets of terms "butch" and "man" and "femme" and "woman" interchangeably. Butch and femme roles are defined within a prison-based moral economy by a set of rules called *snanga, snangananga*, or simply "the game" (see Dirsuweit, 1998, 1999, for a more detailed discussion). While rules for butch and femme roles may exist within certain spatially specific communities outside of the prison, the express codification of *snanga* is unique to women's prisons.[3]

The kingpins, who were invariably butch and the most transgressive of state sactioned norms of femininity, rarely worked and never worked within state-sanctioned activities. Many of the butches I spoke to had few or no visits from family and friends. The only way they can get resources, therefore, is to either do unsatisfying work for a meager salary or smuggle for money, food, and drugs. The spatiality of this counter-economy of the prison is significant: transactions are set up and executed for the most part in the *spiraal*, a large graded central spiral at the heart of the prison. In the center of the *spiraal* is a courtyard and embedded in the walls are a series of thin windows approximately 10 cm wide. Any position in the *spiraal* enables a view across the courtyard through the narrow windows. In this way a small area affords the viewer a line of site which includes most of the *spiraal*. The design ensures that any activity in the *spiraal* is observable from a vantage point which may not be observed.

While the *spiraal* apparently disciplines behavior in its design, it is also the center of prison culture. A vibrant culture persists in the prison despite efforts to alienate, confine, and discipline. The *spiraal* is a central element of this and is the home of a flagrant disregard for its surveillant structure. One prisoner named it "the highway of [the prison]." The *spiraal* is where the underground economy and social life of the prison is played out. It is here that activities, which are sometimes illegal, but always transgressive, take place. Most of the smuggling of food and drugs, looking for partners, gossiping, and prostitution occur in the *spiraal*. It is also an important space for communication with the outside world through the interaction between sentenced and unsentenced prisoners (which is strictly forbidden). Dealing drugs occurs in the *spiraal* and is a central part of the constant conflict around the symbolic tenure of this space.

A prisoner describes how prisoners deal at the steel door leading to the section where the men (biological) are housed.[4]

> And then one of the girls knocks on the door and then she throws three rand through the top. It's like a slot machine. And this piece of paper comes sliding through the opening. A piece of magazine gets folded in a certain way . . . and the quality is good . . . there's always someone at the door. If they hear you're white, they sluk you [take your money].

Institutional authorities are obviously aware of the process. The door has been welded and rewelded repeatedly in an effort to close off any spaces through which packages can be passed. However, despite the efforts of institutional authorities, there was always a small interstitial space where the door had not been completely welded shut, making this door the center of prisoner-based economies. Prisoners have thus appropriated the meaning of the *spiraal* from a place intended to discipline their behavior as they move from one part of the prison to another to a public arena for themselves and everyday activities. In this reclamation of prison space, and by extension body space, prisoners place into question ownership claims of the prison authorities. As such, there is a continual conflict around the control of this part of the prison.

Smuggling is also about the territorialization of space and the *spiraal* is a key site within this process. Finding femmes takes place in the *spiraal*, as does the meeting of butches to exchange gossip and perform a number of economic transactions. For these reasons, men often described the *spiraal* as their terrain. Indeed, they are the most visible transgressors of this space of surveillant control. This sentiment is reflected in a comment made by a senior member of staff that butches were "the real problem" in female prisons. Self-identified male prisoners flout the norms of state-sanctioned discipline by rejecting both work and the feminized normalization inherent therein.

For these men, there is a radical disjuncture (rupture) between the materiality of their bodies and the assumption of masculine signifiers. Women reinscribe their institutionalized bodies as male; they rewrite the biotechnological script of the institution. The interface between the institution and the body is reconfigured particularly at the point of the *spiraal*. Here male-identified butch prisoners cross the thresholds between space and the body in the opposite direction because they reinscribe the space of the *spiraal* with a different meaning that is related to the functioning of the prison counter-economy. This economy in turn has its own gendered signification which is related to masculinity. That this resignification or reinscription of institutional space is transgressive is demonstrated in the responses of male (biological) prisoners and staff members to male-identified butch prisoners: "We were standing around talking to each other and [he] came past and grabbed his crotch and said 'I'm the only man here'." Another femme recounts the responses of male (biological) prisoners:

> X was doing this [traditional] dance with some other women. As they were finishing, his [a butch prisoner] hat fell off and he had shaved his head. These guys [prisoners from the male (biological) prison] just jumped up and started shouting "Sho, what is this?"

In both incidences and in the responses of the institutional authorities butch performances of masculinity are transgressive of heteronormative sexual structures. At the same time masculinity is constructed within masculine privilege.

Butler (1990) argues that transgression lies in the deconstruction of heterosexual economies of signification through the exploitation of the spaces opened up by these economies. For Butler (1990, p. 136), the "the gendered body has no ontological status apart from various acts which constitute its reality." Gender is a performance, a series of ritualized acts which are reinforced by a heterosexual regulatory regime. An act of transgression of this regime, then, is the appropriation and parodic inversion of hegemonic gendered performances. Butch identity becomes heroic in this formulation, since it "resignifies masculinity" by juxtaposing "male" identity on a "culturally intelligible female body" (Butler, 1990, p. 123). In prison, butches go one step further; they actively reject the state-sanctioned performance of femininity in the prison. The body, masculinity, and its relationship to institutional space remain resistant to institutional inscription which occurs through gendered signification entrenched within institutional work regimes.

Violence and the Moral Economy of Gender

Butch identities in the prison are enacted not only in relation to authorities and normative practices, but in relation to femme prisoners. The rules of *snanga*, in addition to shaping the alternative economy of the prison, require butches to take a protective role in relation to femmes. Femmes, on the other hand, are required to fulfill a domestic, home-bound role. In the words of a man interviewed: "The woman is supposed to wash your clothes if they are dirty, she must make food and everything like that. We have to sleep together. She has to wake up in the morning and run the bath so that I can bath and that sort of thing." One of the mechanisms of gaining and maintaining power in the prison was through what can only be described as horrific acts of violence and violent threats often instituted by butches against femmes. These incidents were so pervasive that it was rare for an interview to close without mention of one.

> I think if I had to attack [him, he'd] easily kill me. I'm not scared of that, but because [my girlfriend] is in the same cell, I'm trying to stay calm.

> . . . after he bust [her] nose . . . there was no charge or anything against [him].[5] Because he said that if there is a charge against [him] then he'll kill her. And there was another incident yesterday where he attacked [someone else]. And she had to pack up and move out of the cell because the members insisted that she move on or else he will put a knife in her.

> There was this one prisoner who beat up [his] girlfriend and other people tried to help her and [he] went crazy [with a knife].

> If they can't get back at you physically, they might decide to poison you. There was one woman who no one really liked and one day she started having these cramps. They just put it down to food poisoning, but that was poison.

Protection from this violence was a valued resource. Many relationships in prison are based on an exchange of sex and gifts for protection. Masculine protection is a powerful commodity for exchange. "[She] wants someone to protect her and people around here do push her around because she's small. They do fuck her up. She needs someone with her to protect her. Just to be there . . ." Translated from Afrikaans:

> *First prisoner*: And you'll protect me if anybody else . . .
> *Second prisoner*: Ja-ja, and also in a case like that. If I notice I don't know you but you're talking to [him] just as you please, and look here, I come straight, I say hey wait man, who're you looking at, that's my wife man, don't talk like that to her, man look here, if you don't know how to talk to her, then just leave it. And don't bother her and what-what. And why are you still here, man, if you gonna bother her again you gonna get me. Right, fine, kwaai [trans. that's all in order]. I'm there to cover for us, I've gotta look after her. If she's got a problem, I must be able to solve everything. Yes. That's the way it is, that I must never not be able to solve a problem that she brings to me.

Ironically, violence in prison is often instituted by butches. The rules of *snanga* normalize butch control of femmes. Before being arrested, many butches were also part of gangs in which violence becomes part of daily life. Violence, coercion, and masculine control were the norm for many prisoners and showing this level of control over femmes was central to many of the butches' identities.

> When I came in I was in opnamers [trans. reception] that night and one of the sergeants came in and said I'm going to D Section because of my sentence, but I must be careful because there's a butch there . . . and [he] is going to rape me . . . but there's like a bond [between us]. . . . [He's] got a bad temper, but since [he's] in a single cell and I'm in a big cell things are better. . . . I don't hate [him], I'm very fond of [him].

> Somehow I think like if you take [A], [A] wants to show [he's] the male, so [he] hits all the females . . . most males here are the domineering part for some or other reason. They don't see themselves as equal.

Femmes who refused to complete domestic tasks were often subjected to violence from butches. At the same time, however, there was an extraordinary amount of resistance to butch control.

> *First prisoner*: She was in our cell, but she saw that there are people who are mad like her and she went out within two days because we weren't going to allow such shit.
> *Teresa*: If you're a man you can show power . . .
> *First prisoner*: Of course.
> *Teresa*: But you're all telling me that women are bigger fighters than the men.
> *Second prisoner*: Oh believe me, the women are big fighters.
> *Third prisoner*: I will not challenge Y [a femme].
> *Fourth prisoner*: I would scratch [a butch prisoner] to death.

> One day when I first came to prison one woman told me come here make me coffee. It was in the middle of the cell and they all watched. And I said "are you talking to me" and she said, "yes, come here you" and I said no I don't take orders from you. I said that.

Where a butch was particularly threatening to a femme, cell members rallied around her:

> The people in my cell keep saying to me that she's got no right to treat me like this and that I must stop making her food and things like that. . . . She tried to hit me and they came out and stood around me and started shouting at her. So she left me alone.

This resistance to butch control transgresses the relations of the prisoner-based moral economy and places the *performance* of the butch and femme actors into relief. The rules of *snanga* (the game) are not static or beyond negotiation. Butch control is constantly questioned. Prisoners identify different levels of masculinity related to the roles that butches played outside the prison. Those whose masculine identity was constructed outside the prison or through going in and out of prison over the years elicited far more respect than those who constructed their identity in prison or had children outside of the prison. While a certain level of domestic violence and general violence is accepted from butches/men, when their coerciveness and violence overstepped the moral economy of the prison, both butch and femme prisoners immediately questioned their gender history. Were they in a relationship with a woman before they entered prison or were they in heterosexual relationships? Did they have children? What role did they play in same-sex relationships before coming to this prison? These questions placed the regulatory practice of *snanga* on unsteady ground and as a result the performance of masculinity was the object of constant questioning by femmes and other women in the prison.

However, it was the materiality of the female body that troubled the performance of female masculinities in prison the most.

> I don't understand why this happens because they've got everything like I've got, how can they think that they are something other than a woman. To me it's just a funny thing.

> *Teresa*: you don't seem to be that scared of the men as other femmes.
> *Interviewee*: The only person I am scared of is God. To me whether you're a woman or a man, you're the same. In here I take you, if it comes to that, as a woman. And I'll tell you, "Hey, you wear the same panties as I do." I'm not scared of the hitting and things like that, but I am scared of the outcome. Maybe I stab her or things like that.

Masculine identities in prison, and the moral economy associated with these identities, offers a profoundly important yet paradoxical site for thinking about feminist theorizations of the body. First, we see Butler's notion of transgressive potential operating (see Butler, 1990): female masculinity is profoundly transgressive in the prison in terms of both parodying heteronormative roles and resisting gendered notions of correct behavior within the institution. The transgressive nature of this performance, however, is tempered by the recognition that the power associated with female/butch masculinity reflects and entrenches a set of gendered relations in broader South African society which are based on violence, domination, and masculine privilege. Second, Grosz's concern with the materiality of the body helps to elucidate prison gender dynamics because the right to masculine privilege is ultimately undermined by the materiality of the female body as well as by the multiple efforts of prisoners to resist masculine privilege. Thus, while female masculinity in the prison represents a disturbance within normative regimes, exploring how this performance operates in practice and in space adds depth and nuance to our understanding of the these embodied performances.

Conclusion

There are two interrelated themes within this case study which speak to some of the concerns within feminist geography. The first relates to resistance and transgression; the second to the geographic theorization of embodiment. There is some concern within geography about the manner in which Butler conceptualizes agency.

Gregson and Rose (2000) and Nelson (1999) argue that Bell et al. (1994) have misread Butler's discussion of transgression and imply in their work on lipstick lesbians and skinhead gays that there is some degree of intentionality associated with the assumption of these parodic roles. This intentionality belongs to modernist understandings of identity, which are related to Goffman's invocation of the term as "performance as staged, as played for spectators both behind the scenes and in the auditorium of the marketplace, these studies construe performance as theatrical and dramaturgical, the product of *intentional, conscious agent*" (Gregson and Rose, 2000, p. 436, emphasis added).

Performance instead is the product of interaction and context: Butler is particularly weak at elucidating either, a point which Nelson (1999) and Thrift and Dewsbury (2000) are quick to make. Performance is something organic which becomes transgressive as it collides with broader social forces, and the dynamics within the prison are more understandable within this formulation. Thrift (2000, p. 231) offers a reading of Deleuze's notion of performativity, where performance is: "a symptomology, a diagnostics of signs which isolates a particular possibility of life which helps to make more modes of existence possible. It is a process that operates by means experimental and unforeseen becomings." Dewsbury (2000, p. 475) concurs and extends these comments:

> performativity is excessive, in that it expends unaccountable energies and is affective rather than purely effective. It speaks of a multitude of possible outcomes within an event: everything that actually happened could also have always happened otherwise and simultaneously other happenings occur irrespective of the focus of one subjective orientation.

Performativity deconstructs the usual formulations of causation. It "adopts an openended stance towards human experience, becoming a *stranger to any idea of genetic axis or deep structure*" (Dewsbury, 2000, p. 477).

In terms of this case study, the performance of masculinity cannot be understood in terms of causation and intentionality. There are (transgressive) political consequences in relation to state-sanctioned (and broader social) notions of femininity. The rules of *snanga* allow masculine-identified women to "deny a materiality or material specificity and determinateness to [their] bodies" (Grosz, 1994, p. 190), yet this denial forms the basis of their claim to power and masculine privilege in prison. This intersection of privilege and transgrssion speaks to Johnston's (1998) reading of women body builders in male-dominated gym environments. Drawing on Grosz, she argues that bodies become sexed according to spatiotemporal particularities. The female built body has three possible readings associated to it: as docile and conforming; as transgressive; and as erotic. These readings coexist, forming what Johnston refers to as a double bind of embodiment:

> [female body builders] are caught up in, and constituted by, other (institutional) biopower relating to the sexed spaces of gyms. As body builders use power through the medium of their bodies, certain "contradictions" emerge. Some female body builders could be understood as docile bodies, and others transgress docile bodies. Female body builders can be partly understood within each regime. . . . The femi-

nine/masculine debate of weakness (female) and muscle (male) is played out consistently on and through corporeality.

This reading of an embodiment entwined in several different discursive structures which enable/disable particular readings of the self is of particular pertinence to carceral institutions where relations between subjectivity, the body, and the state are constantly in process with multiple planar surfaces rather than settled political and spatial artefacts.

NOTES

1 Interviews took place twice a week between mid-1996 and the end of 1998. Interviews are reported without positioning the identities of the prisoners to maintain their anonymity.
2 *Panda* and *zula* in the context of the prison refer to daily activities negotiating and making deals within the counter-economy of the institution.
3 Unfortunately no studies have been done on butch–femme communities in South Africa.
4 In the prison, masculine-identified women often refer to themselves as "men," so to differentiate between women who identify themselves as men and the more conventional biological usage of the term I refer to the second usage of the term as men (biological).
5 When referring to masculine-identified women, pronouns are used interchangeably. In this case the prisoners refer to the same prisoner as both he/him and she/her. For ease of understanding, these have been standardized to he/him.

BIBLIOGRAPHY

Bell, D., Binnie, J., Cream, J. and Valentine, G. (1994) All hyped up and no place to go. *Gender Place and Culture*, 1, 31–47.
Butler, J. (1990) *Gender Trouble: Feminism and the Subversion of Identity*. New York: Routledge.
Butler, J. (1993) *Bodies that Matter: On the Discursive Limits of "Sex."* New York: Routledge.
Dewsbury, J.-D. (2000) Performativity and the event: enacting a philosophy of difference. *Environment and Planning D: Society and Space*, 18, 473–96.
Dirsuweit, T. (1998) Sexuality and space: sexual identity in South African mine compounds and prisons. *Development Update*, 2, 107–13.
Dirsuweit, T. (1999) Carceral spaces in South Africa: a case study of institutional power, sexuality and transgression in a women's prison. *Geoforum*, 30, 71–83.
Foucault, M. (1991) *Discipline and Punish*, trans. Alan Sheridan. London: Penguin.
Gregson, N. and Rose, G. (2000) Taking Butler elsewhere: performativities, spatialities and subjectivities. *Environment and Planning D: Society and Space*, 18, 433–52.
Grosz, E. (1994) *Volatile Bodies: Towards a Corporeal Feminism*. Bloomington: Indiana University Press.
Grosz, E. (1995) *Space, Time and Perversion: Essays on the Politics of Bodies*. New York: Routledge.
Hannah, M. (1997) Space and the structuring of disciplinary power: an interpretive review. *Geografiska Annaler*, 79, 171–80.

Johnston, L. (1998) Reading the sexed bodies and spaces of gyms. In H. J. Nast and S. Pile (eds), *Places through the Body*. London: Routledge.

Nelson, L. (1999) Bodies (and spaces) do matter: the limits of performativity. *Gender, Place and Culture*, 6, 331–53.

Thrift, N. (2000) Afterwords. *Environment and Planning D: Society and Space*, 18, 213–55.

Thrift, N. and Dewsbury, J.-D. (2000) Dead geographies – and how to make them live. *Environment and Planning D: Society and Space*, 18, 411–32.

Chapter 25

HIV/AIDS Interventions and the Politics of the African Woman's Body

Kawango Agot

By December 2002, the Joint United Nations Program on HIV/AIDS (UNAIDS) estimated that 42 million people were living with HIV/AIDS worldwide, 29.4 million (70 percent) of whom were in sub-Saharan Africa. The subcontinent was also leading in the number of new infections (70 percent), HIV-related deaths (77 percent), AIDS orphans (92 percent, December 2001 data), and infected women (81 percent) (UNAIDS, 2001, 2002a). AIDS is presently the fourth leading cause of mortality worldwide and the number one killer in sub-Saharan Africa, where at least 16 countries have an adult prevalence of over 10 percent.[1] Over 90 percent of HIV infections in African adults are reported to result from heterosexual intercourse. Hence, in the absence of a vaccine, behavior change interventions still hold the key to addressing the scourge. In light of this, the main HIV prevention strategies promoted by many countries in the region include voluntary HIV testing and counseling, increased condom use, reduction in numbers of sexual partners, and treatment of sexually transmitted infections (Hudson, 1996; Cohen, 2000).

The impact of these interventions, however, has been minimal as HIV continues to spread rapidly in many parts of the continent, notably Southern Africa (UNAIDS, 2002a). In Kenya, the seroprevalence has plateaued in the past three years, even though at a high level of about 13 percent in the general population (Government of Kenya (GOK), 2001a; UNAIDS, 2002b). National scale statistics are somewhat misleading, however, because there are marked variations within the country and among different subpopulations. Of the 43 ethnic communities in Kenya, the Luo of Nyanza Province (see figure 25.1) accounted for the highest HIV prevalence (30 percent) in the country in 2001, despite the fact that the community constitutes only 12 percent of the national population (GOK, 2001a, c).

The Luo community presents a paradox in the HIV/AIDS epidemic. On the one hand, it has accounted for the highest prevalence of the disease in the country since the late 1980s when complete data became available (GOK, 1998, 1999). On the other hand, it is the most researched group in terms of identifying risk factors for

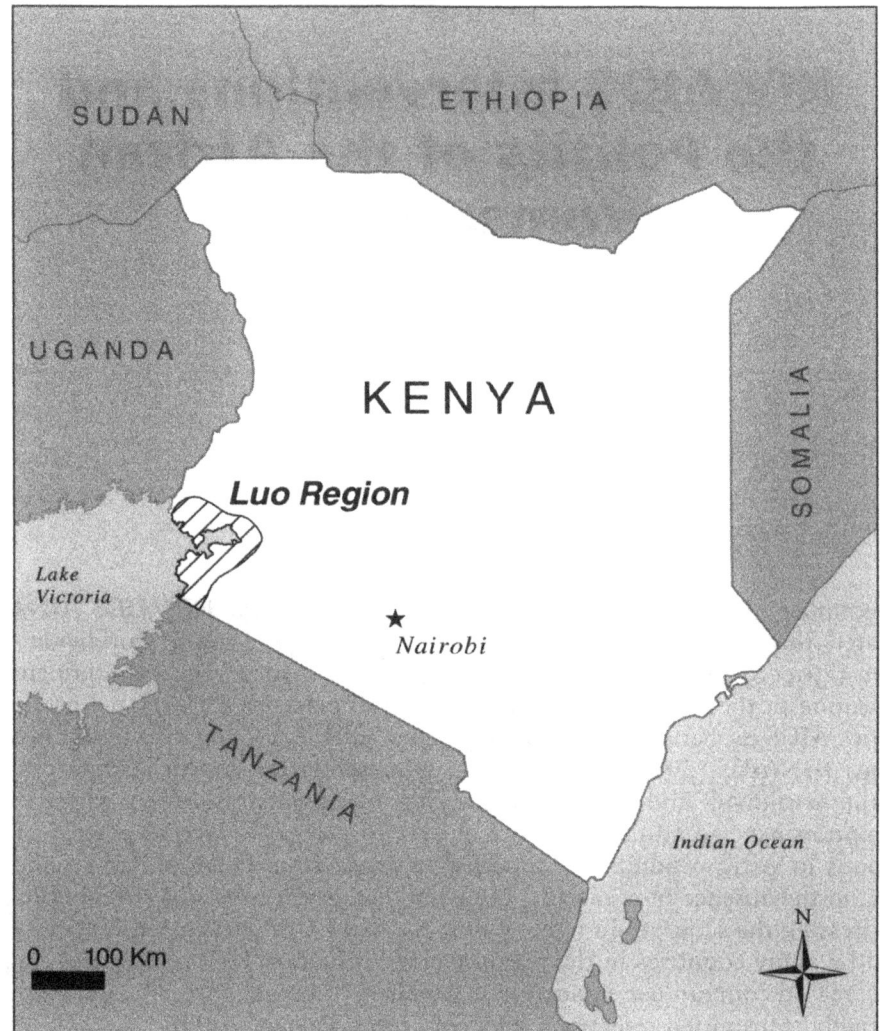

Figure 25.1 The Luo Region of Nyanza Province, Kenya.
Source: Data from ESRI and Kawango Agot.

HIV (Odindo, 1998; Agot, 2001; GOK, 2001b) and the recipient of some of the
most intense interventions in the country to date (GOK, 1998; Otieno, 1999; Agot,
2001). The result of this is 99 percent HIV awareness in the community against a
background of 30 percent seroprevalence. Such discordant findings call to question
the effectiveness of the current interventions.

The government of Kenya, like international health agencies, attributes the
disproportionate prevalence of HIV among the Luo ethnic community in Western
Kenya to, among other factors, the cultural practice of widow inheritance (WI):

> With the advent of AIDS, some of these [cultural] beliefs . . . promote behaviors which
> put individuals at risk of contracting or transmitting HIV. These include different types

of marital union like polygamy . . . [and] leveretic (widow inheritance) relationships. (GOK, 1997, p. 11)

some parts of western Kenya have the highest recorded rates [of HIV] in the country . . . many factors contribute to these . . . cultural practices such as . . . widow inheritance. (GOK, 2001a, p. 44)

To check the spread of HIV in the community, the government has suggested that WI be eliminated.

WI is a practice where a widow is "taken over" by a brother or a cousin to the late husband and where, even though the relationship does not constitute a formal marriage, sex is often an integral part. It is carried out to discourage widows from abandoning their marital homes and their children; to restrain widows from seeking sexual liaisons outside the husbands' clan; to give childless widows an opportunity to get children, especially sons, who would continue the lineage of the deceased; to entitle the widow to social and economic support from the inheritor; and to enable her to participate in specific social events for which ritual sex is a component. Ritual sexual relations are essential, for example, during farming seasons (cultivating, planting, weeding, and harvesting), during ceremonies associated with rites of passage of close family members (birth, marriage, and death), and in the building or repair of homes. The practice of WI is characteristic of patrilineal and patrilocal societies among whom the widow and her children, as well as the family property, belong to the husband, and when he dies, to his family (Agot, 1996, 2001; Mboya 1997; Ntozi 1997; Ogutu 1999).

Given the cultural and social centrality of widow inheritance, members of the Luo community have a different view of the practice from that of the government. "Akinyi" is a 27-year-old widow from Asembo in Luo Nyanza. She lost her husband shortly before the interview. She had the following to say:

My husband died six months ago and I am not talking about something I have heard or something I have been told. I am talking about what I have gone through. I am talking about my own life. How many mothers [polite reference for women, regardless of whether or not they have children] here in this meeting are discussing inheritance from personal experience? [She did not wait for a response.] I just want you people to talk about the issue of inheritance with a lot of caution because those who are still young must be inherited. How do you expect her to initiate farming if she is not inherited? How about harvesting? Building a house? Because most of us widows are young and we were left still living in *simba* [bachelors' hut]. We have children and we cannot do any of these things and "kill" our children with *chira*.[2]

Community and family custom encouraged her to become "inherited" by her late husband's brother, a relationship that would assure continuity of the family line, security, and social acceptability for her and her family. For example, the new relationship established by inheritance allows her to perform sexual rituals seen as crucial to successful harvests, and to keeping the family cleansed and safe from possible misfortunes (*chira*). "Akinyi" decided to be inherited, seeing the well-being of her children as the primary benefit.

Despite its central role in community and family relationships, the cultural practice of WI has been identified and treated by HIV prevention experts as contributing to the spread of HIV among the Luo people of western Kenya. The government of Kenya cites the practice as one of the behaviors fueling the spread of HIV in the community. Yet the repeated and costly efforts undertaken by a variety of international and national agencies to eliminate this practice, as well as other cultural practices deemed risky in the Luo region, have performed dismally. I argue that much of the failure is attributable to their negligence of the sociocultural contexts within which these practices take place. With the help of the international donor community, which channels funds mainly through health ministries, AIDS prevention has been medicalized to the exclusion of other knowledges. Sexuality has been constructed as a biologically driven individual behavior, divorced from culturally significant meanings and social relations, and AIDS interventions have likewise been individually based (Abstinence, Being faithful to one uninfected partner, and Condom use, otherwise known as the ABC of AIDS prevention).

In Africa, however, many sexual practices, such as WI, are embedded in complex cultural matrices. Sexual activities are central to numerous social and cultural rituals and prescriptions. This notwithstanding, many of the HIV/AIDS intervention programs addressing WI in Africa and Kenya more specifically have been built on the assumption that a particular sexual practice can be treated as an isolated trait to be eliminated through intervention efforts. This strategy ignores the compelling and complex undercurrent of factors that shape up this custom, and underlies the dismal performance of HIV prevention programs.

Drawing inspiration from literatures on the body, sexuality, race, and colonialism in feminist geography and related fields (see Schoepf, 1991; Parker et al., 1992; Brown, 1995; McClintock, 1995; Duncan, 1996; Radcliffe and Westwood, 1996; Dyck et al., 2001; Longhurst, 2001), this chapter explores these complex cultural and social matrices of widow inheritance in the Luo community of western Kenya, highlighting women's experiences and interpretations of the practice as well as their perspective on HIV prevention efforts. Based on 15 months of ethnographic research in the Luo region in 1999–2000, my critical exploration of HIV prevention policies targeting widow inheritance demonstrates why the policies have largely failed. Inspired by the work of feminist geographers and postcolonial theory, I link these interventions, and their deployment in particular places such as the Luo region, to longstanding colonial mentalities that otherize African women's bodies and African cultures. Moving from local to transnational, I demonstrate how overreliance on foreign sources to fund HIV/AIDS prevention activities has configured intervention policies in sub-Saharan Africa in ways that have compromised the ability of many countries to meaningfully define their own priorities in combating the epidemic.

The failures of these prevention strategies lie in their Western-centric conceptions of time and space, assumptions that abstract and "otherize" African women's bodies and African cultures. Within this discourse, the fight against AIDS in Africa is often presented as a fight against "cultural barriers" that are seen as promoting the spread of the HIV virus. By treating African cultures as an exotic "other" to Western culture, as well as temporally "backward" and "primitive," the ongoing interventions are targeting changes in behavior (of WI) to the exclusion of social and cultural contexts within which the behavior is practiced. Because these interventions target individual widows outside the social structure of which they (the widows) are part, and in which HIV is acquired or transmitted, they are doomed to failure.

Drawing on women's life stories in the Luo region, I demonstrate that interventions that do not take into account the prevailing social and cultural contexts will continue to be ineffectual, regardless of the amount of time or the resources expended to implement them.

The Meaning of Widow Inheritance in the Luo Region

"Atieno" is a widow in her mid-forties. She is aware that the government is trying to stop people from practicing widow inheritance because of HIV, but she also feels strongly that failure to be inherited would bring misfortunes (*chira*) to her family, particularly to the children. She maintains that there are too many stakeholders in the community who would be concerned about the well-being of her children if she refuses to be inherited:

> Children get support from other family and clan members [telling them] that "You see, it is your mother [who is] killing your children." So the woman gives in and is given "some shepherd full of disease."

> Women now [in this discussion group] are putting a "tick" [an OK] to stop inheritance, but when there is death, then clan members or family members make it impossible for us to refuse to be inherited. They tell your children that, "If your mother is not inherited, then there will be misfortunes in your family." So they [the children] force you, or else they abandon the home and rent a place at a nearby market center [and live there] until you are inherited or until you die. It forces you to look for any man and in the process, collect AIDS and bring [it] home.

Clearly, besides herself, other people in her social support network would also be concerned by her choice to not accept inheritance, including her in-laws, her own parents, her brothers and sisters, her co-wives, and even her own children. In fact, many see failure to follow this cultural prescription as *leading* to *chira*, a misfortune that often takes the form of a wasting disease (see note 2). Many in the community, as I explain further below, confuse *chira* with AIDS, and thus community efforts to avoid HIV and AIDS oppose government ideas of how to limit HIV and AIDS.

The concern expressed by HIV prevention experts about widow inheritance is not completely unwarranted, however misguided: the Luo region of Kenya has the highest seroprevalence rate in the country, at 30 percent (GOK, 2001a). International agencies have flooded the region with money and programs designed to prevent HIV. After more than 15 years of concerted intervention campaigns, however, infection rates remain high, and surveys demonstrate that the majority of men and women in the region still do not understand the basic mechanisms through which HIV is spread, and see no connection between their cultural practices and HIV (Okeyo and Allen 1993; Agot, 2001; see also Ntozi 1997). WI is part of a complex web of social relations and rituals, ones that have very subtle cultural underpinnings and thrive on a delicate labyrinth of sexual detail. My research examining the views of the Luo on eliminating WI as a measure to reduce the spread of HIV shows a more complex cultural matrix than that which is captured in HIV intervention policy discourse.

From Atieno's story, it may not come as a surprise that decontextualized HIV intervention polices have been near universal failures, but the experts employed by international agencies are perplexed at the dismal outcome of sustained efforts and hundreds of thousands of dollars spent to discourage WI. If you talk to people in the Luo region, however, the causes of this failure become clearer: there exist dramatic tensions and contradictions between local peoples and cultures, and the normative and Westernized expectations and strategies of HIV prevention policies. In short, the hubris of policy-makers, and their blindness to local cultural logics and place-specific cultural practices, doom these policies to failure.

To many in the Luo region, failure to be inherited, or being inappropriately inherited, is a violation of a sexual taboo and leads to a wasting disease called *chira*. Similarly, AIDS is seen as resulting from violation of some sexual pact, which also leads to wasting. Thus, AIDS and *chira* are conceived as one and the same thing. As "Awino," a woman in her mid-forties from the Luo region, describes it,

> When a woman is not inherited, this brings *chira*, which is also called AIDS. A woman must be inherited so that she can "open" for her children. If her son wants to marry, to straighten the "things of the home," the mother must be inherited before he [the son] can go ahead with his marriage plans. Failure to observe this tradition will kill the son with *chira* which people will "change" and call AIDS. [In response to my question as to whether what she was describing was *chira* or AIDS, she said] *Chira* is AIDS due to loss of traditions; we Luos call it *chira* and it is the AIDS in your [the research team's] language.

"Achieng," in her fifties, urged:

> Let us treat *chira* before it grows – if we leave it [to grow] until it turns into AIDS, then we won't be able to treat it. [Pointing at young widows.] They don't follow traditions these days; they go for sexual cleansing in towns where they work. Whoever cleanses her [has sexual relations with her first following her husband's death], when he goes back to his house carrying the "shadow of impurity" with him [and sleeps with his wife], can he escape from being infected with AIDS? If this is not AIDS, then what is? How does it look like? Isn't that a big *chira* that cannot be treated? [Referring to the case of a widow rushing back to the city and having sex with an unsuspecting man who does not realize that he is, in fact, inheriting her.]

Besides this conflation of AIDS and *chira*, the practice of WI is also seen within the Luo community as a *protection* against HIV because it is argued that formally inherited widows are more inclined to have fewer sexual partners, and hence would have a lower risk of transmitting or contracting HIV compared to uninherited widows (the argument is further developed below). The following quotes capture the general feeling of the respondents. "AIDS infects those who are 'promiscuous,' not those who are inherited" (from a female respondent in her thirties). In the view of a male respondent in his twenties,

> Let the topic be that of sexual "promiscuity," not inheritance. Inheritance does not bring [lead to infection with] AIDS; "promiscuity" does. Breaching sexual taboos around inheritance [*nyuandruok*] leads to *chira*; outright sexual "promiscuity" [*terruok* or *chode*] is the one that "brings" AIDS. There is a big difference.

In the Luo region, Western-centric assumptions have translated into particular programs, deployed by internationally based and Kenyan agencies, that seek to prevent the spread of HIV by eliminating local cultural practices such as WI. By ignoring the local cultural logics of HIV-relevant practices, these policies are doomed to failure and function to reproduce imperialistic relationships. For example, the government of Kenya, with funding from the West, has earmarked for elimination not only WI, but polygyny as well. However, discussions with and participant observation of thousands of women and men in 1999 and 2000 indicate that these externally conceived prevention strategies are seen as nonsensical and often insulting at the local level.[3] A male respondent in his late fifties, for example, charged that:

> You people who are coming with new and foreign teachings want to interfere with our culture. Why are you and those who sent you [implying the government] targeting the Luo? And if inheritance can bring AIDS, which is a lie, then it is bringing AIDS to everyone who has sex with an infected widow, not just the Luo. It is you educated Luo people who are responsible for bringing to us strange and foreign traditions. Go and tell those who sent you that we will never condemn inheritance [and] it does not matter what! Even if it is announced over the radio how[ever] many times. Let them just say, and continue to say until they get tired. What the problem is, is sexual "promiscuity" by widows, not inheritance.

What is clear is that people are told WI is "bad" but they are never taught the basic infection mechanisms of HIV which would make WI a possible risk practice, nor are they asked their opinion about how best to prevent infection in the context of their lives as individuals or communities. In fact, they are not even involved in discussions on how intervention activities targeting their culture should be approached. A male respondent in his twenties commented that:

> We have not seen anyone coming here to discuss AIDS with us. You [the research team] are the first ones. What we see are big vehicles [on] which are written "Save your family: Discuss AIDS" busy running on the roads but they only discuss it in big hotels. You just see them in Imperial Hotel.[4] You hear about them in Sunset [in] Kisumu. And even if they come to Bondo, they only go to the best hotel in Bondo. They tell those who already know, but they don't have courage to come where we are. We have no Imperial [Hotel] food to give them.

Not only do health officials not reach out to people when setting up policies to eliminate cultural practices, they also treat women as isolated and disembodied rather than socially embedded subjects. They are urged through workshops, the media, and community meetings to refrain from being inherited when widowed, discourses that ignore the social and cultural forces that interact to sustain the practice. "Mikayi" is a 31-year-old mother of two young boys. She has just buried her husband of ten years, and having lived most of her married life in the city, she was hardly acquainted with her in-laws prior to the funeral. She recalls an international workshop she had attended the previous year and the resolutions she made. At the time, feeling very sensitized by the activities at the workshop, she led a group of other equally enthusiastic women in passing a resolution that none of them would succumb to the family pressure to be inherited, should they be widowed. A record of this workshop and this resolution was sent to the donors as an indicator of the willingness of women to denounce the practice, and to demonstrate the "success" of the intervention program.

Today, one year after she made the pledge, she was faced with the reality of being a widow. The workshop excitement had long since subsided, and there were no more charged supporters rallying around her and urging her on. There were no more pledges to be made. Instead, there she was, bereaved, vulnerable, and largely at the mercy of in-laws, most of whom she was meeting for the first time and who had no idea that she made such a pledge. She now wishes that the workshop organizers had invited stakeholders of the tradition (cultural brokers, men, grown children, religious leaders) to work out feasible alternatives to the practice. Perhaps she would have known what to do in reality; perhaps she would have made resolutions that work. With mounting pressure from her in-laws to be inherited, Mikayi decides to escape back to the city before she gives in. She is aware, however, that with this decision, she may not be permitted to visit her parents or married sisters; she may not be allowed to put up a home in the ancestral land of her late husaband; and when her children grow up, she may not participate in their marriage ceremonies. But the potential loss of social acceptance was less threatening to her than the risk of contracting HIV through inheritance.

Mikayi's story is a typical example where African cultural practices are targeted for elimination without bringing into the field all the players. Working and living with women of the Luo region made it clear that only those strategies that start in conversation with local peoples and culture will prove to be effective over the long term. The customary lack of attention to indigenous voices has resulted in policies that largely ignore the social and economic contexts wherein HIV is contracted, and it is only by appreciating what the individuals know (and tell) that we can understand why they chose to follow certain routes and reject others.[5]

HIV/AIDS Interventions in Sub-Saharan Africa: Spatializing Culture into "Us" versus "Them"

I trace the ineffectiveness of HIV prevention strategies in the Luo region to erroneous conceptions of "space" and "time" embedded within HIV policy discourse. Treating Africa as a monolithic, undifferentiated space, one subordinate and backward compared to the West, lies at the heart of misguided HIV and AIDS intervention policies as they are funded and deployed from the United States and Europe. Depictions of HIV/AIDS in sub-Saharan Africa replicate colonial mentalities and sidetrack pressing issues relevant to designing and/or executing effective interventions.

First and foremost, policy discourses and practices treat sub-Saharan Africa as a monolithic entity when in fact the region includes 46 different countries,[6] with at least 900 ethnic communities, close to 2000 subethnic groups, and 300 language families (Treichler, 1999, p. 124). Zambia alone, for instance, has more than 70 languages, while Kenya has 43 ethnic communities with distinct languages (GOK, 2001c). In Kenya, the Luyia and Kalenjin communities comprise nine or more relatively distinguishable subgroups who not only have different linguistic dialects, but among whom certain cultural traditions, such as WI, are carried out differently

(Burudi, 2000; D. O. Isoe, personal communication, May 17, 2001; H. Mahindu, personal communication, May 17, 2001). As such, the conception of sub-Saharan Africa as a single entity obscures salient features specific not just to different countries or different ethnic communities within each country, but even to a single community.

Temporal difference

Euro-American based health policy-makers, political leaders, and funding agencies view AIDS in Africa through a lens of *temporal difference*, perceiving the subcontinent as a backward version of the West. The construction of the so-called "African AIDS" by proponents of "Africa-as-backward" discourse build their position on what Wilton (1997, p. 3) describes as the intersection of a range of pre-existing discursive "packages" about the sub-Saharan subcontinent as a single spatial entity, such as in terms of darkness, barbarism, culturally backward, and sexually wild, insatiable, and exotic (see also Jarosz, 1992; Setel, 1999; Gausset, 2001).

Chirimuuta and Chirimuuta (1989), Schoepf (1991), Jarosz (1992), and Gausset (2001) observe that Western popular and biomedical accounts of AIDS contain numerous examples of racist discourse about African culture. Africa is often designated as the ecological/biological source of AIDS, a scientific discourse that implicitly or explicitly holds "backward and exotic customs" as responsible for the transmission of the disease to humans. Literature on AIDS in Africa consistently characterizes Africans across the board as irredeemably oversexed, abnormal, untamed, dangerous, and different from other peoples elsewhere (read: whites):

> Numerous sexual partners, mobility and prostitution are frequently cited as the causes of heterosexual transmission of AIDS in Africa. Neglecting the political, economic, and social context of disease transmission, discussion of African sexuality and sexual practices is burdened with value-laden assumptions. . . . Blaming the victim encompasses assertions of male virility, female passivity and helplessness, and the view of Africans as "oversexed" and incapable of modifying risky sexual behavior. (Jarosz, 1992, p. 112)

As Agot (2001) and Gausset (2001) contend, African culture is being taken up wholesale to explain, justify, and excuse non-existent, inadequate, or failed intervention campaigns.

Within the "Africans-as-backward" discourse, the fight against AIDS is often presented as a fight against "cultural barriers" that are seen as promoting the spread of the HIV virus. According to Gausset (2001), a Danish anthropologist who conducted a study on HIV/AIDS intervention among the Tonga of Zambia, this attitude is based on a long history of Western prejudices about sexuality in Africa, which focus on its so-called exotic aspects only (such as polygyny, female genital cutting, dry sex, WI), with the assumption that the practices are invariably incompatible with a safer behavior and that their eradication would ensure protection of the people. The author begins his soul-searching article by reporting how, while doing a study in Zambia, a Danish journalist interviewed them about their work and weeks later (mis)reported the interview in a major newspaper in Denmark, which he titled, "The fight against AIDS [in Africa] is the fight against culture," transforming it to fit what he (and the readers) expected the situation to be in Africa regarding AIDS and sexuality.

Gausset cites several examples in scientific literature to argue that placing the blame on "African cultural practices" is part of a widespread discourse of accounting for the epidemic in the region. For example, in what he aptly termed "AIDS and the study of African sexuality," the author maintains that AIDS has made it legitimate once again to study sexuality in Africa, and since lives are at stake, anthropological research on sexuality can no longer be accused of being motivated by exoticism or an interest in "ethnopornography."

Colonial and postcolonial interest in African sexuality and prejudices on the part of Europeans was based on the compilation of anecdotes, rumors, unresearched media reports, and decontextualized data gathered mainly from travelers, missionaries, or colonial administrators (Chirimuuta and Chirimuuta, 1989; Schoepf, 1991; Jarosz, 1992). Much of the current "scientific" data, particularly in the early years of the epidemic, revived the hitherto thinly veiled prejudices. By focusing on cultural practices as barriers to HIV prevention and calling for their elimination, HIV policy discourses ethnocentrically frame local cultures as either irrational (lacking any cultural or social explanation) or immoral. Little effort is made to understand the broader sociocultural context in which these practices are embedded. Gausset (2001, p. 510) points out that "African sexuality" has more often than not been studied only in so far as it was different from "European sexuality" (see also Wilton, 1997). Essentially, presenting AIDS prevention in terms of culturally defined "risk behaviors" diverts attention from more fundamental problems. First, it may transform the fight against AIDS into a fight between cultures, with one culture trying to impose its own conditions on the others. What is at stake then becomes cultural practices instead of AIDS. Second, it may have counterproductive effects of creating an imagined immunity for those who are not practitioners of the perceived "risk behavior." Third, it alienates members of the risk group who are stigmatized as "Other," an alienation rampant among the Luo people who see HIV prevention policy as insulting and nonsensical.

Gausset (2001) refers to this as "the double discourse on [HIV] prevention," in which HIV policy in the West operates under different standards as compared to African AIDS policy. He draws attention to the fact that in the West, when a correlation is found between HIV and certain practices, such as intravenous drug use or anal intercourse, the practices are improved and made safer. When a correlation is observed between HIV and an African cultural practice, the practice is to be eradicated. The author points out that AIDS prevention campaigns in the West do not suggest that subpopulations initially associated with high risk stop their practices; instead, these campaigns seek to make these practices safer. The same understanding is lacking in Africa; the cultural practices which are seen as barriers to AIDS prevention are completely decontextualized, and their importance for people's identities is overlooked because they are implicitly or explicitly treated as part of a "backward," "primitive," or "exotic" continent. Thus, in the West, HIV prevention strategies respect different cultural and sexual behaviors, while attempting to make them safer; in Africa, policy-makers adopt the opposite attitude, seeking to eradicate what are identified as "cultural barriers" to AIDS prevention.

What would it look like to undertake HIV prevention research or policy that did not otherize gendered African bodies and cultures? A growing number of scholars are critically exploring the discursive constructions embedded within HIV/AIDS policy in Africa, constructions that limit the effectiveness of the intervention strategies. Treichler (1999, p. 161), for example, calls on those involved with HIV research to recognize the "rare moments when medicine's narration of the *real* is interrupted long enough to glimpse other narratives; when social and cultural questions are periodically allowed to disrupt the tidy biomedical narrative." In order to accomplish this, researchers or observers must get close enough to local cultural practices and social milieus within which sexual practices are embedded, because "getting a good hard look requires uncovering the (biomedical scientific) rules of everyday practice and attempting to capture the meanings in what is being observed." Unfortunately, there has been a lot of what Treichler (1999) refers to as "textual cleansing and fortification," "smoothing data," and "omitting untidy anomalies" when our research does not conform to our theories (and prejudices) about African cultural practices and HIV.

A classic example of the Western colonial gaze in HIV research in this category would be Gould (cited in Treichler, 1999, pp. 253–4), who, dismissing the researched and documented heterosexual transmission of HIV in Africa, sought familiar "explanation" of the "African AIDS," and listed the following "causes" of "African AIDS": unadmitted homosexual or quasi-homosexual transmission, unadmitted drug use, the practice of anal intercourse as a method of birth control, the widespread use of unsterilized needles, a history of immune suppression and infectious diseases, scarification, clitoridectomy, circumcision (its presence in females and absence in males), and violent, excessive, or exotic sexual practices. In attempting to force explanations that conform to the popular Western perceptions of the "African AIDS crisis," writers like Gould succeed only in contributing to denial of AIDS and slowing efforts to halt the spread of infection, as both African leaders and publics defend African personhood against such stigmatizing attacks. As Schoepf (1991) amply demonstrated in her intensive anthropological investigation of circumstances that led to HIV infection in a sample of women in the Democratic Republic of the Congo, HIV spreads in African populations not because of exotic cultural practices, but because of many situations of everyday life that could be addressed through health policy without denigrating the cultural practices.

Spatial difference

To counter these hegemonic conceptions of African difference I subscribe to the position of *spatial difference*, which decenters "the West" as the normative goal for African societies and cultures. This is a spatial recognition of difference and vulnerability which acknowledges that sub-Saharan Africa may not just be following the West (in the sense of time), but might actually have its own story to tell. As Massey (1999a) argues, spatial differences as a matter of being advanced or backward deny the possibility that there might be alternative stories and trajectories emanating from different regions, and might explain why Africans appear different and unfamiliar to Western observers, as, indeed, Westerners appear different and unfamiliar to Africans. In other words, the position challenges us to go beyond simply the desire to understand or to describe the "Other" culture, and demands that researchers and policy-makers provide space for and give analytical weight to the stories and experiences of diverse embodied subjects affected by the AIDS epidemic (see also Spivak, 1988, cited in Craddock, 2000). These voices and groups of people will then be in a position to contribute effectively to the formulation and implementation of HIV prevention strategies. Only then can we displace Western interpretations that focus on Africa and Africans as backward or as "Other."[7]

In my view, the most important "take home" message about HIV/AIDS *is not* so much to look for spatial or behavioral commonalities that tie together what Treichler (1999) refers to as white or black AIDS, gay or straight AIDS, European or African AIDS, East or West German AIDS, Central or West African AIDS, foreign or native AIDS, guilty or innocent AIDS. *It is not* to look for a familiar explanation or to search for a coherent "rationale" explaining the AIDS that "makes White or Western sense" to the Western professional and technological agencies. Instead, *it is* to learn to listen to the multiple voices of the population groups most directly affected, and, ultimately, to challenge the entire discursive formation of international AIDS discussions that are often applied unthinkingly and, in some sense, imperialistically to diverse cultures – in what Jarosz (1992) would call "flattening places and people."

Conclusion

By ignoring social and cultural contexts of widow inheritance and HIV/AIDS, the intervention planners and practitioners have relied only on the plausibility of a biological mechanism to classify and target the behavior as a "risk" for HIV. In the Luo region, most of those for whom the interventions are designed use cultural conceptions of AIDS as *chira* to frame the association between the practice of WI and the risk of HIV. The result is that while the Kenyan state is campaigning for the elimination of WI as a potential risk behavior for acquiring HIV, the community is campaigning for a revamping of the practice as a potential check against the rapid spread of the disease. Because the intervention planners are insensitive to the ways in which the community is constructing AIDS, their message continues to be at variance with the local knowledge of the people, leaving them ill prepared to tackle pertinent issues about the practice.

I argue that unless the custodians of the custom of inheritance – members of the larger community – are genuinely involved in defining and setting up interventions, singling out the practitioners (widows) independent of their social and cultural contexts will remain ineffectual in addressing the forces that are responsible for the continuation of the risky components of the practice. Even when people understand the message perfectly, they weigh the relative social costs of refusing widow inheritance, and often decide to engage in the longstanding cultural practice anyway, however logical and compelling the evidence may seem to the scientifically trained intervention provider. This is because some of the secondary and tertiary consequences of eliminating the practice may be highly undesirable from the standpoint of many of the Luos, necessitating a careful consideration of the advantages and disadvantages of the practice within the prevailing cultural, economic, religious, and medical contexts.

In addition to demonstrating the need to fundamentally revamp HIV intervention policy, the dynamics in the Luo region shed light on the value of incorporating culture, identity, and deconstructive analysis of the body within medical geography and epidemiology more broadly (Agot, 2001). Feminist approaches that link the local and the global, and ones that highlight intersectionality in the context of women's lives (Mohanty et al., 1991; Katz, 2001; Nagar et al., 2002), are essential perspectives for teasing out the complex effects of health policy interventions on a global scale.

ACKNOWLEDGMENTS

Appreciation goes to Maylian Pak, University of Oregon graduate student, for designing the map of Kenya.

NOTES

1 Botswana (39 percent), Swaziland (33 percent), Zimbabwe (34 percent), Lesotho (31 percent), Zambia (20 percent), South Africa (20 percent), Namibia (19 percent), Malawi (16 percent), Kenya (13 percent), Central African Republic (14 percent), Mozambique (13 percent), Djibouti (12 percent), Burundi (11 percent), Rwanda (11 percent), Côte d'Ivoire (11 percent), and Ethiopia (10 percent) (UNAIDS, 2001, 2002a, b).

2 *Chira* is a wasting disease believed to result from breaching of taboos, including sexual taboos such as declining to be cleansed. Besides wasting, other symptoms include persistent coughing, vomiting, diarrhea, thinning of hair texture, loss of hair, and generalized tiredness and depression. Respondents were conflating *Chira* and AIDS because of the similarities in the signs and symptoms between them.

3 Thousands of people in the Luo region spoke to the team of researchers with whom I worked between 1999 and 2000 (Daniel Oguso of the Organization of African Instituted Churches, Tom Liwindi of Nomiya Church, Erastus Aroko and George Awino of Kisumu District Hospital, Ephraim Odeny and Noah Yogo of Nyanza Provincial General Hospital, John Osodo of the Department of Geography at Moi University, and Victor Onyango and Rosemary Achieng of Impact Research and Development Organization). We met seven days a week, over a 15-month period, with males and females from the Luo region, largely organized around 399 congregations from 68 religious denominations. These meetings were run as focus groups where a head count was taken at each meeting and if more than ten people were present we divided them into smaller groups for discussion. While not everyone spoke at every event, within the small groups the researcher-facilitator opened up conversations about widow inheritance, polygyny, and circumcision. We then shared the small group perspectives back with the large group at the end of the meeting.

4 Imperial and Sunset are four-star hotels in Kisumu, the main city in the area. Reference to them in the rural countryside was their way of saying that workshops are only held in the best venues that are out of reach for most of the rural folk, and that the AIDS war is fought mainly in workshops, not among those infected or at risk of infection.

5 It is important to point out that although the chapter focuses on forces that often make a widow succumb to inheritance, even against her will, I do not intend to portray widows as helpless to resist the tradition, or indicate that adherence to it is not changing. I also do not want to make it appear as if all relatives are out to pressurize the widow to be inherited – some would also try to persuade her to refrain from being inherited, even against her own will (for detailed discussion see Agot, 2001).

6   Angola, Benin, Botswana, Burkina Faso, Burundi, Cameroon, Cape Verde, Central
    African Republic, Chad, Comoros, Congo, Democratic Republic of Congo, Côte
    d'Ivoire, Equatorial Guinea, Eritrea, Ethiopia, Gabon, Gambia, Ghana, Guinea, Guinea-
    Bissau, Kenya, Lesotho, Liberia, Madagascar, Malawi, Mali, Mauritania, Mauritius,
    Mozambique, Namibia, Niger, Nigeria, Rwanda, São Tome and Principe, Senegal,
    Seychelles, Sierra Leone, Somalia, South Africa, Swaziland, Tanzania, Togo, Uganda,
    Zambia, Zimbabwe.

7   My reference to "Western" should not be understood to imply that African researchers,
    professionals, politicians, and those with the power to collect and disseminate informa-
    tion about HIV/AIDS are exempt from deploying ethnocentric, Western interpretations.
    Many of the African leaders and policy-makers involved in prevention programs come
    from very different socioeconomic backgrounds (and cultural backgrounds) than the
    groups targeted for HIV/AIDS intervention policies, and often closely adhere to what I
    am labeling as "Western" interpretations of HIV and AIDS policy.

BIBLIOGRAPHY

Agot, E. K. (1996) *The Impact of Polygyny, Widow Inheritance, and Urban Migration
   on HIV/AIDS Transmission among the Luo Women of Rural Kenya*. Research project
   sponsored by The Organization for Social Science Research in Eastern and Southern
   Africa (OSSREA) for the Seventh Research Competition on Gender Issues. Addis Ababa:
   OSSREA.

Agot, E. K. (2001) Widow inheritance and HIV/AIDS interventions in sub-Saharan Africa:
   contrasting conceptualizations of "risk" and "vulnerability." Unpublished PhD dissertation, University of Washington, Seattle.

Bajos, N. and Marquet, J. (2000) Research in HIV and sexual risk: social relations-based
   approach in cross-cultural perspective. *Social Science and Medicine*, 50, 1533–46.

Barnett, T. and Grellier, R. (1996) Cultural influence on societal vulnerability. In J. Mann
   and D. Tarantola (eds), *AIDS in the World II. Global Dimensions, Social Roots and
   Responses: The Global AIDS Policy Coalition*. New York: Oxford University Press.

Brown, M. (1995) Ironies of distance: an ongoing critique of the geographies of AIDS.
   Environment and Planning D: Society and Space, 13, 159–83.

Burudi, J. M. (2000) Is it wife or widow inheritance? *Daily Nation*, November 12.

Chirimuuta, R. and Chirimuuta, R. (1989) *AIDS, Africa and Racism*. London: Free Association Books.

Cohen, M. S. (2000) Preventing sexually transmitted infection of HIV: new ideas from
   sub-Saharan Africa. *New England Journal of Medicine*, 342, 970–2.

Craddock, S. (2000) Disease, social identity, and risk: rethinking the geography of AIDS.
   Transactions of the Institute of British Geographers, n.s., 25, 153–68.

Delor, F. and Hubert, M. (2000) Revisiting the concept of "vulnerability." *Social Science and
   Medicine*, 50(11), 1557–70.

Duncan, N. (ed.) (1996) *Bodyspace*. London: Routledge.

Dyck, I., Davis Lewis, N. and McLafferty, S. (eds) (2001) *Geographies of Women's Health*.
   New York: Routledge.

Gausset, Q. (2001) AIDS and cultural practices in Africa: the case of the Tonga (Zambia).
   Social Science and Medicine, 52, 509–18.

Government of Kenya (1997) *Sessional Paper No. 4 of 1997 on AIDS in Kenya*. Nairobi:
   Government Printer.

Government of Kenya (1998) *Situation Analysis for the National HIV/AIDS and STDs Control Programme (NASCOP)*. Nairobi: Government Printers.

Government of Kenya (1999) *Strategic Plan for the Kenya National HIV/AIDS and STDs Control Program for 1999–2004*. Nairobi: Government Printers.

Government of Kenya (2001a) *AIDS in Kenya: Background, Projections, Impact, Interventions, Policy*. Nairobi: Government Printers.

Government of Kenya (2001b) *Kenya National HIV/AIDS Strategic Plan – National Aids Control Council 2000–2005*. Nairobi: Government Printers.

Government of Kenya (2001c) *1999 Population and Housing Census: Counting Our People for Development Volume 1*. Nairobi: Central Bureau of Statistics and Ministry of Finance and Planning.

Hill-Collins, A. (1998). *Fighting Words: Black Women and the Search for Justice*. Minneapolis: University of Minnesota Pres.

Hudson, C. P. AIDS in rural Africa: a paradigm for HIV-1 prevention. *International Journal of STD and AIDS*, 7, 236–43.

Jarosz, L. (1992) Constructing the Dark Continent: metaphor as geographic representation of Africa. *Grografiska Annaler*, 74B(2), 105–15.

Kalipeni, E. (2000). Health and disease in Southern Africa: a comparative and vulnerability perspective. *Social Science and Medicine*, 50, 965–83.

Katz, C. (2001) On the grounds of globalization: a topography for feminist political engagement. *Signs*, 26(4), 213–34.

Longhurst, R. (2001) *Bodies: Exploring Fluid Boundaries*. New York: Routledge.

Massey, D. (1984). Introduction: geography matters. In D. Massey and J. Allen (eds), *Geography Matters! A Reader*. Cambridge: Cambridge University Press.

Massey, D. (1999a) Rethinking space and place: spaces of politics. In D. Massey, J. Allen and P. Sarre (eds), *Human Geography Today*. Oxford: Blackwell.

Massey, D. (1999b) The nature of geography: issues and debates. In D. Massey, J. Allen and P. Sarre (eds), *Human Geography Today*. Oxford: Blackwell.

Mboya, P. (1997) *Luo Kitgi gi Timbegi: A Handbook of Luo Customs*. Kisumu: Anyange Press.

McClintock, A. (1995) *Imperial Leather: Race, Gender, and Sexuality in the Colonial Contest*. New York: Routledge.

Mohanty, C. T., Russo, A. and Torres, L. (eds) (1991) *Third World Women and the Politics of Feminism*. Bloomington: Indiana University Press.

Nagar, R., Lawson, V., McDowell, L. and Hanson, S. (2002) Locating globalization: feminist re-readings of the subjects and spaces of globalization. *Economic Geography*, 78(3), 257–84.

Ntozi, J. P. M. (1997) Widowhood, remarriage and migration during the HIV/AIDS epidemic in Uganda. *Health Transition Review*, 7 (suppl.), 125–44.

Odindo, R. (1998) *Overview of intervention in Nyanza Province, Kenya*. The Futures Group International, The HAPAC project, November 25–27.

Ogutu, G. E. M. (1999) *Mila: Prolegomena to the Study of Luo Beliefs and Practices*. Nairobi: Pretthad Office Equipment.

Okeyo, T. M. and Allen, A. K. (1993) Influence of widow inheritance on the epidemiology of AIDS in Africa. *African Journal of Medical Practice*, 1(1), 20–5.

Oppong, J. R. (1998) A vulnerability interpretation of the geography of HIV/AIDS in Ghana, 1986–1995. *Professional Geographer*, 50(4), 437–48.

Otieno, C. (1999) AIDS forum begins today: leaders to discuss out-moded norms. *Daily Nation*, February 19.

Parker, A. (ed.) (1992) *Nationalisms and Sexualities*. New York: Routledge.

Radcliffe, S. and Westwood, S. A. (1996) *Remaking the Nation: Place, Identity and Politics in Latin America*. New York: Routledge.

Schoepf, B. G. (1991) Inscribing the body politic: women and AIDS in Africa. In M. Howe and M. Kaufert (eds), *Pragmatic Women and Body Politics*. Cambridge: Cambrige University Press.

Setel, P. W. (1999) *A Plague of Paradoxes: AIDS, Culture, and Demography in Northern Tanzania*. Oxford: Oxford University Press.

Sibley, D. (1999) Geography and difference: creating geographies of difference. In D. Massey, J. Allen and P. Sarre (eds), *Human Geography Today*. Oxford: Blackwell.

Smith, S. J. (1999) Geography and difference: the cultural politics of difference. In D. Massey, J. Allen and P. Sarre (eds), *Human Geography Today*. Oxford: Blackwell.

Treichler, P. A. (1999) *How to Have Theory in an Epidemic: Cultural Chronicles of AIDS*. Durham, NC: Duke University Press.

UNAIDS (2001) http://www.unaids.org/fact_sheets/files/Africa_Eng.html

UNAIDS (2002a) *AIDS Epidemic Update*, December (http://www.unaids.org/worldaidsday/2002/press/update/epiupdate2002_en.doc).

UNAIDS (2002b) *Epidemiological Fact Sheets on HIV/AIDS and Sexually Transmitted Infections*, 2002 Update (http://www.unaids.org/hivaidsinfo/statistics/fact_sheets/pdfs/Kenya_en.pdf).

Watts, M. J. and Bohle, H. B. (1993) The space of vulnerability: the causal structure of hunger and famine. *Progress in Human Geography*, 17(1), 43–67.

Wilton, T. (1997) *EnGendering AIDS: Deconstructing Sex, Test and Epidemic*. London: Sage.

Chapter 26

British Pakistani Muslim Women: Marking the Body, Marking the Nation

Robina Mohammad

The Body Geographic

Felicity Callard (1998, p. 387) notes how in recent years "'The body'...[has become] a preoccupation in geographical literature." The body after all is, as McDowell and Court (1994) point out, a site. It is a site onto which different power relations are mapped in particular times, places and contexts (Bailey, 1993). As Rabinow (1984, p. 173) notes, "power relations invest it, mark it, train it, torture it, force it to carry out tasks, to perform ceremonies, to emit signs." Butler (1993) contends that marks of gender, ethnicity, or sexuality as facets of identity are fabrications produced in discourse that translate into bodily performances. Performances are part of the production of "fictions" giving a regulatory coherence to identity, achieved "Through acts, gestures and clothes" (McDowell and Court, 1994, p. 732). Thus identity is not natural but "tenuously constituted in time, instituted in an exterior space through a stylized repetition of acts" (McDowell and Court, 1994, p. 732).

In this chapter, I analyse the female body, (hetero)sexuality, and the role of dress in the assertion of a distinctive collective identity. Drawing on research with British Pakistani women, I raise questions about identity construction processes that target women's bodies.[1] I am concerned first with how notions of purity and "community" identity intersect with the spatial containment of women — or attempts to confine women to what I term *transparent* spaces where their dress and behavior can be monitored by family and "community" members. Second, I am interested in the ways that "community" norms of appropriate females dress are imposed upon and negotiated by young British Pakistani women.[2] While linked to the construction of Pakistani identity in Britain, such processes articulate with global Islamization projects to mark the identity, or border, of the "community," on and through women's bodies.

Islamist discourses affirm women's sexuality only to deny women sexuality by positing it as a threat to the Muslim family, the foundation of Muslim identity. In this formulation women's bodies are understood as endangering their own (hetero)sexual purity, which is key to marriage and the formation of Muslim families. As I elaborate further below, the performative construction of space is key to maintaining the boundaries of collective identity and female sexual purity among working-class Pakistani communities in Britain. Certain public spaces are viewed as heightening the danger to women, danger that can be countered in many instances through community norms of appropriate female dress. I discuss these constructions of space in terms of transparent and opaque spaces — the former referring to public spaces within which "community" members can monitor women's behavior and the latter referring to public space that makes monitoring more difficult. I return to the construction and negotiation of transparent and opaque in the next section.

It is young women who are most affected by "community" norms of identity and dress. Younger, mostly unmarried women occupy the weakest social position beneath men and older women, who as mothers have a higher status. They are considered to present a greater danger to collective identity because of their sexualized bodies and relative lack of socialization in "community" norms due to their age. Thus young women, more than older women, are the target of "community" policing strategies to ensure that they perform collective identity as required.

Despite strong pressures, young British Pakistani Muslim women, whose subjectivities have been produced across a matrix of discourses, Islamist, consumerist, secular, and/or orientalist, do not unquestioningly adopt "community" norms. They translate and renegotiate these norms through their performance of collective identity, effecting what Butler (1993) would call a slippage in the repetition of performance. While the "community" seeks to interpellate new generations of British Pakistani women to perform, display, and parade collective identity, particularly through dress and demarcation of appropriate spaces, young women find ways to renegotiate, resist, or reject the "community's" control. Women can refuse the roles allocated to them or reinterpret these roles in ways not intended by the "community."

In the following sections I explore the daily lives, dress, and self-conception of young British Pakistani women. Their stories and experiences demonstrate that Muslim women constantly negotiate various boundaries that are aligned with Muslim collective identity. These boundaries are shaped by multiple binaries, including Islam/west, good/bad, sexed/pure, docile/wild, and victim/agent. Some reject "community" identity and resist regulation by manufacturing an apparent compliance that enables covert resistances separated by distance. At times they may seek to modify their own behavior to conform to the positive position on these binaries. At other times they may redefine the meanings of these terms in order to make them more inclusive. Yet another alternative is to reject these as positive terms and simply reverse the binary. This last group, rather than escape binary thinking, remains firmly within its grip.

Making Identity, Marking Bodies: The Moral Problematization of (Women's) Pleasures

The performance of Muslim womanhood in Britain, and the role of these performances in the construction of Muslim (collective) identity, relate to the rise of religious fundamentalisms and global constructions of Islam. Both Ahmed (1992) and

Kandiyoti (1993) specifically relate the rise of radical Islamism to the historical antagonism between Islam and the West. In ideological terms colonial and post-colonial assaults on Islam have posited it as primitive and "backward," claims problematically read from Islamic practices such as veiling.[3] Economically, the Muslim world has experienced the failure of modernization and development initiatives associated with the West. Yuval-Davies (1992) argues that the decline of secularism from the 1960s onwards and the rise of radically conservative religious fundamentalisms is related to the crisis of modernity, development, and the belief in rationality and progress (see also Afshar, 1994; Sayyid, 1997). Across the Muslim world modernization was promoted on the promise of development and economic prosperity. At the same time development discourses made modernization and progress synonymous with Westernization (Sayyid, 1997). The failure of modernization and the continuation of Western domination in the form of neocolonialisms saw the emergence of a revolutionary Islam that "created an area of [Muslim] cultural resistance around women and the family which came to represent the inviolable repository of Muslim identity" (Taraki, 1995, p. 645). Neither communism nor capitalism has been able to fulfill material, emotional, and spiritual needs, which according to Yuval-Davies (1992) leads to solace in religion.

In the case of British Pakistani communities, identification with Islam promotes a sense of empowerment as it transforms a marginal "community" within the heart of the West into a member of an imagined global community. Pakistanis in Reading (in Southeast England) in economic terms are highly marginalized. The rate of unemployment among Pakistani men is the highest of all ethnic groups. The 1991 Census suggests a figure of around 20.4 percent, compared with 6.8 percent for whites, 7.4 percent for Indians, and 16.4 percent for Bangladeshis and Afro-Caribbeans. Census figures show that there are no Pakistanis in the professional category, and only 15.8 percent in the managerial and technical category. The vast majority of Pakistani men are in skilled manual work. While the rate of economic activity, at 76.4 percent, among Pakistani men is comparable to that for white men, the rate for Pakistani women, at 29.2 percent, is only fractionally higher than the 28.6 per cent for Bangladeshi women, who have the lowest rate for all ethnicities (British Census, 1991).

Clearly multiple exclusions, based on colonial legacies as well as relations of race and class, intensify the need for a distinctive Muslim identity among working-class Pakistanis in Britain. In this context, the "community's" commitment to Islam sits alongside a discursive if not material rejection of Western liberal modernity (Choueiri, 1992). This rejection operates on many levels. For example, the opposition between Islam and the West is evident in the Islamic texts circulating among Pakistanis in Reading. Islamic texts by Sarwar (1980) and Khan (1995), I was informed, were used for the teaching of Islamic education to young Pakistanis. Written in English, they specifically target young people who would not necessarily be literate in a Pakistani language — Urdu, Punjabi, or Mirpuri. Sarwar (1980) and Khan (1995) rely on and reinforce the opposition of Islam to the West as a means of persuading readers of the superiority of Islam. In his foreword, Sarwar (1980, p. 7) draws attention to the dangers of the non-Islamic environment. He explicitly states that the Islamic education of future generations is key to the construction of a Muslim identity, an education that represents a "great challenge" in a non-Islamic society.

According to these texts, the formation of future generations takes place on the mother's lap, within the Muslim family. In a non-Islamic society weight is given to the education and transmission of values through the family, particularly through appropriate mothering. Thus the Muslim family is the foundation of Muslim collectivity (Afshar, 1998). As Foucault (1979) points out, the family is a key site for the production of subjectivity. In this sense it is a unit for the reproduction of the collectivity both biologically *and* culturally. Only marriage is seen to ensure that children who are born to "women are not only biologically but also symbolically within the boundaries of the collectivity" (Yuval-Davies, 1992, p. 285). As such, the guardians of collective identity also become the guarded (Afshar, 1994; Badran, 1995).

Women's (hetero)sexual purity is key to marriage and the Muslim family and these in turn are significant for the production of "community" identity, bringing a focus on women's bodies. As one "community" elder I spoke with argued, "women must be protected, they must be shielded from corruption, because they are in charge of our future generations."[4] They are responsible for the transmission of religion and culture to their children. Thus, both the formation of Muslim families and the role of motherhood demand a sexual regulation of women that confines legitimate sexual activity to marriage. Marriage is the basis of the family and the parental family facilitates women's sexual regulation prior to marriage to ensure their (hetero)sexual purity on entering marriage. (Hetero)sexual purity is key to making, as well as maintaining, a good marriage (Rosario, 2002).

While the construction and maintenance of the Muslim collective identity requires women's (hetero)sexual purity, gendered biological and religious discourses construct male sexuality in such a way that women's (hetero)sexual purity comes to be threatened through encounters with unrelated men (for a detailed discussion of Islamic understandings of women's sexuality see Mernissi, 1975). This scripting of women's bodies is not confined to Islam (see Ortero, 1999). Conservative discourses both secular and religious have posited male sexuality as "naturally" "rampant and unfettered" (see McDowell and Court, 1994, p. 730). As a corollary, women are scripted as "naturally" sexually provocative. In the case of Islamic society, women are associated with *fitna* (chaos) (Badran, 1995).[5] It is in this context that processes for Islamization of collective identity proceed through and legitimate the targeting of women's bodies. The threat to Muslim identity emanating from women's (hetero)sexuality is heightened within particular spatial contexts, a threat which legitimates the spatial containment of women.

The (In)visible Presence under the Eye of Power[6]

> A whole history remains to be written about *spaces-* which would at the same time be the history of powers. (Foucault, 1980, p. 149)

> Discipline proceeds from an organisation of individuals in space. Once established, this grid permits the sure distribution of the individuals who are to be disciplined and supervised. (Rabinow, 1984, p. 17)

Spaces beyond the British Pakistani home are regarded by the "community" as highly dangerous for women's (hetero)sexual purity and thus "community" identity. It is a view that not only draws on Islamist constructions of gendered divisions between public and private, but also is adapted to the diasporic context that challenges identity formation in distinct ways. Radically conservative, Islamist, religio-cultural-nationalist discourses in Islamic states represent space as fundamentally gendered and further divided into public and private, Western and non-Western. Feminist theologian Riffat Hassan (quoted in Rahman, 1996, p. 92) notes how Islamist ideologies relegate "women . . . [to] a sphere and the rest of the world belongs to men." As in other conservative contexts, the sphere of the home is the "private" space that gets naturalized as the domain of women.

In the diasporic context in Britain, working-class Pakistanis faced with the need to maintain and strengthen collective identity negotiate a very different kind of space as compared to Islamist states: they must contend with liberal Western urban landscapes. The "community" seeks to address the impact of these landscapes on women's (hetero)sexual purity, which (as discussed in the previous section) undergirds community identity. This creates many contradictions and paradoxes for women. In Britain state laws do not regulate against the free mixing of men and women in public spaces, as would be the case with an Islamist state. Families and "community" members thus actively seek to monitor and control women's dress, behavior, and presence in public spaces.

The threat from masculine, liberal Western space is heightened by its level of opacity, a term I use to denote those public spaces particularly resistant to surveillance by "community" members. In the town of Reading, the site of my study, it is not so much Pakistani Muslim women's visible presence in transparent public spaces but their presence in insufficiently regulated *opaque* spaces that provides cause for concern. As many social theorists have commented, urban streets allow anonymity and with this the freedom to disappear and escape from the gaze of the Pakistani "community" into the crowd of city life (Wirth, 1938; Simmel, 1971), thus frustrating the "community's" attempts to police women. For the "community," therefore, dangerous spaces are not only the dark alleyways that may characterize general fears about women's safety in public spaces but well lit "unregulated" spaces that might be considered "safe" for women by the wider white "community," such as the high street, certain spaces of the education system, and the labor market.

According to many women respondents, "community" norms construct colleges, universities, and the labor market as spaces that offer dangerous possibilities for sexual immorality, social mobility, and financial independence for women.[7] Both higher education and waged work facilitate assimilation into the wider society. For women, unlike for men, such assimilation is not easily reconcilable with their accepted place within the "community." As the bearers of collective identity, the assimilation of women into white society is more visible and has the potential to affect the "community" more than if a man were to do so. Moreover, when women marry they become part of their husband's family and the religion of the children follows that of their father. Marriage to a white non-Muslim would then also lead women out of the group.

The question of spatial regulation was frequently raised in conversations with younger British Pakistani women. Saima (aged 17) explains, "there is this idea that if you hang around town especially as a girl you are bad." Another 17-year-old, Marina, explained that "girls are not allowed out even to town for personal shopping, for longer than say a couple of hours. Boys are allowed to be out all day, therefore they can go to the cinema etc. without it being noticed [by their families]." Parents can control the school environment to some extent by choosing single-sex or where available Muslim schools (see Joly, 1986; Kelly and Shaikh, 1989). Natasha (also aged 17) notes that even some mixed-sex state schools have in place some form of surveillance of young Pakistani Muslim women as a reassurance for their parents:

> What they [parents] believe is at school you have the supervision of teachers. At our school that is definitely what it is. If they find some of the Asian girls doing things they shouldn't be doing then, because the teachers are so aware of Islamic culture and the Hindu culture and all the other multicultures, they do inform the parents and the parents are aware of that.

Thus state schools (single-sex and even mixed-sex) are preferable to college or university due to perceptions that they provide more adequate supervision (see Mohammad, 1999).

Spaces of education and work not only enable encounters with men but might also require them. Farzana, aged 17, an academically able young woman who was in part-time paid work during the summer vacation, told me that in her job she has to "deal with men and shake hands with them at times [and therefore, she] can't be a good little Muslim girl." Sixteen-year-old Rhea felt that "parents say that you can't work with boys but this is an English country, you can't change that." Parents, particularly those who are less educated, feel that they have little influence over these spaces and seek to assert control over their daughters to prevent and/or minimize access to unsuitable spaces. These controls limit the opportunities available for young women, often guiding them toward marriage and the marital home.

Parental constraints and women's limited access to labor market opportunities seemed to be confirmed by the experiences of a large number of my respondents between the ages of 15 and 30 ($n = 25$), only five of whom were in employment. Of these, only one was in permanent, full-time paid work in a high street retail store. Another who was working as an administrator had gone part-time after having a baby. The remaining three, all of whom were high academic achievers, were working part-time during school vacations. Fifteen respondents were in school or college, the majority studying for GCSE or A-level qualifications. Five of the respondents were at home (parental or marital) full-time.

Yasmine (aged 27), after succeeding in persuading her husband to allow her to undertake paid work part-time, had been unable to find it and was still at home caring for her children. Ruksana (aged 25) had decided against going back to paid work after having a baby. Shela (aged 27) was also at home caring for her children. Finally, two respondents aged 16, Farah and Zubia, were at home (at the time of the research) after being denied parental permission to continue with their studies. They were to remain in the parental home, looking after the home, until their marriage, after which they would move to the marital home.

Many explained to me that in this locality young Pakistani women's entry into the labor market is a recent phenomena.[8] Hitherto, young women were expected to leave school and remain at home until marriage. Women's entry into the labor market brought up new concern for their presence in opaque unregulated spaces. As Sabina reflected, "paid work is becoming the norm among the younger people . . . the older ones [generation] are not too happy about it for some apparent reason because you go out to work and this can happen and that can happen." Similarly, Natasha (aged 17) spoke of "the restrictions laid down by the parents, you can work in such and such a place but you can't work here and there because it is bad." Their comments indicate that there is an important distinction made between work that can be monitored by community members and work that occurs in more opaque spaces.

One respondent, Sharon (aged 20), is still bitter about the way in which she was forced by her father to leave her office job with prospects.[9] Despite the fact that her employers had permitted her to maintain Pakistani dress (one of her father's earlier conditions for her entry in to the labor market), her father would not entertain the possibility of her continuing in this environment. Soon after her appointment he realized that his daughter's office environment contrasted with the space of school. The office job did offer more money but it was also a space that undermined parental supervision and regulation, as well as providing opportunities for promotion. In contrast, her father approved of her work as a shop assistant in a town center store. Unlike the office environment, the shopfloor fixes her both socially (there are fewer possibilities for promotion) and spatially, creating a more transparent and panoptic space. In this space Sharon performs her duties as shop assistant wearing a modified uniform that identifies her as a member of staff, engaging with the customers. Sharon's body is disciplined and self-regulated at all times by the public gaze and the presence of CCTV cameras monitoring the activities of the shoppers and the staff. Moreover, while in the office Sharon could often slip out for half an hour unnoticed, on the shopfloor she may only leave for lunch and her two daily breaks at set, pre-established times. Sharon's father (who lives nearby) can call in at any time during the day on the pretext (if he should need one) of browsing the merchandise and at the same time ensure that his daughter is exactly where she claims to be. In view of the possibility for policing offered by the shopfloor, Sharon's father had adjusted his demands that she wear Pakistani dress in the workplace. Pakistani dress is not only a visual marker of the "community" but also important in countering and containing women's (hetero)sexuality. Alternatively, it is significant in providing a means of negotiating and resisting "community" constraints and/or membership.

Women's Bodies and Dress: Marking the Border, Making the Border

Women . . . the privileged signifiers of difference. (Kandiyoti, 1993, p. 376)

Through their dress women's bodies perform the boundary that marks the imagined, psychic and physical space of the "community." Pakistani Muslim women are posited as the face of Islamic society and required to comply more stringently with Islamist expectations (Yuval-Davies, 1997). Women's dress has been the focus of attention in a variety of Islamist states. Afshar (1998, p. 197) comments that today, "Almost two decades after the Islamic revolution, . . . the only visible sign of

Islamification that remains in Iran . . . is the presence of veiled women." It is also why the process of Islamization in Pakistan proceeded through a focus on women's bodies and attire (Khawar and Shaheed, 1987).

The term *purdah* combines the notion of dress with the idea of private space. Mohammed Anwar (1979, p. 165) notes, for example, how "apart from veiling in the traditional sense, back home [in Pakistan *purdah*] . . . also refers to the restrictions on the physical movement of Pakistani women." *Purdah* signifies dress as a form of coverage that creates a private space secluding women. There are many forms of *purdah*; for example, the *hijab* covers the head and hair, while the *burka* is a two-piece garment that totally covers the body and face. The *chador* covers the head and swathes the body. All three are invested with notions of religiosity, morality, and sexual purity. Islamic texts circulating within this Pakistani "community" in Britain advise that women must refrain from arousing "man's base feelings" by concealing the contours of their bodies in order to safeguard their sexual purity and *izzat* (honor) (Sarwar, 1980, p. 183). Appropriate dress is a means to reduce the social risk posed by Muslim women's entry into the world beyond the protective confines of the home. These garments not only work at the material level by desexualizing the body by covering the hair, and/or hiding its sexually provocative contours, but also symbolize the purity of intentions of the wearer. In the Reading-based "community" few women adopted the Muslim dress. The "community" lays stress on Pakistani national dress, widely understood to be a Punjabi suit, a long tunic top and baggy trousers and matching dupatta (a wrap/scarf, see figure 26.1).[10]

Through the Punjabi suit women's bodies mark community identity among British Pakistanis. It provides adequate coverage while allowing Pakistani women to be readily identifiable in transparent public space. Sharon's father, in the story related in the previous section, initially allowed Sharon to enter paid employment on the condition that she wear Punjabi clothing. She was only able to enjoy this facility in an office environment where employees were not dealing with members of the public. With the recognition that the office environment was a relatively opaque environment that undermined his ability to view and regulate her behavior within it, Sharon's father demanded that she find paid work in a more transparent space such as the shopfloor, even if this meant that she had to swap the Punjabi suit for a uniform. For Pakistani Muslims, wearing Pakistani and/or Muslim dress is clearly linked to the assertion of "community" identity and the maintenance of female purity. Yet this does not prevent some young women from using this dress as a mechanism for resisting community norms: control over the communication of purity facilitates women's negotiations and resistances. For example, for Tara (aged 17), who has adopted the *hijab*, *purdah* is a means of communicating her (hetero)sexual purity in order to negotiate parental permission to attend a mixed-sex college to continue with her studies. But *purdah* is also a means of creating an outward façade of compliance, a point that I return to below.

My own entry into the "community" for the purpose of conducting the interviews also demonstrates the importance of dress and the performance of identity in the British Pakistani "community." My identity as a "Pakistani" was seen by the funding body as well as the "community" as crucial to implementation of the research. In order to undertake the research I had to perform Pakistaniness in a variety of contexts. In other words, confronted by an imagined but real

Within the image:

OPULENT ELEGANCE

Naheed Asim presents her bridal designs in an array of rich silks and luxurious jamavars with intricate embroidery in dabka, bead and thread-work. Here is a sample of her rich traditional collection.

Worked *jamavar* bodice in *zari* complements this plain fuchsia shirt worn with a contrasting green shalwar and a chatapatti dupatta

Figure 26.1 A Punjabi suit for evening wear.
Source: Women's Own (Pakistan), 2001.

"community" border I, like my respondents, had to negotiate and perform group identity through my outward appearance. My experience and appearance, whose visible markers are my Muslim name, skin color, and Punjabi dress (Salwar Khameez), were regarded as a passport into Pakistani "communities."

During interviews and informal conversations, "community" members constantly negated the differences between us, significantly of education and class, to perceive me only in terms of a shifting position on the Western/Pakistani binary (for a detailed discussion of this see Mohammad, 2001). In many respects British Pakistanis were able to position me by reading certain iconographic markers as well as my very basic, general understanding of customs, and fluency in written and spoken Urdu/Punjabi. Remarks like "I can tell you because you'll know what I'm talking about" confirmed my position as an "insider" by acknowledging the presence of otherness "outside."

At the same time, there were many dimensions of my performed identity and experience that positioned me as an outsider to the women I interviewed, a position that often shifted over the course of the interview. Some respondents only opened up to me after they were convinced of my Pakistani credentials, yet in other cases it was my perceived Westernness that my interviewees valorized. For my interviews with first-generation women I asserted my Pakistani identity by dressing in Punjabi dress, while for interviews with second-generation women I often remained in Western dress. In one interaction with a second-generation young woman I was dressed in jeans, while she wore Punjabi dress. In this case my "Western" dress overrode my color and name to exclude me from the position of an "insider," generating discomfort, defensiveness, and apprehension on the part of the young woman. Halfway through the interview she appeared to soften toward me. I realized that she had seen me glancing over at and reading a leaflet that was in Urdu. In contrast, for other young women, my "Westernization," confirmed by my position at that time as a (relatively) independent divorcee, facilitated disclosures that performing the role of a "typical" British Pakistani woman would have foreclosed. Because I was an "insider," respondents felt more able to be critical of the "community" in my presence without feeling that they were being, or appearing to be, disloyal to it. This was evident from the comments, particularly of first-generation women, such as "we shouldn't really say this since it is disloyal [to the 'community'] but since you are one of us it is all right."

It is worthwhile examining the relationship of the researcher's ethnicity to research findings through a comparison with Dwyer's (1999) findings. Claire Dwyer's study of young British Pakistani Muslim women was based on in-depth interviews that took place in schools. The disclosures made to me reflected my respondents' perceptions of my position as Western and/or Pakistani, a position that was central to the issues they raised. Our conversations differed in significant ways from the concerns of Dwyer's (1999) respondents. In my study there was less emphasis on racism, something encountered outside in the white world beyond the "community," and more focus on the constraints of "our culture" produced within and by the "community." It seems that perceptions of Dwyer's position on the Asian/Pakistani, white/West binary, as well as her experiences as a former student and former school teacher, were significant in the themes and tone of the discussions and the disclosures made in her presence by the respondents. Her study suggests that an important issue for young British Muslim women, across class, is the negotiation of "English" spaces through dress, in order to "fit in" and counter racism. By contrast the significant issue for my respondents was the "community" border, its negotiation and resistance through dress.

Negotiating the Border, Mounting Resistances

> With chains of matrimony and modesty
> You can shackle my feet
> The fear will still haunt you
> That crippled, unable to walk
> I shall continue to think.
> (Kishwar Naheed, a contemporary Urdu poet, from Khawar and Farida, 1987)

In the face of parental and "community" restrictions British Pakistani women mount resistances through the deployment of Western discourses and norms, which involves a reassertion and reconfiguration of identity. Sharon had very strong feelings about religion, which she recognized as providing the legitimation for "community" restrictions:

> Religion is manipulated by men to suit their own requirements [to the point that] it has now become one big lie. If there was a God up there surely he would be fair. Why then are women made to cover up from head to toe even in hot climates when men can wander around in shorts? If women's semi-naked bodies are a temptation for men what about the temptation for women from, say, men in shorts. These rules are not the work of God but men.

Resistance was aimed toward not only "community" norms, but also the constraints and racism of British society. Shela (aged 27), married with five children who has experienced racism, affirmed her position as a Pakistani: "English people are different from us. They don't understand us. They think, because they are white they are better than us. They aren't."

Other respondents, like Natasha, contested the meanings of Asianness by insisting: "My brother m'ashallah has got his Asianness but he has also moved on, he hates the idea of sticking to all these negative things we've got." She legitimated this progressiveness on the part of her brother with reference to Islam: "He can move on [because] Islam is all about progression, improvement, not about sticking to your origins." This has certain resonances with Sharon's quote above. These young women may resist Islamist discourse that the "community" draws on to legitimate the controls on women's lives. They may contest Islam or seek to redefine its terms as a way to negotiate the boundary of the "community." For example, Natasha, who wanted to study at university and travel, sought to make these compatible with "community" membership through reference to Islam.

Twenty-year-old Selina did not contest the meanings associated with Englishness (that is, understood as white Western) or a Pakistani position but was trying to establish her place *vis-à-vis* both. Like South Asians in general, the "community" demands that women maintain a visual contrast to white Western women. As Hiro (1991, p. 151) notes, South Asians are concerned that their daughters should "dress modestly [and] must not 'like the cheap English girls' exhibit their anatomy." Selina, however, did not see herself as Pakistani: "I see myself as exactly half English and half Pakistani. When I step outside the [front] door I'm me. If I wear Asian clothes outside the house I don't feel normal." She suggested that the Asian self she was expected to present/perform in the house is not one that recognizes and registers her "Englishness." I found that these women were drawn to me because they believe that my "Westernness" acknowledges theirs. Sharon, also aged 20, made this explicit:

> I only agreed to speak to you because of what I had heard about you being divorced and living on your own. When I met you and your children I thought that they were very English, I thought she's alright and decided to tell you everything. Had you dressed in Punjabi dress and looked Pakistani . . . I would have refused to speak to you.

So while Shela declared that "I value our tradition, I am proud to wear our traditional costume [Punjabi dress]," Sharon wore her resistance on her body as she sat in front of me in my dining room, dressed in a long skirt and long sleeved blouse.

Afshar (1994), drawing on her study of Pakistani Muslim women in Yorkshire, remarks on how historically South Asian women have never revealed their legs even when they have displayed their waists. Thus the skirt, irrespective of whether it reveals the legs, becomes a symbol of the West and Western immorality. When a British Pakistani woman wears a skirt it draws comment from all sections of the "community" that range from outright denouncement to thinly veiled moral disapproval to celebration. Natasha linked the wearing of the skirt to immorality when she observed of British Pakistani girls at her school: "From what you can see the things they do to get the attention of those boys . . . I mean they're smoking, wearing their skirts, flirting around, running around doing things that probably even these English girls wouldn't do." Sharon is fully aware of these perceptions but insists that at school, she: "hated everything Asian. I always hung out with English girls. When we had non-uniform days all the other Pakistani girls would dress in Punjabi dress but I just wore my uniform because I felt it was more me." By rejecting Punjabi dress she sought to resist all that is invested in it: notions of difference, subservience, subordination, docility, and virginity, attributes that are valorized at home but attract racism at school (see also Dwyer, 1999).

At the same time, wearing Western clothes provided a means of performing her difference from the Pakistani "community." In the environment of the interview Sharon felt enabled to perform the role of an "English" girl, most immediately through visual markers such as dress, but also through physical desires, which even when felt are not to be spoken off by "good" Pakistani Muslim women. While she openly rejects the markers of "community" and its controls, other respondents used dress innovatively to negotiate a variety of spaces.

Goodwin (1995) speaks of the "games" (see also Dwyer, 1999) engaged in by elite Muslim women in Pakistan who have been educated abroad, produced within Western value systems, who seek to maintain relatively Western lifestyles back in their home country. These women are forced to perform an outward compliance with local norms through the maintenance of *purdah*, while covertly resisting it. As she writes:

young [elite] woman [in Pakistan] may leave home covered from head to toe, and in the company of a chaperon. Once she arrives at her destination and sheds her *Chador* she may be wearing a miniskirt and she may exchange her chaperon for her boyfriend almost as quickly. . . . Just as in the West, her boyfriend may proposition her, and she may accept, having taken the precaution of stocking up on the contraceptive pill while . . . abroad. And should she have the misfortune to become pregnant, she'll go abroad again, on the pretext of a shopping expedition, to have an abortion. (Goodwin, 1995, p. 68)

Similarly, in the British context, Pakistani women, in an environment that enables resistances, may also resist "community" fictions by maintaining an outward compliance. One respondent referred to a young woman who travels from her home in one locality "in trousers and gets changed into a skirt at work [in another locality]. She thinks [this] is the only way she is able to let the boys know that she's an OK person." Young women struggled to manage their sexuality *vis-à-vis* the performance of the "community" border that often drove them to resist its controls.

Young women understood resistances in dualistic terms, promoted by their formation within a discourse that offers only two possible positions — good or bad. Repressive technologies that work through the emphasis on sexual purity, as in Catholic cultures (see Lee, 1991), posit a virgin/whore opposition. Both Mariam and Sharon sought to embody this binary. They drew on and played with the notion of virginal, docile Muslim woman as a means of negotiating the border of the "community." They sought spaces away from the gaze of the "community" to assert alternative oppositional identities that were not compatible with the Pakistani Muslim "community" that claims them. Sharon explained how "playing the perfect daughter [the virgin] in the home, by dressing in Punjabi dress and covering my hair, [buys me trust and freedom] to be the imperfect daughter [the whore] outside it."

Both Mariam and Sharon (as her "imperfect" version) presented themselves to me as sexual goddesses. This not only supports orientalist eroticization of Muslim women but also confirms the fears of the "community" and legitimates the Islamist understandings of the dangers associated with women's sexuality. Mariam (aged 20) had had a non-consultative arranged marriage with a relative in Pakistan and was awaiting his arrival in Britain. She talked to me about the impact of her body on her current boyfriends (whom her family are unaware of and who are also unaware of each other's existence): "Nadim and Vince, two of the guys I'm seeing, claim that just sitting next to me sends them into ecstasy."

As Foucault (1980) argues, the continuous vigilance against women's sexual activity has the paradoxical result of sexualizing the body. Thus, in a discussion on the subject of virginity and marriage, Sharon declared that "I started wanting guys at the age of ten." She felt the need to explain to me how "a woman's body is far more sensitive than a man's — they only have to touch you here and there and you are off." As the "perfect" daughter she might have agreed to have an arranged marriage with her first cousin in Pakistan, but in keeping with Western notions of love and sex she stated that "as far back as I can remember I have thought that my first one was going to be special, someone who wanted me for my personality, my looks and my body, not a British passport."

I was recruited as an audience to her sexual history through the disclosure of information and photographs of her boyfriends. Sharon explained to me how she selected her boyfriends from different localities and conducted these relationships away from the eyes of the "community" in her home town to minimize the risk of discovery. She believed that her parents and the "community" had no knowledge of their existence but the boyfriends in question were also unaware of each other. McDowell and Court (1994, p. 733) note that the body is controlled:

> through the manipulation of patterns of desire, fantasy, pleasure, and self-image, in which gendered [and ethnicised] notions of appropriate behavior and expressions of sexuality in this wide sense are important mechanisms in the establishment of particular norms of acceptable behavior [and ways of being].

Sharon delighted in narrating to me how her presence made men lose control. She talked, for example, about her expeditions to other localities where she sought to distract and entice men by "wearing jeans and a crop top with my waist showing . . . [this has the desired effect because] every few minutes you get looks, beeps, whistles. You have to give them the vibes." Sharon drew my attention to her body and how good it was looking because she has been dieting and had lost 14 pounds in weight. She narrates how while most Pakistani (both local and from Pakistan) men would expect her to keep it hidden, under "drab" Asian clothes, from the eyes of unrelated men, "one of my boyfriends, [a] Pakistani from Canada [who] lives in Middlesex, believes that if you have a good body show it."

Valentine (1999, p. 334) has commented on links between the appetite for food and carnal desire or pleasures of the flesh found in Christianity. Early Christian practices reflected the struggle of the mind to retain control over "the sinful demands of the flesh" in terms of both a physical and a sexual appetite. A contemporary translation of this struggle takes place when white heterosexual women, in order to enhance their sexual attractiveness and sexual pleasure, seek to comply with the norms and ideals of health and beauty. Here power works less through repression than through "control through stimulation. 'Get undressed — but be slim, good looking, tanned'" (Foucault, 1980, p. 57). Thus women are driven to controlling their physical appetites by following the rules of slimming diets in order to manipulate body weight. A mind and body struggle occurs where the mind seeks to assert control over the physical appetite while the body desires food (Valentine, 1999).

Muslim women are constantly negotiating multiple boundaries that are aligned with Muslim collective identity, including binaries between Islam/west, good/bad, sexed/pure, docile/wild, and victim/agent. At times they may seek to modify their own behavior to conform to the positive position on these binaries. At other times they may redefine the meanings of these terms in order to make them more inclusive. Yet another alternative is to reject these as positive terms and simply reverse the binary. For this last group, rather than escaping binary thinking, they remain firmly within its grip.

Conclusion

Radical Islamism is a response to Western imperialism, neocolonialism and subordination. If, as Goodwin (1995, p. 9) argues, "Women have become symbols of men's Islamic commitment," then the precise image presented and performed by

Muslim women is significant in so far as it communicates masculine control and regulation over them. The control of women's bodies is an expression of possession and territoriality. The fabrication of Islamic identities proceeds by marking women's bodies in terms of both targeting and inscribing.

Identity politics works by bolstering hierarchy (Bondi, 1992) so that those who feel more marginalized *vis-à-vis* the West and/or within the West, men, are able to assert power over those who are weaker still, women. The weakness of women's position leads them to build alliances with men, fathers, husbands, and sons. This is particularly visible in anti-colonial struggles; for example, in the Pakistani nationalist movement (Khawar and Shaheed, 1987). As mothers and mothers-in-law they become co-opted into repressing those who are weaker still: young women. In this way it is the weakest members who are subjected to more direct and visible control.

Young (unmarried) women, those who are identified as being at greater risk of seduction by the West and its values and as such suffer the greatest containment, are also those who are most likely to mount resistances. Their resistances are enabled by oppositional discourses such as Western liberalism, which as in the case of Sharon offer women a way to challenge Islamism. In other instances women draw on alternative interpretations of Islam and or Asianness, as Shela has done to challenge "community" expectations. Resistances did come at the cost of moral disapproval and guilt. But while Sharon and Shela were uneasy about the idea of transgressing, they viewed it as a cost that simply has to be endured.

Their resistance of community norms does produce paradoxes. Sharon, who had denied herself the pleasures of food to achieve her curves, harbored a desire to reveal and showcase her body for the male gaze. Like Selina and Mariam, Sharon contested Islamist fictions through the deployment of secular Western discourses which function less by external control and more through self-production. Sharon experienced "community" expectations as domination but did not experience as repressive the power of sexual desire that held her in its grip, that drew her focus to the space of her own and other (male) bodies and directed her resistances to parental control. Thus, contrary to Valentine's (1999) contention, she did not experience disempowerment by the sexual gaze of male bodies. On the contrary, she found herself in the mirror of this gaze. But her dependency on it for a sense of empowerment and the way in which it in turn impelled her to modify the space of her body, its weight, shape, and tone, means that it is only the balance between domination, self-surveillance, and subjectification that has shifted.

Some British Pakistani women, however, resist the idea of resistance, and embrace community norms. In the case of Shana and Natasha, their compliance with "community" regulation is purchased by their parents' love. As Saima describes it, "if you care about your parents you wouldn't want to do anything to hurt them." Both Shana and Natasha register the seductiveness of "Western" norms and the pressure to have a boyfriend. Natasha declared:

> I have a very caring family and friends of the family. Also [following] my own principles I do intend on saving myself. I'm preserving myself. If I did meet somebody I wouldn't go to the extent of having sex with them. You can go out and about with them which is enough, Also I would make sure that my parents knew, otherwise you feel like a *chor* (a thief) within yourself.

She suggests that covert resistances are a deception of those who are closest to you and whom you love. Other young women, however, comply with community norms because of their stake in collective identity in the context of a racist society. For Shela, compliance stems as much from the pressures of her family as from her sense of exclusion in British culture.

The border marking difference and identity for working-class Pakistani communities in Britain is a thin and constantly negotiated line, marked by controls and transgressions in the daily lives of women. This line sets apart the "good," "pure" woman from those who are "bad" and "wild." For young unmarried Pakistani Muslim women being "good" often involves resisting the pleasures of the flesh in order to maintain the sexual purity necessary to making a marrage and the (re)production of "community." Clothes provide a visual alert of women's positioning on this line. If the *hijab* symbolizes purity and goodness, then jeans or the skirt, particularly if it is short, communicate transgression. Longhurst (2000, p. 458) argues that the expectation of pregnant white women, whose body is sexed/sexual by virtue of its pregnant state, is that they "act modest, coy and demure in relation to their bodies in public places." Young Muslim women's bodies are associated with a burgeoning sexuality, or as the mother of one respondent put it, "a heat, a fire that can burn and destroy." The effects of this perceived burgeoning sexuality must be countered through the presentation of modesty, (hetero)sexual purity, and/or spatial constraints. While Islam is translated by some as "submission to the will of God," there are many varieties of freedom and submission.

NOTES

1   The data used in this chapter are drawn from research that was originally carried out in 1995 and 1996 for a two-part project in Reading, UK, on employment opportunities for Pakistani women, funded by the Department of Employment (via the local council). Initiated by "community leaders" at the Pakistani Community Centre, and undertaken in collaboration with Sophie Bowlby and Sally Lloyd-Evans, this research included participant observation and 50 in-depth interviews with women of different age groups that focused on women's experiences in the labor market. The analysis in this chapter draws primarily from interviews with younger women.

2   I place the term "community" in quotation marks to problematize the cohesiveness or presumed singularity of the British Pakistani population.

3   It is notable that while Westerners saw veiling as oppressive of women, Victorian corsets that physically constrained women's bodies have only recently attracted criticism from feminists (Sayyid, 1997; El Guindi, 1999).

4   Mr Abdul (who is in his fifties) is involved in the administration of the Pakistani Community Centre in Reading. All the names of respondents in this chapter are pseudonyms.

5   As "The Prophet [of Islam] said '[a]fter my disappearance there will be no greater source of chaos and disorder for my nation than women'" (quoted in Mernissi, 1975, p. 13).

6   The Eye of Power is the title of a conversation between Michel Foucault, Jean-Pierre Barou, and Michelle Perrot which was a preface to Bentham's *Panoptique* (see Foucault, 1980).

7   Young Pakistani men often attend college seeking a means to defer the entry into the labor market that is expected of them. The "community" imagines this as a place that young Pakistani men are attracted to in search of sexual adventure with young Pakistani

women. As a mixed-sex college it is regarded as a place that encourages sexual liaisons and permissiveness (Mohammad, 1996, 1999).

8 Young women's participation in the labor market was shifting in part due to changes in legislation that made the ability to support a fiancé from Pakistan a condition of his obtaining a British visa. Marrying young women to spouses born and bred in Pakistan is another way for parents to counter the effects of Western influences on second-generation women (see Mohammad, 1999). Yet paradoxically, new visa requirements made more young women seek out waged labor.

9 The pseudonym Sharon was chosen to reflect this young woman's vision of her own position within the Pakistani "community."

10 It is interesting to note that the Punjabi suit in itself is not a marker of an Islamic identity because it is a popular form of dress across South Asia among Sikh and Hindu, as well as Muslim, women.

INTERVIEWS (PSEUDONYMS)

Mr Abdul (late fifties), interviewed in 1995.
Saima (17 years), interviewed in 1994/5.
Marina (17 years), interviewed in 1995.
Natasha (17 years), interviewed in 1995.
Farzana (17 years), interviewed in 1995.
Rhea (16 years), interviewed in 1995.
Yasmine (27 years), interviewed in 1996.
Ruksana (25 years), interviewed in 1995.
Shela (27 years), interviewed in 1994.
Farah (16 years), interviewed in 1996.
Zubia (16 years), interviewed in 1996.
Sharon (20 years), interviewed in 1994/5.
Tara (17 years), interviewed in 1995.
Selina (20 years), interviewed in 1994/5.
Mariam (20 years), interviewed in 1994.

BIBLIOGRAPHY

Afshar, H. (1988) Behind the veil. In B. Agarwal (ed.), *Structures of Patriarchy: the State, the Community and the Household*. London: Zed Books.

Afshar, H. (1994) Muslim women in West Yorkshire: growing up with real and imaginary values amidst conflicting views of self and society. In H. Afshar and M. Maynard (eds), *The Dynamics of "Race" and Gender: Some Feminist Interventions*. London: Taylor & Francis.

Afshar, H. (1998) *Islam and Feminisms: An Iranian Case Study*. Basingstoke: Palgrave.

Ahmed, Akbar and Hastings, D. (1994) *Islam, Globalisation and Postmodernity*. London: Routledge.

Ahmed Alia (2001) 'Why I chose *hijab*. *Women's Own* (Karachi), February, 82–4.

Ahmed, L. (1992) *Women and Gender in Islam: Roots of the Modern Debate*. New Haven, CT: Yale University Press.

Al-Azmeh, A. (1993) *Islam and Modernities*. London: Verso.

Anwar, M. (1979) *The Myth of Return Pakistanis in Britain*. London: Heinemann Educational.

Badran, M. (1995) *Feminists, Islam, and Nation: Gender and the Making of Modern Egypt*. Princeton, NJ: Princeton University Press.

Bailey, M. E. (1993) Foucauldian feminism: contesting bodies, sexuality and identity. In C. Ramazanoglu (ed.), *Up Against Foucault: Explorations of Some Tensions between Foucault and Feminism*. London: Routledge.

Ballard, R. (1991) The Pakistanis: stability and introspection. In C. Peach (ed.), *The Ethnic Minority Populations of Great Britain, Volume 2*. London: HMSO.

Berrington, A. (1991) Marriage patterns and inter-ethnic unions. In D. Coleman and J. Salt (eds), *Ethnicity in the 1991 Census*. London: HMSO.

Bondi, L. (1992) Locating identity politics. In M. Keith and S. Pile (eds), *Place and the Politics of Identity*. London: Routledge.

Brah, A. (1996) *Cartographies of Diaspora: Contesting Identities*. London: Routledge.

Butler, J. (1993) *Bodies that Matter: On the Discursive Limits of "Sex."* New York: Routledge.

Callard, F. (1998) The body in theory. *Environment and Planning D: Society and Space*, 16, 387–400.

Choueiri, Y. M. (1992) *Islamic Fundamentalism*. London: Pinter.

Dwyer, C. (1999) Veiled meanings: young British Muslim women and the negotiation of difference. *Gender, Place and Culture*, 6(1), 5–26.

Dwyer, R. (2000) The erotics of the wet sari in Hindi films. *South Asia*, 23(1), 143–59.

El Guindi, F. (1999) *Veil: Modesty, Privacy and Resistance*. Oxford: Berg.

Foucault, M. (1979) On governmentality. *Ideology and Consciousness*, 6, 5–22.

Foucault, M. (1980) *Power/Knowledge Selected Interviews and Other Writings 1972–1977*. New York: Pantheon Books.

Ghvamshahidi, Z. (1995) The linkage between Iranian patriarchy and the informal economy in maintaining women's subordinate roles in home-based carpet production. *Women's Studies International Forum*, 18(2), 135–51.

Goodwin, J. (1995) *Price of Honour: Muslim Women Lift the Veil of Silence on the Islamic World*. London: Warner.

Hiro, D. (1991) *Black British White British: A History of Race Relations in Britain*, 2nd edn. London: Paladin.

Joly, D. (1986) The opinions of Mirpuri parents in Saltley, Birmingham about their children's education. Centre for Research in Ethnic Relations, University of Warwick, Paper 2.

Kandiyoti, D. (1993) Identity and its discontents: women and the nation. In P. Williams and L. Chrisman (eds), *Colonial Discourse and Post-colonial Theory: A Reader*. New York: Harvester Wheatsheaf.

Kelly, A. and Shaikh, S. (1989) To mix or not to mix: Pakistani girls in British schools. *Educational Research*, 31(1), 10–19.

Khan, B. (1997) *Sex, Longing and Not Belonging: A Gay Muslim's Quest for Love and Meaning*. Oakland, CA: Floating Lotus.

Khan, M. W. (1995) *Woman between Islam and Western Society*. New Delhi: The Islamic Centre.

Khawar, M. and Farida, S. (1987) *Women of Pakistan: Two Steps Forward One Step Back?* London: Zed Books.

Lee, W. (1991) Prostitution and tourism in South-East Asia. In N. Redclift and T. M. Sinclair (eds), *Working Women International Perspectives on Labour and Gender Ideology*. London: Routledge.

Leidner, R. (1991) Selling hamburgers and selling insurance: gender, work and identity, in interactive service jobs. *Gender and Society*, 5(2), 154–77.

Longhurst, R. (2000) Corporeographies of pregnancy: "bikini babes." *Environment and Planning D: Society and Space*, 18, 453–72.

McDowell, L. and Court, G. (1994) Performing work: bodily representations in merchant banks. *Environment and Planning D: Society and Space*, 12, 399–416.

Mernissi, F. (1975) *Beyond the Veil: Male–Female Dynamics in a Modern Muslim Society*. London: Schenkman.

Modood, T. (1992) British Asian Muslims and the Rushdie affair. In J. Donald and A. Rattansi (eds), *"Race," Culture and Difference*. London: Sage/Open University.

Mohammad, R. (1996) An exploration of the (re)production of an "other" patriarchy: the case of an urban based working-class Pakistani Muslim community. Unpublished master's thesis, University of Bristol.

Mohammad, R. (1999) Marginalisation, Islamism and the production of the "other's" "Other." *Gender, Place and Culture*, 6(3), 221–40.

Mohammad, R. (2001) "Insiders" and/or "outsiders": positionality, theory and praxis. In M. Limb and C. Dwyer (eds), *Qualitative Methodologies for Geographers*. London: Arnold.

Mohammad, R. (2002) The state, governance and visions of womanhood: Spain's voyage to the centre. Unpublished PhD thesis, University of London.

O'Kane, M. (1997) A holy betrayal. *Guardian* (London), 'Weekend' supplement, November 29, 38–44.

Ortega, L. M. (1988) La educación de la mujer en las edades moderna y contemporanea. *Historia*, 16, 41–6.

Ortero, L. (1999) *La Española Cuando Besa*. Barcelona: Plaza y Janés.

Rabinow, P. (1984) *The Foucault Reader*. London: Penguin.

Rahman, F. N. (1996) A feminist theologian on women, Islam and feminism. *Women's Own* (Karachi), August, 91–3.

Rosario, S. (2002) Poor and "dark": what is my future? Identity construction and adolescent women in Bangladesh. In L. Manderson and P. Liamputtong (eds), *Coming of Age in South and Southeast Asia, Youth, Courtship and Sexuality*. Richmond: Curzon.

Sarwar, B. (1996) Women's rights: the actual story. *Libas International*, (9)4, 68–73.

Sarwar G. (1980) *Islam Beliefs and Teachings*. London: The Muslim Educational Trust.

Sayyid, B. S. (1997) *A Fundamental Fear: Eurocentrism and the Emergence of Islamism*. London: Zed Books.

Shaarawi, H. (1986) *Harem Years: The Memoirs of an Egyptian Feminist*. London: Virago.

Simmel, G. (1971) Metropolis and mental life. In D. N. Levine (ed.), *On Individuality and Social Forms: Selected Writings*. Chicago: University of Chicago Press.

Stivens, M. (2000) Becoming modern in Malaysia: women at the end of the twentieth century. In L. Edwards and M. Roces (eds), *Women in Asia*. St Leonards: Allen and Unwin.

Taraki, L. (1995) Islam is the solution: Jordanian Islamists and the dilemma of the "modern woman." *British Journal of Sociology*, 46(4), 643–61.

Thompson, J. (1981) *Islamic Belief and Practice*. London: Edward Arnold.

Valentine, G. (1999) A corporeal geography of consumption. *Environment and Planning D: Society and Space*, 17, 329–51.

Wirth, L. (1938) Urbanism as a way of life. *American Journal of Sociology*, 44, 1–24.

Yuval-Davies, N. (1992) Fundamentalism, multiculturalism and women. In J. Donald and A. Rattansi (eds), *"Race," Culture and Difference*. London: Sage/Open University.

Chapter 27

Transversal Circuits: Transnational Sexualities and Trinidad

Jasbir Kaur Puar

In the past ten to fifteen years, there has been a rapid proliferation of literature on gay, lesbian, bisexual, transgender, and queer identities within transnational and global frames. The work of queer scholars such as M. Jacqui Alexander (1991, 1994, 1997), Yukiko Hanawa (1994), Martin Manalansan (1995), José Muñoz (1995, 1997), and David Eng (2001) seeks to link an earlier generation of work on race and sexuality within national contexts (see Anzaldua, 1987; Berlant and Freeman, 1991; Fung, 1991; Moraga, 1993) to current theorizing of globalization and transationalism. By creating new analytical linkages and exploring new sites of inquiry, this new generation of queer scholarship expands the theoretical scope of queer theory *and* provides a crucial contribution to theorizing globalization and transnational processes.

The work by José Muñoz and others functions as a crucial intervention into queer theories that have largely drawn on literary and psychoanalytic frameworks.[1] Literary and psychoanalytic approaches, while undermining assumptions of essentialized identities, unwittingly posit sexual orientation along a singular axis of identification by inadvertently freezing the fluidity of other positions of race, class, gender, and nationality (see Hennessey, 1994; Cohen, 1997). By exploring the construction and negotiation of queerness through a transnational frame, recent work in queer theory bridges psychoanalytical understandings with the concrete material geographies of daily life on a global scale.

At the same time, this emerging literature makes an important contribution to growing concerns about theorizing globalization, the local–global nexus, glocality, glocalization, and translocality. "Queer globalization" scholarship provides an important disruption of the masculinist configurations of subjectivity, time, space, and place that continue to dominate these conversations. Following Stuart Hall's seminal piece "The local and the global: globalization and ethnicity" (1991), local–global relationships have been fruitfully mined within debates about queer globalization, focused largely on three lines of inquiry. First, recent work examines questions of identification in transnational contexts, i.e. what kinds of sexual identities are available, to whom, where, and when, especially in relation to national, diasporic, and immigrant identities (Morris, 1997; Muñoz, 1997; Lee, 1998; Luibheid, 2002). Second, this scholarship explores transnational organizing, in particular as it relates to NGOs, activist collaborations, conferences, human rights discourses, and movements of capital (Manalansan, 1995; Bacchetta, 2000). Third, a growing number of scholars are engaging questions of queer visibility and consumption debates, i.e. market access versus civil rights: tourism, class politics, representational and media issues (Hennessey, 1994; Chasin 2000; Puar 2002; see also Glen Elder, chapter 38 in this volume).

A central question for many of these scholars is whether globalization produces queerness (queer identities) as part of Western imperialism or whether queerness is produced as an identity of liberation from and resistance to dominant narratives. Local–global nuances have allowed thinking beyond a framework of resistance versus assimilation, but the tension between these two positions in queer globalization studies continues, with increasing repetition and little resolution (and I include my own work in this assessment). Despite trenchant critiques disavowing such oppositions, global–local theorizing frequently reproduces a (seemingly irreconcilable) dichotomy of "local versus global." The constant invocation of "the local and the global" creates a hierarchy that implicitly or explicitly lauds the local as the space of authenticity – the local as democratic, originary, and in the case of queer research, the local as a site of "pure" homosexuality or a site of specific pre-queer identities unavailable elsewhere. Queer theorists, then, frequently reproduce a dichotomy present in the broader literature on globalization. They treat the local as the space of postmodern difference, and of (feminized) resistance, while implicitly conceptualizing the global, the "other" side of this dichotomy, as the space of opportunity. In this formulation the global is either homogenizing and equalizing, the space of sameness, or inversely, it is colonizing and creating greater inequities. In either regard the global is figured as masculine.

In the following discussion on sexuality and Trinidad, I begin to shift from local–global terminology to that of spatiality, scale, circuits, and geopolitics. This shift is inspired by the sexuality and space literature from US, British, and Australian queer geographers: Gill Valentine (1993), Jon Binnie (1995), Larry Knopp (1995, 1996, 1998), Ki Namaste (1996), Glen Elder (1998), Michael Brown (2000), Heidi Nast (2003); as well as collections engaged with space, place, and sexuality such as *Mapping Desire* (Bell and Valentine, 1995), *Body Space* (Duncan, 1996) , *Queer Space: Architecture and Same-Sex Desire* (Betsky, 1997), *Queers in Space: Communities/Public Places/Sites of Resistance* (Ingram et al., 1997), and *Places through the Body* (Nast and Pile, 1998), among others. Contextualizing "queer space" in terms of political economy, gentrification, tourism, urban/rural divides, privatization of public spaces, and hate violence, queer geographers have begun to foreground the relations of built, virtual, and conceptual space to theories of subjectivity, identity, performativity, and power. This work foregrounds multiple local networks that operate within national, subnational, regional, urban, and neighborhood arenas.

While queer theorist geographers have produced pathbreaking work on the relations between space, place, and sexuality, they have had less to say about transnational and globalized arrangements of sexual economies and grids. In other words, the important work on queer geography cited above generally does not analyze global connections *between* queer spaces/places/geographies. In the case of Trinidad, I examine two circuits suggestive of spatial, identity, and power relations between seemingly far-flung queer spaces. First, I explore how Western and modernist conceptions of gay rights and gay identities intersect with local activist politics and identities – a relationship that is not simple or necessarily mutually beneficial. Gay cruise ships, and their perceived right to dock at Caribbean harbors, embody the scalar articulation of US identity politics and colonial legacies with the perceptions and histories of Trinidadian gay and lesbian activists, who cannot necessarily afford to actively support the docking-rights of the cruise ships. Second, I examine the local constructions of Afro-Trinidadian and Indo-Trinidadian sexualities in the context of "Diva," a drag show.[2] These performances, and the audience reaction to them, elucidate the intersection between sexuality, race, and nation in Trinidad. As performances increasingly included in the global gay tourism route, they intersect with changing global political economies and cultural dynamics. Through these two examples, I trace the spatiality of "transversal" transnational sexualities, disrupting the singular "local–global" framing of transnational approaches and expanding the geographies of queer theory.

Exploring "transversal" circuitries sheds lights on unusual or unexpected arrangements of space, scale, and identities, and avoids the dichotomy of repeated invocations of "the local and the global" in transnational theory. Following the lead of Paola Bacchetta (2002), I am interested in thinking through what Bacchetta hails as "transversal queer alliances," which she describes as "connections of solidarity both within and across scale, such as within a local site, from one local site to another, from a local to a regional site, or transnationally, in a myriad of possible arrangements" (Bacchetta, 2002, p. 947). Bacchetta problematizes both the temporal and spatial patterns asserted in dominant configurations of "transnational queerdom": representations of queers outside the USA that reinscribe contemporary narratives of visibility and sexual emergence while effacing earlier, usually pre-internet histories of activism, organizing, and community-building.

These representations of "transnational queerdom," found on the internet, in tourist literature, and in scholarly and activist work, privilege the activities of consumer-queers (queer tourism and scholars of "anthroqueer studies," for example) and activist-queers (involved in global NGO work, for instance, and "national subjects who speak in transnational forums"; Bacchetta, 2002, p. 951). She points out, rightly, that these forms of visibility work to obscure other queerness; in other words "transnational queer representations can be inadvertently paired with other queer effacements in an inseparable representation/effacement configuration" (ibid., p. 947). In complicating the space of the "local" and scrambling the trajectories of the global, the concept of transversal circuits allows me to examine these relationships of representation and allude to the unanticipated spatial and scalar disjunctures within which gay and lesbian activists in Trinidad and the Caribbean struggle.

While some of what follows draws from research trips to Trinidad over a period of five years (1995–2000), I am most explicitly concerned with a period in 1998, from January to March, when I was in Trinidad for the entire length of the Carnival season.[3] The purpose of my presence as an "ethnographer-tourist" in

Trinidad was to evaluate the relationships between globalization, gender, and sexuality.[4] Specifically, the aim of my field research was to query how globalization could be defined in terms of gay and lesbian spatialities and what, in turn, was shaping gay and lesbian spaces in Trinidad in the wake of contemporary processes of globalization.

Certainly, palpable effects of globalization upon gay and lesbian communities seemed to be surfacing in Trinidad at every moment. Gay and lesbian activists were taking part in national, regional, and international organizing networks. The first formal organization for gays and lesbians in Trinidad, LAMBDA, was created in 1994, formed with a model of the "fist in the air activist" paradigm of Act-Up in mind. Commonly referred to in Trinidad as promoting "advocacy politics," LAMBDA was composed predominantly of working-class members and focused on confrontational and publicly visible projects. The Gay Enhancement Association of Trinidad and Tobago (GEATT) created a year later, soon thereafter morphed into the Caribbean Forum of Lesbians, All-Genders, and Gays (C-FLAG); both groups directed their energies toward issues of legislative change. C-FLAG's regional meetings in 1996 (Jamaica) and 1997 (Curaçao, Netherlands Antilles), funded primarily by the Dutch government, engendered comparative discussions of state oppression as well as a sense of broader "cosmopolitan" credibility.[5] The HIV/AIDS epidemic in the Caribbean has generated a tremendous amount of funding and research support from former colonizing countries in the past fifteen years.[6]

Furthermore, by the late 1990s, the internet had enabled global connections that were formerly impossible. An increasing number of gay and lesbian tourists, both "diasporic expatriates" and otherwise, were learning about gay and lesbian community meetings and fetes as well as gay-friendly Carnival masquerades specifically through new websites and e-mail lists created in 1998 for Trinidadian gays and lesbians.[7] Furthermore, a tremendous amount of internet activity, diasporic familial scatterings, and educational ventures had enabled a relatively small but privileged and prominent segment of the gay and lesbian community in Trinidad to experience what they called "gay life" not only in other parts of the Caribbean but also in Miami, New York, Toronto, and London. Finally, Carnival the world over was becoming increasingly coded and identified as a gay and lesbian affair, especially by the gay and lesbian tourist industry, and the case was no different in Trinidad.[8]

Tourism, Globalization, and Sexuality

In February 1998, a curious incident set off a series of conversations about the often tense relationships between the interests and effects of globalization and postcolonial gay and lesbian identities. After the Cayman Islands refused docking privileges to a so-called gay cruise originating in the United States, several other Caribbean governments expressed the intention of refusing the same cruise ship and any that might follow. The local Caribbean media engaged in no editorial discussion or debate about the cruises, but printed press releases from Reuters and other global wire services. Caribbean Cana-Reuters Press reported that in the Bahamas, a cruise with 900 gay and lesbian passengers, arranged by California-based Atlantis Events Inc., had become a "test for the tourist-dependent Caribbean islands after the Cayman Islands refused the ship landing rights" in December (*Trinidad Express*,

1998a, p. 29). Officials from the Cayman Islands, a British territory in the western Caribbean, said gay vacationers could not be counted on to "uphold standards of appropriate behavior" (*Trinidad Express*, 1998a, p. 29). Islanders were apparently offended ten years earlier when a gay tour landed and men were seen kissing and holding hands in the streets.[9] A USA-based gay rights organization called on the British government to intervene. British Prime Minister Tony Blair did so and determined, in the case of the Cayman Islands (dubbed by *Out and About* the "Isle of Shame"), that codes outlawing gays and lesbians, many a legacy of colonial legislation, violated the International Covenant of Human Rights and must be rescinded.[10] United States officials followed suit, insisting that human rights had been violated.

While the controversy focused predominantly on the Bahamas and the Cayman Islands, Trinidadian activists from C-FLAG, GEATT, and Artists Against Aids were outraged that gay and lesbian cruises could be denied docking privileges.[11] Interestingly enough, no gay and lesbian cruises had yet ventured to Trinidad, although it had one of the most active gay and lesbian movements in the Caribbean and the largest (and "parent") Carnival in the Caribbean (Nurse, 1999, p. 677). One explanation for this, perhaps, is that in Trinidad tourism makes up only about 3 percent of the gross domestic product, most of which is generated during the Carnival period (the beaches of Tobago are the other main attraction). Thus, the impact of tourism on Carnival, while growing, still appears to be minimal, since the demands of expatriates are "less intrusive," according to Peter Mason (1998, p. 125). He writes: "This phenomenon, plus the fact that most tourists still come from English-speaking parts of the world with fairly close links to Trinidad, has so far kept the demands of tourism to a manageable level" (ibid.).

Interest in Trinidad as a gay and lesbian tourist site is growing, however, due to the growth of Carnival as a gay and lesbian tourist event, the increasing promotion of cruises and other forms of tourism by the Trinidadian government, and the overall expansion of the global gay and lesbian tourism market. Highlighted in a "Carnival around the world" special issue, the editors of *Out and About* (the leading gay and lesbian travel newsletter) write that "Trinidad's Carnival is the biggest gay event in the region" and claim "The gay community here is relatively uncloseted. . . . Gays play an important role in the social fabric of the country, especially in the arts and in Carnival. . . . The islands are at their gayest, figuratively and literally, during the weeks prior to Ash Wednesday" (December 1996, p. 147). While many diasporic Trinidadian gays and lesbians express reluctance about coming "back home" because of the dearth of gay life in Trinidad, *Out and About, Odysseus: The International Gay Travel Planner*, and *A Man's Guide to the Caribbean 98/99* all list party and dining spots and bars for gay, and mostly male, travelers to Trinidad.

Therefore, Trinidadian gay and lesbian activists had good reason to anticipate that the gay cruises would eventually become an issue in Trinidad as well. I watched in confusion, hopeful on the one hand that the former British colonies would tell Blair and the United States to mind their own business in response to their neo-colonial gestures, but also aware on the other of my ambivalent solidarity with Caribbean activists, many of whom were uncertain about openly supporting the cruises.[12] Some activists, attempting to generate support of the cruise ships through an appeal to the profit motive, did comment that "anti-gay protests could be costly to the tourist economies of the Caribbean, a favorite playground for affluent gays" (*Trinidad Express*, 1998a, p. 29). However, most lesbian and gay organizations in Trinidad decided against issuing an official response, fearing that local publicity and exposure could generate backlash against individuals as well as nascent NGOs that were just barely surviving.

It seemed ironic to me that the UK and the USA advocated protection for cruise ships in the Caribbean, while granting no such absolute rights for the passengers upon their return home. Thus, on the transnational scale, these queer citizens are protected, while, on the national scale, they are not, suggesting another reschematization of scale. Even so, the official actions and statements of the two nations may well allow European and American cruisegoers to leave the Caribbean with a sense of liberal belonging and only a surface understanding of the complex politics of sexuality and the specific postcolonial struggles at issue in the region.

In the meantime, the debates stimulated by the arrival and presence of these ships produced complicated and ambivalent responses from local gay and lesbian populations who feared greater local backlash as a result of the increasing discussion and their more marked visibility. The cruise lines appealed to global gay and lesbian identity politics in order to bring about international intervention so that they might dock. By engaging in such strategies, the cruise lines affected the visibility of gays and lesbians and other non-normative sexualities in the Caribbean. In other words, the visibility of ships often creates a need to "lay low" – that is, for decreased visibility or invisibility – for Caribbean gay and lesbian activists; the case may be even more urgent for those not involved in identity politics. Gay and lesbian populations are caught in an oppositional conflict between postcolonial and former colonizing governments, and in a sense are used as examples or pawns in conflicts that may or may not be about sexuality.

The final irony of scale here, of course, is the presence of a mainstream "globalizing" signifier of gay and lesbian identities, namely of a cruise ship with self-proclaimed (professional) gays and lesbians aboard whose presence can be justified not only in humanitarian or in human rights terms, but also in economic terms as contributors to the local economy. A fairly narrow, and perhaps even conservative, segment of gay and lesbian tourists thus winds up triggering the most contentious political discussions on homosexuality in the Caribbean, in effect becoming symbolic of "radical" gay and lesbian activism. Distinctions drawn around mainstream, normative, or corporate homosexuality (Muñoz, 1997, p. 98) cannot fully absorb the irony that certain forms of "corporate gayness" are fueling the supposedly radical agenda of liberationist human rights projects in the context of gay and lesbian tourism.

The cruise ship can be thought of as one of Bacchetta's "hot sites of power" ("points at which power interacts, settles, and gives off effects") that incites a transnational circuitry. As a hot site, the cruise ship occupies several discrepant meanings through simultaneous scalar registers: an alternative form of travel for European and North American lesbians and gays marginalized by a heteronormative travel industry; a marker of mobile, cosmopolitan queerness for those (e.g. Caribbean gays and lesbians) without access to such spaces; an *ad hoc* space, for the cruisegoers, of global sexual citizenship that temporarily reinstates previously denied national rights; a visible catalyst of contamination and/or an affront to postcolonial governments.[13] Thus transversal circuits display not only queer alliances of scale, but also queer disjunctures: the sites and moments where connections that appear naturalized due to a logic of scalar progression and coherence are disrupted and fissured.

Globalization, Gender, Sexuality, and Drag in Trinidad

> ... to be "visible" in the Caribbean is literally to be on stage, to perform. (Personal communication between Gordon Rohlehr and Tejaswini Niranjana, as quoted in Niranjana, 1999, p. 239)

> In the Caribbean we are all performers. (Antonio Benitez-Rojo, as quoted in Muñoz, 1995, p. 83)

My attention, and the attention of many of my informants, flipped-flopped back and forth between the debates about the cruise ships and the preparations for "Diva" and Carnival.[14] "Diva" started in 1992 not explicitly as a drag show but as an "artistic production" for professional actors. According to the producer, a Chinese-Trinidadian man in his fifties, labelling "Diva" an artistic production was a strategic approach used to circumvent the reluctance of theater owners to host the show. Over the years, however, "Diva" has become increasingly identified as a gay and lesbian community event, drawing increasing numbers of amateur participants as well as expatriates and overseas visitors. Since 1998, it has been advertised on several websites created by and for gay and lesbian Trinidadians, and it appears only a matter of time before "Diva" will be listed in mainstream gay and lesbian tourism publications as a cruising spot for gay men. "Diva" illustrates another transversal circuit of globalization, one that highlights different regimes of gay and lesbian identities and the attendant concerns of race, class, ethnicity, gender, sexuality, and nation as they occur in Trinidad. Colonial histories of cross-dressing and transvestism compete with contemporary globalizing understandings of drag to create debates among producers as well as spectators about whether "Diva" is, or should be, a "gay show."

In 1998, ironically and yet appropriately enough, "Diva" was held in Queens Hall, a central and prominent theater in Port-of-Spain. The performances ranged from the spectacularizing of glamor, to comedic parodies, to tragic depictions of HIV/AIDS, poverty, and sex workers. Lip-syncing to Diana Ross's "Ain't no mountain high enough," three performers in shiny yellow latex bodysuits, sporting huge feathered headdresses and sequined capes, echoed carnivalesque costumes and glamor.[15] In several scenes, participants emphasized similar tropes of beauty and glamor, with heavily sequined ballgowns and cocktail dresses, as was the case in a James Bond *Goldfinger* skit and in an Annie Lennox impersonation. The dramatic performances also included somber depictions of a patient dying of AIDS in front of an AIDS quilt, as well as scenes of domestic abuse and a sex worker being kicked around by her pimp. While the judging still favors conventional glamor drag over pointed social commentary, the show has always been heavily dominated by references to the HIV/AIDS epidemic. The show has never been advertised as a gay show or even a drag show – it has simply been announced as "Diva," a performance guided primarily by a serious artistic and competitive agenda.

Over the years, the performers have tended to be working-class Afro-Trinidadian men. The audience is usually largely middle class, surprisingly even in terms of gender, and racially very mixed; thus there is commonly a notable disparity in the racial and class makeup between the audience and the performers. It also includes many diasporic expatriates who are home for Carnival, as well as tourists and well traveled Trinidadians. The show is not inexpensive by Trinidadian standards: in 1998 both evenings cost a total of 100 TT, the equivalent of US$18. According to the organizers, who claim to know who is in and out of the "community," the audience is always at least half "straight."

The reviews of the contest for the past six years have followed nearly the same format. They commonly ridicule the visuality of drag, regardless of the author or whether they are published in the *Trinidad Express* or the *Trinidad Guardian*, both mainstream daily newspapers. The commentaries focus on what they consider to be the comedic moments of the performance and dismiss much of the serious content. As with the advertisements for the shows, no mention of sexuality or gender is ever made. Explained the producer:

> Last year when we were at the Central Bank, we got a review from someone that wrote from the news . . . *Showtime Magazine*. And she loved everything. She was surprised at the standard, she loved the acts, the lights, everything. But she started talking about the show as it being a gay show. And I had to write back, you know, and say, this is not a *gay* show. Because she had the conception that it was.

Thus, maintaining Diva first and foremost as an artistic production is one strategy used to enable this space for non-normative sexualities.

Most of the seven drag performers I interviewed talked about contemporary "Divas," ranging from Barbara Streisand and Marilyn Monroe, who were parodied in the early shows, to Patti La Belle, Tina Turner, and Toni Braxton. Said a working-class Afro-Trinidadian male who has performed in nearly every "Diva" show: "The first two years, the older actors involved were very aware of the female icons of the cinema . . . Marilyn Monroe, Marlene Dietrich. Now, because they are younger and also *blacker*, they tend to follow Patti La Belle, Toni Braxton." While figures of white womanhood are prominent in earlier "Diva" shows but somewhat absent in more recent ones, the competing definitions of black womanhood reflect the distinctions made between African-American, Afro-Trinidadian, and Indo-Trinidadian femininities. Thus "Diva" is an event where questions of racial performativity and Indian–African relations are highlighted.

A brief historical discussion of race and colonialism helps to situate contemporary racial performances in "Diva." Trinidad's decolonization in 1962 is ironic in that it left two groups of color of nearly equal proportion (Indians at 40.7 percent and Africans at 40 percent) pitted against each other as economic and cultural rivals. The impression of growing racial antagonisms has been termed the "war of cultures" by the media. Indian migration to Trinidad, precipitated in the mid-1800s by the need for indentured labor in the sugar industry, was part of a "first wave" of colonial migrations from India, including migration to Fiji, Suriname, Guyana, and Africa. Although Indians occupied a "fourth tier" underneath white colonists, mixed populations, and former black slaves (for excellent historical analysis see Kale,

1995), in contemporary postcolonial Trinidadian society many Indo-Trinidadians are beginning to access education and other institutionalized privileges. Moreover, they are entering key political, cultural, and social realms in increasing numbers.

Thus, while Afro-Trinidadians have historically dominated the political arena and are culturally associated with the Caribbean, Indo-Trinidadians have recently emerged as powerful challengers to the political and cultural space of the nation. This challenge, precipitated to some extent by the redistribution of wealth due to the oil boom and bust in the 1970s and 1980s, is evident in the growing economic power of the Indian bourgeoisie as well as the election of Basdeo Panday, the first Indo-Trinidadian Prime Minister, in 1995. That year also marked the 150th anniversary of Indian Arrival Day, although this date is not commemorated annually as an official holiday. Currently, attention is being drawn to the globalization of Indian ethnicity occurring throughout the Caribbean through the dissemination of Hindi film and the increasing circulation of aspects of Indo-Trinidadian popular culture, such as chutney music (Niranjana, 1999; see also Khan, 1995). However, the illegitimate, disavowed modernity of Indo-Trinidadians posits them as other *vis-à-vis* Afro-Trinidadians. Indo-Trinidadians appear as either culturally mimicking dominant notions of Trinidadian identity or cast in the realm of the traditional, backward, and primitive.[16] The unmitigated effect of this renders Indo-Trinidadians without claim to a national Trinidadian identity of their own.

In this context, the categorization of who is and is not in "drag" is an important reflection of the relationships between African and Indian ethnicities. Despite the increasing Indianization of Trinidad, however, "Diva" continues to be dominated by Afro-Trinidadians. Every year has seen an Indian act, and 1998 was no different. As the producer comments:

> I've always had an East Indian act. Always had one. I nearly did not have one this year. I always wanted one, I like variety. The gay community in Trinidad has a lot of class and racial differences, and you would find the Chinese, whites, the lighter-skinned Trinis would not be eager to participate in something like "Diva." They would come and look at it.

About halfway through the first show, the "Indian" act was announced, first, by the MC's comments on the problems he was having pronouncing the Indian names, and, second, by the distinct introductory notes of Indian film music that was quite different from the more contemporary "Top 40" pop tunes used in the rest of the acts. A pair of Indo-Trinidadians mimicked the motif of seduction so common in Hindi films. The male figure, dressed in an Indian kurta and pyjamas, pranced after his flighty, pouting partner, who was dressed in a bright pink top and long silk skirt, around trees and through fields. The female figure was barefoot, with long braided hair, an exposed belly, and gold earrings and wrist bangles. One could even imagine the rain so typical of Bollywood films. At the end of the scene, the male figure hoisted the female figure into his arms. This was the only coupling in any performance of "Diva," and the only performance of desire expressed through heterosexual partnering in the two shows.

Throughout, the audience was generally appreciative, but not overly enthusiastic. An undercurrent of chattering increased as the performance continued, and the final applause was lukewarm. During the most comedic moments, the audience did hoot with laughter. These acts were part dance, part acting, part parody. The familiarity of Hindi films to Trinidadian audiences is enabled by regular screenings in theaters as well as by the availability of Indian cable channels and Indian MTV, not to mention by exposure to the rich culture of dance and music made possible through contests held in the south and central areas of Trinidad.

The question remains: were the performances drag? The various answers to this question may illuminate the differences between the visibilities of race *and* sexuality versus the visibilities of race *or* sexuality. The differences between Afro-femininity and how it "gets dragged" and the dragging of Indian femininity are striking.[17] In the audience response surveys that I conducted after the shows, many comments indicated that the Indian performers were regarded as closeted and thus not "really" in drag; rather, they were simply performing an "ethnic" dance. One Afro-Trinidadian male interviewee claimed: "This is an Indian drag queen who is inhibited by fears of people discovering who she really is." An Indo-Trinidadian female judge, lamenting the dearth of Indo-Trinidadian performers, noted: "This is marked as ethnic dance, as Indian dance, while the African is not marked." Another judge, an Afro-Trinidadian man, commented: "It's just a dance. It's a dance to me, to you. The judges don't know what the movements mean. It's not like a Hindi film – there are no subtitles." During my questioning of audience members, I asked repeatedly whether the Indian dance, in the context of the "Diva" contest, was considered to be drag. "He was pretending to be a woman but he does Indian dance anyway," said an Afro-Trinidadian female. "It's an Indian dance because we can classify it like that – it's easy to classify. It's not drag though."

These varied reactions point to several connections between performances of drag and the moments of cultural, racial, and national strategies utilized in them. The characterization of this performance as an Indian one erases Indo-Trinidadians in drag even as it simultaneously enables participation in a Trinidadian national space of drag. The connections between drag and the reterritorializing of national spaces are located when African traditions that are hailed as national traditions, or inversely those national traditions implicitly assumed to be African ones.[18] When is the specter of tradition just barely referenced or not, and who is able to avoid that reference? In relation to Trinidad, the forces of cultural nationalism often prevent queer historical essentialisms. Indian cultural traditions are only marginally available to legitimize non-normative sexualities given that the category of "Indian culture" is so often mobilized by Afro-Trinidadians as one of tradition and backwardness, while by normative Indo-Trinidadians it is also used as a way to claim national space.

I met Sasha and Vik at the cast party after the second night of the "Diva" contest,[19] where I asked if they would be interested in being interviewed for my project. They readily agreed and we arranged a meeting spot at the Grand Bazaar, a relatively new mall. The Grand Bazaar is located at the entrance to the freeway considered the gateway to the "South," a demarcation commonly alluding to the rural, the Indian, the backward spaces of Trinidad from the vantage point of cosmopolitan Port-of-Spain.

We sat in Pizza Hut. Sasha was still in drag. S/he had long painted nails, wore lipstick, and had pinned up his/her long dark hair into a high ponytail. Vik hovered over both of us, getting us drinks and winking at Sasha. We started by talking about the performance and how they felt about the rehearsals and the show. Both Vik and

Sasha were excited about having had the opportunity to perform, and had not felt marginalized by the African-dominated spaces of the show, saying that the audience really appreciated Indian dance.

We spent hours talking about dance in general, about different types of Indian dance, and about the development of Vik's and Sasha's dance school, their business partnership, and the kinds of reactions their families and residential community had about their interest in an alternative career which was not conventional for Indo-Trinidadian men. They had established their dance school nearly six years earlier, and had performed all over Trinidad at Indian weddings and community events, as well as at Trinidadian cultural shows. They had also performed overseas in Guyana, New York, and Miami.

The point is that for the first two hours of the interview, we never once talked about drag, sexuality, homosexuality, gays, lesbians, or gendered roles. I hesitantly read my own assumptions of their sexual relationship through certain moments of affection between the two of them and their narration of a long joint history of living and working together. Having a partnership routed through material business arrangements is a common phenomenon for same-sex liaisons, especially in Indian circles in Trinidad, and may even be facilitated by the concept of arranged marriage that is seen purely as a familial and financial arrangement that benefits everyone. The one fleeting reference to anything remotely related to "Diva" as a space of gender illusion was made when Vikram commented about the Port-of-Spain "community parties" being pleasant though somewhat alienating.

Unlike with the other drag performers I interviewed, who were Afro-Trinidadian, I simply could not bring up the question of sexuality with Vikram and Sasha, largely because they did not appear gay to me in any intelligible way. That they were "closeted" is easy to assume here, except that Vikram and Sasha exist in their hometown of Chaguanas as "openly" as any couple ever could, in a somewhat accepted/tolerated/negotiated transgendered partnering. I struggled to respect their privacy and interpretation on the one hand, and to access the meaning of their relationship in terms I could comprehend on the other. My problematic enthrallment with Sasha and Vikram may well have reflected my desire to produce a "queerer than queer" counternarrative to the homogenizing impulses of metropole-produced queer theory. I was also unable to gain any insight into what Sasha and Vik were thinking about me or if they read me as a lesbian; they asked me only about my family in the United States, my knowledge of Indian dance, and my connections to Indian musicians and performers overseas.

In fact, toward the end of our second hour together, Sasha and Vik started pressing what seemed to me at that time their real agenda: they wanted to know if I had any business contacts on the West Coast who could set them up with a show. In another illustration of jumping scale, in which they indicated that their shows were quite successful in New York and Miami, their emphasis on institutional and economic constraints and opportunities served to foreground the materiality of sexualities. They wanted to know what California was like. What may well have been most enabling for Sasha and Vik were the economic networks they mobilized and within which they moved. This is what I find so interesting, that Sasha and Vik had no investment whatsoever in the process of queer liberation. It is precisely their refusal of a politics around sexuality that was most striking; they appeared completely uninterested in the politicized project of gender bending that often occupies centerstage in USA-based queer theory. Sasha and Vik, and arguably many of the other drag contestants who yearn to be awarded the prize money at "Diva," linked their sexual subjectivity to their work status.

I do not intend here to reductively position the wide range of different kinds of gender, racial, class, and national identifications in such examples. Instead, I want to suggest several conundrums of Bacchetta's "representation/effacement" paradigm that might actually privilege the modality of absence as desirable over presence. If it is visible, is it queer visibility in the ways queer liberation in the United States might define it? It may be in/visible, but is it in/visibility? Not every invisibility involves an assimilationist narrative. Sasha and Vik were more visible in "Diva" as ethnic Indian dancers than as drag performers, or rather they were invisible as gay. Yet they were more visible at home in Chaguanas as a male/female couple than as a gay couple; viewed through the lens of gender rather than sexual orientation, they were invisible as a gay couple. Sasha and Vik's "queerness" is rendered nearly invisible in the context of gay and lesbian identity politics in Port of Spain, yet their invisibility, or perhaps more precisely their partial visibility, may be the most "radical" element of their gendered sexuality.

I will not go so far as to say that the possibilities of Vikram and Sasha as a couple, or of Sasha in drag or as transgendered, are completely invisible *and* accepted without repercussions by a largely Indian community in Central Trinidad. In fact, Sasha and Vik performed the very same acts in "Diva" as they did for Indian weddings and other community functions in South and Central Trinidad. Given the history of female impersonation and cross-dressing in Indian dance as well as in contemporary Bollywood films, the framework of drag may well be irrelevant in these contexts.[20] When I asked about the tradition of Indian cross-dressing in Indian dance, and how it was received at these predominantly heterosexual functions, Vik stated: "I do think maybe they do still have a few negative people, I'm not saying no. The majority of people widely accepted the fact that we do dance together. . . . And they do enjoy seeing boys dress up and dance, so we do the most popular ones, the most acceptable dances."

In this circuit, the globalization of Indian ethnicity via Hindi films and popular culture, as well as diasporic cultural venues, is in conversation with sexuality and race. These moments of sexuality and race are traversed in the movement from the South to the North, from supposed subalternized rural Indian territory to cosmopolitan, urban African territory. Qualifying Vik and Sasha as a male/female couple is too reductive, though perhaps it is precisely this reductive reading that allows them a certain degree of gender fluidity. Similarly ineffectual is the "third gender" status often accorded to *hijras* in India and Native American berdaches.[21] Despite my lack of information or evidence about Sasha and Vik's sexual orientations or their sexual relationship, what remains interesting here is the de/stabilizing of sex/gender binaries within kinship structures, community events, and global labor/work networks.[22] Furthermore, distinctions rendered between drag, drag performers, drag queens, subjects of drag, transsexuals, transgenders, cross-dressing, transvestism, and sex-changes are intrinsically determined as much through racial and class distinctions as they are through distinctions of sexual and gendered practices and subjects.

Globalization 2000: "Circuits of Desire"

In closing, I want to return to the opening dilemma posed by the cruise ship with 900 gays and lesbians from the United States, its presence intertwined with the performances of "Diva" and Sasha and Vik. As a South Asian queer academic based in the United States, I located myself as part of these multiple circuits: complicit with the production of queer theory in the United States and often unable to resist this location as my reference point, I attempted to comprehend the specificities of sexual identities in Trinidad.[23]

What do these circuits say about the uneven and contradictory situations enabled by globalization, particularly in terms of gender and sexuality? In Trinidad, queer sexualities create new communities that in some ways reinvent the spaces of the nation, redefine some citizens of Trinidad, and create new transversal queer alliances across ethnicity, gender, and certainly across scale through regional, transnational, and diasporic connections. I contend that these spaces also create and recreate long-standing divisions among and between different communities, sites, and locations, foregrounding the necessity to examine transversal queer *complicities*, not just alliances.

Furthermore, while the local may not be "dead," as some theorists of globalization who focus on "world without boundaries"[24] might claim, the local as a space of sexual signification continues to reference culture, ethnicity, and so-called native or indigenous sexual practices that are highly unnamable, i.e. unmappable within Euro-American identity frameworks and categories. Their unmappability renders them not sophisticated or "modern enough" to enter the realm of identity politics, nor representatable, a moment of subalternization, even as they are then often mobilized as transgressive sexual practices through postmodern queer politics. The uncomfortable hierarchies resurrected in static local–global formulations with regard to non-normative sexualities could be as follows: the local is the space of the liminal, indigenous, the sexuality that cannot be named, the primitive, and the backwardness of tradition. The global represents the formations of identity that are deemed necessary for certain political movements and moments to take place, and in the case of sexuality, signals the positions of gay and lesbian. Meanwhile queerness is then proffered as a postmodern refusal of identity *par excellence*, often through a reclamation of the space of the indigenous, and without careful contextualization of the spatial interactions of these positions. Thus this reclamation seems at once a harking back to pre-modern pre-identity as much as an attempt at a new postmodern fluidity, one that keeps being captured by modernity.

While I want to insist on the refusal of an imported versus indigenous binary, mapping my own circuits of desire has been a difficult and confusing task. It is precisely upon the erasure of these circuits of globalization that my own desires, in the search for nameable and counter-nameable subjects, have often hinged. In retrospect, it is hard for me to say whether the "refusal of the subject" was indeed the denial of Sasha and Vik as the gay subjects that I could most easily identify, or actually my refusal to allow Sasha and Vik to be the (gay?) subjects that they are (Visweswaran, 1994). If the latter is the case, then I, too, colluded with Afro-Trinidadian assessments that they were not in drag; I, too, viewed the specificity of Sasha and Vik's lives through the lens of romanticized queerness, searching for some kind of sexual liminality that I could not name or see, but still could somehow know.

I have also, with ambivalence, used the terms *gay* and *lesbian* as well as *transgender* to describe people in Trinidad, namings that circulate in tandem with local nomenclature such as "buller" (a reclaimed derogatory slang word for men, its nearest equivalent being "faggot") and the phrase "she goes with a woman," while I used the term *queer* for myself. I have done this in part because *queer* does not yet circulate as a descriptor in Trinidad. However, I am well aware that for some readers this may be seen as a "withholding" of sorts that reinscribes the centrality of queer theory (and myself as a queer theorist) that I have attempted to trouble here. For other readers, using the term generically would have been unforgivably neocolonialist. Though I have resisted offering definitions of these terms as a preface to this material, since the argument made in this discussion renders such definitions counterproductive to my theoretical intent, I have recuperated namings at moments when there appears to be no linguistic escape. All namings are underpinned by tensions between identity positions around race, ethnicity, class, and gender in ways that mark subjects beyond genre and sexual signification. In the context of theorizing about globalization, these namings are often freighted with the difficulty of being untranslatable across social locations.

This chapter set out to look at some specific moments of the globalization of gender and sexuality in the context of Trinidadian identities and the effects of globalization on sexuality. I have focused on Trinidad, not primarily or only as a site, nation-state, or legislative entity, but rather as a series of spatial relationships which decenter any one force to be an overriding determinant of social change, mobility, persecution, or promise. The larger project from which this work derives is also concerned with an exploration of the continuities and discontinuities between colonial and postcolonial legislation, i.e. a study of the British discourses of racialization of the Caribbean and Trinidad before and during the period of decolonization to understand how racial groups in Trinidad were being constructed through sexualization. The development of gay and lesbian activism in Trinidad and its links to international organizing, the negotiation of transsexual and transgender identities in Trinidad, and the practices of consumption, tourism, and cultural production will all continue to alter the ways that gay and lesbian sexualities are understood in Trinidad. The growing participation of "queers of color" organizations as well as postcolonial queers and people of color travelers in these practices and spaces that foreground global and diasporic unity around queerness mean that such discourses are no longer only, though still primarily, in the service of Euro-American universalisms and Enlightenment discourses. Does globalization entail a predictable teleological march towards recognizably gay, lesbian, bisexual, transgender, and queer identities? This is the question that I, and others, continue to explore.

NOTES

1   See the work of Judith Butler, Teresa de Lauretis, and Diana Fuss, among others.
2   "Diva" has taken place every year in Trinidad's capital, Port-of-Spain, since 1992. While annual gay parties, or "fetes," during the holidays and Carnival had become routine, and public events for International AIDS Day and even gay pride had previously been staged in Trinidad, during the time of my visits "Diva" was still considered among the most established and widely recognized public arenas of gay and lesbian interaction.

3   The research in this chapter is based on fieldwork conducted in Trinidad during intermit-
    tent trips from 1994 to 2000. The analysis herein is derived from participant-
    observation in the field as well as from more than 30 interviews with activists from Trinidad
    and other Caribbean countries, "Diva" performers, producers, and judges. I also distrib-
    uted 47 audience response surveys after the "Diva" shows, and organized post-"Diva"
    discussion roundtables. Concurrently I spoke with participants in the urban Port-of-Spain
    Trinidadian gay and lesbian scene, including local residents, Trinidadians from other areas
    who traveled to the capital frequently to attend community events, and tourists.
4   I use the term *ethnographer-tourist* not to minimize or compromise my activities as an
    ethnographer and researcher but to highlight my overlapping positioning and partici-
    pation in tourist circuits in Trinidad. Much has been written on the ethnographer as
    traveler. However, less has been discussed about how a hierarchical distinction between
    traveler and tourist serves to obscure the ways in which ethnographers are tourists in
    the field to varying degrees and are implicated in tourist economies.
5   There are tensions between different activist strategies and the class, racial, and gender
    components of each. Many middle-class gays and lesbians in Port-of-Spain tend to route
    their politics around sexuality through HIV/AIDS activism, and are extremely distanced
    from if not opposed to pushing for legislative changes, while many working-class gays,
    and fewer lesbians, are involved in forming national, regional, and international NGOs
    to address questions of persecution and oppression.
6   The most prominent example of such globalized organizations, the Caribbean
    Epidemiology Centre (CAREC), is located in Port-of-Spain, Trinidad, and is funded by
    various Caribbean islands as well as the Dutch and British governments.
7   See, for example, http://www.search.co.tt/trinidad/gay/
8   Keith Nurse (1999, p. 673) notes that the globalization of Carnival also generates a
    tremendous amount of travel and work opportunities through an overseas Carnival
    circuit that spans the Caribbean, North America, and Europe and involves some of the
    largest gatherings in those locales.
9   In the same article, Bahamian clergymen claimed it was the "power of prayer" that
    steered the ship away from the island, a decision that was claimed to have been made
    due to inclement weather. Clergy said the cancellation was due to "divine intervention"
    (*Trinidad Express*, 1998a, p. 29).
10 The editors of *Out and About* (1998a, p. 27), the leading gay and lesbian tourism
    newsletter, called for a travel boycott against the Cayman Islands, encouraging letter
    writing campaigns to American Airlines, American Express (the "official card" of the
    Cayman Islands), and Norwegian Cruise Lines.
11 See "Isle of Shame" in *Out and About* (1998b) for excerpts from the statement of
    welcome to gay and lesbian travelers eventually issued by the Prime Minister of the
    Bahamas, Hubert A. Ingraham. For contextualization of the tourist industry in the
    Bahamas, see Alexander (1997).
12 Debates continued through the spring, preceded by prison riots in Jamaica over the dis-
    tribution of condoms and continuing pressure from the British to liberalize anti-gay laws.
    Britain had previously abolished the death penalty in several British territories (Anguilla,
    the Cayman Islands, British Virgin Islands, Turks and Caicos, and Monserrat) "despite
    public support for capital punishment in the colonies and throughout the Caribbean." In
    response to Britain's insinuation that it would do the same with regard to laws on homo-
    sexuality, Anguilla's chief minister, Hubert Hughes, stated: "We would like Britain to
    understand that even though we are dependent on British aid, we will definitely not com-
    promise our principles when it comes to Christianity" (*Trinidad Express*, 1998b, p. 30).
13 In this scenario, what I have left unattended is vast: disjunctures within the population
    of the ship, between consumers and laborers perhaps; internal debates within LGBTQ

organizations that represented fissures despite the appearance of a unified stance; the relations and interactions between disembarked cruisegoers and residents of these Caribbean nations.

14 While I focus in this chapter on "Diva," there are several other notable spaces of drag performance in Trinidad. Two examples are those created by drag performer Juana La Cubana, a well known figure in entertainment circles in Trinidad, and in the stage production of "Mark, Maureen, and a Drag Queen" in October 1998.

15 In a longer unpublished version of this chapter I discuss Bakhtin and his conceptualization of the "carnivalesque" to shed more light on the genealogies of the costumes in these drag performances.

16 This effectively projected the modernities of East Indians in Trinidad, among other indentured populations, as what Niranjana (1999, p. 243) calls an "illegitimate modernity," and later, an "artificial modernity," as well as a "disavowed double," because, as she writes, "they had not passed through, been formed by, the story of the [Indian] nation in the making" (Niranjana, 1999, p. 232).

17 Kanhai (1995) has written on how the tensions of decolonization make African and Indian divides more difficult for women as "cultural containers." Kanhai claims that the image of the oppressed Indian "coolie woman" associated with indentureship has led to a preponderance of work on violence against women in Indo-Trinidadian communities. About the "gender control" of Indian women during indentureship and afterwards, Kanhai (1995, p. 9) writes: "Indeed the history of Indian presence in the Caribbean seems to be a chronicle of abusive male control within the community." She notes how the feminist movement in Trinidad is complicit with, and responsible for, the perpetuation of images that inscribe a "tradition"/"modernity" dichotomy between African and Indian women.

18 One of the most important figures constructing African traditions as national ones is the Dame Lorraine, a traditional Carnival character who originally mocked French plantation wives. The Dame Lorraine, a highly performative form of "colonial mimicry" of French Creole whiteness, became a part of carnival processions in 1884 (Bhabha, 1984). The Dame Lorraine can be seen as a covert figure of legitimization, one which functions as a marker of Carnival masquerading and, hence, of a national tradition of cross-dressing and female impersonation (see also Hill, 1972).

19 These are pseudonyms. Vikram calls himself "Vik" for short, and Sasha is a female version of a more masculine Indian name.

20 Due to space constraints here I can only mention this argument. Generally, I want to caution against a decontextualization of histories of female impersonation in Indian dance that often happen through a queer reading that privileges drag in these performances. See Hansen (1992) on female impersonation in Indian dance.

21 The *hijra* in South Asian queer diasporic contexts has become a figure of transgressive sexuality that largely effaces the often non-transgressive (though not "normal" either) status of *hijras* in India. The Native American concepts of berdache and two-spirit have also been applied to contemporary queer liberationist projects in a similar fashion. The figures can be used by diasporic communities in a historically essentialist way as evidence of homosexual traditions within the culture, but they are also used by more mainstreamed gay, lesbian, and queer organizing in similar ways but *without* the requisite attentiveness to issues of racism, immigration, and nationalism (see Nanda, 1993; Patel, 1997).

22 For more detailed studies about these relationships in different contexts, see Prieur (1998) on homosexuality in Mexico and Kulick (1998) on "travestis" in Brazil.

23 This circuit has altered significantly over the years of my research in Trinidad. When I first came to Trinidad in 1994, the few contacts that I made in the gay and lesbian

community were located through word of mouth, primarily from Trinidadian friends in the United States. Information was always cautiously dispensed. Now, fetes that were once invitation-only and known about strictly through word of mouth are advertised on the world wide web. It is also less problematic for me to write about specific places, events, and even people in Trinidad because they have all been "outed" by these websites as well as by the gay and lesbian tourism industry.

24 See Kaplan (1995) for a critique of this version of transnationalism as well as a study of how it operates in relation to consumption.

BIBLIOGRAPHY

Alexander, M. J. (1991) Redrafting morality: the postcolonial state and the Sexual Offences Bill of Trinidad and Tobago. In C. T. Mohanty, A. Russo and L. Torres (eds), *Third World Women and the Politics of Feminism.* Bloomington: Indiana University Press, pp. 133–52.

Alexander, M. J. (1994) Not just (any)*body* can be a citizen: the politics of law, sexuality and postcoloniality in Trinidad and Tobago and the Bahamas. *Feminist Review*, 48 (Autumn), 5–23.

Alexander, M. J. (1997) Erotic autonomy as a politics of decolonization: an anatomy of feminist and state practice. In M. J. Alexander and C. T. Mohanty (eds), *Feminist Genealogies, Colonial Legacies, Democratic Futures.* New York: Routledge.

Angelo, E. and Bain, J. (eds) (1999) *Odysseus: The International Gay Travel Planner.* New York: Odysseus.

Anzaldua, G. (1987) *Borderlands/La Frontera: The New Meztisa.* San Francisco: Spinsters/Aunt Lute.

Bacchetta, P. (2002) Rescaling transnational "queerdom": lesbian and "lesbian" identitary-positionalities in Delhi in the 1980s. *Antipode: A Radical Journal of Geography*, 34(5), 947–73.

Bell, D. and Valentine, G. (eds) (1995) *Mapping Desire.* London: Routledge.

Berlant, L. and Freeman, E. (1991) Queer nationality: the political logic of queer nation and gay activism. In M. Warner (ed.), *Fear of a Queer Planet.* Minneapolis: University of Minnesota Press.

Betsky, A. (1997) *Queer Space: Architecture and Same-sex Desire.* New York: William Morrow.

Bhabha, H. K. (1984) Of mimicry and man: the ambivalence of colonial discourse. *October*, Spring, 125–33.

Binnie, J. (1995) Trading places: consumption, sexuality, and the production of queer space. In D. Bell and G. Valentine (eds), *Mapping Desires: Geographies of Sexualities.* New York: Routledge.

Boodram, K. (1997) Up close at 5th annual transvestite contest: drag queens big night out. *Trinidad Express*, February 10, 26.

Brereton, B. (1989) Society and culture in the Caribbean: the British and French West Indies, 1870–1980. In F. Knight and C. Palmer (eds), *The Modern Caribbean.* Chapel Hill: University of North Carolina Press.

Brown, M. (2000) *Closet Space.* New York: Routledge.

Chang, C. (1998) Chinese in Trinidad Carnival. *The Drama Review*, 42(3), 213–19.

Chasin, A. (2000) *Selling Out: The Lesbian and Gay Movement Goes to Market.* New York: St Martin's Press.

Cohen, C. (1997) Punks, bulldaggers, and welfare queens. *GLQ*, 3, 437–65.

Cordova, S. (1998) *A Man's Guide to the Caribbean, 1998–99.* New York: Centurion Press.

Damron, B. (1998) *Damron Accommodations.* San Francisco: Damron.

Damron, B. (1999) *Damron Men's Travel 2000*. San Francisco: Damron.

Duncan, N. (ed.) (1996) *Body Space: Destabilizing Geographies of Gender and Sexuality*. New York: Routledge.

Elder, G. (1998) The South African body politic: space, race and heterosexuality. In H. J. Nast and S. Pile (eds), *Places Through the Body*. London: Routledge.

Eng, D. (2001) *Racial Castration: Managing Masculinity in Asian America*. Durham: Duke University Press.

Farah, G. (1998) *Trinidad Guardian*, February 8, 19–20.

Fung, R. (1991) *My Mother's Place*. Toronto: VTape.

Hall, S. (1991) The local and the global: globalization and ethnicity. In A. King (ed.), *Culture, Globalization, and the World-System*. London: Macmillian.

Hanawa, Y. (1994) Introduction. *Positions: Circuits of Desire*, 2(1), v–xi.

Hansen, K. (1992) *Grounds for Play: The Nautanki Theatre of North India*. Berkeley: University of California Press.

Hennessey, R. (1994) Queer visibility and commodity culture. *Cultural Critique*, 29, 31–76.

Hill, E. (1972) *The Trinidad Carnival: Mandate for a National Theatre*. Austin: University of Texas Press.

Ingram, B., Bouthillette, A.-M. and Retter, Y. (eds) (1997) *Queers in Space: Communities/Public Places/Sites of Resistance*. Seattle: Bay Press.

Kanhai, R. (1995) The Masala stone sings: Indo-Caribbean women coming into voice. Paper presented to the ISER-NCIC Conference, Trinidad.

Kaplan, C. (1995) "A world without boundaries": the Body Shop's trans/national geographics. *Social Text*, 29, 45–66.

Khan, A. (1995) Homeland, motherland: authenticity, legitimacy, and ideologies of place among Muslims in Trinidad. In P. van der Veer (ed.), *Nation and Migration: The Politics of Space in the South Asian Diaspora*. Philadelphia: University of Pennsylvania Press.

Knopp, L. (1995) Sexuality and urban space: a framework for analysis. In D. Bell and G. Valentine (eds), *Mapping Desires: Geographies of Sexualities*. New York: Routledge.

Knopp, L. (1996) Space(s) lost in George Chauncey's gay New York. *Environment and Planning D: Society and Space*, 14, 759–61.

Knopp, L. (1998) Sexuality and urban space: gay male identity politics in the United States, the United Kingdom, and Australia. In R. Fincher and J. Jacobs (eds), *Cities of Difference*. New York: Guilford Press.

Kulick, D. (1998). *Travesti*. Chicago: University of Chicago Press.

Lee, J. Y. (1998) Toward a queer Korean American diasporic history. In D. Eng and A. Hom (eds), *Q&A: Queer in Asian America*. Philadelphia: Temple University Press.

Lee, S. (1994) TT gays stay in the closet. *Trinidad Guardian*, July 3, 12.

Luibheid, E. (2002) *Entry Denied: Controlling Sexuality at the Border*. Minneapolis: University of Minnesota Press.

Manalansan, M. (1995) In the shadows of Stonewall: examining gay transnational politics and the diasporic dilemma. *GLQ: A Journal of Lesbian and Gay Studies*, 2(4), 425–38.

Martin, C. (1998) Trinidad Carnival glossary. *The Drama Review*, 42(3), 220–35.

Mason, P. (1998) *Bacchanal! The Carnival Culture of Trinidad*. Philadelphia: Temple University Press.

Miller, D. (1994) *Modernity: An Ethnographic Approach. Dualism and Mass Consumption in Trinidad*. Oxford: Berg.

Miller, D. and Slater, D. (2000) *The Internet: An Ethnographic Approach*. Oxford: Berg.

Mirror (1996) 800 lesbians on Caribbean cruise. January 5, 21.

Moraga, C. (1993) Queer Aztlan: the reformation of Chicano tribe. In *The Last Generation*. Boston: South End Press.

Morris, R. C. (1997) Educating desire: Thailand, transnationalism, and transgression. *Social Text*, 52/3, 53–79.

Muñoz, J. (1995) The autoethnographic performance: reading Richard Fung's Queer Hybridity. *Screen*, 21(1/2), 83–99.

Muñoz, J. (1997) "The white to be angry": Vaginal Davis's terrorist drag. *Social Text*, 52/3, 80–103.

Nast, H. and Pile, S. (eds) (1998) *Places through the Body*. New York: Routledge.

Nast, H. (ed.) (2003) Queer patriarchies, queer racisms, international. *Antipode: A Radical Journal of Geography*, special issue.

Namaste, K. (1996) Genderbashing: sexuality, gender, and the regulation of public space. *Environment and Planning D: Society and Space*, 14, 221–40.

Nanda, S. (1993) Hijras as neither man nor woman. In H. Abelove, M. A. Barale and D. M. Halperin (eds), *The Lesbian and Gay Studies Reader*. New York: Routledge.

Niranjana, T. (1999) "Left to the imagination": Indian nationalisms and female sexuality in Trinidad. *Public Culture*, 11(1), 223–43.

Nurse, K. (1999) Globalization and Trinidad Carnival: diaspora, hybridity and identity in global culture. *Cultural Studies*, 13(4), 661–90.

Out and About (1995) Gay Caribbean. 4(1).

Out and About (1996) Carnival around the world. 5(10).

Out and About (1998a) Cayman Islands: boycott update. 7(2), March.

Out and About (1998b) Isle of shame. 7(4), May.

Patel, G. (1997) Home, homo, hybrid: translating gender. *College Literature*, 24(1), 133–50.

Patton, C. and Sanchez-Eppler, B. (eds) (2000) *Queer Diasporas*. Durham, NC: Duke University Press.

Prieur, A. (1998) *Mama's House, Mexico City: On Transvestites, Queens, and Machos*. Chicago: University of Chicago Press.

Puar, J. (2002) Queer tourism: geographies of globalization. Special issue of *GLQ: A Journal of Lesbian and Gay Studies*, 8(1/2).

Trinidad Express (1998a) Gay cruise leaves Bahamas. February 10, 29.

Trinidad Express (1998b) Collision course in church colonies. February 25, 30.

Valentine, G. (1993) (Hetero)sexing space: lesbian perceptions and experiences of everyday space. *Environment and Planning D: Society and Space*, 11, 395–413.

Visweswaran, K. (1994) *Fictions of Feminist Ethnography*. Minneapolis: University of Minnesota Press.

Part V Environment

Chapter 28

Listening to the Landscapes of Mama Tingo: From the "Woman Question" in Sustainable Development to Feminist Political Ecology in Zambrana–Chacuey, Dominican Republic

Dianne Rocheleau

In the rural countryside of the Dominican Republic environmental change has long been tied to livelihoods and landscapes and enmeshed in struggles for social justice and rights to land. In the early 1990s I went with a team of three other researchers to the rolling hills of the Zambrana–Chacuey region in the center of the country to learn about and document the recent community forestry experience of women and men who had been engaged in peasant land struggles against large commercial landowners for decades. Our goal was to see how gender and class had affected their sustainable development and forestry enterprise efforts, and in turn, how these initiatives had changed gendered social relations in the region. We ended up in a dialogue that I call "listening to the landscape," since every feature in this patch-work of farms, forests, gardens, and homesteads was tied to stories of individual lives, families, communities, and social movements.

How exactly does one listen to a landscape? Well, it takes decades of practice and more than a little luck, and, for me, ten years to play it all back and listen again. During the four months from October 1992 to February 1993 I listened attentively to the braided life stories and landscape histories of women and men in the Rural Federation of Zambrana–Chacuey in the Dominican Republic, one of 17 such regional groupings in the national Confederation Mama Tingo. The Zambrana–Chacuey Federation that formed the context for our research on gender and farm forestry (our "sample") was part of a larger land struggle movement spanning two decades, originally led by Florinda Soriano Munoz. Also known as "Mama Tingo," she was an elder woman, a proud farmer, a renowned and revered *campesina* leader, and a martyr slain by armed gunmen during the land redistribution campaign she led in 1974 (Ricourt, 2000). Throughout the stories of the farms and life histories in our study ran the chronicle of the Zambrana–Chacuey Federation and under that the current of a narrative of resistance, resurgence, and complex relations of power spanning centuries. Mama Tingo had led a wave of land struggle that grew to international prominence in the 1970s as part of a braided national movement of liberation theology and land struggles that in turn spawned many similar movements throughout Latin America. The Rural Federation of Zambrana–Chacuey was a regional affiliate of the formal organization that grew out of that movement, though it was not formally publicly founded until 1978. The women and men we spoke to told stories (some recent, some decades past) of facing soldiers, police, and forest guards, as well as tales of encounters with drought, floods, and boom–bust markets for coffee, cocoa, and tobacco.

In the process of listening to the stories, I learned about the co-construction of a regional smallholder patchwork of croplands, forest, and gardens from the mix of ranches, large coffee and cocoa holdings, and forest fragments that predominated in the area from the 1950s through the 1970s. Woven more deeply into that patchwork was a narrative tapestry of species, spaces, people, and power that joined women and men in a tangle of relational webs. This bramble of organic networks included power over (coercion), and power against (resistance), as well as power with (solidarity) at every level, from households and localities to regional, national, and international scales. The strength of roots that tied people to the land, to each other, and to a complex array of species and landscapes varied substantially, with a repertoire of rooting strategies as diverse as those in the ecosystems in which they were embedded – from the tap roots of generations on the same plot, to the tendrils of epiphytes (air plants) holding fast to any surface that offers a resting place. Yet others followed the habit of the spider plant, setting down roots outside fixed and overcrowded containers and dropping small offshoots into new sites; some people moved from one site to another in pursuit of dreams of refuge, and a way of being-in-relation with each other and the land. In Zambrana during the 1970s people from this whole spectrum of situations came together to make common cause in a collective struggle for private property, which they would cobble together into a new commons of viable and secure smallholder communities within a regional agroforest.[1]

Getting There: How the Personal Becomes Professional

It is a feminist proposition that we must first see like ourselves, and know where we stand, in order to engage others in honest dialogue, then look back at ourselves and the world around us, to see like others, with their permission and assistance. I saw the links between the land struggle movement, Mama Tingo, and the landscapes and ecologies of Zambrana in 1992 from a very specific position in a tangled social and ecological web. The early roots of this web spring from childhood and cultures of being-in-relation, of seeing not like a state (Scott, 1999), but like a woman, a feminist (Enloe, 2003), and the granddaughter of a coal miner, a housewife, and two farmers turned factory worker-organizers.

Reading the landscapes of home(s)

My mother cultivated in me a fascination with landscape derived from the long forest and roadside walks of her Appalachian girlhood. During the regular trips we made from our various homes in Connecticut to her home, the Sample Run coal miners' camp nestled in the green ridge and valley topography of western Pennsylvania, she always narrated our progress into forest and mountain country. And finally, once we passed the burning cinder cone of mine waste, redolent of sulfur, I knew the next place we would stop would be where Grandma's flower garden rose out of a small patch of earth in front of Grandpap's workshop. Once there, we walked the same paths she had traveled as a girl, and she pointed and recited the litany of place-markers that anchored memory. There was the creek, the rope swing, the rocks where she had seen snakes, the railroad from the coal mine, and the now redundant outhouses with the ditches draining into the creek. She always showed me the teacher's house on Miners' Row, her best friends' houses, each by name, tied to the stories of their families (who had gone where, who had died, who was still there, with updates at every telling). Sometimes she pointed out the road to the cemetery, reminding me of all the miners there whose tombstones bore the same two dates, for the cave-in and the explosion that nearly claimed her father. Then she always mentioned the union and how it had stood by the miners in hard times. Talking landscapes and livelihoods was a family pastime on Sample Run and not outside the realm of women, whether on forest paths or at kitchen tables. The landscapes we talked about were not devoid of humans; nor were they separate from livelihoods. Mom conveyed an Appalachian sensibility of living in a landscape, coal miners' politics, and a curiosity about my surroundings that made me aware of things I might not otherwise have seen or felt.

My father's contribution was no less formative. He learned all his job skills (mechanic, machinist, carpenter) by asking people about what they did and how they did it. I used to go with him to antique car shows, friends' houses, boat yards, and farms, where he would talk with people about their work and watch them at it, and sometimes join them in some project, get them to teach him. Without realizing it, I became an acute and appreciative observer of people's work, especially craftwork or manual labor, from miners, farmers, barbers, boatbuilders, handymen, factory workers, seamstresses, and bakers (at home or in the workplace). Dad's Québecois parents migrated from farms in Canada and Vermont to factory towns in Massachusetts and then to Hartford, Connecticut. They were blacklisted out of the factories in Hartford for being shopfloor union organizers at the Underwood–Olivetti typewriter factory and spent most of the next three decades in less than minimum wage restaurant jobs. The unofficial union organizer blacklist of the industrialists in Hartford was suspended during the Second World War, and my grandparents built their fair share of airplane parts at Pratt Whitney Aircraft, then went back on the list until the late 1950s.

Running silent and cold through all of this early learning was the whispered other history of my grandmothers and great aunts, forbidden fragments of stolen history that fell like crumbs from the hushed conversations among women. Sons and brothers recited the praises of strong women: "thirteen kids, divorced, working three jobs and no welfare"; "raising her own seven children and healing wounded foster children"; "still together after all these years"; "outlived three husbands, all miners". But the women's whispers marked time with bitter memories – dropped like table scraps of history, they spoke of the pain erased from the public scripts. The moments ran together in a litany of shame and hurt: the shotgun wedding in 1927, the beatings, a lost kidney and near death, the deaths in childbirth and replacement by younger wives, the philandering husbands, a woman obliged to have sex during the onset of labor, the daily abuse, the public humiliation. They spoke of patient long-suffering as well as defiance, of the locked (separate) bedroom doors, the suicide attempt by the divorced mother of 13. Later she would tell me: "they didn't even breed cows that often". She asked the priest for church-subsidized housing and he told her she had "too many children."

Participant observation yielded its insights as well. Patriarchy was not an abstraction and did not depend on an all-star cast of evil men. It (and liberal doses of alcohol) was what brought many of these otherwise good men to practice thoughtless cruelty, first and foremost at home, to visit their own humiliations from the workplace and public space on the women in their midst. I loved them for their way of making farming and factory lives sing with life, their love of work, their dancing, stories, and hearty laughter, but I knew they were also tied up with something real that was not good for women, children, themselves, or most other living things.

So an intense curiosity about work and knowledge, bound tightly with a respect for workers and a concern for their conditions, was part of the invisible baggage that I carried into my academic preparation and later into research. The heroic stories of motherhood and working women, the whispered secrets of women's private lives, their knowledge of home and the world, ran through it all like a subterranean stream of double – no, multiple – consciousness. This childhood apprenticeship gradually merged with my own professional field experience in the Dominican Republic and Kenya from 1979 to 1992 and followed me to Zambrana in 1992. As I went into the field, always someone else's home, to work with farmers, I would watch, ask questions, listen, and try to learn firsthand how people (individually and collectively) made a living. I looked at how they lived and worked on the land, separately and together, under conditions of displacement, eviction, land shortage, droughts, floods, fluctuating markets, and hostile political regimes. And I asked about their knowledge, their perspectives, and the terms and conditions of work, whether "in the home," on family farms, on other people's farms, in plantations, factories, mines, or the military. I also asked about conflicts, struggle, and solidarity, from individual to international scale. While my personal interest always followed these lines my professional focus shifted from physical geography, land use, and environmental degradation to an increasing emphasis on the social relations of environmental change, to an emphasis on gender and environment, and back again to landscapes and ecologies, but through a feminist poststructural lens. Between 1979 and 1993 I traced a path from working mainly with men on questions of environment, with a personal interest in women's lives and gender relations, to a focus on women in sustainable development, to a feminist political ecology and beyond to a feminist poststructural approach to changing landscapes and ecologies.

How the professional becomes personal and political: Sierra, Dominican Republic 1979–81

In 1979–81 I undertook my dissertation research – on watershed systems – in the Sierra of the Dominican Republic. During these two years, when I was also employed on a regional rural conservation and development project in the Central Mountains of the Dominican Republic, I learned most of what mattered to interpret my own empirical physical watershed data by using (at first unconsciously) the ethnographic skills acquired during childhood in everyday life, as we had crossed from Québecois to Appalachian cultures and in and out of various versions of "mainstream" culture, traversing class and culture lines as well as localities in our many moves. While my dissertation proposal did not mention participant observation, it became the central integrative method that joined my biophysical watershed data with the social context of the Sierra. I made sense of the biophysical results from a combination of informal observations, conversation, and formal interviews with people I had come to know in my capacity as a watershed researcher and resident from 1979 to 1981. Informal conversations with rural women and men shaped my understanding of environmental history, resource management, and small farm economics, as well as gendered identities, labor, and land tenure in the Sierra – all relevant to the erosion and runoff rates I had documented.

One of the insights I acquired, incidental to my dissertation research, set me on a new path. It became clear to me that many women were "marooned" in the agrarian landscape, forbidden, by custom, to cultivate land independently of the men in their families. By the late 1970s, many of these men had migrated to New York and were only part-time residents or occasional visitors in their Dominican Republic communities, and many poor or illegal migrants lacked the money or the documents to travel home periodically. One woman named Sarah explained how she had come to break the code against women farming alone, and how difficult the decision had been. Her husband had gone to New York illegally and could not return. At first he sent money, then less, then none. It was assumed that he had another family to support by then. With the help of her children she opened a *conuco* (a mixed crop plot with a diverse repertoire of trees, annual crops, and herbs). She plowed the land at night, by moonlight, to avoid the taunts and stares of her neighbors. She explained that it was something for a man to do, that this opening up of land was a very serious transgression of proper womanhood.

In contrast, Carmen, a rural woman from the same area, lived alone and had had "men friends" who supported her after her husband left. No one seemed to be scandalized by this arrangement. For a woman to cultivate was worse, it seemed, than to be a mistress; better to be a "bad woman" than to be unwomanly. Yet these other men could not cultivate a married woman's land either. It belonged to the absent man and his (male) heirs. In a region rife with male labor migration, gender identity was having a major but invisible impact on land use practice, livelihoods, and environmental degradation.

Class, gender, and migration status jointly marked the landscapes of the Sierra. To read that landscape I had to listen to the women's and men's stories, and to an anthropologist colleague, Nia Georges (1985), then look again. The women's stories would emerge only tangentially in my interpretation of my dissertation field data, but would filter back into case studies of gendered landscapes and livelihoods after my subsequent research (and steep learning curve) in Kenya.

Kenya: women, agroforestry, and feminist critiques of science

Following on the heels of my feminist observations in the Sierra I embarked upon another life changing experience, this time working with, observing, talking with, and explicitly writing about the women's groups in a handful of communities in Machakos District, Kenya. As a geographer and postdoctoral fellow at the International Council for Research in Agroforestry (ICRAF) I found myself in a professional culture of gendered denial. From local to international level, every kind of research and development program – across institutions – was working almost exclusively with men, focused on experiments with imported trees, crops, livestock, and ideas. Yet women were literally everywhere on the landscape and 60 percent of the households were women managed. Thirty percent of the households were women headed, and another 30 percent had absentee male labor migrant heads of household. Women engaged in farm labor, harvested forest products (fuelwood, medicinal herbs, wild foods, and vegetables), herded animals, carried water, sold produce, and staffed small shops at the village marketplace. They participated in community life through everything from building roads, digging communal wells, and fixing gullies, to writing and singing campaign songs for hire in local elections. There were also women (albeit a minority) prominent in local and regional political committees. The women's self-help groups, a complex combined legacy of traditional cooperative labor practices and colonial forced labor brigades for women, were highly visible on the landscape in soil erosion control, grazing land rehabilitation, and public works projects of all kinds. Moreover, there were parallel networks of women's knowledge, authority, and governance operating through the ubiquitous women's self-help groups, as well as through traditional religious practice, elders' councils, herbal medicine, and midwifery. Beginning in 1983 (and still continuing after two decades), I learned several lessons about the history and possible futures of land use in the region based on a gendered perspective on landscapes, labor, and political organization.

My initial efforts to make women and their work visible in agroforestry focused on what Sandra Harding (1986) would call the equity question in the feminist critiques of science, the need to put more women to work as scientists and to see women as proper clients and subjects of research. I eventually traversed the full spectrum of five feminist critiques of science identified by Harding. I quickly moved on to the second feminist critique, the abuse of science by biased practitioners to perpetuate inequalities and oppression. The prevailing research in farming systems at the time was perhaps best encapsulated in the common phrase "the farmer he." Even as some of us made women more visible and documented the predominance of women's labor (60–80 percent) in Kenyan agriculture at that time, many technical and social scientists still insisted that women were "mere laborers," the men actually were "the managers and owners" and as such would be "the ones to talk to." The tools of social science were used to classify the women as "non-owners," and not "primary decision-makers." Eventually, with by now a fully developed sense of "feminist curiosity" (Enloe, 2003), I found myself questioning the possibility of objectivity (Harding's third feminist critique of science) and suspicious of seeking single answers and simple explanations. I began to focus increasingly on what Haraway (1991) and Harding (1986) would call the power of partial perspective, based on situated knowledge.

In my Kenya research, I interviewed groups of men and women separately about crops, trees, forest products, wild foods and medicinal plants, land management, land use, and terms of resource access and control, to get a sense of the distinct bodies of knowledge, interests, and visions of men and women. Over time I increasingly asked people about their analyses, their knowledge, judgment, and vision, rather than simple "information." My relationship to this community in Machakos, while I lived in Nairobi for seven years, and later, as I returned periodically to do research, changed not only how I saw science-as-usual, but also how I saw landscapes and ecologies as gendered terrains (Rocheleau, 1991; Rocheleau and Edmunds, 1997). By the time I had been back in the USA and in the academy for three years, in 1992, the quest was on for ways to demonstrate empirically the multiple social and ecological realities in places like Machakos. I also wanted to take up Harding's challenge to do science differently based on feminist perspectives. I was searching for a way to deal with complex social identities – simultaneous membership in multiple groupings – and to extend that analysis to humans and other beings, in relation, in place, and in networks across places. In short, I was seeking to embody a feminist political ecology in my work, with equal commitments to social, political, and ecological dimensions. I planned a return to Machakos for one year of fieldwork during my pre-tenure sabbatical in 1992–3.

Being there (Zambrana, not Machakos)

My critical environmental history research agenda for my sabbatical year in Kenya was derailed by a politically charged postponement of election dates and ethnic clashes in Kenya fomented by various actors, including some powerful forces within the ruling party. So, I decided, with my family, to instead spend the first semester of sabbatical in the Dominican Republic. What was to have been a short applied research project to write a gender and environment case study – an add-on at the end of my sabbatical year – became a fully fledged research effort during the first semester, complete with setting up house with my family and a graduate student, Laurie Ross, in a small provincial capital, Cotui. My two sons, Rafael and Ramon, and my husband Luis settled into our new neighborhood, school, and their own projects. Laurie and I, along with two Dominican colleagues, Professor Julio Morrobel and Ricardo Hernandez, arranged to conduct our study on a farm forestry project in the rolling hills of nearby Zambrana, with the well established Rural People's Federation of Zambrana–Chacuey and ENDA (Environment Development Alternatives) – a collaborating international non-government organization. ENDA met with us at their Cotui office, and the Federation put us on the agenda in their monthly plenary meeting of the representatives from the member groups – 59 farmers, housewives, and youth associations from 31 communities.

I knew it would be different from previous research sites when we first set foot in the headquarters of the Federation in Zambrana, prior to our presentation at the assembly. The walls were covered with cards from previous meetings of local associations. Each card was covered in turn with words printed in the stark thick letters of ink markers: Blackness, Power, Respect, Woman, Together, Solidarity, Knowledge; or Fear, Alone, Poverty, Ignorance, Humiliation. Paulo Freire clearly lived here, in spirit, and in practice. Having duly noted the mention of race, and the positive association with blackness, neither a commonplace in my prior experience in the Dominican Republic, I was struck by a poster portrait on the wall of cards. The figure on the poster was an apparently African woman with strong cheekbones and the wrapped head scarf that I had come to associate with women farmers I'd known in Kenya. "Who's that?" I asked. "Mama Tingo," they'd said, as if I would know immediately, as if of course everyone would know who *she* was. I didn't, I admitted sheepishly, and there began my first lesson about how people came to be living in these tightly knit networks of smallholder communities in the hills and valleys of Zambrana and Chacuey Districts between Santiago and Santo Domingo.

The story of Mama Tingo was at the root of the history of the Rural Federation of Zambrana–Chacuey. It was all part of a larger national social movement that arose in the 1970s, which was in turn part of a very long history of surviving and selectively taming and negotiating with a long series of political and economic regimes hostile to rural smallholders. With Mama Tingo watching from the wall, I was on notice that this would be a different perspective on gender, race, land, and politics than I had found in the Sierra communities only 50 miles and two hours away where I had worked from 1978 to1981. My own history and my experience in the Sierra and Machakos would train my ear and vision, but neither could offer a blueprint for understanding Zambrana. I would have to start from people's lives, and in particular from women's lives, and follow their footsteps on the land and their webs of relationships, to understand the landscapes and livelihoods of Zambrana.

At the assembly that day we explained that we were funded by USAID (United States Agency for International Development) to write a case study to train foresters, agronomists, and resource managers to work democratically and effectively with women and men across classes, to develop and support sustainable farming, forestry, and land use practices. For our part, we wanted to understand how the Federation members lived on the land, how they (women and men, across localities and across the social and economic spectrum) were dealing with rapid social and environmental changes, and how they saw the future. We hoped to learn from their organizational and land management experience and to provide suggestions to ENDA and the Federation based on what they would tell us, directly and indirectly, through a series of interviews and surveys. What did they want in return? "Write our history. . . . Tell people what happened here. Write a book about the Federation." That seemed both simple and serendipitous. It was not, as it turns out, an easy task, and I am still trying to do an honest job of telling pieces of that story.

Over the course of the next four months we visited and interviewed 31 local associations (farmers, housewives, and youth groups) in 16 communities (out of a total of 59 Federation-affiliated associations in 31 communities, each association composed of roughly 12–30 people from a farming community in a specific locality). We combined ethnographic, standard survey, and feminist methodological approaches, including: participant-observation; group interviews; key informant interviews; life history interviews; community and organizational histories; detailed sketch mapping and land use history with 20 selected households; land use simulation board games with selected groups and households; and a formal survey with a gender-stratified random sample (45) of the more than 700 Federation members in farmers associations and housewives associations, respectively. What we learned complicated rather than simplified the story of this region, along with the social and environmental categories that explained its past and suggested possible futures.

Feminist Political Ecology in Zambrana–Chacuey

In 1993, Zambrana–Chacuey, a hilly farming region, was home to roughly 12,000 people, most of them smallholder farmers engaged in a tenuous mix of subsistence and commercial agricultural production on one-half to two hectares of land per household. The landscape supported a range of land use and cover types, from rice and cattle on large expanses of flatland, to forests, coffee, and cocoa on steep upland slopes and riparian forests along streams and rivers. Farmers cultivated tobacco, citrus and other fruit trees, forested home gardens (*patios*) and *conucos* (diverse plots of cassava, yam sweet potato, taro, and other vegetable, medicinal, and herb crops that reflect the mixed Spanish, African and Caribbean culture), and some of them both planted and harvested trees for timber, woodworking, and charcoal. Almost every household, if it could, included some off-farm wage labor income to complement farm production (Rocheleau and Ross, 1995).

The experience in Zambrana–Chacuey during the 1980s and 1990s exemplified the complex and contradictory nature of simultaneous national trends to strengthen environmental protection and to promote agricultural exports, as well as the attempt to reconcile them under the ideological umbrella of sustainable development. During the 1980s the military-based forestry service mounted an intensive anti-deforestation effort – the *Selva Negra* campaign – complete with armed troops and helicopters. At the same time absentee land speculators and ranchers, as well as agribusiness corporations (Leche Rica citrus and Dole pineapple, prominent among them), sought to acquire more land in the region, in direct competition with smallholders. Meanwhile, smallholder farmers turned to the production of tobacco and or cassava (*yuca*) cash crops on part of their holdings in order to survive the decline in coffee and cocoa prices and the suppression of charcoal and woodworking activities.

In 1984 Zambrana–Chacuey became the site of a collaborative forest enterprise pilot project sponsored by the Federation and ENDA. The initiative combined elements of peasant cooperatives and commercial contract farming models, with widespread planting of timber (an exotic tree, *Acacia mangium*) as a monocroppped cash crop on smallholder farms, followed by the construction of a cooperatively owned sawmill to produce timber for sale. This was the best cash crop going if you could manage to wait six to eight years for the timber to grow and if you could dedicate a plot to it, since it doesn't mix well with other trees and crops. The tree was also set loose upon a complex gendered and class-divided field of power; it was sought after by some women and contested by others based on their specific land tenure and livelihood situations, their interests in food or commercial crops, and their ability to reconcile the tree with their croplands and patio gardens. While there were broad public proclamations of solidarity among all Federation members, my own childhood experience told me that just because women and men share daily work, broad social values, and political involvement does not guarantee similar visions, interests, and equity in community and household economics and politics.

As we began to study the gendered terms of participation, decision-making, and distribution of benefits (Rocheleau et al., 1996; Ross, 1996) of this apparently successful community forestry project, Laurie started to trace the history of the Federation, along with our larger survey. I was also increasingly drawn to the story of the Federation itself, especially the contributions of women, in no small part due to my personal history, and I began to look at the commercial farm forestry trend through a critical feminist lens on ecological, cultural, and political grounds. So the woman question in the forestry project quickly gave way to a long list of feminist questions about sustainable development in general, the forest project in particular, and the broader trends in the landscapes and ecologies of Zambrana–Chacuey. The Federation itself emerged as a major actor in addressing social and gender relations *per se* and also in restructuring the landscapes and ecologies of the region, first through the land struggle and later through the agricultural and sustainable development projects.

In 1993, over 700 people were members of the Federation, which directly served roughly 4000 people through their families and neighbors and provided broader support to the entire population of 12,000 in Zambrana and Chacuey Districts. The organization was rooted in three separate wings of a very broad movement, each with a different and equally rich version of the Federation's history and distinct hopes and fears for its future direction. The three wings were: cooperativist, Catholic liberation theology, and more traditional Catholic Church basic needs advocates. Most of the farmers we met had acquired (or recaptured) their land in the previous 20 years through a series of land struggles using non-violent civil disobedience (by men and women) in the face of armed soldiers and police, jail terms, and campaigns of intimidation. Among the members of the Federation and their families the politics of place was entwined with the practices of solidarity and affinity, across lines of difference (race, class, gender, and political affiliation).

The Rural Federation of Zambrana–Chacuey, as noted earlier, was one of 17 federations within the national Rural People's Confederation of Mama Tingo, named for her after her assassination in 1974. The Federation, in turn, consisted of 59 local member groups (farmers, housewives, and youth associations) in 31 communities. The farmers associations were comprised of mostly married men farmers, with some younger single men and also some women farmers – a mix of women who were the "main farmers" in their "male-headed" households, and others who were single, divorced, or widowed. These groups had waxed and waned over time, with peak periods of activity based on the land struggle campaigns and on market and production support for various cash crops, following boom and bust cycles. The youth associations were less common and generally not as active in the full range of activities of the Federation. The housewives associations were composed entirely of women and were women's mutual support organizations with a broader, more social focus than the farmers associations and more continuity of membership and activity over the years.

From the outset the Zambrana–Chacuey Federation in this area had a base in women's groups and women's politics of place, though that history was often forgotten or subsumed under a more economic or traditional political explanation. Tito Mogollon, one of the original human rights promoters sponsored by the Catholic bishop in the 1970s, noted that he and other organizers originally approached women's groups in the nearby community of Los Cacaus, who were threatened with eviction by the Rosario Dominicano Gold Mine in 1974. Eventually two women's groups, one in each district (Zambrana and Chacuey), formed the nucleus of two new associations, which grew into the Zambrana–Chacuey Federation. The successful non-violent land struggle waged by these groups was rooted in symbols and icons that appealed to long histories in place and the rights of rural people to maintain or regain lands lost to the USA-based sugar corporations, the Trujillo regime, and its wealthy clients. However, the movement also proclaimed the right and the profound need to create space, through land tenure reform, for displaced and landless families from other regions to make new homes and new communities, based on a shared sense of purpose, respect, and mutual support. In an early act of civil disobedience in a neighboring region, a group called La Otra Banda opened the 1970s cycle of land struggles by occupying a 1500 acre plot in a Gulf and Western Corporation sugarcane field. The group cleared the land and planted traditional food crops, underscoring the importance of traditional crops as a political and cultural symbol. That event and its symbolic linkage of culture, politics, and traditional crop species became a touchstone of the Zambrana–Chacuey Federation and many others established in that period.

Women were key actors in the land struggles – they were recognized and visible leaders at local, regional, and national levels. Mama Tingo had made a prominent and public place for women in this struggle through her life and death and she served as an inspiration to women and men and a tangible cherished example of leadership for women in the liberation theology wing of the movement, as well as in political parties and cooperativist organizations. Her story, as a dark-skinned woman farmer who became a local and national leader, contradicted popular images of the peasant farmer (*campesino*) as a heroic, light-skinned rural man, wresting a living from the land. In contrast she embodied the history of rural women and men (*campesinas/campesinos*) of all colors from light to black who had labored in the fields for centuries to produce food and cash crops from Dominican soil.

Throughout the country in the twentieth century women were part of farm operations: in some regions of the country they identified as "farmers," while in others (like the Sierra) women's farm labor was seen as "helping" men in the family. In the central and eastern parts of the country (including Zambrana–Chacuey, and neighboring Yamasa, a focal point of Mama Tingo's campaign) African heritage was clearly expressed in the popularity of Afro-Caribbean religious practices, as well as in distinct land use traditions and a flexible gender division of labor. The legacy of Mama Tingo touched every aspect of the regional Zambrana–Chacuey Federation and its local associations that grew out of the early 1970s land struggle she led, but one of the most significant features of this movement, and the federations it spawned, was the strength and distinctive character of women's leadership. This was certainly evident in the experience of women in the Rural People's Federation of Zambrana–Chacuey.

The history of women in the Zambrana–Chacuey Federation exemplifies the complexity of the Federation and its ability to accommodate difference, its egalitarian ideals and its contradictions. Women were involved from the beginnings of the Federation, through the local housewives associations that constituted two of the founding groups. Women served in leadership positions, including a past vice president and past and current members of the board of directors. Among the women leaders of the Zambrana–Chacuey Federation in 1992, including a past vice president, were herbalists, healers, and midwives, Catholic Church and traditional Caribbean religious leaders, as well as founding members of local women's and farmers associations. Women's "traditional" power in medical, religious, and mystical domains translated into recognized political leadership within the Federation. The same women who had delivered many of the babies in their communities, attended to the sick, and advised local women in personal, marital, and financial matters had also participated at the forefront of land invasions and public demonstrations. They had faced armed soldiers inside and outside of local and provincial jails where Federation members were held in large numbers during major local and regional campaigns that involved the national Confederation Mama Tingo (usually these actions were initiated locally and supported by the larger Confederation). Many of the Federation's women leaders in 1993 were also active in party politics, exemplified in the contrast between one woman active in the leftist Liberation Party and one (surprising only to me it seemed) in the right-wing Reformist Party. Both women leaders were instrumental, along with a handful of men leaders, in negotiating access to resources and support from their respective parties and state officials under various regimes over the course of three decades. Brief moments from encounters with three women leaders exemplify the complex and integrative nature of the broader social and ecological relations that shaped the Federation and the landscapes and the ecologies of Zambrana–Chacuey.

Juana, a former officer of the Federation, an active member of the Liberation Party, and a consistent leader in the local women's association, recounted the role of women in the land struggles beginning in her Zambrana community in 1959 and extending throughout the region into the 1980s. She noted that women sometimes began as scouts and as cooks, bringing food to the men who occupied the large-holders' pastures and plantations. "Then when they arrested the men we would take their place or go to the jail to surround it, we would call for others from the Federation, and from further away, from the [national] Confederation Mama Tingo. We filled the jail in Cotui and then in San Francisco de Macoris."

Another leader from Chacuey, Cristobalina, recalled the many campaigns from land struggles to later road closings and demonstrations for schools, roads, and clinics. She remembered facing armed soldiers outside the local jail on multiple occasions. A one-woman short course in feminist poststructuralism, she explained her identity as a woman, and Catholic Church leader, and alluded to her other traditional Afro-Caribbean religious duties. She was a midwife, a herbalist, a healer, a businesswoman (crop merchant), a landowner, a mother, a leader in her local farmers and housewives associations, and a founder and former elected board member of the Zambrana–Chacuey Federation. She was also a member of the right-wing Reformista party, which intitally shocked me. When I asked her about feminism, she recounted how she had taught her daughter: "I taught her to say 'mine'."

Yet another elder woman, Maria, a locally beloved and revered religious leader and single mother of several (grown) children, expressed the intensely personal politics of some women as farmers, gardeners and agroforest dwellers : "I am a farmer . . . I am a women of the land . . . " In response to this her friend and neighbor Alfonso laughed and affectionately called her "Mama Tingo." Later as we visited her home and the surrounding agroforest, Julio Morrobel (of the research team) recognized a tied bundle of grass suspended outside her door and we asked her about it. She explained that she would sometimes light it as incense to ward off evil. "I make this during Holy Week on Good Friday. . . . I spend the day in silence, walking through the forest, taking one leaf of each kind of plant, then I tie them and place them here." So the farm and the forest and the worlds of spirits and the hearth met at the doorstep to Maria's home.

Women members of local housewives and farmers associations, as well as women married to farmers association members, also played a major role in shaping the landscape as farmers, as gardeners, as nursery and livestock keepers, and as gatherers and managers of fuelwood, medicinal herbs, and water supplies. During the course of the 1980s the women of the Zambrana–Chacuey Federation (as Federation board members, and as members of local housewives associations, and/or farmers associations) led the Federation in an increasing involvement in development, moving from campaigns for land and marketing support to infrastructure and basic services and beyond into a long-term sustainable development partnership with ENDA. The latter progressed from a research and development initiative on women's ethnobotany and herbal medicine, to a broader collaboration on agriculture, forestry, and local enterprise development (Ross, 1996) that continues until the present.

As we followed women's footsteps, their activities, their memories, and their social connections across the landscape, we encountered a multiplicity of unfolding ecologies. This journey blurred the maps of property and land use, and the categories of forest/farm, wild/domestic, farming/gathering, and the distinctions of private, common, and public lands and resources, as well as the lines between household, community, and political entities. The outcome of both the land struggles and the research and development efforts of the 1970s and 1980s had major implications for diversification of livelihoods, landscape, and the restoration of diverse species and cultivars to the landscape under smallholder agroforestry systems. The people who gained land through these campaigns were almost always tree planters and many fostered rich agroforests in areas formerly used as pastures, cropland, or cocoa and coffee plantations by large holders.

Rather than being content with this profusion of social data I found myself absolutely "needing to know" about the biodiversity of these patches and ribbons of forest in a regional landscape quilted by the Federation members into a distinctive and gendered socioecological formation, a regional agroforest rooted in community, a shared history of struggle, and visions of a possible agrarian future. As we proceeded with the sketch maps and survey listing of tree and crop species it became apparent that the patio gardens constituted a polka-dot forest, a major component of a biodiverse regional agroforest. The fact that patio gardens were largely women's domains was an exciting finding for a project focused on gender, but I was also delighted to find that the seeds of forest past and forest future were basically wrapped around people's homes, that the highest biodiversity was found close to – not removed from – the focal point of human habitation. This was true for indigenous forest tree species as well as for overall species diversity, including crops.

So Zambrana–Chacuey was not only home to a regional organization derived from an inspiring social movement led by both women and men, it was also home to an officially invisible regional agroforest – a patchwork of trees, crops, homesteads, gardens, livestock, and wildlife. This diverse agroforest formation constituted the ecological context for the people, who both shaped and adapted to distinct biotic assemblages. The people of the area existed within landscapes of cohabitation with a variety of beings-in-relation, a network that entwined culture, politics, and biotic elements. Based on empirical observation and social and ecological analysis, we came to see this landscape as home to a regional agroforest, rather than a formerly intact entity violated and diminished by humans. It is a biotic community (and, arguably, a cyborg forest) forged in the relations of different groups of people with each other and with various other life forms, non-living landscape features and changing technologies (see Haraway, 1991, on feminism and cyborgs). The agroforest(s) of Zambrana–Chacuey constituted a cyborg forest, a gendered construct that reflected the political courage as well as the calloused hands and watchful eyes of Mama Tingo, not only in the dramatic land takings but in the everyday remaking of the regional ecologies of home. So Mama Tingo and her legacy in Zambrana–Chacuey brought me to the threshold of a new agenda, the study of nature/culture in emergent ecologies (Rocheleau et al., 2001) and hybrid geographies (Whatmore, 2002) in combination with political ecology.

NOTE

1   Agroforest refers to the purposeful combination of trees, crops, and animals in managed ecosystems to enhance production as well as conservation, for economic as well as ecological ends.

BIBLIOGRAPHY

Enloe, C. (2003) *Bananas, Beaches and Bases: Making Feminist Sense of International Politics*. Berkeley: University of California Press.

Georges, E. (1985) The causes and consequences of international labor migration for a rural Dominican sending community. PhD dissertation, Department of Anthropology, Colombia University.

Georges, E. (1990) *The Making of a Transnational Community: Migration, Development and Cultural Change in the Dominican Republic*. New York: Colombia University.

Haraway, D. J. (1991) *Simians, Cyborgs and Women: The Reinvention of Nature*. London: Routledge.

Harding, S. (1986) *The Science Question in Feminism*. Ithaca, NY: Cornell University Press.

Ricourt, M. (2000) From Mama Tingo to globalization: the Dominican women peasant movement. *Women's Studies Review*, 9, 1–10.

Rocheleau, D. (1991) Gender, ecology and the science of survival. *Agriculture and Human Values*, 8(1), 156–65.

Rocheleau, D. (1995) Maps, numbers, text and context: mixing methods in feminist political ecology. *Professional Geographer*, 47(4), 458–67.

Rocheleau, D. and Edmunds, D. (1997) Women, men and trees: gender, power and property in forest and agrarian landscapes. *World Development*, 25(8), 1351–71.

Rocheleau, D. and Ross, L. (1995) Trees as tools, trees as text: struggles over resources in Zambrana–Chacuey, Dominican Republic. *Antipode*, 27(4), 407–28.

Rocheleau, D., Ross, L., Morrobel, J. and Malaret, L. (2001) Complex communities and emergent ecologies in the regional agroforest of Zambrana–Chacuey, Dominican Republic. *Ecumene*, 8(4), 465–92.

Rocheleau, D., Thomas-Slayter, B. and Wangari, E. (eds) (1996) *Feminist Political Ecology: Global Perspectives and Local Experiences*. London: Routledge.

Ross, L. (1996) What happens when a grassroots organization meets an international NGO. Master's thesis, Program in International Development and Social Change, Clark University, Worcester, MA.

Scott, J. (1999) *Seeing Like a State: How Certain Schemes to Improve the Human Condition Have Failed*. New Haven, CT: Yale University Press.

Whatmore, S. (2002) *Hybrid Geographies: Natures Cultures Spaces*. Thousand Oaks, CA: Sage.

Chapter 29

Gender Relations beyond Farm Fences: Reframing the Spatial Context of Local Forest Livelihoods

Anoja Wickramasinghe

Background

Generations-long forest-centered human livelihoods in South Asia have gone through several transitions, often resulting in physically detached but materially, perceptionally, and emotionally attached practices and traditions. The role played by the forest and the place given to the forests in livelihood systems varies widely across space. Various groups, some of them external, with different interests have designated stretches and patches of forests under many labels. Designations such as "The Strict Nature Reserves," "Heritage Sites," "Man and Biosphere Reserves," "Conservation Forests," and "National Parks," for example, are evidence of the global interest in Third World forests more than of local interests.

There has also recently been a paradigm shift inquiring into the relations between society and forests. In Sri Lanka and India recent research on "people and the forest," the "anthropogenics" of forestry, non-timber forest-product gathering, and local conservation efforts has generated interest in the multiple social, cultural, and also material linkages with the forests (Wickramasinghe, 1995, 1997a, 2003a, p. 105, 2003b, p. 101; Wickramasinghe et al., 1996; Saxena, 1997; Ramakrishnan et al., 1998). The valuation of non-timber forest resources has contributed directly to this paradigm shift (see Panayotou and Ashton, 1992; deBeer and McDermott, 1996).

A longstanding social and economic harmony between local people and their forests has enabled the people, particularly those in the forest peripheries, to deal with their territorial resources in a sustainable way. In this traditional social and local context there has been a strong need to place forest resources within the livelihood system – rather than seeing forest resources as collections of use-specific products and life forms. The links between forest resources and livelihood are organized through various roles and activities performed by men and women in gender-specific patterns: the links between people and the forest resources are influenced strongly by their needs, culture, lifestyle, perceptions, livelihood – and gender. It is in this context that "gender relations" become central in the process of linking the forest resources with livelihood. ("Gender" here refers to the relations of power between women and men which are revealed in a range of practices, ideas, and representations, including the division of labor, roles, and resources between men and women, and ascribing to them different abilities, attitudes, desires, personality traits, and behavioral patterns (Agarwal, 1994).) Gender has been studied most in connection with reproduction and production; much less attention has been paid to gender relations in resource-based livelihood systems. How gender relations extend beyond the domains of households and farms is a rich area of interest for geographers, especially those working in landscape geography, resource geography, and the disciplines of forestry and feminist studies.

Several research questions are raised in the effort to analyze gender relations "beyond farm fences." The first is: what kinds of gender-specific needs do forest resources satisfy? The second is: what factors influence men and women in their choice of resources to satisfy those needs? And, finally, are men's and women's resource perceptions related to their gender?

The Distinctiveness

The social and cultural roles of women and men have resulted in most "well-being and welfare" activities being assumed to be women's work, and most production and resource-based work being perceived as men's work. This simple binary assumption has continued despite growing evidence that, looked at as a whole, livelihood systems are complexly gendered: livelihood systems incorporate specific places and spaces for men and women whose work roles and responsibilities are shaped by particular social and cultural forces.

Across South Asia, women's resource work extends the space of women beyond the household to include many ecosystems that they occupy and deal with. Agarwal (1994), in her work on gender and land rights in South Asia, has examined the link between land and gender relations, and the legal and social context of land ownership. Referring to Sen (1981), Agarwal discusses the factors influencing the bargaining strength of women and men within the family *vis-à-vis* subsistence needs. These factors include: the private ownership and control over assets, especially arable land; employment and other income earning means; access to communal resources (village commons and forests); access to external social support systems; and access to support from the state or from NGOs. According to this analysis these factors strengthen a person's survival ability outside the family. Despite the contextual variations prevailing across the South Asia region, which she has discussed referring to cases, it has been recognized that the village commons and state forests are important for two primary reasons. The first is the high dependence of the rural households on these for a wide variety of items essential for daily use; and the second is that their products are primarily gathered by women and children. The level of dependence being determined by the local context, or the contextual variations, is an indication of the livelihood security of women in particular.

In Bangladesh, patriarchy shapes a distribution of power and resources within families where power and control of resources rest with men, and women are powerless and dependent on men (Cain 1979); here, the material base of patriarchal control is secured through interlocking elements of the kinship system, political systems, and religion. In Sri Lanka, where disparity between men and women is reported to be relatively narrow when compared with other countries in the neighborhood, agrarian reforms and forestry policies have failed to assure women's rights to their livelihood sources (Wickramasinghe, 1994, 1997b, p. 37). Despite such local and national differences, women's constrained access to resources is similar across geographic areas and societies, and to the extent that women engage in village-based, ecosystem-based, and family-based resource management they do so for the security of their own, their family's, and their community's livelihoods. For instance, Bhadra (1997), in her case study conducted in Chalnakhel in Nepal, reports that farms and forests are the space of women to produce crops and earn a living. This author's field notes in the same community show that women work from dawn to dusk throughout the year, attending to multiple tasks of managing crops, livestock, shrubs, and trees. This work spatially extends between fields, forest, and cattle sheds. Tasks such as carrying organic residues and cow dung to enrich the fields and carrying fuelwood and fodder from the forest are crucial to sustain the integrity of the livelihood systems.

Beyond Farming

Most studies of women's contributions to production have focused on their roles in agricultural crop production. As a result there is an increasing recognition that women are dealing with land, farming for food and subsistence, and sustaining family farms. Various new methodologies have also been introduced to enumerate women's labor in agriculture – especially "time budget" studies to account for the time allocated on various activities, including unpaid work and the irregular enrolment and division of time between production and reproduction. Motivated by global concerns over women's contribution to economic development, during the UN women's decade of 1975–1985 a tremendous amount of information on women's agricultural labor was generated and disseminated. Dixon (1982), for instance, based on ILO estimates and FAO censuses of about 82 countries, estimated that the proportion of women in the agricultural labor force was about 42 percent. For sub-Saharan Africa the regional average was 46 percent; for North Africa and the Middle East, 31 percent; and the highest, 48 percent, was recorded for Asia. However, contextual differences in the nature of this labor and the nature of the goods and services provided by men and women in a given society that differ across space, at various levels from households to ecosystems, regions, countries, and broad geographical regions, have never been fully explored.

Much of the research on women and agriculture rests on several assumptions. Most of the research focused on household production pays little attention to the integrated elements of the farming systems, and in particular has been guided by the assumption that agriculture excludes silviculture. However, we do know that each household's income is derived from a great variety of sources that constantly change in response to available opportunities according to the season, the state of the market, and even the time of the day – and includes the use of a changing mix of resources.

The second assumption tends to be that agricultural crops are the main supports for meeting household subsistence requirements. The multiple production systems of tropical farmers, which include home gardens, highland and lowland fields, and also common property consisting of more than one unit, have seldom been separately examined – instead, study of these has been included under the umbrella of "household farming."

The widely adopted household model considers the household as a single functional unit where capital and labor are organized and divided, and consumption of output takes place. The consequence of this has been the greater weight given to the patriarchal social system and to masculinized livelihood sources. Women's spaces and sources of livelihood, such as forests and common properties, are not theoretically or materially integrated in conventional "household" studies. Household analyses leave out diverse sources of livelihood sustenance, and also mask women and their contributions.

Typically, household models narrow down gender relations in households to the point where only domesticity and reproduction roles of women are emphasized. Studies on household resource allocation on farm production and production of services have failed to produce a full scenario of livelihood systems, and have failed to examine the ways in which households achieve livelihood security by integrating multiple sources in the village and communal ecosystems.

Integrative Approach

Studies in South Asia (including my own research described below) are now importantly revealing the complexity of livelihood systems. Most farm households throughout Asia live on extremely small plots of land and forest materials are a crucial part of their livelihood system. They suffer from a lack of basic assets – the land to produce essential items such as fuelwood, materials for porridge making, condiments, vegetables, etc. – within the limited space that they own. The allocation of space for forest resource exploitation spatially stretches beyond the boundaries of households and communities; the allocation of labor stretches beyond farm fences to satisfy multiple livelihood needs.

Instead of a "household" approach, a "livelihood" approach allows us to examine the complexity of multiple aspects of resource use, and the interrelated issues of gender. It can be argued that "livelihood," being a more human-activity-based frame, provides a broader framework for analyzing gender in relation to local contexts. The relevance of livelihood analyses to gender and geography is its ability to accommodate wider perspectives. The intra- and extra-household linkages that are inherent in livelihood studies are essential to reveal gender beyond the unit of household and beyond farm fences. Livelihood studies also include consideration of non-private resources: men's and women's livelihood sources stretch out family property and private property and encompass common property. Access to and use of resources under customarily recognized *de facto* rights is a key condition that contravenes gender-related subordination often associated with the distribution of private resources.

Household models particularly underemphasize work that takes place beyond the household, especially work that takes place in commonly owned and spatially distant forests. Various types of forest resources that people use to satisfy livelihood needs (food, fuelwood, fodder, construction, and production work) are left outside the paradigm of "forestry" and are typically not incorporated into household analyses.

Forests that are common or state-owned property in many countries are conceptually (and sometimes materially) placed outside the livelihood systems of people and established as a space of masculine domination. The significance of gender relations in the management of forests and their resources has never been accommodated, because "people" are seen as *de facto* users, not authorized resource users. Such uses are generically labeled under various categories, such as "multiple use," "non-timber uses," "traditional knowledge." The conceptual understanding of forest and forest resources has been placed in the paradigm of the scientists, conservationists, and the global community. The discipline of forestry and forest science, dominated by men, sustains these rigid boundaries. The masculine domination results in narrowing down the broader paradigms of livelihood to the households and farms, thus impinging on the analysis of gender relations "beyond farm fences."

Livelihood models are especially useful for considering the role of spatially dispersed forest resources in sustenance patterns. Being holistic, such an approach avoids the limitation of separating production from reproduction, *de facto* use of livelihood sources from the *de jure* property use. Forests are marked with much broader values that include materials, therapeutic values, services, and the amenity values. The livelihood model, in its incorporation of integrated and multifaceted contexts, also incorporates perspectives of feminist geography.

Resource Perceptions

Various aspects of socialization, including gender, culture, lifestyle, and livelihood, influence how people perceive resources. Janis Alcorn's (2000) definition of resource perceptions related to ethnobotany is especially relevant for inquiring into gender relations. In Alcorn's framework, "resource perception" refers to the process of assigning a particular resource role, or "use," to a plant by evaluating that plant's possible utility and the consequences of using it. This humanistic approach to resource definition allows for flexibility and complexity: the users determine the resource role of plants, their utility values, and the resulting effects. This approach allows for cross-cultural and gender-sensitive assessment, since users are heterogeneous groups, driven by various expectations placed on them by society.

My research in the traditional communities in the southern periphery of Sri Lanka's Adam's Peak Wilderness – the area commonly known as "Samanala Adaviya" – demonstrates the importance of gender-specific resource perceptions. The forests in this region are rich in goods of livelihood importance, therapeutic values, and "service" functions (including amenity values). Forest resources regenerate through natural processes without depending on external input; the confidence placed on the continuity of forest resources by women is greater than that of their male counterparts. Strong gender differences manifest themselves between the "regenerating" resources that women deal with and the end-harvest of timber that men deal with.

In my survey, 86 percent of women perceive resources of the forest primarily as a replenishing storage: the fruits, pods, leaves, resin, twigs, flowers, rhizomes, fungi, etc. are found year after year if not season after season. The resource roles of the above-canopy strata of multistory forest are hierarchical, not merely in terms of material output but also in terms of security and the well-being of the ones below.

Table 29.1 Dominant gendered perceptions regarding various forest life resources, Samanala Adaviya, Sri Lanka

Life form	Livelihood significance	
	Women	Men
Trees	Producers of multiple biomass, regenerators, nurturers/supporters	Woody parts, timber in particular
Shrubs	Producers of multiple biomass, supporters	Small timber, fencing materials
Lianas/vines	Raw materials for utensil making	Binding materials
Bamboo	Raw materials for making utensils, and thatching	Construction
Rhizomes	Food and medicinal	Food
Fungi	Food and medicinal	Food
Pandanas	Raw materials for making mats and other utensils	
Herbs/ground flora	Medicinal and food	Medicinal and food

Women's resource perception in this respect is holistic. Trees, shrubs, lianas/vines, bamboo, etc., which occupy the strata above the ground cover, are made to take care of the herbs, pandanas, fungi, and rhizomes below (see table 29.1). Producing the biomass of multiple uses while nurturing and supporting others, including the next generation stocks, is perceived as a unique function. Men's greater concern over the harvest of timber for construction and the forests' industrial values, as was reported in 92 cases, demonstrates the influence of gender relations on resource perceptions.

Gender Relations through "Needs" and "Resources"

The differences in men's and women's perceptions of the forest are grounded by the gendered division of livelihood responsibilities, by generations-long experience, and passed on knowledge which has made women concentrate more on the intergenerational sustainability of society and forests. Throughout the history of forest resource utilization in Sri Lanka (and possibly universally), women have contributed overwhelmingly to deciding the "utility value" of forest resources. The process of converting raw materials into useful forms and items has evolved in relation to needs; needs to sustain lives and needs to satisfy the gender-specific expectations placed on them by society. Food culture has evolved in relation to forest resources in the domain of women, who have become the pioneers in converting biomass into useful commodities. For example, it is women who know how to mix various forest-based ingredients to make food delicacies, increase palatability, eliminate unpleasant taste, reduce harmful elements in one food by adding others, etc.; therefore, women's contributions in this regard have resulted in shifting the paradigm of forestry from mere protection toward livelihood security.

Table 29.2   Gender specific activities in securing forest products

Utility by category	Materials/forest output	Activities by gender	
		Men	Women
Carbohydrate	Seed, fruit, tubers	Harvesting – upper story products	Harvesting, gathering, processing, digging
Vegetable	Leaves, twigs, fungi, rhizomes, seeds, fruit, pods, flowers, tubers	Harvesting – upper story products	Harvesting, gathering, processing
Porridge	Leaves, roots	Harvesting	Harvesting, gathering, processing
Fruit	Fruit	Harvesting	Harvesting, gathering, processing
Beverages	Stems, leaves, rhizomes, flowers	Harvesting	Harvesting, gathering, processing
Condiments	Pods, seeds, bark, leaves, fruit	Harvesting	Harvesting, gathering, processing
Sweetening	Sap, honey	Harvesting	Processing
Medicinal	Leaves, herbs, stems, bark, pods, fruit, rhizomes, root	Harvesting, gathering, preparing, treating	Harvesting, gathering, processing, preparing, treating
Binding materials	Lianas, barks, leaves	Harvesting	Harvesting, processing
Fuelwood	Branch wood, deadwood	Harvesting	Gathering, headloading, processing, stacking
Timber/construction	Trees, branches, bamboo, reeds	Harvesting	Stacking
Raw materials	Leaves, rattan, bamboo, pandanas, reed, timber	Harvesting, producing heavy utensils, furniture	Harvesting, processing, producing utensils (mats, baskets, winnowing fans, etc.)

Source: Information from Wickramasinghe (1995, 2002).

The activities performed by men and women in relation to the forest reveal the structure of gender relations extended beyond the household and the farmlands. All forms of life available in the forest, such as trees, shrubs, lianas, pandanas, reeds, and rhizomes, serve one need or another, or in some cases multiple needs, that men and women are (differently) expected to satisfy (see table 29.2). The procuring activities are diverse, depending on the methods of securing materials, the technology used, and also the gender-specific social ideology related to the work involved. Some activities, such as climbing trees are exclusively masculine, while gathering forest products off the ground or harvesting products of the lower story and ground layers are considered to be suitable roles for women. The gendered assignments of men to "heavy work" and women to labor-intensive "light" work are reflected in the work performed in forest interiors. On the same grounds, it is women's labor that fulfills the frequently recurring work and the ongoing collection of forest products for subsistence. By contrast, men's labor is more occasional and sporadic, focusing on occasionally occurring heavy work such as harvesting tree produce and timber, and the gathering of goods in bulk for the market.

The processing of forest produce takes place at home and is often combined with the domestic chores of women. This is also influenced by the purpose of forest gathering and the end use. The studies conducted in Labunoruwa (Wickramasinghe, 1997c, p. 89) and in the Adam's Peak Wilderness (Wickramasinghe, 1995) reveal that the majority of the forest products gathered by women are meant to satisfy practical needs, while men's engagements are more focused on the produce sold at the market. The natural biomass entering the household, including foods of various qualities, edible varieties of plants, roots, fruits, and nuts, and wild plants or parts with medicinal value, are collected by women. Ethnoforestry practices, including the use of plants, have revealed that forest products are found in every diet and the local food culture and traditional herbal medicines are integral elements in the local livelihood. Some 68–82 percent of the work related to the gathering of biomass of food quality as well as processing and utilizing are exclusively performed by women.

Women spend 18–22 hours in foraging per week and they often add about four or five hours during the peak seasons. Men's contribution to foraging is marginal, and they enter the forest for harvesting rather than gathering what is needed and available through the entire phenological cycle of the forest. Except where socially accepted masculine tasks are involved – for instance, climbing trees to collect the florescent sap of the fishtail palm (*Caryota urens*) and bee honey – women dominate gathering, processing and utilizing of the multiple forest products. Women, being regular and frequent visitors to the forest, always find at least a couple of varieties of food and medicine, if not several, to bring back to enrich the household food bowl. In Sri Lanka, women collect 119 varieties from the forest, including 20 varieties used as a source of carbohydrate, 38 green vegetables, 15 varieties for making porridge, 28 varieties of fruit, ten varieties for making beverages, and eight varieties of condiments; 280 varieties of medicinal products were also on their list (Wickramasinghe, 2002). Women's work in the forests is thus related to many linkages: adding value to forest resources; food security and biodiversity; disease prevention and curation.

Choice of Resources

Forests provide multiple activity choices for both men and women; however, resource extractions are not similar. Women mainly extract materials for subsistence. Domestic and subsistence activities being interrelated, women's resource choices are focused on livelihood for themselves, their families, and their communities. Their labor transforms raw products into consumable goods (food, utensils) through their experience and knowledge. The fundamental feature noted here is the division of labor by the output that they generate, not by biomass alone.

The choice of forest resources depends on the users' needs, social parameters, including culture, perceptions, and value system, and resource ownership, access to resources, and traditional knowledge. Men and women are active decision-makers in assessing what is to be produced and gathered, what sources are to be used and by whom, how work is to be attended. These decisions take place in their day-to-day lives, influenced by tradition and gender-shaped practices. Within the vast tracts of forests, men and women locate their work, and their perception of resources, in spatially particular niches and habitats.

Women's choices about resources are rather complex, because one visit to the forest allows them to bring a headload of fuelwood, a bunch of greens, a sack of seeds, pods, and fruit, whatever they find readily available on the forest floor or ready to be harvested. This differs from the men's visits to the forest, which have a predetermined goal rather than searching for opportunities. Some 60–80 percent of men's visits to the forest are made for specific purposes: to tap fishtail palms, harvest construction timber and rattan, or collect resins and medicinal plants. Men's activities in the forests are thus focused on gathering resources with cash potential, and for occasional needs such as construction.

Similarly, gender influences the intensity of men's and women's links with specific habitats and niches. Except for fuelwood, in all other cases the locations of communities rich in specific varieties/species are known to women. They walk to the "hot-spots," the communities dominated by one single variety, which are known as "gonna," or "arana," or "yaya" when they decide to get large quantities. Resource gathering during random walks is diverse and often no excess is left for off-harvest consumption. Women, being influenced by their subsistence needs and their responsibility for family food and health, know and use a diverse and wide range of resources. The consequence is that women encounter and occupy the forest regularly and frequently, much more so than men. Ranking of forest resources by women reveals that they see the forest as a reservoir of resources rich in food, medicine, and fuelwood. For men it is primarily a source of goods with market potential and food supplements.

Conclusion

The points raised in this chapter challenge the assumption that "women's work" is primarily limited to domestic chores and farming. A full profile of women's work in rural areas, forest peripheries in particular, and also in village ecosystems that include forestscapes, shows that forest resources are central to livelihood systems. Biomass that regenerates through natural processes and recurs across the forests is fundamental for livelihood security. The human capital invested in procuring and converting raw products into consumable output should be repositioned to a central place in considerations of how livelihood needs are met.

The premise about the linkages of forest resources and livelihood security is rather complex. It challenges the popular assumption about women's dependence on farming for subsistence, and also challenges the paradigm of "conservation through protection." Both men and women rely on forest resources for their livelihood security, though their engagement and resource dependency are not the same. Women strongly depend on the forest for food supplements and substitutes, fuelwood, and medicinal products, and their dependence stems from their responsibility for family food security. Men are associated more with products used to supplement income and timber for construction and building, which have a direct bearing on the gender and masculinity of the tasks.

The forest, being a source of livelihood, is a genderscape. It includes niches and habitats of gender-specific contacts. The natural and manipulated forests have the capacity to satisfy social expectations placed on them. The functions of the forest that ecologists distinguish are enriched by multiple livelihood functions innovated more by women. Recognition of such functions allows the integration of culture, which has been extended by men and women into the forest ecosystems. In this respect, the ethnoforestry practices add value to the forest and disclose the roots to ethnocentrism for conservation. Many forms of vegetation, be they trees, shrubs, herbs, lianas, vines, fungi, or rhizomes, are in this culture.

The reframing of livelihood systems by focusing on forest resources and including the life-long practices of women demonstrates the gender-centralism of forestry. Feminist analysis of women's use of resources and ecosystem contacts challenges forest management policies biased toward conservation through protection. This creates another shift in the forestry paradigm. The first shift was from scientific management and forest protection toward conservation, and the second was toward participatory management. The challenge now is to revisit national policies and see whether the potential of local resources for maintaining food security, food culture, and traditional knowledge are marginalized in favor of global interdependence. The second challenge is to re-examine whether the forests supporting local livelihood are alienated and kept in compartments by the state to promote global partnerships.

BIBLIOGRAPHY

Agarwal, B. (1994) *A Field of One's Own: Gender and Land Rights in South Asia.* Cambridge: Cambridge University Press.

Alcorn, J. B. (2000) Factors influencing botanical resource perception among the Huastec: suggestions for future ethnobotanical inquiry. In P. E. Minnis (ed.), *Ethnobotany: A Reader.* Norman University of Oklahoma Press.

Bhadra, C. K. (1997) The land, forestry and women's work in Nepal. In A. Wickramasinghe (ed.), *Land and Forestry: Women's Local Resource-based Occupations for Sustainable Survival in South Asia.* Kandy: CORRENSA.

Brydon, L. and Chant, S. (1993) *Women in the Third World: Gender Issues in Rural and Urban Areas.* Aldershot: Edward Elgar.

Cain, M. (1979) *Class, Patriarchy and the Structure of Women's Work in Rural Bangladesh.* Population Council Working Paper No. 43. New York: Population Council.

Cloud, K. (1985) Women's productivity in agricultural systems: considerations for project design. In C. Overholt, M. B. Anderson, K. Cloud and J. E. Austin (eds), *A Case Book: Gender Roles in Development Projects.* West Hartford, CT: Kumarian Press.

deBeer, J. H. and McDermott, M. (1996) *The Economic Value of Non-timber Forest Products in Southeast Asia*, 2nd rev. edn. Amsterdam: Netherlands Committee for IUCN.

Dixon, R. (1982) Women in agriculture: counting the labour force in developing countries. *Population and Development Review*, 8(3), 539–66.

Frankenberger, T. R. (1996) Measuring household livelihood security: an approach for reducing absolute poverty. Paper presented at the Anthropology Meetings, Baltimore, March 27–30.

Lee, R. (1982) Politics, sexual and non-sexual, in an egalitarian society. In E. Leacock and R. Lee (eds), *Politics and History in Band Societies.* New York: Cambridge University Press.

Panayotou, T. and Ashton, P. S. (1992) *Not by Timber Alone: Economics and Ecology for Sustaining Tropical Forests*. Washington, DC: Island Press.

Ramakrishnan, P. S., Saxena, K. G. and Chandrashekara, U. M. (eds) (1998) *Conserving the Sacred: For Biodiversity Management*. New Delhi: UNESCO and IBH Publications.

Sarin, M. (1995) Joint forest management in India: achievements and unaddressed challenges. *Unasylva*, 46, 30–6.

Saxena, N. C. (1987) *Commons, Trees and the Poor in the Uttar Pradesh Hills*. London: ODI Social Forestry Network.

Saxena, N. C. (1997) Women and afforestation programmes in India: policy issues. In A. Wickramasinghe (ed.), *Development Issues across Regions: Women, Land and Forestry*. Kandy: CORRENSA.

Sen, A. K. (1981) *Poverty and Famines: An Essay on Entitlement and Deprivation*. Delhi: Oxford University Press.

Townsend, J. and Momsen, J. (1987) Towards a geography of gender in the Third World. In J. Momsen and J. Townsend (eds), *Geography of Gender in the Third World*. London: Hutchinson.

White, B. (1976) Population, involution and employment in rural Java. *Development and Change*, 7, 280.

Wickramasinghe, A. (1994) *Deforestation, Women and Forestry: The Case of Sri Lanka*. Utrecht: International Books.

Wickramasinghe, A. (1995) *People and the Forest: Management of the Adam's Peak Wilderness*. Colombo: Sri Lanka Forest Department, Forestry Information Services.

Wickramasinghe, A., Pérez, M. R. and Blockhus, J. M. (1996) Nontimber forest product gathering in Ritigala forest (Sri Lanka): household strategies and community differentiation. *Human Ecology*, 24(4), 493–519.

Wickramasinghe, A. (1997a) Anthropogenics related to forest management in Sri Lanka. *Applied Geography*, 17(2), 87–110.

Wickramasinghe, A. (1997b) Rural women's problems as issues for agrarian reforms. In A. Wickramasinghe (ed.), *Development Issues across Regions: Women, Land and Forestry*. Kandy: CORRENSA.

Wickramasinghe, A. (1997c) Women harmonizing ecosystems for integrity and local sustainability in Sri Lanka. In A. Wickramasinghe (ed.), *Land and Forestry: Women's Local Resource-based Occupations for Sustainable Survival in South Asia*. Kandy: CORRENSA.

Wickramasinghe, A. (2002) Ethnoforestry practices in the context of resource management and livelihood security. Paper presented at the Annual Research Sessions, University of Peradeniya, Peradeniya, Sri Lanka, October 30.

Wickramasinghe, A. (2003a) Conservation innovations of peripheral communities: case study of Adam's Peak Wilderness. *Journal of the National Science Foundation of Sri Lanka*, 31(1/2), 105–23.

Wickramasinghe, A. (2003b) Adam's Peak Sacred Mountain Forest. In C. Lee and T. Schaaf (eds), *The Importance of Sacred Natural Sites for Biodiversity Conservation*. Paris: UNESCO Division of Ecological Science.

Chapter 30

The New Species of Capitalism: An Ecofeminist Comment on Animal Biotechnology

Jody Emel and Julie Urbanik

On July 5, 1996, at the Roslin Institute in Scotland, Dolly the sheep was born. She was the first mammal to be successfully cloned using adult cells, and the world met her with a wave of surprise, interest, and immediate concern (Kolata, 1998). Those expressing concern were mostly interested in what this new technology might mean for humans, not animals. Would this cloning success open the way for a new age of designer babies or cloned servants? Could one actually have oneself cloned? Could scientists now clone Hitler or Jesus? Who would control cloning technologies and applications? Since Dolly, the debates have been framed by concern about the human possibilities and costs of this biotechnology – with the most recent outcry prompted by an announcement by a religious sect, the Raeliens, in December 2002 that they had cloned the first human and named her Eve (Clonaid, 2002). While the Raeliens have provided no proof of their feat, the issues around human cloning remain at the center of public debates.

In the meantime, biotechnology companies around the world continued to press forward with animal cloning research. Dolly remained their cause célèbre until January 8, 2001, when Noah was born. Advanced Cell Technologies, Inc. (ACT) of Worcester, Massachusetts, had successfully cloned an extinct guar cow by using a domestic cow as the surrogate mother (ACT, 2001). This announcement, right here in our hometown, surprised us for two reasons. First, animal biotechnology of this scale happening in Worcester was previously unheard of. Second, the scientific community seemed to welcome the birth of Noah in a way that clearly showed animal biotechnology did not provoke the serious debate and introspection that human biotechnology does.

As ecofeminist geographers steeped in the knowledge of laboratory science and its historical relationship to the domination of nature and women (Haraway, 1991, 1997; Birke, 1994), we realized we needed to try to understand exactly what Dolly and Noah might mean for humans, animals, and the planet. Since Worcester is our "backyard," our "place," so to speak, and mindful of the environmental and progressive political exhortation to "think globally and act locally" (or, as in some versions, to think locally and act globally), we began by visiting ACT here in town.

The biotech park is located directly across the street from the University of Massachusetts Medical Center and sits up on a hill overlooking Lake Quinsigamond. There are three buildings of red brick and reflective bluish tinted windows. There are few signs directing visitors and we wandered unquestioned through the first building, where we saw biohazard signs on glass windows opening up to labs populated by, in this case, all women working in protective paper suits busily tending a variety of tubes and machines. There was no one in the hallways and all the doors leading off the hallways were locked. The one person we did encounter had never heard of ACT and recommended that we try another building.

ACT was in the second building and again we walked into an open foyer with no visible security. On the first floor was Cyagra, Inc., which is the commercial arm of ACT. On the second floor was ACT itself. A secretary dressed in jeans and a T-shirt gave us all of the company's promotional and scientific published literature on cloning and xenotransplantation. It was a tiny office and no one paid much attention to us as we gazed at photographs on the walls of cloned animals, which, we realized later, were posed outside in green fields and not in the concrete laboratories where they were created. Since that initial visit, ACT officials have repeatedly refused to do an interview with us.

In this chapter, we want to share what we have learned about animal biotechnology as it had developed by the fall of 2002, primarily in the United States (and, specifically, in Massachusetts). It is our position that animal biotechnology has important implications for nature–society relationships because it is actively altering the shape of three main geographic subfields. First, animal biotechnology is altering local and global political economies as biotech firms lobby for less regulatory oversight, obtain total ownership via patent holdings, and expand the capitalist reach to include the commodification of the genetic inheritance of the planet. Second, animal biotechnology brings a new set of problems to geographers studying global environmental change, sustainable futures, and biodiversity stabilization because the risks/hazards associated with producing (and releasing) genetically modified organisms have not been adequately studied. Third, animal biotechnology is altering the way we understand both humans and animals. Not only is animal biotechnology another way to industrialize animals as a resource for humans – at great cost to the individual animals – but this technology is altering our understanding of concepts such as evolution, domestication, breeding, wild, and the very definition of what a "species" is.

We begin our discussion by outlining our theoretical position as materialist ecofeminists because this theory prods us to look in certain places and make certain phenomena visible that otherwise are unexplored in the human–nature–space triumvirate. We then provide a brief overview of the animal biotechnology industry before delving more deeply into our three areas of focus: political economy and patents, risks/hazards, and animal identity. It is our hope that this chapter will serve to raise awareness of the spaces and places of biotechnology and their impacts upon the bodies of individual animals and humans as well as our planet.

An Ecofeminist Geographer's Lens

Ecofeminism is a radical politics working at the intersection of feminisms and environmentalisms. The basic framework holds that contemporary "Western" use of nature is related to forms of domination or exploitation such as sexism, racism, classism, speciesism, and heterosexism (Sturgeon, 1997; Warren, 2001). According to Mary Mellor the Western hegemonic concept of nature–society relations maintains the notion of a "public world constructed on the false promise of an independently functioning individual, with the nurturing, caring and supportive world hidden, unpaid and unacknowledged" (Mellor, 1997, pp. 239–40). She is speaking primarily of women's private and unacknowledged labor, but we (the authors) contend that this is also true of animals' labor, sacrifice, and bodily contribution to the economy and to science (see also Wolch and Emel, 1998).

While, at one time, ecofeminists could be assailed for an overemphasis on the cultural, they have never ignored the materiality of the economy. In fact, a major part of the ecofeminist project is to explain how Western society has estranged itself from its own material embodiment in nature. Carolyn Merchant (1980) demonstrated how the industrial revolution, as a product of capitalism and patriarchy, contributed to this estrangement, while Val Plumwood (1993) illustrates how the philosophical lineage from Aristotle through the Enlightenment has produced persistent hierarchical dualisms such as "man/woman," "culture/nature," "human/non-human," where one side was valued over the other. This "1/0" binary thinking, as Ariel Salleh (1997) calls it, developed into a dualism of transcendent or immanent (Mellor, 1997) ways of being in the world. Mary Mellor claims that an ecofeminist materialist account is one of immanence – a recognition of our embodiment and direct dependence on nature – while Western science and economic structures are transcendent and believe that through "progress" humans can diminish and/or forget their physical embodiment and connection with the rest of the natural world.

Animal biotechnology extends and reinforces the 1/0 binary as it proclaims the necessity of manipulating non-human beings for the sake of human "progress." According to ecofeminist Carol Adams, "Animal researchers say we are just giving the public what they need. Yet what their research focuses on often is 'protection' from something we are doing to ourselves or remedying the results of what we have done" (Adams, 1994, p. 46). A materialist ecofeminist position calls for a "politics of human–nature relations" (Mellor, 1997, p. 188) that simultaneously disrupts notions of false transcendence and recreates a politics with nature as a subject or agent. Animal biotechnology is best understood as a product of transcendent thinking. The promise is that it will be used to make people live longer (xenotransplantation and pharming), to help the environment (cloning endangered animals), or to feed the world (agricultural biotechnology). The flaw in this position, from a materialist ecofeminist perspective, is that we animal humans do not transcend our animal minds and bodies, our "nature," our brothers and sisters in the primate family, and other such linkages. The usefulness of the transcendent position is that it allows us to ignore the fact that billions of animals are used as invisible labor and lose their own lives in the name of human scientific and economic progress. Human animals are dependent upon other animals, come together in affectionate expression with some non-human animals, and share only fuzzy boundaries with animal others (i.e. boundaries collapse between the me and not-me). The arguments of animal biotechnology are not congruent with a position of immanence or an ecofeminist politics of human–nature relations.

These narratives of "progress" offered up by biotechnology firms, built as they are on the legitimacy of concepts such as "human health," "human longevity," "feeding the world," are understood by ecofeminists as concepts intricately intertwined with institutional structures that promote corporate profits as well (Warren, 2001). The gendered, raced, classed, and species benefits and costs from "development" and "scientific progress" are central questions for ecofeminist geography. The postcolonial ecofeminist Vandana Shiva (2000, p. 1) illustrates that "what the industrial economy calls 'growth' is really a form of theft from nature and people." She asks:

> What is our responsibility to other species? Do the boundaries between species have integrity? Or are these boundaries mere constructs that should be broken for human convenience? The call to "transgress boundaries" advocated by both patriarchal capitalists and postmodern feminists cannot be so simple. It needs to be based on a sophisticated and complex discrimination between different kinds of boundaries, an understanding of who is protected by what boundaries and whose freedom is achieved by what transgressions. (Shiva, 2000, p. 57)

Ecofeminist geographers attempt to understand humans, nature, and non-human beings in a synthetic political, economic, and ethical analysis (Seager, 1993; Emel, 1995; Wolch and Emel, 1998). As ecofeminists we are interested in an "earth democracy, across cultures and species" (Salleh, 1997, pp. ix–x), and as geographers we are interested in uncovering the spaces and places that either contribute to or hinder the potential for this democracy. We believe animal biotechnology is one site of contention that is hindering an earth democracy. A materialist ecofeminist position not only examines the socioeconomic aspects of animal biotechnology but must also include an exploration of the risks to humans and "nature" in general from the putative "advances" offered by the biotechnological manipulation of animals.

An Overview of Animal Biotechnology

Using bacteria or yeast to make food items such as beer, wine, cheese, and bread might be considered to be a form of "biotechnology," one that humans have employed for thousands of years. However, using existing organisms to do a job they "naturally" do (such as cause fermentation or mold on cheese) is substantially different than contemporary biotechnology, which aims to manipulate the genetic structures of living organisms. This "new" form of biotechnology structurally alters an organism to make it produce or act in a way desirable to the "scientist." Such biotechnology has been used commercially for industrial purposes and in plant agriculture for many years. Most public debate and concern thus far has centered on plant biotechnology in terms of labeling, selling, and consuming genetically modified foods (Rifkin, 1998; Priest, 2001; Tokar, 2001). It is our intention to bring *animal* biotechnology into the fold of this larger biotechnology debate.

Genetic modification – whether of plants or of animals – is often presented by biotech advocates as merely an extension of familiar "selective breeding" practices that humans have deployed over several centuries. This is a disingenuous argument, and modern biotechnology is very different from traditional methods of domestication. Traditional domestication or selective breeding practices are not appropriate analogues for contemporary biotechnologies that target specific genes and change their functioning in the course of a single generation, that clone animals, or that insert genes from one species into another (transgenics). The term "genetic modification" itself is defined by the UK-based Genewatch as the "alteration of the genome of an organism by inserting genes from another organism, or altering genes which are native to that organism . . . genetic modification may also involve disruption of native genes" (Rutovitz and Meyer, 2002).

While the methods of genetic modification vary and can seem to be quite complicated to the uninitiated, the end products can be easily categorized into four main types of commercial activity: the production of biological and medical goods; drug production through genetic manipulation of host organisms (pharming); increased agricultural output; and cultural uses. Of these four, biological/medical products (such as Oncomouse, a mouse altered to develop cancer) are already firmly entrenched in the marketplace, while pharming is rapidly becoming the next mass marketed category of animal biotechnology. Table 30.1 provides a context for understanding where and how animals are being altered or cloned for capital gain in the United States.

Pharming can be summed up by the following: "The production of a foreign protein in the milk of transgenic animals was first demonstrated in 1987 when scientists at the National Institute of Health (NIH) in the USA reported the production of human tissue plasminogen activator in mouse milk. Since then, at least 29 human therapeutic proteins have been produced in transgenic animals, most of them in milk, but some in blood, urine, or sperm" (Rutovitz and Meyer, 2002, p. 39). The Canadian company Nexia Corporation, which garnered wide attention for its trademarked product Biosteel, has created a different route for pharming. Biosteel is made from female goats spliced with a spider gene to produce silk in their milk and is touted by Nexia as the next great material for the medical, military, and fashion industries (http://www.nexia.com).

To aid the development of biological medical products, manufacturers like DuPont and patent holders like Harvard University currently use an estimated 91–95 strains of transgenic mice in medical research. One of the largest suppliers of transgenic mice in the USA, Charles River Labs of Boston, which has earned the name "The Mouse Company," saw its revenues soar to $465 million in 2001 – of which 40 percent came directly from its "animal business" (Aoki, 2002). The Swiss pharmaceutical company, Novartis, whose $115 billion market capitalization places it among the world's top six pharmaceutical companies, has recently leased lab space in Boston for which it will invest about $750 million for development (Diesenhouse, 2002). Novartis is interested in xenotransplantation (growing organs in animals that will then be transplanted into humans) and moved to the USA to "escape the complexities of Europe's regulatory systems for the biopharmaceutical industry and to have access to United States' talent and consumers" (Diesenhouse, 2002). The estimated world market for xenotransplantation is expected to reach $6 billion by 2010

Table 30.1 Commercial uses of animal biotechnology

Commerical purpose	Sample products/uses	Animals used	Sample of associated US and Canadian companies
1 Pharming: therapeutic drug production for human consumption	Blood coagulation, anemia associated with cancer, insulin-like growth hormones, cancer treatments, tissue sealant	Cattle, chickens, goats, pigs, rabbits, sheep	Genzyme (MA), AviGenics (GA), GeneWorks (MI), DNX Inc. (NJ), Infigen (WI), Virginia Polytechnic Institute (VA)
2 Biological and medical products	Xenotransplantation (organ transfer between species), Biosteel (spider silk produced in goat milk), transgenic mice/rats for medical research	Pigs, goats, mice, rats	Charles River (MA), Advanced Cell Technologies, Inc. (MA), Circe Biomedical (MA), Diacrin (MA), Immerge Therapeutics (MA), Nexia Corp (Canada), producers of trademarked Biosteel
3 Animal agriculture	Cloning high value stud animals, faster reproduction of factory farmed animals, genetically "enhancing" existing farm animals	Cattle, chickens, pigs, horses, sheep, fish	Advanced Cell Technologies (MA), Aqua Bounty Farms (MA), Cyagra Corp. (KS), Genetic Savings and Clone (TX), Animal Cloning Sciences (CA), Amgen (CA), ABS Global (WI), Infigen Inc. (WI), EmTran (PA)
4 Cultural	Cloning endangered and/or extinct animals, cloning family pets	Cats, the guar cow, other pets, possibly pandas, cheetahs, the Tasmanian wolf	Advanced Cell Technologies, Inc. (MA), PerPetuate (MA), Genetic Savings and Clone (TX), Transgenic Pets (NY), XY Inc. (CO)

and companies like Novartis working to clone pigs for transplantation are planning on cornering a very large part of it (Fano, 2001).

The manipulation of animals for various commercial purposes seems to know no bounds. Horse breeders and owners can have their favorite horse cloned for $10,000 (Animal Sciences Corp., 2002); Animal Cloning Sciences (ACS) "expects to be able to produce cloned embryos at under $1000 each, which should result in a gross profit per clone of $9000. ACS expects to be able to produce 10 clones per day. At this capacity ACS should gross $23,000,000 per year" (ACS, 2002). Worcester's Cyagra, Inc. will clone prized cattle for a $10,000 deposit and $9000 upon delivery of the cloned animal. Cyagra is "in the cloning business to give Mother Nature a hand producing duplicates of the very best and most profitable animals in the dairy and beef industries while eliminating costly trial and error breeding" (http://www.cyagra.com). In Israel researchers have developed a featherless chicken to "stay cooler" and "save on plucking costs" (Bennet, 2002).

The market for companion animals or endangered animals is expected to increase rapidly. For example, pet owners can now have their pet's DNA banked for $700 up front and $90 each year for storage until the cloning technology becomes more reliable and economically feasible (PerPETuate, 2002). "Genetics Savings and Clone" is working to cover all the bases: you can get your pet's DNA banked, your livestock cloned, and according to the website is working on cloning wildlife and assistance/rescue dogs (http://www.savingsandclone.com). When ACT announced that it had cloned the first endangered animal, Noah the guar cow, CEO Michael West said that "the successful birth represents a new way of preserving diminishing genetic diversity in endangered animal populations" (Levine, 2001). Though Noah lived for only two days and died of dysentery (supposedly unrelated to the cloning process), it gave hope to many that cloning technology will be the savior of endangered species. ACT spent $200,000 in 690 attempts to create embryos, which resulted in 30 viable cow eggs, eight pregnant cows, and only Noah being carried to term (Levine, 2001).

The Political Economy of Patents

Patents are at the heart of the biotechnology debate. In nearly all societies, individual animals are "owned" – as companions, food, and entertainment. But does the meaning of "ownership" change if a *patent* is given for an *entire* breed of animal? Is patenting an organism truly a new form of property ownership – a "gargantuan shift in the way society values organisms," as one observer suggests (Mitchell, 1997) – or is it simply an extension of the way animals are currently valued as commodities?

In the past two decades the US Patent and Trademark Office (USPTO) has issued hundreds of animal biotechnology patents (USPTO, 2002). Section 101 of the US patent regulations states that: "Whoever invents or discovers any new and useful process, machine, manufacture, or composition of matter, or any new and useful improvement thereof, may obtain a patent therefore, subject to the conditions and requirements of this title" (US House, 2002). The "composition of matter" clause in this section has proved to be the defining language in the biotechnology patent debates.

Diamond v. *Chakrabarty* (1980) is the beginning of the controversy over patenting living non-plant organisms (Mitchell, 1997; Rifkin, 1998; Tokar, 2001). A. Chakrabarty, then an employee of General Electric, had altered the *Pseudomonas* bacterium to enable it to degrade crude oil. In 1976, he applied for a process patent and a "composition of matter" patent for the "new" bacterium itself. Initially the USPTO and the USPTO Board of Appeals rejected the "composition of matter" patent, ruling that the bacterium was a product of nature and not an invention of man *and* that the bacterium was a living organism, which would also preclude it from being patented. But in 1979, the US Court of Customs and Patent Appeals reversed the decisions of the USPTO, claiming that "the fact that the bacterium was alive was without legal significance" (Mitchell, 1997, p. 41) and that, in fact, the bacterium was created by Chakrabarty and therefore deserved patent protection. The USPTO appealed the reversal to the US Supreme Court and in a five to four decision in 1980 the court upheld the reversed decision and claimed that "everything under the sun made by man" could potentially be patented and the relevant distinction was not between living and non-living things but between products of nature and human made inventions (Wilson, 2001).

In 1988, the first patent protection was extended to multicellular living organisms, namely a genetically engineered oyster. Less than a year later, patent number 4,736,866 was issued to Philip Leder and Timothy Stewart of Harvard University for their "oncomouse." The patent actually covers *any* mammal whose genome has been altered to produce cancerous neoplasms. While there were protests against the patent, case law had already established that multicellular organisms could be patented so there was not much room for legal protest.

A 1990 article by Lisa Raines entitled "Public policy aspects of patenting transgenic animals" remains useful for understanding the debate around patenting genetically modified animals. In brief, those who support patenting claim that resisters are hindering the *progress of science* and trying to hurt US economic interests; resisters believe science is putting profits before planetary safety and cultural/ethical considerations.

Arguments in favor of patenting include: the exorbitant cost of conducting animal biotechnology research/development and the need to guarantee exclusivity to the "creator" to allow funding to continue; the claim that there is no fundamental difference between "owning" a patented animal or an unpatented one (which is done in every facet of our culture); the medicinal and nutritional benefits of genetically modified animals make further research necessary; and, finally, the need to maintain the US competitive edge in the global marketplace.

Arguments against the patenting of animals are tied into beliefs about patenting any kind of lifeform. Activists such as Jeremy Rifkin (1998) are concerned about violating "species integrity," while religious groups believe patenting life and conducting biotechnology research violates "creation's inherent structures and boundaries." Animal rights groups argue that patenting animals devalues the lives of individual animals and the process of animal biotechnology causes unnecessary pain and suffering to the experimental animals (Humane Society of the United States, http://www.hsus.org, accessed October 15, 2002). Still others claim that patenting of life is the final excess of capitalism and is wrong because it gives control over plants, animals, and even people to profit-driven multinational corporations.

Vandana Shiva points to an important contradiction within the biotech industry itself with regard to patenting. The industry claims that there is "substantial equivalence" between genetically modified products and natural ones, while at the same time claiming patent rights because these GMOs are "novel" and different (Shiva, 2000, p. 109). For Shiva, "this ontological schizophrenia is a convenient construct to create a regime of absolute rights and absolute irresponsibility" (Shiva, 2000, p. 109). A similar contradiction exists for animal experimentation – animals are seen as different enough from humans that they fall outside of moral consideration but they are enough like us that experiments on them will help us. Shiva points out several cases where the biotech industry has been able to maneuver back and forth between these two positions to further its economic expansionism.

Risks/Hazards

There are two major sites of contention around the risks/hazards associated with animal biotechnology. The first is the issue of food safety and the second is concern over releasing genetically modified animals into the environment.

The US government has taken the position that genetically modified foods are no different from regular foods and they see no need to implement a labeling system to alert consumers to the presence of modified products:

> In this case, whether intentionally or not, the absence of a single, specific federal policy or standard concerning the labeling of genetically engineered products and processes has the effect of keeping the public debate narrowly focused on those aspects that most clearly fall under existing regulatory policy. . . . In effect, the decision as to whether or not to buy genetically engineered foods cannot be made by consumers on the basis of concerns about such things as environmental impact, regulatory adequacy, economic justice, farm policy, ethical considerations (including animal welfare concerns and religious reservations), or a cautious stance with respect to potential health impacts that might be subtle and are not yet proven. (Priest, 2001, p. 88)

Currently, genetically modified animals have not entered the food supply (NRC, 2002) but the US Food and Drug Administration (FDA) is considering action to rescind its hold request (which is all that is preventing them from reaching the market at this point) to farmers not to sell cloned animals (Humane Society of the United States, 2002). The FDA is basing its decisions on a 2002 National Research Council (NRC) report, which found cloned animals to be safe for human consumption because they retain all the same genetic material. The NRC admits that there is some risk associated with genetically modified animals being consumed as food because of potential allergies and disease transmission. They believe this risk is minimal, but argue for more scientific safety tests before allowing the animals into the market.

With respect to the release of genetically modified animals into the environment, many activists suggest that we should be concerned not just about risks to human health in consuming such animals as food, but also about the risks to whole ecological systems. Evidence for widespread environmental problems associated with the release of genetically modified animals is provided by a recent study at Purdue University funded by the US Department of Agriculture Biotechnology Risk

Assessment Program (Tally, 2002). Scientists found through modeling exercises that the release of transgenic fish with a higher likelihood of mating success drives a wild population extinct in as few as 40 generations. They believe that non-domesticated transgenic animals will be more of a risk to the environment because the "wild" animals are more vulnerable to huge displacements in genetic structures.

The cloning of endangered and/or extinct species provides an interesting extension of this discussion of risk. In addition to the Guar calf at ACT, other institutions, such as the Australian Museum in Sydney, are currently working on cloning extinct animals. The Australian Museum is working with Tasmanian Tiger cloning – after discovering a pup preserved in ethanol, which allowed the DNA to remain viable enough to extract for duplication (Australian Museum, 2002). Cloning extinct and/or endangered animals may soon become a reality. Conservationists raise a basic question about saving (or recreating) extinct species when there is no cognate habitat to return the animal to. For example:

> Already $1 million ($US) have been invested in preparations for cloning the Indian cheetah. It's money some conservationists think would be better invested in saving existing species and their habitats. No cheetahs are left in India, and the genes for the project will have to be imported from another country. (Levine, 2001)

If there is no habitat for the animal to return to, and the medical costs of the cloning technology are so high, who actually benefits? Many conservationists argue that money spent on flashy cloning technologies is diverting funds from zoos and other conservation agencies, and from work on actual habitat restoration. Advocates believe that cloning technology can be the last resort for many species, can help to increase the genetic pool, and can reduce the already overpopulated zoo populations because it would allow targeted breeding. In terms of bringing back extinct animals, the question of habitat becomes even more crucial. Would we be bringing back the animal just to watch it die because the habitat has changed, the diseases are different, or too many complications arise from cloning? How would cloning endangered/extinct animals change our cultural understandings of what is wild? No matter what the answers to these questions, it must be recognized that the case of cloning endangered animals clearly demonstrates that humans are pushing forward in animal biotechnologies in terms of release into the environment without taking adequate precautionary measures.

Animal Welfare

In the rapidly evolving world of cloning technologies, research on animals has been overshadowed by a media focus on human cloning and human stem cell therapy (Priest, 2001). The media focus on humans has detracted from the public's ability to reflect on animal research and has worked to the advantage of animal biotechnology because animals themselves remain invisible and as long as the public outcry focuses on humans the work of researchers manipulating the genes of animals continues unabated. Animal welfare laws in the USA, especially with regard to scientific research, are well behind those of Europe (Rogers, 2001; Rutovitz and Meyer, 2002). This is because, by and large, the American public accepts the "scientific" view that animals *must* be used in research to help humans. Animal researchers have successfully convinced the public that animals are like humans (at least enough for them to be medically or commercially useful) – but that they do not warrant the same ethical consideration. As long as science is able to convince the public that what they are doing to animals is "for the good of humanity" then the public seems largely to accept animal experimentation.

In the USA, groups like the Humane Society of the United States (HSUS) and People for the Ethical Treatment of Animals (PETA) have argued that animal biotechnologies continue human exploitation of animals. For example, HSUS writes that: "We are not aware of major food safety issues, but we condemn cloning as yet another move away from regarding animals as animals, and yet another development that will favor large corporations over small ones – as well as further narrowing the genetic base of our livestock, which might readily cause a problem in case of disease" (Humane Society of the United States, 2002).

Animal rights activists in the USA, however, have not been as thorough as some groups in Europe. For example, Genewatch UK has recently published a report (*Genetically Modified and Cloned Animals: All in a Good Cause?*, Rutovitz and Meyer, 2002) that examines all aspects of animal biotechnologies from an animal welfare perspective. Their results are not comforting. Their research includes chapters on pharming, lab animals, agriculture, xenotransplantation, cloning, ethics, and animal welfare. In each case they provide a history of the technology and the invasive and painful results for the animals used, and provide numerous alternatives that do not use animals at all. They conclude that:

> The genetic modification of animals is an assault on the integrity of another species. Not only can genetic modification cause considerable suffering to the animals involved, but it changes our relationship with the natural world and contributes to the commodification of animals by using them as we wish and for maximum commercial gain. The presumption in every case should be against such interventions in the absence of extremely compelling reasons for them. (Rutovitz and Meyer, 2002, p. 76)

Genewatch's report further finds that animal biotechnologies are reversing the 1990s trend of reducing the number of animals in scientific experiments. The biotechnology industry views animals as "equipment" or "products," not as sentient beings. Joan Dunayer documents how animal experimenters often use proprietary claims to justify their work: "Nude mice did not exist in nature," one vivisectionist states. "We have created them for our use and are, therefore, justified in using them accordingly" (Dunayer, 2001).

A feminist view of animal welfare recognizes the sentience of many non-human animals, the commonalities between human animals and other animals, the mutual dependencies between humans and other animals, and the human responsibility to witness the existence of other animals. A feminist view of animal welfare requires a transparency of commodity chains – life in the feedlot, death in the slaughterhouse – and the wealth generation that is unevenly generated as a result of the latter. Animal biotechnology continues and escalates this violation of the existence rights of non-humans, who, like children and women in many situations, have no voice to resist or participate in the directions their lives take.

Conclusion

Ecofeminist Mary Mellor claims that "getting the relations between humans right will not resolve the ecological imbalance" because the source of much of the conflict between humans is the unacknowledged problem of dependency on nature and the beings that populate it (Mellor, 1997, p. 183). We make the same claim for animal biotechnology. Arguments about animal biotechnology healing and feeding humans, or saving animal species themselves, hide the socioeconomic structures that sanction their exploitation and manipulation. This socioeconomic system is self-sustaining: the scientific–university–private firm biotechnology nexus can "stay in business" only so long as what they are doing is seen as justifiable. They are able to make it socially justifiable by not presenting the whole picture – a picture which includes disclosure of profits, alternatives to consumption, the debates over patenting life forms, alternatives to using animals, and ultimately an alternative to the scientific mindset that technology equals progress and must be pursued at all costs to "benefit" humanity.

BIBLIOGRAPHY

ACS (2002) *Investor Information*. Animal Cloning Sciences (http://www.animalcloning-sciences.com/investors.htm). Accessed October 20, 2002.

ACT (2001) ACT announces birth of first cloned endangered species – Noah is born. Advanced Cell Technology Inc. (http://www.advancedcell.com/pr_01-12-2001.asp). Accessed October 11, 2002.

Adams, C. J. (1994) *Neither Man nor Beast: Feminism and the Defense of Animals*. New York: Continuum.

Aoki, N. (2002) The mouse company roars. *Boston Globe*, May 27, Business.

Australian Museum (2002) Project to clone the extinct Tasmanian tiger (http://www.amoonline.net.au). Accessed May 28, 2002.

Bennet, J. (2002) Cluck! Cluck! Cluck! Chickens in their birthday suits. *New York Times*, May 24.

Birke, L. (1994) *Feminism, Animals and Science: The Naming of the Shrew*. Buckingham: Open University Press.

Clonaid (2002) The world's first human clone baby is born (http://www.clonaid.com/english/pages/press.html). Accessed December 28, 2002.

Diesenhouse, S. (2002) From candy to chromosomes in Cambridge. *New York Times*, October 9, Business.

Dunayer, J. (2001) *Animal Equality: Language and Liberation*. Derwood, MD: Ryce Publishing.

Emel, J. (1995) Are you man enough, big and bad enough? Wolf eradication in the United States. *Environment and Planning D: Society and Space*, 13(6), 707–34.

Fano, A. (2001) If pigs could fly, they would: the problems with xenotransplantation. In B. Tokar (ed.), *Redesigning Life? The Worldwide Challenge to Genetic Engineering*. London: Zed Books.

Haraway, D. (1991) *Simians, Cyborgs and Women: The Reinvention of Nature*. London: Routledge.

Haraway, D. (1997) *Modest witness@second millennium: Femaleman meets Oncomouse*. London: Routledge.

Kolata, G. (1998) *Clone: The Road to Dolly and the Path Ahead*. New York: William Morrow and Co.

Levine, T. (2001) Noah's birth brings a flood of questions. *Alternatives Journal*, 27(2), 4.

Mellor, M. (1997) *Feminism and Ecology*. New York: New York University Press.

Merchant, C. (1980) *The Death of Nature*. San Francisco: HarperCollins.

Mitchell, C. B. (1997) Patenting life: an examination of some ethical implications of biopatents. Thesis, doctor of philosophy, University of Tennessee, Knoxville.

National Research Council (2002) *Animal Biotechnology: Identifying Science-based Concerns*. Washington, DC: National Academy of Sciences.

PerPETuate (2002) Cost and payment page (http://www.perpetuate.net/cost.asp). Accessed May 21, 2002.

Plumwood, V. (1993) *Feminism and the Mastery of Nature*. London: Routledge.

Priest, S. H. (2001) *A Grain of Truth: The Media, the Public, and Biotechnology*. Lanham, MD: Rowman and Littlefield.

Raines, L. J. (1990) Public policy aspects of patenting transgenic animals. *Theriogenology*, 33(1), 129–49.

Rifkin, J. (1998) *The Biotech Century: Harnessing the Gene and Remaking the World*. New York: Jeremy P. Tarcher/Putnam.

Rogers, C. P. (2001) Solution or stumbling block? Biological engineering and the modern extinction crisis. *Georgia Journal of International and Comparative Law*, 30(141), 141–59.

Rutovitz, J. and Meyer, S. (2002) *Genetically Modified and Clone Animals: All in a Good Cause?* Derby: GeneWatch UK.

Salleh, A. (1997) *Ecofeminism as Politics: Nature, Marx and the Postmodern*. London: Zed Books.

Seager, J. (1993) *Earth Follies: Coming to Feminist Terms with the Global Environmental Crisis*. New York: Routledge.

Shiva, V. (2000) *Stolen Harvest: The Hijacking of the Global Food Supply*. Boston: South End Press.

Sturgeon, N. (1997) *Ecofeminist Natures: Race, Gender, Feminist Theory and Political Action*. London: Routledge.

Tally, S. (2002) Model shows bioengineering could mean extinction of species. *Agriculture Online News*.

Tokar, B. (ed.) (2001) *Redesigning Life? The Worldwide Challenge to Genetic Engineering*. London: Zed Books.

US House of Representatives (2002) *US Code 35, Chapter 10, Section 101: Inventions Patentable* (http://uscode.house.gov/title_35.htm). Accessed June 14, 2002.

US Patent and Trademark Office (2002) *General Information Concerning Patents* (http://www.uspto.gov/web/offices/pac/doc/general/index.html). Accessed September 22, 2002.

Warren, K. (2001) *Ecofeminist Philosophy: A Western Perspective on What It Is and Why It Matters*. Lanham, MD: Rowman and Littlefield.

Wilson, K. A. (2001) Exclusive rights, enclosure and the patenting of life. In B. Tokar (ed.), *Redesigning Life? The Worldwide Challenge to Genetic Engineering*. London: Zed Books.

Wolch, J. and Emel, J. (eds) (1998) *Animal Geographies: Place, Politics, and Identity in the Nature–Culture Borderlands*. London: Verso.

Chapter 31

Siren Songs: Gendered Discourses of Concern for Sea Creatures

Jennifer Wolch and Jin Zhang

Introduction

One need not look far for evidence of gender divisions in thinking about animals, or to find deeply engrained differences in animal practices. Anthropological chestnuts such as "man the hunter" may have been tossed into the fire and thankfully gone up in smoke, but certainly most traditional societies had clear gender divisions of labor when it came to hunting animals, their husbandry and slaughter, and use of animal products for food, clothing, or tools (Clutton-Brock, 1981). Such divides have been eroded but not eliminated; in most parts of the world, women are still less likely to be hunters than men, certain tasks associated with the keeping and slaughter of animals are often gendered, and certain animal foods are considered appropriate food for either men or women but not both.

Moreover, new sorts of divisions have arisen, especially around attitudes toward animals and animal practices in the First World, particularly North America, Europe, Australia, and New Zealand. Animal welfare and rights organizations, for example, were historically and still tend to be dominated by women, especially at the member and staff levels, women are more apt to avoid eating animals than men, and their attitudes toward the treatment of animals in a variety of contexts tend to be more supportive of animal protection and welfare.

Geographers, including feminist geographers, had not been major contributors to the research literature on human–animal relations (much less their gendered quality) since the decline of Sauerian geographies of domestication and animal husbandry in the mid-twentieth century (Wolch et al., 2002). This was despite the fact that nature–society relations had become, once again, central to the discipline (Turner, 2002), despite the fact that animals, presumably, are a major component of "nature" (although they are notorious boundary-crossers). Within the past decade, however, research on "animal geographies" has made something of a comeback (Wolch and Emel, 1998; Philo and Wilbert, 2000). This emerging body of work includes work on animals and human identity (Anderson, 1995; Philo, 1998; Whatmore, 1999), interactions and networks (Tuan, 1984; Whatmore and Thorne, 1998, 2000), representation (Davies, 2000; Woods, 2000), the economy (Furuseth, 1997; Ufkes, 1998; Robbins, 1998; Thorne, 1998), landscape and place (Brownlow, 2000; Matless, 2000; Waley, 2000; Yarwood and Evans; 2000), race and racialization (Anderson, 1997, 2000; Elder et al., 1998; Griffith et al., 2002), cities (Gullo et al., 1998; Wolch, 1998; Griffiths et al., 2000), and ethics (Whatmore, 1997; Wescoat, 1998; Lynn, 1998; Jones, 2000).

Feminist geographers in particular have begun to explore the nexus of gender and human–animal relations (Emel, 1995; Michel, 1998; Howell, 2000; van Stripriaan and Kearns, 2002; Wolch et al., 2000). The present chapter seeks to extend this nascent feminist research on animal geographies, through an analysis of gendered discourses about marine wildlife offered by men and women living in southern California's coastal zone. This metropolitan region is one of the most diverse in the nation, with high shares of immigrants from many parts of the world, creating a setting in which attitudes are apt to be heterogeneous, and sometimes in conflict. Marine creatures are an appropriate focus of study, because coastal residents are likely to have frequent experience with marine wildlife animals – for instance, seeing birds, seals, or dolphins while at the beach, fishing in coastal waters, whale watching – and thus they are a visible part of everyday life and the local cultural imagination.

In the next section, we review the literature on gender and attitudes toward animals, highlighting gender differences in a variety of human–animal interaction realms, providing an overview of feminist approaches to gender differences, and noting work done by geographers. Then, we describe a recent survey of attitudes toward marine wildlife in urban southern California, which asked respondents to reflect on how marine animals should be treated, how they felt about controversial animal practices that involve harm to animals, and whether or not they ever felt stigmatized by either their ideas on animals or their animal practices. Our results are presented next, focusing on the gender differences in responses, and how gender and race cross-cut to create a more complex series of discourses.

Gender and Attitudes toward Animals

Academic studies of human–animal relations have sought to test dominant cultural assumptions about the gendered nature of thinking about animals, and animal practices. Most of this work has been conducted by psychologists and sociologists, using mainstream research methods (such as surveys), and has focused on attitudes toward the human use of animals (as pets, food, research tools), and human–animal interactions (responses to companion animals, reactions to wildlife management strategies, etc.). Such literature has not typically offered explanations for the consistent findings of gendered attitudes, however. In contrast, feminist work on gender and animals has provided theoretical approaches to understanding gendered attitudes, mostly predicated (following Gilligan) on the assumption that women are more apt to develop an ethics of care than are men. Animal geographers have embraced feminist frameworks, and developed analysis of such topics as wolf eradication, pet keeping, wildlife rehabilitation, and animal welfare campaigns. Despite the

flowering of work on gender and human–animal relations, however, there are critical absences, especially around issues of gender and attitudes across race/ethnic
groups, and attitudes toward controversial animal practices often associated with
specific race/ethnic communities.

Attitudes toward animals

Early research on attitudes toward animals began in the late 1970s, with the work
of Stephen Kellert. With his colleagues, Kellert developed a typology of attitudes
and surveyed a national sample of the US population (Kellert, 1978, 1984; Kellert
and Berry, 1980). Findings from Kellert's survey and subsequent studies revealed
that gender, age, income, rural versus urban background, religion, and race/
ethnicity were some of the demographic features that explained attitudinal variation (O'Donnell and van Druff, 1987; Bowd and Bowd, 1989; Driscoll, 1992). Since
then, research has focused on specific dimensions and types of attitudes; for
example, ideas about how to treat lab animals (Pifer et al., 1994; Plous, 1996), or
attitudes toward a specific policy action such as the culling of a deer herd (O'Donnell and van Druff, 1987). Studies have also identified a broadened range of attitudes, reflecting societal concerns over environmental degradation and loss of
biodiversity that have intensified since the 1970s.

Findings with respect to gender differences in values and attitudes have been
remarkably consistent. Although women may be more negativistic about animals
that are dangerous or unpleasant, they tend to be more biocentric – concerned about
animal welfare and protection – and ecocentric, or oriented toward ecological or
environmentalist values. They are also more apt to disapprove of biomedical and
psychological research involving harm to animals, and to favor animal rights and
avoid meat-eating.

For example, Kellert and Berry (1987) found that women tended to be more
humanistic and moralistic than male respondents, who generally had stronger utilitarian and dominionistic views about how people should interact with animals.
This attitudinal difference is reflected in studies of broader value orientations, where
women have expressed more bio- and ecocentric, and less utilitarian, values toward
the natural environment (Steger and Witt, 1989; Vaske et al., 2001), or are more
oriented toward altruism, which in turn influences environmental attitudes (Dietz
et al., 2002).

A large number of studies have investigated the role of gender in shaping attitudes toward various uses of animals, as well as activism around animal protection
and rights. With regard to the use of laboratory animals, survey research reveals
that women are more apt to oppose animal research than men. Plous (1996), for
example, found that opposition to the use of animals in psychological research and
testing was stronger among female than male respondents to a survey. Similarly,
Driscoll's (1992, 1995) studies of species preferences and attitudes toward animal
research found that although most respondents supported animal research, women
were more likely to be opposed, while Navarro et al. (2001) report that among
Spanish psychology students, women less frequently expressed support for animal
research. Likewise, Eldridge and Gluck (1996) assessed gender differences, finding
that women were not only more likely to support animal protection, but also had

higher probabilities of opposing animal research, being concerned about the suffering of research animals, and questioning the scientific validity of animal research. They were also more willing to make sacrifices, such as giving up meat in their diets, or forgoing the medical benefits of animal-based research, in order to protect animals. And Pifer (1996) found that feminist attitudes, as well as gender, were powerful predictors of opposition or support for animal research. Such results appear to cut across many geographic lines; in a study of attitudes toward animal research in 15 nations, Pifer et al. (1994) discovered that in all 15 countries, women were significantly more apt to disapprove of research involving animal subjects.

Women also tend to be more strongly bonded with companion animals. A study by Kafer et al. (1992) revealed that women scored higher than men on the Pet Relationship Scale, which measures the strength of affection, the extent to which people saw companion animals as an equal member of the family, and the frequency of physical contact and interaction. In turn, those with stronger attachments to companion animals were more likely to oppose hunting and animal research.

Women are less apt to favor lethal methods of wildlife management, or to favor hunting. Although they may express heightened concern for the environmental risk associated with wild animals this does not necessarily translate into desires to eradicate the sources of such risk. Zinn and Pierce (2002), for example, explored values, gender, and concern about potential mountain lion attacks among a large sample of Colorado metropolitan residents. They found that women were more apt to express concern about a cougar attack, indicating heightened perceptions of environmental risk, a result consistent with earlier studies suggesting that women tended to be more concerned with local environmental risks than global risks (Davidson and Freudenberg, 1996). They were less utilitarian and hence less likely to favor the killing of cougars that happened into residential areas, however.

Women are also more likely to be vegetarians, and to support animal rights, according to a variety of historical studies and surveys (Elston, 1987; Sperling, 1988; Leneman, 1997). In Victorian Britain, for instance, vegetarianism, animal rights, and suffrage were interconnected movements (Landsbury, 1985). Santos and Booth (1996) studied meat avoidance among British undergraduates and found that women were more likely to avoid meat than were male students. With respect to animal rights, Plous (1991), Herzog (1993), Nibert (1994), and Peek et al. (1996) all found that women were more likely to express pro-animal rights attitudes than men, and become involved in the animal rights movement. Kruse (1999), in a study of attitudes from the 1994 US General Social Survey, similarly found greater support for animal rights among women, linking such support to their lower prevalence of holding Darwinian views of nature that suggest constant struggles for survival within the natural realm, and the need for humans to control nature.

Feminist perspectives

Feminists (and particularly ecofeminists) have also argued for gender differences in attitudes toward animals, as well as for the need for a specific feminist perspective on the position of animals in society (Adams, 1990; Gaard, 1993, 2001; Adams and Donovan, 1995). Positions on gender differences are often linked to Gilligan (1993), who suggested that moral development in girls was oriented toward valuing and maintaining networks of relationships with others. This conditioning, in turn, could be expected to lead to a more situationist ethics and explain differences in attitudes and practices. Moreover, feminist theories of sexual oppression have informed the gendered nature of animal practices such as hunting, instigating explorations of the sexualized nature of male violence against animals (Kheel, 1995), and, in turn, led to arguments about the ways in which animals become an "absent referent" in discourses and practices of male violence against women (Adams, 1990).

More general feminists' positions on the relationships between women and animals are often at odds with each other. Spiritual feminists, accepting biological determinism, typically assert that women are more in tune with nature, and thus have stronger connections with the non-human animal world. Liberal and socialist feminists, in contrast, have historically argued that women use intellect and reason like men and hence not like animals, with liberal feminists striving for rights and equality of opportunity, and socialist feminists asserting the need for more fundamental changes in political economic structures and institutions. Others, such as French (1985), Spelman (1982), and Donovan and Adams (1995), suggested that in Western societies male dominatory philosophy has indeed involved casting women, children, and animals as part of nature and thus in opposition to reason, while rejecting the narrow and anthropocentric nature of many liberal and socialist feminist agendas and instead embracing a broader and more radical cultural feminist agenda that stands in opposition to oppression of all life forms.

Such views are echoed by ecofeminist philosophers (for example, Donovan, 1990; Warren, 1990; Plumwood, 1993). These ecofeminists have critiqued mainstream environmental ethics as well as deep ecology as being linked to patriarchal projects and, in the case of deep ecology, for its holism that eradicates human–animal and animal–animal differences that are crucial to acknowledge. Animal rights philosophies, such as those launched by Singer (1975) and Regan (1983), in contrast, have also been rejected due to their inherent individualism as well as their anthropocentric emphasis on human–animal similarities (especially in the realm of cognitive performance) as a basis for making judgments about respect or rights. A major thrust of these philosophers has been to juxtapose such approaches, and articulate the lineaments of a feminist philosophy of care.

Feminist research on human–animal relations has been largely theoretical or confined to case studies, except in the case of psychologists who have explored the relevance of Gilligan's arguments. Wang (1999), however, attempted to empirically investigate some hypotheses derived from feminist theory using survey research methods. Grounding his survey on a basic ecofeminist framework that linked the logic of domination of women to that of animals, he found that among his sample of respondents, there was a correlation between attitudes toward women and attitudes toward the environment. In particular, authoritarianism had a consistent positive link to patriarchial positions for both men and women, and among women, to anthropocentrism in attitudes toward nature.

Galvin and Herzog (1992) surveyed college students to assess Gilligan's notions of gender differences in moral development. Women and those who expressed higher levels of idealism had greater concern for animal welfare. Belief in universal as opposed to relative or situational ethical rules was not linked with concern for animals. These findings provide support to the argument that women are more likely to develop a situational ethic of care that extends to animals.

Kalof (2001) also developed an empirical, feminist-inspired analysis of "discourses of concern" for animals. Drawing on Kellert's typologies, she surveyed attitudes toward animals and other aspects of their orientations (political, sexual, food preferences, environmental activism, etc.) among demographically diverse college students, and performed Q factor analysis to characterize the discourses of concern that emerged. This produced six discourses or attitude clusters, and the associated characteristics of their adherents. One cluster emphasized moralism, love for animals, and animal rights, whose adherents were Latino or white women who self-identified as political moderates, and had high levels of altruism. Cluster 2 was anthropocentric and utilitarian and supportive of killing animals for human use or pleasure, yet also supportive of animal welfare; most adherents were white men with low altruism scores and conservative or moderate political values. Cluster 3 was utilitarian also, but with a moralistic emphasis, and adherents were urban women of color with high altruism scores and liberal–moderate politics. Cluster 4 was negativistic but also environmentalist and protective of wildlife populations; respondents here were older men and women, conservative to moderate politically, with high altruism scores. The fifth cluster emphasized pets and love for animals, but respondents were also negativistic and not necessarily environmentalist in orientation; respondents were young suburban Asian women. The sixth and final cluster was centered on wildlife protection, but was also negativistic, and supported hunting and recreational choices even if they harm animals; its adherents were young African Americans and a white women with a high altruism score and a rural background. Although working with a small sample size, this analysis showed that attitudes are multilayered, and people often have views that run counter to simpler ideas about attitude structure that emphasize the centrality of one or two attitudinal constructs. Moreover, such multilayered discourses or attitude clusters are gendered, and linked to levels of altruism – again echoing Gilligan and feministic arguments about women and the ethics of care.

Gender and animal geographies

Geographers have contributed to the literature on gender and human–animal relations, primarily using feministic geographic frameworks. In so doing, they have broadly illuminated questions of attitudes and how gender relations (as well as race) shape discourses about animals and animal practices – and vice versa. Emel (1998), for example, presented a powerful ecofeminist analysis of wolf eradication in the American West, connecting animal representations to constructions of human masculinity in processes of place domination. Wolf representations emphasized so-called savagery, lack of mercy, unfair habits of pack hunting, and cowardice – all contravening norms of masculinity in the American frontier centered on virility and prowess, sporting honor, and willingness to kill in the name of chivalry, morality, progress, and civilization. Emel showed how these sorts of images not only devastated wolves and other animals, but were analogous to racist and sadistic treatment of people falling below European-American males on the hierarchy of beings.

In another consideration of the gendered character of nineteenth-century human–animal interactions, this time in an urban context, Howell's (2000) study of dog-stealing in Victorian London showed both dogs and bourgeois women as victims of a patriarchal society, confined to domestic captivity, but vulnerable to the actions of lower-class men, the venal public world of commerce, and dangers lurking in the city's poorest districts. Borrowing from Virginia Woolf's satirical tale of the theft of Elizabeth Barrett's dog, written from the dog's point of view, he articulates a political geography of dog-stealing characterized by class antagonisms and exploitation of rich by poor, and deeply ingrained practices of "domestication" itself – both of dogs and of women confined by Victorian ideals of femininity, obedience to male authority, and middle-class domesticity.

Considering a very different time and place, Michel (1998) explored golden eagle rehabilitators and wildlife educators in San Diego's metro fringe. There, conventional planning around endangered species and habitat conservation relies on scientific discourse and legitimacy and excludes alternative arguments based on the connections people feel with wild animals that have been injured as the result of urbanization. Grounded in struggles to save injured eagles and starving eagle chicks, and to nurture responsibility and consideration for animals among children, eagle rehabilitation and wildlife education for children constitute an ecofeminist, personal politics of both animal and human social reproduction that asserts the agency of wildlife in defining pathways to human–animal coexistence and shared places. This practice allows children, and by extension their parents, into the world of the golden eagle and the birds' fight for survival, helping them to voice their views to the larger public through letter-writing campaigns and special events – in this way helping to recast the nature of grassroots environmental activism in southern California.

Gender relations and animals are also at the forefront of van Stipriaan and Kearn's (2002) study of billboards as an influential aspect of Auckland's built environment that shapes ideas about gender and local urban culture. They considered the case of a notorious Royal New Zealand Society for the Prevention of Cruelty to Animals spay-neuter campaign, showing a dog "dressed" in a woman's wig, sexy glasses, a frilly skirt, and high heels, wearing lipstick and smoking a cigarette, and urging Auckland residents to "De-sex your bitch." Van Stipriaan and Kearns argued that the campaign and the controversy it generated destabilized human–animal relations and highlighted the gendered nature of animal representations in popular culture – thus indicating that animals not only have a bodily presence in the city but also play important representational roles in local cultural formations.

Building on postcolonial and critical race theory, as well as feminist thought, Wolch and her co-workers investigated attitudes toward animals among women of color to help to understand the dynamics of racialization in culturally diverse world cities under conditions of globalization (see also Elder et al., 1998). As international migration brings together heterogeous people, "out-of-place" animal practices risk being interpreted as trangressions of species boundaries. In a focus group study, for example, Griffith et al. (2002) discovered a willingness to tolerate dog eating among Filipinas in Los Angeles that seemed to reflect their own experience as a marginalized group in American society, and their sensitivity to racialization based on color and culture. Filipina respondents hesitated to condemn other groups whose animal practices, while alien or distasteful, were rooted in their particular culture, instead adopting a position of cultural relativism that served to protect their own cultural practices. Similarly, central city African American women queried about their attitudes toward animals provided responses that also reflected their doubly marginalized status as women of color. They segmented the animal world into three categories: "food," "pet," and "wildlife." "Food" animals were simply necessary for survival; people had to distance themselves from their unfortunate fate. Pets and wild animals, in contrast, demanded compassion – people should help wildlife in distress, just as people should help each other regardless of color, hinting at their solidarity with animals as brethren due to their outsider status (Wolch et al., 2000).

Animals can also become links to a past identity and so cushion the shock of the new urban environment among immigrant women, just as they also play a role in assimilation. Lassiter and Wolch (2004), for instance, found that for Latina immigrants, keeping animals like chickens in the backyard is one way to retain connection to rural landscapes they left behind, but over time the culture of pet-keeping – and the assimilation that it implies – seems to take over. When animals once relegated to the barnyard or backyard are welcomed into the house, traditional human–animal boundaries become destabilized and new forms of relationship become possible.

Summary

Taken together, conventional social scientific, feminist, and human geographic research on gender and human–animal relations has produced several key findings and insights. First and foremost, most studies find that human attitudes toward animals are gendered, although many other factors influence ideas about how people should think about, and interact with, non-human others. In particular, race/ethnicity and immigrant status appear to intersect with gender to produce distinctive discourses of concern for animals, sometimes leading to a form of cultural relativism in which traditional practices involving harm to animals are nonetheless defended as a means of resisting mainstream cultural domination. Second, gender differences may be rooted in the ways that women and girls are socialized and relate to others, making them more likely to develop interdependent networks of friendship and solidarity with others, including animals. For some, this leads to animal rights activism and a politics of care for non-human animals. Third, realms of patriarchy, male domination, and violence can operate in parallel for women, people of color, and animals, and dominatory relations with animals contribute to the contours of masculinity. The results can be both literal and figurative or discursive, with representations of women and animals shaping everyday urban landscapes.

Despite these emergent understandings, we are still left with important questions. Much of the social science research on attitudes has used either non-representative samples (university students, for example) or national surveys that lack any grounding in place or shared daily experience with forms of wildlife. Feminist animal geographers exploring attitudes have conducted place-based research, but relied primarily on qualitative methods such as focus groups that offer rich insights but do not capture systematically the extent to which attitudes toward animals are gendered (or raced) in a particular urban-bioregional population. No research has attempted to understand attitudes toward the wide range of animal practices that involve animal pain, suffering, or death; most studies focus on one or two practices only, such as the use of laboratory animals, or companion animal euthanasia, despite evidence of mounting concern about a much broader spectrum of practices. Nor have researchers delved into questions of human–animal relations and feelings of belonging or marginalization. In a world of intensifying place-based cultural mixing and hybridity, such research would seem vital to grasping emergent nature–culture networks and conflict.

In what follows we explore the complex, and complexly gendered, discourses of concern expressed by culturally diverse residents in a major world city-by-the-sea – Los Angeles – about one crucial realm of the animal kingdom: marine wildlife.

Exploring Discourses of Concern about Marine Creatures in Los Angeles

The coastal zone of the Los Angeles metropolitan region is one of the most populous and culturally diverse urban settings in the United States, and is one of California's most important economic and environmental resources. Yet this highly valuable asset is increasingly threatened by development and human encroachment into dwindling coastal wildlife habitats. There is tremendous economic pressure to develop the few remaining parcels of open space, fueling controversy among environmentalists, developers, and local government. Moreover, the burgeoning population places heavy demands on coastal resources, including recreational facilities, such as beaches and boardwalks, and ecological attractions, such as tidepools, kelp forests, and coastal marshes. In addition, conflicts over issues pertaining to human interactions with marine wildlife are becoming more commonplace.

Past research indicates that in general, US residents are convinced that ocean life is crucial to human survival and the planet's ecology, and believe in protection of marine environments (Kidd and Kidd, 1998). But despite the importance of southern California's coastal zone, the harms that its marine wildlife faces, and the potential for cross-cultural conflict, little prior research has focused on culture-based attitudes toward marine wildlife and habitats. Further, there is a paucity of information regarding how residents differentiated by gender, race/ethnicity, class, and other key dimensions think about and utilize coastal resources in Los Angeles, how they feel about the animal practices of others, and how their practices and perceptions might impact the coastal zone and marine wildlife in the long term.

To better grasp human–marine wildlife relations in the coastal zone, especially as they vary by cultural background and gender, we conducted a telephone survey of Los Angeles county residents, based on previous attitudinal research. Our survey design was based on previous attitudinal research, focus group findings (see, for example, Griffiths et al., 2000; Lassiter and Wolch, 2000; Wolch et al., 2000), and pilot survey experience (Whitley, 1998). Administration of the survey itself was conducted by Responsive Management Incorporated, a public opinion polling and survey research firm specializing in fisheries, wildlife, natural resource, outdoor recreation, and environmental issues.

The survey was designed to determine how demographic traits, socioeconomic status, personal background features, and past or present geographic and cultural context might shape attitudes toward marine wildlife in the Los Angeles coastal zone. The survey instrument was comprised of questions and statements surrounding respondents' demographics, beach utilization and activities, knowledge about marine wildlife and the coastal zone, stance on policy issues, attitudes toward how people should interact with marine animals, and the nature of attitudinal change over time. In addition, to gauge respondents' attitudes toward controversial, cross-cultural practices – or tolerance toward what are often considered controversial interactions with animals – statements regarding culturally sensitive practices involving animals and nature were included in the survey. They were also queried as to whether they felt looked down upon – or stigmatized – for their own animal practices.

Attitudinal responses were grouped as either anthropocentric or bio/ecocentric, and further classified into ten attitudinal categories. Anthropocentric attitudes included:

- *Utilitarian–dominionistic*: principal concern for the mastery or control of animals and nature.
- *Utilitarian–stewardship*: foremost interest in the practical value of animals and the natural environment.
- *Negativistic*: fundamental interest in avoidance of animals due to indifference, dislike, or fear of animals.
- *Aesthetic*: primary interest in the physical attraction or beauty of animals and nature.
- *Animal welfare*: principal concern for the right and wrong treatment of animals and nature.
- *Spiritualistic/supernatural*: fundamental interest in the supernatural properties of animals and nature.

Bio/ecocentric attitudes included:

- *Environmental–naturalistic*: primary interest in direct contact with wildlife in undisturbed, natural settings.
- *Environmental–stewardship*: principal concern for ecological characteristics of wildlife and natural habitats.
- *Animal rightist*: foremost concern for the rights and well-being of individual animals.
- *Coexistence*: primary interest in the harmonious coexistence between humans and animals.

The overall sample was roughly even with respect to gender, but was divided between those who were relatively well educated and affluent, and those who had less education and far lower incomes. Most were under age 45. Whites and Latinos comprised 70 percent of the sample and were represented in almost equal numbers. African American and Asian Pacific Islanders were also similarly represented, comprising 12 and 10 percent, respectively. The remainder of the sample fell into the "other" category. This roughly reflects the general demographic composition of the county, although African Americans and Asian Pacific Islanders were oversampled. A majority of survey respondents identified themselves as Christian, and more than half were born in the USA. Mexico was the most common country of origin among non-US born respondents, followed by China. The vast majority of all respondents had lived in the USA for longer than two years.

Discourses of Concern: The Role of Gender

Striking gender differences emerged in our analysis of attitudes – which, following Kalof (2001), we also term "discourses of concern." This is despite the fact that very few gender differences emerged within the sample in terms of time spent at the beach, the incidence of seeing particular types of animals along the coast, or activities engaged in while beachgoing (although men were more apt to fish, while women were more likely to sunbathe, swim or wade, or visit with friends). Also, there were few significant differences in knowledge levels. Women were less likely to identify Pacific cormorants as endangered (they are not), but less knowledgeable about which fish are not safe to eat if caught off the southern California coast. In addition, there were few demographic differences between men and women, although predictably men reported somewhat higher incomes and more education.

Basic attitudes toward marine wildlife

As shown in table 31.1, women were significantly less likely to express utilitarian attitudes, and more likely to agree with statements in support of animal welfare and the spiritual value of marine creatures.

For example, women tend to oppose the idea that recreational fishing was acceptable regardless of whether the catch was consumed, and were less apt to agree that lobster fishing was analogous to growing apples, or to consider that the most important reason to protect fisheries was to insure sufficient food for humans. Although they were less negative about seagulls, they were more likely to be put off by kelp and seaweed, and jellyfish or crabs. With respect to aesthetic discourses, there was only one significant difference, although in all cases women were more likely to agree with statements emphasizing aesthetic values. Support for animal welfare was stronger among women; for example, they were significantly more likely to agree that the use of barbed hooks and killing whales were cruel acts. And although they were less apt to think that eating freshly caught fish gives people more energy, they were more likely to express the view that seeing wild marine animals gave them a magical feeling.

Women were also more likely to support bio/ecocentric discourses about marine wildlife. Support for environmental–naturalistic statements was mixed, in that women were more apt to support coastal wetland protection for seabird habitats, but less likely to agree that nature should "take its course" if a whale was beached or another animal became stranded, despite the fact that they expressed more support for allowing wild animals to live their lives without interference from people. Women also expressed stronger sentiments in favor of stewardship for the benefit of animals and nature; they were significantly more likely to agree that oil spills should be prevented in order to save sea birds, and that overfishing should be avoided in order to insure that there was enough food for animals that depended on the ocean's bounty for survival. They also advocated animal rights more strongly than men, agreeing more often that the fates of individual animals mattered to them, and that marine animals should not be kept in captivity because they had a right to be free. They were also less apt to agree that the idea of animals having legal rights was absurd (although not significantly so). Lastly, although no coexistence

Table 31.1 Gender differences in attitude constructs

Attitude	Variable	Gender	Mean	t-test for equality of means	p-value
Anthropocentric					
Utilitarian–dominionistic	I think that recreational fishing is fine, regardless of whether you eat the fish you catch.	Female Male	−0.38 0.09	−4.414	0.000
Utilitarian–dominionistic	Restaurants shouldn't serve swordfish if their numbers are significantly declining.	Female Male	1.27 1.25	0.364	0.716
Utilitarian–dominionistic	Populations of sea lions should be reduced if they eat too many fish that people eat.	Female Male	−0.61 −0.58	−0.324	0.746
Utilitarian–dominionistic	Since mile-wide fishing nets are so efficient, they should be used even though they cause ecological damage.	Female Male	−0.99 −0.87	−1.226	0.221
Utilitarian–stewardship	As long as the lobster population is healthy, commercial lobster fishing is no different than harvesting apples each year.	Female Male	0.43 0.63	−2.156	0.031
Utilitarian–stewardship	The most important reason to protect areas where fish mature and reproduce is to insure that people will have enough fish to eat in the future.	Female Male	0.63 0.93	−3.173	0.002
Utilitarian–stewardship	It is okay for sharks and other marine animals to be used for food and medicines so long as the animals are not endangered.	Female Male	1.06 1.19	−1.560	0.119
Negativistic	I find seagulls to be a real nuisance.	Female Male	−0.80 −0.68	−1.281	0.201
Negativistic	Seaweed and kelp are dangerous to swimmers.	Female Male	0.08 −0.03	1.087	0.277
Negativistic	When I go to the beach, I don't go in the water because there might be unpleasant animals like jellyfish or crabs there.	Female Male	−0.43 −0.76	3.140	0.002
Aesthetic	If I were to visit a marsh or wetland, it would be to watch the colorful birds and other wildlife that live there.	Female Male	1.46 1.42	0.660	0.509
Aesthetic	One of the most striking things about whales is their grace and beauty.	Female Male	1.60 1.55	0.772	0.441
Aesthetic	I don't like the idea of mounting fish on the wall as trophies.	Female Male	0.50 0.34	1.545	0.123
Aesthetic	If I had to choose, I'd rather snorkel than surf because snorkeling allows me to see beautiful fish.	Female Male	1.24 1.08	2.151	0.032
Animal welfare	Catching fish with barbed hooks is cruel.	Female Male	0.90 0.30	6.130	0.000
Animal welfare	Killing whales is a cruel act.	Female Male	1.45 1.22	2.783	0.006
Animal welfare	Keeping smart animals like seals and killer whales in aquariums is cruel.	Female Male	0.52 0.34	1.800	0.072
Spiritualistic/supernatural	I avoid some kinds of animals because they bring bad luck.	Female Male	−1.50 −1.52	0.322	0.748

Table 31.1   *continued*

Attitude	Variable	Gender	Mean	t-test for equality of means	p-value
Spiritualistic/ supernatural	It gives your body more energy to eat fish that's just been caught fresh.	Female Male	0.39 0.52	−1.275	0.203
Spiritualistic/ supernatural	Seeing wild animals like dolphins in the surf would give me a magical feeling.	Female Male	1.45 1.20	3.149	0.002
Bio/ecocentric					
Environmental– naturalistic	When stranded animals wash up on the beach, we should let nature take its course and not intervene.	Female Male	−0.33 −0.25	−0.755	0.450
Environmental– naturalistic	It's unfortunate to see whales beach themselves but that's "nature's way."	Female Male	0.53 0.83	−3.290	0.001
Environmental– naturalistic	It's never OK for people to interfere with wild animals, who should be free to lead their lives without interference from people.	Female Male	0.66 0.45	2.042	0.042
Environmental– naturalistic	If I were to support the protection of coastal marshes or wetlands, it would be to allow seabirds to live in their natural habitat.	Female Male	1.59 1.52	1.203	0.229
Environmental– stewardship	It is important for sea lions to exist in Southern California because that's where they've historically lived.	Female Male	1.47 1.43	0.658	0.511
Environmental– stewardship	Creatures like sand worms and marsh mice are not ecologically important.	Female Male	−0.55 −0.72	1.888	0.059
Environmental– stewardship	The most important reason to prevent oil spills is because local populations of sea birds could be wiped out.	Female Male	1.32 1.07	2.948	0.003
Environmental– stewardship	If we decide to protect coastal marshes, it should be because that's where many young fish populations grow up.	Female Male	1.27 1.29	−0.278	0.781
Environmental– stewardship	The most important reason to avoid over-fishing is to make sure there's enough food left in the oceans for other animals.	Female Male	1.42 1.22	2.779	0.006
Animal rights	The fates of individual animals matter to me, not just what happens to endangered species.	Female Male	1.55 1.37	2.706	0.007
Animal rights	The idea of marine animals, like whales or dolphins, having legal rights just like people do is absurd.	Female Male	−0.16 −0.02	−1.318	0.188
Animal rights	We should not keep marine animals in aquariums because they have the right to be free.	Female Male	0.68 0.42	2.621	0.009
Coexistence	Sea lions shouldn't be removed from beaches just to make room for people.	Female Male	1.13 1.02	1.143	0.253
Coexistence	Although the beach is the seagull's natural habitat, when I'm there I don't want them around me because they are messy.	Female Male	−0.69 −0.67	−0.251	0.802
Coexistence	It's OK when pelican steal fish from commercial fishermen because pelicans have to eat too.	Female Male	1.40 1.29	1.600	0.110

statements elicited significantly gendered responses, women were more apt to agree to such statements than men.

The survey also asked respondents about three policy issues, all of which have received attention from the local media recently. This issue revolved around illegal collection of tidepool animals, use of dolphin-safe tuna fishing methods, and the development of tidal wetlands. There were no significant differences in position regarding policy approaches to tidepool collecting or dolphin-safe tuna; most men and women supported either fines or public education campaigns to dissuade collectors, and over 70 percent called for making dolphin-safe tuna fishing techniques mandatory. But women were significantly more likely to call for protecting coastal wetlands from development, regardless of economic impact (48.7 percent of women versus 42.6 percent of men).

Tolerance toward controversial animal practices and feelings of stigma

Women's discourses of concern for animals emerged as particularly distinct from those of male respondents when certain controversial animal practices – many of them culturally coded – were discussed. In order to gauge attitudes toward controversial practices – or tolerance – a series of yes/no statements regarding culturally sensitive practices involving animals and nature were included in the survey. Prior to each statement, respondents were asked to keep in mind that other cultures treat animals differently. They were then queried as to whether "it was OK" that people engage in certain culture-specific practices, such as participation in horse-tripping events at Mexican-style rodeos, sacrificing animals for religious purposes, as in the Santeria religion, or raising calves in confinement for veal (a largely Anglo-European practice).

Women were almost invariably less tolerant than men of controversial practices, as shown in figure 31.1. For example, women were more likely to condemn whaling, religious animal sacrifice, eating sea turtles, dog-eating, donating pets to research labs, attending or participating in bullfights, dogfights, horse-tripping, or calf-roping, eating factory farmed meat or keeping veal calves in crates, or cropping the ears or tails of dogs. Tolerance was especially limited (under 15 percent) with regard to animal sacrifices, dog eating, and dog fighting – the latter two practices predictably unpopular in a place where most members of the dominant culture consider dogs as members of the nuclear family (Beck and Katcher, 1996). Only on a couple of items were there no differences – littering on the beach, participating in cockfighting (both practices of which a vast majority of both men and women disapproved), and spending money on pets, something that between two-thirds and 70 percent of respondents were willing to indulge. Despite the heightened concern for animal welfare and rights among women that is reflected in these responses, we should note that over 50 percent of both men and women approved of keeping animals alive until they were ready to be eaten (as is often practiced in Asian restaurants), and eating factory farmed meat.

Survey participants were also asked if they felt looked down upon – or stigmatized – by their own animal practices. More than 40 percent (42.2 percent) of all respondents felt looked down upon, or stigmatized, and about one-fifth felt that people looked down on them for their belief that animals have rights like people.

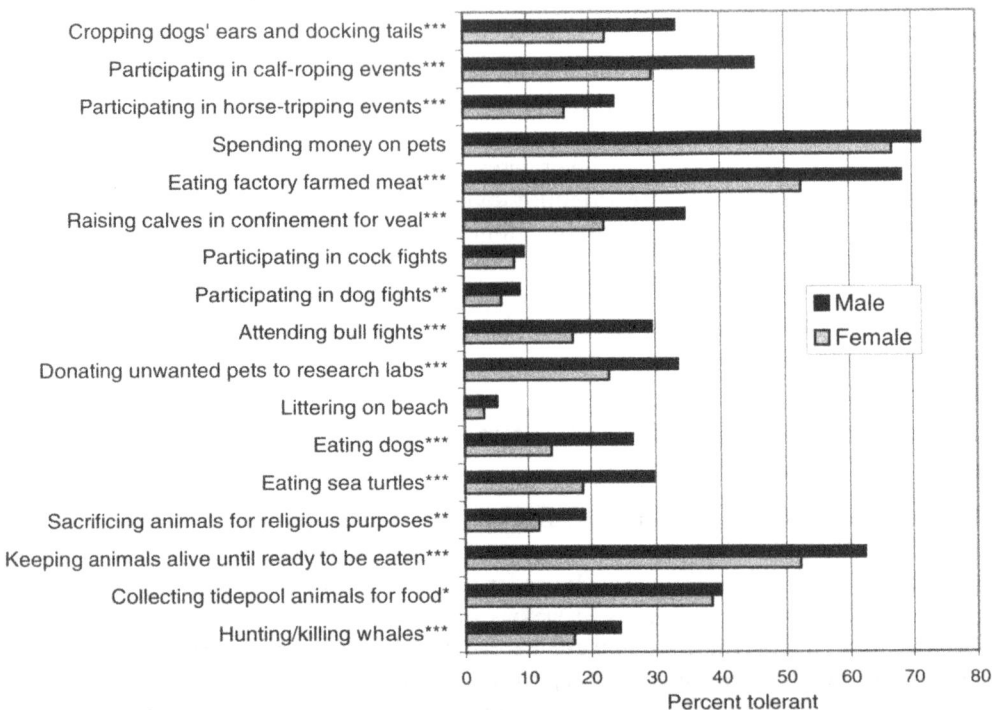

Figure 31.1 Gender differences in tolerance toward controversial animal practices. Tests significant at *10, **5, and ***1 percent levels.

Eleven percent felt that others took exception to the kinds of animals they ate, and almost 10 percent because of the amount of money they spent on their pets. Approximately 6 percent of respondents felt that others disapproved of the sorts of animals they kept at home, and the way they treated or trained their animals.

Although the difference was not statistically significant, women were somewhat more likely to feel looked down upon than men (table 31.2). Tellingly, they were significantly more apt to report stigma because they thought animals had rights. Men, in contrast, felt more stigmatized because they hunted or fished (although percentages here are very small).

Race and Discourses of Animal Concern

Women's attitudes varied not only from men's, but dramatically by race/ethnicity, revealing the interplay of race/gender and their mutually constitutive role in shaping discourses around marine animals. Indeed, there was a stronger pattern of race/ethnic differences than gender differences, at least in terms of numbers of attitude questions that were significantly different. Moreover, there were consistent race/ethnic differences in the extent to which men and women of different race/ethnic backgrounds were comfortable about controversial animal practices. In some instances, however, race-specific gender differences were even more pronounced that overall gender contrasts.

Table 31.2 Perceived social stigma (percentages)

Do you ever feel that people look down on you or think you are strange because of the . . .	Female (n = 423)	Male (n = 427)
I never feel that way	56	58.8
Kinds of animals you eat	10.4	12.2
Sorts of animals you keep at home	6.6	6.3
Way you treat or train your animals	6.1	7.3
Fact that you don't really like animals	3.3	2.1
Fact that you think animals have rights like people**	23.6	18
Money you spend on your pets	11.1	8
Fact that you hunt*	1.9	4
Fact that you fish*	1.4	4

Tests significant at *10 and **5 percent levels.

Race/ethnic contrasts in women's attitudes

On 28 of the 35 attitude questions, for example, there were statistically significant differences among women according to their race/ethnicity. Indeed, there were more such differences within the female subsample than among the male group, where attitudes tended to be more homogeneous in areas of utilitarian–stewardship and environmental–naturalistic attitudes. The most dramatic differences among women's discourses were between Latinas and Asian Pacific Islander women. Latinas were far more likely to express strong support for animal rights and welfare, spiritualism, aesthetic values, and environmental stewardship. Moreover, they were less utilitarian–dominionistic, and much more likely to support intervention to help wildlife. For example, 81.9 percent of Latinas strongly agreed with the statement that "the fates of individual animals matter to me, not just what happens to endangered species," versus only 38.9 percent of Asian Pacific Islander women. Similarly, only 19.4 percent of Latinas strongly disagreed that the use of barbed fishhooks was not cruel (an animal welfare statement), compared to 65 percent of Asian Pacific Islander women. Latina coexistence attitudes were much stronger too; almost 85 percent expressed strong agreement that it was OK for pelicans to "steal" food from fishers in order to survive, compared to only 25 percent of Asian Pacific Islander women. There were a couple of contradictions, however; Latinas were more apt than Asian Pacific Islander women to strongly agree that sharks and other marine animals could be used if not endangered, and to embrace lobster fishing if their populations were healthy – indicating stronger utilitarian–stewardship attitudes. Thus, despite their relative biocentrism and support for animal welfare, Latinas expressed something of a utilitarian streak as well.

On many items, white and African American women fell in between the two poles of Latinas and Asian Pacific Islander women, but typically they were closer to the Latina end of the discourse spectrum. However, there were exceptions, when both groups fell at the ends of the discourse spectrum. For instance, white women were far less negativistic about dealing with beach creatures such as crabs or jellyfish (only 13.7 percent strongly agreed that they wouldn't enter the water because of possible contact), than African American women (39.7 percent). And white women were the least utilitarian–dominionistic: only 4.1 percent strongly agreed that sea lion populations should be reduced if they ate too many fish (compared to 29.8 percent of Asian Pacific Islander women), and only 19.4 percent found the use of mile-wide nets acceptable fishing practice, compared to 62.3 percent of women in the Asian Pacific Islander subsample who were in strong agreement with this approach to fishing. Interestingly, white women were the least apt to strongly agree that interference with wild animals was appropriate (only 20.5 percent), compared to almost three-quarters of Latinas who favored intervention to help animals. African American women were least likely to embrace animal rights discourses around legal rights for animals. The share expressing strong agreement that such rights were "absurd" was 36.2 percent among African American women, compared to only 14.4 percent for white women.

Were race/ethnic differences among women's discourses reflected among the male respondents? Results here were fascinating. Considering only those questions on which there were significant gender differences in response patterns, we found that the pattern of race/ethnic differences was not always the same (and in fact was usually different) across genders. In particular, white men were sharply less apt to embrace animal rights and welfare views than women. Only 55.4 percent strongly agreed that the fates of individual animals mattered to them, compared to 70.5 percent of white women; only 21.7 percent strongly felt that the use of barbed fish hooks was cruel, compared to 44.5 percent of white women; and only 57.3 percent strongly agreed that killing whales was cruel, compared to almost three-quarters of white women. White women were far less utilitarian, with only 17.1 percent strongly agreeing that lobster fishing was fine as long as the population was healthy, compared to 40.8 percent of white men; and they agreed more strongly with environmental stewardship attitudes: for example, 61 percent of white women strongly agreed that the most important reason to prevent oil spills was to protect seabird populations, compared to 38.2 percent of white men.

Some other intriguing results pertained to Latinos and African Americans too. Among Latinas, strong agreement that marine animals should not be kept in confinement because they had rights to be free was common (73 percent, compared to only about half among Latino men). African American women had stronger aesthetic attitudes than their male counterparts, with only about a third of the latter indicating that they would snorkel in order to see beautiful sea creatures, compared to over half of African American women.

There were, however, no statistically significant differences in women's attitudes toward the particular policy issues we queried (dolphin-safe tuna, tidepool animal collecting, and protecting wetlands). In general, whites and Asian Pacific Islanders were the least apt to support dolphin-safe tuning fishing methods, and whites the most apt to suggest simply ignoring illegal tidepool collection (although across all groups, support both for dolphin-safe tuna fishing and protecting tidepool animals was very strong). Asian Pacific Islander women were most likely to favor developing wetlands; African American women, in contrast, were the most apt to opt for protecting wetlands no matter what.

There were no systematic race/gender differences in these policy-related attitudes. On some issues, men of specific race/ethnic groups were more likely to support policy protection than their female counterparts, but not on others. On only one of

Table 31.3 Women's tolerance for controversial animal practices by race/ethnicity (percentages)

Percentage tolerant of	White (n = 146)	African American (n = 58)	Latina (n = 160)	Asian Pacific islander (n = 36)
Hunting/killing whales***	23.3	19	8.1	19.4
Collecting tidepool animals for food***	45.9	51.7	26.3	44.4
Keeping animals alive until ready to be eaten***	55.5	67.2	46.3	47.2
Sacrificing animals for religious purposes**	15.8	15.5	6.9	13.9
Eating sea turtles***	19.2	31	10.6	30.6
Eating dogs**	19.2	13.8	5	27.8
Littering on beach	2.1	3.4	3.1	2.8
Donating unwanted pets to research labs	21.2	24.1	21.3	30.5
Attending bull fights**	16.4	27.6	13.8	16.7
Participating in dog fights	5.5	8.6	4.4	11.1
Participating in cock fights	6.2	8.6	6.9	19.4
Raising calves in confinement for veal*	25.3	34.5	15.6	16.7
Eating factory farmed meat***	57.5	69	38.1	75
Spending money on pets***	82.2	72.4	51.9	58.3
Participating in horse-tripping events***	7.5	19	20.6	27.8
Participating in calf-roping events*	37.7	32.8	23.1	27.8
Cropping dogs' ears and docking tails***	32.3	32.8	12.5	11.1

Tests significant at *10, **5, and ***1 percent levels.

the policy issues did race/ethnicity become statistically significant among men: dolphin-safe tuna fishing, where Latinos were far more likely to support the enforcement of dolphin-safe methods than were other men of other race/ethnic groups.

Race/ethnic variations in tolerance of controversial animal practices

Although female respondents were on average less tolerant of controversial animal practices than their male counterparts, race/ethnic differences in discourses were also strong. Women's responses about the acceptability of various practices differed significantly by race/ethnicity on 13 out of 17 items (compared to only 11 out of 17 among men). Overwhelmingly, Latinas were the least tolerant – on 10 out of these 13 items, the share expressing approval of a controversial practice was the lowest (table 31.3). The differences between Latinas and other women were often large, ranging from around 15 percentage points (whaling) to over 36 percentage points (factory farming).

On a few items, Latinas and Asian Pacific Islander women were similarly reluctant to approve, and in both instances the practices in question were "mainstream" US customs. For example, cropping the ears or tails of dogs was far less acceptable to Latinas and Asian Pacific Islander women (12.5 percent for Latinas, 11.1 percent for Asian Pacific Islander women), as was raising calves in confinement (15.6 percent for Latinas, 16.7 percent for Asian Pacific Islander women). In comparison, almost a third of white women found tail docking and ear cropping acceptable, as did 22.8 percent of African American women, while over a third of the latter found no problem with veal calf confinement, and a quarter of white women were similarly unbothered.

White women were the least tolerant on only one item: horse-tripping. Here only 7.5 percent approved, compared to acceptance rates of 19–28 percent for the other groups (Asian Pacific Islander women being most tolerant, but Latinas being close behind). Given the close bond between (especially) Anglo women and horses in the USA, as well as the horse's iconographic status in Western culture, this is not surprising. African American women were the most tolerant on four items but were never the least tolerant among the four subgroups of women. Asian Pacific Islander women, in contrast, were the most tolerant, or second most tolerant, of two culturally coded practices: eating turtles and dogs. They were, however, also the most tolerant of factory farming and horse-tripping – practices linked to mainstream US food systems and Mexican rodeo, respectively. They were the least tolerant on only one item: cropping/docking of canine ears and tails.

On specifically marine-related practices, women were uniformly less tolerant; however, there were strong variations, with Asian Pacific Islander and African American women being more tolerant of eating sea turtles, for example, and Latinas being more tolerant of tidepool collecting (figure 31.2).

Although women of all four race/ethnic groups were less tolerant than their male counterparts, gender differences were far greater for some groups than others. For example, Latinos, like their female counterparts, were by far the least tolerant of controversial animal practices, and although their tolerance levels were higher than those of Latinas, the gender gaps were not very large and often were smaller than the average gender gap for tolerance questions. African American and Asian Pacific Islander gender differences were more uneven (sometimes more pronounced, but often less than the average gap).

White men and women, on the other hand, differed sharply on a majority of the tolerance questions: women were far less tolerant (except when it came to spending money on pets, which they favored more often than men). Some contrasts were extraordinary: less than a fifth of white women approved of eating dogs or turtles, compared to over 40 percent of white men; only about a quarter approved of keeping veal calves in confinement, compared to almost 50 percent of white men; and 55.5 percent of white women agreed with keeping animals alive until ready to be eaten, compared to 77.1 percent of white men. These white gender gaps were much larger than average, in some cases by almost an order of magnitude.

These findings suggest that that popular conceptions about the relative dominance of whites among animal welfare and rights supporters may be erroneous – although whites may be more likely to become activists on animal-related issues. Moreover, animal rights and welfare support among whites may be mostly a result of white women's attitudes, rather than those of white men. Latinos in general, and Latinas in particular, expressed far less tolerance toward practices involving harm to animals, regardless of cultural coding. Even horse-tripping, closely associated with Mexican cultural practice, failed to win approval from more than a fifth of Latinas – lower than among Asian Pacific Islander women, about the same as among white men and African American women, and substantially lower than among African American or Asian Pacific Islander men.

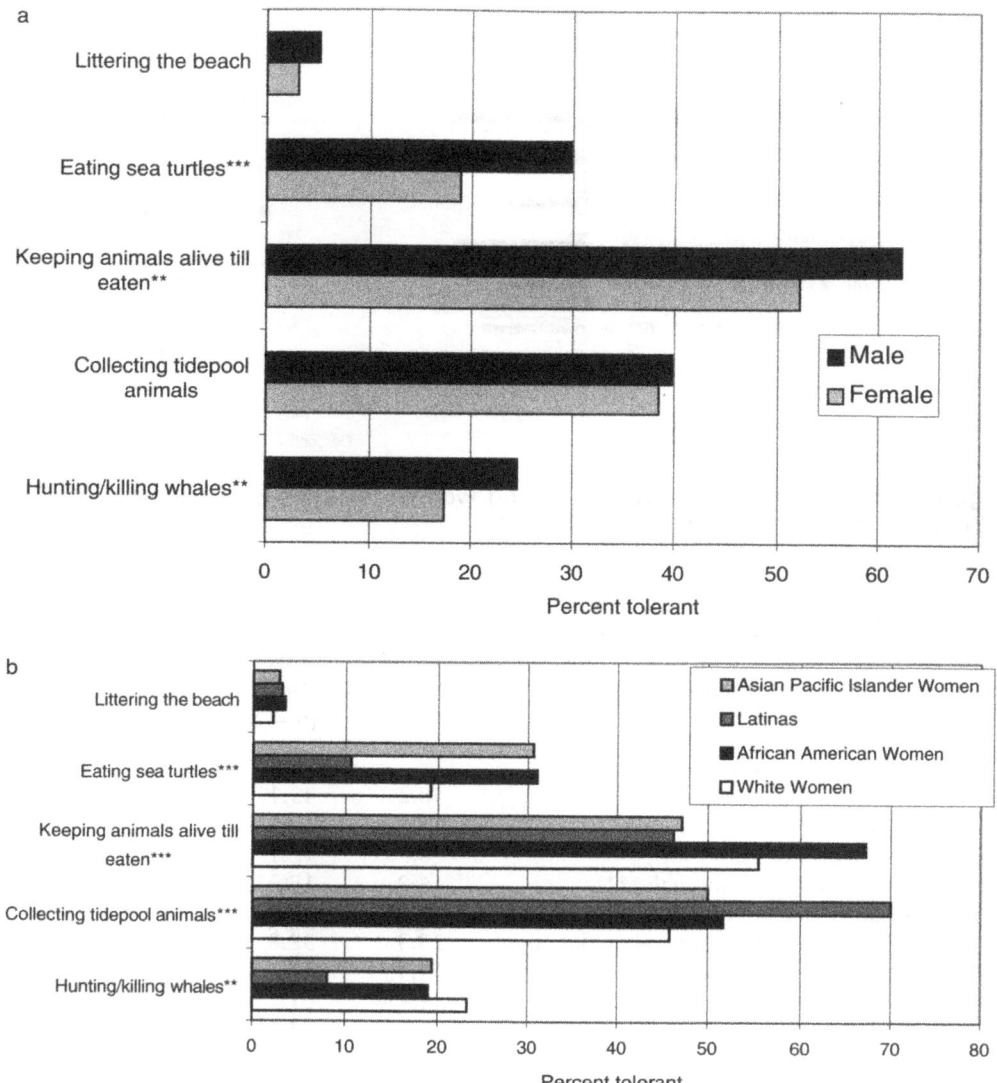

Figure 31.2 (a) Gender differences in tolerance toward controversial marine practices. (b) Race/ethnic differences in tolerance toward controversial marine practices. Tests significant at *10, **5, and ***1 percent levels.

Race and stigma

Race/ethnic differences in feelings of stigma among women were far more common than amongst men (figure 31.3). Although in neither case were overall reports of stigma significantly different across race/ethnic lines, women differed significantly across race/ethnic group on six out of the eight specific stigma questions.

In general African American women were the most likely to express feelings of stigma, especially with respect to the types of animals eaten, training approaches and treatment of pets, spending money on pets, and the fact that they don't like

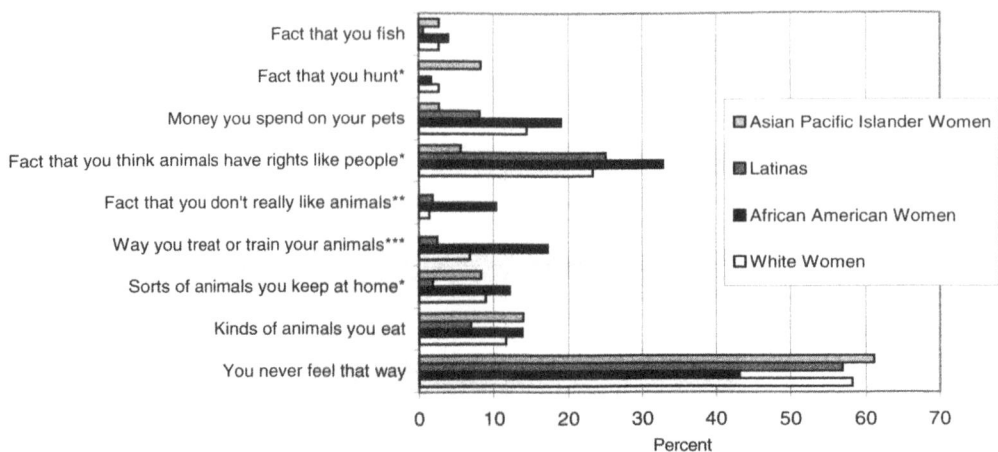

Figure 31.3 Race/ethnic differences among women in feelings of stigma. Tests significant at *10, **5, and ***1 percent levels.

Table 31.4 Women's perceived social stigma by race/ethnicity (percentages)

Do you ever feel that people look down on you or think you are strange because of the ...	White (n = 146)	African American (n = 58)	Latina (n = 160)	Asian Pacific islander (n = 36)
I never feel that way	58.2	43.1	56.9	61.1
Kinds of animals you eat	11.6	13.8	6.9	13.9
Sorts of animals you keep at home*	8.9	12.1	1.9	8.3
Way you treat or train your animals***	6.8	17.2	2.5	0
Fact that you don't really like animals**	1.4	10.3	1.9	0
Fact that you think animals have rights like people*	23.3	32.8	25	5.6
Money you spend on your pets	14.4	19	8.1	2.8
Fact that you hunt*	2.7	1.7	0	8.3
Fact that you fish	2.7	4	0.6	2.8

Tests significant at *10, **5, and ***1 percent levels.

animals. The biggest contrast, however, was around whether respondents felt stigmatized because they thought animals had rights. Here, almost a third of African American women reported having been made to feel strange because of such convictions, compared to only 5.6 percent of Asian Pacific Islander women (table 31.4).

Men, in contrast, only differed across race/ethnic group with respect to hunting. On most items, men of particular race/ethnic background were less apt to report feeling stigmatized than their female counterparts. In particular, on the general question of whether respondents had ever been made to feel strange on account of their animal practices, 40.9 percent of African American men answered in the affirmative – about the same as white and Latino men, and somewhat higher than Asian Pacific Islander men – compared to 56.9 percent of African American women.

Discussion

Our research into the gendered nature of attitudes toward marine wildlife – or discourses of concern – reinforces past findings about gender as well as race/ethnic differences in nature–society relations, and raises a number of fascinating questions for further research. Women's discourses of concern consistently expressed greater sympathy with bio/ecocentric viewpoints, especially around environmental–stewardship and animal rights, and embraced perspectives in support of animal welfare and spiritual values. Moreover, women were on average far less tolerant of controversial animal practices than men, and were more likely to feel that the ways in which they interacted with, or thought about, animals caused them to be looked down upon, particularly around the question of whether they thought animals had rights. Lastly, discourses of concern, as well as expressions of tolerance and feelings of stigmatization, varied strikingly by race, with race/ethnic differences in some instances outweighing gender contrasts.

Taken together, these findings break new conceptual as well as empirical ground. Although results generally support past research on gendered attitudes toward animals, they provide a far more detailed assessment of conventional attitude categories and shed light on more nuanced categories (such as spirituality and coexistence) not considered in past studies. Moreover, we have extended the exploration of attitudes, to try to better understand gendered and raced dimensions of the morality of nature–society relations.

Although past work has investigated practices such as experimentation on laboratory animals, finding that women are less tolerant, we found that women respond differently to a wide range of specific human behaviors that inflict pain, suffering, and/or death on particular types of animals, and that their responses interact with race, with the "race gap" in some cases being larger than the "gender gap" in tolerance. Along with their greater biocentrism, the intolerance of Latinos, in particular, is compelling evidence of an emergent and powerful Latino environmentalism, while the results for African Americans belie past assertions of lower concern about the environment generally, and animals specifically.

The question of stigma, or "othering" as a result of animal practices, has not been addressed before. But this is clearly important given recent research on the ways in which animal practices may fuel racialization, an implication furthered by the fact that African American women were most apt to feel stigmatized. Intriguingly, their strongest perception of stigma arise from their beliefs that animals have rights – a belief that African American women overall were the least likely to hold (itself perhaps due to deep distrust of any comparisons between their situation and that of animals, given a long and terrible history of dehumanization). This suggests both in-group and out-group pressures on those committed to animal rights, and reveals the complexity of dynamics between racialization and ideas about animals.

How can we understand the gendered and raced nature of discourses of concern, as revealed here? And what sorts of non-essentialist feminist geographic research might help in developing such an understanding? Ecofeminist theory suggests that the situated character of women's knowledge works to build broader and deeper social networks that are more apt to span the species divide. Building on the work of Whatmore and Thorne (1998) and others, actant network theory (ANT) might prove useful to emplacing women and sea creatures (or other animals) in networks that heighten women's solidarity with non-humans. Such an analysis could help us to understand the extent to which discourse variations are related to networks, or, following Gilligan, are predicated on deeply gendered socialization processes and a resulting ethics of care among women. Alternatively, drawing on Adams's work on animals as the "absent referent" embedded in speech acts and patriarchal treatment of women, women as marginalized social actors may relate to animals as quasi-women, and react more forcefully against their victimization – more often than not at the hands of men who hunt, fish, rope, trip, fight, and slaughter.

The crucial role that race plays, however, cautions against any framework rooted solely in gendered developmental psychology or patriarchal structures and practices. Culture – and in all likelihood culture contact – clearly matters a great deal. The discourses of Latinos and Asian Pacific Islanders in our study, most of whom were immigrants, are apt to be shaped by attitudinal norms in their places of origin, but also influenced by new surroundings and nature–society conventions. The latter may make newcomers feel more like outsiders, and prompt them to either align their perspectives with the new setting or resist any incursions on traditional thinking and practice as a means to combat marginalization. Thus the fact that Latinos – both men and women – were so intolerant of controversial animal practices (even practices such as charreada or horse-tripping, culturally coded as Mexican) cannot tell us whether such attitudes reflect desires to assimilate, distinctive "modern" biases characteristic of those who decide to immigrate, or deep-seated cultural norms (indigenous, settler?) carried across the border where their light is shone over alien nature–society relations. Thus cross-cultural ethnographic and archival work on attitudes toward animals is critical.

The relative tolerance of whites – especially men – is provocative given popular perceptions of animal welfare and rights as being white projects, and stands in contrast to the fact that whites were not necessarily the most anthropocentric respondents. Do whites trying to be sensitive to cultural heritage and racially coded representation and embodied practices try to avoid condemning the sorts of interactions with animals characteristic of other cultures, while at the same time holding disparate views about how they themselves should behave? Qualitative research with white populations could help us to understand this seeming contradiction.

Finally, feminist and anti-racist geographers could build on both the classic works on stigma (by Goffman, for example), as well as recent research on disability and the body by Wilton (1998) on abjection and exclusion, to better understand the stigma associated with animal practices. Are relationships with animals simply part of a larger dynamic by which many aspects of everyday life – real and perceived slights, verbal abuse, discrimination – become causes for racial anxiety and tension? Another possibility is that animals – with all the implications for purity and danger, defilement and redemtion that they hold – play a more specific role in defining both gendered and racialized identities, even in contemporary society, a role that might be excavated by a feminist animal geography.

Many men may not "yield to the song of the siren nor the voice of the hyena, the tears of the crocodile nor the howling of the wolf" (Chapman et al., 1999). The ears of women seem slightly more attuned to these songs and voices – the roars and cheeps, chitters and barks, growls and croaks – suggesting the possibility of mutual learning and dialogue, and willingness to at least talk. Feminist geographers have much to learn about nature–society relations from listening in.

ACKNOWLEDGMENTS

Funding for this research was provided by the US Sea Grant Program, whose support is gratefully acknowledged. The authors also thank Unna Lassiter and Marcie Griffith, who participated in the development and analysis of the survey utilized in this study, and Mark Duda and Responsive Management Inc. for survey design assistance and administration. Correspondence should be directed to: Jennifer Wolch, Department of Geography, University of Southern California, 3620 S. Vermont Ave., Los Angeles, CA 90089-0255 (e-mail: wolch@usc.edu).

BIBLIOGRAPHY

Adams, C. J. (1990) *The Sexual Politics of Meat: A Feminist–Vegetarian Critical Theory.* New York: Continuum.

Adams, C. J. and Donovan, J. (eds) (1995) *Women and Animals: Feminist Theoretical Explorations.* Durham, NC: Duke University Press.

Anderson, K. (1995) Culture and nature at the Adelaide Zoo: at the frontiers of "human" geography. *Transactions of the Institute of British Geographers,* n.s., 20, 275–94.

Anderson, K. (1997) A walk on the wild side: a critical geography of domestication. *Progress in Human Geography,* 21, 463–85.

Anderson, K. (2000) "The beast within": race, humanity, and animality. *Environment and Planning D: Society and Space,* 18, 301–20.

Beck, A. and Katcher, A. (1996) *Between Pets and People.* West Lafayette, IN: Purdue University Press.

Bowd, A. D. and Bowd, A. C. (1989) Attitudes toward the treatment of animals: a study of Christian groups in Australia. *Anthrozoös,* 3, 20–4.

Brownlow, A. (2000) A wolf in the garden: ideology and change in the Adirondack landscape. In C. Philo and C. Wilbert (eds), *Animal Spaces, Beastly Places: New Geographies of Human–Animal Relations.* London: Routledge.

Chapman, G., Jonson, B. and Marston, J. (1999) *Eastward ho.* Manchester: Manchester University Press. originally published in 1605.

Clutton-Brock, J. (1981) *Domesticated Animals from Early Times.* Austin: University of Texas Press.

Davies, G. (2000) Virtual animals in electronic zoos: the changing geographies of animal capture and display. In C. Philo and C. Wilbert (eds), *Animal Spaces, Beastly Places: New Geographies of Human–Animal Relations.* London: Routledge.

Davison, D. J. and Freudenberg, W. R. (1996) Gender and environmental risk concerns: a review and analysis of available research. *Environment and Behavior,* 28, 302–39.

Dietz, T., Kalof, L. and Stern, P. C. (2002) Gender, values and environmentalism. *Social Science Quarterly,* 83, 351–64.

Donovan, J. (1990) Animal rights and feminist theory. *Signs: Journal of Women in Culture and Society,* 15, 350–75.

Donovan, J. and Adams, C. J. (1995) Introduction. In C. J. Adams and J. Donovan (eds), *Women and Animals: Feminist Theoretical Explorations.* Durham, NC: Duke University Press.

Driscoll, J. W. (1992) Attitudes toward animal use. *Anthrozoös*, 5, 32–9.

Driscoll, J. W. (1995) Attitudes toward animals: species ratings. *Society and Animals*, 3, 139–50.

Elder, G., Wolch, J. and Emel, J. (1998) La practique sauvage: race, place, and the human–animal divide. In J. Wolch and J. Emel (eds), *Animal Geographies: Place, Politics, and Identity in the Nature–Culture Borderlands*. London: Verso.

Eldridge, J. J. and Gluck, J. P. (1996) Gender differences in attitudes toward animal research. *Ethics and Behavior*, 6, 239–56.

Elston, M. A. (1987) Women and anti-vivisection in Victorian England. In N. A. Rupke (ed.), *Vivisection in Historical Perspective*. London: Routledge.

Emel, J. (1995) Are you man enough, big and bad enough? Ecofeminism and wolf eradication in the USA. *Environment and Planning D: Society and Space*, 13, 707–34.

Emel, J. (1998) Are you man enough, big and bad enough? Ecofeminism and wolf eradication in the USA. In J. Wolch and J. Emel (eds), *Animal Geographies: Place, Politics, and Identity in the Nature–Culture Borderlands*. London: Verso.

French, M. (1985) *Beyond Power*. New York: Summit.

Furuseth, O. W. (1997) Restructuring of hog farming in North Carolina: explosion and implosion. *Professional Geographer*, 49, 391–403.

Gaard, G. (ed.) (1993) *Ecofeminism: Women, Animals, and Nature*. Philadelphia: Temple University Press.

Gaard, G. (2001) Ecofeminism on the wing: perspectives on human–animal relations. *Women and Environments International Magazine*, 52/53, 19–22.

Galvin, S. L. and Herzog, H. A. (1992) Ethical ideology, animal rights activism, and attitudes toward the treatment of animals. *Ethics and Behavior*, 2, 141–9.

Gilligan, C. (1993) *In a Different Voice: Psychological Theory and Women's Development*, 2nd edn. Cambridge, MA: Harvard University Press.

Griffith, M., Wolch, J. and Lassiter, U. (2002) Animal practices and the racialization of Filipinas in Los Angeles. *Society and Animals*, 10, 221–48.

Griffiths, H., Poulter, I. and Sibley, D. (2000) Feral cats in the city. In C. Philo and C. Wilbert (eds), *Animal Spaces, Beastly Places: New Geographies of Human–Animal Relations*. London: Routledge.

Gullo, A., Lassiter, U. and Wolch, J. (1998) The cougar's tale. In J. Wolch and J. Emel (eds), *Animal Geographies: Place, Politics, and Identity in the Nature–Culture Borderlands*. London: Verso.

Herzog, H. H. (1993) The movement is my life: the psychology of animal rights activism. *Journal of Social Issues*, 49, 103–91.

Hogan, L., Metzger, D. and Peterson, B. (eds) (1998) *Intimate Nature: The Bond between Women and Animals*. New York: Ballantine.

Howell, P. (2000) Flush and the *banditti*: dog-stealing in Victorian London. In C. Philo and C. Wilbert (eds), *Animal Spaces, Beastly Places: New Geographies of Human–Animal Relations*. London: Routledge.

Jones, O. (2000) (Un)ethical geographies of human–non-human relations: encounters, collectives and spaces. In C. Philo and C. Wilbert (eds), *Animal Spaces, Beastly Places: New Geographies of Human–Animal Relations*. London: Routledge.

Kafer, R., Lago, D., Wamboldt, P. and Harrington, F. (1992) The pet relationship scale: replication of psychometric properties in random samples and association with attitudes toward wild animals. *Anthrozoös*, 5, 93–105.

Kalof, L. (2001) The multi-layered discourses of animal concern. In H. Addams and J. Proops (eds), *Social Discourse and Environmental Policy*. Cheltenham: Edward Elgar.

Kellert, S. R. (1978) *Policy Implications of a National Study of American Attitudes and Behavioral Relations to Animals*. Washington, DC: Dept of the Interior, US Fish and Wildlife Service.

Kellert, S. R. (1984) Urban American perceptions of animals and the natural environment. *Urban Ecology*, 8, 209–28.

Kellert, S. R. and Berry, J. K. (1980) *Knowledge, Affection, and Basic Attitudes towards Animals in American Society, Phase III.* New Haven, CT: US Dept of the Interior and Fish and Wildlife Service.

Kellert, S. R. and Berry, J. K. (1987) Attitudes, knowledge, and behaviors toward wildlife as affected by gender. *Wildlife Society Bulletin*, 15, 363–71.

Kheel, M. (1995) License to kill: an ecofeminist critique of hunters' discourse. In C. J. Adams and J. Donovan (eds), *Women and Animals: Feminist Theoretical Explorations.* Durham, NC: Duke University Press.

Kidd, A. H. and Kidd, R. M. (1998) General attitudes toward and knowledge about the importance of ocean life. *Psychological Reports*, 82, 323–9.

Kruse, C. (1999) Gender, views of nature, and support for animal rights. *Society and Animals*, 7, 179–98.

Lansbury, C. (1985) *The Old Brown Dog: Women, Workers, and Vivisection in Edwardian England.* Madison: University of Wisconsin Press.

Lassiter, U. and Wolch, J. (2002) Socio-cultural aspects of attitudes toward marine animals: a focus group analysis. *California Geographer*, 42, 1–24.

Lassiter, U. and Wolch, J. (2004) From barnyard to backyard to bed: attitudes toward animals among Latinas in Los Angeles. In G. Hise and W. Deverell (eds), *Land of Sunshine: The Environmental History of Greater Los Angeles.* Pittsburgh: University of Pittsburgh Press.

Leneman, L. (1997) The awakened instinct: vegetarianism and the women's suffrage movement. *Women's History Review*, 6, 271–87.

Lynn, W. S. (1998) Animals, ethics and geography. In J. Wolch and J. Emel (eds), *Animal Geographies: Place, Politics, and Identity in the Nature–Culture Borderlands.* London: Verso.

Matless, D. (2000) Versions of animal–human: Broadland, c.1945–1970. In C. Philo and C. Wilbert (eds), *Animal Spaces, Beastly Places: New Geographies of Human–Animal Relations.* London: Routledge.

Michel, S. (1998) Golden eagles and the environmental politics of care. In J. Wolch and J. Emel (eds), *Animal Geographies: Place, Politics, and Identity in the Nature–Culture Borderlands.* London: Verso.

Navarro, J. F., Maldonado, E., Pedraza, C. and Cavas, M. (2001) Attitudes toward animal research among psychology students in spain. *Psychological Reports*, 89, 227–36.

Nibert, D. A. (1994) Animal rights and human social issues. *Society and Animals*, 2, 115–24.

O'Donnell, M. and van Druff, L. (1987) Public attitudes and response to wildlife and wildlife problems in an urban–suburban area. In L. W. Adams and D. L. Leedy (eds), *Integrating Man and Nature in the Metropolitan Environment: Proceedings of the National Symposium on Urban Wildlife.* Columbia, MD: National Institute for Urban Wildlife.

Peek, C. W., Bell, N. J. and Dunham, C. C. (1996) Gender, gender ideology, and animal rights advocacy. *Gender and Society*, 10, 464–78.

Philo, C. (1998) Animals, geography and the city: notes on inclusions and exclusions. In J. Wolch and J. Emel (eds), *Animal Geographies: Place, Politics, and Identity in the Nature–Culture Borderlands.* London: Verso.

Pifer, L. (1996) Exploring the gender gap in young adults' attitudes about animal research. *Society and Animals*, 4, 37–52.

Pifer, L., Shimuzu, K. and Pifer, R. (1994) Public attitudes toward animal research: some international comparisons. *Society and Animals*, 2, 95–113.

Plous, S. (1992) An attitude survey of animal rights activists. *Psychological Science*, 2, 194–6.

Plous, S. (1996) Attitudes toward the use of animals in psychological research and education: results from a national survey. *Psychological Science*, 7, 352–8.

Plumwood, V. (1993) *Feminism and the Mastery of Nature.* New York: Routledge.

Proctor, J. (1998) The spotted owl and the contested moral landscape of the Pacific Northwest. In J. Wolch and J. Emel (eds), *Animal Geographies: Place, Politics, and Identity in the Nature–Culture Borderlands*. London: Verso.

Regan, T. (1983) *The Case for Animal Rights*. Berkeley: University of California Press.

Robbins, P. (1998) Shrines and butchers: animals as deities, capital, and meat in contemporary North India. In J. Wolch and J. Emel (eds), *Animal Geographies: Place, Politics, and Identity in the Nature–Culture Borderlands*. London: Verso.

Santos, M. L. S. and Booth, D. A. (1996) Influences on meat avoidance among British students. *Appetite*, 27, 197–205.

Singer, P. (1975) *Animal Liberation*. New York: New York Review of Books.

Spelman, E. (1982) Woman as body: ancient and contemporary views. *Feminist Studies*, 8, 109–31.

Sperling, S. (1998) *Animal Liberators: Research and Morality*. Berkeley: University of California Press.

Steger, M. A. and Witt, S. (1989) Gender differences in environmental orientations: a comparison of publics and activists in Canada and the US. *Western Political Quarterly*, 42, 627–50.

Thorne, L. (1998) Kangaroos: a non-issue. *Society and Animals*, 6, 167–82.

Tuan, Y.-F. (1984) *Dominance and Affection: The Making of Pets*. New Haven, CT: Yale University Press.

Turner, B. L. (2002) Contested identities: human–environment geography and disciplinary implications in a restructuring academy. *Annals of the Association of American Geographers*, 92, 52–74.

Ufkes, F. M. (1998) Building a better pig: fat profits in lean meat. In J. Wolch and J. Emel (eds), *Animal Geographies: Place, Politics, and Identity in the Nature–Culture Borderlands*. London: Verso.

van Stipriaan, B. and Kearns, R. A. (2002) Bitching about a billboard: advertising, gender and canine (re)presentations. Unpublished manuscript, Department of Geography, University of Auckland.

Vaske, J. J., Donnelly, M. P., Williams, D. R. and Jonker, S. (2001) Attachment and demographic influences on environmental value orientations and norms. *Society and Natural Resources*, 14, 761–76.

Waley, P. (2000) What's a river without fish? Symbol, space and ecosystem in the waterways of Japan. In C. Philo and C. Wilbert (eds), *Animal Spaces, Beastly Places: New Geographies of Human–Animal Relations*. London: Routledge.

Wang, A. Y. (1999) Gender and nature: a psychological analysis of ecofeminist theory. *Journal of Applied Social Psychology*, 29, 2410–24.

Warren, K. (1990) The power and the promise of ecological feminism. *Environmental Ethics*, 12, 125–46.

Wescoat, J. (1998) The "right of thirst" for animals in Islamic law: a comparative approach. In J. Wolch and J. Emel (eds), *Animal Geographies: Place, Politics, and Identity in the Nature–Culture Borderlands*. London: Verso.

Whatmore, S. (1997) Dissecting the autonomous self: hybrid cartographies for a relational ethics. *Environment and Planning D: Society and Space*, 15, 37–53.

Whatmore, S. (1999) Hybrid geographies: rethinking the human in human geography. In D. Massey, J. Allen and P. Sarre (eds), *Human Geography Today*. Cambridge: Polity Press.

Whatmore, S. and Thorne, L. B. (1998) Wild(er)ness: reconfiguring the geographies of wildlife. *Transactions of the Institute of British Geographers*, 23, 435–54.

Whatmore, S., and Thorne, L. B. (2000) Elephants on the move: spatial formations of wildlife exchange. *Environment and Planning D: Society and Space*, 18, 185–203.

Whitley, L. N. (1998) Cultural diversity and attitudes toward marine wildlife. Masters thesis, University of Southern California, Los Angeles.

Wilton, R. (1998) Disability, identity and exclusion: community opposition as boundary maintenance. *Geoforum*, 29, 173–85.

Wolch, J. (1998) Zoöpolis. In J. Wolch and J. Emel (eds), *Animal Geographies: Place, Politics, and Identity in the Nature–Culture Borderlands*. London: Verso.

Wolch, J., Brownlow, A. C. and Lassiter, U. (2000) Constructing the animal worlds of inner-city Los Angeles. In C. Philo and C. Wilbert (eds), *Animal Spaces, Beastly Places: New Geographies of Human–Animal Relations*. London: Routledge.

Wolch, J. and Emel, J. (eds) (1998) *Animal Geographies: Place, Politics, and Identity in the Nature–Culture Borderlands*. London: Verso.

Wolch, J., Emel, J. and Wilbert, C. (2002) Re-animating cultural geography. In K. Anderson, M. Domosh, S. Pile and N. Thrift (eds), *Handbook of Cultural Geography*. London: Sage.

Woods, M. (2000) Fantastic Mr Fox? Representing animals in the hunting debate. In C. Philo and C. Wilbert (eds), *Animal Spaces, Beastly Places: New Geographies of Human–Animal Relations*. London: Routledge.

Yarwood, R. and Evans, E. (2000) Taking stock of farm animals and rurality. In C. Philo and C. Wilbert (eds), *Animal Spaces, Beastly Places: New Geographies of Human–Animal Relations*. London: Routledge.

Zinn, H. C. and Pierce, C. L. (2002) Values, gender, and concern about potentially dangerous wildlife. *Environment and Behavior*, 34, 239–56.

Chapter 32

Geographic Information and Women's Empowerment: A Breast Cancer Example

Sara McLafferty

In communities across the United States, women are actively working to improve health for themselves, their families, and their communities. Women's health activism rests on a recognition that local health problems are often deeply connected to the local environment – to the quality of air and water, to the presence and impacts of environmental contaminants, to the availability and quality of health, social, retail, and recreational services, and to the place-based relations among people and institutions. Digital geographic information and geographic information systems (GIS) technologies are playing increasingly important roles in describing and understanding the links between health and the local environment. Drawing upon a case study of women's breast cancer activism in Long Island, New York, this chapter looks critically at the use of geographic information and GIS in research and activism around environmental health. I explore how geographic information underpinned women's quest for knowledge about the local breast cancer problem; how women framed "community GIS queries" about the links between breast cancer and the environment; and how a federally sponsored health GIS for the region is in turn reshaping research and activism. Feminist analysis points out the "social construction" of GIS – how GIS transformed women's breast cancer activism, and how politics and power relations affected the design and implementation of GIS.

Geographic information refers to digital information that is tied to locations on the earth's surface. Residential addresses of people with breast cancer, air pollution levels across a metropolitan area, hospital locations, and poverty statistics by zip code are all examples of geographic information. Geographic information systems (GIS) are computerized systems for managing, mapping, and analyzing geographic information. Increasingly, public health researchers, policy-makers, and advocacy groups are utilizing GIS to explore the causes and consequences of ill-health, to evaluate access to health care, and to examine the links between environmental quality and health (Cromley and McLafferty, 2002). At the same time, researchers are beginning to look critically at the broader roles and impacts of geographic information by examining how GIS is changing society and how social relations are affecting the structuring of GIS (Pickles, 1995; Sheppard et al., 1999).

Recent literature on GIS and society emphasizes that in both its design and its use, GIS is a *social* practice (Sieber, 2000). "The evolving relationship between society and GIS can take many forms depending on local context and circumstances" (Sheppard et al., 1999, p. 814). Social and political processes affect the design of GIS, including issues of data availability, scale, and access, as well as types of queries and spatial analysis tools. Although GIS was traditionally viewed as a positivist tool that offered fixed representations of Cartesian spaces, current work moves far beyond this, arguing that GIS can incorporate both qualitative and quantitative information, thus creating multiple and even feminist representations (Kwan, 2002; Pavlovskaya, 2002). Researchers are also beginning to look critically at the use of GIS. Questions of power lie at the heart of GIS use: how is power negotiated via GIS – by whom and for whom (Ellwood, 2002)? While GIS has enormous power to disseminate useful information, it also creates a "technocratic elite" that can create barriers to empowerment (Clark, 1998; Ghose and Huxhold, 2001). It is critically important to understand the types of knowledge created and privileged by GIS in particular contexts and the community impacts of such knowledge. Drawing upon my earlier work (McLafferty, 2002), this chapter explores such issues in relation to women's breast cancer activism.

The chapter is divided into three sections. The first section presents a brief discussion of breast cancer, emphasizing current understanding of the disease and its connection to the environment. The next section describes a case study of women's breast cancer activism in Long Island, New York. It highlights the shifting roles of geographic information and GIS in breast cancer activism. In the final section, I explore how varying conceptions of knowledge and power are represented in the breast cancer GIS case study.

Breast Cancer

Breast cancer is the most common cancer in women in the United States. One out of every eight women born today will develop breast cancer in her lifetime, and a substantial fraction of those women will die from the disease. The incidence of breast cancer varies greatly among racial and ethnic groups in the USA. White women have the highest incidence rate, followed by African-American, Hispanic and Asian women. However, African-American women have the highest death rate from breast cancer and the highest incidence of the disease at young ages (Kreiger, 2002).

The causes of breast cancer are complex and poorly understood. The disease has a significant genetic component: 10–20 percent of women with breast cancer have a close relative with the disease, and a specific genetic mutation accounts for roughly half those cases. Breast cancer risk is also related to menstrual and reproductive history (NCI, 2002). Early menarche and late menopause increase the risk of breast cancer, whereas early age at first pregnancy is associated with reduced risk. Epidemiological studies reveal a socioeconomic gradient in breast cancer risk. Incidence is higher among women with high levels of education and income, leading some to call breast cancer "a disease of affluence"; but the socioeconomic gradient partly reflects better detection and access to mammography rather than true differences in incidence (Kreiger, 2002). Obesity and a high-fat diet have also been linked to heightened breast cancer risk, but these effects too are complicated by differences in detection.

Most breast cancer cases in the United States (>60 percent) are not explained by known risk factors (NCI, 2002). This gap in knowledge has stimulated interest in the role of environmental contaminants such as organochlorines, polyaromatic hydrocarbons, and electromagnetic fields. Some environmental contaminants have estrogen-like effects and thus may promote tumor development; others are known to cause breast cancer at high doses (NCI, 2002). Research on environmental contaminants has produced complex and often contradictory findings (Calle et al., 2002). We still do not know which, if any, environmental factors are important in breast cancer development, and how and to what extent environmental factors interact with genetic, behavioral, and social factors in affecting disease risk.

Mapping and GIS are emerging as important tools in untangling the connections between environmental hazards and breast cancer risk. Such systems are useful in characterizing the environmental contaminants and exposures that people face in their everyday lives and for exploring the presence and geographical locations of breast cancer clusters. GIS has been embraced and promoted by breast cancer activists who see the technology as a way of "grounding" breast cancer research in local neighborhood environments. How has breast cancer activism affected the use of GIS in research and policy analysis, and how, in turn, has GIS affected breast cancer activism? The following case study of women's breast cancer activism shows how geographic analysis and mapping can be simultaneously empowering and disempowering in efforts to improve community health.

Case Study

In the mid-1980s, women in Long Island, NY, a suburban area located east of New York City, became concerned about high rates of breast cancer in their communities. Cancer statistics showed that women in Long Island had among the highest rates of death from breast cancer in the nation (National Cancer Institute, 1987), and women activists wanted to know why. They worried about possible environmental causes: agricultural chemicals that might have seeped into the ground water and contaminated drinking water supplies; electromagnetic fields from overhead power lines and generating facilities; and hazardous chemicals dumped and emitted by the large number of industrial and laboratory facilities scattered throughout the region.

Working with county politicians, local activists asked the state health department to investigate the causes of breast cancer in Long Island. The state prepared an analysis based on data from the state's cancer registry, ordered by zip code. The report's main conclusion was that socioeconomic and demographic factors were to blame, including high income and education levels and Jewish ethnicity (Fagin, 2002a).

Many women were unsatisfied with the state's response. They felt that it blamed them for their illness. Women began to organize and mobilize, forming local breast cancer coalitions to raise awareness and spearhead political action. Some of these coalitions decided to conduct their own investigations to contest the state's assertions about breast cancer. Women wanted to know more. What was the real prevalence of breast cancer in their communities? Were there geographical clusters of breast cancer? Was breast cancer connected to hazards in the environment? Was breast cancer being adequately diagnosed and treated?

A community-based breast cancer GIS

One of these breast cancer coalitions, the West Islip Breast Cancer Coalition (WIBCC), conducted a door-to-door survey of breast cancer prevalence in the town of West Islip (www.1in9.org). Volunteers combed the community, asking women a series of simple questions about breast cancer prevalence and various risk factors: family history, age, residential address, reproductive history, use of oral contraceptives, and length of residence at the current address. Over 9000 women responded, roughly 60 percent of women in the community. This was an extraordinarily high response rate for an unscientific, voluntary survey. WIBCC began preparing "pin maps" to show the extent of breast cancer in the community, but this was a daunting task given the large number of survey responses. They then asked our research group at Hunter College, along with researchers at SUNY Stonybrook, to incorporate the survey data in a GIS for mapping and analysis.

We created a rudimentary breast cancer GIS for the WIBCC. With the tax parcel map as a base, the GIS included the survey data, census data, streets, water mains, and locations of hazardous facilities. Each survey was geocoded by residential address to the corresponding tax parcel. This made it possible to use survey information on length of residence and family history and other risk factors in exploring geographic patterns of breast cancer in the community. At the coalition's request, we created dot maps of breast cancer prevalence based on the survey information (Timander and McLafferty, 1998). Locations of women with breast cancer and without the disease were displayed on the maps. The maps attracted a lot of attention, showing vividly the wide extent of breast cancer prevalence (Fagin, 2002b).

We also encouraged WIBCC to propose "community queries": simple hypotheses about breast cancer that could be explored using GIS. For example, some women thought that breast cancer cases might be clustered at the ends of water mains, where contaminants in the water supply might accumulate. To examine this query, the breast cancer survey data and water mains were overlaid, and the "ends" of the water mains were identified (Gilman, 1997). We calculated breast cancer prevalence among long-term residents who lived at the ends of water mains and among those who lived elsewhere in the community. A comparison of prevalence rates showed that prevalence was not higher at the ends of water mains, and the finding remained when we controlled for known risk factors like family history of breast cancer.

We examined several other community queries using the data and spatial analysis tools available in the GIS. Although none of these analyses found a connection between breast cancer and environmental hazards, members of the WIBCC expressed satisfaction in having community concerns investigated. It was empowering for the coalition to have some control of the breast cancer data and how they were analyzed. This and other community-based efforts created a vision of a breast cancer GIS for the region.

The Long Island Breast Cancer Study Project

Activism often occurs at several geographical scales simultaneously. While these local efforts were occurring, coalitions and activists throughout the region lobbied at the national scale to convince the federal government to do something about breast cancer. A group of activists from Long Island went to Washington, DC, to call for increased federal funding of breast cancer research and more attention to environmental causes. They asked that a GIS be created for Long Island to explore the association between elevated breast cancer risk and a wide range of environmental and social hazards.

With the help of Al d'Amato, the powerful senator from Long Island, the activists' goals were put into legislation. Enacted in 1993, Public Law 103-43 directed the National Cancer Institute to "conduct a case–control study to assess biological markers of environmental and other potential risk factors contributing to the high incidence of breast cancer in Nassau and Suffolk counties in New York" (PL103-43, sec. 1191). The legislation also specified "the use of a geographic system to evaluate the current and potential exposure of individuals" to diverse environmental hazards. Termed the Long Island Breast Cancer Study Project (LIBCSP), the study received approximately $30 million in federal funding for the 1993–2000 fiscal years. PL103-43 was unprecedented in its specificity. It stipulated not only the geographic area of interest (Nassau, Schoharie, and Suffolk counties in New York and Tolland county in Connecticut), but also the methodology (case–control study and "a geographic system").

In 1999, the National Cancer Institute awarded a $5 million contract to a private firm for developing and implementing a GIS to support breast cancer studies in Long Island. The GIS is web-based and aims at providing researchers and community residents with a tool for examining geographic associations between breast cancer and environmental and social conditions. With more than 400 data layers, the GIS brings together information on population sociodemographic characteristics, air and water pollution, hazardous facilities, transportation, health services, land use, and so on. The GIS will mesh these disparate data in an easy-to-use format. Breast cancer data from the state's cancer registry will be available at the zip code level. There is even a process for incorporating information from residents on current or historical land uses and environmental hazards, but it is unclear how the information will be verified and who will ultimately decide which community information to include. A prototype GIS for the LIBCSP currently exists on the internet at www.healthgis-li.com.

Access to the GIS and its data layers and spatial analysis tools will be tightly regulated, because of concerns about the privacy and confidentiality of health data and security-related concerns for environmental and infrastructure data. Although the GIS was intended to give both the public and researchers access to geographic information for Long Island, the design of the system privileges researchers. The public will only be allowed to view data that are "publicly available . . . and not subject to privacy restrictions." Most of the publicly available data, such as breast cancer data by zip code, are already available on the internet. Access to more detailed information requires a password and is limited to researchers. To get the password, researchers must go through an approval process that involves submitting an approved research protocol. Thus, even for researchers the design of the GIS and the approval process appear to discourage exploratory analysis. Instead they require "scientific" studies with well defined hypotheses and research methods. Access to geographic tools for analyzing the information is also restricted. The public will only be allowed to view data and prepare simple maps, not to analyze geographic associations. In contrast, approved researchers will have access to a suite of spatial analysis tools. In summary, although the design of the GIS and the regulations governing access continue to evolve, this $5 million GIS is several steps removed from the grassroots activism that initiated it.

In recent years, the breast cancer movement on Long Island has begun to splinter and fade. Many of the activists moved on, some died, and the movement is losing momentum, although some local coalitions remain strong. A series published in the Long Island paper *Newsday* chronicles the decline of breast cancer activism and the many disappointments of the women involved (Fagin, 2002a). One activist is quoted as saying: "The fight against breast cancer that we started in the early 1990s is dead in the water" (Fagin, 2002a). Activists are frustrated by the long delays: research findings from the federally funded studies of pollution and breast cancer are just beginning to appear in the literature, almost ten years after they were funded, and the GIS, at the time of writing, is still unavailable. Preliminary results from the research studies have been disappointing, showing little or no association between pollutants and cancer. Although some of the studies have important methodological limitations, their widely publicized findings are demoralizing to breast cancer activists. An article in the *New York Times* questioned the whole basis of activism with the headline: "Breast cancer on Long Island: no epidemic despite the clamor for action" (Kolata, 2002). Asserting that breast cancer incidence is not exceptionally high on Long Island, the article portrays the LIBCSP as a huge waste of taxpayer dollars and an example of sincere, but misguided, community activism.

Geographic Information and Women's Activism

The breast cancer case study – a grassroots effort transformed into a multimillion dollar GIS – illustrates the changing and often uneasy relationship between GIS and women's activism for community health. Throughout the 1990s, geographic information and GIS were central to the growth and development of grassroots activism. Under the LIBCSP, however, GIS became a destabilizing force that may have weakened, or at least failed to advance, the breast cancer movement. These changes reveal important *social* dimensions of GIS – how the technology is structured and used, for what and by whom.

Concepts of knowledge and power underpin the evolving definition of GIS and its (dis)connection to the communities that initiated it. Feminists argue that knowledge is situated and contested. There are many types of knowledge, each contingent on the position of the knower in relation to the subject of knowledge (Haraway, 1991). By comparison, traditional scientific knowledge rests on a "myth" of objectivity, a rationalist assumption that the knower, along with her or his beliefs, values, and biases, is detached from the subject (Rose, 1993).

The role of geographic information in the breast cancer example epitomizes these contrasting perspectives on knowledge and knowing. From the very beginning, geographic information was crucial to women's understanding of breast cancer in their communities. Women's knowledge was grounded in their own experiences with breast cancer and the experiences of friends and neighbors. By drawing on published breast cancer data and information from community surveys, women were able to situate personal experiences in a wider community/regional context. Women's understanding of the disease process for breast cancer was also highly grounded. In their GIS efforts and their appeals to the National Cancer Institute, activists viewed breast cancer as intimately tied to the local environment. Not just a medical condition, the disease was wrapped up with environmental hazards, activity patterns, land use, and health care services. GIS-based community queries grew out of local experiences and concerns. Attitudes toward maps and geographic information also reflect a grounded understanding of breast cancer. In the West Islip case, activists recognized intuitively the need for breast cancer information at a fine geographic scale, a scale that would allow detailed mapping and analysis of environmental associations. They collected their own data to fulfill that need, and the maps they created were critically important in mobilizing local residents around breast cancer.

In contrast, the GIS developed by the National Cancer Institute represents a more scientific, "masculinist" view of knowledge: knowledge comes from science and scientists, and from the procedures that make up the scientific method. In the NCI's GIS, only approved researchers will have access to detailed GIS data layers, and a research proposal will be required for approval. Geographic information is to be used for scientific studies and not for community-driven queries. In designing the GIS, NCI expressed concern that untrained users, such as community residents, might jump to the wrong conclusions based on geographic information. Although this is a valid concern, NCI has made little attempt to bridge the gap between scientific understanding and community concerns. GIS can be structured so that community residents can access data for flexible, user-defined geographic areas (Cromley et al., 2004) or submit their own geographic queries for investigation by GIS analysts, as occurred in the West Islip case.

Power and empowerment are also central to the evolving breast cancer GIS. Empowerment is a multidimensional concept that describes the processes by which individuals or groups acquire power. Ellwood (2002) identifies three components of empowerment: *distributive change*, which involves improvements in access to goods and services; *procedural change*, in which marginalized groups gain a voice in decision-making; and *capacity building*, which enhances a group's ability to take action on its own behalf. The multifaceted nature of empowerment is important because it means that there are many ways to gain or lose power. Events can be both empowering and disempowering at the same time by inducing different types of change. Geographic information and GIS have varying effects on empowerment depending on how the technologies are structured and used, where and by whom (Harris et al., 1995; Clark, 1998).

Geographic information and the computerized systems for analyzing that information were both empowering and disempowering for the breast cancer movement, and the balance shifted over time. In the early years, GIS-based maps of breast cancer assisted in capacity-building and distributive change. Many activists commented that the maps immediately attracted attention. They made the breast cancer issue "real" by tying it to a familiar community map and making the wide geographic extent of the disease clearly evident (Carlin, 2003). Public support for the movement grew and new participants were pulled in. At the same time, a regional organization of local breast cancer coalitions began to emerge as pieces of the breast cancer map were patched together. GIS also provided a vision of how to conduct research into the environmental causes of breast cancer. Based on experiences in communities like West Islip, activists began to see GIS as a key tool for linking disease information with data about environmental contamination and exposures. They lobbied successfully for federal funding of GIS-based breast cancer research – a distributive change that poured millions of dollars into breast cancer research on Long Island.

As the nexus of power shifted from grassroots groups to state and federal health authorities – from the local to the national scale – GIS became more a disempowering technology for breast cancer activists. Because of its design, the NCI's GIS essentially separates activists from the issues that concern them. The main problems stem from the rules governing access to data and spatial analytic tools and the lack of a community-oriented interface for investigating local queries. The GIS neither facilitates community access to detailed health and environmental data nor provides a mechanism for responding to community queries. Even the simplest GIS-based community query requires a full-blown research proposal. One activist commented: "People thought they were going to be able to tap into this GIS and discover why they got breast cancer. It's not going to happen" (B. Balaban, quoted in Fagin, 2002b). As a result, the hard-won $5 million GIS is becoming a white elephant that symbolizes wasted local activism.

Despite the changing balance of power, breast cancer advocacy continues in Long Island. Women activists are contesting images of the breast cancer movement depicted in the media and pursuing their own GIS-based investigations. Some breast cancer coalitions have gone back to developing their own community-based GIS and are doing innovative local mapping projects (Carlin, 2003). In addition, similar kinds of GIS are being developed in other regions of high breast cancer incidence like Cape Cod, MA (Paulu et al., 2002). Conducted in collaboration with local experts, these community-driven GIS present a vision of the technology as emancipating and empowering. Such a vision is consistent with Kwan's (2002) argument that GIS presents possibilities for critical activism and "feminist visualization." It also suggests the potential for community-based GIS that focus on local health and social concerns and that involve local residents and grassroots organizations as active participants (Ghose, 2001).

Feminists view power as situated and gendered. How power is negotiated via GIS depends critically on the context in which the technology is developed and implemented. The sociopolitical context of Long Island clearly influenced the role of GIS in breast cancer activism. A suburban area with an affluent, educated population, Long Island provided a very favorable environment for grassroots GIS development. Many activists were well educated and had connections with local university researchers. Although not trained in GIS, the activists had a strong intuitive sense of what mapping and environmental overlays could contribute to breast cancer analysis. In addition, breast cancer coalitions had significant financial and political resources. These were crucial for developing community-based GIS and pushing for federal legislation. Most communities – especially low-income communities that lack political and economic clout – face major barriers to effective GIS implementation for community health (Ghose and Huxhold, 2001).

Changes in the scale of activism and the role of federal health agencies led to a redefinition of GIS and corresponding changes in community activism. As federal involvement increased, GIS was redefined in biomedical terms that involved "detaching" the disease from the communities affected by it. Despite their relatively high levels of education and income, women activists in Long Island lost power in controlling the direction and form of GIS development, and the breast cancer movement lost momentum. Thus, the negotiation of power via GIS was influenced by changes in the scale and context of GIS development.

Conclusion

The Long Island case study reveals the complex and at times contradictory roles of GIS and geographic information in women's efforts to improve community health. Clearly, GIS has the potential to propel activism and to support progressive improvements in community health; but the technology can also be disempowering by separating activists from the issues of concern. These varying constructions of GIS reflect the privileging of certain types of knowledge – in one case a form of feminist, grounded knowledge and in the other scientific, rationalist types of knowledge. Differences in power, participation, and sociopolitical context influence the structuring of GIS. GIS is more than just a technology. It encompasses the people and organizations that use, create, and interact with the technology and whose lives are affected by it. Understanding how GIS affects women's activism for community health and how such activism in turn shapes the construction and use of GIS offers exciting prospects for geographic inquiry.

BIBLIOGRAPHY

Calle, E., Franklin, H., Henly, S., Savitz, D. and Thun, M. (2002) Organochlorines and breast cancer risk. *CA: Cancer Journal for Clinicians*, 52, 301–9.

Carlin, S. (2003) Community breast cancer mapping (http://phoenix.liu.edu/~scarlin/).

Clark, M. (1998) GIS: democracy or delusion? *Environment and Planning A*, 30, 303–16.

Cromley, E., Cromley, R. and Ye, Y. (2004) Online reporting and mapping of spatially-aggregated individual records selected by user. *Cartographica*, 39(2).

Cromley, E. and McLafferty, S. (2002) *GIS and Public Health*. New York: Guilford Press.

Ellwood, S. (2002) GIS in community planning: a multidimensional analysis of empowerment. *Environment and Planning A*, 34, 905–22.

Fagin, D. (2002a) Tattered hopes: a $30 million federal study of breast cancer and pollution on Long Island has disappointed activists and scientists. *Newsday*, July 28.

Fagin, D. (2002b) Still searching: a computer mapping system was supposed to help unearth information about breast cancer and the environment. *Newsday*, July 30.

Ghose, R. (2001) Use of information technology for community empowerment: Transforming geographic information systems into community information systems. *Transactions in GIS*, 5(2), 141–63.

Ghose, R. and Huxhold, W. (2001) Role of local contextual factors in building public participation GIS: the Milwaukee experience. *Cartography and Geographic Information Science*, 28, 195–208.

Gilman, B. (1997) Spatial clustering of breast cancer along dead-end water mains in West Islip, NY. Unpublished MA thesis, Department of Geography, Hunter College.

Haraway, D. (1991) *Simians, Cyborgs and Women: The Reinvention of Nature*. New York: Routledge.

Harris, T., Weiner, D., Warner, T. and Levin, R. (1995) Pursuing social goals through participatory geographic information systems. In J. Pickles (ed.), *Ground Truth: The Social Implications of Geographic Information Systems*. New York: Guilford Press.

Kolata, G. (2002) Breast cancer on Long Island: no epidemic despite the clamor for action. *New York Times*, August 23.

Kreiger, N. (2002) Is breast cancer a disease of affluence, poverty or both? The case of African-American women. *American Journal of Public Health*, 92, 611–13.

Kwan, M. (2002) Feminist visualization: re-envisioning GIS as a method in feminist geographic research. *Annals of the Association of American Geographers*, 92, 645–61.

McLafferty, S. (2002) Mapping women's worlds: knowledge, power and the bounds of GIS. *Gender, Place and Culture*, 9(3), 263–9.

National Cancer Institute (1987) *Atlas of Cancer Mortality among Whites: 1950–80*. Washington, DC: US Department of Health and Human Services, DHHS Publication No. (NIH) 87-2900.

National Cancer Institute (2002) *Genetics of Breast and Ovarian Cancer* (http://www.cancer.gov/cancerinfo/pdq/genetics/breast-and-ovarian#Section_25).

Paulu, C., Aschengrau, A. and Ozonoff, C. (2002) Exploring associations between residential location and breast cancer incidence in a case–control study. *Environmental Health Perspectives*, 110(5), 471–8.

Pavlovskaya, M. (2002) Mapping urban change and changing GIS. *Gender, Place and Culture*, 9(3), 281–9.

Pickles, J. (ed.) (1995) *Ground Truth: The Social Implications of Geographic Information Systems*. New York: Guilford Press.

Rose, G. (1993) *Feminism and Geography: The Limits of Geographical Knowledge*. Minneapolis: University of Minnesota Press.

Sheppard, E., Couclelis, H., Graham, S., Harrington, J. and Onsrud, H. (1999) Geographies of the information society. *International Journal of Geographical Information Science*, 13, 797–823.

Sieber, R. (2000) Conforming (to) the opposition: the social construction of geographical information systems in social movements. *International Journal of Geographical Information Science*, 14, 775–93.

Timander, L. and McLafferty, S. (1998) Breast cancer in West Islip, NY: a spatial clustering analysis with covariates. *Social Science and Medicine*, 46, 1623–35.

Chapter 33

Performing a "Global Sense of Place": Women's Actions for Environmental Justice

Giovanna Di Chiro

The current era of corporate globalization marked by increasing capital mobility, global economic restructuring, and the attendant unrestrained flows of production systems, technologies, people, and pollution across international borders has introduced tough challenges for environmental justice activists worldwide. Recognizing the significance of these broad-scale economic and social transformations, activists have responded by expanding their concepts of community identification and forging new local/global modes of political mobilization. While significant scholarly attention has been paid to the well known ensemble of powerful states, corporate elites, and international trade and financial institutions, accounts of the transnational activities and movements of *community*-based actors and organizations are much less visible.[1] In this chapter, I highlight the complex ways that "local" activists and organizations reconstruct their social identities and environmental politics to confront the changing social, economic, and ecological circumstances engendered by globalization. I examine one instantiation of the post-NAFTA transnational networking practices of women environmental justice activists, centering in particular on the work of Teresa Leal, a Mexican Opata Indian who works with marginalized communities living along the USA – Mexico border in the twin cities of Nogales, Arizona and Nogales, Sonora. On a daily basis, Leal traverses multiple borders – political, geographic, cultural, epistemic – and shapes a transnational advocacy network of women activists devoted to environmental justice on the border.

For many years, Leal's attention has focused on the health and welfare of her family, community, and native land – the biocultural region comprising southern Arizona and northern Sonora, which is bifurcated by the official, national boundary dividing the cities of *Ambos Nogales* (both Nogales). Yet Leal's ecological vision for achieving environmental justice extends beyond her local environment in the Sonoran Desert, encompassing a broader reach and assembling a larger "community." In this chapter, I examine how the movements, actions, strategies, epistemic systems, and passions of women activists working on the border reveal the systemic interconnections between humans and the environments in which they live and on which they depend. By focusing in on the transnational actions of one of these activists, Teresa Leal, I explore how her sense of community and "place" is shaped by a collision at the border of the circuits of capital, its "global assembly line," and the toxic flows of industrial pollution through the environments and bodies of the inhabitants of the Sonoran Desert.

As I will present, this community/environment concept defines a new *ecosystem* – one that includes people and their interrelationships with the social and physical landscapes, and with the circulating air, water, and industrial poisons that travel across international, geographic, watershed, neighborhood, and corporeal borders. I examine how women activists like Leal construct a "global sense of place" (Massey, 1994, pp. 146–56) and a new concept of a healthy environment by articulating scientific knowledge about ecological stability with local knowledge about the impacts of global environmental change in their communities and the bodies of their children and families. This emergent "global sense of place" represents one of the critical innovations in the field of environmental justice, and is *enacted*, or "performed," by a growing international network of women activists from diverse racial/ethnic, class, and national backgrounds. Although not explicitly adopting a feminist identity, this globalizing environmental justice network is grounded in a heterogeneous, grassroots-shaped, feminist environmentalist politics.

Locating Environmental Justice Praxis in the Age of Globalization

The US environmental justice movement emerged in the 1980s with a strong "politics of place" opposing the disproportionate siting of industrial facilities emitting environmental toxins and hazardous wastes in communities of color (Bullard, 1990; Bryant and Mohai, 1992; Gottlieb, 1993; Hofrichter, 1993). The history of the production of these spatial inequalities traces the complex histories and geographies of industrial capitalism, including labor migration and job discrimination, "uneven" urban development, Fordist production systems, housing, zoning and real estate practices, suburbanization, and the exploitation of class and racial divisions (Smith, 1990; Szasz, 1994; Hurley, 1995; Harvey, 1996). Part of this history was the effectiveness of the "not in my backyard" (NIMBY) phenomenon, which enabled the more affluent, politically connected segments of the population to express successfully their opposition to a polluting facility located in their neighborhoods and to insist it be sent elsewhere. According to Robert Bullard, " 'Somewhere Else, USA' often ends up being located in poor, powerless, minority communities" (Bullard, 1992, p. 85). This historical process of the unequal spatial distribution of hazardous exposure and the consequent unequal protection of poor and minority communities within the state regulatory apparatus was identified by activists as "environmental racism" and since the mid-1980s has been a central organizing feature energizing the oppositional politics of the environmental justice movement (Bryant and Mohai, 1992; Bullard, 1994, 1999; Pulido, 1996).

The struggle against environmental racism has foregrounded the fact that some landscapes and the people who live in them are deemed more worthy of attention and protection than others. This nexus of racial/ethnic identity, place, and unequal protection has succeeded in galvanizing political action and in demonstrating that the concerns of mainstream environmentalism have most often been limited to those landscapes of interest to white, middle-class communities. The selective focus on the connections between place and social identity, however, may be politically detrimental to the success of the environmental justice movement. According to some social geographers, most social movements emphasize a "place-bound" identity, which leaves them vulnerable to the "fragmentation which a mobile capitalism and flexible accumulation can feed upon" (Harvey, 1996, p. 302). Limiting the organizational attention to protecting the "local community," critics argue, cannot tackle the superior spatial power of capitalist production, since polluting corporations, taking advantage of the present climate of "free trade" and permeable borders, can easily relocate to places even more marginalized and even less empowered to resist. The threat of relocation in an industry's "race to the bottom" strategy presents a dual dilemma for environmental justice activists – either they are faced with the untenable trade-off of "jobs versus the environment" or they risk securing their own community's protection from environmental harm at the expense of another's (Faber, 1998; Clapp, 2001). In recognition of this quandary, many environmental justice activists reject the self-preservationist NIMBY stance and instead argue for a reconstructed, more *inclusive* "politics of place."

Women's "Place" in Transnational Environmental Justice Movements

Social movement scholars have long observed that, on a global scale, women are consistently the first to become aware of and take action on environmental problems in their communities (Di Chiro, 1992; Seager, 1993; Sturgeon, 1997; Merchant, 1996; Stein, 2004). Most analyses of this sociological phenomenon seek explanations of the observed "woman – environment" relationship and develop critiques of how, or the ways in which, women are "closer" to nature. While there are many theories that expound on varying interpretations of this supposed "closeness," the most salient analysis for the purposes of this chapter examines the ways that women "encounter" nature and the conditions of their environments through the sphere of "women's work" in sustaining the home, family, and community. Feminist analyses of the politics of "separate spheres" have shown that women's *place* in many societies is restricted to the "home," and that women's environmental consciousness springs from the specific roles and responsibilities associated with this spatial assignment (Domosh and Seager, 2001). For example, women in most cultures around the world are responsible for the food security and healthcare of their families and communities. As Mona Domosh and Joni Seager (2001, p. 191) write:

> [Women] are in many cases, responsible for creating the means of subsistence – globally, it is often women's work in the waged workforce and in the family fields that provides the basics for family survival. And in all economies, women take raw materials and income, whether provided by themselves or by others, and fashion it into food, clothing, housing, health care and child care. In environmental management terms, we could say that women make the primary consumer and resource-use decisions for their families and their communities.

Watching their children and neighbors becoming ill from exposure to contaminated air, water, or food from a local toxin-emitting facility, and experiencing the threats to their economic security through the destruction of their local resource base by clear-cut logging, pesticide-dependent export agriculture, or polluting export-processing factories, has mobilized women all over the world to become environmental activists to safeguard their homes, families and communities. In so doing, they have built, step by step, one community struggle at a time, the local, national, and transnational networks that comprise the global environmental justice movement. Paradoxically, as feminists scholars have noted, women activists need to *leave* their homes and local communities in order to protect them (Krauss, 1998). Resisting the social, cultural, and economic structures of power that "put women in place," women environmental justice activists defy gender, race, and class stereotypes, leave their "places," and infiltrate the male-dominated spheres of environmental science, law, and politics to defend the environmental security of their communities.

Recognizing the importance of this "place-based" commitment, however, does not preclude their adopting a *trans*local standpoint in the quest for an inclusive environmental *justice* that ensures that the externalities of industrial "progress" do not end up in the "backyards" of marginalized communities in the USA, or anywhere else in the world (Schlosberg, 1999; Cole and Foster, 2001; McDonald, 2002). The knowledge that toxic pollutants, when released into the environment, circulate throughout the ecosystem, and the awareness that the age of corporate globalization has greatly facilitated the mobility of industrial capitalism's hazardous production and disposal practices, have mobilized many women (and men) activists to develop new translocal, transnational, and "global" political formations that acknowledge the interconnectedness of ecological, political, and economic systems (e.g. Yearley, 1996; Keck and Sikkink, 1998; Peña, 2001; Di Chiro, 2003).

For environmental justice activists like Teresa Leal, this has meant "coming out of our little trenches" and building alliances across cultures, landscapes, and borders of all kinds. The development of such cross-border environmental activist identities from the margins is consistent with what sociologist John Brown Childs has called "transcommunality," the process of political identity formation that "maintain[s] particularistic rooted affiliations, while creating broad constellations of inclusive cooperation that draw from multitudes of distinctly rooted perspectives" (Childs, 1998, p. 145). Transcommunal cooperation is a method that entails face-to-face contact and mutual trust that is built up through "shared practical action" in which people from different "emplacements of affiliation" can work together around common future goals. According to Leal and other women activists, the extent to which environmental justice activists are able to create these sorts of intercultural, transcommunal identities and alliances will foreshadow the success or failure of the environmental justice movement and, indeed, the future of the planet and its inhabitants.

A geography of environmental inequality produced by the communities that are suffering the injustices of economic globalization inspires a reconstructed sense of place, in fact a *global* sense of place that recognizes the *different* ecological predicaments of marginalized peoples and landscapes around the globe while embracing an intercultural understanding of *common* struggles and shared futures. Grassroots activists in the environmental justice movement, most of whom are women of color, offer a set of theories and practices for an ecologically grounded politics of place that reshapes the universalizing tendencies of "spaceship earth" discourse, an oft-quoted global environmental metaphor that tends to hide from view the widely divergent social and environmental conditions of the people who inhabit different "berths" on the spaceship (Chatterjee and Finger, 1994; Conca and Dabelko, 1998). In the following narrative of transborder environmental justice organizing, I argue that women activists produce a politics of place that articulates the "militant particularism" of environmental/geographic specificity, together with a hope-filled cosmopolitanism embodied in the quest for environmental justice and "earth democracy" (Ichiyo, 1993; Shiva, 2003). To bring to light this local/global environmental justice methodology, I sketch an ethnographic account of a day-long "toxic tour" hosted by Teresa Leal that I attended during the summer of 2001.[2]

"Toxic Touring" on the Border

On a blisteringly hot summer day in June 2001, I joined a group of women environmentalists on a road trip across the USA–Mexico border to see the social and ecological impacts of the manufacturing and trade policies spawned by the 1994 authorization of the North American Free Trade Agreement (NAFTA).[3] With Teresa Leal as our "tour guide," our group left the sleepy, border town of Nogales, Arizona, and entered the bustling, crowded streets of Nogales, Sonora, driving further south into the upper reaches of the Santa Cruz River Basin, a regional watershed that extends from Nogales, Sonora, to Phoenix, Arizona.

Crossing the international border in our car went smoothly enough, although the 14-foot-high, two-mile-long steel walls flanking the city belied this perception of a permeable boundary.[4] At one time, Ambos Nogales enjoyed a relaxed border ambience, in which Mexican and Indian "day-crossers" casually passed through the flimsy chain-link fence to take a job, to visit family, or to do some shopping. In the early 1990s, after the INS instituted a border crack-down in California and Texas, immigrants flooded to Ambos Nogales, which soon became a people-smuggling hot spot crawling with the ubiquitous green and white Ford Explorers favored by *la migra* (Davidson, 2000). The recent deployment of high-tech surveillance systems and military maneuvers enhancing the policing of the 2000-mile-long USA–Mexico border has transformed Ambos Nogales into a "low-intensity conflict" zone and one of the more reinforced entry points along the international frontier (Dunn, 1996).

Border crossings have been regular features in Teresa Leal's life. When Teresa was a young girl, her mother crossed the international border from Nogales, Sonora, into the United States to insulate her daughter from the turmoil of Mexico's revolutionary land reform movements of the 1940s and to shield her from the radical influences of her Trotskyite father. This was to no avail, however. Her father regularly stole across the frontier to see his daughter, bringing stories of his battles for workers' and peasant farmers' rights, which would later inspire the political commitments shaping Teresa's life work, so much so that her mother lamented that her radical leanings were "in her genes." This afternoon, as our car wound its way through Nogales's traffic in a futile attempt to avoid crevices and potholes, Teresa recounted another memorable border crossing: her voyage to Rio de Janeiro in 1992 to attend the United Nations Conference on Environment and Development

(UNCED). Traveling on the Panamerica Highway in a car packed with fellow activists from the street theater group Centro Libre de Experimentación Teatral Artistica (CLETA), Teresa crossed 12 international borders in two weeks to participate in the Earth Summit's NGO Forum to "let people know what is happening at the US–Mexican border."[5] In any case, our most recent crossing was only one in a long history of Teresa's movements to and fro, to gather with family and friends, to earn her livelihood, or to forge political alliances for social and environmental justice.

Upon crossing the border into one of the hubs of Mexico's "free trade zone," we motored alongside the dry streambed of the Nogales Wash, one of the primary tributaries that empties into the north-flowing Santa Cruz River.[6] Perennially dry since the mid-1950s due to years of groundwater overdraft accompanying the development boom on both sides of the border, the Nogales Wash has become a convenient dumpsite for garbage, sewage, and all manner of industrial waste (Udall Center, 1993; Ingram et al., 1995). Teresa commenced her narrative history of environmental destruction in the Sonoran Desert ecosystem by chronicling the unusually high incidences of cancer, neurological disease, miscarriage, and birth defects faced by the low-income Latino and Native American communities living in the Santa Cruz River Basin. As we would learn, the protection of the ephemeral and "sacred" water resources of this desert ecosystem swirls at the center of the struggle for environmental justice along the border.

Our party drove along the international highway through the center of town on our way to Nogales's industrial park district, which houses 100 industrial facilities, or *maquiladoras*, foreign-owned "export-processing" plants erected on the Mexican side of the border, originating in the mid-1960s as an economic development strategy to encourage foreign investment and create jobs to bolster Mexico's flagging economy (Kearny and Knopp, 1995; Kopinak, 1996). With their minimal tax and tariff liabilities, loose environmental restrictions, and the open-door atmosphere afforded by NAFTA, the maquiladoras play a pivotal role in Nogales's economic "health," even though, according to many environmental, public health, and labor activists, they also contribute to the deteriorating health of the workers, local residents, and surrounding desert environment.[7]

Improving the health and livelihoods of Nogales's residents, however, may not mean taking to the streets to shut down the factories. "We know they'll just go elsewhere and do the same things to other people," says Teresa, "and then the people here will be left without jobs."[8] Supporting a strategy that brings the community and the corporations to the table as equal partners, Teresa argues for a middle ground that may require crossing political and ideological "borders" to take seriously the region's need for both economic development and environmental security. She explains, "there are many ways to get things done, and this might mean building non-traditional partnerships and thinking outside our own little sectarian corridors."[9]

Local Knowledge, "Good Gossip," and Political Transformation

En route to our first "tour" stop, the Samson/Samsonite maquiladora, our plan was to meet up with members of *Comadres*, a binational grassroots organization of women seeking to improve the working and living conditions of women factory workers and their families. Teresa established Comadres in the early 1970s after being approached by women workers who knew of her community work and had listened to her weekly radio program, "Hablando de Mujeres," which provided information about women's and children's health, domestic violence, and the importance of protecting water quality in the *colonias*, the squatter settlements encircling the city where many of the maquiladora workers live.

One Samson/Samsonite luggage seamstress, Panchita, had sought Teresa's help after she contracted mysterious skin rashes that resulted in bleeding sores on her legs and arms. Fearing the materials she was working with were harmful, Panchita asked Teresa to translate the instructions label stamped on a bolt of the textile she used every day sewing the liners of Samsonite suitcases. Reading that the material was made of fiberglass, Teresa informed Panchita that "sewing through that material created microscopic little shards of glass that fly through the air and into her clothes and body. That's why she had all these rashes" (quoted in Adamson, 2002, p. 48). Teresa recounted that the doctor, rather than investigating the serious skin problem, admonished Panchita, saying, "You don't clean your house; you live in the dump and that's why you have mange. Because your hygiene is so bad" (ibid., p. 49). Although she was poor and lived in humble conditions in the colonias, Panchita knew she and her children were not dirty, and so she began to attend meetings of Comadres and talked with other women with similar stories.

Comadres aims to empower women to stand up for their rights for a clean and healthy environment and to exchange valuable information about preventing chemical toxicity in the workplace, purifying the water used in their homes, accessing public health and environmental services, and developing income-generating strategies, such as weaving and sewing co-ops, to supplement their poverty-level wages. Through their associations with the Comadres network, the women workers share their experiences of environmental illnesses and blend this knowledge with the technical expertise of medical doctors and environmental scientists. In this way, they produce a hybrid environmental expertise that emerges from their everyday lives and their encounters with professional knowledge.[10] This aggregate environmental health perspective – the product of anecdotal *and* scientific knowledge – comprises "women's talk," argues Teresa, but it is much more than trivial "gossip."

> As Latina women [Comadres have] always been . . . put down by the machos; [we] are considered *mitoteras*. Mitoteras comes from *mitote*, which means "gossip." Mitoteras are gossipy. But we are changing the meaning of "gossip" from negative gossip to positive gossip. We put out the good mitote, or good gossip. We get out the news about what's going on in the villages, or colonias, and in the factories. We share information. (Adamson, 2002, p. 47)

As they learn about the widespread environmental impacts of the maquiladora operations and become aware of the epidemiological data documenting a dizzying array of health problems suffered by the people living adjacent to the polluting facilities, the Comadres are transformed into ecopolitical agents who shift their attention from individual self-help efforts to organizing strategies that address wider systemic problems. These "macro-Comadres," Teresa explains, are [those] that help not only individuals, but also share the results of their efforts with their community. They're scavengers; they find materials that can be used to help build shelters for workers who have just come to the community. They scavenge food, clothes. The people who are seeking aid see what the Comadres are doing and go to them for help. . . . They're the ones challenging the system, challenging the government, and trying to stop the railroad tanker cars filled with toxic materials that roll through our communities. They're yelling about the fact that there's no water, no electricity, no police, no safeguards. (Quoted in Adamson, 2002, pp. 47–8)

Clean, potable water is a limited resource in the colonias and it becomes essential to maintain the purity of the water that is delivered weekly by *pipas*, private vendors who procure clean water from distant wells and, for a small fee, transport it by truck to families living in the colonias (Ingram et al., 1995, pp. 76–8).[11] Upon discovery that the 55-gallon drums women workers were scavenging from the maquiladoras or the local dump to store water had previously contained hazardous chemicals, the Comadres organized training sessions to teach people who had no choice but to use the barrels how to line them prior to a water delivery, not with concrete, tar, or lead-based paint, but with heavy polyethylene plastic.[12]

Evidence of the success of the Comadres' popular education efforts to protect the water supply dots the hilly landscape of Nogales's poorest barrios: virtually every dwelling – small houses constructed of discarded plastic sheets, wooden shipping pallets, and the occasional cinder-block wall – is equipped with one or two water barrels stationed next to the road. Teresa describes how the Comadres designed and distributed through word-of-mouth networking a water-labeling system to differentiate between the barrels filled with salvaged gray water, which could be used for gardening or washing the house, and barrels that are reserved exclusively for drinking, bathing, and cooking. "It's a simple system, but it works," she says. "The women cover the top of the 55-gallon drums that are meant for clean water with metal or wooden boards," a sign to the pipas that only those should be replenished with the precious commodity.[13] "So, you see the beauty of word-of-mouth strategies, of mitote, of gossip," Teresa exclaims. "Once it starts going, you can't stop it. That's how Comadres uses the power of 'gossip'" (quoted in Adamson, 2002, p. 49).

Organizing from the ground up, women's organizations like Comadres become what Naples (1998) calls "grassroots warriors," effecting small-scale transformations that empower maquiladora workers to understand their environmental and labor rights and to see how these entitlements guaranteed by Mexican law have been even further curtailed under NAFTA's neo-liberal policies. One woman maquiladora worker commented: "I used to think that the factory was more important than I was, because that's how they make you feel. . . . [Now], I realize that I have rights and that I have to make them worth something at work, at home, and in the streets" (quoted in Light, 1999, p. 2). As a member of Comadres, Teresa joins with women workers to recruit and educate other maquiladora workers about the everyday, personal impacts of global economic restructuring, to organize practical campaigns to protect the health of their families, and to take the small steps necessary to build support for a women's labor union (Kamel and Hoffman, 1999; Louie, 2001; Naples and Desai, 2002).

Sacred Waters and Toxic Effluent

Arriving at the entryway to the Samsonite factory, we waited in the car as Teresa inquired about a meeting with Panchita and other Comadres who worked at the plant. A lovely, peach-tinted stucco façade decorated the external walls of the Samsonite maquila, which was elegantly landscaped with fountains, ornamental palms, and sweet-smelling flowers. The loading zone at the rear of the factory painted a very different portrait of the maquila's operations: a small, attached shed labeled *Químicos Peligrosos* (dangerous chemicals) stores the powerful glues and solvents used to assemble the pricey suitcases. The women who work all day on the assembly line, hovering over vats of glue, are known as *las loquitas* (the crazy ones), a double meaning referring both to their willingness to take the job and to the glue's deleterious effect on their brain cells.

The security guard informed us that no workers were allowed on breaks that day due to an unscheduled "inventory." Irritated at the dubiousness of this explanation, we drove instead to the home of María Luisa, another comadre who had sought Teresa's advice in dealing with the Nogales Health Department after the death of her two young daughters – the four-year-old from pancreatic cancer and the 16-year-old from leukemia. The family, suspecting that the rebar used to fortify the house's foundation had been salvaged from a shipment tainted with depleted uranium, asked Teresa to help them demand health department officials to inspect the house for radiation. Finding no "unusual levels of radiation," the health inspectors offered no explanation of the cause of the children's cancers. Although María Luisa's three surviving children remain healthy, they also grew up in the same house and so she continues to worry about their fates.

María Luisa invited us into her home, a small stucco bungalow perched alongside the Nogales Wash. We sat in her tidy front room and listened to a painful story of the illnesses and deaths of the two girls, and the public health professionals' medical exegesis of their untimely demise: "It was probably a genetic defect transmitted through the females."[14] María Luisa, still grieving six years later, resigned herself to "the will of God," and held tightly to Teresa's hands. We shared our condolences and exited our host's front door to an unobstructed view of the Nogales Wash, the desiccated channel that serves as an industrial dumpsite, an open sewer, and the local water source for the city of Nogales, Sonora. Teresa sighed, obviously frustrated at the explanatory pre-eminence of either genetic reductionist theories of disease or religious ones: "The family has been drinking this polluted water for years, but you have to start where people are, you can't impose your ideas on them and tell them they're crazy. In this overpumped, overused desert wash, people are more obsessed with the quantity of water available to them, and this can sometimes make them overlook the quality."[15] Despite the unassailable physical evidence of a potential health hazard provided by the proximity of the polluted streambed to people's homes and to the city's wells, Teresa said, "the health department is unwilling to accept a causal relationship."[16]

Leaving María Luisa's house and the crowded urban center, we drove past the Nogales Wash and the make-shift shanties people had erected along its banks, and proceeded 18 kilometers west to the mountainous terrain that is home to the new Nogales Sonora Sanitary Landfill. We parked along a hillside road just north of the landfill, grateful for the peace and quiet and the panoramic view of the Sierra Madre mountains, but this bucolic scene would be short-lived. Barreling up the steep road appeared one fully loaded dump-truck after another, conveying undifferentiated domestic refuse and hazardous waste from the city to this out-of-the-way sacred grove, now an out-of-sight disposal site. After years of binational grassroots pressure succeeded in shutting down the former landfill, a toxic stew notorious for bouts of spontaneous combustion and located a few kilometers south of the border, city officials selected this distant locale to solve their waste disposal problems (Ingram et al., 1995; Davidson, 2000). To escape the clouds of dust and grit kicked up by the constant trail of trucks, as well as the stench of garbage, our group set out into the sage brush and scrub oak, heading down a precipitous canyon, in search of the Venero San Antonio an artesian spring flowing in a southerly direction toward the Mambutu River.

As a young mother, Teresa brought her children to the San Antonio Spring, telling them stories passed on by her grandparents of her indigenous ancestors who had lived in this magnificent territory long before the Spanish conquistadores claimed the land. The miracle of a steady source of underground fresh water in the Sonoran Desert always awed Teresa, and she hoped to transmit this respect for the natural environment to her offspring. In recent years, however, rather than embarking on leisurely excursions to these beloved mountains, she finds herself struggling hard to protect their precious watercourses. Discovering that her ancestral canyonlands would become the dwelling place for the new "sanitary" landfill, Teresa embarked upon a water testing program to ascertain the pre-landfill quality of the surface and underground water. After the installment of the landfill in 1995, Teresa worked with scientists and community activists from the Grupo Ecologista Independiente to determine if the hazardous chemicals and liquid wastes from the dump were seeping into the groundwater and contaminating the spring. Teresa regularly hitchhiked to the site and trekked the ten-mile route tracing the spring from its headwaters near the landfill to its confluence with the Mambutu River, collecting water samples along the way. Through her itinerant water sampling methods, she hopes to build a water quality record that can provide early warnings and evidence of contaminants escaping the landfill in violation of Mexican environmental law and polluting the Mambutu, a river often touted as pristine and, therefore, poised as a candidate to replace the dwindling and contaminated water resources of the Santa Cruz.

Teresa and the *ecologistas* with whom she collaborates have sent samples to the Sonoran Water Laboratory for general water quality assessments and have learned that "the pre-landfill samples showed that the springwater was very pure." "But," she continues, "after the landfill went in, the samples have come back with descriptions like 'it's not potable' or there's residue of oils, gasoline, or industrial acetone."[17] When Teresa presented this evidence to the Sonoran ministry for environmental quality, she was told "but the Mambutu is a south-flowing river, so it will not impact Nogales at all."[18] In spite of the knowledge of this geomorphology, immense above-ground water pipes pump water from the Mambutu to Nogales's water distribution system to supplement the subsiding water table around the border, belying this simplistic assessment of the region's hydrological processes. Ignoring the ecosystem comes at a risk, Teresa argues: "They think the landfill is so far away and won't affect Nogales, Sonora." But, through human disregard and manipulation of the desert's subtle yet inexorable water cycles, the industrial wastes from the maquiladoras are returned to the city, circulate through the environment into the rivers and washes, and eventually find their way into the bodies of the people. "And, we know the Santa Cruz River flows from south to north, so the chemicals go over the border into Arizona."[19]

Cross-border Actions

Returning to Nogales, Sonora, we once again crossed the Nogales Wash as we made our way to the site of the old landfill, a massive scar in the desert landscape in striking distance of the international border. In the early 1980s, Teresa had targeted the old landfill as a site of action because of the local waste management authority's negligent regulatory practices and the increasing health and environmental dangers associated with the dump. Knowing that the dump provided a dangerous livelihood as landfill scavengers for mothers and children who lived in Bella Vista, the nearby colonia, Teresa set up Mi Nueva Casa (My New House), a drop-in center only yards from the international fence that offered food, drug treatment, and literacy classes.

Teresa's forays to the dump also entailed a water monitoring agenda. She and other activists had long suspected that high levels of rare cancers – multiple myeloma, pancreatic cancer, childhood leukemia – and autoimmune diseases such as lupus were caused by the seepage of poisons from the landfill into the groundwater and eventually into the city wells. Meanwhile, the landfill was growing beyond its fenced borders, and semi-trucks brimming with noxious cargo were filling up the lagoons, gullies, and washes surrounding the engorged site, which happened to lie a few hundred meters from an elementary school. Working together with the schoolchildren, parents, and teachers, Teresa organized a grassroots campaign appealing to Mexico's child safety laws that prohibit the siting of a potentially harmful entity within 2500 meters from a school, daycare center, church, etc. Aware that this law was originally intended to protect children from exposure to bars, adult bookstores, liquor stores, or pharmacies, Teresa asserted that going to school next to a hazardous landfill was an even greater danger to the welfare of children.

Generating awareness and knowledge about the environmental and human costs of the landfill, a mixed, binational community of street children, students, parents, teachers, environmentalists, and workers mobilized forces in what Teresa called "urban anarchy at its finest":

> We got masks and buckets and we brought water from the filled-up lagoon and dumped in on the steps of the mayor's office – we washed his steps with this water – and it was just awful. The fathers got pieces of old, junk cars and scrap materials at the dump and, at night, they would drag those pieces and put them across the road where the dump trucks would come in the morning. They would stand on top of the scraps during the day, in very macho style, just challenging the drivers to come and get them out of the way. The truck drivers would say, "well, we couldn't deliver the garbage because of these people." So, it started escalating. It was a very citizen movement and, finally, we got [the landfill] closed down. Two weeks later the mayor had to hire dump trucks from Hermosillo to remove the landfill; it took them two weeks day and night![20]

We drove alongside the decommissioned landfill's rusty fence, the solitary barrier separating it from the city limits, and stared out at the sacrificial scene before us: a vast, dirt-capped landmass littered with a field of methane off-gassing vents. Perched just outside the fence sat the Torreon water tank, an old well, Teresa bemoaned, that was still being used by several barrios in the area. The shallow – 50 feet or less – wells sunk along the Nogales Wash, she explained, have long been contaminated, through seepage from the landfill and dumping from the factories, with high levels of TCE (trichloroethylene), PCE (tetrachloroethylene), trichloroethane, chloroform, and various toxic metals such as lead, mercury, chromium, and copper (Berry and Simms, 1994; Davidson, 2000).

In the years following the heterogeneous grassroots mobilizations that succeeded in shutting down the landfill, Teresa had joined forces with a binational community-based organization called LIFE (Living Is For Everyone). LIFE was founded in the early 1990s by residents of Carrillo Street, a pleasant Latino subdivision in Nogales, Arizona, that had been built on the grounds of Camp Little, a US Army garrison established in 1910 to defend the town from possible invasions spilling over from the Mexican Revolution. Located one mile north of the border and a few blocks from the Nogales Wash, the Carrillo Street community had for several years shown signs of an emergent disease cluster of uncommon illnesses, including multiple myeloma, pancreatic cancer, anencephaly, and lupus. To collect evidence in support of this hypothesis, a group of neighbors led by Susan Ramirez, Anna Acuña, and Jimmy Teyechea launched their own "popular epidemiological" health studies of their neighborhoods.[21] They had each been collecting informal data on the rising numbers of illnesses in their community and "talked about what could be causing the diseases. Could it be, as Susan believed, the water? Or, was it, as Anna believed, the air?" (quoted in Davidson, 2000, p. 58). Gathering at Acuña's home, a group of neighbors produced a map using colored pins signifying particular diseases (red for blood cancers, black for lupus, orange for ovarian cancer), which graphically revealed that most of the victims' homes were clustered together on streets straddling the wash. The LIFE group contested the conventional argument that their illnesses were caused by either "genes or lifestyle or God or fate," (ibid., p. 59) and became increasingly convinced that their members and families had become sick due to polluted air drifting northward from the constant plastic and refuse fires at the Nogales, Sonora, landfill, and poisoned groundwater from the landfill, the maquiladoras, and the dilapidated Nogales International Wastewater Treatment Plant (NIWTP).

Committed to pursuing their epidemiological research, and encouraged by the support of two University of Arizona researchers, Doctors Joel Meister and Larry Clark, the LIFE organizers recruited Teresa Leal to help to generate a more comprehensive health survey along the Ambos Nogales border. Teresa recalls the origin of this binational health study:

> [LIFE] thought the natural conclusion was that if there were high levels of cancer, neurological and respiratory diseases on the US side, there should be an equal amount, if not worse, on the Mexican side. And they were running into stone walls everywhere in the medical community, so they invited me to shadow [their study] ... so we could do coverage on both sides.[22]

Explaining that her interest in conducting community-based epidemiological research was "in the name of searching for better quality information," Teresa elaborated on the virtue of blending the often discordant epistemic systems of the scientific method with local, experiential observation:

> It's our human nature to be opinioned and we have our views, we have faith in our instincts, which makes us suspicious, or inquisitive, and so we need to do more than just feel thoughtful . . . we also have to be open to the possibility that there are broader parameters than what we in our own mind have determined. I think that's the virtue of science, that it can take an objective view. And, sometimes that objective view corrects things that have just been handed down as half-truths. I think that science does have the power to center us, and that's not a bad idea. But, it's not always so powerful, that some other information or view cannot share. So, I think we have to be humble about the fact that we don't know everything and that we get more powerful the day that we admit that we don't know everything. It makes us think more clearly. In the name of searching for better quality information, we come up with synergisms that are much better, much more powerful, and it's nobody's possession, nobody's property.[23]

Publication of the results of the binational community-sponsored health studies in Ambos Nogales, and the post-NAFTA scrutiny by the international environmental justice community of the worsening social and environmental conditions in the free trade zones, brought some long-awaited attention to the region: *USA Today* did a story titled "Nightmare on the border," the news program *A Current Affair* aired a show about the rising death toll on the border, and, in 1993, Arizona Governor Fife Symington and Senator John McCain visited the homes of LIFE members and allocated funding for a $100,000 health study in Nogales, Arizona (Davidson, 2000, p. 70). Both US and Mexican governments began to pay attention to the environmental and health problems escalating along the border by tracking hazardous waste shipments from the maquiladoras, licensing waste haulers, and providing limited funding to support the factories to institute pollution prevention systems in their manufacturing and waste management practices (Clifford and Sheridan, 1997). Although these are signs of improvement, Teresa argues, "globalization is just making things worse, and there are no provisions in NAFTA for environmental remedial clean-up along the border."[24]

Clean Water, Contingent Alliances

Anxious for some more agreeable scenery, our tour party left the dump and recrossed the international border into Nogales, Arizona, on our way to Teresa's home, a small apartment just blocks from the line, whose multiple uses include a home base for children and grandchildren, a busy office, and a safehouse for battered women.[25] Entering her sunlit apartment, its walls painted a dazzling yellow, we witnessed Teresa move into action. Her answering machine was crammed with a remarkable array of messages, among them an interview request from NPR, an environmental justice colleague wanting to firm up their plans to attend the United Nations Conference on Racism, Xenophobia, and Related Forms of Discrimination in Durban, South Africa, and a message from a member of the Sierra Club's Grand Canyon Chapter, updating Teresa on the status of their joint lawsuit against the

International Boundary Water Commission (IBWC), The City of Nogales, Arizona, and the US Environmental Protection Agency (EPA).

Binational strategies proved successful in pushing the Mexican government to shut down the overfull, uncontrolled landfill in Nogales, Sonora, and in raising awareness and generating scientific and government attention to the environmental justice issues at the border. Teresa decided to get involved in yet another cross-border action when she learned of the decades-long negligence of various binational regulatory authorities charged with protecting the water quality of Ambos Nogales. Apparently shirking their duties to monitor and upgrade the Nogales International Wastewater Treatment Plant (NIWTP), the IBWC, the EPA, and the City of Nogales, Arizona, were listed as defendants in the suit charging violations of the Clean Water Act of 1972.[26] When the District Court decided that the EPA has "absolute discretion to ignore Clean Water Act violations at the Nogales Plant,"[27] the plaintiffs (Teresa Leal and the Sierra Club), with representation from the Arizona Center for Law in the Public Interest, challenged this decision in the US Court of Appeals, arguing that, in accordance with the legislative history of the Clean Water Act, the EPA has "a *non*-discretionary mandate to act upon consistent violations at the International Wastewater Treatment Plant in order to avoid contaminating the river and to efficiently clean the residual waters that come into the waste water treatment plant as best they can."[28] In late 2001, the defendants signed a consent decree in which "the IBWC and the city of Nogales agreed that our complaints are real and that between now and 2005 they will work on them. We have the power to intervene, and if we feel it's not being done right, we have the power to seek explanation."[29]

Allying with the Sierra Club, an organization that "in working to preserve natural landscapes, often ignores or denies that some people need to live off the land, and depend for their survival on natural resources . . . a blind spot [that] does not set us up to be 'good neighbors'" (quoted in Adamson, 2002, p. 56), was a decision Teresa mulled over at length. "But," she argues, "in a world increasingly affected by globalism, we cannot afford to work against each other. [In] this lawsuit . . . the Sierra Club and I are working together for a common goal. . . . We're coming from different perspectives and yet we are working towards a common goal" (ibid.), and "if that's what it takes to move these agencies to do a better job, if the decay of our river basin is out of control, then we have to do all that is in our power to help."[30]

Transnational organizing has become a staple of the environmental justice movement, and Teresa's activist history is no exception. As the co-chair of the Southwest Network for Environmental and Economic Justice, (SNEEJ), a network of 85 organizations focusing on environmental health, workplace safety, sustainable jobs, and cultural survival, Teresa works with her colleagues to build alliances with organizations and indigenous tribal councils that span many national borders. As an organization supporting leadership development in low-income communities of color, SNEEJ organizes training workshops that address environment, health, and work in a global framework. Teresa explains:

> We are working on what we call a "just transition," which is not about "free trade," but "just trade." We try to prepare workers for globalization so they will not be victims of globalization. The real problem with globalization is that it threatens people's cultures and identity, and an identity is necessary for people to consider that they have

something to fight for . . . and care for. Our workshops teach people about . . . sustainable economic development in their communities. (Quoted in Adamson, 2002, p. 51)

After traveling to and participating in the WTO demonstrations in Seattle, Teresa was convinced that SNEEJ members needed to "come out of our little trenches" and "join forces with other environmentalist groups, with the turtle people, the whale people, with the monks from Tibet, with the Raging Grannies" (ibid., p. 53). For Teresa, eroding the borders that have separated environmental groups from women's groups from human rights groups and from labor organizations generates a grassroots-based globalization "from the ground up" and broadens the capacity to fight the destructive forces of globalism (Brecher et al., 2000). She argues that the long history of sectarian divisions, whether in the form of international boundaries, environmental and medical expertise, or grassroots organizing based on identity politics, corrodes the potential for cross-border community alliances. Furthermore, she contends:

> The air is for all of us; the water is for all of us. Shit and pollution, toxic substances, do not ask for permission to come into your house; they do not need a passport to cross the border. . . . We can't say, oh that person has cancer because they're poor. No. Cancer hits everybody . . . toxins can find their way into everyone's house, whether they're rich or poor. . . . There is plenty of research to suggest that POPs [persistent organic pollutants] are released into the environment by industrial processes, and by the spraying of pesticides, etc. These toxins are flowing through intercontinental airways. Do you know what that means? They're spreading all over the planet! Why don't we stop it? Because corporate profiteering is paramount; corporate heads don't seem to be able to live without exorbitant profits. So different groups need to come together to fight the corporations. No one group can do it alone. (Quoted in Adamson, 2002, pp. 54–5)

Beyond Geographic and Epistemological Border Zones: Seeking "Coalescence"

In recent years, Teresa joined with other activists to form the Cuenca (watershed) Network, a coalition of different groups interested in the preservation of the Santa Cruz River Basin. She explains:

> My concern is the holistic approach to saving the river, both for ample water supply and quality of water. On the Mexican side of the river, the users like the *ejidatarios* (collective farmers and ranchers), need to be participating; they need to be included and they're not being included as we speak. We could get a biological corridor strategy going that transcends the border, that takes in cultural concerns and makes a constant effort to be bilingual and bicultural. We have to stop creating limits to what we can do.[31]

The Cuenca Network has brought her into contact with the Friends of the Santa Cruz River, the Sustainable Border Group, and Amig@Naturales, as well as university researchers in Arizona and Sonora to discuss protocol design for an inventory of "the natural wildlife of the river to create a holistic plan . . . so that we can better understand what the river is about" (quoted in Adamson, 2002, p. 56). Two graduate students working at the Biosphere II station outside of Tucson helped to develop a website on the diverse cross-border "coalescences" to protect the health of the Santa Cruz River and its human and non-human desert inhabitants.[32]

Cuenca networking creates a global sense of place and relies upon many forms of border crossing by its members in order to understand and articulate the many different affinities and antagonisms that flow throughout the Santa Cruz Watershed and surface in the Ambos Nogales border region. For Teresa, the border is a "meeting place" of a tangled network of uneven social and economic relations, environmental circumstances, and ecosystemic limitations, all of which are constructed on a far larger scale than the place itself. In this sense, Teresa's ecological imaginary retains the idea of specificity and uniqueness of place without appealing to a static, nativist formulation of "community," because it "allows a sense of place which is *extroverted*, which includes a consciousness of its links with the wider world, which integrates in a positive way the global and the local" (Massey, 1994, p. 155, italics added).

Teresa's most recent campaign (together with the binational women's organization Las Sinfronteras), to buy a small house on the Sonoran side of the border that would serve as both an environmental education center and a safehouse for women maquiladora workers who are exposed to violence on the job and in the home, illustrates this articulation of the global and the local. "I remember [at the Earth Summit] in Rio, the incipient message was that women's rights were environmental rights," Teresa recalls. "Now, we're getting ready for Rio-plus-10 in Johannesburg in September [2002]. OK, let's see how we fared."[33] The terrible environmental conditions at the border and the miserable living and working conditions of the maquiladoras' predominantly female labor force cannot be disarticulated, she argues, and neither is faring very well. The violence against young women in the global assembly line, whose livelihoods and survival needs are tied to the factories, parallels the wholesale poisoning of the water and air of the Sonoran Desert by transnational corporations supplying low-priced products for upscale markets around the globe. The struggle for environmental justice on the border necessitates paying attention to these interlocking layers of injustice, and devising locally rooted and globally informed strategies to confront them.

Like all dwellers in arid landscapes, Teresa and her comadres express a heightened awareness of the sacredness of water, the lifeblood of daily existence. It would not be overstating the issue to say that for those people living in the Sonoran Desert, survival depends on the safekeeping of water. Protecting the local watershed, however, requires a larger-than-local water politics that spans borders of all kinds: national, racial, gendered, economic, linguistic, ecological, technological, spiritual, and epistemic. The "popular" knowledge that grounds struggles for environmental justice on the border is more than local, more than anecdotal, more than just personal experience. It is the outcome of shared observation, careful research, and the forging of synchretic assemblages of "experts" of all stripes. As Teresa argues, our partial and limited worldviews cannot hope to address environmental problems of such magnitude; it is the "coalescence of people with differing views that is really inspiring."[34]

While inspired by the *uniqueness* and particular circumstances of the peoples and landscapes that comprise the desert ecosystem in which they live, women activists like Teresa Leal enact an ecological feminist consciousness that constructs a grassroots-driven cosmopolitanism challenging the rapacious reach of twenty-first-century corporate globalization. Committed to the social, economic, and ecological integrity of *everybody's* backyard, the diverse communities nestled along the dry streambeds of Ambos Nogales cross many borders to "coalesce" in a spirit of environmental justice for all.

NOTES

1   There has been some recent attention to the transnational political stakes of community groups who join "transnational advocacy networks" to bring attention to the local impacts of global processes and to gain leverage on the international stage. For example, see Keck and Sikkink (1998) and Naples and Desai (2002).

2   This article incorporates work previously published in an earlier work, titled " 'Living is for everyone': border crossings for community, environment and health," *Osiris*, 19 (2004).

3   The North American Free Trade Agreement is a trade treaty signed by Mexico, Canada, and the USA that eliminates tariffs and reduces non-tariff barriers on the circulation of goods, raw materials, and services among the three countries. Launched in 1994, it created the largest "free trade zone" in the world (encompassing approximately 405 million people) and was designed to foster increased trade and investment among the partners. Although the agreement stipulates that trading partners must respect local labor and environmental regulations, analysts have shown that corporations object to these regulations as excessive "non-tariff" barriers to free trade. See Deere and Esty (2002).

4   I was struck by how the border means nothing and everything to local inhabitants. Crossing it is a daily, unremarkable happening for many people, like Leal, a resident alien, who have family and business on both sides of the border. It is also a constant presence in their lives, however, and especially for those undocumented Mexicans, Indians, and Central American immigrants who do not benefit from easy access and mobility. The border for many is arbitrary and some argue that the "border" is everywhere in the USA, in the many barrios that house the immigrants who work in factories, homes, farms, and gardens throughout the country. The official border is a militarized zone marked by high security, sophisticated surveillance technologies, and menacing young border patrol agents driving around in SUVs. It is a staged exhibition of disciplining and resistance.

5   Author's phone interview with Teresa Leal, February 12, 2002.

6   Once a perennially flowing river lined with majestic cottonwood trees, and now a parched "cut in the earth," the Santa Cruz River has sustained decades of uncontrolled pumping of its surface and groundwaters and years of unregulated dumping of toxic chemicals, copper mine tailings, and untreated sewage, which have contaminated its already depleted water resources. For a more detailed study of the environmental justice activism in the Santa Cruz River Basin, see Di Chiro (2003). For an exhaustive history of the Santa Cruz River, see Logan (2002).

7   For discussions of the health problems along the border resulting from the maquiladoras, see Berry and Simms (1994).

8   Author's interview with Teresa Leal, Nogales, Sonora, June 11, 2001.

9   Ibid.

10 For a similar analysis focusing on white working-class women in the United States, see Krauss (1998, pp. 129–50).

11 The price of delivered water ranges from 4.5 to 5.0 pesos ($1.32–1.47) per 200 liters of water.
12 Author's interview, Teresa Leal, Nogales, Arizona, June 13, 2001.
13 Author's interview with Teresa Leal, Nogales, Sonora, June 12, 2001.
14 Author's interview with María Luisa Garcia, Nogales, Sonora, June 12, 2001.
15 Author's interview with Teresa Leal, Nogales, Sonora, June 12, 2001.
16 Ibid. For an extensive study of the status of groundwater resources in Arizona and New Mexico see *Sacred Waters: The Life-Blood of Mother Earth*, ed. The Southwest Network for Environmental and Economic Justice and the Campaign for Responsible Technology (Albuquerque, New Mexico, 1997).
17 Phone interview, Teresa Leal, February 12, 2002.
18 Ibid.
19 Ibid.
20 Ibid.
21 Popular epidemiology refers to community-based research into the etiology of environmental illnesses that is conducted by members of the contaminated community itself. See Brown (1992).
22 Author's interview, Teresa Leal, Nogales, Arizona, June 13, 2001.
23 Ibid.
24 Author's interview, Teresa Leal, Nogales, Arizona, June 13, 2001.
25 Presently, due to lack of funding, Teresa has been forced to relinquish this apartment, which, she laments, has left the women who need shelter "houseless." Author's phone interview with Teresa Leal, June 16, 2003.
26 The lawsuit was heard in two federal courts: The United States District Court, *Sierra Club and Teresa Leal* v. *Robert Ortega, Commissioner, IBWC et al.* (CIV 00-184-TUC-RCC), and the United States Court of Appeals for the Ninth Circuit, *Sierra Club and Teresa Leal* v. *Christine T. Whitman, EPA, Robert Ortega, IBWC, and Marco A. Lopez, Mayor of Nogales, Arizona* (00-16895).
27 Letter to Teresa Leal from Vera Kornylak, Staff Attorney for Arizona Center for Law in the Public Interest, June 6, 2001.
28 Author's interview, Teresa Leal, Nogales, Arizona, June 13, 2001.
29 Author's phone interview, Teresa Leal, February 12, 2002.
30 Author's interview, Teresa Leal, Nogales, Arizona, June 13, 2001. Leal reports that at present, "the IBWC, the EPA and Nogales, Arizona continue to defer on next steps while the river and ecology continue to erode." Author's phone interview with Teresa Leal, June 16, 2003.
31 Author's interview, Teresa Leal, Nogales, Arizona, June 13, 2001.
32 See http://www.geocities.com/woborders/2nogales1
33 Author's phone interview, Teresa Leal, February 12, 2002.
34 Ibid.

BIBLIOGRAPHY

Adamson, J. (2002) Throwing rocks at the sun: an interview with Teresa Leal. In J. Adamson, M. M. Evans and R. Stein (eds), *The Environmental Justice Reader: Politics, Poetics, and Pedagogy*. Tucson: University of Arizona Press.

Berry, T. and Simms, B. (1994) *The Challenge of Cross-border Environmentalism*. Albuquerque, NM: Resource Center.

Brecher, J., Costello, T. and Smith, B. (2000) *Globalization From Below: The Power of Solidarity*. Cambridge, MA: South End Press.

Brown, P. (1992) Popular epidemiology and toxic waste contamination: lay and professional ways of knowing. *Journal of Health and Social Behavior*, 33(3), 267–81.

Bryant, B. and Mohai, P. (eds) (1992) *Race and the Incidence of Environmental Hazards*. Boulder, CO: Westview Press.

Bullard, R. (1990) *Dumping in Dixie: Race, Class and Environmental Quality*. Boulder, CO: Westview Press.

Bullard, R. (1992) Environmental blackmail in minority communities. In P. Mohai and B. Bryant (eds), *Race and the Incidence of Environmental Hazards: A Time for Discourse*. Boulder, CO: Westview Press.

Bullard, R. (ed.) (1994) *Unequal Protection*. San Francisco: Sierra Club Books.

Bullard, R. (1999) Leveling the playing field through environmental justice. *Vermont Law Review*, 23, 453–78.

Chatterjee, P. and Finger, M. (1994) *The Earth Brokers: Power, Politics, and World Development*. London: Routledge.

Childs, J. B. (1998) Transcommunality: from the politics of conversion to the ethics of respect in the context of cultural diversity – learning from Native American philosophies with a focus on the Haudenosaunee. *Social Justice*, 25(4), 160–72.

Clapp, J. (2001) *Toxic Exports: The Transfer of Hazardous Wastes from Rich to Poor Countries*. Ithaca, NY: Cornell University Press.

Clifford, F. and Sheridan, M. B. (1997) Borderline efforts on pollution. *Los Angeles Times*, June 30.

Cole, L. and Foster, S. (2001) *From the Ground Up: Environmental Racism and the Rise of the Environmental Justice Movement*. New York: New York University Press.

Conca, K. and Dabelko, G. D. (eds) (1998) *Green Planet Blues: Environmental Politics from Stockholm to Rio*, 2nd edn. Boulder, CO: Westview Press.

Davidson, M. (2000) *Lives on the Line: Dispatches from the US–Mexico Border*. Tucson: University of Arizona Press.

Deere, C. L. and Esty, D. C. (eds) (2002) *Greening the Americas: NAFTA's Lessons for Hemispheric Trade*. Cambridge, MA: MIT Press.

Di Chiro, G. (1992) Defining environmental justice: women's voices and grassroots politics. *Socialist Review*, 22(4), 93–130.

Di Chiro, G. (2003) Beyond ecoliberal "common futures": toxic touring, environmental justice and a transcommunal politics of place. In D. Moore, J. Kosek and A. Pandian (eds), *Race, Nature, and the Politics of Difference*. Durham, NC: Duke University Press.

Di Chiro, G. (2003) Steps to an ecology of justice: women's environmental networks across the Santa Cruz River watershed. In V. Scharff (ed.), *Seeing Nature Through Gender*. Lawrence: University of Kansas Press.

Domosh, M. and Seager, J. (2001) *Putting Women in Place: Feminist Geographers Make Sense of the World*. New York: Guilford Press.

Dunn, T. (1996) *The Militarization of the US–Mexico Border, 1978–1992: Low Intensity Conflict Doctrine Comes Home*. Austin: University of Texas Press.

Faber, D. (ed.) (1998) *The Struggle for Ecological Democracy: Environmental Justice Movements in the United States*. New York: Guilford Press.

Gottlieb, R. (1993) *Forcing the Spring*. Washington, DC: Island Press.

Harvey, D. (1996) *Justice, Nature and the Geography of Difference*. Cambridge, MA: Blackwell.

Hofrichter, R. (ed.) (1993) *Toxic Struggles: The Theory and Practice of Environmental Justice*. Philadelphia: New Society Publishers.

Hurley, A. (1995) *Environmental Inequalities: Class, Race, and Industrial Pollution*. Chapel Hill: University of North Carolina Press.

Ichiyo, M. (1993) For an alliance of hope. In J. Brecher et al. (eds), *Global Visions: Beyond the New World Order*. Boston: South End Press.

Ingram, H., Laney, N. and Gillilan, D. (1995) *Divided Waters: Bridging the US–Mexico Border.* Tucson: University of Arizona Press.

Kamel, R. and Hoffman, A. (eds) (1999) *The Maquiladora Reader: Cross-border Organizing since NAFTA.* Philadelphia: Mexico-US Border Program, American Friends Service Committee.

Kearny, M. and Knopp, A. (1995) *Border Cuates: A History of US–Mexican Twin Cities.* Austin, TX: Eakin Press.

Keck, M. and Sikkink, K. (1998) *Activists Beyond Borders: Advocacy Networks in International Politics.* Ithaca, NY: Cornell University Press.

Kopinak, K. (1996) *Desert Capitalism: Maquiladoras in North America's Western Industrial Corridor.* Tucson: University of Arizona Press.

Krauss, C. (1998) Challenging power: toxic waste protests and the politicization of white, working class women. In N. Naples (ed.), *Community Activism and Feminist Politics: Organizing across Race, Class, and Gender.* New York: Routledge.

Light, J. (1999) The long, slow road to organizing women maquiladora workers. *CorpWatch*, June 26, 1–4.

Logan, M. (2002) *The Lessening Stream: An Environmental History of the Santa Cruz River.* Tucson: University of Arizona Press.

Louie, M. C. Y. (2001) *Sweatshop Warriors: Immigrant Women Workers Take on the Global Factory.* Cambridge, MA: South End Press.

McDonald, David (ed.) (2002) *Environmental Justice in South Africa.* Athens: Ohio University Press.

Massey, D. (1994) *Space, Place, and Gender.* Minneapolis: University of Minnesota Press.

Merchant, C. (1996) *Earthcare: Women and the Environment.* New York: Routledge.

Naples, N. (ed.) (1998) *Grassroots Warriors: Activist Mothering, Community Work, and the War on Poverty.* New York: Routledge.

Naples, N. and Desai, M. (eds) (2002) *Women's Activism and Globalization: Linking Local Struggles and Transnational Politics.* New York: Routledge.

Peña, D. (2001) Globalization of the environment and local communities of resistance. In J. Ageyman *et al.* (eds), *Just Sustainabilities: Development in an Unequal World.* Cambridge, MA: MIT Press.

Pulido, L. (1996) *Environmentalism and Economic Justice: Two Chicano Struggles in the Southwest.* Tucson: University of Arizona Press.

Schlosberg, D. (1999) *Environmental Justice and the New Pluralism: The Challenge of Difference for Environmentalism.* Oxford: Oxford University Press.

Seager, J. (1993) *Earth Follies: Coming to Feminist Terms with the Global Environmental Crisis.* New York: Routledge.

Shiva, V. (2003) Earth democracy. *Tikkun*, 18(1), 43.

Smith, N. (1990) *Uneven Development: Nature, Capital and the Production of Space.* Oxford: Blackwell.

Stein, R. (ed.) (2004) *New Perspectives on Environmental Justice: Gender, Sexuality and Activism.* New Brunswick, NJ: Rutgers University Press.

Sturgeon, N. (1997) *Ecofeminist Natures: Race, Gender, Feminist Theory, and Political Action.* New York: Routledge.

Szasz, A. (1994) *Ecopopulism: Toxic Waste and the Movement for Environmental Justice.* Minneapolis: University of Minnesota Press.

Taylor, D. (1997) Women of color, environmental justice, and ecofeminism. In K. J. Warren (ed.), *Ecofeminism: Women, Culture, Nature.* Bloomington: Indiana University Press.

Udall Center (1993) *Water in Nogales: Survey of Use, Issues and Concerns.* Report Prepared by the Udall Center for Studies in Public Policy, University of Arizona, Tucson.

Yearley, S. (1996) *Sociology, Environmentalism, Globalisation.* London: Sage.

Part VI State/Nation

Chapter 34

Feminist Political Geographies

Eleonore Kofman

Today political geography is a vibrant subdiscipline making important contributions both to understanding contemporary affairs and to the development of geography as a whole. . . . However, the subdiscipline has still yet to meet the challenge of feminist geography whose concerns for power in place and space from a gender perspective have only appeared intermittently in contemporary political geography. (Taylor, 2000, pp. 596–7)

In the quote above Peter Taylor raises the lack of impact that gender issues and feminist geographies have had on political geography. It is a question that is increasingly being addressed by mainstream (Toal and Shelley, 2002) and feminist geographers (Staeheli, 2001; Staeheli et al., 2004). Political geography appears not to have been able to shake off its conservative tradition, one that can be traced back to the heritage of the "founding fathers" of political geography such as Mackinder and Ratzel at the end of the nineteenth century. They wrote the grand narratives about the strategic struggles conducted by dominant states over a world space which at the end of the nineteenth century had been entirely carved up. While women were mostly excluded from the discipline and its institutional structures, there existed dissident figures, such as the anarchist geographers Peter Kropotkin (Breitbart, 1981) and Elisée Reclus (Dunbar, 1981), whose geographic understandings could provide another lineage for a political geography engaged with emancipation, power, and forms of domination operating at different levels and including racism and sexism.

Responding to these divergent genealogies of political geography during the past two decades, feminist political geography has challenged prevailing conceptions of and offered alternatives to the political, its sites, spaces, and practices.[1] Politics is about how people exercise power through both material and discursive practices (Peake, 1999) within specific places and across all scales of analysis (Kodras, 1999). This extends beyond the inclusion of women and gender issues and embraces a rethinking of what political geographies could be about, the different relationships between the political and power (Brown and Staeheli, 2003), and the different forms of political activities in which people are engaged.

However, there has been limited recognition of feminist political geographers by both feminists and male political geographers. It should be said that many feminists actually do political geography but do not identify with political geography. Thus the number of feminist political geographers who see themselves as political geographers (though not necessarily exclusively), participate in its institutional structures, and/or publish in the journals associated with the subdiscipline remains small. Lynn Staeheli (2001) has analyzed some of the reasons for the gap between the two constituencies and the lack of intellectual traffic between them. One important factor shaping this outcome is that the key journals in the subdiscipline are not seen to be welcoming. There have been few articles in *Political Geography* on feminism, gender, sexual and race relations, or identity politics (Kodras, 1999), although feminist political geographers have published articles on other topics, such as citizenship and national identity. Responses to a survey conducted by Staeheli (2001) also mentioned the feeling that political geography dealt with the big issues, eschewing the community and the everyday, both theoretically and methodologically, and preferring the realm of elites, policy documents, and more recently the media.

Despite the limited participation of feminist geographers in political geography, which has largely remained masculinized in its theoretical approaches and concerns, feminists have contributed to revitalizing and producing radical political geographers. I show this in relation to traditional topics (state, nationalism, and more recently citizenship), as well as themes analyzed at scales above (globalization, geopolitics) and below (everyday life) the state. Recent interest in the social construction of scale has been one of the contributory factors in raising a greater awareness of the explicitly political nature of much feminist geography.

Feminist Contributions to Radical Political Geographies

The journal *Antipode*, though often left out of histories of political geography (Taylor, 2002), contained the earliest analyses of women and geography prior to the development of feminist organizations in the discipline (see Hayford, 1974; Helms, 1974). Hayford traced the historical shift from political power exercised through the household to the public sphere, as well as women's diminished role, as a result of the development of capitalism. Helms critiqued the treatment of older women in an economic and social system which marginalized them and denied them social rights. Building on this early work, "feminist geography" became increasingly visible toward the end of the 1970s, roughly at the same time as the renewal of political geography. Nevertheless, when the subdiscipline became institutionalized through the creation of study groups and the journal *Political Geography*, there were few women or feminists involved.

Broadly speaking, during the 1980s feminist geographers forged few links with feminist political studies. The first areas to be taken up by feminists were the local state (Fincher, 1987), urban politics (Bondi and Peake, 1988), and electoral geography (Drake and Horton, 1983; Peake, 1984). This minimal engagement with political geography stems from the fact that feminism in geography largely made an impact through social geography, and it was in sociology that feminist perspectives progressed most rapidly within the social sciences. By the end of the 1980s, however, the important influence of feminist international relations scholarship (see Enloe, 1989) inspired political theorizing by a growing number of feminist geographers in the 1990s and beyond (see Kofman and Youngs, 2003).

In 1990, *Political Geography Quarterly* published a special issue on "gender and political geography," with Eleonore Kofman and Linda Peake as guest editors. The special issue was the first to draw out more comprehensively the ways that gender is implicated in fundamental issues explored by political geography. At the time, however, there were few feminist political geographic studies to illustrate the themes raised in that volume. Thus, the special issue sought to chart a gendered agenda for political geography by sketching out how feminists in related disciplines of politics and international relations had challenged the conception of politics and questioned dichotomies based on the private/public and the formal/informal.

Contributors to that volume opened up the notion of politics, its forms, and the sites of its practices from a geographic and feminist perspective. For example, they analyzed the traditional topic of the state in its patriarchal and sociosexual forms (Pateman, 1989; Connell, 1990) and drew attention to the relevance of welfare and reproduction to the maintenance of state power, as opposed to a singular focus on sovereignty and borders. As importantly, the special issue charted important trajectories of feminist geography into the realms of international politics, conflicts, and participation in national liberation movements at a time when a gendered historiography of geography (Domosh, 1991a, b) and the role that gender, race, and sexuality played in imperialism, especially of the British Empire (Blunt and Rose, 1994; McClintock, 1995), had not yet been written.

As interest in gendering political geography grew in the early to mid-1990s, questions about the state virtually withered away as a subject of analysis. Nevertheless, related concerns about national identity and citizenship, which continued to be located within a state framework, came increasingly to the fore. During the early 1990s, critical interrogations of scale (Taylor, 1982; Smith, 1993) drew attention to levels other than the nation-state, and feminist political geographers subsequently turned in particular to the local and everyday (Marston, 2000). However, with the exception of feminist development geographers, feminist geographers have tended to eschew the global.

Given the importance of state and global levels, the remainder of this chapter examines these issues in more depth. The section immediately following this one traces current and future work on the state, nation, and citizenship in feminist geography, followed by a section devoted to scales above and below the state: the global, on the one hand, and everyday life and spaces, on the other. These emerging approaches in feminist geography serve as antidotes to the emphasis on elites and formal political institutions.

State, Nation, and Citizenship

In response to criticisms levelled against the state-centrism of political geography, poststructuralist and postmodern theorists argued against the need for theorizing it. It was suggested that the state could be more appropriately conceptualized as a set of arenas and collections of practices (Pringle and Watson, 1992). Yet there was some insightful engagement with the state as a gendered institution and constellation of practices. Connell (1995, p. 73) argued that the state is a masculine institution in that its organizational practices are structured in relation to the reproductive arena. Its different gendered regimes (Connell, 1990) help us to understand sociosexual arrangements and practices in a wide range of arenas and in different types of states. Despite these important insights, however, it is important to think about the state transnationally. Western feminists have paid little attention to the experience of Third World women under postcolonial states and the strategies available to them in their struggles and negotiations with the state (Rai, 1996, p. 5).

Proclamations of the transcendence of the state are particularly prevalent today within studies of globalization and transnationalism. Taking their cue from a simple reading of technological time–space compression and economic activities, these perspectives emphasize the decline in traditional aspects of state sovereignty. The declared demise of the state does not take into account the complexity and unevenness of a wide range of economic, social, and political processes that articulate with state institutions at various scales. They also ignore the concrete gendered impact of state policies and practices. Thus, while critics demonstrated the shortcomings of state-centrism, taking the state down from its pedestal should not lead to relegating it to a secondary role not worthy of sustained analysis. The state, though neither monolithic nor consistent, remains the most organized institution in society (Knuttila and Kubik, 2000). It is a site of coordination and the playing out of diverse and often opposing strategies, with which feminists interact in different ways and with varying outcomes (Chappell, 2000).

Feminist political geographers are beginning to return to considerations of gendering the state, a trend illustrated by a session of that title at the 2001 Annual Meeting of the Association of American Geographers (with a selection of papers published in Desbiens et al., 2004). A more considered analysis of the state in globalization is reviving interest in the role of the state in "managing, shaping, regulating and supporting" complex, overlapping, and often contradictory circuits of capital and people (Mitchell et al., 2003). At the same time the state has divested itself of many of its responsibilities for social reproduction and in doing so reconfigured welfare provision.

Though not explicitly theorizing the state, many feminist scholars (Curthoys, 1993) and geographers have nevertheless maintained indirect concern with it and its policy outcomes through studies of citizenship in different societies (Smith, 1989; Staeheli and Cope, 1994; Fenster, 1997; McEwan, 2000; Kofman, 2003a; Nelson, 2004). Marshall (1950), the best known theorist of post-war citizenship, conceived of it as a status bestowed on those who were full members of a community, which served to reduce inequality. For him, citizenship evolved and expanded in relation to three types of rights – civil, political, and social. These cover both the formal aspect of rights and obligations which members of a community enjoy and the substantive rights they exercise, i.e. it comprises status and practice (Lister, 1997). It expresses the power relations and structures between groups and the outcomes of processes determining who does, and does not, belong to the community, which is itself not fixed.

Feminist scholars have drawn attention to the fact that citizenship was constituted on the basis of presumed independence of the individual embodying male norms and attributes, especially the primacy of formal work and politics. The engendering of citizenship has primarily drawn attention to inequalities of access to resources across private and public spheres (Walby, 1994; Lister, 1997). Often, however, citizenship studies in geography and related disciplines have a highly Western-centric undertone. Western feminism assumes the existence of welfare and has tended to emphasize social citizenship (McEwan, 2000) as well as political participation.

Through studies of women's political activism, geographers have advanced our understanding of the complexity and fluidity of the relationship between private and public spaces and the different ways such spaces of citizenship can be constructed (Staeheli, 1996; Fincher and Panelli, 2001). They caution us to think carefully about the meaning of "private" and "public." Transgressive acts and spaces can be extremely effective tools in pursuit of political aims. What are normally considered to be private acts (breast feeding or kiss-ins) can be performed in public spaces (Staeheli, 1996). Identities associated with the private and familial sphere may be deployed in public, e.g. women in their role as mothers protesting against state policies. Activists can deliberately seek to blur the public/private divide. Private spaces can be used to develop and orient campaigns in the public sphere, while public spaces can be the location for campaigns to change family lives (Fincher and Panelli, 2001). Activism itself may be about the demand for the right to public visibility, as with migrant women (Yeoh and Huang, 1998). For all these studies by feminist geographers and others, citizenship is an important way to frame processes of inclusion and exclusion in the public sphere.

Civil rights, for Marshall the first set of rights to be won, have probably received less attention from feminists, but they actually encompass a series of rights that are particularly pertinent. These include freedom of speech and association, as well as the ability of an individual to conclude contracts and dispose of his or her own body. Issues of marriage, divorce, abortion, fertility, rape, and domestic violence, to name just a few significant ones, fall into this category; they are closely connected with discussions about embodying geography (see below). As contracts, such as those of marriage and employment (Pateman, 1989) became codified in the nineteenth century, women's rights, especially those of married women, often regressed during the apogee of private patriarchy (Walby, 1990). These legal structures subsequently became a focal point for struggles for gender equality. Today gays and lesbians have increasingly contested their exclusion from contracts hitherto reserved for heterosexual couples, such as adoption, marriage, and pensions. It should be noted that there are debates among lesbians and gays (Bell, 1995) about the desirability of a rights-based citizenship which may impose conformity to dominant heterosexist norms and practices, e.g. marriage.

In the context of transnational networks, new regional arrangements and discourses of international human rights, there have been continuing debates about the different scales at which citizenship claims and rights are negotiated. In particular, some scholars (Turner, 1993; Soysal, 1994) have questioned the suitability of the nation-state to accommodate citizenship rights, especially in multicultural societies. Yet despite an increasingly multilevelled governance, states retain considerable power to shape the institutions and discourses of citizenship (Kofman, 2003b). Processes of citizenship include and exclude differently positioned subjects and articulate gender with other dimensions of identity and inequality, such as "race," ethnicity, religion, sexuality (Bell, 1995; Valentine, 1996), age, and disability (Chouinard, 2001). Citizenship therefore involves a sense of belonging to or exclusion from a collectivity, whether it be at the local, national, or international scale.

For migrants, formal citizenship remains significant and serves to differentiate national and foreign populations. Immigration and settlement policies constantly redraw boundaries delineating those who belong and the rights they may exercise. Indeed, states have reasserted their right to political and cultural sovereignty in relation to the movement of people (Hyndman, 2001; Kofman, 2003b) and must therefore be seen as a key element in facilitating and constraining people's mobility, which varies according to class, nationality, gender, race, and immigrant status (see Silvey in this volume). Though seemingly neutral, the categories used to classify different types of immigration (labor, highly skilled and less skilled, family, refugee) are gendered in their assumptions and outcome. Skilled labor is conceptualized as male, family as female and dependent upon males. Immigration policies also demonstrate the interplay of the state with economic and social actors across different scales.

Kim England (2003) notes that in retrospect what had started out as a project on care of children in the home, albeit by female migrant domestic workers, could become an example of how the state is deeply implicated in the construction of racialized and sexualized identities and in regulating and shaping the supply of labor from the Third World. It thus clearly shows the interaction of labor and chains of care across different scales and the fluidity between the private and the public (Hochschild, 2000). Those employed in reproducing the labor of wealthier states, especially within the domestic sphere, whether it be in Singapore (see Yeoh in this volume), the Middle East, North America, or Europe, enjoy at best a partial citizenship (Parenas, 2001), and one that limits their own rights to family life (see Pratt in this volume). Increasingly, feminist geographers are beginning to make the connections between the movement of embodied workers, state policies, and citizenship. For Chang and Ling (2000), the claim in the technomuscular discourse of globalization that the state is being hollowed out needs to be balanced by the renewed vigor of the state in shaping the labor intimacy regime reliant on the global circulation of female domestic labor. Both labor contracts and racialized and sexualized bodies are regulated by the complicit patriarchal state and employers.

Another important theme related to the state is that of nations and nationalism. Feminist geographers have argued that gender, sexuality, and nation are socially constructed (Jackson and Penrose, 1993) and reveal the intimate relationships expressed through the nation (Radcliffe and Westwood, 1996; Nast and Pile, 1998; Mayer, 2000), often configured as a family (Kofman, 1993; McClintock, 1993), and where each has his or her place. The (national) male is portrayed as the heroic defender of the nation and protector of women against external violation (Sharp, 2000) and internally from foreign men, as in the ideology of the National Front in France (Kofman, 1993). Women, as the biological and cultural reproducers of the national collectivity (Yuval-Davis and Anthias, 1989), are not equal in the nation but play a symbolic role (Sharp, 2000). In violent confrontations and war, rape of women (and to a lesser extent men) may be used to emasculate the men of the enemy nation through dishonoring and penetrating their women, the upholders of its traditional virtues, as happened in Bosnia-Herzogovina in the 1990s (Mayer, 2004). Nationalist sentiment is expressed both through the gendered iconography of monuments (Johnson, 1997) and landscapes (Nash, 1993) and through the mundane practices of everyday life that unthinkingly reproduce distinct national identities (Sharp, 2002). It is thus embedded and embodied in institutions, everyday life, symbolic practices, and representations.

As several chapters in this volume (by Domosh, Hannah, and Hyndman) highlight, epochs of insecurity reinforce the power of the state as defender and protector of the nation as haven. More attention is being paid to the nationalist rhetoric of dominant states and the formation of coherent national identities around the articulation of external threat and evil, especially in the definition of American national identity (Sharp, 2002). This Cold War discourse has been revived in the post September 11 period to embrace an internal enemy and stiffle dissent (see chapters in this volume by Hannah and Hyndman). Too often, and especially during periods of heightened and strong nationalist sentiments, hegemonic, heterosexual, and militarized masculinities (Connell, 1995) are propelled to the fore.

Given the fascination with the body and embodiment in geography and the social sciences (Moss and Dyck, 2002), it is not surprising that a political geography of the body is emerging around the political connections between the body, power, and spaces. Influenced by Foucault's writings, the body can be seen as a political field, the object and target of power and manipulation (Teather, 1999, p. 10). Thus bodies are the site of oppression and resistance of both the individual and the collectivity.[2] They are being used strategically to assert power over, control, and weaken the adversary in different spaces, as in the case of rape, whether it be in the home or camps in times of conflict. In relation to domestic labor this may extend to controlling the sexuality and imposing pregnancy tests on female domestics, as in Singapore and Taiwan. Mayer (2004) illustrates the use of rape in the Bosnian conflict and the way it shifted scale between the home, the local, and the national. Often those with least power are deploying their bodies as a form of resistance and to draw attention to their cause: for example, Palestinians who have destroyed their bodies in the face of Israeli occupation; prisoners on hunger strike; or refugees, as in Australia and the UK, who have mutilated themselves in protest against inhumane treatment. Some states (Australia, Canada, the USA and the UK) now recognize gender persecution as grounds for asylum; many of these claims center on treatment and comportment of the body, e.g. genital mutilation, rape, appropriate behavior and dress in public spaces (Crawley, 2001).

Extending Feminist Political Geography to Other Scales

While feminist geographers continue to deepen their inquiry on the state, citizenship, nation, and national identity, outlined above, they are also extending the scope of feminist political geography at other scales. Here I touch upon three key themes relevant to recent work in feminist political geography: globalization and the relationship of feminist geographers to the South; a feminist rereading of geopolitics; and theorizations of scale.

The view of the world as a single interconnected entity can be traced back in political geography to the work of Peter Taylor (1985) on the world system which began to emerge in the fifteenth century with the explorations and colonizations of powerful European states. Since then markets have penetrated all areas of the globe, transforming traditional societies. In the past few decades the pace of change has quickened and interdependence intensified such that globalization has become since the 1990s a major conceptual tool in understanding the contemporary world and its time–space compression (Harvey, 1989). While its leading protagonists are male elites (Hooper, 2000) who fill the managerial and technological corridors driving the so-called knowledge economy, their needs (cleaning, restaurants, sex) are increasingly supplied by female labor from the periphery, an issue examined by Samarasinghe in this volume. As Sassen (2000) rightly argues, though evicted from the narratives of globalization, their labor is essential and increasingly constitutes a strategy of survival and a counter-geography of globalization.

While globalization may appear to offer a paradigm that opens horizons onto the world outside the North American and European axes, in reality the interpretation of economic, sociocultural, and political change emanates from the North. Feminist geography of the North usually stays close to home, reflecting a lack of interest in the construction of knowledge and inequalities of the South. In some ways this parallels the tendency of many postcolonial studies in geography to trace diffusion from the West and the effect of colonialism and imperialism on metropolitan centers (Clayton, 2002). As he comments, "the critique of colonialism can become a seductive but sanitized Western intellectual pastime . . . that barely connects with the practical predicaments of formerly colonized people and places." We know much about women's exclusion from the colonial and exploratory enterprise and their attempts to overcome it, but uncover little of the "historical geographies of the colonized world" (Yeoh, 2001).

Feminist geographers, particularly those working in development geography, provide an exception to postcolonial theorizing in geography. Many of these scholars trace the material impacts of neocolonial processes on women's lives, labor, and politics and in the process have collaborated with colleagues (not necessarily academic) in the South (Peake and Trotz, 1999; Townsend et al., 1999; Laurie and Calla, 2004). They are producing political geographies that link different places analytically with studies based on situated knowledges (Katz, 2000) and anchored in landscapes, inhabited by individuals and groups engaged in local (Nagar, 2000; Nelson, 2003) and translocal politics.

A second dynamic area within feminist political geography is feminist critiques and engagement with critical geopolitics. During the early 1990s a promising avenue of global thinking had seemed to be in the offing with critical geopolitics, which sought to problematize "epistemological assumptions and ontological commitments of conventional geopolitics, to question state-centric premises and deconstruct the power relationships underpinning the geopolitical practices of states, though almost always the dominant ones" (O'Tuathail, 1996). Among both its practitioners (Dodds, 2001) and feminist geographers (Kofman, 1996; Sharp, 2000; Hyndman, 2001), critical geopolitics has been submitted to increasing critique.

On the whole (but see Dalby, 1994; Sparke, 1996) critical geopolitics is strong on deconstructing dominant conceptualizations but has contributed little to investing its political landscapes with gendered figures (Kofman, 1996; Hyndman, 2001) or reconstructing alternatives (Hyndman in this volume). Textual critiques of critical geopolitics privilege the words and texts of the elites (Sharp, 2002). The disem-bodied and privileged geopolitical world-views of traditional geopolitics have not been replaced by studies which demonstrate how geopolitics works in everyday life (Thrift, 2000), nor does it explore how the global interacts with the local (Secor, 2001). Nor have ethnographic studies demonstrated how, for instance, foreign poli-cies emerge through the interaction of government departments, gender divisions of labor and particular textual practices. Critical geography foregrounds the practices of elites, politicians, and to a lesser extent popular culture (Sharp, 2000), but with a few exceptions fails to engage with oppositional movements in what Routledge (1996, 2003) calls an anti-geopolitics. This approach links up most closely with what a feminist geopolitics might look like.

For Routledge (1996), the study of social movements contributes another way of approaching critical geopolitics through turning attention away from the machina-tions of the state. He explores how different social movements challenge state-cen-tered notions of hegemony, consent, and power and challenges its colonization of the "political" (Routledge, 1996, p. 509). Contemporary social movements have extended the conception of the political to include issues of gender, ethnicity, and autonomy. They frequently embody popular practices so that these different ways of becoming expand the notion of the political domain to include everyday prac-tices and knowledges that are articulated as counter-hegemonic positions (ibid., p. 512). Today participation in protests, especially against neoliberal globalization, contributes to geopolitical revisionings of the world. The internet has greatly facil-itated the possibility of gathering protestors together in a short time and was used very effectively in the anti-Iraqi war demonstrations coordinated throughout the world (late 2002 to March 2003). Yet as Laurie and Calla (2004) caution us, the celebratory focus on social movements as a source of women's informal political activities may close off a wider analysis of politics in relation to other institutions, such as the state and other political arenas.

Many feminist (Marston, 2000; Dowler and Sharp, 2001) and other geographers (Routledge, 1996; Thrift, 2000; Flint, 2002) have suggested that the everyday offers a way of grounding political geography and geopolitics and an understanding of how the international, the national, and the banal are all reproduced in everyday life. It may seem surprising that a well known political geographer of the traditional kind, Norman Pounds (1972, p. vii), wrote that "people act politically every day of their lives, and their actions are no less susceptible of political analysis than those of the decision-makers in the nation's capital." There is of course no intrinsic reason why political geography cannot attend to the everyday life embedded in statescraft and the global events, as feminist international relations has done (Enloe, 1989; Sylvester, 1998). The slogan "the personal is political" drew attention to the polit-ical dimension and power relations of the spaces of the household (Millet, 1972). These calls to incorporate the everyday will need to go further and reflect on the nature of the everyday (Gardiner, 2000) and how it enables us to synthesize differ-ent processes, both formal and informal, across different scales.

Finally, feminist political geography has been invigorated by work questioning the scales at which social science analysis is conducted, the significance of scale for social and political action and organization, and the ways in which different scales could be connected (Smith, 1999; Marston, 2000; Hyndman, 2001; Staeheli, 2001).

Political geographers (Smith, 1993; Taylor, 1999) and feminist geographers (Marston, 2000) recognize the significance of the home and household for wider economic, social, and political processes. Marston argues that the mainstream literature on scale has overlooked social reproduction and consumption and that the social reproduction of the household by late nineteenth- and early twentieth-century urban middle-class women played a major part in the smooth running of capitalism and the remaking of the state. Taylor (1999) too traced the creation of domesticity, modernity, and changing gender relations through the home of the Dutch, British, and American hegemons. American modernity, which was exported as an integral element of its superior culture (see Domosh in this volume), replaced the masculine idea of the home as haven with a more feminine, though more isolated, space. Opening up the scales of social reproduction and consumption represents a nexus of exchange between feminist and political geographies.

This theorization of scale could also move in the opposite direction. Jennifer Hyndman (2001) transposes concepts normally associated with a finer scale to a coarser one. She points to the War Crimes Tribunal for the former Yugoslavia, where an act such as rape, usually associated with the local, has been recognized for the first time in history as a weapon of war and a crime against humanity. People's bodies become sites of aggression similar to territory and the body as a geopolitical site is politicized (Mayer, 2004). Thus the public/private divide is applied to the transnational scale. In the critical debates on wider notions of security to include people and not just states, this could embrace feminist geography's work on fear and security in local spaces.

Conclusion

As I have shown in this chapter, feminist geographers have much to contribute to reconceptualizing and extending political geography. It has been at the forefront of a radical political geography that questions the relationship between people, places, and power at different scales. Yet many feminist geographers have not seen themselves as doing or identifying with something called political geography (Staeheli, 2001). Political geography was considered to be preoccupied with the state, especially in its more public manifestations, and unconcerned with issues of identity, difference, and the articulation between different identities. Foregrounding scale (Staeheli, 1994; Marston, 2000) and understanding that agents move between and operate across scales have served to locate what may have appeared to be local and domestic phenomena, e.g. migrant domestic workers, in a wider, and explicitly political, context. Moreover, the transgression of scales and spaces also draws attention to the specifically political dimensions of agency, processes, and institutions.

There has been an intellectual return of the state, but now embedded in a series of overlapping and interconnecting scales. At the same time, we need to analyze the global scale which many feminist geographers have not seriously confronted. It would be ill-advised to retreat to the local as the more approachable and neighborly scale and thereby leave unquestioned gender inequalities and activism at the global scale. In this endeavor a sisterly regard for how feminist economics and international relations have handled the global, the national, and the everyday would be useful.

NOTES

1 Despite the quote at the beginning of the chapter, it is interesting that Taylor (2002) does not consider feminist approaches as an alternative radical political geography.

2 There is a history of women's movements, such as the suffragettes at the beginning of the twentieth century, using their bodies to press their claims. In the UK, they threw themselves at the King's horses, tied themselves to railings and went on hunger strikes.

BIBLIOGRAPHY

Bell, D. (1995) Pleasure and danger. The paradoxical spaces of sexual citizenship. *Political Geography*, 14(2), 139–53.

Blunt, A. and Rose, G. (1994) *Writing Women and Space: Colonial and Postcolonial Geographies*. New York: Guilford Press.

Bondi, L. and Peake, L. (1988) Gender and the city: urban politics revisited. In J. Little, L. Peake and P. Richardson (eds), *Women in Cities*. London: Macmillan.

Breitbart, M. (1981) Peter Kropotkin: the anarchist geographer. In D. Stoddart (ed.), *Geography, Ideology and Social Concern*. Oxford: Blackwell.

Brown, M. and Staeheli, L. (2003) Are we yet there? Feminist political geographies. *Gender, Place and Culture*, 10(3), 247–55.

Chang, K. and Ling, L. (2000) Globalization and its intimate other: Filipina domestic workers in Hong Kong. In M. Marchand and A. Sisson Runyan (eds), *Gender and Global Restructuring. Sightings, Sites and Resistances*. London: Routledge.

Chappell, L. (2000) Interacting with the state. *International Feminist Journal of Politics*, 2(2), 244–76.

Chouinard, V. (2001) Legal peripheries: struggles over disabled Canadians' places in law, society and space. *The Canadian Geographer*, 45(1), 187–92.

Clayton, D. (2002) Critical imperial and colonial geographies. In K. Anderson, M. Domosh, S. Pile and N. Thrift (eds), *Handbook of Cultural Geography*. London: Sage.

Connell, R. (1990) The state, gender and sexual politics. Theory and appraisal. *Theory and Society*, 19(8), 507–44.

Connell, R. (1995) *Masculinities*. Berkeley: University of California Press.

Crawley, H. (2001). *Refugees and Gender. Law and Process*. Bristol: Jordans.

Curthoys, A. (1993) Feminism, citizenship and national identity. *Feminist Review*, 44, 19–38.

Dalby, S. (1994) Gender and critical geopolitics: reading security discourse in the new world order. *Environment and Planning D: Society and Space*, 12, 595–612.

Dodds, K. (2001) Political geography III: critical geopolitics after ten years. *Progress in Human Geography*, 25(3), 469–84.

Domosh, M. (1991a) For a feminist historiography of geography. *Transactions of the Institute of British Geographers*, n.s., 16, 95–104.

Domosh, M. (1991b) Beyond the frontiers of geographical knowledge. *Transactions of the Institute of British Geographers*, n.s., 16, 488–90.

Dowler, L. and Sharp, J. (2001) A feminist geopolitics? *Space and Polity*, 5, pp. 165–77.

Drake, C. and Horton, J. (1983) Comment on editorial essay: sexist bias in political geography. *Political Geography Quarterly*, 2, 329–35.

Dunbar (1981) Elisée Reclus, an anarchist in geography. In D. Stoddart (ed.), *Geography, Ideology and Social Concern*. Oxford: Blackwell.

England, K. (2003) Towards a feminist geography? *Political Geography*, 22, 611–16.

Enloe, C. (1989) *Beaches, Bananas and Bases: Making Feminist Sense of International Politics*. London: Pandora.

Fenster, T. (1997) Ethnicity, citizenship, planning and gender: the case of Ethiopian immigrant women in Israel. *Gender, Place and Culture*, 15(2), 177–89.

Fincher, R. (1987) Space, class and political processes: the social relations of the local state. *Progress in Human Geography*, 11, 496–515.

Fincher, R. and Panelli, R. (2001) Making space: women's urban and rural activism and the Australian state. *Gender, Place and Culture*, 8(2), 129–48.

Fleming, M. (1979) *The Anarchist Way to Socialism. Elisée Reclus and Nineteenth-century Anarchism*. London: Croom Helm.

Flint, C. (2002) Political geography: globalization, metapolitical geographies and everyday life. *Progress in Political Geography*, 26, 391–400.

Gardiner, M. (2000) *Critiques of Everyday Life*. London: Routledge.

Harvey, D. (1989) *The Condition of Postmodernity*. Oxford: Blackwell.

Hayford, A. M. (1974) The geography of women: an historical introduction. *Antipode*, 6(2), 1–19.

Hays-Mitchell, M. (1995) Voices and visions from the streets: gender interests and political participation among women informal traders in Latin America. *Environment and Planning D: Society and Space*, 13(4), 445–70.

Helms (1974) Old women in America: the need for social justice. *Antipode*, 6(2), 26–32.

Hochschild, A. R. (2000) Global care chains and emotional surplus value. In W. Hutton and A. Giddens (eds), *On the Edge. Living with Global Capitalism*. London: Jonathan Cape.

Hooper (2000) Masculinities in transition. The case of globalization. In M. Marchand and A. Sisson Runyan (eds), *Gender and Global Restructuring: Sightings, Sites and Resistances*. London: Routledge.

Hyndman, J. (2001) Towards a feminist geopolitics. *The Canadian Geographer*, 45(2), 210–22.

Jackson, P. and Penrose, J. (eds) (1993) *Constructions of Race, Place and Nation*. London: University College Press.

Johnson, N. (1997) Cast in stone: monuments, geography and nationalism. In J. Agnew (ed.), *Political Geography Reader*. London: Arnold.

Katz, C. (2000) On the grounds of globalization: a topography for feminist political engagement. *Signs: Journal of Women in Culture and Society*, 26(4), 1213–34.

Knuttila, M. and Kubik, W. (2000) *State Theories: Classical, Global and Feminist Perspectives*. London: Zed Books.

Kodras (1999) Geographies of power in political geography. *Political Geography*, 18, 75–9.

Kofman, E. (1993) National identity and sexual and cultural differences in France. In M. Kelly and R. Bock (eds), *France: Nation and Regions*. Southampton: University of Southampton Press.

Kofman, E. (1995) Citizenship for some but not for others: spaces of citizenship in contemporary Europe. *Political Geography*, 14(2), 121–38.

Kofman, E. (1996) Gender relations, feminism and geopolitics: problematic closures and opening strategies. In E. Kofman and G. Youngs (eds), *Globalization: Theory and Practice*. London: Pinter.

Kofman, E. (2003a) Political geography and globalization as we enter the twenty-first century. In E. Kofman and G. Youngs (eds), *Globalization: Theory and Practice*. London: Continuum.

Kofman, E. (2003b) Civil rights and citizenship. In J. Agnew, K. Mitchell and G. Toal (eds), *A Companion to Political Geography*. Oxford: Blackwell.

Kofman, E. and Peake, L. (1990) Into the 1990s: a gendered agenda for political geography. *Political Geography Quarterly*, 9(4), 313–36.

Kofman, E. and Youngs, G. (eds) (2003) *Globalization: Theory and Practice*. London: Continuum.

Laurie, N. and Calla, P. (2004) Development, post-colonialism and feminist political geography. In L. Staeheli, E. Kofman and L. Peake (eds), *Feminists Mapping Politics: Feminist Perspectives on Political Geography*. New York: Routledge.

Lister, R. (1997) *Citizenship: Feminist Perspectives*. London: Macmillan.

McClintock, A. (1993) Family feuds: gender, nationalism and the family. *Feminist Review*, 44, 61–80.

McClintock, A. (1995) *Imperial Leather: Race, Gender and Sexuality in the Colonial Context*. London: Routledge.

McEwan, C. (2000) Engendering democracy: gendered spaces of democracy in South Africa. *Political Geography*, 19, 627–51.

Marshall, T. H. (1950) *Citizenship and Social Class*. Cambridge: Cambridge University Press.

Marston, S. (1990) Who are the people? Gender, citizenship and the making of the American nation. *Environment and Planning D: Society and Space*, 8(4), 449–58.

Marston, S. (2000) The social construction of scale. *Progress in Human Geography*, 24(2), 19–42.

Mayer, T. (ed.) (1994) *Women and the Israeli Occupation: The Politics of Change*. London: Routledge.

Mayer, T. (ed.) (2000) *Gender Ironies of Nationalism: Sexing the Nation*. London: Routledge.

Mayer, T. (2004) Nation, gender and boundaries: feminist political geography and the study of nationalism, In L. Staeheli, E. Kofman and L. Peake (eds), *Feminists Mapping Politics: Feminist Perspectives on Political Geography*. London: Routledge.

Millet, K. (1972) *Sexual Politics*. London: Abacus.

Mitchell, K., Marston, S. and Katz, C. (2003) Introduction: life's work: an introduction, review and critique. *Antipode*, 35(3), 415–42

Moss, P. and Dyck, I. (2002) Embodying social geography. In K. Anderson, M. Domosh, S. Pile and N. Thrift (eds), *Handbook of Cultural Geography*. Oxford: Blackwell.

Nagar, R. (2000) Mujhe jawab do! (Answer me!): women's grass-roots activism and social spaces in Chitrakoot (India). *Gender, Place and Culture*, 7(4), 341–62.

Nash, C. (1993) Remapping and renaming: new cartographies of identity, gender and landscape in Ireland. *Feminist Review*, 44, 39–57.

Nast, H. and Pile, S. (eds) (1998) *Places through the Body*. New York: Routledge.

Nelson, L. (2003) Decentering the movement: collective action, place and the sedimentation of radical political discourses. *Environment and Planning D: Society and Space*, 21, 559–81.

Nelson, L. (2004) Transnational topographies of gender and citizenship: Purhépechan Mexican women claiming political subjectivities. *Gender, Place and Culture*, 11(2), 163–87.

O'Tuathail, G. (1996) *Critical Geopolitics*. London: Routledge.

Parrenas, R. (2001) Trangressing the nation-state: the partial citizenship and "imagined (global) community" of migrant Filipina domestic workers. *Signs: Journal of Women in Culture and Society*, 26, 1129–54.

Pateman, C. (1989) *The Disorder of Women: Democracy, Feminism and Political Theory*. Cambridge: Polity Press.

Pcake, L. (1984) How Sarvlik and Crewe fail to explain the Conservative victory of 1979 and electoral trends in the 1970s. *Political Geography Quarterly*, 3, 161–7.

Peake, L. (1999) Politics. In L. McDowell and J. Sharpe (eds), *Glossary of Feminist Geography*. London: Arnold.

Peake, L. and Trotz, A. (1999) *Gender, Ethnicity and Place: Women and Identities in Guyana*. London: Routledge.

Pounds, N. (1972) *Political Geography*. New York: McGraw Hill.

Pringle, R. and Watson, S. (1992) "Women's interests" and the post structuralist state. In M. Barrett and A. Phillips (eds), *Destabilizing Theory. Contemporary Feminist Debates*. Cambridge: Polity Press.

Radcliffe, S. and Westwood, S. (eds) (1993) *Viva: Women and Popular Protest in Latin America*. London: Routledge.

Radcliffe, S. and Westwood, S. (eds) (1996) *Remaking the Nation: Place, Identity and Politics in Latin America*. London: Routledge.

Rai, S. (1996) Women and the state in the Third World: some issues for debate. In S. Rai and G. Lievesley (eds), *Women and the State. International Perspectives*. London: Taylor and Francis.

Routledge, P. (2003) Anti-geopolitics. In J. Agnew, K. Mitchell and G. Toal (eds), *A Companion to Political Geography*. Oxford: Blackwell.

Routledge, P. (1996) Critical geopolitics and terrains of resistance. *Political Geography*, 15(6/7), 509–31.

Sassen, S. (2000) Women's burden: counter-geographies of globalization and the feminization of survival. *Journal of International Affairs*, 53(2), 503–24.

Secor, A. (2001) Towards a feminist counter-geopolitics: gender, space and Islamist politics in Istanbul. *Space and Polity*, 5(3), 191–211.

Sharp, J. (1996) Gendering nationhood: a feminist engagement with national identity. In N. Duncan (ed.), *BodySpace: Destablizing Geographies of Gender and Sexuality*. London: Routledge.

Sharp, J. (2000) Re-masculinising geo-politics? Comments on Gerard O'Tuathail's *Critical Geopolitics*. *Political Geography*, 19, 361–4.

Sharp, J. (2002) Gender in a political and patriarchal world. In K. Anderson, M. Domosh, S. Pile and N. Thrift (eds), *The Handbook of Cultural Geography*. London: Sage.

Sharp, J. (2003) Feminist and postcolonial engagements. In J. Agnew, K. Mitchell and G. Toal (eds), *A Companion to Political Geography*. Oxford: Blackwell.

Smith, F. (1999) Discourse of citizenship in transition: scale, politics and urban renewal. *Urban Studies*, 36, 167–87.

Smith, F. (2001) Refiguring the geopolitical landscape: nation, transition and gendered subjects in post-Cold War Germany. *Space and Polity*, 5(4), 213–35.

Smith, N. (1993) Homeless/global:scaling places. In J. Bird (ed.), *Mapping the Futures*. London: Routledge.

Smith, S. (1989) Society, space and citizenship: a human geography for the "new times"? *Transactions of the Institute of British Geographers*, 14, 144–56.

Soysal, Y. N. (1994) *Limits of Citizenship: Migrants and Postnational Membership in Europe*. Chicago: University of Chicago Press.

Sparke, M. 1996. Negotiating national action, free trade, constitutional debates and the gendered geopolitics of Canada, *Political Geography*, 15, 615–40.

Staeheli, L. (1994) Empowering political struggle: spaces and scales of resistance. *Political Geography*, 13(5), 387–91.

Staeheli, L. (1996) Publicity, privacy and women's political action. *Environment and Planning D: Society and Space*, 14, 601–19.

Staeheli, L. (2001) Of possibilities, probabilities and political geography. *Space and Polity*, 5, 177–89.

Staeheli, L. and Cope, M. (1994) Empowering women's citizenship. *Political Geography*, 13, 443–60.

Staeheli, L., Kofman, E. and Peake, L. (eds) (2004) *Mapping Gender, Making Politics: Feminist Perspectives on Political Geography*. New York: Routledge.

Sylvester, C. (1998) Handmaiden tales of Washington power: the abject and the real Kennedy White House, *Body and Society*, 4, 39–66.

Taylor, P. (1982) A materialist framework for political geography. *Transactions, Institute of British Geographers*, 7, 15–34.

Taylor, P. J. (1984) Introduction: geographical scale and political geography. In P. Taylor and J. House (eds), *Political Geography: Recent Advances and Futures Directions*. London: Croom Helm.

Taylor, P. J. (1985) *Political Geography: World-economy, Nation-state and Locality*. London: Longman.

Taylor, P. J. (1999) Places, spaces and Macy's: place–space tensions in the political geography of modernities. *Progress in Human Geography*, 23(1), 7–26.

Taylor, P. J. (2000) Political geography. In R. J. Johnston, D. Gregory, G. Pratt and M. Watts (eds), *Dictionary of Human Geography*, 4th edn. Oxford: Blackwell.

Taylor, P. J. (2002) Radical political geographies. In J. Agnew, K. Mitchell and G. Toal (eds), *A Companion to Political Geography*. Oxford: Blackwell.

Teather, E. (1999) Introduction: geographies of personal discovery. In E. Teather (ed), *Embodied Geographies: Spaces, Bodies and Rites of Passage*. London: Routledge.

Thrift, N. (2000) It's the little things. In K. Dodds and D. Atkinson (eds), *Geopolitical Traditions. A Century of Geopolitical Thought*. London: Routledge.

Toal, G. and Shelley, F. (2002) Political geography: from the long 1989 to the end of the post cold-war peace (http://mailer.fsu.edu/~psteinbe/pgsg,pgcgia.doc).

Townsend, J. (ed.) (1999) *Women and Power*. London: Zed Books.

Turner, B. (ed.) (1993) *Social Theory and Citizenship*. London: Sage.

Valentine, G. (1996) An equal place to work? Anti-lesbian discrimination and sexual citizenship in the European Union. In M. D. Garcia Ramon and J. Monk (eds), *Women of the European Union: The Politics of Work and Daily Life*. London: Routledge.

Valentine, G. (2003) Sexual politics. In J. Agnew, K. Mitchell and G. Toal (eds), *A Companion to Political Geography*. Oxford: Blackwell.

Walby, S. (1990) *Theorising Patriarchy*. Oxford: Blackwell.

Walby, S. (1994) Is citizenship gendered? *Sociology*, 28(2), 379–95.

Yeoh, B. (2001) Historical geographies of the colonised world. In B. Graham and C. Nash (eds), *Modern Historical Geographies*. London: Longman.

Yeoh, B. and Huang, S. (1998) Negotiating public space: strategies and styles of migrant female domestic workers in Singapore. *Urban Studies*, 35(3), 583–602.

Yuval-Davis, N. and Anthias, F. (1989) *Woman–Nation–State*. London: Routledge.

Chapter 35

Gender, Race, and Nationalism: American Identity and Economic Imperialism at the Turn of the Twentieth Century

Mona Domosh

The hyperpatriotism that followed the terrorist attacks on the United States in 2001 is a poignant and often painful reminder of the power of nationalism to unite and divide. Constituting this hyperpatriotism is a nationalistic rhetoric that has fashioned the United States into a nation of purported equal and organically united citizens – a "homeland" – in need of defense from "evil" and "uncivilized" barbarians (Kaplan, 2003). The discursive and material division of the world into "us" and "them" is certainly not new, but the virulence of this recent manifestation of nationalism makes it all the more critical to take notice of it again. As Jan Pettman (1996, p. 48) argues, "nationalism constitutes the nation as above politics, and so disguises the politics of its making. This is the extraordinary power of the nation as that thing which people will kill and die for." Nationalisms, then, are "dangerous" (McClintock, 1997), and disclosing the "politics of [their] making" is of critical importance.

This "deconstruction" of nationalisms has been one of the primary tasks of relatively recent feminist political scholarship. Working largely within a postcolonial framework, and with insights drawn from political and feminist theory and criticism, scholars have shown how nationalism is gendered, raced, and sexualized (see, for example, Spivak, 1987; Stoler, 1991, 2001; McClintock, 1995; Pettman, 1996). In other words, they have shown how nationalism operates through and is constituted by the ideologies, discourses, and practices of culturally constructed hierarchies based on gender, race, and/or sexuality. This should not be particularly surprising. As "imagined communities," in the now famous words of Benedict Anderson (1991), nations are not naturally formed but are culturally imagined. The sense of belonging together implied in the word nation is predicated on a division between what is "us" and what is "them," a distinction that has often been based on gendered, raced, classed, and sexualized hierarchies. For example, scholars have shown how the Spanish constructed the native peoples of South America as effeminate and homosexual during the time of conquest in the seventeenth and eighteenth centuries in order to legitimize their imperial conquest (Trexler, 1995), while Anne

McClintock (1995) offers a persuasive account of how English national identity during the nineteenth century was formulated around ideals of a white patriarchal family, in distinction to the culturally constructed image of effeminate men and masculine women who inhabited the areas of the world under British control. Feminist geographers too have explored the interlinked roles that gender, race, class, and sexuality have played in national identity formation and the legitimation of imperialism (e.g. Ware, 1992; Blunt, 1994, 1999; Blunt and Rose, 1994; Bell, 1995; Kenny 1995; McEwan, 1996; Morin, 1998).

What I explore in this chapter is the role that gendered and racialized ideologies played in a type of imperialism that is not generally analyzed as such, an imperialism based primarily on economic expansion, not political or military control. My focus is on how American national identity formations were used to legitimize the massive economic expansion beyond the borders of the United States in the last decades of the nineteenth and the first decades of the twentieth centuries. Although it is different from direct military and political control, I am referring to this export of American mass-produced commodities as an imperial process because it was enabled by and in turn produced power over space and over people, and by so doing, it created new spatial arrangements, refashioned geopolitics, and carried with it and in turn shaped novel cultural forms and geographical knowledges (Wilkins, 1970; Rosenberg, 1982; O'Rourke and Williamson, 1999; Jacobson, 2000).

And this American economic imperialism was formulated through and legitimized by a set of interrelated gendered and racialized discourses. For example, as I discuss in this chapter, American companies often used visual and verbal images in their advertisements that drew heavily on Victorian notions of separate spheres and whiteness to show the "civilizing" effects of their products. This served to legitimize their sale at home and overseas, as it associated consumption with the highest "stage" of the civilizational hierarchy. One of the key tropes in these commercial representations was the figuring of the nation – the United States – in the form of a proper, white, Victorian family. Singer and McCormick corporations, for example, showed how their products – sewing machines and farming equipment – fostered women's proper domestic roles and men's proper productive roles, thus creating the ideal, patriarchal, white, American family.

This use of the image of the family in nationalistic and imperial discourses, of course, is not unique to the case of American commercial expansion; it can be seen in a diverse array of settings and time frames. It is not difficult to see why. If the nation is often defined in distinction to those not belonging, those who do belong are often discursively figured as a family, people tied together by "natural" bonds of blood or race. In this way, the nation is imagined as a home, as a domestic and secure space, in distinction to the foreign and threatening spaces beyond. Yet the idea of figuring the nation as family and home relies on sets of interrelated gendered and racialized meanings that are highly problematic. As scholars have pointed out (McClintock, 1995, p. 97; Kaplan, 2003), the idea of the nation as family helps to secure the "naturalness" of social hierarchies *within* nations by associating them with the purported "naturalness" of the Western family: the patriarch with subordinate wife and children, all supposedly forming an organic unity with the same inherent interests. At the same time, the family trope legitimizes a view of the "natural" hierarchy that exists *between* nations – colonial states are often spoken of as "children" within the "family" of nations, while imperial nations are figured as their parental protectors. And the family metaphor enables a country to present two useful yet somewhat contradictory images of itself – the feminine home, the heartland, in need of defense, and the masculine state, aggressively defending the national hearth – that are seen as united under the common banner of the "family."

Feminist geographers have built on this work, focusing on the complexities of the gendering of nationalisms within particular locales, drawing on the metaphors of family and domestic space to deconstruct notions of nationalism, and developing constructs that situate gender, race, and sexuality within analyses of imperial regimes. For example, in her analysis of Irish nationalism, Catherine Nash (1994) shows how the country's social and economic context shaped its gendered forms of nationalism. In the late nineteenth century Irish nationalists looked to the Celtic western portions of Ireland, and to the iconography of the peasant women who lived there, to create an image of an "authentic" and "wild" Irishness that could be distinguished from those parts of the country under British influence. Yet that imagery was potentially unsettling to the reigning socioeconomic system. The association of women with an "authentic" Irish wildness created an iconography of promiscuous and fertile women, a symbolic imagining that conflicted with Catholic notions of proper womanhood, and with a socioeconomic system that relied on family farming and the regulation of women's sexuality. In the first decades of the twentieth century, Irish nationalists reformulated the image of the West of Ireland as a Gaelic, masculine frontier, thus serving to distinguish it from the overly civilized British, while resituating the feminine within the constrained boundaries of a proper wife on the family farm.

The gendering of the nation is also related to a country's particular geopolitical context. In countries such as Ireland that were formulating anti-colonial movements, images of an "authentic" heartland were important tools in legitimizing political and military actions. In other situations, as Sarah Radcliffe's (1996) work on Ecuador shows, what is important is to present a country as modern and Western, in order to gain symbolic and economic participation in the global marketplace. In contemporary Ecuador, then, nationalism is figured around the modern, bourgeois, "liberated," independent woman, while the peasant women who produce much of the agricultural wealth of the country are erased from public forms of representation. What I explore in this chapter is the particular geopolitical context of the United States at the end of the nineteenth century, when national identity was figured partly by contrasting America's more "civilized" form of conquest through consumption to the European model of conquest through military control. Because of the long historical association of consumption with femininity, this situation placed women and the feminine in a prominent discursive position in the representations of American commercial imperialism.

By focusing on American nationalism and economic expansion at the turn of the century, I hope to highlight both the *pervasiveness* of gendered and racialized ideologies to the cultural construction of national identity and imperial expansion, and the *elasticity* of these ideologies. In other words, part of the power of these ideologies is that they could be used to promote economic as well as political and military imperialism, but in different ways. By highlighting the flexibility of these discourses and their applicability to economic imperialism, I also hope to suggest, by implication, the need to analyze the gendered and racialized discourses that underlie contemporary economic expansion and globalization. Let me start my analysis with a discussion of American national identity at the turn of the twentieth century, before turning to how that nationalistic discourse was used to legitimize economic imperialism.

The Woman with the Machine in the Garden: American National Identity at the Turn of the Twentieth Century

The idea of America as a place essentially different from Europe and as the embodiment of a pastoral ideal dates back to Shakespeare's *The Tempest*, but it wasn't until the late eighteenth and early nineteenth centuries that the idea was articulated as a potent political and social force (Smith, 1957; Marx, 1973). Jefferson's 1785 *Notes on Virginia* is generally credited as the piece of writing that pulled together the various strands of pastoralism with the notion of American exceptionalism to clearly articulate the ideal of the American garden. That ideal can be expressed in spatial form: a farmhouse set within agricultural fields, surrounded by modified nature (not wilderness) and removed from any signs of urbanization or industrialization. The figure that dominates is the independent white farmer and patriarch who by virtue of his status as an agricultural land-owner is a rational citizen of the nation. He is the "true American . . . whose values are derived from his relations to the land" (Marx, 1973, p. 130). And those things he valued were democracy, rationality, egalitarianism, and self-sufficiency.

This imagery took on great symbolic significance throughout the nineteenth century; it became the "master symbol" (Smith, 1957, p. 138) of American identity. Its metaphors of fecundity and growth served as elements in American prose and poetry, and its imagery figured prominently in American visual arts. But more to the point here, the image of the garden with its virtuous white farmer was useful in promoting agricultural settlement in the West – useful, that is, in American imperial dominance over Native Americans. As John Agnew (1993, p. 212) argues, Jefferson's strategy for territorial expansion was "to 'conquer without war' – to pursue the objectives of American policy by economic and other peaceable means of coercion and consent," and the garden as a spatial form of national identity was a potent ideological means of coercion. The garden was a middle landscape – it was neither urban nor wild and uncivilized. And since it was this landscape that was thought to create the "true" American, the wild had to be transformed into the civilized garden, tended by the white patriarch farmer and his dependent wife.

As enduring and powerful as this image was, it was challenged by the economic and social forces that were shaping most Americans' lives in the second half of the nineteenth century – industrialization, urbanization, technological advances, the rise of consumer culture. As Leo Marx argued (1973), the garden could accommodate certain forms of the machine, but it was remade in the process. The garden was adapted to a nation whose goals were now about the productivity that only the machine could bring, but a nation still committed to thinking of itself as a civilized paradise of the middle landscape. The result was a sentimental form of the pastoral ideal – the ideal served a rhetorical and ideological purpose, but it no longer carried any real weight. It had become by the mid to late nineteenth century "an increasingly transparent and jejune expression of the national preference for having it both ways" (Marx, 1973, p. 226).

In its new guise, as a garden that accommodated machines, the pastoral ideal's plasticity became more apparent. One particularly potent form, which appeared in popular magazines, novels, and advertising, stressed domestic consumption. In this "new" garden, the imagery retained the sense of rural harmony that defined the pastoral ideal, but it changed key aspects of that ideal. In the imagery of the "old" garden, a balance was maintained between industrialization (*over*civilized) on the one hand, and the savagery of the wilderness (*under*civilized) on the other, through the attachments of men to good agricultural lands. Access to land and to the "unspoiled American landscape" allowed farmers to nurture their morality; the land, in the words of Leo Marx (1973, p. 131), "disseminates germs of virtue" that kept farmers from the two extremes of industrialization and savagery. This imagery, then, was certainly appropriate to legitimizing America's expansion westward, conquering wilderness, and turning it into the garden for the white patriarchal family.

America's industrial corporations, however, were not particularly interested in land acquisition; instead, they were concerned with the expansion of markets for their products. An imagery that spoke of the virtues of consumption, therefore, began to replace an imagery that centered on land. In the words of Jackson Lears (1994), "fables of abundance" shifted their emphasis from the farm, to the factory and the domestic machines at home. In this new, consumption-oriented garden, the balance between the overcivilized and the undercivilized could be maintained through the figure of the late nineteenth-century white middle-class woman – according to Victorian gender ideology, she was the "innately" moral keeper of domestic virtues. She could live in a city, participate in the new industrial economy by serving as the family's consumer, yet still keep the American family in balanced harmony through her "essential" commitment to domesticity, while her husband left home to engage in the acquisition of wealth and power. This new garden imagery of American national identity was therefore much more malleable than its previous incarnation: it could accommodate both the urban and the rural; consumption and production. For example, figure 35.1, a *c.*1890s trade card advertising Household Sewing Machines, depicts an array of consumer goods, including machines, on the one hand, while also relaying an image of the slow-paced, family-oriented values of rural life on the other. The emphasis on proper domesticity and the home helps create a middle landscape – one filled with commodities (therefore not *under*civilized), but commodities used wisely and morally (therefore not *over*civilized).

By creating a different garden image whose central "virtuous" relationship was that between a white middle-class woman and her products, the pastoral ideal continued the ideological work of its previous incarnation: it "enabled the nation to continue defining its purpose as the pursuit of rural happiness while devoting itself to productivity, wealth and power" (Marx, 1973, p. 226). With women buying and using mass-produced goods to keep the domestic space of home in good order, and men using machines to produce in factories, farms, and offices, the idealized middle ground of American national identity could continue, albeit in different form. This new "garden" also served two related purposes. It was of course a particularly useful ideological tool in promoting mass consumption, necessary to the Fordist economic structure that was beginning to dominate the United States (see Sklar, 1988; Harvey,

Figure 35.1 Trade card for the Household Sewing Machine Company, c.1890. Collection of the author.

1990). And the United States could use the new garden to legitimize its commercial expansion overseas – American culture could be bought, for example, along with a new sewing machine, and American corporations spoke of the necessity of selling overseas in order to help to spread their "superior" form of culture, of civilization. The virtues of civilization would come not from access to land, but from access to new products. As a promotional book about the McCormick farm machinery company proclaims: "the harvester is the best barometer of civilization. . . . Whoever operates a harvester must not only be intelligent: he must be free" (Casson, 1908, p. 159).

The Discourse of Civilization and American Nationalism

As the quote about the McCormick harvester suggests, American products at the turn of the twentieth century were associated at some level with more than mere utility – these products, according to these booster accounts, brought with them qualities of intelligence and freedom, part of American "civilization." The term civilization had a very particular meaning in the late nineteenth- and early twentieth-century United States – it coalesced into one word a powerful ideological discourse that was fundamental to understanding turn-of-the-century America. If "the woman (or man) in the garden with the machine" formed a two-dimensional, spatial imagery of American national identity, what gave that imagery energy and dynamism and propelled it beyond national borders was the ideology of civilization. Let me briefly outline this discourse before turning to a case study of how it was used to underpin American economic imperialism.

As Kay Anderson (2000) has argued, drawing a distinction between the "savage" and the "civilized" has been a primary way of organizing social and economic power hierarchies in the West since at least the Ancient Greeks. Yet this distinction, and its concomitant narrative of "improvement," took on particular salience in the mid-nineteenth century as it merged with Darwinian evolutionary discourse and modern forms of racism born out of European imperialism. In other words, a complicated discourse of racism that sorted the world's regions into a hierarchy based on the "race" of its inhabitants was twinned with Darwinian notions that posited a temporally progressive view of social and animal life. This temporal and spatial ordering of the world merged with classical notions of savagery and civilization to form an incredibly powerful set of ideas that served to legitimize Western imperial expansion. According to this ideology, the world's cultures were on an inexorable journey of "development" out of savagery through various stages and onto civilization. Only the white race, however, was thought to have reached the final stage of civilization, and part of the Christian mission of the white race was to "enlighten" other cultures and help them reach the stage of civilization (through colonization). Some cultures, however, were thought to be unable to reach the final stage, and they were doomed to extinction, vestiges of a remote past.

Gendered as well as racialized ideologies were central to this discourse of civilization (Bederman, 1995; Newman, 1999). The various hierarchical stages of humanity could be identified not only by "race" but also by degrees of sexual differentiation. Civilized societies were defined as those characterized by a strict gender division of labor: women were passive, men active; women were the purveyors of society's morals and spirituality, men were the producers and protectors of society. In other words, the Victorian ideology of separate spheres was completely integrated into the discourse of civilization, and vice versa. If separate spheres defined civilized societies, then savage ones were defined in opposition – men and women's roles did not diverge, so that savage men were thought to be effeminate, emotional, unable to support women and children, while savage women were masculinized, aggressive, engaged in manual labor. As Bederman (1995, p. 25) suggests, "the pronounced sexual differences celebrated in the middle class's doctrine of separate spheres were assumed to be absent in savagery, but to be an intrinsic and necessary aspect of higher civilization."

Gender served another related purpose. Prevalent notions of Christian millenialism called for white, civilized societies to work toward the ultimate goal: the vanquishing of evil and the ultimate rule of Christ over the righteous. Ideas of evolution merged with millenialism, creating a belief that evolutionary forces could act as the hand of god – instead of Christ's mighty hand, the survival of the fittest would lead to the rule of the superior races. And it was considered the duty of civilized people to spread this civilization, just as it was the duty of good Christians to convert heathens. For civilized, manly men, this meant taking up arms (for those races destined not to survive), or converting others through colonization. For womanly women, this meant teaching civilization through more domestic means: showing savage women how to keep house, sharing with them the domestic arts, etc. As Bederman (1995, p. 26) says, this is what was meant when people spoke of the "advancement of civilization."

The "advancement of civilization," then, is what many would have referred to as the reason for nineteenth-century imperial expansion, the primary means of forcibly spreading civilization. The advancement of civilization was also particularly useful as legitimation for the United States' economic expansion – selling certain kinds of products like sewing machines was thought to free women from manual labor, thus making them more "civilized." It also connected to national identity formation: the women (or the men) with their machines in the garden were performing their civilized roles; that is, the man was engaged in productive labor, the manly patriarch, while the woman was engaged in domestic work, reproducing the civilized family. In this way, we can begin to see how the complex set of ideas contained within the discourse of civilization fueled expansion overseas for Western powers, *and* was a potent tool in promoting American economic imperialism. Although part of the American myth of self is a belief in its exceptionalist national character (Agnew, 1983; Kaplan, 1993; Adas, 2001), we can see that its commercial imperialism participated in an ideological formation very similar to, though not identical with, what underlay European territorial expansion. Let me now turn to several case studies to show how American national identity, fueled with the rhetorical and ideological power of "civilization," was used to legitimize economic imperial expansion.

The Woman (and the Man) with Machines in the Garden (Both Domestic and Foreign)

The Singer Sewing Machine Company and the McCormick Farm Machinery Company were two of America's dominant international companies in the late nineteenth century. As early as 1863, Singer was selling over 40 percent of its machines beyond American borders (Davies, 1976), while McCormick extended its markets into the agricultural regions of Russia, Mexico, Argentina, and India.[1] Both companies were constantly dealing with fierce competition, and found mass advertising a powerful means to increase their market share both at home and overseas, and both used images of American national identity to promote their products.

Singer used a variety of advertising and marketing strategies throughout the late nineteenth century: door-to-door canvassing, displays at major world expositions, and advertisements printed on trade cards, and in magazines, catalogues and brochures. Before the technological innovations of the turn of the twentieth century that allowed for affordable large-size images to be printed in magazines, and long before such images could be replicated in color, trade cards provided the one means of transmitting visual information in color and in mass quantities (Garvey, 1996). Singer's set of "nation" trade cards were considered by the company to be particularly successful in promoting sales in general (Davies, 1976). These cards drew on a motif common to many sewing machine cards – that of the proper, "civilized" white woman using a sewing machine in her parlor. Figure 35.1, as we have seen, depicts an advertising card for the Household Sewing Machine company, and the image is rife with tokens of middle-class propriety: the woman is dressed impeccably, working on her domestic duty of sewing, while also overseeing her son; the parlor is outfitted with all the elements of proper American middle-class life, such as potted plants, a grandfather clock, a fireplace mantle adorned with knick-knacks, cuddly cats, and even a bird in a lovely cage. All of these items are symbolic tokens of the domestic sphere and bespeak of the "highest" stage of civilization, and together they represent the new American pastoral, a middle ground where industrial products fit well into domesticated surroundings.

Singer drew on this set of associations when it released its first set of "nation" cards in 1892 (several others were added in succeeding years). Each card was titled with the name of a particular country where Singers were sold. The front of each card was a color lithograph depicting "native" people using the machines, while the flip side contained text describing the country. The people on the cards are shown wearing clothes considered typical of their country, clothes that were presumably sewed on the Singer machine. For example, figure 35.2 depicts two "natives" of Korea, dressed in their everyday clothes, using of course the Singer. This was a fairly standard format of Singer's sets of nation cards, a woman or sometimes a man using the machine, sometimes alone, sometimes posed with other groups of "natives."

A comparison between this card and figure 35.1 reveals interesting similarities and differences. The Korean woman is certainly participating in a form of American civilization – she is doing her proper domestic work of sewing, and is posed as if she is in a patriarchal relationship, with her purported husband nearby and above her. Devoid of other details it is difficult to discern the exact qualities of the "garden" she might inhabit, although the beautiful floor speaks of domestic attention. Yet she really isn't completely civilized – there are no children present to complete the patriarchal scenario (none of the nation cards depicts children), nor pets, nor tokens of "proper" home decoration. The text on the back of the card makes more clear the "message" of the image: "The natives show unmistakable evidence of descent from the half savage and nomadic tribes of Mongolia and northern Asia and the Caucasian people from western Asia. They are a thriftless race, preferring rather to be governed than to govern. Western civilization, however, is beginning to work wonders throughout the peninsula; and ere long, it is to be hoped, modern science and education will predominate." The implication here is that the Singer is doing the work of civilization, helping women with their domestic chores while at the same time bringing new technology and science to the "half savage." Purchasing and using a Singer machine, then, was a civilizing act for the Korean woman – she was on her way to becoming civilized. Using the Singer was also a "civilizing" act for the American woman. By using the machines she was participating in her proper domestic role while keeping up-to-date and modern, *and* she was fulfilling her feminine role as the civilizer of society, as she could associate her use of Singer machines with their "civilizing" work abroad. In this case, then, the imagery of the woman with the machine in the garden is used to legitimize the sale of Singer machines both at home and overseas – the product was both proof that one was civilized (for the American woman) and a means of attaining it (for the Korean woman).

McCormick Farm Machine Company also drew on images of American identity to promote their products as bearers of civilization. But given the nature of McCormick's products – farm equipment – the company used images of men using machines in the garden, particularly the farmer sitting atop his McCormick machine dutifully performing "civilized" masculinity; that is, engagement in productive activities with the use of modern machinery. On the cover of its 1894 catalogue (figure 35.3), for example, the relationship between a white, patriarchal farmer and American nationalism is made explicit: as the text reads, the farmer (sitting atop

Figure 35.2 Trade card for the Singer Sewing Machine Company, c.1894. Collection of the author.

Figure 35.3 Cover of the McCormick Farm Machinery Company catalogue, 1894. Warshaw Collection of Business Americana, Archives Center, National Museum of American History, Smithsonian Institution.

the McCormick in the bottom image) is the key to national prosperity, to the "balance" of power, and the greatness of the nation. On the first page of the catalogue, just inside this cover, is a poem with accompanying images that depicts life "before McCormick invented the reaper" and "after our latest achievement" (see figure 35.4). Before the McCormick, women had to work in the fields, and men's agricultural work was laborious – these were "savage" times, "no joy, no sweetness his." After the McCormick, the patriarchal American pastoral is achieved, with machines working the productive fields, and husband and wife occupying the cultivated and civilized parlor, watching out the window as "sweet Progress drew back" the "curtain folds." So the McCormick makes possible civilization in the form of the American pastoral – its mechanized form of farming brings prosperity to the nation at the same time that it allows for men to assume their civilized role of patriarch and women to tend to the domestic sphere.

McCormick's depiction of the use of its machines abroad shows clearly the "progress" that its machines were making possible. An 1887 image from one of its catalogues, for example, shows the machines being used in Asiatic Russia to great effect – by using five machines at once the process is orderly and fast, producing neat bundles of grain placed at exact intervals (see figure 35.5). This is certainly quite an "advancement" over how grain was harvested *before* the McCormick, as the company was always quick to point out in the text that accompanied these images – before the machines were in use places like Russia were living in semi-barbarism, using hand scythes to cut grain, with large peasant women providing a good deal of the labor. Yet, of course, this image suggests that the Asiatic Russians are not completely civilized or Americanized – even though they are using modern machinery and saving women from hard labor, they have not fulfilled the American pastoral. First and foremost, these men appear to be laborers, not yeoman farmers; that is, they do not own this land but are working for someone else, presumably the man sitting on horseback surveilling them. And even though women are not working the fields, it is not clear from this image what sort of domesticity is awaiting these men when they go home. This type of imagery was characteristic of McCormick's depiction of its overseas sales areas – rarely if ever is the American domestic ideal of patriarchal family depicted in these images. Presumably, however, given the "progress" that McCormick machines bring with them, it would be just a matter of time and continual consumption and use of American machines until the Russians reached the stage of civilization.

Concluding Remarks

My analysis of these select images begins to show how American commercial expansion in the late nineteenth and early twentieth centuries was represented through and in turn legitimized by images of American national identity. American products were represented as the bearers of and symbols for American civilization, and that civilization took the form of a heteronormative, patriarchal white family using the products of American industrialization in a modified garden. McCormick drew on images of the virtuous yeoman farmer to show how their products brought civilization through the liberation of women from manual labor and the productivity that their machines made possible. Singer drew on images of the "woman with the machine

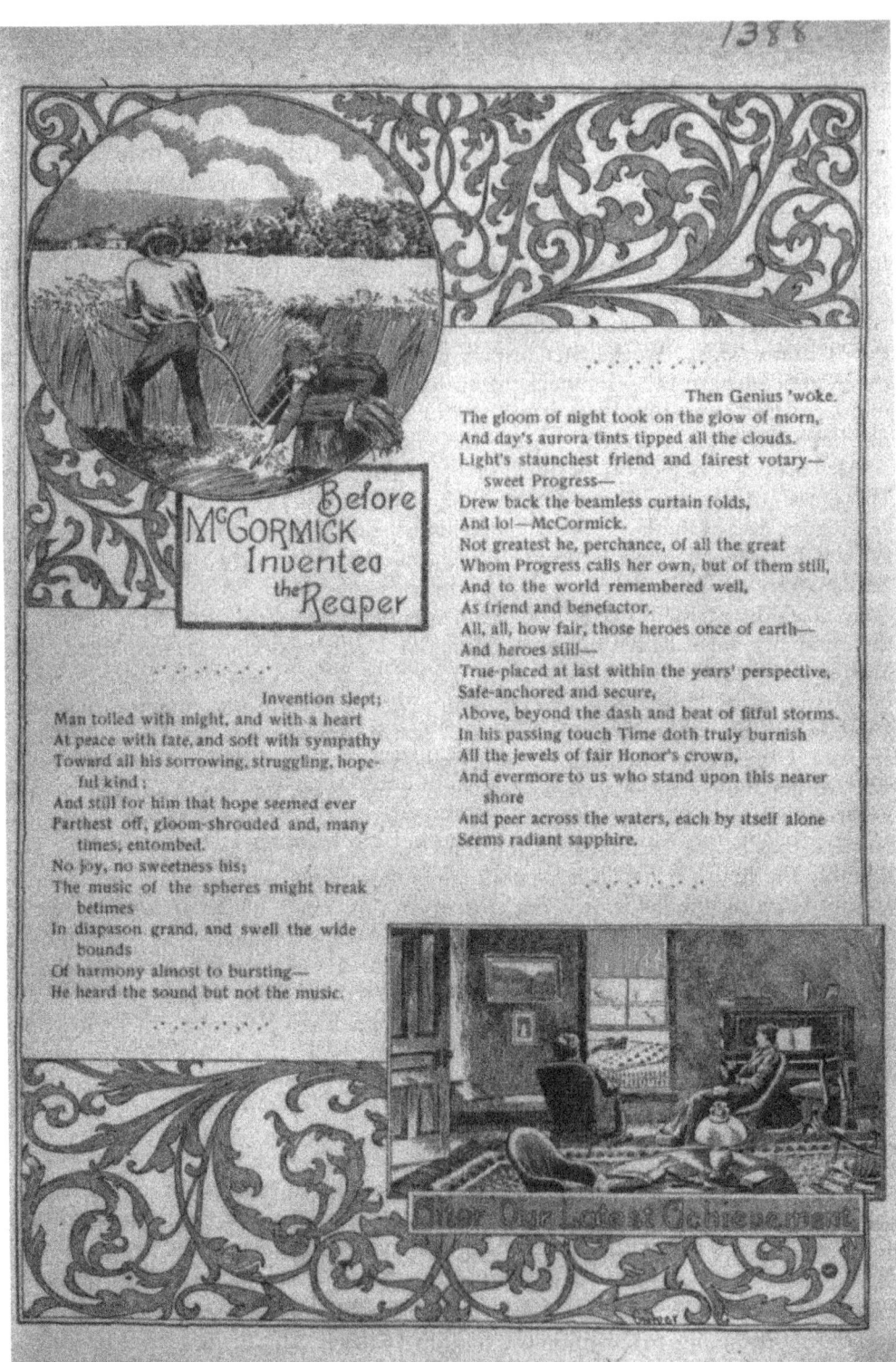

Figure 35.4 "Before McCormick Invented the Reaper," McCormick Farm Machinery Company catalogue, 1894, inside front cover page. Wisconsin Historical Society, McCormick Collection, Image ID 5448.

THE McCORMICK IN ASIATIC RUSSIA.

From a Photograph taken on the Estate of Ivan Pleshanow, 190 miles east of Samara, Russia, who operates five McCormick Harvesters and Binders, all drawn by Camels.

Figure 35.5 "McCormick in Asiatic Russia," McCormick Farm Machinery Company catalogue, 1887, p. 29. Wisconsin Historical Society, McCormick Collection, Image ID 5454.

in the garden," tending to her domestic duties of social reproduction and assisted by modern machines. In both cases, foreign peoples are presented in the process of becoming civilized through the use of these products; their complete conversion to civilization was presumably just a matter of time, and correct consumption.

American economic imperialism, then, was fueled by gendered and racialized ideologies similar to those that, according to Anne McClintock and others, underpinned European imperialisms. The purported naturalness of a hierarchy between nations and races paralleled the purported naturalness of the patriarchal family, and vice versa. Singer's "nation" cards, for example, lined up on a fireplace mantel of an American Victorian parlor, represented a family of unequal nations. Yet the particularities of a system of imperialism based primarily on consumption shaped gendered and racialized ideologies in particular ways. Because economic imperialism brought with it not colonies but customers – that is, the "others" over which it exerted power were not political subjects but economic ones – its ideological formations allowed for more flexible "subjects." American companies used foreign peoples in their advertisements to suggest the civilizing benefits their products brought to consumers. By so doing they suggested that other peoples and other "races" were civilizable, able to become "white" through correct consumption. America's prosperity and identity, based as it was on mass production and consumption, and imagined as the white patriarchal family using modern machines but within a domesticated garden, was, ideologically at least, up for sale.

In reality, that sharing of prosperity was rarely, if ever, accomplished. Integral to American nationalism since the late nineteenth century, as we have seen, has been an expansive rhetoric of sharing the American dream, *and* a restrictive view of who actually gets into the "garden." As consumers, Singer's Korean "natives" and McCormick's Asiatic Russians were presented as potentially civilizable, but never as "equal" citizens. Perhaps it has been this ongoing powerful discrepancy between an American rhetoric of economic advancement – what Michael Watts (2002) has called the "discourse of development" – and the realities of structural uneven development that is at the root of much of today's anti-American and anti-globalization sentiment. In that case, critical and reflexive historical feminist geographies of nationalism may offer one path to understanding contemporary global anxieties and strife.

NOTE

1 My analysis of these two companies is based primarily on archival information: corporate documents, letters, catalogues, and brochures. Both companies left their archives at the State Library of Wisconsin, Madison. I have chosen to limit my analysis here to several select images that I found to be characteristic of these company's advertising strategies.

BIBLIOGRAPHY

Adas, M. (2001) From settler colony to global hegemon: integrating the exceptionalist narrative of the American experience into world history. *American Historical Review*, 106, 1692–720.

Agnew, J. (1993) The United States and American hegemony. In P. Taylor (ed.), *Political Geography of the Twentieth Century*. London: Belhaven Press.

Agnew, J. (1983) An excess of "national exceptionalism": towards a new political geography of American foreign policy. *Political Geography Quarterly*, 2, 151–66.

Anderson, B. (1991) *Imagined Communities: Reflections on the Origin and Spread of Nationalism*. London: Verso.

Anderson, K. (2000) "The beast within": race, nature and animality. *Environment and Planning D: Society and Space*, 18, 301–20.

Bederman, G. (1995) *Manliness and Civilization: A Cultural History of Gender and Race in the United States, 1880–1917*. Chicago: University of Chicago Press.

Bell, M. (1995) A woman's place in "a white man's country": rights, duties and citizenship for the "new" South Africa. *Ecumene*, 2, 129–48.

Blunt, A. (1994) *Travel, Gender, and Imperialism: Mary Kingsley and West Africa*. New York: Guilford Press.

Blunt, A. (1999) Imperial geographies of home: British domesticity in India, 1886–1925. *Transactions of the Institute of British Geographers*, 24, 421–40.

Blunt, A. and Rose, G. (eds) (1994) *Writing Women and Space: Colonial and Postcolonial Geographies*. New York: Guilford Press.

Casson, H. (1908) *The Romance of the Reaper*. New York, Doubleday, Page & Co.

Davies, R. B. (1976) *Peacefully Working to Conquer the World: Singer Sewing Machines in Foreign Markets, 1854–1920*. New York: Arno Press.

Garvey, E. G. (1996) *The Adman in the Parlor: Magazines and the Gendering of Consumer Culture, 1880s to 1910s*. New York: Oxford University Press.

Harvey, D. (1990) *The Condition of Postmodernity*. Oxford: Blackwell.

Jacobson, M. F. (2000) *Barbarian Virtues: The United States Encounters Foreign Peoples at Home and Abroad, 1876–1917*. New York: Hill and Wang.

Kaplan, A. (1993) "Left alone with America": the absence of empire in the study of American culture. In A. Kaplan and D. Pease (eds), *Cultures of United States Imperialism*. Durham, NC: Duke University Press.

Kaplan, A. (2003) Homeland insecurities: transformations of language and space. *Radical History Review*, 85, 82–94.

Kenny, J. (1995) Climate, race and imperial authority: the symbolic landscape of the British hill station in India. *Annals of the Association of American Geographers*, 85, 694–714.

Lears, J. (1994) *Fables of Abundance: A Cultural History of Advertising in America*. New York: Basic Books.

McClintock, A. (1995) *Imperial Leather: Race, Gender and Sexuality in the Colonial Conquest*. New York: Routledge.

McClintock, A. (1997) "No longer in a future heaven": gender, race, and nationalism. In A. McClintock, A. Mufti and E. Shohat (eds), *Dangerous Liaisons: Gender, Nation, and Postcolonial Perspectives*. Minneapolis, University of Minnesota Press.

McEwan, C. (1996) Paradise or pandemonium? West African landscapes in the travel accounts of Victorian women. *Journal of Historical Geography*, 22, 68–83.

Marx, L. (1973) *The Machine in the Garden: Technology and the Pastoral Ideal in America*. New York: Oxford University Press.

Morin, K. (1998) British women travelers and constructions of racial difference across the nineteenth-century American West. *Transactions of the Institute of British Geographers*, 23, 311–29.

Nash, C. (1994) Remapping the body/land: new cartographies of identity, gender, and landscape in Ireland. In A. Blunt and G. Rose (eds), *Writing Women and Space: Colonial and Postcolonial Geographies*. New York: Guilford Press.

Newman, L. M. (1999) *White Women's Rights: The Racial Origins of Feminism in the United States*. New York: Oxford University Press.

O'Rourke, K. and Williamson, J. (1999) *Globalization and History: The Evolution of a Nineteenth-century Atlantic Economy*. Cambridge, MA: MIT Press.

Pettman, J. (1996) *Worlding Women: A Feminist International Politics*. New York: Routledge.

Radcliffe, S. (1996) Gendered nations: nostalgia, development and territory in Ecuador. *Gender, Place and Culture*, 3, 5–21.

Rosenberg, E. (1982) *Spreading the American Dream: American Economic and Cultural Expansion, 1890–1945*. New York: Hill and Wang.

Sklar, M. (1988) *The Corporate Reconstruction of American Capitalism, 1890–1916: The Market, The Law, and Politics*. Cambridge: Cambridge University Press.

Smith, H. N. (1957) *Virgin Land: The American West as Symbol and Myth*. New York: Vintage Books.

Spivak, G. C. (1987) *In Other Worlds: Essays in Cultural Politics*. New York: Methuen Press.

Stoler, A. L. (1991) Carnal knowledge and imperial power: gender, race, and morality in colonial Asia. In M. di Leonardo (ed.), *Gender and the Crossroads of Knowledge: Feminist Anthropology in the Postmodern Era*. Berkeley: University of California Press.

Stoler, A. L. (2001) Tense and tender ties: the politics of comparison in North American history and (post) colonial studies. *Journal of American History*, 88, 829–65.

Trexler, R. (1995) *Sex and Conquest: Gendered Violence, Political Order, and the European Conquest of the Americas*. Ithaca, NY: Cornell University Press.

Ware, V. (1992) *Beyond the Pale: White Women, Racism, and History*. London: Verso Books.

Watts, M. (2002) Alternative modern: development as cultural geography. In K. Anderson, M. Domosh, S. Pile and N. Thrift (eds), *Handbook of Cultural Geography*. London, Sage.

Wilkins, M. (1970) *The Emergence of Multinational Enterprise: American Business Abroad from the Colonial Era to 1914*. Cambridge, MA: Harvard University Press.

Chapter 36

Virility and Violation in the US "War on Terrorism"

Matthew G. Hannah

Two Stories

Five days after the September 11 attacks, the *New York Times* ran a front page article by David Sanger and Don van Natta Jr entitled "Four days that transformed a President, a presidency and a nation, for all time" (Sanger and van Natta, 2001). Near the beginning a before-and-after dualism is set up:

> On Monday night, [Bush] was laughing over dinner with his brother Jeb at a seaside Florida resort, posing for pictures with the restaurant staff and dodging questions from reporters about looming battles over the vanishing budget surplus. By this morning, with downtown Washington locked down by the military, he was conducting a war council at Camp David and demanding that countries around the world, starting with the Arab world, declare whether they were allies in the war on terrorism.

> As he rode Marine One from Andrews Air Force Base to the White House on Tuesday evening, Mr Bush watched the smoke billowing from the jagged gash in the Pentagon and seemed to recognize how profoundly his young presidency had been transformed.

The young, not yet manly Bush is recognizable in the unsteadiness of his first public appearances after the disaster, and in his willingness to let Vice President Cheney and his staff determine where he would fly to remain safe. But even while still allowing himself to be ferried about, there are signs of an emerging hardness, as he reportedly tells Cheney, "That's what we're paid for, boys. We are going to take care of this. When we find out who did this, they are not going to like me as President. Somebody is going to pay." After a meeting in a bunker in Nebraska (according to Condoleeza Rice, presumably now an honorary "boy"), Bush resolves not to allow himself to be kept away from Washington any longer and insists that his next address to the nation be made from the Oval Office. The day after returning to the White House, Bush then takes further steps toward assuming the mantle of full manhood, personally making the decision in council to use the term "war," and then giving a speech at the damaged Pentagon, where "It was the first time he spoke without notes, and he seemed far more comfortable." The rest of the article fills in the picture with further anecdotes chronicling the President's "fast track" acquisition of the sort of manly aura necessary to inspire the confidence of the nation.

On November 9, 2001 another article appeared on the front page of the *New York Times*, this one with the title "To the stranger, a wild land, strangely awesome." The piece strings together musings and experiences dramatizing Afghanistan as a rugged, "uncivilized" country, and the relation between the landscape and its inhabitants. The reporter, Dexter Filkins, writes:

> The natural world looms large in Afghanistan and its landscape seems bound up in all its parts. The faces of its people, now captured in a thousand photographs, seem merely the human reflection of the country's geography: all crags and fissures, desiccated and rough. To picture the war being fought here, imagine fighting in the Grand Canyon or Escalante National Monument, or perhaps even on the moon. (Filkins, 2001, p. 1)

As geographers well know, such perceptions of non-European peoples as essentially linked to the features of their environments have a long and ignoble history in the service of colonialism and imperialism on many different continents. Viewing the "dusky natives" as connected to nature has made it a more palatable proposition to conquer territory than if we viewed them, for example, as world citizens who happen to live in another place. We can claim (and we often have over the past two centuries) to be on a "civilizing" mission that will benefit the natives (see Jennings, 1976; Berkhofer, 1978, p. 30; Blaut, 1993). These days, of course, "civilization" is a word that can't be used indiscriminately. In the *New York Times* article, the word "technology" serves as a substitute. "Like a desert plant," Filkins claims, "technology clings to the soil here with shallow roots, struggling to take hold, and then a wind comes and blows it away." He writes, "In Farkhar, there is no mail delivery, no streetlights. No telephones, no telephone poles. Farkhar darkens with the sunset; most of the villagers, lacking anything else to do, go to sleep." But if someone sets up a TV or a satellite telephone, lines quickly form, life becomes interesting. Much as in descriptions of American Indian tribes during the nineteenth century, Afghans are here characterized fundamentally in terms of lack: it is difficult to imagine them doing anything but twiddling their thumbs in boredom, perhaps fighting each other to pass the time (and to express their innate savagery), but nevertheless remaining inseparable from their craggy landscapes.

The articles glossed above are only two of the more prominent installments in a pair of traditional American discourses that have already animated hundreds, perhaps thousands, of articles, newscasts, official statements, and commentaries since September 11: the American culture of masculinity and the American frontier myth. This chapter is an attempt to explain how these two closely related traditions have shaped the public discourses around US foreign policy since the September attacks, and in particular to argue that their intertwined logics can help us to understand the prominence of themes of *violation* and *penetration* in official policy pronouncements and media commentary. I suggest that post-September 11 American policy, at least as it has been presented to the American public, represents a deepening of the imperial project that draws heavily on rhetoric and traditional narratives of "manly" behavior.

By "imperial project" I mean to refer both to policy-making proper and to strategies for securing assent to those policies. Although I offer the outlines of an argument that the American culture of masculinity did in fact play some part in *directly motivating* policy responses to September 11, this claim remains speculative and circumstantial. And at a certain level, it is less important than the ideological role of manhood ideals in *selling* the "war on terror." Whatever the "real" reasons were behind the invasion of Afghanistan, there is no question that a mobilization of destructive manhood ideals was centrally important in allowing the Bush administration to proceed with the support of most Americans.

My argument can be situated between two types of study. In a brilliant use of primary policy documents (transcripts of White House meetings), Robert D. Dean has demonstrated definitively that the Johnson administration's unwillingness to pull out of the Vietnam conflict in the late 1960s can be traced in large part to the same manhood ideals I discuss here (Dean, 2001). Lacking access to such overwhelming causal evidence, I cannot show, for example, that it was masculinity more than oil that drove the invasion of Afghanistan. However, public policy pronouncements and media discourse do more fully support the kind of conclusions reached by Kristin L. Hoganson regarding the role of masculinist assumptions in securing public support for the Spanish–American and Philippine–American wars a century ago (Hoganson, 1998). Commenting on the unifying effects of manhood ideology in selling those adventures, Hoganson writes, "On the one hand, gender served as a cultural motive that easily lent itself to economic, strategic and other justifications for war. On the other, gender served as a coalition-building political method, one that helped jingoes [a term originally coined for supporters of those two wars] forge their disparate arguments for war into a simpler, more visceral rationale that had a broad appeal" (Hoganson, 1998, p. 8). Today's jingoes, too, are unified to an important degree by a gender ideology mobilized to promote aggressive reactions to world events.

The present study needs to be situated, finally, in the context of feminist political geography. Jennifer Hyndman's definition of the aims of "critical geopolitics" perfectly captures my purposes as well: critical geopolitics is "a taking apart of normalized categories and narratives of geopolitics. [It] is about questioning assumptions in a taken-for-granted world and examining the institutional modes of producing such a world *vis-à-vis* writing about the world, its geography and politics" (Hyndman, 2001, p. 213). When the "normalized categories and narratives of geopolitics" are essentially gender categories and narratives, as is so clearly the case here, critical geopolitics becomes indistinguishable from feminist geopolitics. In general, my aim is to expose the purportedly natural, "unmarked" pattern of the American response to September 11 as contrived and very much "marked" by the masculine (and related, though subordinate, racist) coloration of the "war on terror."

Another sense in which the analysis offered here furthers a feminist agenda in political geography has to do with scale. Anna Secor argues that:

> In offering a critique of public-private dualisms and assumptions about scale and politics, feminist research need not simply replace the global with the local as the "authoritative" scale of analysis. On the contrary, feminist approaches show how the

(immanently political) categories of public and private, global and local, formal and informal, ultimately blur, overlap and collapse into one another in the making of political life. (Secor, 2001, p. 193)

While I don't pursue the theoretical implications, I do show in the final section of the chapter that the "penetrative" or "violative" impulses of the discourse surrounding the "war on terror" imprint the results of aggressive geopolitics across all scales, from that of the "international community" to that of the individual bodies held at Guantanamo Bay.

American Manhood Ideals and the Frontier Myth

In America, almost since the first stirrings of the industrial revolution, men have generally understood and experienced our manhood as something that *needs to be proven, and proven repeatedly*. Michael Kimmel argues that this imperative emerged during the nineteenth century, as the traditional pillars of a sense of manhood, land ownership and independent artisanry, became available to fewer and fewer American men (Kimmel, 1996; the brief cultural history to follow is condensed from Kimmel, 1996; Rotundo, 1993). The onset of industrialization brought a widespread perception of new economic opportunities, and gave birth to the resilient and still-important ideal of the "self-made man." At the same time, the journey of many men from a familiar rural life, a life led primarily among family, friends, and acquaintances (and perhaps known enemies), to a busy life in the newly bursting urban centers, where most daily contacts were often with strangers, brought about a new emphasis on external presentation, on impressing others in the course of the search for employers, partners, or investors. The audience for such self-presentation was (and remains) *other men*, the coveted commodities in their possession, respect, trust, confidence.

As the century progressed, however, and as the American industrial capitalist system matured, it became clearer that the dream of the self-made man could only be realized by a few. In addition to the difficult economic situation, women were entering traditionally male public life in ever greater numbers, and insisting on a larger role in the organization of American society generally. The pressure to prove one's manhood did not abate, but, as economic proofs were so hard to come by, American men were forced to turn to other strategies. In the decades surrounding 1900, many of the basic features of the culture of masculinity that survive to the present day were put in place. Men turned in large numbers to recreational sports, to body-building and prize-fighting, they organized and joined homosocial organizations (joining college fraternities, but also Moose, Elk, Odd-Fellows lodges, becoming Shriners and Masons, etc., etc.). Since the late nineteenth century especially, proving one's manhood has also meant proving that one is *not feminine*. The intensified stigmatization of gay men, and actual violence against gays or perceived "weaklings," became and have remained part and parcel of an evolving suite of behaviors by which men try to signal and establish their masculinity (Kimmel, 1996; see also Rotundo, 1993). Angus McLaren notes that it was during this period that many Western nation-states first criminalized homosexuals as a type of *person* (see McLaren, 1997, p. 219, also pp. 137–57).

As in many other societies throughout history, *fighting a war* has a long pedigree in the USA as one very important way to certify masculinity. Teddy Roosevelt played a crucial role in explicitly proposing the outward projection of violence through war as one response to the contemporary crisis of manhood. In so doing, he tapped into another extremely important mother-lode of cultural memory, the American *myth of the frontier*. This myth has been tied up both with ideals of masculinity and with our national identity since shortly after the Revolution (Slotkin, 1998a, b, 2000; sources for the following gloss also include Drinnon, 1980; Limerick, 1987; Nelson, 1998). According to America's frontier myth, our national identity was forged out of the strenuous demands placed on us by the task of conquering, and wresting a living from, the vast, primeval North American wilderness. Our fore-fathers ventured out beyond where it was safe, exposing themselves to danger, fight-ing Indians for the right to possess the land, hunting wild game, domesticating the natural world (of which the Indians were only the most dangerous representatives), making the continent safe for civilization.

In doing all this, American men (until very recently, women's only acknowledged roles in the story were to bear children, feed men, and be protected by them) so to speak "burned off" all the excess frippery and feminine culture associated with Europe. No longer "dandies," no longer subtle thinkers, no longer inhabitants of plush, perfumed sitting room furniture, American men had founded a culture of honesty, of hard work, of self-reliance, of physical toughness, decisiveness, and moral simplicity. For real American men, there was little that couldn't be solved by toughness and straightforwardness, no reason to think national or international problems couldn't be addressed by the same sort of simple wisdom and forceful action that got the horse through the gate. It should be obvious that there is a lot of overlap between the image of manhood forged on the frontier and the more urban ideal of the self-made man.

A readiness to get violent has been a central feature of the frontier manhood ideal throughout. So, too, has a presumed right to push onward past legal, official bound-aries into "virgin" territory, to take what can be taken and the law be damned. Richard Drinnon argues that this aggressive expansionism did not by any means stop when our forefathers reached the limits of the North American continent, but also helped to propel American military and economic interventions across the Pacific and around the globe (Drinnon, 1990). The evolution of the protagonists in this myth can be traced from Cooper's *Leatherstocking* through Custer, Owen Wister's *Virginian* and Teddy Roosevelt, to the Lone Ranger, John Wayne, Clint Eastwood, and Sylvester Stallone's *Rambo* (Drinnon, 1990; Slotkin 1998a, b, 2000). The archetypal situations in which these and other mythical heroes have taught American men they must prove themselves are Indian fighting and the showdown between gunslingers.

The evolving American culture of masculinity has been a culture of perpetual crisis, in the sense that men cannot very easily cease to worry (at least subconsciously) about their manhood, and must remain ready at all times to behave so as to certify and recertify it in front of other men (see Beneke, 1997, for a psychoanalytically grounded interpretation of this compulsion, and Robinson, 2000, for a discus-sion of white male "crisis" since the 1960s). In this general context, the already-powerful frontier myth encourages men to link their individual efforts at establish-ing masculinity with *national* military actions, and thus both to "masculinize" the meaning of "America" and to "Americanize" the meaning of "masculinity" (Nelson, 1998). The frontier myth also encourages the identification of "American" with "white" (i.e. of Northern and Western European descent), as has been clear from the numerous attacks against Middle-Eastern Americans over the past two years.

Nevertheless, my primary focus here remains masculinity because, in the context of the "war on terror," it is the elephant in the room that nobody is talking about. Among other things, I contend that the masculinist fusion of the national and personal helps to account for the broad but counterintuitive consensus among American public figures that the beginning of a war is an *inappropriate* time for critically questioning foreign policy decisions. It might be expected that critical discussion would *intensify* at the prospect of large-scale destruction, loss of life, and international destabilization, but this has not been the pattern. The familiar calls for "unity behind the President" are to a significant degree expressions of *masculine* identification. No individual man, and certainly no national administration, is *forced or required* to tap into these patterns, but they are so prevalent in so many of our cultural institutions (from child-rearing to education to television and cinema) that they mark out what is certainly the easiest path for many men to follow. A Gallup poll in the fall of 2001 demonstrated that following September 11, many American women accepted these basic stories about how American men and the national government should behave (Jones, 2001). It should not be surprising that two of the most frequently discussed books on American foreign policy to appear since September 11, Robert D. Kaplan's *Warrior Politics* (2002) and Max Boot's *The Savage Wars of Peace* (2002), are centrally concerned with tapping into and updating the intertwined myths of masculinity and the frontier.

All of this should not be taken to imply that other cultures in other parts of the world happily lack destructive myths of conquest or needlessly violent masculine ideals. Cynthia Enloe convincingly argues that it is impossible fully to understand the complexities of international politics without devoting attention to the variety of *different* cultures of masculinity inflecting various strands of militarism in all corners of the globe (Enloe, 1990, p. 13, 1993, pp. 71–101). Interestingly, in so far as masculinity has been explicitly thematized in the American media since September 11, the focus has been exclusively on *foreign* or *marginalized American* cultures of masculinity, not on the role masculine ideals play in the decisions and justifications of the actions taken by the US government and other powerful actors. Robert Kaplan's best-selling book *Warrior Politics*, a sloppily argued but conveniently timed call for the USA to get tough on the international stage and stop apologizing for its dominating behavior, is chock full of references to the dangerous hordes of "young males" in poor parts of the world ripe for recruitment into violent movements (Kaplan, 2002). But gender is conspicuously absent (or is suddenly naturalized) in these accounts when it comes to discussions of the heavy-handed behavior of American presidents and generals.

Terrorism and the Logic of the Showdown

Testeria: Crippling condition found in males, sometimes dangerously pathological. Accounts in part for the ability of the male ruling class to efficiently, calmly, and maturely carry out planetary catastrophe. Male inventions like war, capitalism, totalitarianism, industrialism, and other atrocities are only possible if millions of efficient, calm, mature male people are diligently repressing their healthy, humyn emotions. Since the turn of the century, over 50 million humyn beings have been slaughtered in war by psychiatrically normal male people. (Loesch, 1992)[1]

This definition of "testeria" was clearly intended to recast the "typically male" suppression of compassion and other human emotions as irrational and destructive, and thereby to denaturalize and problematize what has become the normal course of world affairs in a world still run chiefly by men. In my view this definition of testeria is too narrow, because it neglects the destructive effects of the *precipitate expression* of "typically male" emotions such as anger and aggression. Both the frontier myth and the American culture of masculinity charge aggression with positive meaning as a "natural" feature of male behavior. Whether or not the Bush administration and the media barons were acting "testerically" in the calm, calculating sense evoked in the dictionary definition, the support they generated among the American public rested in large part on the stoking of a "hotter," more emotional form of testeria. (My use of the term is of course cultural, not biological: women, too, can act testerically if they have internalized the masculinist standards of male conduct analyzed here.) Having set the cultural stage in the foregoing section, I attempt in this section to explain the process by which this hotter strain of testeria was induced.

The American culture of masculinity encourages us to conceive of conflicts in terms of tests of strength or prowess between two combatants playing by the same rules, as in the stereotypical western gunfight. The gunfight crystallizes and shimmers with the concentrated pressure of American manhood ideals more intensely than any other scenario of male action. If we[2] flip back our duster-coat to reveal our big, shiny revolver, and stare down the villain with our steely gaze, he won't make the fatal move for his own weapon. If he does, our steady nerves and cool hands will cut him down in the blink of an eye. The cultural icons we picture when imaging the showdown (John Wayne, Clint Eastwood, etc.) are familiar. The basic problem, the source of the widespread testeria that followed, is that the September 11 attacks, like most terrorist attacks, short-circuited the logic of the showdown. Much has been made in the press of the emotional unwillingness of many Americans in the months following 9/11 truly to accept that things couldn't be made right again, that we couldn't undo what had been done. It is my contention that this unwillingness to accept, and the enraged frustration that came with it, can only be fully explained with reference to specifically masculine myths. The basic structure of the showdown, of attempts to prove manhood in combat, is one of confrontation, followed by escalation, followed by a climax of either actual violence or the facing down of the weaker man by the stronger. It is crucially important in this mythical scenario that violence not happen right away, because men prove their manhood not simply or even primarily by *being* violent but, more importantly, by demonstrating unflinching resolve and courage *despite the danger of a violent exchange*. In other words, the showdown has to last a little while so the men have a chance to prove that they are brave, and for that to be possible, *each gunslinger has to be aware of the danger posed by the other*.

Measured against this mythical showdown scenario, the terrorist would indeed seem to be "a faceless coward," as Bush put it (quoted in Sanger and van Natta,

2001). But "cowardice" is not really the right word: it requires that someone be "cowed." The attackers of September 11 never gave the US military a chance to "cow" them. They bypassed the whole escalation scenario, never appeared for a preliminary confrontation or gave the warning that would have allowed the administration to show its fearless resolve. They simply produced horrible violence right away, horrible violence that will not be erased or made right even if the American military kills them all in a bloody war that shows the world once again the strength of the USA. The short-circuiting of the showdown, I submit, is one important factor behind the rage felt especially by many American men, and may help to explain the first name given to the US war of revenge in Afghanistan: "Operation Infinite Justice." With the chance to prove our manhood already irretrievably lost, the task of restoring our manly dignity indeed becomes "infinite."

The short-circuiting of the showdown also helps to explain the often surprising depth of fear engendered by the "terrorist threat" as composed not merely of physical fear but also of gender-based insecurity. It is possible to think of relations between masculine ideals and warfare as occupying a spectrum, or perhaps an evolution, with conventional wars at one end (in which conditions allow the showdown logic to operate), guerrilla warfare in the middle (already frustrating to conventional desires for open confrontation, but still involving the occasional possibility of direct tests of strength: see Boot, 2002, p. xv), and terrorism at the other end (which most thoroughly negates the conditions for manly confrontation). This is one place where racial and gender ideologies intersect: centuries of stereotypes of Indian combat provide the vocabulary for present-day accusations of terrorist "cowardice."

The gendered frustrations inherent in the present situation are a trigger for the "savage war" option in American policy, the quest for "extinction" of the enemy. Historian Richard Slotkin was one of a number of scholars asked to offer their thoughts on September 11 in a special supplement to the *Chronicle of Higher Education* that appeared two weeks later. Slotkin sees the USA opting for

> the myth of "savage war," based on the oldest US myth, the myth of the frontier. The myth represents American history as an Indian war, in which white Christian civilization is opposed by a "savage" racial enemy: an enemy whose hostility to civilization is part of its nature or fundamental character; an enemy who is not just opposed to our interests but to "civilization itself." The myth also provides a recipe for countering the threat, a model of heroic action that will bring victory and resolve the crisis. The hero of this myth is the wielder of extraordinary violence: He can win only by fighting fire with fire, evil with evil, and he must fight until the enemy is exterminated or utterly subjugated. In war with such an enemy, nothing less than total victory is acceptable. (Slotkin, 2001, p. B11)

Bush's language in his first address to the nation after September 11 fits this pattern fairly well:

> Our grief has turned to anger, and anger to resolution. . . . Our war on terror begins with al Qaeda, but it does not end there. It will not end until every terrorist group of global reach has been found, stopped and defeated. . . . With every atrocity, they hope that America grows fearful, retreating from the world and forsaking our friends. . . .

But this country will define our times, not be defined by them. I will not forget this wound to our country or those who inflicted it. I will not yield; I will not rest; I will not relent in waging this struggle. (Bush, 2001)

In assuming the mantle of the avenging hero, Bush identifies his "savage war" with the masculine project of recuperating injured masculine honor, and lends his own personal mission national redemptive significance.

"Punitive" Expeditions

[S]ymbols of right and wrong manhood have . . . become lodged in our political consciousness and in the decision-making culture of our great institutions. These symbols make certain choices automatically less acceptable, and in doing so they impoverish the process by which policy is made. We are biased in favor of options we consider the tough ones and against those we see as tender; we value toughness as an end in itself. We are disabled in choosing the wise risk from the unwise, and tend to value risk as its own form of good. In this manner we are hurt by the cultural configuration of manhood. (Rotundo, 1993, p. 291)

Ideas of masculinity have to be perpetuated to justify foreign-policy risk-taking. To accept the Cold War interpretation of living in a "dangerous" world also confirms the segregation of politics into national and international. (Enloe, 1990, p. 13)

Max Boot, at the time editorial features editor at the *Wall Street Journal*, gained notoriety almost overnight for an October 2001 article in the conservative *Weekly Standard* entitled "The case for American empire: the most realistic response to terrorism is for America to embrace its imperial role" (Boot, 2001). He was working at the time on a full-length book, an unapologetic chronicle of episodes of American empire spanning the past two centuries. Like Kaplan, Boot urges us to stop whining and face the fact that we have been for a long time, and remain, an imperial power aggressively pursuing its interests worldwide. Also like Kaplan, Boot adapts the pattern identified by Enloe above to the post-Cold War world, strongly distinguishing the domestic political realm, in which democracy and participation can play a major role, from international politics, where a "dangerous world" demands quick executive decisions by a small military-executive elite. Most importantly, his argument allows us to pinpoint more exactly that aspect of American foreign decision-making that can be attributed to our culture of masculinity rather than to any identifiable strategic necessity. The historical evidence amassed by Boot points to a more or less typical course of American armed intervention in the nineteenth century:

After killing some natives, the Americans seldom stayed long; nor did they usually involve themselves in local politics. What did they achieve? Sometimes a trade treaty; at other times, simply the satisfaction of having instilled fear of the Stars and Stripes. . . . The US strategy, if that is the right word for such a haphazard enterprise, might best be characterized as "butcher and bolt" – a bit of slang popular in Britain's Indian Army to describe punitive expeditions against troublemaking tribes, expeditions designed not to occupy territory but to "learn 'em a lesson." (Boot, 2002, p. 38)

In the twentieth century and up to the present day, our interventions have come more often to include extended stays on foreign soil, the assumption of administrative duties, the supervision of elections, etc. But the often strategically irrelevant "punitive" thread has continued, and in the post-9/11 world, has enjoyed a revival.

Boot traces this "punitive" thread in American "small wars" originally to the culture of nineteenth-century Navy officers: "The Navy's way of doing things created a starched, stiff-necked officer corps – imperious, hot-blooded, quick to take offense, and above all, brave, sometimes suicidally so" (Boot, 2002, p. 40). To Boot's credit, while he views this testerical hot-headedness as quasi-"natural," even in some cases admirable, he does not claim that it has any *intrinsic* strategic usefulness to American foreign policy. This tendency to excessive "punitive" retaliation even for imagined slights would of course be strengthened exponentially after a real attack such as that of September 11. In the last substantive section of this chapter, I briefly discuss some important geographic dimensions to the discourse surrounding the "war on terror." I submit that, whatever purely "strategic" purposes they may serve, the common *emotional* thread binding these dimensions together and making them largely acceptable to the American public is a ferocious "punitive" impulse to retaliatory violation brought on by a deeply felt injury to (explicitly masculine) American honor.

Infinite Justice and Violation at Different Scales

The description of "savage war" borrowed above from Richard Slotkin clearly entails a fundamental refusal to respect boundaries of all sorts (legal, national, material, or moral). How else could a tireless international hunt of the sort that Bush announced in his address to the nation be pursued? In what follows I try to give a more concrete geographic sense of what this means, with respect to international boundaries, sovereignty rights, and individual bodily integrity.

In his September 20 address to the nation, Bush had put the rest of the world on notice: "Every nation, in every region, now has a decision to make. Either you are with us, or you are with the terrorists. From this day forward, any nation that continues to harbor or support terrorism will be regarded by the United States as a hostile regime" (Bush, 2001). The UN General Assembly met shortly after the September 11 attacks, and condemned them in no uncertain terms. However, the implications of Bush's strict "good guy or bad guy" distinction were cause for alarm, especially among some of the world's poorer countries. A number of representatives to the General Assembly worried about how they would be able to acquire the resources to make sure they would not be harboring terrorists in the future. Speaking for Barbados, Ambassador June Clarke agreed that no country, however small, was immune from terrorism: "small countries are particularly vulnerable because they frequently do not have the logistical and intelligence assets to effectively track the activities of terrorists and other agents of transnational crime." The Acting Permanent Representative of Botswana, Leutlwetse Mmualefe, said many developing countries faced enormous social and economic problems and were "in dire need of resources and technical assistance to help us upgrade our capacities to effectively participate in the global coalition against terrorism." He stressed that conflict-ridden regions of the world had proven to be easier breeding grounds and havens for terrorist activities (UNGA, 2001). In other words, Bush's stark division of the world into good and evil demanded a degree of border integrity and internal territorial surveillance far beyond the resources of many nations.

US policy has put poor countries in a position of long-term vulnerability to American intervention, has rendered large parts of the world available in principle to violation at our whim. This is a new and serious weakening of the notion of national sovereignty and national boundaries. Its newness derives not from the actual violations of sovereignty that may follow (Max Boot shows just how "traditional" illegal violation is as a tool of US foreign policy), but because its explicit espousal as a legitimate general policy option for the world's strongest power and self-styled "good guy" spells trouble. Attuned to the newly explicit and aggressive American assertion of a right to play global bully, Robert Kaplan crows:

> The concept of "international law" promulgated by Hugo Grotius in seventeenth-century Holland, in which all sovereign states are treated as equal and war is justified only in defense of sovereignty, is fundamentally utopian. The boundaries between peace and war are often unclear, and international agreements are kept only if the power and self-interest are there to sustain them. (Kaplan, 2002, p. 118)

In other words, the surface of the Earth is our field of action, and we no longer put much stock in the claims other nations may have to the exclusive use and control of their own territory. According to these pundits, we can also now be frank about the narrowness of the interests we pursue when throwing our weight around; we can dispense with the laborious charade of trying to argue that our interventions are always in the service of universal human values.

But in the new dispensation, the Earth's surface is not merely our "field"; it is also our "oyster," as the saying goes, and in our pursuit of international terrorists, we have already begun to focus our technology on cracking the shell and digging for what we want. This "depth" dimension of our imperial reach comes out clearly in reportage on US attempts to root al Qaeda out of the caves of Afghanistan. As Andrew Revkin reported in the *New York Times* of December 3, 2001, "The Pentagon is hurriedly developing new earth-penetrating weapons even as American forces are striking dozens of suspected underground hide-outs of al Qaeda and the Taliban with specialized tunnel-blasting bombs and missiles." The new push to improve such weapons began in 1991 after the Persian Gulf War, when Iraq's "subterranean activities" were revealed. "It accelerated later in the 1990s as Libya, Iran, North Korea, terrorist camps in Afghanistan and other adversaries shifted activities underground to avoid attack and detection by satellites and aircraft." Revkin quotes Clark A. Murdock, former deputy director of long-range planning for the Air Force: "The crown jewels increasingly are buried, and that's why people are thinking about these things." Among the weapons being developed:

- the "Deep Digger" cannon, which "is said to eat into rock or reinforced concrete with a series of blasts, using secondary explosions to remove the resulting rubble quickly";
- the AGM-86D, a "refurbished deep-penetrating version of the Air Force's formerly nuclear-tipped aircraft-launched cruise missile";
- a "hard-nosed bomb designed to penetrate and incinerate buried caches of chemical or biological weapons without releasing contamination into the air";

- a warhead "carried on a missile or a bomb [that] would attack tunnels by piercing a reinforced door, then skipping several times down a passageway – keeping track of the bounces and detonating only when it was deep inside";
- a "thermobaric" weapon, developed specifically for use against North Korea, whose effectiveness "lies in a new mix of highly explosive materials that when released and detonated in a tunnel create a long-lasting wave of high pressure that kills anyone or destroys equipment throughout a maze of passageways" (Revkin, 2001).

These weapons are certainly not mere figments of the American culture of masculinity. They have their origins in a program of global military dominance inseparable from the wider imperial project identified by Boot, Kaplan, and others. But I would suggest that in concert with the penetrative measures being considered at both larger and smaller geographical scales, they help to constitute "Operation Infinite Justice" as a masculine psychopolitical program of violative revenge. A "savage war" of "infinite justice" certainly has nothing to do with the idea of rehabilitative punishment, which is not only aimed at reintegrating the perpetrator into "normal" social life but also in principle *proportional to the crime*. Infinite justice is punishment *out of all proportionality* to the crime, and the demand for it points to psychological needs that in fact have little to do with the notion of balance implicit in the image of the "scales of justice." Again, I suggest that this excessive, vengeful anger was provoked in significant measure by the fact that the terrorist attacks short-circuited the masculine showdown logic, that they placed American national manhood in a compromised, unredeemable position from the outset. At the heart of the administration's (and many Americans') reaction to September 11 was an insistence that *somebody pay dearly for this*, and that the magnitude of payment exceed (in an almost ritual way) that of the original injury. This interpretation is also suggested by the fact that recourse to torture of suspected terrorists in US custody became an explicit topic of discussion in late October and early November 2001.

Anna Secor rightly insists that feminist political geography not abandon the international scale in favor of the local, but strive to show how they interdigitate (Secor, 2001, p. 193). In the aftermath of September 11, it was clear that the bodily and the geopolitical scales could invest one another directly. An article by Jim Rutenberg in the November 5 *New York Times* entitled "Torture seeps into discussion by news media" begins with a quote from *Newsweek* columnist Jonathan Alter: "In this autumn of anger, even a liberal can find his thoughts turning to . . . torture" (quoted in Rutenberg, 2001). According to Rutenberg, Mr Alter reported subsequently that he had not received a "significant flood of e-mail messages or letters" objecting to the mention of such extreme measures. Other news commentators quoted in the article, while hesitant to endorse torture as such, justified at least considering it by arguing that information obtained through torture could end up saving lives in the future.

An October 21 *Washington Post* article by Walter Pincus had detailed the frustrations felt by FBI agents at the stonewalling of the suspected terrorists in their custody. At that time the Bureau was considering a number of alternative strategies, among them "drugs or pressure tactics, such as those employed occasionally by

Israeli interrogators, to extract information. Another idea is extraditing the suspects to allied countries where security services sometimes employ threats to family members or resort to torture" (Pincus, 2001). The government has thus far not openly admitted pursuing such suggestions. However, the images of kneeling prisoners under strict sensory deprivation at Guantanamo Bay suggest a physical vindictiveness not justified merely by the desire to prevent communication among prisoners (Amnesty International, 2002). The main thing about these prisoners is that *we have them*, and until we have bigger fish in our nets, it is their bodies that will have to bear the brunt of our desperate insistence on making someone pay.

In the discursive field opened up by this early, tentative, and explicitly emotional broaching of the subject, it has become possible in the intervening years to discuss torture in cool and apparently rational terms as a US policy option. A cover article by author Mark Bowden featured in the generally liberal *Atlantic Monthly* (Bowden, 2003) begins by positing a distinction between "torture" and "coercion" (the latter leaving no lasting *physical* damage), a distinction rejected by the Geneva Convention, the Universal Declaration of Human Rights, and all international humanitarian organizations. Much of the article consists of a long survey of available interrogation methods, inviting the reader to listen in (presumably with rapt admiration) to the "fishing stories" of legendary interrogators in Israel and New York City. Bowden dismisses the flat no-torture position of groups such as Amnesty International as a product of the "civilian sensibility," which "prizes above all else the rule of law" (Bowden, 2003, p. 70). The "warrior sensibility," by contrast, "requires doing what must be done to complete a mission. By definition, war exists because civil means have failed" (ibid.).

Quite apart from the very problematic assumption that all available civil means for combating terrorism have been exhausted (for that matter, even attempted in any serious way), the masculine underpinnings to this "rational" position are not difficult to detect. Despite Bowden's attempts to argue that all of this has nothing to do with emotions, he is clearly advocating the "savage war" option identified by Slotkin (see above), this time at the level of individual bodies. Bowden's conclusions:

> The Bush Administration has adopted exactly the right posture on the matter. Candor and consistency are not always public virtues. Torture is a crime against humanity, but coercion is an issue that is rightly handled with a wink, or even a touch of hypocrisy; it should be banned but also quietly practiced. . . . It is wise of the President to reiterate US support for international agreements banning torture, and it is wise for American interrogators to employ whatever coercive methods work. It is also smart not to discuss the matter with anyone. (Bowden, 2003, p. 76)

Conclusion

The "crown jewels" the US administration is after in its "war on terror" may take the form of organizations hidden inside the boundaries of a foreign country, they may take the form of leadership cadres hidden in networks of underground caves, they may take the form of information hidden inside the bodies of individual terrorism suspects. No matter. At all of these scales, we are contemplating or already using the physical force necessary to violate boundaries and reach our targets. This multilevel program of penetration is not merely (if it is at all) a neutral commitment to locate guilty parties; it is also a vindictive, symbolic exercise of grimly enraged conquering power, one in which innocent civilians, whether in Afghanistan or Iraq, are acceptable "collateral damage." The "real reasons" why Bush and many media commentators have insisted on prevailing in such a definitive, violative manner may not come to light for years. But it is in any case certain that their success in winning the support of many Americans can be traced to a virulent outbreak of testeria nourished in the soil of the American frontier myth and our culture of masculinity.

NOTES

1   I am grateful to Peggy Luhrs for introducing me to the term "testeria."
2   The politics of "we" are of course very tricky. I have chosen to identify myself here with American males, as my own social integration was saturated with the logics described here, and it would be naive of me to pretend that I have been able to expunge all traces of these ideologies from my own thinking. More problematically, I have also used "we" when referring to Americans in the political context after 9/11. Many on the American left insist that the Bush Administration is not "us," and that it is important to signal to the rest of the world that "we" are not represented by its actions. I accept this, but use "we" anyway to evoke a sense of our (Americans') continuing urgent responsibility to change the situation that has led to this disidentification.

BIBLIOGRAPHY

Amnesty International (2002) USA: treatment of prisoners in Afghanistan and Guantanamo Bay undermines human rights, May 11 (http://www.amnesty.org/web/content.nsf/pages/gbrsep11crisis).

Beneke, T. (1997) *Proving Manhood: Reflections on Men and Sexism*. Berkeley: University of California Press.

Berkhofer, R. F. Jr (1978) *The White Man's Indian: Images of the American Indian from Columbus to the Present*. New York: Vintage Books.

Blaut, J. (1993) *The Colonizers' Model of the World*. New York: Guilford.

Boot, M. (2001) The case for American empire. *The Weekly Standard*, October 15.

Boot, M. (2002) *The Savage Wars of Peace*. New York: Basic Books.

Bowden, M. (2003) The dark art of interrogation. *Atlantic Monthly*, 292(3), 51–76.

Bush, G. W. (2001) Address to a joint session of Congress and the American people, September 20 (http://www.whitehouse.gov/news/releases/2001/09).

Dean, R. (2001) *Imperial Brotherhood: Gender and the Making of Cold War Foreign Policy*. Amherst, MA: University of Massachusetts Press.

Drinnon, R. (1980) *Facing West: The Metaphysics of Indian Hating and Empire Building*. New York: Schocken Books.

Enloe, C. (1990) *Bananas, Beaches and Bases: Making Feminist Sense of International Politics*. Berkeley: University of California Press.

Enloe, C. (1993) *The Morning After: Sexual Politics at the End of the Cold War*. Berkeley: University of California Press.

Filkins, D. (2001) To the stranger, a wild land, strangely awesome. *New York Times*, November 9.

Hoganson, K. (1998) *Fighting for American Manhood: How Gender Politics Provoked the Spanish–American and Philippine–American Wars*. New Haven, CT: Yale University Press.

Hyndman, J. (2001) Towards a feminist geopolitics. *Canadian Geographer*, 45, 210–22.

Jennings, F. (1976) *The Invasion of America: Indians, Colonialism and the Cant of Conquest*. Williamsburg: W. W. Norton.

Jones, J. (2001) Men, women equally likely to support military retaliation for terrorist attacks, October 5 (http://www.gallup.com/poll/releases).

Kaplan, R. (2002) *Warrior Politics: Why Leadership Demands a Pagan Ethos*. New York: Random House.

Kimmel, M. (1996) *Manhood in America: A Cultural History*. New York: Free Press.

Limerick, P. (1987) *The Legacy of Conquest: The Unbroken Past of the American West*. New York: Norton.

Loesch, J. (1992) Quoted in Testeria. In C. Kramarae, P. Treichler and A. Russo, *Amazons, Bluestockings and Crones: A Feminist Dictionary*. New York: Pandora Press.

McLaren, A. (1997) *The Trials of Masculinity: Policing Sexual Boundaries, 1870–1930*. Chicago: University of Chicago Press.

Nelson, D. (1998) *National Manhood: Capitalist Citizenship and the Imagined Fraternity of White Men*. Durham, NC: Duke University Press.

Pincus, W. (2001) Silence of 4 terror probe suspects poses dilemma. *Washington Post*, October 21.

Revkin, A. (2001) US making weapons to blast underground hide-outs. *New York Times*, December 3.

Robinson, S. (2000) *Marked Men: White Masculinity in Crisis*. New York: Columbia University Press.

Rotundo, E. A. (1993) *American Manhood: Transformations in Masculinity from the Revolution to the Modern Era*. New York: Basic Books.

Rutenberg, J. (2001) Torture seeps into discussion by news media. *New York Times*, November 5.

Sanger, D. and van Natta, D. Jr (2001) Four days that transformed a President, a presidency and a nation, for all time. *New York Times*, September 16.

Secor, A. (2001) Toward a feminist counter-geopolitics: gender, space and Islamist politics in Istanbul. *Space and Polity*, 5, 191–211.

Slotkin, R. (1998a) *The Fatal Environment: The Myth of the Frontier in the Age of Industrialization, 1800–1890*. Norman: University of Oklahoma Press.

Slotkin, R. (1998b) *Gunfighter Nation: The Myth of the Frontier in Twentieth Century America*. Norman: University of Oklahoma Press.

Slotkin, R. (2000) *Regeneration through Violence: The Mythology of the American Frontier, 1600–1860*. Norman: University of Oklahoma Press.

Slotkin, R. (2001) Myths provide society with a functioning memory system. *Chronicle of Higher Education*, September 28, B11.

United Nations (2001) Daily highlights, October 5 (http://www.un.org/news/dh/20011005.htm).

Chapter 37

Feminist Geopolitics and September 11

Jennifer Hyndman

As I sit down to write this introduction, the first "anniversary" of September 11 is being marked by several weeks of media coverage. The *New York Times* continues to publish "portraits of grief," the biographies of people killed in the attacks on the World Trade Center and the Pentagon, including the police officers and firefighters who attempted to rescue those stranded. September 11, 2002 offers much more to think about in light of the "war on terror." Afghanistan remains highly insecure, despite the removal of the Taliban in November 2001. Afghan President Hamid Karzai has survived assassination attempts; already one member of the cabinet has been killed. Many people in the country still face the conditions of starvation present a year ago, and less than one-third of the $1.8 billion in international aid pledged to Afghanistan for this year has materialized. Afghan civilians have been killed by the aerial bombings once too often, forcing the USA to apologize publicly in July 2002 for killing most members of a wedding party. Added to this is the seemingly hidden cost of the war against terrorism on human and civil rights for Arab Americans, documented and undocumented immigrants from majority Muslim countries, and those captured and kept as "enemy combatants" on the US base at Guantanamo Bay, Cuba, including several minors. Has there been *any* constructive political change or justice served since September 11, 2001?

This chapter is not simply an analysis of the events of September 11. It aims to reconstitute conventional understandings of security and to construe them in ways that are broadly in keeping with a feminist politics. The events and aftermath of September 11 provide a basis on which to explore geopolitics from a highly embodied, situated, yet globalized location. While I do aim to *analyze* the events and aftermath of September 11, I also want to position myself politically in relation to the violence they embody. Nothing justified the killing of innocent unarmed people on September 11. Nothing justifies the retaliatory killing of innocent people elsewhere. The perpetrators of such destruction and violence should be brought to justice, though I lament the fact that no civilian venue has yet been identified.[1] I do not condone US and British bombing in Afghanistan, which has killed uncounted civilians and imperiled the lives of hundreds of thousands more. Such "collateral damage," as these deaths are euphemistically referred to, is unlikely to ever be prosecuted. Terror in the USA on September 11 has been met with increased terror in Afghanistan, and elsewhere, since October 7, 2001 and continuing well into 2002.[2] The uniqueness of this event in the USA has been confused with the ordinariness of such agony resulting from ongoing terror and senseless violence in other parts of the world.

The events and aftermath of September 11 ineluctably ended the already precarious distinction between domestic, sovereign space and interconnected global space where transnational networks, international relations, multilateral institutions, and global corporations operate. If it existed, any comfortable distinction between domestic and international, here and there, us and them, has ceased to have meaning after that day. One reaction to these horrific acts of violence has been to seal the border, but this is short-sighted if none of the 19 men who committed the atrocious acts of violence crossed an international border that day. Both the USA and Canada are countries of immigrants whose labor is critical to a prosperous economy. On September 11, 2001 most of the hijackers had entered the USA legally. In exploding *domestic* aircraft, carrying US civilians and themselves, they imploded any notion that political borders contain political conflict. Feminists have long argued that private–public distinctions serve to depoliticize the private domestic spaces of "home" compared to more public domains (Domosh and Seager, 2001). The attacks of September 11 certainly exposed the limitations of "domestic" space, somehow bounded and separated from the processes and politics of economic, cultural, and political integration.

In this climate of fear, public consent has been mobilized to reconstitute the country as a bounded area that can be fortified against outsiders and other global influences. In this imagining of nation, the USA ceases to be a cosmopolitan constellation of local, national, international, and global relations, experiences, and meanings that coalesce in places like New York City and Washington, DC. Disparate processes of globalization underscore the estimated one million immigrants who move to the USA annually or the huge flows of foreign investment in global production that flow the other way. The nation-as-state is increasingly defined by a "security perimeter" and the strict surveillance of borders. Before September 11, I contend that the nation-state was considered a more relational, cosmopolitan, and dynamic geographical notion of place.[3] The meeting of social, economic, political, and cultural relations at different scales in a given location produces place (Massey, 1997). In contrast, bounded thinking about discrete political borders has had concrete implications for bodies marked as "other," airline security, immigration and visitor visa regulations (in the USA and Canada), and customs control, especially at the land border between the USA and Canada. But it has more reactionary, if less tangible, implications for American politics, US immigrants, and questions of how "civil" society should be. Anti-terrorist legislation, anti-immigrant rhetoric, and public support for a military response are as much an expression of outrage and insecurity as they are evidence of government resolve and a heightened need for increased security in the face of the heinous attacks of September 11. What has been disturbing, however, is the shrinking space and number of venues available for open dialogue about the attacks and responses to them. I attempt to reclaim some of this space by calling for what might be thought of as a *feminist geopolitics*, a more accountable and embodied notion of politics that analyzes the intersection of power and space at multiple scales, one that eschews violence as a legitimate means to political ends (Hyndman, 2001).

In what follows, I outline the theoretical underpinnings of what I refer to as feminist geopolitics, tracing its antecedents in critical geopolitics and feminist critiques of international relations theory. Employing this analytical framework (see Hyndman, 2003), I construct a political space beyond the binary logic of "either/or" and advocate more embodied ways of seeing and knowing by examining the casualties in the wars of/on terrorism.

Feminist Geopolitics

For the purpose of this chapter, I define "feminist" as analyses and political interventions that address the inequitable and violent relationships of power among people and places based on real or perceived differences. Implicit in this definition is an understanding that masculinist power relations imbue the social, economic, and political arrangements that frame livelihoods across the world. Yet, while gender remains a central concern of feminist politics and thought, its primacy over other social, political, and economic locations is not fixed across time and place. Feminist geopolitics attempts to develop a politics of security at multiple scales, including that of the (civilian) body. It decenters state security, the conventional subject of geopolitics, and contests the militarization of states and societies with a "world system" perspective. It seeks embodied ways of seeing and material notions of protection for people on the ground. Feminist geopolitics is not a new theory of geopolitics or a new ordering of space (Hyndman, 2001). It is an analytic and politics that is contingent upon context, place, and time. Just as place is constituted at multiple scales, so too is geopolitics.

Within geography, Eleonore Kofman imagines a feminist geopolitics as one which would incorporate feminist analyses and gender into an extant set of geopolitical practices.

> The most successful incorporation of feminist insights and gender issues into geopolitics would dismantle and democratize geopolitics such that it no longer involved the personnel of statecraft located with the most repressive echelons of the state. Real groups would then begin to figure in the landscapes and maps of the global economy and power relations. Geopolitics would open out into a broader context which we could call global political geography, in which comparative analyses and the local, however that is defined, would also be included. (Kofman, 1996, p. 218)

Kofman's description of feminist geopolitics is uniquely situated within both political and feminist geography. It aspires to a less punitive version of the state-centric realist geopolitics. It also tacitly identifies a gap in the geographic literature: that the scale at which security is generally conceptualized precludes collective concerns, civil groups, and individual protection. A feminist geopolitics might be viewed at once as a critical approach and a contingent set of political practices operating at multiple scales that include, but are not restricted to, the nation-state (Hyndman, 2001).

Feminist geopolitics owes a significant debt to feminists in political science who have developed substantial critiques of international relations theory (Enloe, 1989, 2000; Peterson 1992, 1996; Whitworth, 1994; Pettman 1996). These authors, among others, have provided sustained and incisive critiques of international relations (IR) theory. Feminist critiques of security, for example, challenge the tacit territorial assumptions of states by asking whether states actually render their populations secure (see Peterson, 1992). Many of these analyses, however, have failed to go beyond the neorealist narratives of international relations.

Critical geopolitics as a subfield of political geography that emerged in the late 1980s aims to fill this gap. Critical geopolitics provides a useful departure point for making sense of the responses to September 11. Critical geopolitics is a less a theory of how space and politics intersect than a taking apart of normalized categories and narratives of geopolitics. It is about questioning assumptions in a taken-for-granted world and examining the institutional modes of producing such a world *vis-à-vis* writing about the world, its geography and politics.

This subfield of political geography is about suspending modernist assumptions of pregiven centers from which politics and knowledge are constructed and meaning imposed (Dalby, 1991). It analyzes the discursive practices by which scholars spatialize international politics in a single world characterized by specific groups of people, places, and stories. Within geography, "Critical geopolitics is one of many cultures of resistance to Geography as imperial truth, state-capitalized knowledge, and military weapon. It is a small part of a much larger rainbow struggle to decolonize our inherited geographical imagination so that other geo-graphings and other worlds might be possible" (Ò Tuathail, 1996, p. 256). This struggle to create other possible worlds overlaps with the project of feminist geopolitics.

While critical geopolitics is useful for a feminist geopolitical analysis, its deconstructive impulses are to my mind necessary but insufficient in generating change or preventing violence (see Nelson, 1999). Critical geopolitics decenters the nation-state and exposes the investments that our dominant geopolitical narratives embody, but it doesn't put Humpty Dumpty back together again, so to speak. Nor does it question why Humpty is always falling off the wall. We are left with well interrogated categories, but no clear way forward in practice (Hyndman, 2001).

A feminist imaginary invokes a universe of politically possible interventions, actions, and alliances, but its subject is neither postmodern nor universal. The distinction between a universe of potential modes of engagement (my notion of feminist geopolitics) and universalist notions of what engagement should look like (in which modernity is singular and fixed across space) is another way to get at this difference. I employ this feminist geopolitical imaginary broadly as a grid of intelligibility to analyze the militarized response to the attacks and the aftermath.

The feminist theory and politics of transnational feminist studies also situates a feminist geopolitics (Bachetta et al., 2001). In the wake of September 11, a group of USA-based feminist scholars posted a statement that captures the feminist geopolitics imaginary of which I speak:

> many women in Afghanistan are starving and faced with violence and harm on a daily basis not only due to the Taliban regime but also due in large part to a long history of European colonialism and conflict in the region. The Bush administration's decision

to drop bombs at one moment and, in the next, care packages of food that are in every way inadequate to the needs of the population offers a grim image of how pathetic this discourse of "civilization" and "rescue" is within the violence of war.

We are concerned about the ways in which the "war against terrorism" can be used to silence and repress insurgent movements across the globe. We also emphasize how racism operates in the naming of "terrorism". . . . The production of a new racial category, "anyone who looks like a Muslim" in which targets of racism include Muslims, Arabs, Sikhs, and any other people with olive or brown skin, exposes the arbitrary and politically constructed character of new and old racial categories in the US.

This statement interrogates the dominant ways of seeing the "war on terror" and its solution. Feminist geopolitics allow for "new ways of seeing, theorizing, and practicing the connections between space and politics" (Murphy Erfani, 1998).[4] Contributing to emerging debates in feminist geopolitics, I am concerned about the consequences of "terror" for unsuspecting citizens on both ends of this continuum of violence.

Either/or–Neither/nor: Feminist Geopolitics in Practice

In his televised address on September 20, President Bush drew a clear line between the two sides in "the war against terrorism": "if you are not with us, you are against us." On October 7, Osama bin Laden stated that the world is divided into two regions – one of faith and another of infidelity (Hensmen, 2001). Such binary thinking has become part of the dominant geopolitical narrative, garnering support for both sides, and leaving little space in between for those who fail to identify with either side. This narrative device relies on what Chantal Mouffe (1995) has called "the constitutive outside." Subject constitution and legitimacy require definition, something to define one's project *against*. Not only are such binaries logically questionable, they are also politically weak, reproducing the dominant geopolitical narrative as the only political option. "President Bush's ultimatum to the people of the world – 'If you are not with us, you're against us' – is a piece of presumptuous arrogance. It's not a choice that people want to, need to, or should have to make" (Roy, 2001a). In the aftermath of the attacks on September 11, the creation of an antagonist and an enemy territory (i.e. Afghanistan) has been crucial to the US government's response. Evidence that 15 of the 19 men on the four hijacked flights that crashed were Saudi nationals, and that Al Qaeda is a transnational network with operatives in at least 34 countries, including Germany, the USA, and Canada, has been glossed over. A target was sorely needed and quickly identified.

Feminist political sociologist Cynthia Cockburn (2000) rejects the logic of "either/or" militarization in the context of NATO attacks on the Federal Republic of Yugoslavia in 1999. As forces of the Milosevic government attacked ethnic Albanians in Kosova/o, two options were presented: *either* NATO attacks the Serbs (including some civilian targets) *or* ethnic Albanians in Kosova/o will be annihilated. Cockburn contests the militarized either/or ultimatum with a logic of her own, noting that "neither/nor" was an option that received little attention (Cockburn, 2000). The same feminist logic, I contend, applies to the events and aftermath of

September 11: neither is the killing of thousands of innocent civilians in New York, Washington, and Pennsylvania warranted, nor is the killing of thousands of innocent civilians in Afghanistan. The space to voice this kind of dissent has, however, been highly restricted in North America since the attacks, a point to which I will return.[5]

Few Degrees of Separation

A brief political geography of the Taliban and Osama bin Laden's early years also breaks down the convenient shorthand of "either/or." Certainly, an analysis of the history of all US foreign policy and its consequences is neither possible nor appropriate here, but a brief survey of US foreign policy as it relates to the formation and activities of the Taliban and Osama bin Laden is surely justified. In this context, mutually exclusive spaces of "here" and "there" and political dyads of "us" and "them" allow us to see the world more clearly, but less honestly. Michel Chossudovsky (2001) attempts to unsettle the dominant geopolitical narrative that emerged after the attacks of September 11, highlighting instead the connections between the USA, the Taliban, and Osama bin Laden. He notes that the largest covert operation in the history of the CIA was launched in 1979 in response to the invasion of Afghanistan by the Soviet Union. Working together with the Pakistan Inter Services Intelligence (ISI), the CIA actively encouraged political instability in Afghanistan and beyond, with the idea that Muslim states could eventually defeat the Soviet Union. Between 1982 and 1992, some 35,000 radicals from 40 Islamic countries joined Afghanistan's fight. Tens of thousands more went to study in madrassas (Koranic schools) in Pakistan, as refugees fleeing the fighting in Afghanistan (Rashid, 1999).

Pakistan's ISI was the go-between between the CIA and the Mujahadeen (which included Osama bin Laden) in Afghanistan. The CIA's support was covert and indirect, so as not to reveal its own geopolitical investments. While the Cold War began to fade and Soviet troops withdrew in 1989, the Islamic jihad based out of Pakistan did not, and the civil war in Afghanistan continued unabated. *Jane's Defense Weekly* (cited in Chossudovsky, 2001) reported that half of Taliban manpower and equipment originated in Pakistan under ISI. What is more disturbing is that the Clinton Administration appears to have known about links between the ISI and Al Qaeda, including the former's use of Al Qaeda camps in Afghanistan to train covert operatives for use in a war of terror against India. The "evildoers," as they have been called by US President Bush, were once US allies. One can speak more accurately, then, about degrees of separation between the USA and the Taliban than about historic enmity, longstanding hatreds, or the absence of political ties.

Just as Cold War geopolitics connected the USA with Central Asia, the Taliban, and indirectly Osama bin Laden, tracing the geopolitics of oil interests goes some distance in explaining the United States' hands-off approach to the Taliban and its treatment of women in Afghanistan, one which parallels US treatment of Saudi Arabia, its most important oil ally before the invasion of Iraq. When the Taliban won control of Kabul in 1996, Washington said nothing. In December 1997, Taliban leaders met with US State Department officials in Washington and visited Houston, Texas, to meet with UNOCAL oil executives (Pilger, 2001). "At that time the

Taliban's taste for public executions and its treatment of Afghan women were not made out to be crimes against humanity" (Roy, 2001b). Assured access for an oil pipeline from the Caspian Sea oil and gas reserves to the Indian Ocean would decrease US reliance on Middle Eastern sources (Chossudovsky, 2001). Turkmenistan, which borders on Afghanistan, holds the world's third largest gas reserves and an estimated six billion barrels of oil reserves. The desire for political stability in this region has a subtext.

The political economy of oil, underwriting an earlier war in the Persian Gulf, combined with US efforts to destabilize the USSR not so long ago, illustrate the geopolitical designs of superpower on this region. Such designs shape what we see and hear in the mainstream media. Conventional state-centered and resource-driven geopolitics promote a dominant geopolitical narrative that ensconces the "us"/"them" binary.[6] Such politics obfuscate minor voices and non-militarized responses to the attacks, and muffle dissent where it finds expression. After September 11, a dominant geopolitical narrative generated all-knowing maps of meaning that have been disseminated through the mainstream media. These god's-eye cartographies of peopleless places, mostly in Afghanistan, have mobilized consent for more violence in subtle ways. They enable military maneuvers to proceed, despite significant opposition to attacks that would harm innocent civilians.[7]

Such maps construct particular sightlines that enable one to see "enemy" positions and movements via remotely sensed satellite data, but omit images of the Afghan civilians killed by the American and British attacks. Acts of omission are as much acts of commission in this context, and more accountable maps are in short supply. Matt Sparke (2001) has asked what other maps we might draw: "maps that might trace where bin Laden's financial support has come from over the years; maps that might show how he has been 'harbored' by other states that the US would be much less inclined to bomb; and, maps that show how the dead terrorists themselves were once 'harbored' in states across America itself." Maps that forge links and recognize extant networks among political actors resist "either/or" reasoning and have the potential to enhance their accountability.

Embodied Vision and Visible Bodies

Despite the valuable interventions of critical geopolitics, one of its shortcomings is the highly disembodied mode of its critique (Sharp, 2000; Sparke, 2000). In contrast, embodied vision is ontologically committed and admittedly partial in perspective. Such commitments do have the potential to subvert prevailing ways of understanding "the problem," actions that might have concrete effects on the lives of people who are players in such events. In other words, feminist geopolitics enables political intervention or change to occur, albeit not in any universal or global way. Instead, such actions are strategically essential(ist) while being historically and geographically contingent. Since September 11, 2001 I have read the short, often moving, biographies of hundreds of the people killed, in the *New York Times*. Until the spring of 2002, an updated body count of the people lost in the World Trade Center, in the Pentagon, and on the flights that never arrived at their destination was published every Sunday.[8] As an audience, the human face of these horrific acts of violence in the USA was everywhere apparent. It took a long time, however, before the same paper began to publish photos of civilians who had lost family members, homes, and livelihoods to the bombings in Afghanistan. The *New York Times*, among other mainstream media, took even longer to cover any statistics about how many civilians have been killed in that country by US military planes equipped with smart and not-so-smart bombs (Bearak, 2002). The audible silence around the equally preposterous deaths of a people already ravaged by war and, in many regions, starvation is remarkable.

Reports of civilian deaths in Afghanistan took time to filter back to the US mainstream media, but once they did, an alarmingly visible landscape of death and destruction emerged. Barry Bearak chronicles attacks on five towns and villages, "mysteries" that remain unresolved in which large numbers of civilians were killed. With information from other reporters, his article discusses the questionable use of cluster bombs, some of which fail to detonate on contact and are littered "live" around the countryside of Afghanistan.[9] While the tragedies at both of ends of this violence are not disproportionate, in terms of lives lost, the patriotic values placed on them (or not) vary tremendously. Body counts provide a reality check: violence kills civilians and is unwarranted wherever it occurs. Yet where is the space beyond retribution and the "either/or" logic of militarization?

Neil Smith (2002, p. 635) argues that the "need to nationalize September 11 arose from the need to justify war." Smith argues that the World Trade Center catastrophe was a profoundly local and also global event, yet it was produced as a *national* tragedy. This politically strategic rescaling of September 11 serves at once to limit public and media expression that questions government tactics of combating terrorism and forces citizens to see these tactics as a matter of *national* interest and security.

The Privatization of Women in Public War

Women inside Afghanistan were virtually invisible when US-led bombing of Afghanistan began on October 7, 2001. Taliban rule between 1996 and 2001 ensured that unaccompanied women were not found in public space. Despite the "fall of Kabul" to the Northern Alliance on November 13, 2001, the spectacle of celebrations in mainstream Western media showed mostly men and boys celebrating on the streets. The odd well placed woman can be seen, uncovering her face or returning to paid work, but both represent highly charged notions of ethnocentric "progress" published in these media. The media did "celebrate" women's liberation from Taliban oppression after Kabul fell, but the emancipation of Afghan women did not become a political objective until after September 11, when the social geography of Afghanistan became intelligible to the North American public.

Both President George W. Bush and First Lady Laura Bush began to voice the importance of women's rights in Afghanistan, but not until after Kabul was captured in November 2001. Mrs Bush was the first wife of a US president to deliver the whole of the weekly radio address, on the theme of rights for Afghan women. "Only the terrorists and the Taleban forbid education to women. Only the terrorists and the Taleban threaten to pull out women's fingernails for wearing nail polish," Mrs Bush said. "The plight of women and children in Afghanistan is a matter of deliberate human cruelty carried out by those who seek to intimidate and control," she noted (BBC, 2001). The Taliban had violated women's rights since it took control of Afghanistan in 1996, but only after the USA took control of Afghanistan did their plight receive such high profile attention. In his State of the Union address on January 29, 2002, the President said:

> We have a great opportunity during this time of war to lead the world toward the values that will bring lasting peace. . . . We have no intention of imposing our culture. But America will always stand firm for the non-negotiable demands of human dignity: the rule of law; limits on the power of the state; *respect for women* [emphasis added]; private property; free speech; equal justice; and religious tolerance.

Freeing Afghan women became a convenient aim of the war on terror. The emancipation of similarly affected Saudi women has never been a political objective of US government.

Promotion of the greater civility of the Northern Alliance over the Taliban has been conducted with such enthusiasm that one cannot help but be skeptical. The Revolutionary Association of Women of Afghanistan (RAWA, 2002, p. 122) and Kathy Gannon (2001) both note that the Northern Alliance has no better record in its treatment of women than does the Taliban. These (Western) claims of the liberation of (Eastern) women accentuate the ways in which the "woman question" has been mobilized for specific political agendas, and how real women's bodies get caught in these webs. "There are such clearly racialized, Orientalized discourses about the Taliban women's horrific oppression, circulated by those wanting to rationalize the invasion, and aimed at locating the invaders as the liberators."[10] The visceral and carceral price paid by Afghan women for the "war on terrorism" and the war against the West by the Taliban that preceded it force a reconceptualization of geopolitics in which the scale of the body – the space on which war is effectively waged – is central.

Without Conclusion

The war on terrorism continues, now in Iraq. In this context, feminist geopolitics represents a space beyond the binaries of either/or, here/there, us/them. As an ethnographic, rather than a strategic, perspective it does not promote an oppositional stance in relation to particular political principles or acts. Instead, as an analytic, it attempts to map the silences of the dominant geopolitical position(s) and undo these by invoking multiple scales of inquiry and knowledge production. Scrutinizing the prevailing nation-state-centered discourse of the war on terrorism is critical for recognizing the international and global dimensions of the terror perpetrated on September 11 and seeing the terror invoked on Afghan civilians, in the name of justice.

In February 2002, Daniel Pearl, a *Wall Street Journal* reporter, was abducted and then killed in Pakistan. His tragic death was no doubt fuel for the fire in the war on terrorism, but the words of his wife upon his death are instructive:

> Revenge would be easy, but it is far more valuable in my opinion to address this problem of terrorism with enough honesty to question our own responsibility as nations and as individuals for the rise of terrorism . . .

[I hope] I will be able to tell our son that his father carried the flag to end terrorism, raising an unprecedented demand among people from all countries not for revenge but for the values we all share: love, compassion, friendship and citizenship, far transcending the so-called clash of civilizations. (Globe and Mail, 2002)

While the language of patriotism is clear in Ms Pearl's commentary, her dismissal of Samuel Huntington's clash of civilizations thesis makes it clear that her husband's death should not become a justification for more violence, couched in the language of national security.

Dominant geopolitical narratives shape what we see and hear in profound ways. However, networks of opposition to the state-sponsored attacks on Afghanistan generate snippets of hope that dissident voices have not been silenced. On September 15, 2001, the House of Representatives voted on a resolution permitting the President to use "all necessary force against those nations, organization or persons he determines planned, authorized, committed, or aided the terrorist attacks" (Ibbitson, 2001). Defying unprecedented unity in Congress, all but one member voted for this resolution. The sole voice of dissent was lodged by Barbara Lee, congresswoman for the district of Berkeley and Oakland. She has since received numerous death threats and hate mail. Her dissent does not mean, of course, that women are categorically more peace-oriented than men. Many other US congresswomen supported the resolution. Her position has to do with a confluence between her politics and the geography of the district she represents. Berkeley City Council was the first municipal government in the USA to pass a resolution criticizing the military campaign against Afghanistan. Its history of peace marches and anti-war activism is well known. Lee has taken a courageous stand in a climate of patriotism that has been intolerant of criticism of the US government.[11] Her stance has been applauded by feminists from countries whose experience of terrorism span decades for her ability to connect terrorism in the USA with terrorism elsewhere (Cat's Eye, 2001). There are few degrees of separation between here and there, us and them, either/or. A feminist geopolitics aims to trace the connections between geographic and political locations, exposing investments in the dominant geopolitical rhetoric, in the pursuit of a more accountable and embodied geopolitics that contests the wisdom of violence targeted at innocent civilians, wherever they may be.

NOTES

1   In November 2001, President Bush signed an order creating special military tribunals where foreigners charged with terrorism could be tried in secret (Dao, 2001). Mr Bush would personally determine who is tried, and cases would be heard before juries of military officers, not jurors (*Globe and Mail*, 2001). Verdicts may include the death penalty, and the tribunals include no appeals process. The authority of such tribunals to facilitate due process and mete out justice under the rule of law is, accordingly, very limited.

2   As noted, the perpetrators of the crimes committed on September 11 should be captured and brought to justice. I contend, however, that intelligence networks and targeted covert police operations informed by such networks would minimize the terror experienced by civilians who otherwise risk becoming casualties.

3   The popular, if fictionalized, imaginings of the nation and its vulnerable borders were televised in the season's premiere of the US political teledrama, *The West Wing*. In the

episode, the White House is shut down due to a security breach, with specific reference to a terrorist threat at the Ontario–Vermont border. The threat attests to a widespread feeling at the time that the Canadian border was linked to the attacks of September 11, and while no evidence to date has shown this to be the case, the USA has pledged to triple border patrols along this border. Given the focus on geopolitics in this chapter, this story is amusing because the Ontario–Vermont border is itself a fiction; Vermont is contiguous only with Québec, but this is irrelevant. The amorphous "Canada" is the weak link in the security perimeter propagated further by the show. As in the highly satirical *South Park* film of 1999, "Blame Canada" is as obvious as it is ludicrous. The *South Park* television series also produced a cutting critique of the US response to September 11 in its season opener.

4   Murphy Erfani (1998, pp. 5–6) argues that classical geopolitics posits a neutral mind detached from the body which observes space "objectively." Her own feminist geopolitics focus on the interstices of mind and body as an inseparable borderland location of the corporeal subject.

5   There are numerous examples to illustrate the policing of dissent in the North American context alone. On September 11, while watching the Canadian Broadcasting Corporation's live television coverage, I witnessed a reporter interview a US passenger who had been diverted to a Canadian airport due to the attacks. When asked his opinion on the attacks, the passenger said that the killings were tragic and that he had sympathy for the families affected, but that US foreign policy was bound to produce consequences and that this was one result. Another passenger standing beside him began hitting him with his carry-on luggage, and the camera panned away.

6   The travesties of the Gulf War should be a reminder of the opportunism and expediency with which alliances are forged. Neither Kuwait nor Iraq was a democratic country, but Kuwait's "promise" of democracy was sufficient ground to warrant forging an alliance for the sake of future democracy (and oil). As Dan Hiebert reminded me, "the war became about fighting for the right of a dictator to save 'his' country from another dictator so eventually democracy could be introduced. Of course all those democratic reforms were quickly forgotten at the end of the war" (personal correspondence, December 11, 2001).

7   In Canada, a poll taken in September before the bombing began found that 73 percent of Canadians (compared to 80 percent of Americans) favored joining the United States in its battle against terrorism. When asked if they approved of such a response if innocent civilians were killed, the approval rate dropped to 43 percent for Canadians (McCarthy, 2001). According to the poll, men are far more supportive of the war against terrorism than women, with some 79 percent of men supporting a war on terrorism compared to 68 percent of women.

8   In the *New York Times* on February 24, 2002, "Officials estimate that as of Friday, 3,062 people had died, or were missing and presumed dead, as a result of the attacks on Sept. 11, not including 19 hijackers."

9   He also notes that approximately 60 percent of the 18,000 bombs, missiles, and other ordnance used between October 7, 2001 and February 10, 2002 were precision-guided, as compared to fewer than 10 percent during the first Gulf War.

10   Personal correspondence from Rachel Silvey, October 7, 2002. I thank Rachel for these and other insights conveyed upon reading an earlier draft of this chapter.

11   In the USA, the geopolitical heartland of free speech, President Bush employed a media strategy just after the bombing began that dissuaded American television news organizations from broadcasting pretaped statements from Osama bin Laden, arguing that he might be using television news to deliver coded messages to his supporters (Stanley, 2001).

BIBLIOGRAPHY

Bacchetta, B., Campt, T., Grewal, I., Kaplan, C., Moallem, M. and Terry, J. (2001) Transnational feminist practices against war (http://www.geocities.com/carenkaplan03/transnationalstatement.html). Accessed July 25, 2003.

BBC (2001) Laura Bush decries Taleban "brutality", November 17 (http://news.bbc.co.uk/1/hi/world/americas/1662358.stm). Accessed July 28, 2003.

Bearak, B. (2002) Unknown toll in the fog of war: civilian deaths in Afghanistan. *New York Times*, February 10.

Campbell, M. (2002a) Thousands of Afghans likely killed in bombings. *Globe and Mail*, January 3.

Campbell, M. (2002b) Afghan civilian death toll high, 2nd estimate. *Globe and Mail*, January 19.

Cat's Eye Collective (2001) America attacked or America attacks? *The Island* (Colombo, Sri Lanka), September 27.

Chossudovsky, M. (2001) Who is Osama bin Laden? Dept of Economics, University of Ottawa/Centre for Research on Globalisation (CRG), Montréal (http://globalresearch.ca/articles/CHO109C.html). Accessed September 16, 2001.

Cockburn, C. (2000) Women in black: being able to say neither/nor, *Canadian Woman Studies*, 19(4), 7–11.

Dalby, S. (1991) Critical geopolitics: discourse, difference, and dissent. *Environment and Planning D: Society and Space*, 9, 261–83.

Dalby, S. (1994) Gender and critical geopolitics: reading security discourse in the new world order. *Environment and Planning D: Society and Space* 12, 595–612.

Dao, J. (2001) More US troops in bin Laden hunt; hide-outs bombed. *New York Times*, November 19.

Domosh, M. and Seager, J. (2001) *Putting Women in Place: Feminist Geographers Make Sense of the World*. New York: Guilford Press.

Enloe, C. (1989) *Bananas, Beaches and Bases: Making Feminist Sense of International Politics*. Berkeley: University of California Press.

Enloe, C. (2000) *Maneuvers*. Berkeley: University of California Press.

Gannon, K. (2001) The Taliban may leave, but burqas will stay. *Globe and Mail*, November 17.

Globe and Mail (2001) Dangerous tribunals, editorial. November 19.

Globe and Mail (2002) The murder of Daniel Pearl, editorial. *Globe and Mail*, February 23.

Hensmen, R. (2001) The only alternative to global terror. *South Asia Citizens Wire*, Dispatch 2, October 24.

Hyndman, J. (2001) Towards a feminist geopolitics. *Canadian Geographer*, 45(2), 210–22.

Hyndman, J. (2003) Beyond either/or: a feminist analysis of September 11th. *ACME: An International e-Journal of Critical Geographies*, 13(3), 1–13.

Ibbitson, J. (2001) The defiant dove of Capitol Hill. *Globe and Mail*, November 5.

Kofman, E. (1996) Feminism, gender relations and geopolitics: problematic closures and opening strategies. In *Globalization: Theory and Practice*. London: Pinter.

McCarthy, S. (2001) Canadians reject war if civilians put at risk. *Globe and Mail*, September 22.

Massey, D. (1997) A global sense of place. In T. Barnes and D. Gregory (eds), *Reading Human Geography: The Poetics and Politics of Inquiry*. New York: Wiley.

Mouffe, C. (1995) Post-Marxism: democracy and identity. *Environment and Planning D: Society and Space*, 13(3), 259–65.

Murphy Erfani, J. (1998) Globalizing Tenochtitlán? Feminist geopolitics: Mexico City as borderland. In D. Spener and K. Staudt (eds), *The US–Mexico Border: Transcending Divisions, Contesting Identities*. Boulder, CO: Lynne Rienner.

Nelson, L. (1999) Bodies (and spaces) do matter: the limits of performativity. *Gender, Place and Culture*, 6(4), 331–54.

Ò Tuathail, G. (1996) *Critical Geopolitics*. Minneapolis: University of Minnesota Press.

Peterson, V. S. (1992) Security and sovereign states: what is at stake in taking feminism seriously. In V. S. Peterson (ed.), *Gendered States: Feminist (Re)visions of International Relations Theory*. Boulder, CO: Lynne Rienner.

Peterson, V. S. (1996) Shifting ground(s): epistemological and territorial remapping in the context of globalization(s). In *Globalization: Theory and Practice*. London: Pinter.

Pettman, J. J. (1996) *Worlding Women: A Feminist International Politics*. London: Routledge.

Pilger, J. (2001) Hidden agenda behind war on terror. *The Mirror* (London), October 29.

Rashid, A. (1999) The Taliban: exporting extremism. *Foreign Affairs*, November/December.

Revolutionary Association of Women of Afghanistan (2002) The people of Afghanistan have nothing to do with Osama and his accomplices. Movement Statements on September 11 and its aftermath. *Inter-Asia Cultural Studies*, 3(1).

Roy, A. (2001a) The algebra of infinite justice. *Guardian* (London), September 29.

Roy, A. (2001b) War is peace. *Outlook Magazine*, October 29.

Sharp, J. P. (2000) Remasculinising geo-politics? Comments on Gearoid O'Tuathail's *Critical Geopolitics*. *Political Geography*, 19(3), 361–4.

Smith, N. (2002) Editorial: scales of terror and the resort to geography: September 11, October 7. *Environment and Planning D: Society and Space*, 19, 631–7.

Sparke, M. (2000) Graphing the geo in geo-political: *critical geopolitics* and the re-visioning of responsibility. *Political Geography*, 19, 373–80.

Sparke, M. (2001) Maps, massacres, and meaning. Unpublished paper, Department of Geography and Jackson School of International Studies, University of Washington.

Stanley, A. (2001) President is using TV show and the public in combination to combat terrorism. *New York Times*, October 11.

Weber, C. (1994) Shoring up a sea of signs: how the Caribbean Basin Initiative from the US Invasion of Grenada. *Environment and Planning D: Society and Space*, 12, 547–60.

Whitworth, S. (1994) *Feminism and International Relations: Towards a Political Economy of Gender in Multilateral Institutions*. Basingstoke: Macmillan.

Chapter 38

Love for Sale: Marketing Gay Male P/Leisure Space in Contemporary Cape Town, South Africa

Glen S. Elder

Introduction

> Yet the most interesting aspects of the gay dance world are the collective ones. The circuit is a sprawling experiment in a virtual community, unbounded by geography. It matters little whether you are dancing in Sydney or Mykanos, Ibiza, Atlanta, Paris or Palm Springs. The community you socialize in, its culture and range of personal relationships, its norms and connections, transcends place. (Nimmons, 2002, p. 159)

Over the past decade, the emergence of recognizable, interconnected, and seemingly interchangeable gay male leisure spaces has not gone unnoticed. Most recently *Antipode* (vol. 34, no. 5, 2002) devoted a special issue in which a mostly feminist analysis was used to review contemporary understandings of the "global gay culture." In this collection, geographer Jasbir Puar succinctly critiques work focusing on this phenomenon in the following way:

> Gay venues in North America and Europe are understood to be producing primarily positive, liberatory disruptions of heterosexual space, unexamined in terms of racial, class, and gender displacements. The assumed inherent quality of space is that it is always heterosexual, waiting to be queered or waiting to be disrupted through queering, positing a singular axis of identity which then reifies a heterosexual/homosexual split that effaces other kinds of identities – race, ethnicity, nationalism, class, and gender. (Puar, 2002, pp. 935–6)

In this chapter, I will use Puar's feminist critique as a point of departure to examine the emergence of gay male consumption space in Cape Town, South Africa. Of particular interest to me is her characteristically feminist method, which questions how these new spaces are implicated in global patterns of racial, class-based, and gender displacements.

My interest in gay male spaces in Cape Town stems in part from the fact that South Africa's post-apartheid cities have come to cater to a growing local and international gay male market. This process has been generated in part from the celebrated gains made by South Africa's homosexual movement during the writing of the national constitution in 1996. References to the constitutional protection of gay rights abound in gay and lesbian travel promotional material. New constitutional protections, combined with the growth of gay and lesbian tourism, have transformed the urban landscape of many South African cities. Since 1996, some urban centers have actively inscribed the gains around sexual orientation in space; so-called "gay neighborhoods" in Cape Town and Johannesburg have "emerged" with purposeful intent.

As geographers have shown, the ghettoization of gay and lesbian experience, because of homophobia, was a logical outcome of a systemic spatial ordering (Knopp, 1987, 1990; Valentine, 1993a, b; Brown, 1997). This distinctly Western European and North American urban experience related to the confluence of particular economic, social, political, health, and safety-related conditions in the North.

There are quite a distinct set of economic, social, political, and health processes converging in the South African context; South Africa's raced and sexed history is remarkably different from that of most Northern urban settings. Given these differences, there is no compelling reason why the spatialized "gay ghetto" should emerge there. In other words, the South African case presents us with a spatial conundrum of sorts. Why has there been a *concerted effort* to map yet a new kind of sexually segregated space onto a landscape that is simultaneously moving away from its racially segregationist past?

Cape Town in South Africa is thus a particularly fascinating context within which to examine the recent and strident efforts that market a particular kind of gay culture and generate particular urban landscapes. By way of a close reading of the promotional material about "Gay Cape Town," especially the much vaunted "Pink Map," I hope to show that this particular contemporary spatial narrative in South Africa actually creates a segregated space of exclusion.

South Africa and International Tourism

South Africa's post-apartheid economic policy has been characterized by a fervent drive to attract foreign direct investment. One such avenue for attracting foreign exchange has been the fairly well established tourist infrastructure built to entertain trapped white South Africans during the apartheid era, when global sanctions prevented extensive travel abroad (Dodson, 2000).

It is not surprising then that within this context an emerging interest in the geography of tourism in South Africa has also begun to take hold. Dodson (2000, p. 418) argues, for example, that: "Certainly South Africa, source and destination of significant numbers of international migrants as well as tourists, is caught up in this rearrangement of the global human mosaic, with its complex reworkings of society and space, place and landscape." In a call for a cultural geography of tourism in South Africa, Dodson (2000) goes on to suggest that the creation of places of consumption/leisure in places like South Africa does not necessarily erase local differences as a great deal of early literature about consumption spaces suggests, but masks local distinctiveness:

Thus geographical *de-differentiation* at one level cannot erase the geography of material inequality and social difference at another, more fundamental level. For if the convergence of consumption, leisure, and tourism is both the product and the preserve of the new middle class, it is at the same time a powerful agent of social differentiation, exclusion, and marginalization. In an inescapable paradox, the superficial homogenization of culture and landscape is the very means by which inequitable access to opportunities and facilities is reproduced. Perhaps nowhere is this more true than in post-apartheid South Africa. (Dodson, 2000, p. 419)

Dodson's astute observation can be extended to help an understanding of the emerging geography of gay and lesbian travel in Cape Town. As she rightly observes, the apparent homogenization of spaces reserved for leisure in South Africa is only a superficial move. So too in the evolving gay and lesbian travel geography, we see an apparent homogenization of gay space in efforts to create interchangeable neighborhoods that produce experiences of place similar to Sydney, Amsterdam, or London. What this space creates is a myth of "community," while also masking the life of gay and lesbian people and the material inequalities of globalization.

We now turn to the process involved in the creation of a dedifferentiated space and how, within the context of evolving gay and lesbian geographies, this process occurs.

The Making of "Pink" Cape Town

"Dubbed the 'mother city,' Cape Town has the continent's most developed gay and lesbian scene and eclipses all other South African cities when it comes to the number of gay and lesbian venues (there are well over 100). It has consistently made it to the top ten of international gay and lesbian travel destinations."[1] Throughout the mid-1990s, it is estimated that at least half a million to upwards of three-quarters of a million tourists visited the Cape Town metropolitan area in any given year (Travel Industry, 2000). South African tourism is presently in the final stages of its international marketing campaign (that does not include gay tourism *per se*), having made the largest global investment to date: R400 million (just over US$60 million). The national campaign has let loose a regional battle between differentially endowed provinces. Within that contest, Johannesburg, Cape Town, and more recently Durban have begun to seek a comparative advantage.

Enter the gay tourist. In 1993, Strub Media Group produced a glossy report that claimed gay American households average $63,100 compared to $36,500 for all households. The report went on to assume that given gay men and lesbians have no children, their disposable income is even higher than their average income would suggest. The astounding longevity of a 1993 USA-based report has gone a long way in fueling marketing efforts targeting a yet to be measured economic entity: the gay market.

Gay people became, in the popular imaginary, playboys of the 1990s. According to the report, while renowned for their gourmet capabilities in the kitchen, apparently 80 percent of gay men also eat out more than five times a month. Apparently, gay men and lesbians also buy more CDs, use their American Express cards more, and generally spend more money on the good life than their straight counterparts. The longevity and profligate power of this study, despite self-evident questions about its actual validity, is interesting. For example, the results of the report were based on responses from a mail survey to subscribers of a magazine that catered to readers who were white, urban, white collar, and mostly male. While more recent reports have gays and lesbians in the United States better educated but without higher incomes (Yankelovich Partners, 2000), the Strub Report still shows up consistently. However, the results of this survey and others like it have become the means by which local tourist associations and city administration seek to "make a case" for investing in the so-called "gay travel" industry.

Lamenting Cape Town's formal embrace of a gay travel niche market, Cape Town Tourism, the for-profit agency that markets the city, noted that one event, the Mother City Queer Project in Cape Town, generated more than R50 million alone, almost half of the R106 million that the Argus Cycle Race managed to generate (Heard and Ludski, 2001). Salivating, promoters argue that gay tourism in the USA is a $331 billion market while Sydney, situated similarly to Cape Town, raises R651 million from its famous Mardi Gras.

Rehashing the Strub Media Group's report again, Sheryl Ozinsky, Cape Town Tourism Manager, noted in March 2001 that "Gay visitors are lucrative, they have double income and no kids, they spend double or triple the regular visitor, they are also exciting visitors, they have *joie de vivre*, they enjoy doing things like sport, clubbing, and they add flavor to the city" (Heard and Ludski, 2001, p. 7). Cape Town Tourism enthuses: "Cape Town has recently been voted the fifth most popular gay destination in the world after the Canary Islands, Ibiza, Los Angeles, San Francisco, and Amsterdam" (Blignaut, 2001).

In contrast to the celebration of gay tourist spaces, contrary anecdotal evidence and research suggests that conspicuous, consumptive outness in some parts of the city of Cape Town has made local gay and lesbian lives in other parts of Cape Town, particularly among its black and colored gay and lesbian identified youth, more life-threatening than ever (Reuters, 2001a, b). Homophobic violence, sexual assault, and deadly STDs continue to stalk townships unchecked. Eprecht (2002) argues that an interconnected HIV/AIDS crisis, crippling structural adjustment programs, and a resurgence of right-wing Christian and Islamic fundamentalism have combined to produce a state-sanctioned homophobia in several southern African states outside of South Africa. Taking a similar position, Altman (2001, p. 98) argues that:

> The strong hostility for some African political and religious leaders toward homosexuality as a "western import" is an example of psychoanalytic displacement, whereby anxieties about sexuality are redirected to continuing resentment against colonialism and the subordinate position of Africa within the global economy.

Increased homophobia and homophobic violence for gays and lesbians in South Africa is eclipsed by the marketing efforts to promote gay tourist sites, discussed in what follows. A cursory glance at the Pink Map Gay Guide to Cape Town reveals a city divided into categories that will please the Phantom gay tourist who apparently likes opera. An alluring white drag queen holds the city's "queer world" in the palm of her hand (see figure 38.1). A list of the information contained in the map appears at the bottom of the page: accommodation, cafes and restaurants, travel info, shopping, services, pubs, clubs, entertainment, transport, adult shops, steam baths, and, curiously, "keeping up."

Figure 38.1 Pink Map (2000). Maps are updated and reprinted annually. This map is used with permission of A & C Maps, Cape Town.

Marketeers' fantasies about "gay life" have organized the city into convenient zones: Shopping, Entertainment, and Accommodation. It is worth noting that this coding of the city does not always provide a contiguous set of spaces, but generally speaking most venues fall within the city center or what is sometimes called the city bowl. "Accommodation" lists over 15 sites of accommodation, many with names that evoke an evocative geography of elsewhere. The Amsterdam Guest House and the Mediterranean Villa are examples.

It is clear that the urban space of Cape Town has been remapped to present a city at once exotic and comfortably familiar to the international (gay?) traveler. By employing profoundly spatial rhetorical strategies, Cape Town Tourism and other publications have sought to refashion a particular time–space nexus in such a way that the particularities of that actual moment are lost and its material consequences at the very least hidden. Three themes are identified and discussed below in some detail. The first seeks to rhetorically conflate time and space, lulling the tourist into a supposed "colonial" aesthetic and style, free of the deeply oppressive structures that helped to produce that stylistic moment. Second, the refashioning of Cape Town is also an exercise that seeks to remake parts of the city into male space. Threats of violent crime and a lack of safe venues for women have come to create a day-and-night geography of pleasure in the city that we might characterize as homo-masculine in nature. Finally, the politics of the anti-apartheid struggle and the spaces of those struggles have been revised to produce a less threatening vision of anti-racist struggles that do not disrupt the consuming patterns of white masculine metropolitan privilege.

Travels in Space/Time

In addition to the Pink Map, the marketing of Cape Town to gay and lesbian travelers has occurred in publications targeting the US market. *Out and About: Essential Information for Gay and Lesbian Travelers* featured South Africa and Cape Town in 1998. This feature article followed a South African Tourism sponsored tour of venues designed for gay and lesbian travel writers in mid-1997. The *Out and About* article describes accommodations in the following manner, encouraging gay tourists to travel in time as well as space: "Enter through [the] gates and you feel as though you have stepped back to the glory days of the British Empire.... If you are nostalgic for the atmosphere, the Mt Nelson will fulfill your Colonist fantasies." Here, pretensions of colonial grandeur lure the tourist, time replaces space, and we recall McClintock's observation that:

> According to the colonial version of this trope, imperial progress across the space of empire is figured as a journey backward in time to an anachronistic moment of pre-history. By extension, the return journey to Europe is seen as rehearsing the evolutionary logic of historical progress, forward and upward to the apogee of the Enlightenment in the European metropolis. (McClintock, 1995, p. 40)

While the hotel in question comes to stand in for "colonial grandeur" (and not the colonized space), movements across space become movements through time in the contemporary postcolonial context and in Cape Town's evolving gay landscape.

Accommodation, however, is not alone in this regard. A restaurant situated in the Castle and occupied by several colonial administrations since the mid-1600s is described in such a way that "colonial taste" is literally consumed. "Enjoy Cape Malay and traditional food in an African-Colonial atmosphere in the oldest building in South Africa."[2] Not to be outdone, "*The Cape Colony Restaurant* offers the city's most memorable dining experience. . . . The recently renovated dining room is a perfect expression of Old World Colonial Opulence, and remarkably affordable at today's exchange rate" (*Out and About*, 1998, p. 39). By acknowledging that indeed the space has been renovated, the brochure winks at the leisure traveler. No, this is not "original" but a postcolonial game in which movement across and through landscape becomes an artificial movement back in time to a solipsistic and foppishly refined colonial world.

The making and sometimes remaking of colonial space into a consumable commodity requires that the traveler either lack knowledge of, or suspend knowledge about, the deeply oppressive colonial histories of the Cape Colony. Recall, of course, that accessing this dreamy world and denying knowledge about its multiple scales of oppression is a privilege granted most readily to white metropolitan subjects who glide seamlessly and without question into the history of colonial racism and sexism.

The invitation to "go back in time" is also an invitation to impose on colonial territories a feminized vision of landscape that renders its taking a matter of course and destiny (McClintock, 1995; Trexler, 1995). While the conflation of time and space to create a desired effect is reflected in most leisure travel brochures about the city of Cape Town, it is the unquestioned notion of the colonial in the context of gay and lesbian literature that is troubling.

The Remaking of Male Space

A more distinctive mark of gay travel brochures is the effort to mark territories with "gay" signifiers. Fantasies about gay men in particular produce a variety of interesting brochures that mark spaces wherein "gay specific actives might be performed." The remaking of particular locales at particular times into homomasculine space renders, among other spatial implications, previously hidden and secluded (gay) beaches (often raided by apartheid era police) an open secret. Indeed Michael Brown has examined the spatialization of the closet in *Closet Space* (2000). In it he shows how the marking of gay space in urban Auckland occurs in ways that rely on the secret of the closet, readable only to gay men. Recall that "wink-wink, nudge-nudge" language is part of the epistemology of the suffocating closet underpinning modern Western thought (see Sedgwick, 1990). The "open secretness" of the following account and the implied sexual activity creates a similarly oppressive stricture on the travel landscape of metropolitan Cape Town:

> *Sandy Bay* is a nude beach frequented by many gays and lesbians. . . . Follow the signs (and the crowds) down to the walking path entrance, which you will recognize by the soft drink vendors parked there. The walk to the beach is about fifteen minutes, so bring along anything you might need. The gay section is at the far end . . .

Graaff's Pool. "Graaff's" is more of a nude sunbathing platform than a swimming pool. You'll recognize it by the concrete steps and platform extending into the surf on the beach in Sea Point. This men-only nude sunbathing spot has a straight side and a gay side; it's not hard to tell the difference. (*Out and About*, 1998, p. 39)

A guide to pubs, clubs, and entertainment once again directs the traveler to 12 locations. The photograph inserted on this map shows a group of mostly white male, presumably gay revelers:

The Bronx is solely a video pick-up bar, catering mostly to men, mostly young and mostly white. Upstairs through an outdoor patio is *Angels*, a dance spot popular with lesbians and coloreds [*sic*, the regional term for those of mixed race]. . . . If you're still not tired after all that dancing, unwind with a sauna at *Steamers*, a place for men to cool down after, um, heating up. For those in search of seamy more than steamy, explore the down-and-dirty drag shows at *The Brunswich Tavern*. Watch the trannies, watch the trade, and, by all means, watch your wallet. Sadly, the closest thing to a lesbian bar is *Café Erte* . . . its lesbian following fluctuates and seems based more on reputation than fact. (*Out and About*, 1998, p. 39)

A list of activities and venues are described in literature and on websites that mark places as gay. The suggestion that many of these places are also sites where sex might be sought is never far from the surface. Increasingly, these efforts to create gay spaces have also come to produce a landscape upon which the norm is understood to be white and male, or homomasculine. Particularly, the "special" references to people of color, lesbians, and transgendered people (marking these groups as different from white, gay, male), as well as the belief that all readers would frequent the bath houses of Cape Town, produce intentionally gay, white, and masculine spaces.

Several bath houses in Cape Town are presented as possible nodes on an evolving urban landscape of homomasculine desire. New and catering to a particular class of gay men, expensive by local standard and exclusionary, these establishments throughout the city stand in stark contrast – in Cape Town and elsewhere – to the history of bath houses in gay male culture.

As George Chauncey (1994) reminds us, at the beginning of the twentieth century and for much of the first half of it:

The safest most enduring, and one of the most affirmative of the settings in which gay men gathered in the first half of the twentieth century was the baths . . . by World War I several of them had become institutions in the city, their addresses and distinctive social and sexual character known to almost every gay New Yorker and to many gay Europeans as well. (Chauncey, 1994, p. 207)

Expanding on the concept of the bath house, Tattleman (1997, p. 394) examines the institutions in history and argues that:

The principle of the bathhouse was that you brought nothing inside with you. Ideally, the bathhouse tried to erase the boundaries that divide people: clothing was removed, and issues of class were left at the lockers. By stripping bare, new experiences became possible. When you left, you took those things you had learned from participating in the bathhouse, by communicating your body, back out into the world (p. 394).

By contrast, a so-called Cape Town "leisure" club, or bath house, described itself thus:

> Situated in the heart of Cape Town's gay area, the [bath house] *offers what you'd expect* [emphasis added] from a steambath: an enormous steamroom, sauna, 2 spa baths, 3 video lounges, maze, showers, darkrooms and cabins. But also 2 bars, a restaurant, private cabins, with or without its own TV and video channel, luxurious double volume lounges, fireplace and a sundeck with the most spectacular views over the harbour, downtown and Table Mountain.[3]

This bath house promotion, in fact, suggests a reversal of Chauncey's and Tattleman's cultural logic as it has been translated into a global postmodern market for gay male travel. Here, the promotional material suggests that the informed tourist go into the facility with a particular set of preconceived ideas. Promotional materials also have pictures of twenty-something, well defined white men making quite clear that the "expected" clientele, once again, is white, middle class and male.

In order to create the sense of a "gay city," Cape Town's commercial promoters have created a sense of a city awash in gay and purportedly lesbian sites. In reality, the promotional material appears, mostly, to appeal to a male market. The list of bars, photographs, and the inclusion of several steam bath locations on the map suggest that the target market is not gay, but middle-class, gay male identified, and white. The effort to produce a "gay-safe space" that is not gay, safe, or space in a contiguous sense is a troubling slight of hand (see figure 38.2).

Conclusion

A globalized tourism process has unleashed a node of leisure spaces, internationally, in cities as distinct as Sydney, Rio, London, Montreal, Hong Kong, and Cape Town. The process of dedifferentiation of gay male consumption space, however, is a scam. Real local differences are purposefully masked in this process and a feminist geographic analysis has gone part of the way in exposing this process.

Despite the emancipatory marketing rhetoric employed to promote such spaces, I would argue that "the closet" remains in place as only a very particular form of same sex desire is articulated. Other same sex models, like monogamous couples, all forms of lesbian desire, non-white and interracial same-sex coupling, remain hidden forms of desire in South Africa and other places.

In addition, the process of dedifferentiation in Cape Town has inadvertently created a landscape which may well be more unsafe for gay tourists. An effort to describe and map Gay Cape Town as a contiguous space when the ground truth suggests something more akin to geographically far flung clusters of activity requires all gay consumers to travel through spaces that do not offer the ambiguous safety of the Western gay ghetto. In fact the spatialized form of the "ghetto" familiar to the traveling tourist is nowhere to be seen in Cape Town at all.

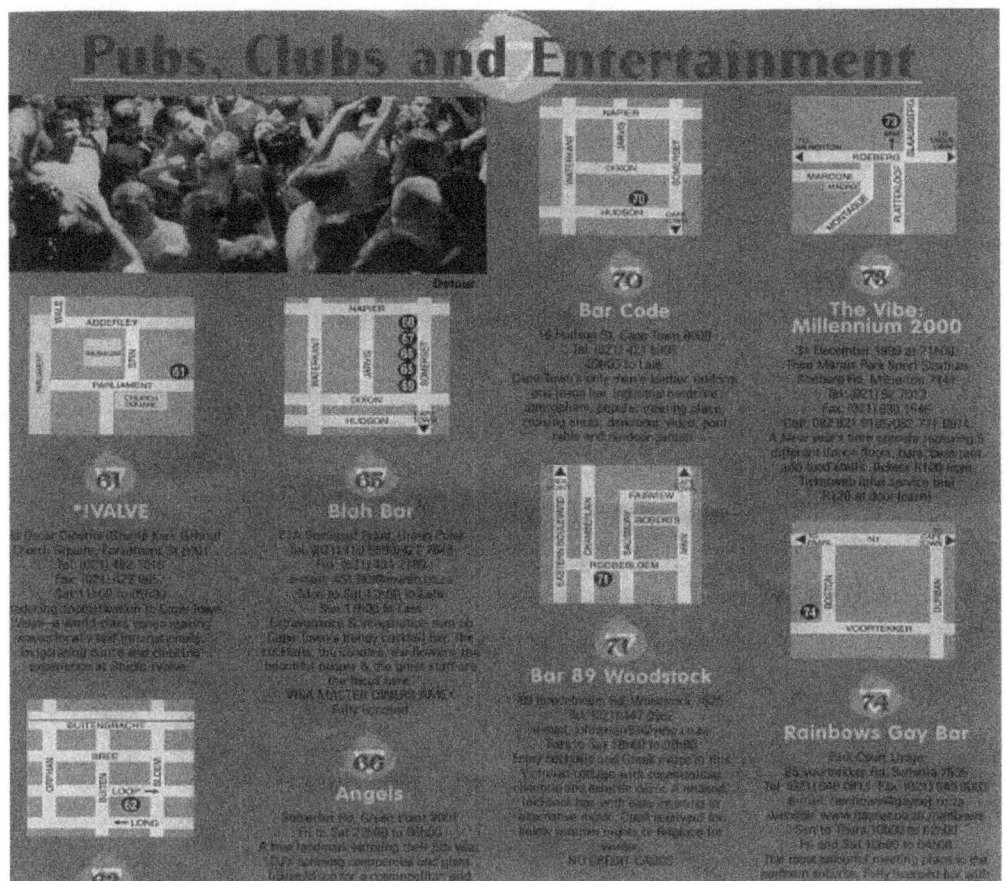

Figure 38.2 Map depicting various locations of bars and dance clubs throughout the city of Cape Town.
Source: Out and About, 1998. Reproduced with permission.

On January 31, 2003, the *Mail and Guardian* newspaper reported the following:

> Police and members of Cape Town's gay community now believe the gory massacre at the Sizzler's massage parlour in Sea Point last week was a message from Cape Town gangs who felt their turf was being invaded. . . . Only one victim survived the massacre. Seven people were found shot dead in the little white house . . . after the survivor stumbled to a local garage and raised the alarm. . . . In November 1999 a pipe bomb rocked the gay Blah Bar, injuring nine; in August 2000 a car bomb exploded outside Bronx, a gay club; and in June 2000 two people were injured in an explosion at New York Bagels. None of these cases has been solved.

A number of unsolved and increasingly grizzly attacks on a disproportionately high number of well known gay locales in Cape Town, over several years, begs questions about the wholesale marketing of these spaces. The high levels of visibility accorded these spaces and their overtly commodified character, I would suggest, has inad-

Figure 38.3 Cartoon depicting connections between violent crime in Cape Town and tourism.
Source: The Sowetan Newspaper, November 30, 1999. Reproduced with permission.

vertently created a "metonymic space" (Brown, 2000) where consumption stands in for gayness. Constitutional guarantees not withstanding, Cape Town remains in many ways a thoroughly homophobic local context. The marketing of these spaces and places as "easily recognizable gay spaces" (for both patrons and homophobes) is not without cost. In a city still recovering from the ruinous psychological, economic, and social effects of apartheid engineering it is not surprising that gay space might become a metonym for opulence and hedonism and a "soft" target in an urban class struggle, most often described in terms related to crime.

NOTES

1 http://www.gaynetcapetown.co.za (November 11, 2000).
2 Pink Map–Western Cape 2000, Gay Guide to Cape Town.
3 http://www.steamers.co.za (accessed July 13, 2001).

BIBLIOGRAPHY

Altman, D. (2001) *Global Sex*. Chicago: University of Chicago Press.

Blignaut, C. (2001) Pink and proud of it. *FairLady Magazine*, March 28, 35–7.

Brown, M. (1997) *RePlacing Citizenship: AIDS Activism and Radical Democracy*. New York: Guilford Press.

Brown, M. (2000) *Closet Space: Geographies of Metaphor from the Body to the Globe*. London: Routledge.

Chauncey, G. (1994) *Gay New York: Gender, Urban Culture and the Making of the Gay Male World, 1890–1940*. New York: Basic Books.

Dodson, B. (2000) Are we having fun yet? Leisure and consumption in the post-apartheid city. *Tijdschrift voor Economische en Sociale Geografie*, 91(4), 412–25.

Eprecht (2002) "What an abomination, a rottenness of culture": reflections on the gay rights movement in southern Africa. *Revue canadienne d'études du développement/Canadian Journal of Development Studies*, 22, 1089–108.

Heard, J. and Ludski, H. (2001) Pink dollars quietly help to put tourism in the black. *Business Times* (South Africa), March 18, 7.

Knopp, L. (1987) Social theory, social movements and public policy: recent accomplishments of the gay and lesbian movements in Minneapolis, Minnesota. *International Journal of Urban and Regional Research*, 11, 243–61.

Knopp, L. (1990) Some theoretical implications of gay involvement in an urban land market. *Political Geography Quarterly*, 9(4), 337–52.

McClintock, A. (1995) *Imperial Leather: Race, Gender and Sexuality in the Colonial Context*. London: Routledge.

Nimmons, D. (2002) *The Soul Beneath the Skin*. New York: St Martin's Press.

Out and About: Essential Information for Gay and Lesbian Traveler (1998) *Out and About*, 7(3), 33–48.

Puar, J. (2002) A transnational feminist critique of queer tourism. *Antipode*, 34(5), 935–46.

Reuters News Agency (2001a) Cape Town draws gay tourists and controversy. Reuters, March 13.

Reuters News Agency (2001b) Cape Town: The new Sodom. Reuters, September 1.

Segwick, E. (1990) *Epistemology of the Closet*. Berkeley: University of California Press.

Strub Media Group (1993) *The Gay and Lesbian Market Potential*. New York: Strub Media Group.

Tattleman, I. (1997) The meaning at the wall: tracing the gay bathhouse. In G. Brent Ingram, A. Bouthillette and Y. Retter (eds), *Queers in Space: Communities, Public Spaces, Sites of Resistance*. Seattle: Bay Press.

Travel Industry (2000) *World Year Book, 43*. Spencertown, NY: Travel Industry.

Trexler, R. (1995) *Sex and Conquest: Gender Violence, Political Order, and the European Conquest of the Americas*. Ithaca, NY: Cornell University Press.

Valentine, G. (1993a) (Hetero)sexing space: lesbian perceptions and experiences of everyday spaces. *Environment and Planning D: Society and Space*, 11, 395–413.

Valentine, G. (1993b) Negotiating and managing multiple sexual identities: lesbian time–space strategies. *Transactions of the Institute of British Geographers*, 18, 237–48.

Yankelovich Partners (2000) *The Changing Demographics of America*. New York: Yankelovich Partners.

Chapter 39

Women's Struggles for Sustainable Peace in Post-conflict Peru: A Feminist Analysis of Violence and Change

Maureen Hays-Mitchell

Introduction

Throughout the 1980s and early 1990s, political violence wracked the Andean country of Peru as the insurgent movement Sendero Luminoso clashed with the Peruvian state. By the attenuation of armed violence in 1993, over 23,000 violent attacks had occurred, 70,000 Peruvians were dead, over 6000 people had disappeared, more than half a million persons had been displaced from their homes, thousands of innocent Peruvians were imprisoned, material damages exceeded US$21 billion, and nearly half the national territory was designated an emergency zone (IPEDEP, 1999; CVR, 2003). To this day, both the structural inequities underlying the conflict and strategies for an integrated reparation remain unaddressed. Within this context, communities themselves have been forced to undertake the process of social reconciliation and economic reconstruction.

As victims of violence, perpetrators of hostilities, and advocates for peace, women played key roles in Peru's armed struggle. Impoverished women, in both urban shantytowns and rural villages, were an especially leading force in (a) the transition toward open resistance to Sendero Luminoso and (b) the movement to hold the Peruvian state accountable for abuses committed by its armed forces (Coral, 1998; Del Pino, 1998; CVR, 2003; interview with Coral, 2001). Nowhere was this more true than in Ayacucho, the epicenter of the violence, where there is an expression that "women ended the war" (see figure 39.1). Today, women continue to play critical roles in the struggle to rebuild Peru's devastated economy and reconcile its strained society.

Despite the role that poor women and their associations played in ending the conflict, and their ongoing concern with constructing what they openly refer to as a "culture of peace," women and their organizations have not been integrated substantively into the official process of rebuilding their country. That is, although non-governmental organizations work directly with women and particular state agencies

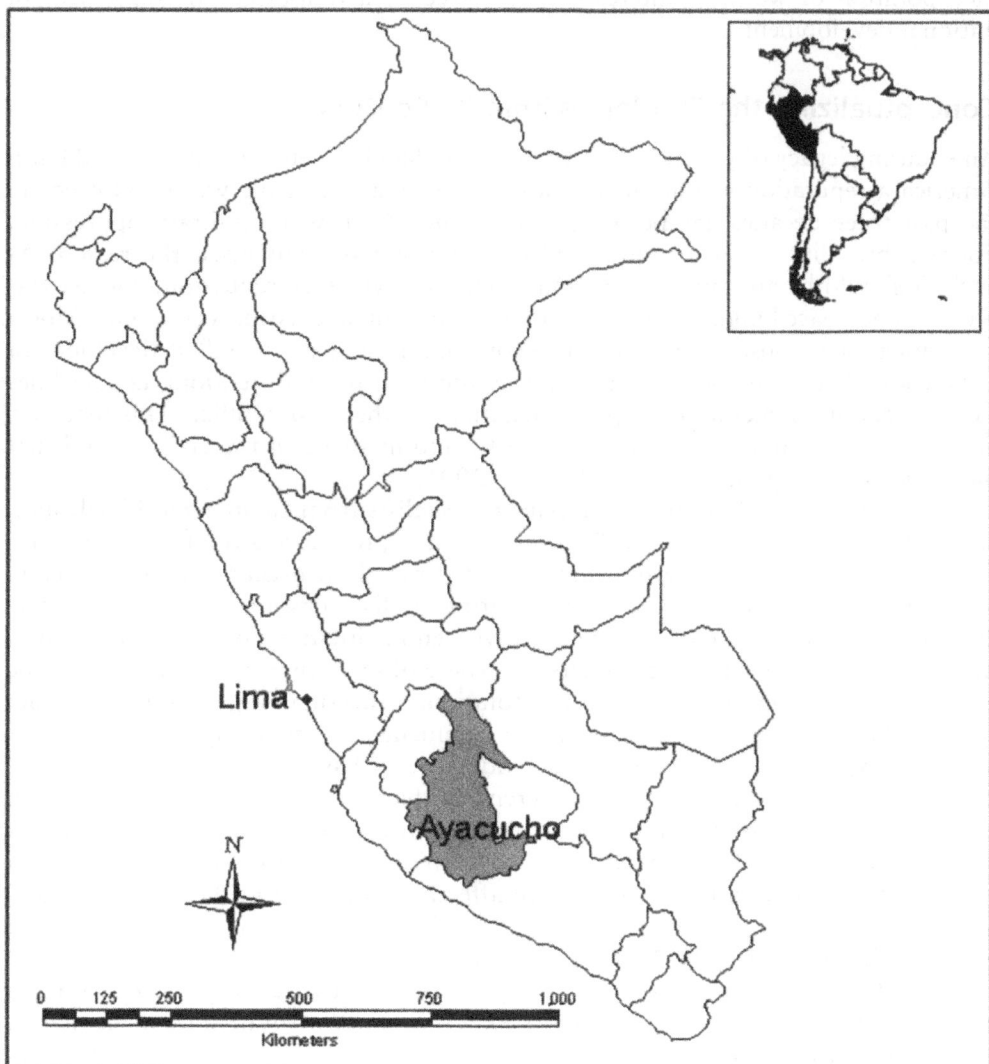

Figure 39.1 Ayacucho Department, Peru.

are designated as women's-interest groups, official reconstruction efforts tend to ignore the needs, interests, and potential contributions of popular-sector women. I posit that this contradiction within post-conflict Peru is the result of the neoliberal restructuring program that frames Peru's reconstruction efforts. When pressed on the matter, officials within the Peruvian government attribute their deficient efforts to the fiscal constraints of the structural adjustment program that was imposed upon Peru in 1990 by the international financial community, and that continues to dictate the course of economic development in post-conflict Peru. Such comments by officials betray the masculinist bias of the neoliberal paradigm upon which the prevailing global political-economic order is premised. Within this paradigm, the incorporation of women's interests is perceived as a non-essential component of national development.

Conceptualizing the Problem within Its Context

An enduring legacy of violence, that spans more than five centuries, has earned Latin America a reputation as one of the most violent regions of the world. Violence in the past three decades has been especially acute. Civil wars, military oppression, political brutality, social turmoil, and civil unrest have convulsed the region. As such, it should be an important region in terms of violence reduction, conflict resolution, and peace building. However, these goals remain elusive. As Caroline Moser and Fiona Clark observe, policies of reconciliation and post-conflict development throughout Latin America typically ignore the need to analyze along gender lines the impacts of conflict and the peace negotiations that end conflict. This tendency is unjust to the women and men affected by and involved in the civil turmoil that has afflicted the region (Moser and Clark, 2001).

In May 2000, scholars and development specialists from Guatemala, El Salvador, Colombia and Peru convened in Bogotá, Colombia, to address the failure to establish sustained peace in their homelands.[1] They noted an urgent need for concrete information on the situation of women during conflict, despite an expanding literature that focuses on the diverse roles of women in situations of political violence. In contrast, scholars and practitioners elsewhere observe that little is written of the gendered dimension of the conflict resolution, peace-making, and post-conflict process. There remains, as Karam (2001) comments, an especially urgent need for information, documentation, analysis, and training based on actual peace negotiations that have taken place in different conflict situations around the world. Notwithstanding, participants in the Bogotá conference concluded that, without comprehensive and reliable information concerning women's involvement in armed groups, the nature and levels of violence affecting women, and the survival strategies women employ in times of conflict, conflict resolution was doomed in the violence-plagued societies of Latin American.[2]

The subtle tension in these views suggests the need to conceptualize women's involvement in armed conflict in a more expansive and integrative manner; that is, a conceptualization that speaks to the interrelated, interdependent, and place-specific nature of women's involvement in armed conflict. To be precise, the approach must contextualize, integrate, and view through a gendered prism: (a) circumstances that precede the outbreak of hostilities; (b) women's diverse and often conflicting roles during conflict; and (c) women's experiences with and/or contributions to processes of conflict resolution and post-conflict reconstruction. To this end, I suggest conceptualizing women's involvement in armed conflict along a spatiotemporal continuum that is historically and culturally grounded, place-specific, integrative, flexible, and fluid.

In this regard, the Peruvian conflict is instructive. Despite the destructive effects and the enduring legacy of civil war, women are reshaping traditional roles and acquiring social visibility. According to Isabel Coral (1998, p. 374), Director of PROMUDEH (El Ministerio de Promoción de la Mujer y del Desarrollo Humano, The Ministry of Women's Advocacy and Human Development) and scholar-activist from Ayacucho, in the course of Peru's civil war,[3]

[Women] achieved a transition from "traditional" roles that situated them as invisible, toward explicit and visible affirmation as protagonistic social actors. In this valuable experience shaped by women there emerged important changes in gender relations. Women drew themselves toward spaces of political power and decision making to press their needs and perspectives, they relocated themselves in spaces of productive economic activity and work, they partially restructured familial relationships of authority and self-esteem.

In referring to this generative or empowering aspect of the war, Coral argues that women emerged as decisive agents in struggles over politics, survival, and reconstruction (interview with Coral, 2001). In her words, "Something was achieved in the area of gender equity and affirmation in the 1980s and 1990s [in Peru]. 'Normalcy' will not bring a return to the point of beginnings . . . the process of 'awakening' and actuation have proved fundamentally irreversible" (Coral, 1998, p. 374).

In this chapter, I examine the transition in women's status within Peruvian society that Coral implies is a consequence of the direct and involuntary involvement of women in the civil war. To this end, I consider the impact of the recently concluded civil war on the rural women of Ayacucho – the epicenter of the violence – for whom the war was most devastating in both material and emotional ways.[4] Initially caught in the crossfire between state forces and Sendero Luminoso, poor indigenous women in rural Ayacucho (i.e. *campesinas*) were, indeed, victims of brutal violence.[5] In the absence of state protection and/or international support, they were forced to rely upon their own resources to protect and ensure the survival of themselves and their families. As the war progressed, so did their capacity to organize and, eventually, resist. These unlikely women built a broad-based and representative movement that is widely credited with leading to the dimunition of violence in 1993.

Today, however, their organization is in turmoil, and they are largely closed out of the official reconstruction effort that denies their achievements and expects them to return to pre-war roles, despite the massive changes in their lives and region provoked by the civil war.[6] Their needs and interests are relegated to NGO-sponsored community programs and conferences; they are perceived as non-essential components of national development. Moreover, their energies are co-opted to fill the void created by state retrenchment from essential social programs, as required by the structural adjustment program to which Peru is bound.[7] Nevertheless, it is these women of humble background who are unofficially credited with bringing a cessation to the violence, and for whom the war's empowering aspect has been most remarkable.

My particular interest lies in understanding the role this cohort of women played in achieving preliminary peace and their potential contribution to the post-conflict reconstruction of their country. Not only is Peru's official process of post-conflict development largely uninformed by women's perspectives, but it fails to incorporate a substantive gender perspective. In the absence of an actual peace proposal, the topics of peace, reconciliation, and even reconstruction tend to be subverted within the national agenda. Instead, under the pretense of "gender-neutrality," the state promotes a masculinist model of economic and political development, bounded by extreme fiscal restraint and dominated by neoliberal discourse. Although much responsibility for this rests with the Peruvian state, factors within the international community contribute significantly to this situation. Consequently, local communities themselves have been left to undertake, with the help of diverse NGOs, the process of economic reconstruction and, in many cases, painful social reconciliation.

In what follows, I reject the masculinist premise upon which neoliberal reform (and, by extension, Peru's process of post-conflict reconstruction) is based. I maintain that, to achieve a lasting peace, the women who are directly affected by conflict must be included as full participants in every aspect of conflict resolution and post-conflict reconstruction. Further, women's perspectives and needs, informed by this and other constituencies of women, should be a basic component of any peace agreement and post-conflict reconstruction. Even humanitarian assistance organizations – increasingly called upon to mediate conflict resolution – have failed to adopt and apply a gender perspective consistently and to integrate local and international women into decision-making positions (Karam, 2001).

This is particularly tragic because the emergence of women activists in conflict areas gives cautious optimism for future transformative change (Mertus, 2001). Appreciating the complexity, intricacies, and uniqueness of Peru's unlikely movement, wherein humble women halted violence, may bring us one step closer to developing a conceptual framework capable of understanding the gendered dimension of conflict and the prerequisites of effective reconstruction within a context of sustainable peace. My goal is to provide justification to strengthen the role of women's associations in the peace-building and reconstruction process in post-conflict Peru and potentially elsewhere. To this end, I analyze the phases through which indigenous women in Ayacucho passed as the Peruvian civil war engulfed their lives, as well as the obstacles they encounter today in their struggle to sustain Peru's tentative peace.[8]

Feminist geography is uniquely suited to contribute to conflict resolution and peace-building. As a human geographer, I embrace the integrative and unifying heritage of my discipline, as well as its grounding in a tradition of hands-on fieldwork. Moreover, I appreciate the uniqueness of places and the significance that individuals ascribe to particular places. As a feminist geographer, I am sensitive to the inequities in power and the violent relationships that often infuse those spaces and inflict suffering upon many who occupy them. I am deeply troubled by injustice and human suffering, and am committed to affecting change through my work. In their analysis of gender in countries affected by conflict, Hyndman and de Alwis (2002, p. 2), contend that "feminist analysis provides a more powerful lens with which to examine the place of women and men, and a more compelling position from which to transform relations that provoke or perpetuate violence, hate, and inequality." They suggest that a comprehensive feminist analysis is especially important in conflict zones because political discourse shapes, and is shaped by, gender relations. In their analysis of the Sri Lankan conflict, Hyndman and de Alwis outline a feminist analytic that is a tool both for understanding social, economic, and political relations, and for changing them. For the purpose of this chapter, similarly placed within a context of violence and turmoil, I invoke their definition of "feminist" as "analyses and political interventions that address the unequal and often violent relationships among people based on real or perceived social, economic, political, cultural and sexual differences" (Hyndman and de Alwis, 2002, p. 3). Human geographers committed to feminist praxis are well positioned to bring rigorous analysis that is at once integrative and historically and culturally grounded to bear upon specific societies afflicted by violence and injustice.[9]

Women, War, and Reconstruction

Interstate and intrastate conflict affects women in varying ways. Notwithstanding, Karam (2001, p. 3–7) identifies some recurring features that illustrate how drastically and dramatically women's lives and futures can change during conflicts. For example, women can be mobilized as soldiers in patriarchal militaries. They can find themselves as heads of households with limited resources. They can be subjected to increased medical and social vulnerability, greater security risks in disintegrating polities, increased sexualized violence, and war-structured sexual work. And they can become disadvantaged refugees. Jeanne Vickers (1993, p. 18) adds:

> [Women], and those they cared for, may be killed or injured in ethnic fighting or civil disturbances. . . . Their houses may be damaged, or they may flee from home in fear of their lives. Dwindling food supplies and hungry children exacerbate tensions. And so, to the loss of husbands, sons and brothers who are killed in battle, is added the longer-term suffering of further deprivation.

Extending this point, as mothers, wives, lovers, or daughters of "the disappeared," women are left to negotiate family disruptions – and, in the case of several Latin American countries including Peru, to demand accountability.

The Bogotá meeting highlighted various ways in which the presence of armed conflict can jeopardize community interaction and organization, even as communities struggle to assume responsibility for security and well-being (see Moser and Clark, 2001, p. 34). For example, warring parties can misinterpret community actions to organize for survival as mobilization against them. Although the mobilization of men is always viewed suspiciously, women's organizations are also exposed and vulnerable. Throughout Peru, women leaders were forced into hiding by threats against them; they were attacked and even killed.[10] In some cases, women's organizations disband, as occurred in many parts of Peru, including Lima. When this occurs, women lose both their influence in local affairs and a ready source of mutual support and reciprocity that is often crucial for family survival (Moser and Clark, 2001). In other cases, especially in Ayacucho, violence directed at grassroots women and their organizations only hardens their resolve.

Like conflict and war, post-war reconstruction presents itself differently to women and men. Women are expected to revert to pre-war gender roles. However,

> Many women will have become widows and single parents, dependent on their own earning power to provide for themselves and their children. In the absence of jobs of the kind they can do, training they can get access to, capital, credit and land, many women fall more deeply into the poverty they knew before war began. (Cockburn, 2001, p. 26)

In the course of civil war in Peru, more than 20,000 women were widowed and 500,000 children stricken with post-trauma stress syndrome (CVR, 2003, p. 343). For this population, returning to their pre-war status is impossible.

Moreover, wartime stigmas persist. Time after time, displaced women within the Ayacuchana community in Lima recounted to me that, although they had fled the terrorist and state-induced violence of their homeland, they were branded as "terrorists" wherever they sought safe haven. Several years after the civil war, they continue to experience arbitrary arrest and employment discrimination based solely upon their residence within an urban enclave to which migrants from Ayacucho were relegated upon their arrival in Lima.[11] Clearly, even after peace, sectors of women remain socially, politically, and economically marginalized – and still vulnerable to violence. Recognizing and understanding that women and men can remain vulnerable in post-conflict situations is fundamental to making them agents in their own survival, recovery, and growth.

Because violence, peace, and reconstruction have been seen predominantly as male domains, women and gender issues have been excluded from discussions and interventions for conflict and peace (Moser and Clark, 2001). A gender lens, indeed a feminist lens, makes a difference to what we see in war, peace, and reconstruction:

> gender consciousness calls, first, for a sensitivity to *gender difference*. It invites us to see how women and men may be positioned differently, have different experiences, different needs, different strengths and skills; and how in different cultures these differences have different expressions. Second, it invites us to notice *gender power relations* – to see how they shape institutions like the family, the military, the state; how they intersect with relations of class and ethnicity; how power, oppression and exploitation work in and through them. . . . It invites us to act for transformative change. (Cockburn, 2001, p. 28)

Although Cockburn refers to "gender consciousness" in this passage, she invokes a contextualized feminist analytic. Whereas gender analysis initially set out to advance understandings of the ways in which gender identities and gender relations are produced, its dilution in development discourse has undercut its utility as a rigorous analytical concept (Parpart, 2000; Hyndman and de Alwis, 2002). In contrast, feminist analysis of conflict allows us to move beyond merely identifying the different ways in which armed violence affects women and men to, as Cockburn says, "act for transformative change." In this case, feminist praxis challenges us to engage the disparate power relations (gender and others) that are at work in processes of peace-building, reconstruction, and development in post-conflict Peru.

Campesina Women and Political Violence in Peru: A Transformative Journey

In 1982 when the armed conflict began in the remote and impoverished Department of Ayacucho (Huamanga),[12] the existing social fabric and civil institutions were incapable of responding effectively to the armed conflict. Nor did these movements and institutions include women in their efforts to stem the tide of conflict. Isabel Coral, former Director of PROMUDEH and Co-Director of CEPRODEP (Centro de Promoción y Desarrollo Poblacional), suggests that because regional social movements were largely authoritarian, exclusionary, and confrontationist, they offered neither the conditions nor the opportunities to integrate women. Coral, a leader and activist, is deeply respected in Ayacucho for her capacity to analyze and articulate their experience of mobilization during the civil war.[13] She notes that while formal political parties dogmatically incorporated the topic of women as part of their discourse, they were not able to establish policies and strategies aimed at promoting women's progress and participation. Hence, feminist discourse and debate – largely an urban phenomenon – had little impact at the regional or rural level (Coral, 2001; interview with Coral, 2001).

A contradiction soon emerged, however, between the dogmatic and radical political projects of Sendero Luminoso, and the authoritarian and repressive traditions of government forces (Coral, 2001; interview with Coral, 2001). Some suggest that the generalized chaos created by the conflicting missions of Sandero Luminoso and the Peruvian state opened space for an alternative; that is, in this context movements arose that began to resist and transform violent situations (Martínez i Álvarez, 1999; Coral, 2001; interviews with Coral, Gamarra, Reynaga, all 2001). In this opening, campesina women developed a movement that, although local, was linked to global concerns. Under these conditions, women were able to transcend their confinement to traditional roles and consolidate themselves as social and political actors (Coral, 2001). As we will see, "[This] process [not only] confined the spaces of war and broadened democratic spaces [but] generated changes in gender relations in private life, in the economy and in politics" (Coral, 2001, p. 156).

Trapped in crossfire and chaos, the rural civilian population was especially vulnerable. Coral explains that because men were initially the principal target of both sides in the war, women were left to confront the new situation. The defense of life reordered their roles within society. Women within this population took up the challenge of building alternative organizational networks as the basis of personal, family, and community survival. With time, immediate family survival broadened to incorporate the integral preservation of family and community life and eventually the defense of human rights (Coral 1999, 2001; interview with Coral, 2001). In order to shield their families from the escalating violence, women devised spatial strategies of diffusion and networking. For example,

> One of the first actions undertaken by a woman was to organize her family's dispersal. A woman would stash her husband and the adolescent children at greatest risk from both sides in zones of refuge as far away as possible. . . . She would try to relocate her middle children in homes of relatives, godparents (compadres), or friends in more secure communities. She would leave family elders at home in order not to lose her family's stake in the community. . . . Such women, carrying their smallest children with them, would become itinerant wanderers – moving back and forth between these different spaces to look after the wellbeing of their relatives and to coordinate economic and family activities. [It was, in effect,] a process of dispersal and subsequent coordination . . . many families experienced four successive displacements during the war. (Coral, 1998, p. 355)

Women's mobilization first took the shape of furtive meetings of relatives of victims of violence. They met to provide moral support and share information. Through these networks of solidarity, Ayacuchana women educated one another about mechanisms and procedures for pressing their demands. With time, they began to identify tasks and delegate responsibilities. They organized themselves geographically in order to coordinate searches for the missing, exchange information on the dead, the disappeared, and the detained, and pressure military and police posts for the release of seized relatives. They quickly learned to utilize legal mechanisms as well as national and international human rights groups. And they learned how to project themselves in the media (Coral, 1998; Martínez i Álvarez, 1999; interview with Coral, 2001). Once again, in the eloquent immediacy of Isabel Coral:

> Profoundly affected and sensitized, the women became the principal protagonists in the defense of human rights. They were spurred not only by the painful process of burying the dead, seeking the disappeared, and trying to free prisoners, but also by the desire to preserve the physical integrity and lives of those who remained with the women. . . . Women from these battered and violated sectors, with no more resources than their will and creativity, burst onto the public stage to assume the giant task of defending human rights. Their work was not easy, because they did not know what rights they had or what mechanisms existed for exercising such rights. They had to discover all of this through the life experience of confronting a previously unknown world. (Coral, 1998, pp. 355–6)

The political violence not only represented a human rights crisis, it generated an economic collapse that threw most households into crisis. Absences of men were massive and growing; women were thrust into the role of head of household. While this implied burdens of work and responsibility, it also implied greater protagonism (Coral, 1998; interviews with Coral, Reynaga, Salcedo, all 2001). The profound imbalance between income and consumption needs pressed women into income-generating activities. These were often collective projects in the shape of food services, production workshops, and communal gardens. With time, these organizations grew more stable – primarily as *clubes de madres* (mothers' clubs). More formal organization lent legitimacy to women's claims and facilitated negotiating with the state as well as developing links with humanitarian organizations (Martínez i Álvarez, 1999).

Coral contends that campesina women organized themselves in two ways: (a) as a women's movement and (b) by linking up with other new social actors, such as the displaced persons movement and community self-defense committees (*rondas*). They accumulated significant organizational experience and the capacity to seize the initiative. In the context of terrifying reprisal in the late 1980s, the grassroots initiative of women was extended progressively to local and subregional levels.[14] Eventually, this led to the development of extensive networks that represented the entire affected population.[15] The process of coordinating the women's organizations culminated at a regional scale in 1988 with the founding of the Federation of Mothers' Clubs of the Province of Huamanga – the immediate province of Ayacucho/Huamanga city. At its inception, 270 mothers' clubs formed the federation. Immediately, eight other provinces in Ayacucho replicated this process and established federations. In November 1991, the First Departmental Congress of Mothers' Clubs was held; representatives of 1200 clubes de madres attended, and FEDECMA (Federación Departamental de Clubes de Madres de Ayacucho, Departmental Federation of the Mother's Clubs of Ayacucho) – an overarching organization that joined all provincial level federations at the departmental level – was founded (Coral, 1998, p. 358, 2001; Martínez i Álvarez, 1999; FEDECMA, 2001; interview with Coral, 2001).

Coral stresses that the grassroots women's movement seemed to advance against the current of the war and, as other social and political organizations faltered, it created the most representative and largest organization in the department of Ayacucho. Although resistance was a condition of survival, the "movement" seemed to grow more because it was based on a consensual objective: to settle the armed conflict.[16] By 1995, FEDECMA included eleven provincial federations, 1400 clubes de madres, and 80,000 affiliated women in rural and urban settings (FEDECMA Archives). Initially, FEDECMA's unifying objectives were (a) the defense of human rights and (b) the struggle for economic survival. It was characterized by an extensive network of female leaders that spanned the entirety of Ayacucho Department. They conveyed information and communicated responses back to their affiliates (Coral, 1998, 2001; interview with Coral, 2001).

My conversations with many of the founding leaders of FEDECMA – all campesina women of modest social and economic circumstances – left me humbled and astounded as they described the conditions under which they initially mobilized. They walked long distances over grueling terrain and in dangerous circumstances to convey messages; and they and their family members were harassed, beaten, humiliated, and subjected to death threats. One leader recounted to me of being apprehended by a band of senderistas, one of whom wore a sweater belonging to the leader's mother – a signal that her mother had been killed, presumably for her daughter's activities, and a veiled threat that the same could happen to her. When pressed about how they maintained their resolve, they explained that they developed a range of strategies – individual and collective – to protect themselves from danger.

Initially, Sendero Luminoso dismissed the women's organizations as inconsequential. However, once they instituted the provincial federation, Sendero actively sought to crush the movement through a variety of heavy-handed tactics. In the end, however, it was impossible for Sendero to halt or control the movement once it had been consolidated.[17] In August 1988, the Federation collaborated in a nationwide march for peace. Coral is convinced that this mobilization marked a turning point in the war. It brought together, at the national level, diverse institutions and organizations from across the political and social spectrums. In Ayacucho, half of the participants were women, and they carried placards stating in Quechua: "Those who kill should die" and "Fear is Gone" (FEDECMA, 2001). When Sendero attempted to disperse the crowd by exploding dynamite and seizing microphones, it was women who forced them to retreat (FEDECMA, 2001; interviews with Aime, Reynaga, both 2001). This was an extraordinarily difficult time. Recent conversations with women who were at the center of these events several years ago evoked moments of open weeping.

In assuming critical political functions, the women's movement became the accepted interlocutor of civil society (Coral, 1999, 2001; interview with Coral, 2001). This new presence within the regional political scene generated a newfound identity – that of being a rural, indigenous woman.

[Rural, indigenous women] were more informed, more independent, more secure; they felt capable of doing many things; they could speak out, express their opinions, and make decisions. They had more self-esteem . . . women were now considered important, above all because of the strength that they showed. (Coral, 2001, p. 158)

From this newly established space of credibility, Coral explains that women set about broadening democratic openings. They organized emergency programs to bring assistance to vulnerable populations. In order to staunch the rural exodus and limit its social impact, they channeled scarce resources to communities that had been raided or were at risk. They directed the rebuilding of community organizations that had disbanded. Moreover, they broadcast a bilingual (Quechua and Spanish) radio program (Coral, 1999, 2001; FEDECMA, 2001; interviews with Coral, Reynaga, both 2001). Within community organizations, they made synonymous the meanings of "woman" and "activist."

By making the most of this space, women generated new conditions in which they could undertake their own offensive aimed at resolving the armed conflict. With time, they were able to attenuate the impact of violence, manage the chaos and bewilderment of the vulnerable population, and isolate Sendero Luminoso so successfully that it was ejected from nearly all the territories it had occupied (Del Pino, 1998; Coral, 2001; interview with Coral, 2001). It would seem that on the basis of experience and organization, Ayacuchana women had obtained legitimacy within regional public opinion as credible social actors. Grassroots campesina women had established the most viable and representative organization within the southern Andes during the Peruvian civil war (Coral, 1998; Del Pino, 1998; CVR, 2003; interview with Coral, 2001).

Gendering Conflict Analysis: Feminist Lessons from Ayacucho

Within the Peruvian conflict, different types of violence – political, economic, social – coexisted and overlapped at different scales and in different time frames. Regardless of its form, violence and conflict eroded physical, human, natural, and social well-being – with differing effects on men and women. Introducing a gender analysis into conflict resolution can break down oversimplified understandings that portray men as the actors and women as the victims in conflict situations; it recognizes that men's and women's experiences and actions during times of political violence and social upheaval are determined by gender roles and identities assigned by society (Moser and Clark, 2001). For example, conventional wisdom tells us that "men make war and women make peace" (see Elshtain, 1987; Lentin, 1997). Yet in no conflict have women not been an integral part. Throughout history, women have fought fiercely for both war and peace. Peru is no exception. As *senderistas*, women took up arms against the state; in *rondas campesinas*, they used traditional farming tools to defend their communities against Sendero; and as members of FEDECMA, they defied weaponry to fight non-violently for peace. Despite this, until very recently, most literature on women in war and conflict has presented women as passive victims rather than as active agents.[18]

In Ayacucho, women found themselves at the center of political violence. Within that space, they occupied diverse roles and created distinct fates for themselves. Following Moser and Clark (2001), to portray these women singularly as victims would fail to identify the opportunities that conflict may have created for them. They observe that,

> Just as the gendered causes, costs, and consequences of violence and conflict are frequently marginalised in international and national debates, so too are the gendered

nature of conflict resolution and the associated humanitarian aid and development. In all cases, the diversity of experiences that women and men have in conflict and peace-building are largely ignored, and their multiple identities obscured by simplistic representations in conflict and peace. These deny men and women their agency as both victims and actors of armed conflict and building sustainable peace. (Moser and Clark 2001, p. 29)

Although we must not depreciate the suffering of men during Peru's political violence, we must be mindful that, in this as in all war, suffering was afflicted upon women and men in very distinct and gendered ways. Peruvian scholar and activist Guilia Tamayo (2000, p. 14) observes that "It is true that men and women share a set of circumstances during armed conflict that exposes them to particularly adverse conditions and to the abuse of their human rights. However, there are certain gender-based risks, dangers and disadvantages, which particularly and disproportionately affect women."

The stories and insights related to me by impoverished women in Ayacucho continually pointed to the gendered nature of outcomes as a result of the Peruvian civil war. For example, mobility was gendered. Men were more able to flee to provincial cities and the departmental capital. Women, encumbered by dependents, were more often forced to stay put. There, they assumed the tasks otherwise done by men in addition to their typical workload. They were charged with the survival of the family, and often placed themselves in grave danger as they ferried food and/or supplies to their menfolk in hiding. Many were forced to produce "taxes" in the form of food and services (including sexual) to occupying forces on both sides. For example, many concurred with this description of life at the height of the violence. "Women had to see to the security of the family, the children, the husband far away. It all revolved around her because the man's presence was gone. The defense of life. . . . The abuse of women, physical and moral abuse, abuse in every sense of the word. Still, we had to concern ourselves with life – tired, frightened, angry." As their workloads increased, they were exposed to ever higher levels of debilitation.

Similarly, their recollections of the crimes committed in Peru's war were gendered. Men and women were abused in different ways; often they died different deaths. More than one woman in Ayacucho implied to me that, although the deaths that men suffered were tragic, the fate of women who survived was worse. One recounted, "To survive is to suffer terribly; we suffer hunger, lack of water, firewood. . . . For nothing, men died. My husband's death was horrific; they made me watch him suffer, scream, die. Now his life is ended; we survive, suffer. . . . We die slowly, one by one, alone." In other words, surviving women were destined to suffer on their own (a) the trauma of losing their mates (often tortured and killed in their presence) and (b) the ongoing violence of the war and its aftermath. Indeed, their lived reality lends credence to Cynthia Cockburn's eloquent observation on the debilitating convergence of gender and class within war:

war and terror have the effect, sometimes deliberate, sometimes incidental, of rending the fine fabric of everyday life, its interlaced economies, its material systems of care and support, its social networks, the roofs that shelter it. This affects women, who in most societies have a traditional responsibility for the daily reproduction of life and community, in ways that are both class and gender specific. The poorest are least able to escape the war zone or buy protection. (Cockburn, 2001, p. 21)

Within Peru, the initiative for resistance took root in the most marginalized sectors of rural and urban society. These were the sectors most directly affected by the violence, and most reluctant to actively engage it. Women were at the forefront of many of these initiatives. For them, resistance was a precondition for survival and the preservation of life itself. Repeatedly, women survivors of the conflict cut to the quick when I asked them why they opted for resistance beyond mere survival. They would simply state the Quechua expression "Kausayta paqarachispam difindiniku"; that is, "Because we give life, we defend it." For them, the choice was clear: "Resist or die." In this sense, the civil war in Peru served as a site for change and empowerment. Coining the concept of "victimized self," Roberta Julian argues that by taking action against victimization, women actually diminish it (Julian, 1997). The experience of Ayacuchana women is a case in point. Despite the intent of the Peruvian state that society return to normal – that is, that women return to their previous situations – the generative aspect of the civil war continues to empower women beyond the cessation of armed conflict. One member of the newly formed Comité de Vigilancia en Vilcashuáman explained, "Because we have survived, we women have the capacity to change things. . . . I will never suffer like that again." By taking action against their victimization, these humble women achieved peace – and empowerment along the way.

Conclusion: The Masculinization of Development, Feminization of Poverty, Bankruptcy of Reconstruction in Post-conflict Peru

If Peru's peace is to be sustained, the influence of economic distress as a causal factor of political violence and its persistence under post-conflict conditions needs to be addressed. The political violence that erupted in the 1980s was an expression of the internal inequality that had long characterized Peruvian society.[19] A combination of internal and external factors exacerbated this structural violence throughout the 1970s and 1980s. Like most developing countries, Peru was hard hit by the abrupt rise of oil prices and interest rates on the world market, coupled with recession and protectionism in more developed countries, and matched by the steady decline in the prices of its commodities. As Peru's indebtedness grew, economic liberalization and structural adjustment, forced upon it by the international financial community, aggravated longstanding internal inequities. Since 1980, Peru has adopted three neoliberal reform programs, the most stringent in 1990. Although political violence diminished dramatically in 1993, the structural inequities underlying it persist and, in combination with the distressing effects of the current structural adjustment program, are fomenting social discontent.[20]

Economic austerity, associated with neoliberal restructuring, has affected all Peruvians. However, it has been a particularly gendered phenomenon. The majority of the poor in Peru are women and dependent children.[21] As adjustment policies have reduced wages, eliminated food subsidies, and inflated prices, the purchasing power of women as food providers has deteriorated. Cutbacks in public expenditures in health care and education have led to diminished care and training for poor women and their families, while it has increased their burden as the primary healthcare providers and educators within families. Throughout Peru, women have been increasingly forced to balance greater amounts of wage work with higher levels of subsistence, domestic, and community production in meeting household needs, thus intensifying their double – indeed triple – burden.[22]

Addressing the debilitating consequences of debt and debt-induced restructuring is essential to creating conditions for sustainable peace. Mindful of the international dimension of the economic crises that grip most conflict-ridden developing countries such as Peru, it is incumbent upon the international financial community to invest itself in creating conditions, not only for economic recovery, but for sustainable development as well. These sensitivities should lead us to challenge national governments, international financial institutions, and development organizations to guarantee that, with the cessation of violence, the precious space momentarily opened for change in Peru is not lost. Instead, in the spirit of feminist praxis, this moment must be seized to (a) secure genuine and lasting transformations in power relations, (b) guarantee that the emerging society reflects women's visions and repays what they have given, and (c) ensure that gender and class relations are not allowed to revert to the pre-war situation (Cockburn, 2001). Lest we forget, in conflict as in all aspects of life, women form half of society – an essential half that represents the ideas, experiences, and hopes of all sectors of society. To ensure that women and their dependants (both young and elderly) are not only combatants and/or victims of conflict, the specificities of women's interests, experiences, and needs must be at the center of official processes of peace-making, reconciliation, and post-conflict development. Without this, lasting peace will remain elusive in countries such as Peru.

ACKNOWLEDGMENTS

The fieldwork upon which this analysis is based was funded by a Picker Fellowship awarded by Colgate University in March 2000. I especially wish to thank Isabel Coral of PROMUDEH; Yanet Palomino of CEPRODEP-Lima; Blanca Zanabria and Vladimiro Hurtado of CEPRODEP-Ayacucho; Edith Villarreal, Ana María Yáñez, and Celina Salcedo of Movimiento Manuela Ramos; Jefrey Gamarra of IPAZ; Gumercinda Reynaga of La Red Nacional de la Promoción de la Mujer; Teodora Aime of FEDECMA; Nori Condor of La Universidad Nacional San Cristóbal de Huamanga for valuable field assistance, especially in translation and interpretation among monolingual Quechua-speaking women; and the Ayacuchana women who graciously shared their stories and insights with me and who must remain anonymous.

NOTES

1 World Bank sponsored conference "Género, Conflicto y la Construcción de la Paz Sostenible: Experiencias de América Latina," Bogotá, Colombia, May 8–9, 2000. Conference proceedings are available from the World Bank and are synthesized in Moser and Clark (2001).

2 See conference proceedings, synthesized by Moser and Clark (2001).

3 PROMUDEH is a state institution nominally designed to support women's rights and development. It was created during the interim presidency of Valentín Paniagua between the resignation in November 2000 of President Alberto Fujimori (amid allegations of

human rights abuses and corruption) and the election in April 2001 of President Alejandro Toledo. Upon Toledo's inauguration in July 2001, the influence of PROMUDEH was greatly diminished and Isabel Coral eventually resigned as its director.

4 More than 40 percent of the deaths and disappearances of the civil war were concentrated in the Department of Ayacucho (CVR, 2003).

5 I use the widely accepted term "campesina" to refer to rural women of indigenous origin. The Truth and Reconciliation Committee (CVR, 2003) posits that there was a significant relationship between poverty and social exclusion and the probability of becoming a victim of violence during the armed conflict. It found that the violence was biased geographically, ethnoculturally, socioeconomically, and by gender (CVR, 2003, pp. 316–17).

6 To refer to an official policy of post-war reconstruction might be disingenuous. For nearly two decades, the Peruvian state refused to acknowledge the internal conflict as a civil war. Although numerous NGOs use the term "reconstruction" in their work, official state agencies invoke the term "development."

7 For more thorough analysis of the gender-specific impact of neoliberal reform on women in Peru, see Hays-Mitchell (1997, 1999, 2002).

8 It is important to note that women in Peru also served as combatants within Sendero Luminoso. However, their experience is not part of this particular analysis.

9 This analysis is one part of ongoing investigations into the gendered implications of economic and political development in Latin America. In 2001, I conducted fieldwork in Ayacucho (or Huamanga) Department and the displaced Ayacuchana (or Huamangüina) community in Lima. During this time, I interviewed grassroots women who had been directly involved in the conflict, many of whom continue their activism today, in addition to a younger generation of grassroots women who are actively involved in the post-conflict reconciliation and reconstruction process. Moreover, I interviewed and accompanied in their work several middle-class, activist women within Ayacucho who were critical in the effort to provide organizational assistance to the grassroots women in the latter stages of the conflict. Finally, I interviewed Peruvian intellectuals with expertise on the civil war and reviewed pertinent archival data. For this reason, I must say that my research – original as it is – is based upon the labors of others, to whom I owe a debt of gratitude.

10 The most widely publicized case being the public assassination in February 1992 by Sendero Luminoso of María Elena Moyano, vocal women's activist and deputy mayor of Villa El Salvador, the largest shantytown in Peru.

11 See Hays-Mitchell (2001).

12 "Huamanga" is the local name, grounded in the indigenous Quechua language, used to refer to Ayacucho Department and its capital city. Use of the term "Huamanga" predates the arrival of the Spanish in the southern Andes. In contrast, "Ayacucho" is the name imposed by the Spanish conquistadores and colonial administrators.

13 The information upon which this section is based is drawn in large part from the work of, and conversations with, Isabel Coral. I owe her an enormous debt of gratitude. Isabel graciously welcomed me into her office and generously shared published and unpublished manuscripts, answered all my questions, and clarified my vision. This section represents a weaving together of Isabel Coral's recounting and analysis of the mobilization of poor rural women in Ayacucho. It is drawn from tape-recorded interviews and published and unpublished works, corroborated and supplemented by information and insights provided by other social activists, development practitioners, academics, and campesina women in Ayacucho.

14 By the early 1990s, over 400 rural communities had been razed or abandoned, mostly at the hands of Sendero Luminoso in reprisal for resistance (CVR, 2003).

15 This stands in stark contrast to the previous experience of women's movements at the national level, in which rural women were largely ignored (see Blondet, 1995).

16   The most distressing dimension of this strategy was reconciliation with the state – represented as it was by the armed forces that had inflicted untold atrocities on the rural population. It was an uneasy alliance based on shared interests.

17   It is interesting to note that their counterparts in the national capital of Lima were beaten down and largely disbanded in the urban offensive of 1992.

18   Sadly, the co-optation and systemization of "gender analysis" by many development and humanitarian organizations, in a way that ignores the specificities of distinct places, has unwittingly contributed to this.

19   See Gorriti (1999), among others.

20   See, for example, Koonings and Kruijt (1999) and Rotker (2002).

21   Instituto Cuánto (2000).

22   For more complete discussion, see Hays-Mitchell (1997, 2002).

INTERVIEWS

Aime, T. Approximate age 50–55 years. Founding President of FEDECMA (Federación Departmental de Clubes de Madres de Ayacucho, Departmental Federation of the Mother's Clubs of Ayacucho). Interviews with author, April 30 and May 8, 2001.

Coral Cordero, I. Approximate age 45–50 years. Series of interviews with author, January–February, April–May 2001.

Gamarra, J. Approximate age 45–55 years. Director of IPAZ (Instituto de Investigación, Promoción, Desarrollo y Paz en Ayacucho. Series of interviews with author, January–February, April–May 2001.

Reynaga, G. Approx age 45–55 years. President of La Red Nacional de la Promoción de la Mujer, Ayacucho. Interview with author, 2001.

Salcedo, C. Approx age 45–55 years. Director of Movimiento Manuela Ramos, Ayacucho. Interviews with author, February 8 and April 24, 2001.

BIBLIOGRAPHY

Blondet, C. (1995) El movimiento de las mujeres en el Perú 1960–1990. In *Perú 1964–1994: Economía, Sociedad y Política*. Lima: Instituto de Estudios Peruanos.

Cockburn, C. (2001) The gendered dynamics of armed conflict and political violence. In C. Moser and F. Clark (eds), *Victims, Perpetrators or Actors? Gender, Armed Conflict and Political Violence*. London: Zed Books.

Comisión de la Verdad y Reconciliación (2003) *Informe Final*. Lima: CVR (http://www.cverdad.org.pe/ifinal/index.php).

Coral Cordero, I. (1998) Women in war: impact and responses. In S. Stern (ed.), *Shining and Other Paths: War and Society in Peru*. Durham, NC: Duke University Press.

Coral Cordero, I. (1999) Organizaciones sociales: de víctimas a actores protagónicos en la construcción de la paz. Unpublished document, Lima.

Coral Cordero, I. (2000) Social organizations: from victims to actors in peace building. In C. Moser and F. Clark (eds), *Victims, Perpetrators or Actors? Gender, Armed Conflict and Political Violence*. London: Zed Books.

Del Pino, H. and P. (1998) Family, culture, and "revolution": everyday life with Sendero Luminoso. In S. Stern (ed.), *Shining and Other Paths: War and Society in Peru*. Durham, NC: Duke University Press.

Elshtain, J. (1987) *Women and War*. New York: Basic Books.

Federación Departmental de Clubes de Madres de Ayacucho (2001) Archives.

Gorriti, G. (1999) *The Shining Path: A History of the Millenarian War in Peru.* Chapel Hill: University of North Carolina Press.

Hays-Mitchell, M. (1997) Development vs empowerment: the gendered legacy of economic restructuring in Latin America. *Yearbook of The International Congress of Latin Americanist Geographers*, 23, 119–31.

Hays-Mitchell, M. (1999) From survivor to entrepreneur: gendered dimensions of microenterprise development in Peru. *Environment and Planning A: International Journal of Urban and Regional Research*, 31, 251–71.

Hays-Mitchell, M. (2001) Danger, fulfillment, and responsibility in a violence-plagued society. *Geographical Review*, 91(1/2), 311–21.

Hays-Mitchell, M. (2002) Coping with austerity: a gendered perspective on neoliberal restructuring in Peru. *Gender and Development*, 10(3).

Hyndman, J. and de Alwis, M. (2002) Beyond gender: a feminist analysis in countries affected by conflict. Unpublished article.

Instituto Cuánto (2000) *Perú en Números.* Lima: Instituto Cuánto.

Instituto Peruano de Educación en Derechos Humanos y Paz (1999) Weaving ties of trust and commitment to build human rights and democracy in Peru. Unpublished document, Lima.

Julian, R. (1997) Invisible subjects and the victimized self: settlement experiences of refugee women in Australia. In R. Lentin (ed.), *Gender and Catastrophe.* London: Zed Books.

Karam, A. (2001) Women in war and peace-building: the roads traversed, the challenges ahead. *International Feminist Journal of Politics*, 3(1), 2–25.

Koonings, K. and Kruijt, D. (eds) (1999) *Societies of Fear: The Legacy of Civil War, Violence and Terror in Latin America.* London: Zed Books.

Lentin, R. (1997) *Gender and Catastrophe.* London: Zed Books.

Martínez i Álvarez, P. (1999) *Kausayta Paqarachispam Difindiniku.* Lima: Centro de Promoción y Desarrollo Poblacional.

Mertus, J. (2001) Grounds for cautious optimism. *International Feminist Journal of Politics*, 3(1), 99–105.

Moser, C. and Clark, F. (2001) Gender, conflict, and building sustainable peace: recent lessons from Latin America. *Gender and Development*, 9(3), 29–39.

Parpart, J. (2000) Rethinking participation, empowerment, and development from a gender perspective. In J. Freeman (ed.), *Transforming Development: Foreign Aid for a Changing World.* Toronto: University of Toronto Press.

Rotker, S. (ed.) (2002) *Citizens of Fear: Urban Violence in Latin America.* New Brunswick, NJ: Rutgers University Press.

Tamayo, G. (2000) La mujer en situaciones de conflicto, pos-conflicto y zonas militares: experiencias de América Latina. Unpublished paper presented at World Bank sponsored conference Género, Conflicto y la Construcción de la Paz Sostenible: Experiencias de América Latina, Bogotá, Colombia, May 8–9.

Vickers, J. (1993) *Women and War.* London: Zed Books.

Index

Printed and bound by CPI Group (UK) Ltd, Croydon, CR0 4YY

17/12/2025

14794971-0005